REFRIGERATION AND AIR-CONDITIONING

Dr. D.S. Kumar

"Dedicated to my wife, Mrs. Manjit Kaur, for her endless patience, continuous moral support and encouragement. Her cheerful compliments and inspiring words lifted my spirits on countless occassions. She has enriched my life and my gratitute to her is boundless"'

BY THE SAME AUTHOR

1 Fluid Mechanics & Fluid Power Engineering

2 Heat & Mass Transfer

3 Engineering Thermodynamics (Principles and Dynamics)

4 Applied Thermodynamics

5 Basics of Mechanical Engineering

6 Engineering Mechanics: Statics and Dynamics

7 Automobile Engineering

8 Mechanical Engineering (Objective)

REFRIGERATION AND AIR-CONDITIONING

By

DR. D.S. KUMAR

B.Sc. Engg. (Hons.), M.Sc. Engg, Ph.D, FIE

Formerly :

Professor Mech. Engg, Punjab Engg. College, Chandigarh,
Director Academic Affairs, Punjab Technical Univ., Jalandhar,
Principal S.U.S. College of Engg. & Technology, Tangori (Mohali) and
Advisor, Manav Rachna Educational Institutions, Faridabad

S.K. KATARIA & SONS®

Publisher of Engineering Books

4885/109, Prakash Mahal, Dr. Subhash Bhargav Lane,
Opposite Delhi Medical Association, Daryaganj, New Delhi-110002 (India)
Tel.: +91-11-43551243, Mobile: +91-9871775858, 7678425209
E-mail: katariabook@gmail.com, katariabooks@yahoo.com
Website: www.skkatariaandsons.com
(e-books, subscriptions and bulk buying also available on our website)

Published by:

S.K. Kataria & Sons®

4885/109, Prakash Mahal, Dr. Subhash Bhargav Lane,
Opposite Delhi Medical Association, Daryaganj, New Delhi-110002 (India)
Tel.: +91-11-43551243, Mobile: +91-9871775858, 7678425209
E-mail: katariabook@gmail.com, katariabooks@yahoo.com
Website: www.skkatariaandsons.com

First Edition: January 2017, Reprint: 2026

Price: ₹ 795/-

Laser Typesetting by : Plus Computers

Printed at: Al-Aziz Binding House

PREFACE TO THE FIRST EDITION

Over the last few years, there has been a large scale development in commercial refrigeration and air-conditioning, and accordingly a lot of emphasis is being placed on teaching of this subject as an independent course rather than being a part of a general course on thermal engineering. This aspect is reflected by the introduction of formal tution in refrigeration and air-conditioning at undergraduate and postgraduate studies of the various universities and technological institutes.

The author has taught this course several years and this monograph is essentially based upon the lectures delivered by him. From the experience gained through useful class discussions and feedback, the lecture notes were revised to improve their clarity and necessary explanatory notes were added during each teaching semester. The subject matter has thus been thoroughly tested in the classroom and found useful.

The text presents an amalgam of numerous ideas and influences and is the outcome of the distillation of the available textbooks written by distinguished authors. Obviously then a question arises to "Why another book on Refrigeration and Air-conditioning". The answer to this question is reflected by the following salient features incorporated in this book :

- Precise and logical presentation of the basic concepts and principles
- Text written in simple, lucid and easily understandable language
- Full use of line diagrams made to supplement the text and to explain a particular phenomenon as clearly as possible.
- Solutions provided to a wide variety of problems of standard comparable to those set for engineering degree, AMIE, GATE and Engineering Service Examinations
- Consistent use of SI units and notations throughout the text
- Inclusion of section on Multiple Choice Questions and short answer questions keeping in view the recent trend of such questions being asked in the various University and competitive examinations.

The author strongly believes that this treatise on Refrigeration and Air-conditioning will contribute effectively to the existing literature in the field, and will be a useful text for the students of engineering curriculum.

Acknowledgements: On completion of the book "Refrigeration and Air-conditioning" I wish to acknowledge my thanks and gratitude to all those who have been associated with this work directly/indirectly. Acknowledgement is due and is hereby made to all the authors whose works have been freely consulted in the preparations of this text. I owe an enormous debt of gratitude to my colleagues and students for much valued assistance in the form of discussion and feedback. Above all, I with to place on record my sincerest and apologies to my near and dear ones who willingly endured certain hardships as a result of my pre-occupation with this work. A special mention in this respect is reserved for my wife, Mrs. Manjit Kaur whose support, patience, encourangement and smiling face always keeps my spirits high and purpose clear.

My efforts in writing this book will be amply rewarded if the readers send their constructive suggestions and objective criticism with a view to enhance the usefulness of the book. Further, I would greatfully acknowledge the contribution of the readers by way of bringing to my attention the errors and printing mistakes that might have inadvertently crept into the book.

D.S. Kumar

World Spa, Gurgaon
January, 2017

CONTENTS

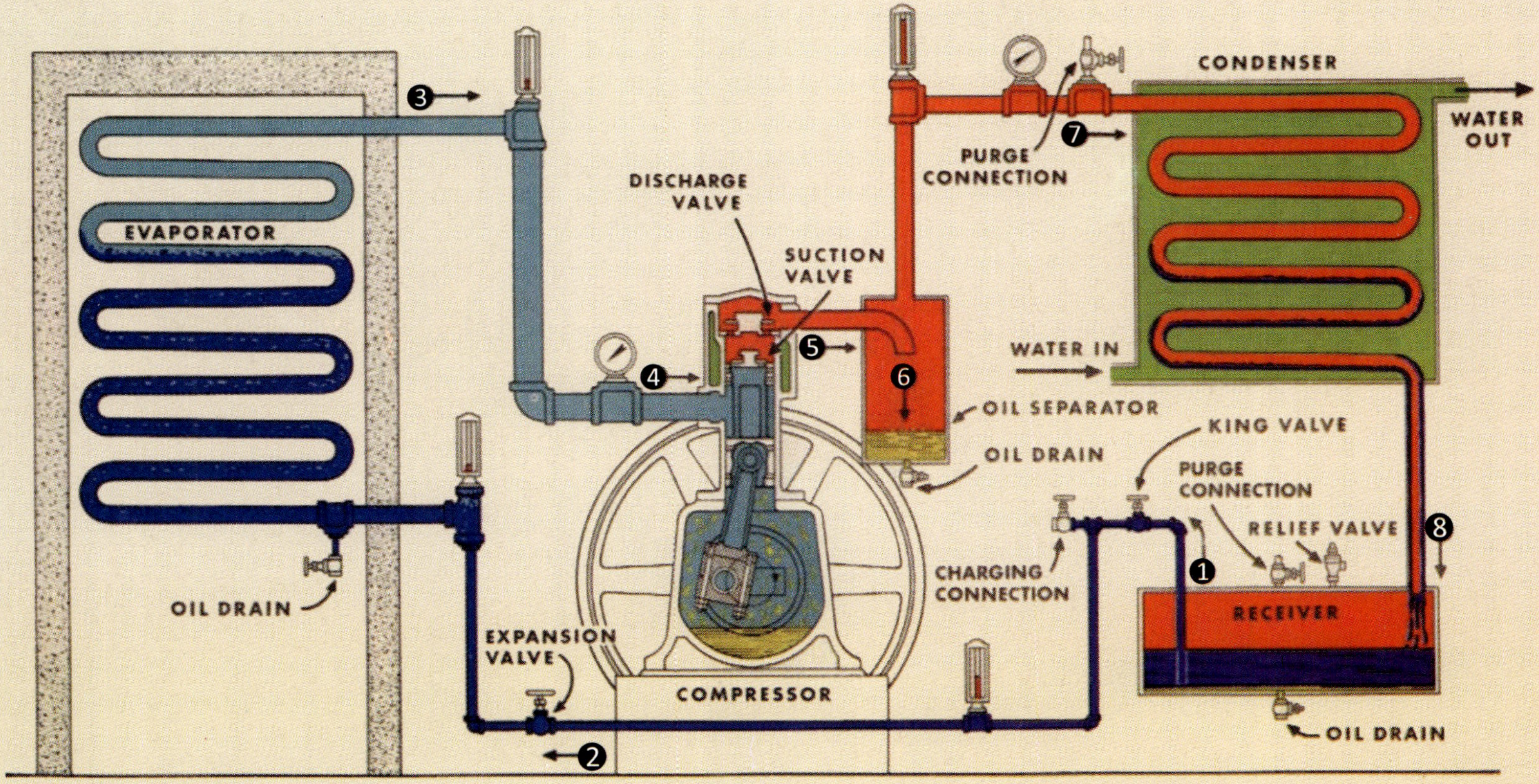

1 Liquid refrigerant leaving receiver
2 Liquid refrigerant passing through expansion valve
3 Low pressure gaseous refrigerant leaving evaporating coils
4 Low pressure gaseous refrigerant entering compressor
5 High pressure gaseous refrigerant leaving compressor
6 High pressure gaseous refrigerant entering oil separator
7 High pressure gaseous refrigerant entering condenser
8 Liquid refrigerant returning to receiver

Legand

- Liquid Refrigerant
- Low Pressure Gaseous Refrigerant
- High Pressure Gaseous Refrigerant
- Oil
- Water

Chapter – 1

Basic Concepts and Definitions

Learning objectives : A study of the subject matter presented in this chapter will enable the students/readers to

- *recapitulate the basic concepts and the laws of thermodynamics*
- *characterise the heat and work interactions in the different processes constituting Carnot cycle*
- *understand and make enthalpy and entropy calculations of a pure substance during phase change*
- *understand the physical picture and laws controlling the different modes of heat exchange*

1.1. THERMODYNAMIC SYSTEM, SURROUNDINGS AND BOUNDARY

A ***thermodynamic system*** represents a prescribed and identifiable (fixed) quantity of matter under consideration to analyse a problem; to study the changes in its properties due to exchange of energy in the form of heat and work. The system may be a quantity of steam, a mixture of vapour and gas or a piston-cylinder assembly of an IC engine and its contents.

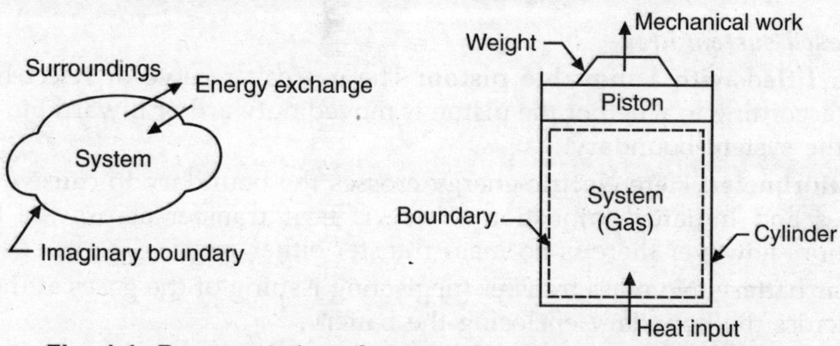

Fig. 1.1. Representation of system, boundary and surroundings

For the description of a thermodynamic system, some of the following quantities need to be specified:

(*i*) quantity as well as composition of the matter

(*ii*) measurable properties such as pressure, temperature and volume of the system

(*iii*) energy of the system.

The combination of matter and space, external to the system, that may be influenced by changes in the system is called ***surroundings*** or ***environment***. The thermodynamic system and surroundings are separated by an envelope called ***boundary*** of the system. The boundary represents the limit of the system and may be either real or imaginary, and may change shape, volume, position and/or orientation relative to the observer. For example, an elastic balloon which is initially spherical in shape may get deformed into a cylindrical shape or some other geometrical shape or it may get squeezed to a reduced volume during a certain process. As such the boundary of the gas enclosed in the balloon would not retain the same shape and size. Further, the boundary may be diathermic or adiabatic depending upon whether it allows or not exchange of energy in the form of heat. The walls or boundaries which do not allow heat transfer to take place across them are named ***adiabatic***. In contrast, the walls that do allow heat interaction across them are called ***diathermic.***

The surroundings can also be regarded as a system, and hence thermodynamics is largely concerned with interaction between systems.

Thermodynamic systems can be classified into:

(*a*) Closed system : A closed system can exchange energy in the form of heat and work with its environment but there is no mass transfer across the system boundary. The mass within the system remains the same and constant, though its volume can change against a flexible boundary. Further, the physical nature and chemical composition of the mass may change. Thus a liquid may evaporate, a gas may condense or a chemical reaction may occur between two or more components of the system.

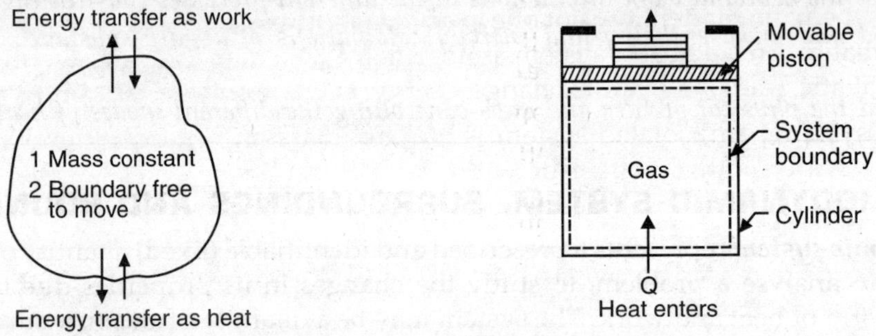

Fig. 1.2. Closed system

Examples of closed system are :

(*i*) **Cylinder fitted with a movable piston:** The gas can receive or reject heat, expand or contract according to whether the piston is moved outward or inward but no matter (gas) crosses the system boundary.

(*ii*) **Bomb calorimeter:** Here electric energy crosses the boundary to cause a spark between electrodes and initiate combustion. There is heat transfer across the boundary after combustion; however there is no mass transfer either way.

(*iii*) **Motor-car battery:** No mass transfer (neglecting issuing of the gases at the battery poles) occurs across the boundary enclosing the battery.

(*iv*) **Pressure cooker, Kitchen refrigerator, Ice cream freezer** (manual crank operation) etc.

(*b*) Open system : An open system has mass exchange with the surroundings along with transfer of energy in the form of heat and work. The mass within the system does not necessarily remain constant; it may change depending upon mass inflow and mass outflow.

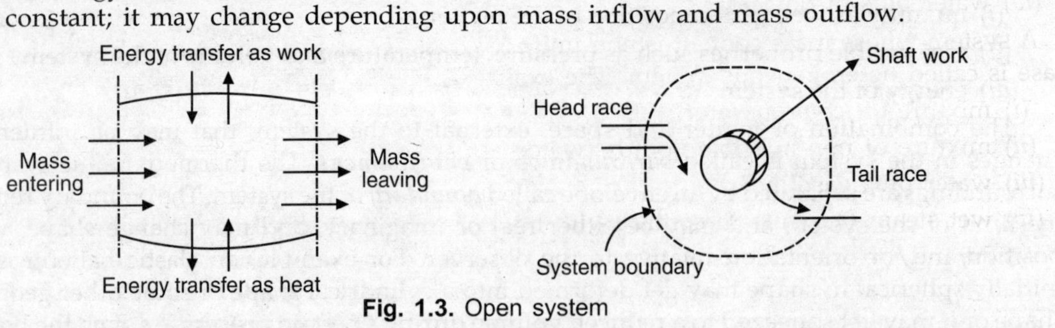

Fig. 1.3. Open system

Examples of an open system are:

(*i*) **Water wheel:** It is a device that converts potential energy of water into mechanical work. Water enters the wheel from head race side and leaves it to the tail race from the other end, and as such the mass crosses the system (wheel) boundary. The work output due to rotation of the wheel also crosses the system boundary.

(*ii*) **Motor car engine:** The engine initially draws charge (mixture of air and petrol) and finally exhausts the spent-up gases to the surrounding atmosphere; the mass flow occurs across the system (engine) boundary.

(*iii*) **Steam generator (Boiler):** Boiler is a device which converts water from liquid to vapour phase. Here contents of the system change; water flows into and steam flows out of the system.

Most of the engineering devices are open systems involving flow of fluids through them.

(*c*) Adiabatic system : There exist walls or boundaries which do not allow heat transfer to take place across them; no matter how hot one member is compared to the other. Such boundaries are named adiabatic. An adiabatic system is thermally insulated from its environment; it is enclosed by adiabatic walls and can exchange energy in the form of work only.

(*d*) Isolated system : An isolated system is of fixed mass and energy; it exchanges neither mass nor energy with another system or with the surroundings. An isolated system has no interaction with the surroundings; it neither influences the surroundings nor is influenced by it. When a system and its surroundings are taken together, they constitute an isolated system. The universe can be considered as an isolated system and so is the fluid enclosed in a perfectly insulated closed vessel (thermos flask).

(*e*) Homogeneous and heterogeneous systems : Based on phase change, a system may be classified as a homogeneous system and a heterogeneous system. A phase represents a quantity of matter that is uniform throughout in physical structure and in chemical composition. Physical uniformity implies that the matter is all gas, or all liquid or all solid. Uniformity of chemical composition means that the chemical composition does not vary from one part of the system to another. An iron piece, a liquid contained in a vessel, a gas enclosed within a container, and a mixture of gases represent one-phase systems. A system consisting of liquid and gas is a two-phase system of a liquid phase and a gaseous phase. Likewise a mixture of solid, liquid and gas constitutes a three-phase system.

A system consisting of a single phase is called homogeneous system. The examples are:

(*i*) ice, water, dry saturated steam
(*ii*) mixture of ammonia in water
(*iii*) mixture of air and water vapour
(*iv*) water plus nitric acid

A system whose mass content is non-uniform throughout, *i.e.*, it consists of more than one-phase is called heterogeneous system. The examples are:

(*i*) mixture of ice and water
(*ii*) mixture of non-miscible liquids (water + mercury)
(*iii*) water plus gasoline
(*iv*) wet steam (vapours in contact with liquid being evaporated)

1.2. THERMODYNAMIC PROPERTY: INTENSIVE AND EXTENSIVE PROPERTIES

A thermodynamic property refers to the characteristics which can be used to describe the condition or state of a system, *e.g.*, temperature, pressure, chemical composition, colour, volume, energy etc. The salient aspects of a thermodynamic property are:

(*i*) it is a measurable characteristic describing a system, and helps to distinguish one system from another,

(*ii*) it has a definite unique value when the system is in a particular state,

(*iii*) it is dependent only on the state of the system; it does not depend on the path or route the system follows to attain that particular state,

(*iv*) its differential is exact.

If P is a thermodynamic property with dP representing its differential change, then the integral between initial state 1 to final state 2 of the system will have only value given by

$$\int_1^2 dP = (P_2 - P_1)$$

Since a thermodynamic property is a function of the state of a system, it is referred to as a **point function** or a **state function**.

There are two kinds of thermodynamic properties namely intensive and extensive.

(*a*) ***Intensive property*** is independent of the extent or mass of the system. Its value remains the same whether one considers the whole system or only a part of it. Examples are pressure, temperature, density, composition, viscosity, thermal conductivity, electrical potential etc.

(*b*) ***Extensive property*** depends on mass or extent of the system ; its value depends on how big a portion of the system is being considered. Examples include energy, enthalpy, entropy, volume etc.

Let a system with value of a certain characteristic/property P be divided into a number of parts. If P_1, P_2, P_3 etc denote the value of that property for various parts of the system, then

For an intensive property : $P = P_1 = P_2 = P_3 = \ldots$

For an extensive property : $P = P_1 + P_2 + P_3 + \ldots$

(*c*) ***Specific property :*** An extensive property expressed per unit mass of the system. Examples include specific volume, specific energy, specific entropy, specific enthalpy etc.

specific volume $v = V/m$

specific energy $e = E/m$

where V, E and m are volume, energy and mass of the system respectively.

Like extensive properties, specific properties too are additive.

(*d*) ***Molar property :*** Ratio of an extensive property to the mole number (or the property per unit mole). For example

$$\text{Molar vol } v = \frac{\text{total volume}}{\text{mole number}} = \frac{V}{n} \text{ m}^3/\text{mol}$$

Check for a property : A given expression $dP = M\,dx + N\,dy$ would be a thermodynamic property if its differential is exact, *i.e.*, if

$$\left[\frac{\partial M}{\partial y}\right]_x = \left[\frac{\partial N}{\partial x}\right]_y$$

1.3. EQUILIBRIUM: MECHANICAL, CHEMICAL, THERMAL AND THERMODYNAMIC

Equilibrium is the concept associated with the absence of tendency for spontaneous change in the value of any macroscopic property of the system when it is isolated from its surroundings.

(*a*) ***Mechanical equilibrium*** (Equality of pressure) : Condition or state in which there is no unbalanced force within the system and nor at its boundaries. Mechanical equilibrium implies uniformity of pressure, *i.e.*, there is only one value of pressure for the entire system. If pressure gradients exist within the system, internal changes (diffusion) take place to wipe out the unbalance and attain the state of mechanical equilibrium.

(*b*) ***Chemical equilibrium*** (Equality of chemical potential) : A system in mechanical equilibrium may undergo a spontaneous change of internal structure due to chemical reaction or diffusion (transfer of matter). For example, there will be spontaneous change in the properties of a mixture of oxygen and gasoline once it is ignited. Chemical equilibrium represents that condition or state of the system when all chemical reactions in it have ceased and there is no mass diffusion.

(*c*) ***Thermal equilibrium*** (Equality of temperature): Condition or state in which the temperature of the system is uniform.

Let a system in mechanical and chemical equilibrium be separated from its surroundings by a diathermic wall (heat flow is allowed). If the surroundings are at a temperature higher than that of the system, then heat will flow towards the system and with the passage of time it will become hotter than what it was before. Subsequently both will attain the same temperature and the heat interaction will cease. When such a stage is reached, *i.e.*, there is no change taking place in the condition of system and surroundings, they are said to be in thermal equilibrium with each other. A system is thus said to be in thermal equilibrium if there exists uniformity of temperature throughout the system or between the system and its surroundings.

It is to be emphasised that the fluids comprising the systems in thermal contact are not reactive chemically. For example, the thermal contact between acid and water, although both at the same temperature, would result in rise in temperature due to the release of heat of dilution, and would not conform to the definition of thermal equilibrium.

(*d*) ***Thermodynamic equilibrium***: A system which is simultaneously in a state of mechanical equilibrium, thermal equilibrium and chemical equilibrium is said to be in a state of thermodynamic equilibrium.

1.4. STATE, PATH, PROCESS AND CYCLE

Consider a system constituted by gas enclosed in the piston cylinder assembly of a reciprocating machine. Corresponding to position of the piston at any instant, the condition of the system will be prescribed by pressure, volume and temperature of the gas. When all such properties have a definite value, the system is said to exist at a definite state. *State is thus condition of the system identified by thermodynamic properties.* With properties as co-ordinates, the state of the system in thermodynamic equilibrium can be represented by a point. (Fig. 1.4)

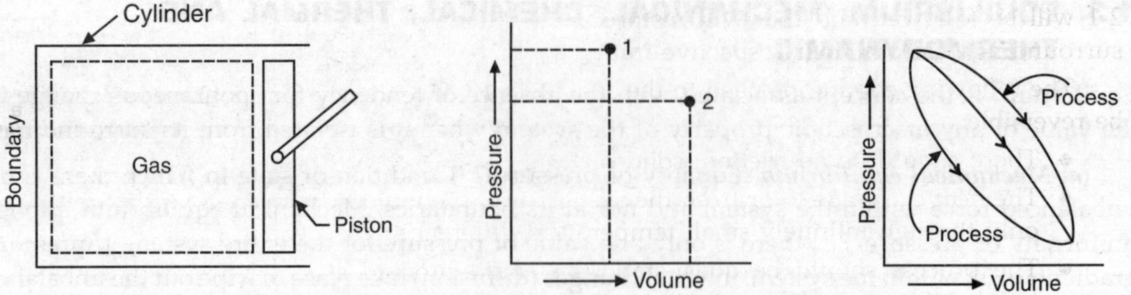

Fig. 1.4. Representation of state, a process and system paths

When the piston moves outwards, the properties of the system change (pressure decreases and volume increases). Any such operation in which properties of the system change is called a ***change of state.*** The locus of the series of states through which a system passes in going from initial state to its final state constitutes the ***path***. A complete specification of the path is referred to as a ***process***. Apparently a *thermodynamic process represents a transition in which a system changes from one state to another*. A process is simply a change in the state of the system and there are infinite ways for a system to change from one state to another state. Reference Fig. 1.4., *A* and *B* represent the terminal states of the piston and the paths *A1B*, *A2B*, and *A3B* traced by the system from state *A* to state *B* represent the processes. Certain processes are identified by special names, *e.g.*, an isochoric process is a constant volume process, an isobaric process is a constant pressure process, and an isothermal process is a constant temperature process.

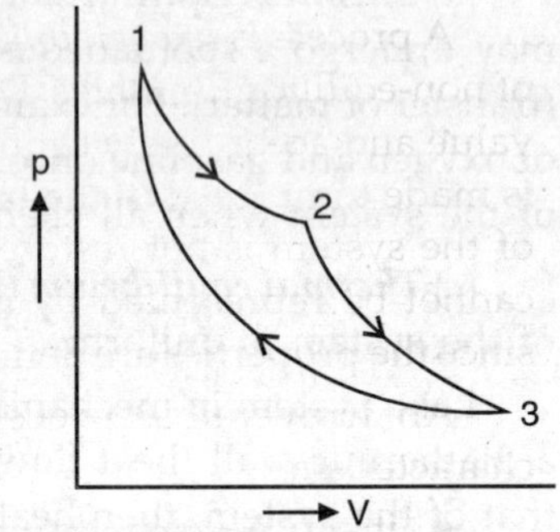

Fig. 1.5. A three process thermodynamic cycle

When a system in a given state undergoes through a series of processes such that the final state is identical with the initial state, a ***cyclic process*** or a ***cycle*** is said to have been executed. An essential feature of the cycle is that the intital condition of the system is restored after the processes. The change in the value of any property of the system for a cyclic process is zero. Mathematically

$$\oint dP = 0$$

where P is any property and symbol $\oint$ represents integration around a cycle.

Note : In an IC engine, the condition of the exhaust gases is not the same as the condition of the charge sucked in. As such, these engines execute only a mechanical cycle and not a thermodynamic cycle.

A cyclic process can be represented on *p-v* diagram (Fig. 1.5.) if the various states and paths are known. The cycle 1-2-3-1 comprises three processes 1-2, 2-3 and 3-1.

1.5. REVERSIBLE AND IRREVERSIBLE PROCESSES

A thermodynamic process is ***reversible*** if the system passes through a continuous series of equilibrium states. An equilibrium state of the system can be located on any diagram drawn for properties (such as pressure-volume diagram) at any instant during the process. The reversible process is then shown by a line drawn through the points representing the equilibrium states. Fig. 1.6. shows a reversible process on A-B plot whose path is 1-2. If the process is reversed, path

2-1 will be followed and that will restore the system as well as surroundings back to their respective initial states.

The following conditions need to the satisfied for a process to be reversible:

- There should be no friction; solid or fluid
- The heat exchange to or from the system, if any, should be only through infinitely small temperature difference
- The process should be quasi-static, it should proceed at infinitely slow speed. For this, the pressure difference between the system and surroundings must be infinitely small.

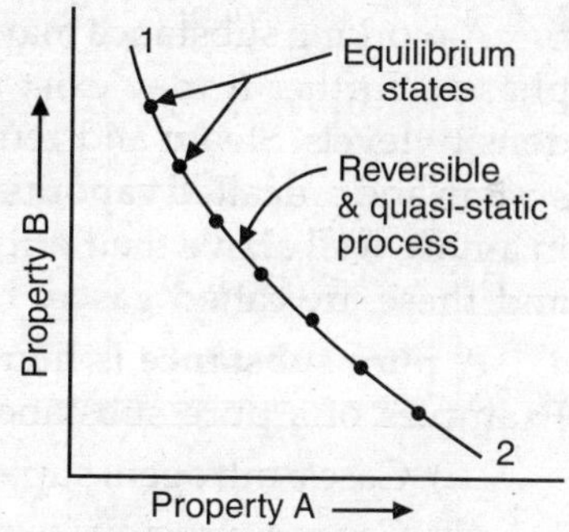

Fig. 1.6. Quastatic and reversible process

A process is ***irreversible*** if the system passes through a sequence of non-equilibrium states. During such a process, the properties of the system donot have a unique value and accordingly cannot be located on any property diagram. When an irreversible process is made to proceed in the backward direction, the original state of the system is not restored. (Fig 1.7). An irreversible process cannot be represented by a full line on diagram of properties since the properties are definite only at the initial and final states.

An irreversible process is identified by the following characteristics:

- it can be carried out in one direction only,
- it occurs at a finite rate,
- it cannot be reversed without causing permanent changes in the surroundings,
- the system is never in equilibrium state at any instant during an irreversible process.

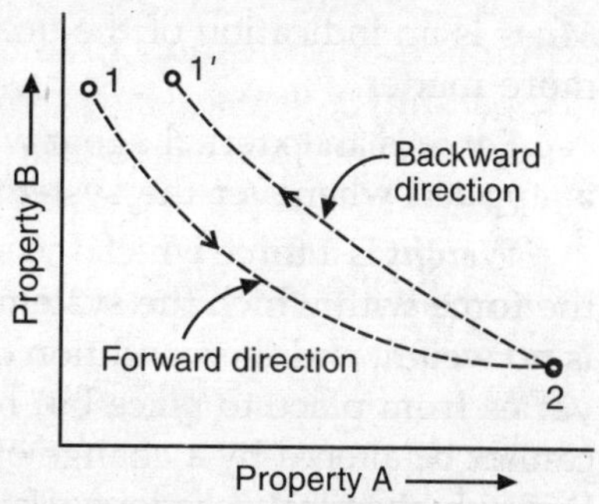

Fig. 1.7.

All natural processes take place spontaneously at finite speeds through finite discontinuities between the system and its environment and as such are irreversible. Processes are rendered irreversible as a result of degradation of energy by such factors as friction, turbulence, diffusion, throttling and combustion etc.

1.6. WORKING SUBSTANCE

A working substance refers to the fluid employed in thermodynamic devices to serve as a medium for the transport of energy between the system and its surroundings. A fluid, which may be gas, vapour, liquid or any non-chemically reactive mixture of these constitutes the matter normally considered in engineering thermodynamics. Fluids can be conveniently made to absorb, store and release energy. Further, they can be easily transported from place to place through pipes.

- Combustion products of hydrocarbons obtained by igniting air-fuel mixture form the working substance in petrol and diesel engines, gas turbines and jet propulsion.
- Water vapour is the working substance in steam power plants. The water is vaporised to steam in the boiler, the steam is fed to the turbine which is set into rotary motion and mechanical work is produced. The spent-up steam (steam leaving the turbine) is condensed to liquid water and is returned to the boiler.
- Refrigerants (NH_3, SO_2, Freon-12 etc.) are the working substances used in refrigerators, air-conditioners and ice-making plants.

A working substance may exist in solid, liquid and gaseous phase or in a combination of these phases. Further it may exist in different conditions, *i.e.*, at different pressure, temperature and density levels. Steam and refrigerants change their phase during processes occurring within the system and are called **vapours**. The substance such as air and combustion products of hydrocarbons in air are well above their critical temperature. In these substances no change in phase takes place and these are called **gases.**

A **pure substance** is homogeneous, and has the same chemical and physical compositions. Examples of a pure substance are:

(*i*) Gases: nitrogen; superheated steam

(*ii*) Gas mixtures: air; oxygen plus gaseous fuels

(*iii*) Combination of liquid and vapour of the same matter: water plus steam; liquid oxygen plus vaporised oxygen

1.7. MASS, FORCE AND WEIGHT

Mass is an indication of the quantity of matter present within a system. The more mass means more matter.

Force is an external agent which tends to change the speed or direction of a system. A force is applied whenever the system needs to be accelerated or decelerated.

Weight is a force which the system exerts due to gravitational acceleration. Weight is actually the force with which the system is attracted towards the earth. Where there is no gravity, there is no weight and this condition of the system is known as weightlessness. The weight of a system varies from place to place but mass remains constant. The mass of a given collection of matter cannot be altered by a change of location, a deformation of shape, a change of temperature, and through chemical reactions which the matter may undergo.

The weight (W) of system equals the product of mass (m) and local gravitational acceleration. The mass of a system remains constant whereas weight varies with change in value of gravitational acceleration from one place to another. The value of g at sea level is 9.8066 m/s^2 and is generally taken as 9.81 m/s^2.

1.8. SPECIFIC WEIGHT, MASS DENSITY AND SPECIFIC VOLUME

* **Specific weight (w)** of a substance is its weight per unit volume

$$w = \frac{W}{V}$$

where W is the weight of the substance occupying volume V. The weight of the substance is the force with which it is attracted to the centre of earth. It is product of mass and local gravitational acceleration, *i.e.*, $W = mg$. Since weight is expressed in newton, the unit of measurement of specific weight is N/m^3.

* **Density (ρ)** is a measure of the amount of working substance contained in a given volume and is defined as the mass per unit volume.

$$\rho = \frac{m}{V}$$

where m is the mass of the substance having volume V. The volume is a measure of the space occupied and m is a measure of the ability of the particles of the substance to resist acceleration. The units of density correspond to those of mass and volume.

The weight W and the mass m of a substance are related to each other by the expression $W = mg$. Dividing this expression throughout by volume V of the working substance, we obtain

$$\frac{W}{V} = \frac{mg}{V} \quad \text{or} \quad w = \rho g$$

The above identity reveals that specific weight w changes with location depending upon gravitational pull.

* **Specific volume (v)** represents the volume per unit mass of the substance; specific volume is the reciprocal of mass density

$$v = \frac{V}{m}; \quad v = \frac{1}{\rho}$$

The specific volume has the units m^3/kg.

The concept of specific volume is found to be practically more useful in the study of flow of gases and vapours.

* **Specific gravity (s)** refers to the ratio of specific weight (or mass density) of a fluid to the specific weight (or mass density) of a standard fluid. For liquids the standard fluid is water at 4°C, and for gases the standard fluid is taken either air at 0°C or hydrogen at the same temperature. Specific gravity is dimensionless and has no units.

A statement that specific gravity of mercury is 13.6 implies that mercury is 13.6 times heavier than water.

1.9. WORK, ENERGY AND POWER

Work is done when the point of application of force moves in the direction of force. The amount of work equals the product of force and displacement in the direction of force.

$$\text{Work done} = \text{force} \times \text{distance}; \quad W = FL$$

where F is the force, W is the work done and L is the distance moved. The unit of work is joule (J) which is defined as the work done when a force of one newton is exerted through a distance of one metre in the direction of the force.

$$1 \text{ J} = 1 \text{ Nm}$$

Energy is defined as the capacity to produce a change from existing conditions, *i.e.*, capacity to exert force through a distance and do work. Energy cannot be seen; however its presence can be felt by properties of the system. When energy is added to or subtracted from the system, there occurs a change in one or more properties of the system. The energy of a system may take the form of

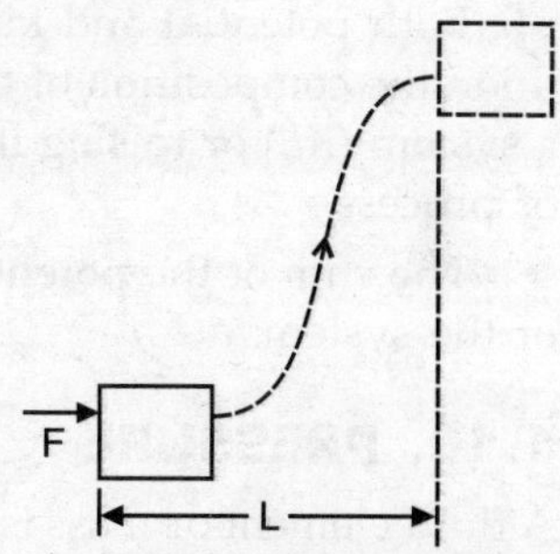

Fig. 1.8. Concept of work as used in mechanics

(*i*) ***Stored energy,*** *i.e.*, energy contained within the system boundaries. Examples are kinetic energy, potential energy and internal energy.

(*ii*) ***Energy in transit,*** *i.e.*, the energy which crosses the system boundaries. Examples are heat energy, mechanical energy and electrical energy.

Whatever may be the form of the energy, its unit is Joule (J) or kilo joule (kJ). The energy per unit mass is the specific energy, the unit of which is kJ/kg. The industrial unit of energy is kilowatt hour (kWh) which is equivalent to work of one kilowatt for one hour.

$$1 \text{ kWh} = 3.6 \times 10^6 \text{ J} = 3600 \text{ kJ}$$

Power is the rate of work or energy transfer. The unit of power is watt (W), kilowatt (kW) or megawatt (MW).

$$1 \text{ W} = 1 \text{ J/s} = 1 \text{ Nm/s}$$

$$1 \text{ kW} = 10^3 \text{ W} \text{ and } 1 \text{ MW} = 10^6 \text{ W}$$

Potential Energy and Kinetic Energy

Potential energy is the energy possessed by a system due to its position or elevation; it represents the ability to do work against the body weight in lifting it from the surface of earth.

When the height of mass changes from elevation z_1 to z_2, the change in potential energy (ΔPE) equals the work done on the system in raising it from lower elevation to higher elevation and is given by

$$\Delta PE = mg(z_2 - z_1)$$

Potential energy of a system is taken zero when its mass rests upon the earth.

Kinetic energy is the energy possessed by a system by virtue of its motion.

If the system experiences a change in velocity from V_1 to V_2, then,

$$\text{Change in kinetic energy, } \Delta KE = m\int_{V_1}^{V_2} V\,dV = \frac{1}{2}m\left(V_2^2 - V_1^2\right)$$

If kinetic energy of the system at rest with respect to the earth is arbitrarily assigned a zero value, then

$$\text{Kinetic energy, } KE = \frac{1}{2}mV^2$$

where V is the velocity of the system relative to the earth.

Both potential and kinetic energies are external forms of energy since they do not depend upon the composition of the fluid comprising the system. Further, the work done in accelerating a system (KE) or raising it (PE) gets stored and can be later recovered by reversing the direction of process.

The sum of the potential energy and kinetic energy of a system is called **mechanical energy** of the system.

1.10. PRESSURE

A fluid element or mass is essentially acted upon by two categories of forces: body forces and surface forces. ***Body forces*** on fluid elements are caused by agencies such as gravitational, electric or magnetic fields. The magnitude of these forces is proportional to the mass of the fluid. ***Surface forces*** represent the action of the surrounding fluid on the element under consideration through direct contact. A force acting on portion dA of a surface can be resolved into two components; one normal and other tangential to the area dA. The normal force per unit area is called ***intensity of pressure*** (or simply pressure). From continuum point of view, the pressure at a point may then be defined as

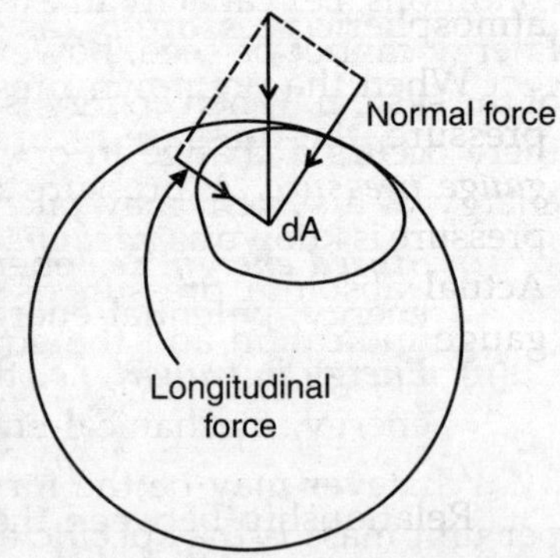

Fig. 1.9. Concept of pressure

$$p = \underset{A \to 0}{Lt} \frac{F}{A} = \frac{dF}{dA}$$

where dF is the force acting normal to the infinitesimal area dA. If the total normal force F acts uniformly over the entire area A, then $p = F/A$. Pressure is usually expressed in N/m^2 or pascal (Pa).

$$1 \text{ pascal (Pa)} = 1 \text{ N/m}^2$$

i.e., pascal represents the force of one newton acting on an area of 1 m^2.

The unit of pascal is very small. Engineers express pressure in kilopascal (kPa) and Megapascal (MPa)

$$1 \text{ kPa} = 10^3 \text{ Pa}; \ 1 \text{ MPa} = 10^6 \text{ Pa}$$

Some other commonly used units of pressure measurement are:

$$1 \text{ bar} = 10^5 \text{ N/m}^2 = 10^5 \text{ Pa}$$

$$1 \text{ standard atmosphere (atm)} = 1.01325 \text{ bar} = 101325 \text{ Pa}$$
$$= 760 \text{ mm of mercury} = 760 \text{ Torr}$$

The pressure exerted by 1 mm column of mercury is named as Torr. Usually pressure below atmospheric pressure is expressed in Torr

$$750 \text{ mm of mercury} = 1 \text{ bar} = 100 \text{ kPa}$$

The following terms are generally associated with measurement of pressure.

(*i*) ***Atmospheric pressure*** (p_{at}): This is the pressure exerted by the envelope of air surrounding the earth's surface. The standard atmospheric pressure is defined as the pressure produced by a column of mercury 760 mm high; the specific gravity of mercury being 13.5951 kg/m^3 and the acceleration due to gravity being 9.80665 m/s^2. From hydrostatic equation

$$p_{at} = \rho g h = (13.5951 \times 1000) \times 9.80665 \times \frac{760}{1000}$$
$$= 101325 \text{ N/m}^2 = 1.01325 \times 10^5 \text{ Pa} = 1.01325 \text{ bar}$$

(*ii*) ***Absolute pressure*** (p_{abs}) : Pressure has been defined as the force per unit area due to interaction of fluid particles amongst themselves. A zero pressure intensity will occur when molecular momentum is zero. Such a situation can occur only when there is a perfect vacuum, *i.e.*, a vanishingly small population of gas molecules or of molecular velocity. Pressure intensity measured from this state of vacuum or zero pressure is called absolute pressure.

(*iii*) ***Gauge pressure*** (p_g) and ***vacuum*** (p_{vac}) : Instruments and gauges used to measure fluid pressure generally measure the difference between the unknown pressure p and the existing atmospheric pressure p_{at} as shown in Fig. 1.10.

When the unknown pressure is more than atmospheric pressure, the pressure recorded by the instrument is called ***gauge pressure***. A pressure reading below the atmospheric pressure is known as ***vacuum, rarefaction*** or ***negative pressure***. Actual absolute pressure is then the algebraic sum of the gauge indication and the atmospheric pressure

$$p_{abs} = p_{at} + p_g$$
$$p_{abs} = p_{at} - p_{vac}$$

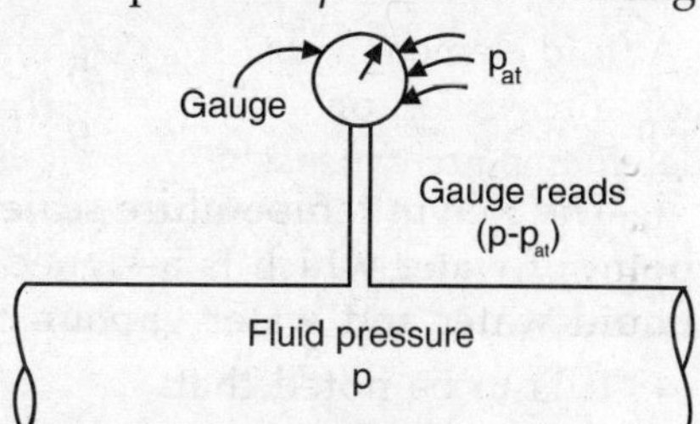

Fig. 1.10. Gauge pressure

Relationship between these pressure terms has been illustrated in Fig. 1.11.

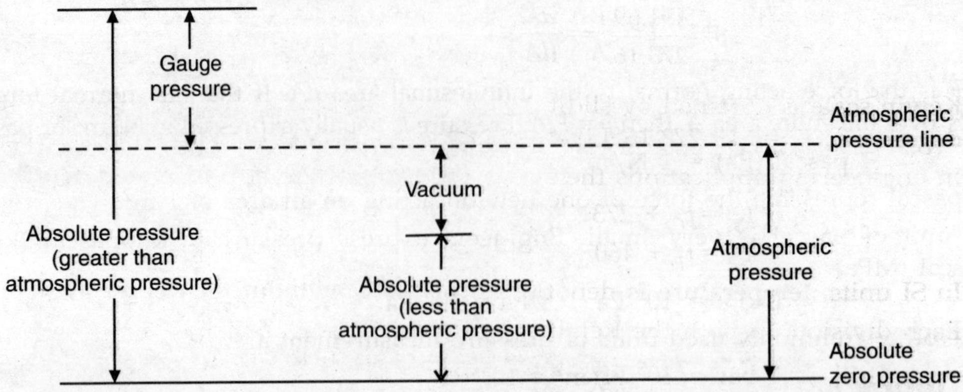

Fig. 1.11. Relationship between absolute, gauge, and atmospheric pressure

1.11. TEMPERATURE

Temperature of a system is that property which determines the ability of the system to transfer energy in the form of heat to any other system in thermal contact. Two systems would be equal in temperature if no change occurs in any property when they are brought into communication.

Two systems which are in thermal equilibrium with a third system are in thermal equilibrium with each other. This is the essence of Zeroth law of thermodynamics. The Zeroth law, like other thermodynamic laws, is based on experience and does not have any mathematical proof. The Zeroth law provides the basis for temperature measurements.

Temperature scales are defined by assigning numbers to the ice point, to the steam point and to the equal spaced marks between them. The ***ice point*** is the equilibrium temperature of ice with water at standard atmospheric pressure and it is assigned a value of 0°C on Celsius scale (also called centigrade scale) and 30°F on Fahrenheit scale. The ***steam point*** is the equilibrium temperature of pure water with its own vapour at standard atmospheric pressure and it is assigned a value of 100°C on Celsius scale and 212°F on Fahrenheit scale. Between the steam and ice points, there are 100 equal parts of Celsius scale and 180 equal parts on Fahrenheit scale. The temperature measured on the Celsius and Fahrenheit scales are correlated by the expression

$$\frac{t_C}{100} = \frac{t_F - 32}{180}$$

or $$t_C = \frac{5}{9}(t_F - 32)$$

The Kelvin temperature scale has been defined with reference to a single point-the triple point of water which is assigned a value of 0.01°C. At the triple point of water: the ice, the liquid water and water vapour coexist in equilibrium state.

It is to be noted that:

(*i*) That Kelvin scale has a value of 273.16 at 0°C and a value of 373.16 at 100°C; it is sometimes called the absolute Celsius scale and is denoted by K. There is an equivalent Fahrenheit scale, called the Rankine scale R where the triple point of water is 491.69°R and ice point is 491.67°R. the Kelvin and Rankine scales are related as

$$\frac{T_R}{T_K} = \frac{491.69}{273.16} = \frac{180}{100} = \frac{9}{5}$$

(*ii*) Kelvin scale is obtained by sliding each value of the Celsius scale upward by 273.16 degree, *i.e.*, $T_K = t_C + 273.16$. Likewise in Fahrenheit and Rankine scale; $T_R = t_F + 459.67$

(*iii*) In engineering applications the above correlations are approximated as

$$T_K = t_C + 273$$
$$T_R = t_F + 460$$

In SI units, temperature is denoted simply by K without the degree symbol.

(*iv*) Each division on both the Kelvin and Celsius scale is equal.

1.12. WORK

The thermodynamic definition of work relates to the concept of system and its surroundings.

(*i*) Work refers to the energy transferred (without transfer of mass and with the help of a mechanism) because of a property difference, other than temperature, that exists between the system and its surroundings.

Work is the energy transfer by reasons other than temperature difference.

(*ii*) Work refers to the interaction between system and surroundings whose sole effect, external to the system, could be reduced to the lifting of mass against the gravitational force.

Consider a system formed by a certain amount of gas contained in a piston cylinder arrangement. Let the piston move outward through a small distance dx ($dx \to 0$) during a small interval dt. Since the piston moves through a small distance, the pressure p acting on the face of the piston can be assumed to remain constant. A single pressure gauge will then indicate the system pressure p, which also prevails over the entire system boundary. The infinitesimal work done by the system is

$$\delta W = \text{force} \times \text{distance moved}$$
$$= (\text{pressure} \times \text{area of piston}) \times \text{distance moved}$$
$$= pA\,dx$$

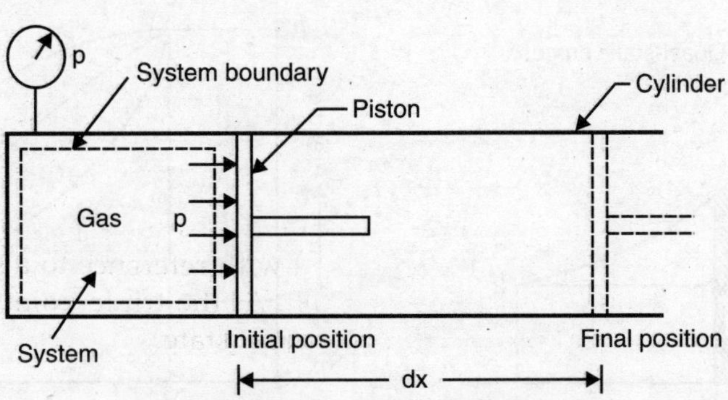

Fig. 1.12. Piston cylinder arrangement

It is to be noted that $A\,dx$ represents the infinitesimal volume dV swept by the piston.

$$\therefore \qquad \delta W = p \times (A\,dx) = p\,dV$$

If the piston moves through a finite distance, the work done by the piston can be evaluated by integrating the above expression between the initial and final states.

Reference Fig. 1.13, the expansion of the gas in the piston cylinder assembly can be reduced to the raising of a mass m against gravity by manipulating surroundings external to the system.

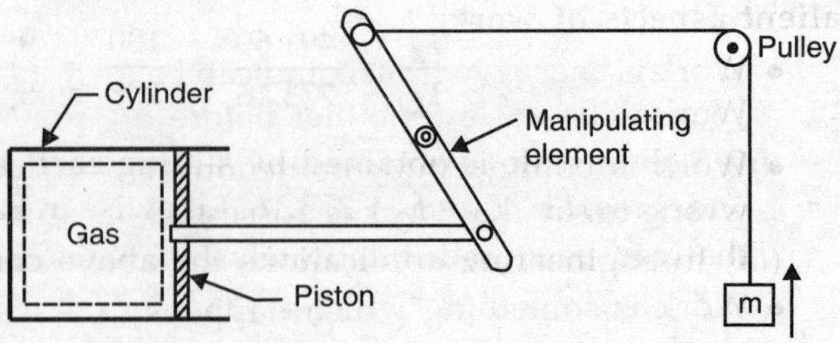

Fig. 1.13. Work done equivalent to raising of mass

1.13. DISPLACEMENT WORK : *p dV*

The expansion of the system (gas) by outward movement of the piston can be represented on p-V diagram like 1-2 (Fig 1.14). The shaded area represents the work done due to small movement of the piston and the total work done by the gas during expansion process is

$$W = \int_1^2 p\, dV$$

Inspection of the p-V plot (Fig. 1.15) shows that mere specifying the end states 1 and 2 does not determine the area (or work); the nature of the curve needs to be known also. The curve may be arched upwards or it may sag downwards, and the area under the curve will vary accordingly. For the same initial and final states, the work done by the system in following the paths A, B and C is different. Obviously work is a path function and not a point function. Accordingly the work transfer across the system boundaries is not classified as a thermodynamic property.

The integral $\int_1^2 p\, dV$ can be evaluated, and hence computations of work done made, only if we know the relation between p and V or the path of the process.

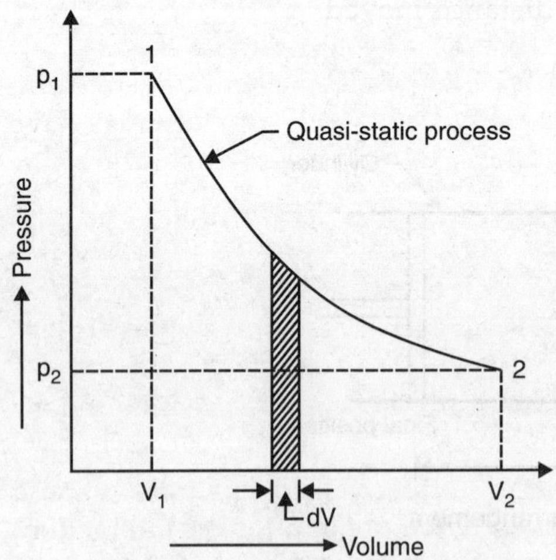

Fig. 1.14. Quasi-static p dV work

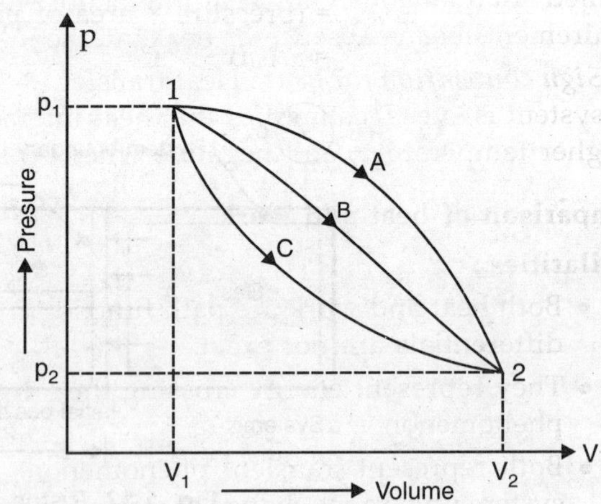

Fig. 1.15. Work-a path function

Salient aspects of work:

- Work is energy in transit and can be identified only when the process is being executed. Work does not exist either before or after the interaction.
- Work is boundary phenomenon; it is an interaction across the boundary. It will be wrong to say that work is possessed or stored by the system. It is the transferred energy that can be stored indefinitely by the system and not work.
- Work is something which happens to a system but it is not the characteristic of the system.

Being an inexact differential, work interaction is represented as δW instead of dW. The identity $\int_1^2 p\,dV$ for work done is valid only if

- the system is closed
- the process is quasi-static, *i.e*, frictionless and reversible.
- there are no viscous effects within the system
- the effects due to gravity, electric current and magnetization are negligible

$$W \neq \int p\,dV \text{ for a non-quasi-static and irreversible process.}$$

Sign convention for work : When work is done by the system it is arbitrarily taken to be positive, and when work is done on the system it is taken to be negative.

1.14. HEAT

Heat is a form of energy that is transferred (without transfer of mass) across the boundaries of a system because of temperature difference between the system and its surroundings, and is always from a higher temperature to a lower temperature. No heat interaction is possible between the system and its surroundings if both are in thermal equilibrium, *i.e.* at the same temperature. Two conditions, namely temperature difference and communication, must be fulfilled simultaneously for heat interaction between two systems to occur; absence of either requirement means absence of heat interaction.

Sign convention for heat : Heat transfer to the system is taken +ve while heat transfer from the system is –ve. This implies that heat interaction is positive when the surroundings have a higher temperature, and negative when the surroundings have a lower temperature.

Comparison of heat and work

Similarities :

- Both heat and work are path functions; they are not properties of the system and their differentials are not exact.
- They represent energy crossing the system boundary; and hence they are a boundary phenomenon.
- Both represent transient phenomenon; these energy interactions occur only when a system undergoes change of state. They don't exist either before or after the interaction.
- Both are something which happen to a system but they are not the characteristics of the system.
- Concepts of heat and work are associated not with a 'store' but with a 'process'.

Dissimilarities :

- There cannot be work transfer in a stable system. No such restriction exists for heat interaction.
- Heat is energy interaction due to temperature difference only; work is by reasons other than temperature difference.
- Sole effect external to the system could be reduced to the raising of mass against gravitational force in case of work. However, during heat transfer other effects are also observed.
- Heat is a low grade energy whereas work is a high grade energy.

1.15. SPECIFIC HEAT AND HEAT CAPACITY

Whether a system has received or rejected heat is indicated by a change in its temperature; the temperature grows with the receipt of heat and diminishes with heat rejection. The quantity of heat received or rejected by a system during the process of heating or cooling is measured in terms of the product of mass, specific heat and temperature changes.

$$Q = m\,c\,(t_2 - t_1)$$

where Q is the heat gained by the system in kJ; m is the mass of the substance in kg; $(t_2 - t_1)$ is the temperature rise in degree centigrade or kelvin; and c is an experimental factor called ***specific heat*** and expressed in kJ/kgK.

The specific heat may be defined as the quantity of heat that enters or leaves a unit mass of the substance when it experiences 1 degree change in temperature.

The product $(m \times c)$ is separately termed as ***heat capacity*** or ***water equivalent*** of the system.

It has been predicted from experimental results that the value of specific heat depends upon

– substance comprising the system;
– condition under which the change of state occurs; and
– state of the system at the time of heat interaction.

The specific heat for liquids and solids varies primarily with temperature and does not differ much with nature of the process. In case of gases, specific heat is also dependent on the path of the thermodynamic process which may be carried out isochorically (at constant volume), isobarically (at constant pressure) or polytropically. Accordingly, the specific heats for liquids and solids are quoted without any mention of the process involved in heat exchange. But for gases, it is the usual practice to quote two specific heats namely specific heat at constant pressure and specific heat at constant volume.

The ***specific heat at constant volume*** (c_v) is defined as the amount of heat added at constant volume to change the temperature of unit mass of the gas by 1 degree. Likewise for heat addition at constant pressure, the quantity of heat required for raising the temperature of unit mass of the gas through 1 degree temperature difference is called ***specific heat at constant pressure*** (c_p). These specific heats are properties of the gas, and are really not constant but vary both with pressure and temperature. However, for moderate changes in pressure and temperature, they are assumed constant.

For water: $c = 4.186$ kJ/kg-deg

For air : $c_p = 1.005$ kJ/kg-deg

$$c_v = 0.718 \text{ kJ/kg-deg}$$

The ratio $\frac{c_p}{c_v}$ is given the symbol γ and is called adiabatic exponent. For air $\gamma = 1.4$.

1.16. IDEAL GAS

An ideal (or perfect) gas has no intermolecular forces of attraction or repulsion, does not change its phase during a thermodynamic process and obeys a set of common rules governing change of its properties.

- Boyle's law provides a reciprocal relationship between pressure and volume of a gas at constant temperature.

$$p \propto \frac{1}{v} \quad \text{or} \quad pv = \text{constant} \qquad \text{(at constant temperature)}$$

- Charles law provides direct relationship
  (*a*) between volume and temperature of a gas at constant pressure

$$v \propto T \quad \text{or} \quad \frac{v}{T} = \text{constant} \qquad \text{(at constant pressure)}$$

  (*b*) between pressure and temperature of a gas at constant volume

$$p \propto T \quad \text{or} \quad \frac{p}{T} = \text{constant} \qquad \text{(at constant volume)}$$

The ideal equation of a state is given by

$$pv = RT \quad \text{or} \quad pV = mRT$$

where v is the specific volume, V is the total volume and R is the gas constant for the particular gas.

If p is in N/m^2, V is in m^3, m is in kg and T is in degrees kelvin, the units of the gas constant R work out to be Nm/kgK or J/kgK.

The gas constant R, though constant for a particular gas, has different values for different gases. For air R = 287 J/kgK.

In practice, no gas obeys the ideal gas equation rigidly. It is essentially valid at low pressure and high temperatures where the density is sufficiently low. An ideal gas has no forces of inter molecular attraction. It obeys the law $pv = RT$. The specific heat capacities are not constant but are functions of temperature.

The gas constant multiplied by molecular mass (MR) is a constant for all gases. This constant is called the ***universal*** or ***molar gas constant,*** and it is equal to

$$R_0 = MR = 8314 \text{ J/kg mol K}$$
$$= 8.314 \text{ J/mol K} = 8.314 \text{ kJ/kg mol K}$$

The value of gas constant R for any gas can be calculated by dividing the universal gas constant by molecular mass of gas. For example, for nitrogen (molecular mass 28)

$$R_{N_2} = \frac{R_0}{M_{N_2}} = \frac{8314}{28} = 296.9 \text{ J/kgK}$$

- **Avogadro's hypothesis :** Equal volumes of different gases at the same pressure and temperature contain equal number of molecules.

The mass of 22.41 m^3 of hydrogen at NTP (atmospheric pressure 1.01325 bar and 0°C) is equal to its molecular mass in kg. For example

22.41 m^3 of hydrogen at NTP has mass of 2 kg

A 22.4 m^3 of volume is known as a mole. Obviously mole is a bigger unit of volume.

- The Van der Waal's equation of state is prescribed as

$$\left(p+\frac{a}{v^2}\right)(v-p)=RT$$

where a and b are the constants for the gas in question, v represents the volume per unit mass and R is the gas constant.

1.17. FIRST LAW OF THERMODYNAMICS

It is special case of the law of conservation of energy and stipulates that:

Energy can neither be created nor destroyed but only converted from one form to another. Total energy of an isolated system, in all its forms remains constant. For any closed system taken through a cyclic process, the network delivered to surroundings is equal to the net heat taken from the surroundings. That is

$$\oint \delta Q = \oint \delta W$$

- The energy of a system comprises of potential energy, kinetic energy, internal energy and terms accounting for electricity, magnetism and surface tension. *Internal energy* is the microscopic energy, *i.e.*, the energy stored in the molecular motions. It is the energy stored in the molecular structure due to heat and work interactions. The change in energy of a system depends on the end states and is independent of the path of process. It is a point function and an exact differential. Thus energy is a property of the system. In a isolated system, the energy of the system remains constant.
- Consider the non-flow energy equation: $\delta Q = \delta W + dU$
  - (*a*) δQ, δW and δU have to be in consistent units
  - (*b*) Heat supplied to gas, work done by gas and increase in internal energy are taken as +ve quantities.
  - (*c*) Heat received by gas, work done on gas and decrease in internal energy are taken as –ve quantities.
- A perpetual motion machine of first kind is a hypothetical machine which produces continuous supply of work without absorbing energy from the surroundings.
- No machine can produce energy without corresponding expenditure of energy. As such a perpetual motion machine of the first kind is impossible.
- The internal energy of a gas is function of temperature only (Joule's law) and for any reversible process

$$\text{Change in internal energy } (U_2 - U_1) = m\, c_v\, (T_2 - T_1)$$

- Enthalpy is the property of the system defined by

$$H = U + pV$$

- The following useful correlations are helpful in the solution of problems:

$$c_v = \left(\frac{dU}{dT}\right)_v; \quad c_p = \left(\frac{dh}{dT}\right)_p; \quad \frac{c_p}{c_v} = \gamma \quad \text{and} \quad (c_p - c_v) = R$$

Further for any polytropic process: pV^n = constant

$$\frac{T_2}{T_1} = \left(\frac{p_2}{p_1}\right)^{\frac{n-1}{n}} = \left(\frac{V_1}{V_2}\right)^{n-1}$$

For a brief summary of non-flow expansion process, refer Table A given below:

Table A: Summary of Non-flow Expansion Process 1–2

Process	*Work output*	*Heat received*
1. Isochoric (Constant volume)	$W_{1-2} = 0$	$Q_{1-2} = U_2 - U_1 = mc_v (T_2 - T_1)$
2. Isobaric (Constant pressure)	$W_{1-2} = p \int dV$	$Q_{1-2} = H_2 - H_1 = mc_p (T_2 - T_1)$
3. Isothermal (Constant temperature)	$W_{1-2} = p_1 V_1 \log_e \frac{V_2}{V_1}$	$Q_{1-2} = W_{1-2}$
4. Adiabatic	$W_{1-2} = \frac{p_1 V_1 - p_2 V_2}{\gamma - 1} = \frac{mR(T_1 - T_2)}{\gamma - 1}$	$Q_{1-2} = 0$
5. Polytropic (pV^n = Const)	$W_{1-2} = \frac{p_1 V_1 - p_2 V_2}{n - 1} = \frac{mR(T_1 - T_2)}{n - 1}$	$Q_{1-2} = W_{1-2} + (U_2 - U_1) = \frac{\gamma - n}{\gamma - 1} \times W_{1-2}$

Change in enthalpy and heat interactions are different quantities. For the solitary case of constant pressure change in enthalpy equals the heat interaction

$$Q_{1-2} = H_2 - H_1 = m\, c_p (T_2 - T_1)$$

1.18. STEADY FLOW ENERGY EQUATION

The mass flow rate $\dot{m}$ is the quantity which passes a reference point per unit time; its value is expressed in kg/s

$$\dot{m} = \rho AV = \frac{AV}{v} \quad \text{where } V \text{ is the flow velocity}$$

For steady flow (no accumulation or diminution of matter within the machine)

$$\dot{m} = \frac{A_1 V_1}{v_1} = \frac{A_2 V_2}{v_2}$$

A *flow process* constitutes an open system through which the working fluid enters and leaves from the surface of the system.

Flow work or *flow energy* refers to the work which a fluid does when it moves in or out of any passage. It represents the product of pressure and specific volume of the flowing fluid.

The general energy equation for flow through a machine is written as

$$m_1\left(u_1 + p_1v_1 + \frac{V_1^2}{2} + gz_1\right) + Q = m_2\left(u_2 + p_2v_2 + \frac{V_2^2}{2} + gz_2\right) + W_s$$

For steady flow $m_1 = m_2$ and by definition $h = u + pv$

Then,

$$m\left(h_1 + \frac{V_1^2}{2} + gz_1\right) + Q = m\left(h_2 + \frac{V_2^2}{2} + gz_2\right) + W_s$$

The steady flow energy equation can be written on the basis of unit mass or on the basis of unit time.

On unit mass basis:

$$m\left(h_1 + \frac{V_1^2}{2} + gz_1\right) + Q = m\left(h_2 + \frac{V_2^2}{2} + gz_2\right) + W_s$$

- Each term inside the brackets is expressed per unit mass and has units of energy
- Normally : Q and W_s are in kJ
  m is in kg; h is in kJ/kg
  V is in m/s ; z is in m and g is in m/s^2

$$m\left(h_1 + \frac{V_1^2}{2\times 10^3} + \frac{gz_1}{10^3}\right) + Q = m\left(h_2 + \frac{V_2^2}{2\times 10^3} + \frac{gz_2}{10^3}\right) + W_s$$

On unit time basis:

$$\dot{m}\left(h_1 + \frac{V_1^2}{2\times 10^3} + \frac{gz_1}{10^3}\right) + \dot{Q} = \dot{m}\left(h_2 + \frac{V_2^2}{2\times 10^3} + \frac{gz_2}{10^3}\right) + \dot{W}_s$$

where all terms represent energy flow per unit time (kJ/s)

1.19. SECOND LAW OF THERMODYNAMICS

1.19.1. Kelvin-Planck Statement

"It is impossible to construct an engine that operates in a cycle and produces no effect other than work output and exchange of heat with a single heat reservoir"

The statement implies that no heat engine can be developed that receives a certain amount of heat from a high temperature source and converts that into an equivalent amount of work. *i.e.*, $W = Q$. The thermal efficiency of such a engine $\eta = W/Q = 1$ or 100 percent. Fig. 1.16(*a*) represents the schematic arrangement of a heat engine that exchanges heat with a single heat source and is 100 percent efficient. Such a system satisfies the principle of energy conservation (1st law) but violates the Kelvin statement of second law. Obviously Kelvin-Planck statement tells that no heat engine can have thermal efficiency equal to 100 percent.

The only alternative for continuous power output from a heat engine is that a portion of the heat received must be rejected to a heat reservoir at low temperature (heat sink). This engine receives Q_1 units of heat, rejects Q_2 units of heat and converts $(Q_1 - Q_2)$ units of heat into work per cycle. All possible heat engines conform to this representation (Fig. 1.16*b*).

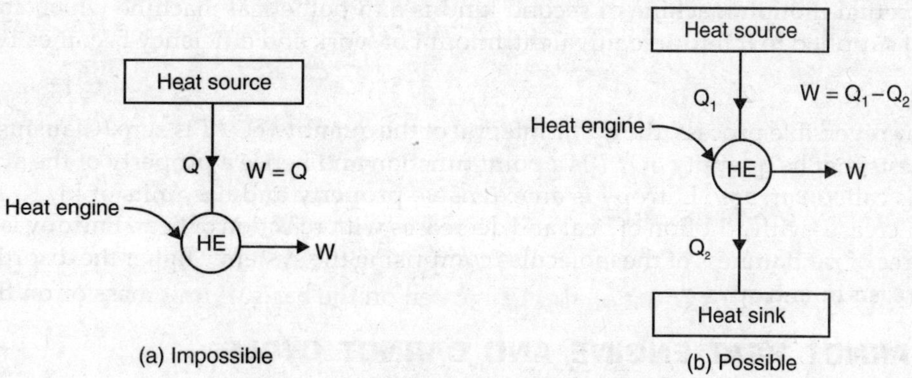

Fig. 1.16. Schematic representation of a heat engine

1.19.2. Clausius Statement

"It is impossible to construct a device that operates in a cycle and produces no effect other than the transfer of heat from a system at low temperature to another system at high temperature".

The statement implies that heat cannot flow of itself from a system at low temperature to a system at high temperature. The schematic arrangement that is prohibited by Clausius statement is shown in Fig 1.17(*a*). The coefficient of performance of such an arrangement equals: $COP = Q/W = Q/0 = \infty$. Obviously the Clausius statement tells that *COP* of a heat pump/refrigerator cannot be equal to infinity.

The only alternative for the transfer of heat from low temperature to high temperature level is that some external work must be supplied to the machine as shown in Fig. 1.17(*b*).

Whereas the Kelvin-Planck statement is applied to heat engines, the Clausius statement concerns heat pumps and refrigerators. Both the Kelvin-Planck and Clausius statements are negative statements, they have no mathematical proof. The law is based on experimental observations, and to-date no observation has been made that contradicts the law and this aspect is taken as sufficient evidence of its validity.

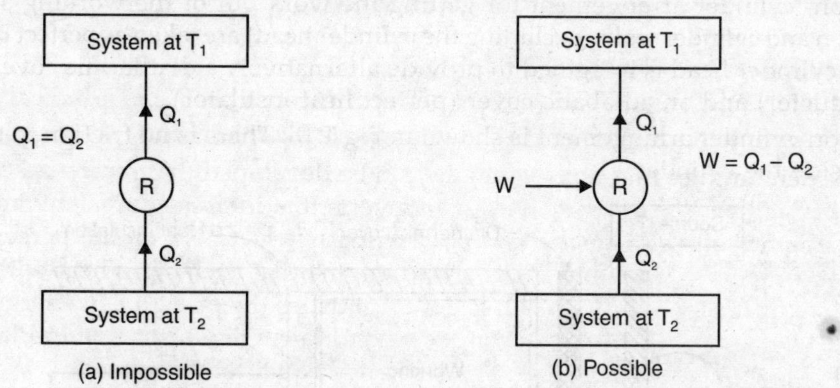

Fig. 1.17. Schematic representation of a heat pump/refrigerator

Though worded differently, both these statements are interlinked and complimentary to each other. Any device that violates Clausius statement leads to violation of Kelvin-Planck statement and vice versa.

A perpetual motion machine of second kind is a hypothetical machine which converts the entire heat supplied to it into an equivalent amount of work and efficiency becomes 100 percent.

Entropy

For any reversible process, the cyclic integral of the quantity $\delta Q/T$ is zero (Clausius theorem). This suggests that he quantity $\delta Q/T$ is a point function and hence a property of the system. This property is called ***entropy***. Entropy is an extensive property and has units of kJ/K. Entropy of a system increases with addition of heat and decreases with rejection of heat. Entropy is a measure of the degree of randomness of the molecules comprising the system. Higher the disorder, greater is the increase in entropy.

1.20. CARNOT HEAT ENGINE AND CARNOT CYCLE

A heat engine is a thermodynamic device used for continuous production of work from heat when operating in a cyclic process. Both heat and work interactions take place across the boundary of this cyclically operating device.

A heat engine is characterised by the following features :

- reception of heat Q_1 from a high temperature source at T_1
- partial conversation of heat received to mechanical work W
- rejection of remaining heat Q_2 to a low temperature sink at temperature T_2
- cyclic/continuous operation and
- working substance flowing through the engine.

A carnot cycle is a hypothetical cycle consisting of four distinct processes : two reversible isothermal processes and two reversible adiabatic processes. The cycle was proposed in 1824 by a young French engineer, Sadi Carnot.

The essential elements needed for making an analysis of this cycle are :-

- a working substance which is assumed to be a perfect gas
- two heat reservoirs; the hot reservoir (heat source) at temperature T_1 and the cold reservoir (heat sink) at temperature T_2
- piston- cylinder arrangement for getting the work out of the working substance. The piston and cylinder walls (excluding the cylinder head) are taken as perfect heat insulators. The cylinder head is imagined to provide alternatively a diathermic cover (perfect heat conductor) and an adiabatic cover (perfect heat insulator).

The piston-cylinder arrangement is shown in Fig 1.18. There is no friction to the movement of piston inside the cylinder.

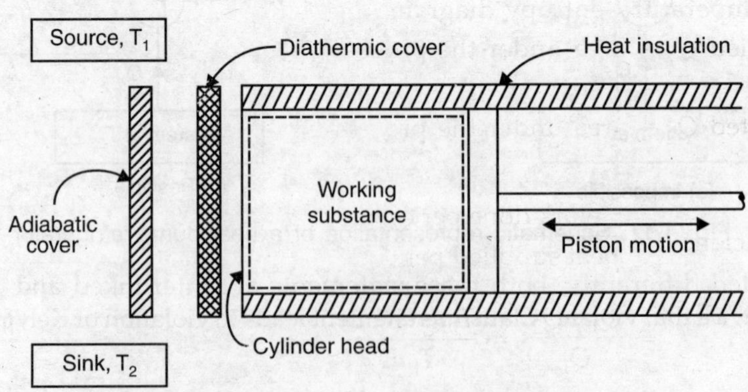

Fig. 1.18. Essential elements for a Carnot cycle

The sequence of operation for the different processes constituting a Carnot cycle is :

Isothermal expansion ($a-b$) : The heat is supplied to the working fluid at constant temperature T_1. This is achieved by bringing the heat source in good thermal contact with the cylinder head through diathermic cover. The gas expands isothermally from state point a (p_a, V_a) to state point b.

Adiabatic expansion ($b-c$) : At the end of isothermal expansion (state point b), the heat source is replaced by adiabatic cover. The expansion continues adiabatically and reversibly upto state point c (p_c, V_c). Work is done by the working fluid at the expense of internal energy and its temperature falls to T_2 at state point c.

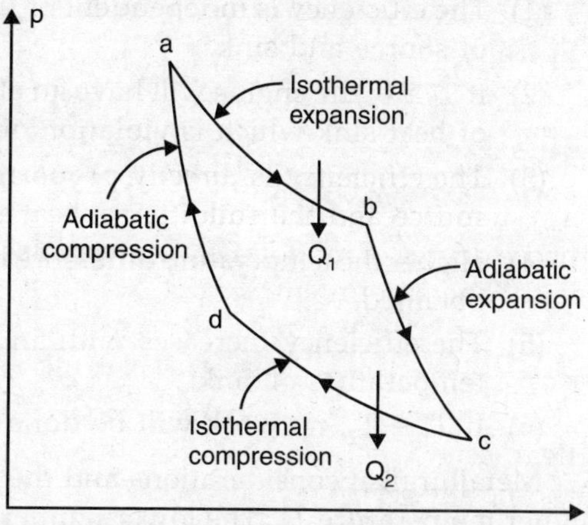

Fig. 1.19. Carnot cycle on P-V plot

Isothermal compression ($c-d$) : After state point c, the piston starts moving inwards and the working fluid is compressed isothermally at temperature T_2. The constant temperature T_2 is maintained by removing the adiabatic cover and bringing the heat sink in contact with the cylinder head. The compression continues upto state point d and this position of the piston is so chosen that it lies on the path of reversible adiabatic drawn from state point a. The working fluid loses heat to the sink and its amount equals the work done on the working fluid.

Adiabatic compression ($d-a$) : At the end of isothermal compression (state point d), the heat sink is removed and is replaced by adiabatic cover. The compression now proceeds adiabatically and reversibly till the working fluid returns back to its initial state point a. Work is done on the working fluid, the internal energy increases and temperature is raised to T_1.

Since all the processes that constitute a Carnot cycle are reversible, the Carnot cycle is referred to as a reversible cycle. Further, a cyclic heat engine working on the Carnot cycle is called Carnot engine and its thermal efficiency is given by

$$\eta = \frac{\text{net work output}}{\text{heat input}} = \frac{W_{net}}{Q_1}$$

For a closed cycle, the net work obtained is equal to difference between the heat Q_1 supplied by the source and the heat Q_2 rejected to the sink.

From the temperature–entropy diagram (Fig. 1.20)

Heat supplied Q_1 = area under the process line $a - b$

$= T_1\, ds$

Heat rejected Q_2 = area under the process line $c - d$

$= T_2\, ds$

$$\text{Efficiency} = \frac{\text{work done per cycle}}{\text{heat supplied per cycle}}$$

$$= \frac{Q_1 - Q_2}{Q_1} = \frac{T_1\, ds - T_2\, ds}{T_1\, ds} = \frac{T_1 - T_2}{T_1}$$

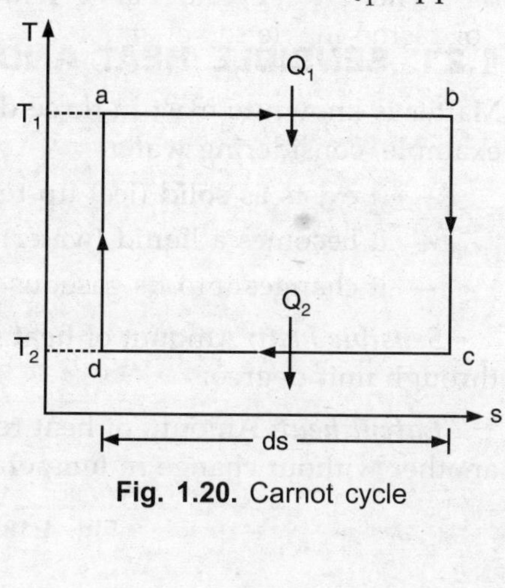

Fig. 1.20. Carnot cycle

Following conclusions can be made with respect to efficiency of a Carnot engine :

(1) The efficiency is independent of the working fluid and depends upon the temperatures of source and sink.

(2) If $T_2 = 0$, the engine will have an efficiency of 100 percent. However, that means absence of heat sink which is violation of Kelvin-Planck statement of the second law.

(3) The efficiency is directly proportional to temperature difference $(T_1 - T_2)$ between the source and the sink.

(4) Higher the temperature difference between source and sink, the higher will be the efficiency obtained.

(5) The efficiency increases with an increase in temperature of source and a decrease in temperature of sink.

(6) If $T_1 = T_2$, no work will be done and efficiency will be zero.

Metallurgical considerations and the high cost of temperature resisting materials limit the higher temperature T_1. The lower temperature T_2 is limited by atmospheric or sink conditions.

Carnot cycle gives the maximum possible thermal efficiency which can be obtained for the given temperature limits. The Carnot engine, however, is a hypothetical device and it is not possible to devise it due to the following reasons :

(*i*) All the four processes have to be reversible. This necessitates that the working fluid must have no internal friction between the fluid particles and no mechanical friction between the piston and cylinder walls.

(*ii*) The heat absorption and rejection have to take place with infinitesimal temperature differences. Accordingly, the rate of energy transfer will be very low and the engine will deliver only infinitesimal power.

(*iii*) For attaining isothermal operation, the piston movement is required to be very slow. However, the piston must move fast for the adiabatic process. A variation in the speed of the piston during different processes of a cycle is rather impossible.

(*iv*) There is insignificant difference in the slopes of isothermal and adiabatic lines. Consequently the p–V plot is greatly extended both in the horizontal and vertical directions. The cylinder then involves greater pressures and volumes, and as such becomes bulky and heavy.

This hypothetical device, however, serves as a yardstick or standard of perfection against which the performance of any practical heat engine can be compared.

1.21. SENSIBLE HEAT AND LATENT HEAT

Matter is known to exist in three different forms namely solid, liquid and vapour or gas. For example, considering water

– it exists as solid (ice) up to 0°C

– it becomes a liquid (water) at 0°C and remains so upto 100°C

– it changes into its gaseous phase (steam) at 100°C

Sensible heat: Amount of heat required to raise the temperature of unit mass of substance through unit degree.

Latent heat: Amount of heat required to convert unit mass of substance from one state to another without change in temperature.

Consider 1 kg of ice at – 10°C (below freezing point) and refer Fig. 1.21 for the changes in its characteristics when heat is supplied to it.

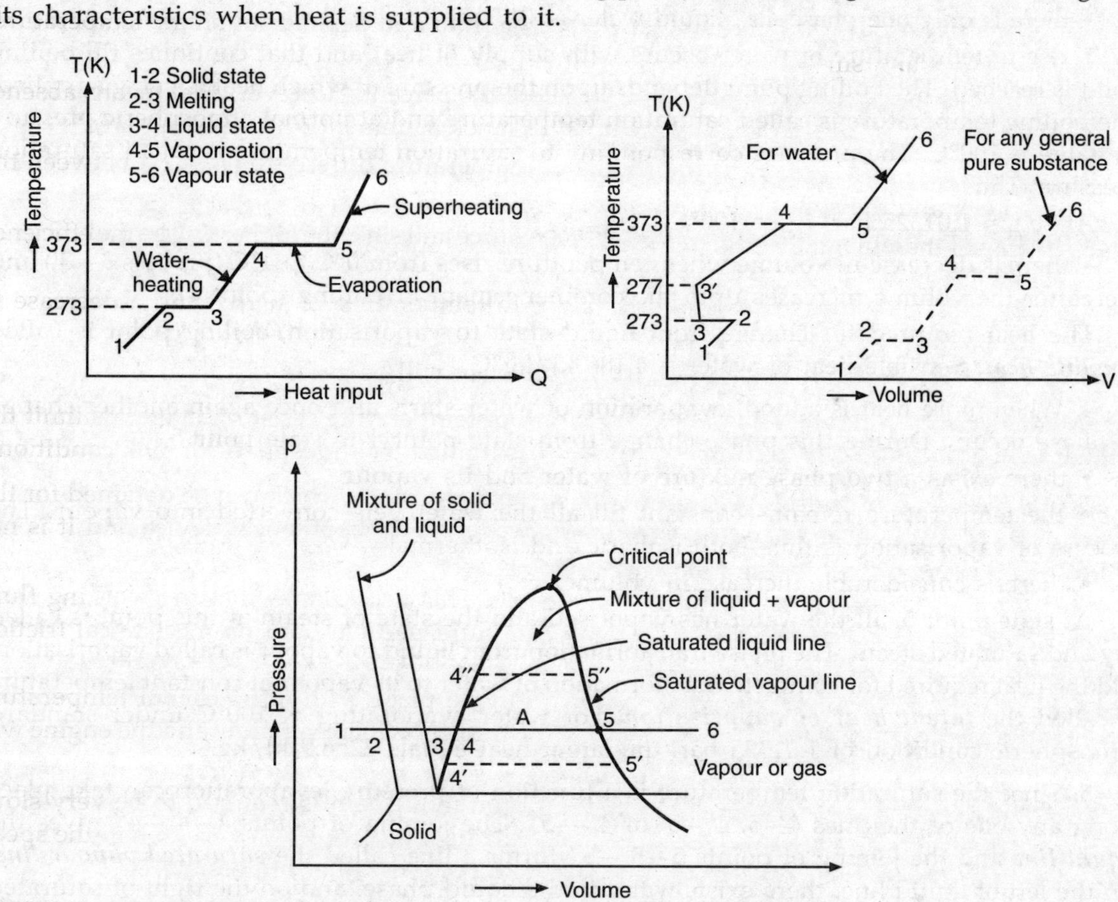

Fig. 1.21

1. The solid ice becomes warmer and the rise in temperature can be detected by sense of touch measured by thermometer. The temperature of the given ice would rise from – 10°C to 0°C and the heat absorbed by ice during this process is called ***sensible heat.*** 1 kg of ice takes about 2.1 kJ of heat to raise temperature through 1°C.

Sensible heat of ice = 2.1 kJ/kg°C

2. With further supply of a heat to solid ice at 0°C, melting starts (2 – 3). During the process of melting (phase change from solid to liquid),

- there exists a mixed phase of ice and water
- the temperature of mixed phase remains at 0°C
- there is nominal decrease in volume ; liquid water at 0°C is heavier than ice. However, for normal substances there is increase in volume when change occurs from solid phase to liquid phase.

This transformation of phase from solid to liquid is called the ***melting*** or ***fusion of ice,*** and the quantity of heat required for complete transformation of ice into water at constant temperature of 0°C is called ***latent heat of fusion.*** At normal atmospheric pressure of 1.01325 bar (1 atm), it equals 335 kJ/kg.

3. When the entire ice has melted (point 3)

• there is only one phase, *i.e.*, liquid water at 0°C.

• rise in temperature of water occurs with supply of heat and that continues till boiling point is reached. The boiling point depends upon the pressure at which heat is being supplied. The boiling temperature is called saturation temperature and at normal atmospheric pressure its value is 100°C. The pressure corresponding to saturation temperature is called saturation pressure. Thus

At p_{sat} = 1.01325 bar, t_{sat} = 100°C

• there is decrease in volume when temperature rises from 0°C to 4°C (process 3–3') and thereafter the volume increases upto the commencement of boiling (point 4).

The heat required for change from liquid state to vaporisation/boiling point is called ***sensible heat.*** Sensible heat of water = 4.186 kJ/kg°C

4. When more heat is added, evaporation of water starts and once again another change of phase occurs. During this phase change from state point 4 to state point 5,

• there exists a two phase mixture of water and its vapour

• the temperature remains constant till all the water gets converted into vapour. The process of vaporisation is thus both isobaric and isothermal

• there is considerable increase in volume.

At state point 5, all the water has vaporised and the state of steam at this point is called dry and saturated steam. The phase transformation from liquid to vapour is called vaporisation, and the heat required for complete transformation of water to its vapour at constant temperature is called the ***latent heat of vaporisation.*** For water evaporating at 100°C under standard atmospheric condition of 1.01325 bar, the latent heat equals 2256.9 kJ/kg.

5. Since the saturation temperature is a function of pressure, evaporation can take place along any one of the lines 4–5, 4′–5′ or 4″–5″. The joining of points 4–4′–4″ gives the ***liquid line*** and the joining of points 5–5′–5″ forms a line called the ***saturated vapour line.*** On the left of liquid line, there exists water in the liquid phase, and on the right of saturated vapour line the water is said to exist in the superheated phase. Between state points 4 and 5 there is a mixture of steam and water, *i.e.*, wet saturated steam with varying dryness fraction. A shift in the state point A towards the vapour line makes the steam drier and drier.

The saturated liquid curve is steeper compared to the dry saturated vapour line because the volume of liquid changes at much slower rate with increase in temperature.

6. The horizontal distance between saturated liquid line and the dry saturated vapour line becomes less and less with increase of pressure and saturation temperature. The point at which this distance becomes zero is called the ***critical point.*** Critical point represents the highest temperature and pressure above which a liquid cannot exist. At the critical point,

• the liquid is directly and fully converted into dry vapour without going into the phase of evaporation, *i.e.*, latent heat is zero

• the liquid and vapour are indistin-guishable

• the properties of saturated liquid and saturated vapour are identical.

The temperature and pressure at the critical point are called ***critical temperature*** t_c and ***critical pressure*** p_c.

For water, the critical values are :

critial temperature t_c = 374.14°C ; critical pressure p_c = 220.9 bar and critical volume v_c = 0.003155 m^3/kg.

7. Once the steam attains dry and saturated state, its temperature and volume start increasing with further supply of heat. The steam so obtained is called ***superheated steam*** and it behaves like a perfect gas.

8. Since supply of heat has been at constant pressure, the heat input can be taken equal to change in enthalpy. During different stages of the process of steam formation from ice at – 10°C, the enthalpy changes are as follows:

(*i*) Enthalpy changes from – 10°C to 0°C (process 1 – 2)

$$(dh)_{1-2} = c_i\, dt \text{ where } c_i \text{ is the specific heat of ice}$$

$$= 2.09\,\{0 - (-10)\} = 20.9 \text{ kJ/kgK}$$

(*ii*) Enthalpy change during melting of ice at 0°C (process 2 – 3)

$$(dh)_{2-3} = \text{latent heat of fusion of ice} = 335 \text{ kJ/kgK}$$

(*iii*) Enthalpy change during heating of water from 0°C to the boiling/saturation temperature 100°C (process 3 – 4)

$$(dh)_{3-4} = c_w\, dt \text{ where } c_w \text{ is the specific heat of water}$$

$$= 4.186\,(100 - 0) = 418.6 \text{ kJ/kgK}$$

(*iv*) Enthalpy change during vaporisation at 100°C (process 4 – 5)

$$(dh)_{4-5} = \text{latent heat of vaporisation}$$

$$= 2256.9 \text{ kJ/kgK}$$

(*v*) Enthalpy change during heating of dry saturated steam upto say 250°C. (process 5-6)

$$(dh)_{5-6} = c_{ps}\, dt$$

where c_{ps} is specific heat of superheated steam

$$= 2.1\,(250 - 100) = 315 \text{ kJ/kgK}$$

9. Depending upon condition of steam, the heat or enthalpy values are defined as:

(*i*) ***Wet steam*** : When the vapours contain particles of liquid evenly distributed over the entire mass of vapour, it is called wet saturated steam. Wet steam is characterized by its dryness fraction which is defined as the ratio of mass of dry saturated vapour to the total mass of the mixture. That is

$$\text{Dryness fraction, } x = \frac{\text{mass of dry saturated vapour}}{\text{mass of mixture}}$$

$$= \frac{m_g}{m} = \frac{m_g}{m_f + m_g}$$

where m is the mass of mixture, m_g is the mass of dry saturated steam and m_f is the mass of water vapour (saturated liquid) in suspension with the saturated dry steam.

Total heat of wet steam represents the quantity of heat required to convert 1 kg of water at 0°C into wet steam at constant pressure.

$$h_{\text{wet}} = h_f + x\, h_{fg}$$

(*ii*) ***Dry saturated steam*** : The total heat of dry saturated steam represents the quantity of heat required to convert 1 kg of water at 0°C into dry saturated steam at constant pressure. The enthalpy of dry saturated steam increases with pressure.

(*iii*) ***Superheated steam*** : The total heat of superheated steam represents the quantity of heat required to convert 1 kg of water at 0°C into superheated steam at constant pressure.

The value of specific heat of steam at constant pressure c_{ps} depends upon the degree of superheat and the pressure of steam generation. Its average value is taken from 2.0 to 2.1 kJ/kgK.

Solid—liquid—vapour equilibrium

Fig. 1.22 shows the phase (p versus T) diagram for water in which regions of solid, liquid and vapour are indicated. Its salient features are:

1. The single phase regions are separated from each other by saturation lines. Along the sublimation line $1-T$ there is equilibrium between the solid and vapour phase, *i.e.*, the solid and vapour phases co-exist. The solid and liquid phases co-exist in equilibrium along the fusion line T−2. There is equilibrium between the liquid and vapour phase along the evaporation line $T-C$.

2. The fusion line for water has a negative slope ($\partial p/\partial T < 0$) indicating that melting point of ice decreases with increasing pressure. However, for most substances, the slope of fusion curve is positive.

3. The evaporation line ends at the critical point where there is no distinct change from the liquid phase to the vapour phase, *i.e.*, the liquid and vapour are indistinguishable.

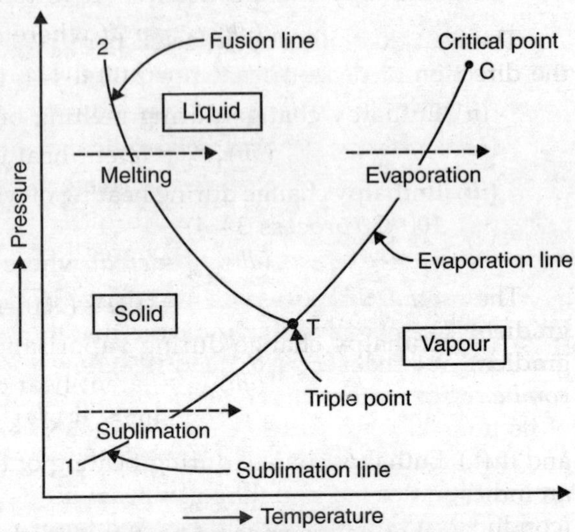

Fig. 1.22. Phase equilibrium diagram on p-T co-ordinates

4. The sublimation, fusion and evaporation lines meet at point T, called *the* ***triple point***. The triple point may thus be defined as the point at which all the three phases - solid, liquid and vapour co-exist in equilibrium. The triple point exists at a definite pressure and temperature. For water, triple point values are

$$t_{tp} = 0.01°\text{C} \text{ and } p_{tp} = 0.611 \text{ kPa}$$

5. Sublimation refers to the heating process where solid water or ice gets directly transformed to saturated vapour phase without first melting into water. The reverse of sublimation is referred to as ablimation.

1.22. MODES AND BASIC LAWS OF HEAT TRANSFER

The literature on heat transfer generally recognizes three distinct modes of heat transmission; *conduction, convection* and *radiation*. These three modes are similar in that a temperature differential must exist and the heat exchange is in the direction of decreasing temperature. Each method has, however, different physical picture and different controlling laws.

Conduction

Thermal conduction is a mechanism of heat propagation from a region of higher temperature to a region of low temperature within a medium (solid, liquid or gaseous) or between different mediums in direct physical contact. Conduction does not involve any movement of macroscopic portions of matter relative to one another. The thermal energy may be transferred by means of electrons which are free to move through the lattice structure of the material. In addition, or alternatively, it may be transferred as vibrational energy in the lattice structure. Irrespective of the exact mechanism, the observable effect of conduction is an equalization of temperature.

Conduction is primarily a molecular phenomenon requiring temperature gradient as the driving force. Experimental evidence does indicate that steady-state one-dimensional flow of heat by conduction through a homogeneous material is given by the ***Fourier law***.

$$Q = -kA\frac{dt}{dx}$$

$$q = \frac{Q}{A} = -k\frac{dt}{dx}$$

The heat flux q (heat conducted per unit time per unit area) flows along normal to area A in the direction of decreasing temperature. The units on each term are :

Q : rate of heat flow, kJ/hr

A : area perpendicular to the direction of heat flow, m^2

dx : thickness of material along the path of heat flow, m

dt : temperature difference between the two surfaces across which heat is passing, degree Kelvin K or degree centigrade C.

The ratio dt/dx represents the change in temperature per unit thickness, *i.e.*, the temperature gradient. The negative sign indicates that the heat flow is in the direction of negative temperature gradient, and that serves to make heat flow positive. The proportionality factor k is called the ***heat conductivity*** or ***thermal conductivity*** of the material. Thermal conductivity of a material is one of its transport properties. Others are the viscosity associated with the transport of momentum, and the diffusion coefficient associated with the transport of mass. Thermal conductivity provides an indication of the rate at which heat energy is transferred through a medium by the diffusion (conduction) process. For a prescribed temperature gradient and geometric parameters, the heat flow rate increases with increasing thermal conductivity.

Assuming $dx = 1$ m ; $A = 1\ m^2$ and $dt = 1°$,

we obtain $Q = k$

Hence thermal conductivity may be defined as the amount of heat conducted per unit time across unit area and through unit thickness, when a temperature difference of unit degree is maintained across the bounding surfaces.

The units of thermal conductivity are worked out from the Fourier law written in the form:

$$k = -\frac{Q}{A}\frac{dx}{dt}$$

Thus;

$$[k] = \frac{\text{kJ}}{\text{hr}}\ \frac{1}{\text{m}^2}\ \frac{\text{m}}{\text{deg}} = \text{kJ/m-hr-deg}$$

The unit kJ/m-hr-deg could also be specified as J/m-s-deg or W/m-deg and this is actually done while quoting the numerical values of thermal conductivity.

From Ohm's law

$$\text{current } (i) = \frac{\text{voltage potential } (dV)}{\text{electrical resistance } (Re)}$$

and from Fourier's law

$$\text{heat flow rate } (Q) = \frac{\text{temperature potenial } (dt)}{\text{thermal resistance } (dx/kA)}$$

Obviously there is a one-one correspondence between the flow of electric current and heat. That is

- electric current (amperes) is analogous to thermal heat flow rate (kJ/hr).
- electric voltage (volts) corresponds to thermal temperature difference (degree Kelvin).
- electric resistance (ohms) is analogous to quantity dx/kA. This quantity is called ***thermal resistance.***

Thermal resistance, $R_t = (dx/kA)$, is expressed in the units s-deg/J or deg/W. The reciprocal of thermal resistance is called *thermal conductance* and it represents the amount of heat conducted through a solid wall of area A and thickness dx when a temperature difference of unit degree is maintained across the bounding surfaces.

Convection

Thermal convection is a process of energy transport affected by the circulation or mixing of a fluid medium (gas, liquid or a powdery substance). Convection is possible only in a fluid medium and is directly linked with the transport of medium itself. Macroscopic particles of a fluid moving in space cause the heat exchange, and thus convection constitutes the ***macroform*** of the heat transfer. The effectiveness of heat transfer by convection depends largely upon the mixing motion of the fluid.

With respect to origin, two types of convection are distinguished; forced, and natural or free convection.

In ***natural*** or ***free convection,*** the circulation of the fluid medium is caused by buoyancy effects, *i.e.*, by the difference in the densities of the cold and heated particles. Consider heat flow from a hot plate to atmosphere. The stangant layer of air in the immediate vicinity of the plate gets thermal energy by conduction. The energy thus transferred serves to increase the temperature and internal energy of the air particles. Because of temperature rise, these particles become less dense (and therefore lighter) than the surrounding air. The lighter air particles move upwards to a region of low temperature where they mix with and transfer a part of their energy to the cold particles. Simultaneously the cold air particles descend downwards to fill the space vacated by the hot air particles. The circulation pattern, upward movement of the warm air and the downward movement of the cool air, is called the ***convection currents.***

In ***forced convection,*** the flow of fluid is caused by a pump, fan or by atmospheric winds. These mechanical devices provide a definite circuit for the circulating currents and that speeds up the heat transfer rate. Examples of forced convection are : flow of water in condenser tubes, fluid passing through the tubes of a heat exchanger, cooling of internal combustion engine; air conditioning installation and nuclear reactors.

Regardless of the particular nature; the appropriate rate equation for the convective heat transfer between a surface and an adjacent fluid is prescribed by Newton's Law of cooling.

$$Q = hA\,(t_s - t_f)$$

where, Q is the convective heat flow rate, A is the area exposed to heat transfer, t_s and t_f are the surface and fluid temperature respectively. The heat transfer co-efficient h depends upon the thermodynamic and transport properties (*e.g.*, density, viscosity, specific heat and thermal conductivity of the fluid), the geometry of the surface, the nature of fluid flow, and the prevailing thermal conditions.

Convection mechanism involving phase changes leads to the important fields of boiling (evaporation) and condensation.

Radiation

Thermal radiation is the transmission of heat in the form of radiant energy or wave motion from one body to another across an intervening space. Unlike heat transfer by conduction and convection, transport of thermal radiation does not necessarily affect the material medium between the heat source and the receiver. An intervening medium is not even necessary and the radiation can be affected through vacuum or a space devoid of any matter. Radiation exchange, in fact, occurs most effectively in vacuum. A material present between the heat source and the receiver would either reduce or eliminate entirely the propagation of radiation energy.

The most vivid evidence of radiation heat transfer is that represented by solar energy which passes through inter-stellar space (conditions close to that for perfect vacuum) on its way to the earth space. Solar radiation plays an important part in the design of heating and ventilating system. Heat transfer by radiation is encountered in boiler furnaces, billet reheating furnaces and other types of heat exchange apparatus. The design and construction of engines, gas turbines, nuclear reactors and solar collectors is also significantly influenced by the radiation heat transfer.

The basic rate equations for radiation heat transfer are based on Stefan-Boltzman law :

$$E_b = \sigma_b AT^4$$

where E_b is the energy radiated per unit time, T is the absolute temperature of the surface, and σ_b is the Stefan-Boltzman constant.

$$\sigma_b = 5.67 \times 10^{-8} \text{ W/m}^2 \text{ K}^4$$

The expression $E_b = \sigma_b AT^4$ is essentially valid for an ideal radiator or a black body-suffix b designates a black surface. The radiant energy emitted by a surface is less than that for an ideal emitter and is given by

$$E = \epsilon \sigma_b AT^4$$

where, ϵ is a radiative property of the surface and is called ***emissivity***; its values depends upon surface characteristics and temperature. It indicates how effectively the surface emits radiations compared to an ideal or black body radiator. Normally a body radiating heat is simultaneously receiving heat from other bodies as radiation. Consider that surface 1 at temperature T_1 is completely enclosed by another black surface 2 at temperature T_2. The net radiant heat transfer is

$$Q = \sigma_b A_1(T_1^4 - T_2^4)$$

Likewise, the net rate of heat transfer between the real surface (called gray surface) at temperature T_1 to a surrounding black surface at temperature T_2 is

$$Q = \sigma_b A_1 \epsilon (T_1^4 - T_2^4)$$

The net exchange of heat between the two radiating surfaces is due to the fact that one at the higher temperature radiates more and receives less energy for its absorption. An isolated body which remains at constant temperature emits just as much energy by radiation as it receives.

1.23. HEAT FLOW BETWEEN SURFACE AND SURROUNDINGS : COOLING AND HEATING OF FLUIDS

When a moving fluid comes into contact with a stationary surface, a thin boundary layer develops adjacent to the wall and in this layer there is no relative velocity with respect to surface. In a heat exchange process, this layer is called stagnant film and heat flow in the layer is covered both by conduction and convection processes. Since thermal conductivity of fluids is low, the heat flow from the moving fluid to the wall is mainly due to convection. The rate of convective heat transfer between a solid boundary and adjacent fluid is given by the Newton-Rikhman law :

$$Q = hA(t_s - t_f)$$

where t_f is the temperature of the hot moving fluid, t_s is the temperature of the wall surface and A is the area exposed to heat transfer. The convective coefficient (film coefficient) h depends upon the thermodynamic and transport properties (*e.g.*, density, viscosity, specific heat and thermal conductivity) of the fluid, the geometry of the surface, the nature of fluid flow and the prevailing thermal conditions. The resistance of the stagnant boundary layer is included in the convection coefficient. The dimensions of h are W/m^2-deg. The factor $1/hA$ represents the thermal resistance of the film heat transfer process.

The heat transfer through a wall separating two moving fluids involves (*i*) flow of heat from the fluid of high temperature to the wall, (*ii*) heat conduction through the wall and (*iii*) transport of heat from the wall to the cold fluid (Fig. 1.23).

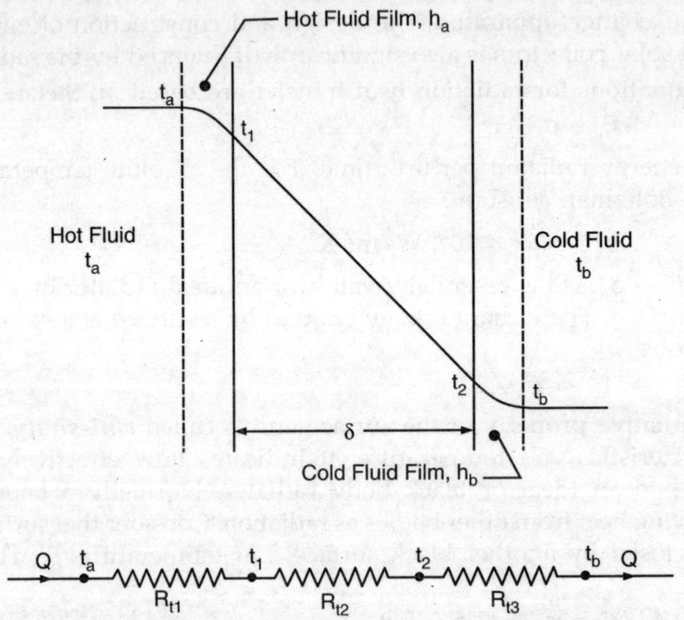

Fig. 1.23. Heat conduction through a wall separating two fluids

Under steady state conditions, the heat flow can be expressed by the equations :

$$Q = h_a\, A(t_a - t_1)\text{s}$$

$$Q = \frac{kA}{\delta}\ (t_1 - t_2) = h_b\, A(t_2 - t_b)$$

where h_a and h_b represents the convective film coefficients, k is the thermal conductivity of the solid wall having thickness δ. These expressions can be presented in the form :

$$t_a - t_1 = \frac{Q}{h_a A}\ ;\quad t_1 - t_2 = \frac{Q\delta}{kA}\ ;\quad t_2 - t_b = \frac{Q}{h_b A}$$

Summation of these equalities yields

$$(t_a - t_b) = Q\left[\frac{1}{h_a A} + \frac{\delta}{kA} + \frac{1}{h_b A}\right]$$

$$\therefore \quad Q = \frac{(t_a - t_b)}{\dfrac{1}{h_a A} + \dfrac{\delta}{kA} + \dfrac{1}{h_b A}} \qquad ...(i)$$

The denominator $(1/h_aA + \delta/kA + 1/h_bA)$ is the sum of thermal resistances of different sections through which heat has to flow.

Quite often the heat flow through a composite section is written in the form

$$Q = UA\,(t_a - t_b) \qquad ...(ii)$$

where U is the ***overall heat transfer coefficient***. It represents the intensity of heat transfer from one fluid to another through a wall separating them. Numerically, it equals the quantity of heat passing through unit area of wall surface in unit time at a temperature difference of unit degree. The coefficient U has dimensions of W/m^2-deg.

From expressions (*i*) and (*ii*),

$$\frac{1}{U} = \frac{1}{h_a} + \frac{\delta}{k} + \frac{1}{h_b}$$

Apparently the overall heat transfer coefficient is reciprocal of unit thermal resistance to heat flow.

The overall heat transfer coefficient depends upon the geometry of the separating wall, its thermal properties and the convective coefficient at the two surfaces. The overall heat transfer coefficient is particularly useful in the case of composite walls, such as in the design of structural walls for boilers, refrigerators, air conditioned buildings, and in the design of heat exchangers.

1.24. HEAT EXCHANGERS

Heat exchanger is a process equipment designed for the effective transfer of heat energy between two fluids ; a hot fluid and a coolant. The purpose may be either to remove heat from a fluid or to add heat to a fluid. Notable examples are :

(*i*) boilers (evaporators), superheaters and condensers of a power plant.

(*ii*) automobile radiators

(*iii*) evaporator of an ice plant and milk-chiller of a pasteurizing plant

(*iv*) water and air heaters or coolers.

The heat transferred in the heat exchanger may be in the form of latent heat (*e.g.*, in boilers and condensers) or sensible heat (*e.g.*, in heaters and coolers).

1.24.1 Logarithmic Mean Temperature Difference

During heat exchange between two fluids, the temperature of the fluids change in the direction of flow and consequently there occurs a change in the thermal head causing the flow of heat. Fig. 1.24 (*a*) represents the temperature conditions existing in surface condenser or feed water heater. The hot fluid is steam and the cold fluid is water. Here the temperature of steam remains constant but the temperature of water is progressively rising. During heat exchange in evaporation of water into steam (Fig. 1.24 (*b*)), the water evaporates at constant temperature and the temperature of hot gases continuously decreases in flowing from inlet to outlet.

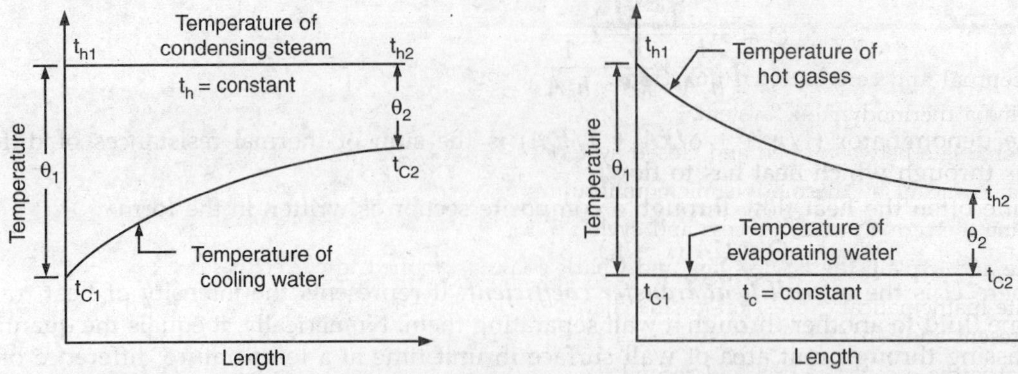

Fig. 1.24. Temperature changes of mediums during (a) condensation (b) evaporation

In many instances, both the fluids experience a change in temperature while flowing through the heat exchange equipment. In a parallel flow system, the thermal head (temperature potential) causing the flow of heat is maximum at the inlet. However, the thermal heat goes on diminishing along the flow path and is minimum at the outlet. In a counter flow system, both the fluids are in their coldest state at the exit.

The rate of heat transfer is calculated from the expression

$$Q = UA\,\theta_m \text{ where}$$

$$\text{logarithmic mean temperature difference } \theta_m = \frac{\theta_1 - \theta_2}{\log_e \dfrac{\theta_1}{\theta_2}}$$

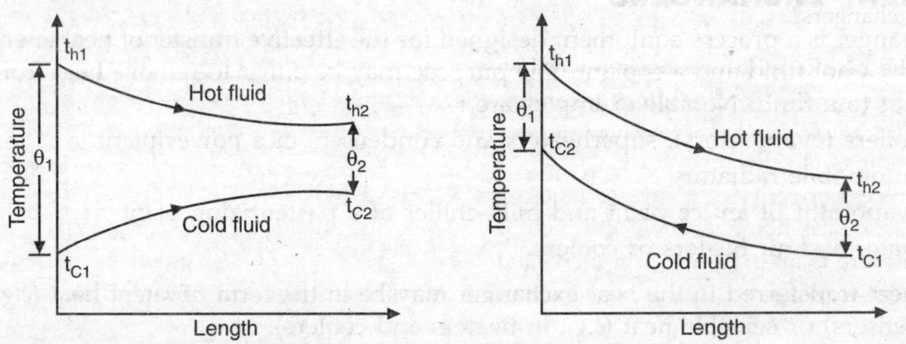

Fig. 1.25. Temperature changes of mediums (fluids) during (a) parallel flow (b) counter flow

For parallel flow arrangement (Fig. 1.25)

$$\theta_1 = t_{h_1} - t_{c_1} \quad \text{and} \quad \theta_2 = t_{h_2} - t_{c_2}$$

For counter flow arrangement (Fig. 1.24)

$$\theta_1 = t_{h_1} - t_{c_2} \quad \text{and} \quad \theta_2 = t_{h_2} - t_{c_1}$$

The logarithmic mean temperature difference for counter flow unit is greater than that of parallel flow system and accordingly the counterflow heat exchanger can transfer more heat than a similar parallel flow heat exchanger. Conversely, a counter flow heat exchanger needs a smaller heating surface for the same rate of heat transfer.

REVIEW QUESTIONS

A. Conceptual and conventional questions.

1. Define a thermodynamic system.
2. Differentiate between open and closed systems: Give examples to illustrate your answer.
3. What is meant by thermodynamic equilibrium?
4. Define the terms : state, process and cycle.
5. State and explain the Boyle's law and Charle's law as applied to perfect gases.
6. Write mathematical expression for the characteristic equation of a gas. What do the different symbols represent and what are their units?
7. What is the essence of first and second laws of thermodynamics?
8. Define enthalpy. How does it differ from internal energy?
9. Why is the concept of enthalpy introduced in thermodynamics?
10. Define the terms saturation temperature and saturation pressure.
11. What is the main feature of triple point? State the values of pressure and temperature at triple point of water.
12. Discuss the process of change of phase of water at constant pressure. Show the various stages on temperature-heat input, pressure-volume and temperature-volume diagrams.
13. (*a*) What is Fourier's law of heat conduction?
    (*b*) Differentiate between natural convection and forced convection.
    (*c*) How is the heat transfer coefficient defined? What is its dimension?
14. What is a heat exchanger? Why are counterflow heat exchangers considered superior to parallel flow heat exchangers?

B. Fill in the blanks with appropriate word/words

1. The system and its surroundings taken together constitute an .... system.
2. The fluid employed in thermodynamic devices for the transport of energy between the system and surroundings is called the ....
3. Any process or series of processes whose end states are identical is called a ....
4. A .... is homogeneous, and has the same chemical and physical compositions.
5. Temperature is an .... property independent of the size and mass of the system.
6. Heat and work are .... interactions which bring about a change in the .... of the system
7. The product of mass and specific heat of a substance is termed as .... or ....
8. For an ideal gas, the .... and .... are functions of temperature alone.
9. The value of gas constant *R* for any gas can be calculated by dividing the universal gas constant by .... of the gas.
10. Heat supplied to a system equals the work done in case of a non-flow process carried out ....
11. The change in the enthalpy of a system equals the heat supplied under constant .... conditions.
12. The .... law of thermodynamics puts a limit on the conversion of a given amount of heat into work.
13. All .... have the same efficiency when working between the same ....
14. A mixed phase of liquid and gas is called a ....
15. The latent heat of vaporisation is .... at the critical point.
16. Triple point is the point where .... exist together.
17. The material medium between the heat source and receiver is not affected during the heat transmission by ....

Answers

1. isolated 2. working substance 3. cycle 4. pure substance 5. intensive 6. energy, properties 7. heat capacity, water equivalent 8. internal energy, enthalpy 9. molecular mass 10. isothermally 11. pressure 12. second 13. reversible engines 14. vapour 15. zero 16. three phases of solid, liquid and gas 17. radiation

C. Multiple choice questions.

1. A thermodynamic system refers to
   (*a*) any defined region in space (*b*) a specified mass in fluid flow
   (*c*) a specified region of constant volume (*d*) a prescribed and identifiable quantity of matter
2. Choose the open thermodynamic system
   (*a*) manual ice cream freezer (*b*) centrifugal pump
   (*c*) pressure cooker (*d*) automobile storage battery
3. Which one of the following represents a closed system ?
   (*a*) bomb calorimeter (*b*) steam generator
   (*c*) universe (*d*) exhaust stroke of an IC engine
4. Which one of the properties given below is an intensive property of the system ?
   (*a*) composition (*b*) volume (*c*) kinetic energy (*d*) entropy
5. Which amongst the followings is not a property of the system ?
   (*a*) pressure (*b*) composition (*c*) thermal conductivity (*d*) heat
6. Choose the correct statement
   (*a*) temperature is an extensive property (*b*) mass remains same in an open system
   (*c*) the system boundaries may expand or collapse
   (*d*) an isolated system allows exchange of energy in the form of heat only
7. Identify the false statement in the context of heat and work
   (*a*) they exist only during interaction between the system and the boundary
   (*b*) are path functions and exact differential
   (*c*) may or may not occur simultaneously
   (*d*) occur due to some unbalanced potential which tends to promote a change of state
8. As differentials, heat and work would be described mathematically as
   (*a*) inexact (*b*) exact (*c*) discontinuity (*d*) point function
9. Identify the wrong statement concerning thermodynamic work
   (*a*) $W = \int p\,dv$ for a reversible process only (*b*) $W = 0$ if $dv = 0$
   (*c*) $W \neq \int p\,dv$ for a free expansion process (*d*) is a path function and inexact differential
10. Which one of the following gases will have the maximum value of gas constant R ?
   (*a*) nitrogen (*b*) carbon dioxide (*c*) sulphur dioxide (*d*) oxygen
11. Consider the following statements
   An ideal gas
   1. has no intermolecular forces of attraction or repulsion
   2. does not change its phase during a thermodynamic process
   3. obeys a set of common rules govering change of its properties

   Which of these statements is/are correct ?
   (*a*) 1 and 2 (*b*) 2 and 3 (*c*) 3 alone (*d*) 1, 2 and 3

12. Molar volume is equal to .... at normai temperature and pressure
(a) 29.27 m^3 (b) 22.4 m^3 (c) 42.7 m^3 (d) 84.8 m^3

13. First law of thermodynamics refers to conservation of
(a) mass (b) momentum (c) energy (d) force

14. The internal energy of an ideal gas is a function of its absolute temperature only. This statement refers to
(a) Avogadro's law (b) Maxwell law (c) Joule's law (d) Regnault's law

15. Heat is being supplied to air in a cylinder fitted with a frictionless piston held by a constant weight. The process is
(a) isothermal (b) adiabatic (c) isobaric (d) isochoric

16. Work output from a system is at the expense of internal energy in a non-flow process carried out
(a) at constant pressure (b) at constant volume
(c) adiabatically (d) polytropically

17. In a Carnot cycle, the rejection of heat is
(a) at constant pressure (b) at constant volume
(c) at constant temperature
(d) partly at constant pressure and partly at constant volume

18. The efficiency of a Carnot cycle engine depends on
(a) speed of the engine (b) working fluid; whether gas or vapour
(c) temperature at which heat is supplied and that at which it is rejected
(d) temperature of ambient air

19. Three engines A, B and C operating on Carnot cycle respectively use air, steam and helium as the working fluid. If all the engines operate with in the same high and low temperature limits, then which engine will have the highest efficiency ?
(a) Engine A (b) Engine B
(c) Engine C (d) All engines will have the same efficiency

20. Vapour is a
(a) pure substance (b) perfect gas (c) mixed phase of liquid and gas
(d) substance homogeneous and invariable in chemical composition

21. At critical pressure (225 bar)
(a) entropy is zero
(b) latent heat of vaporisation is zero
(c) saturated vapour becomes a gas
(d) sublimation occurs from a solid to vapour phase

22. Match the two lists given below :

A. Critical	1. All three phases–solid, point liquid and vapour coexit in equilibrium
B. Sublimation	2. Phase changes from solid to liquid
C. Triple point	3. Properties of saturated liquid and saturated vapour are identical
D. Melting	4. Heating process where solid directly gets transformed to gaseous phase

23. The essential condition for the transfer of heat from one body to another is
(a) both bodies must be in physical contact
(b) heat content of one body must be more than that of the other
(c) one of the bodies must have a high value of thermal conductivity
(d) there must exist a temperature difference between the bodies

24. Consider the following statements :
The Fourier heat conduction equation

$$Q = - k A \frac{dt}{dx}$$

presumes
1. Steady state conditions
2. Constant value of thermal conductivity
3. Uniform temperature at the wall surfaces
4. One-dimensional heat flow.

Which of these statements are correct ?
(*a*) 1, 2 and 3 (*b*) 1, 2 and 4 (*c*) 2, 3 and 4 (*d*) 1, 3 and 4

25. Choose the correct statement with respect to a counter flow heat exchanger :
(*a*) both the fluids at inlet are in their coldest state
(*b*) both the fluids at exit are in their hottest state
(*c*) both the fluids at inlet are in their hottest state
(*d*) one fluid is hottest and the other is coldest at inlet

Answers

1. (*d*)	2. (*b*)	3. (*a*)	4. (*a*)	5. (*d*)	6. (*c*)	7. (*b*)	8. (*a*)
9. (*a*)	10. (*d*)	11. (*a*)	12. (*b*)	13. (*c*)	14. (*c*)	15. (*c*)	16. (*b*)
17. (*c*)	18. (*c*)	19. (*d*)	20. (*c*)	21. (*b*)	22. (*A-3, B-4, C-1, D-2*)	23. (*d*)	
24. (*d*)	25. (*c*)						

Chapter – 2

Refrigerators and Heat Pumps

Learning objectives : A study of the subject matter presented in this chapter will enable the students/readers to

- *state the difference between heat engine, refrigerator and heat pump*
- *know about the historical development of different methods available for producing cooling effect*
- *appreciate the various typical applications of refrigeration*
- *understand the parameters used to assess the performance of a refrigeration system*

2.1. REFRIGERATION: WHAT IS IT?

Science of providing and maintaining temperatures below that of surroundings

(American Society of Refrigeration Engineers)

Refrigeration and air conditioning is the fascinating branch of science which deals with the chilling or freezing of a substance by removing some of its heat. This artificial withdrawl of heat produces within the substance or within a space a temperature below the general temperature of its surroundings. Refrigeration essentially means continued abstraction of heat from a substance (perishable foods, drinks and medicines etc.) at low temperature level and then transfer this heat to another system at high potential of temperature. To accomplish this, mechanical work must be performed to satisfy the second law of thermodynamics.

Air conditioning refers to the simultaneous control of temperature, humidity, cleanliness and air motion within a confined region or space.

2.2. HEAT ENGINE, REFRIGERATOR AND HEAT PUMP

A heat engine is a thermodynamic device used for continuous production of work from heat when operating in a cyclic process. Both heat and work interactions take place across the boundary of this cyclically operating device. Essentially a heat engine takes heat from the combustion of fuel and converts part of this energy into mechanical work.

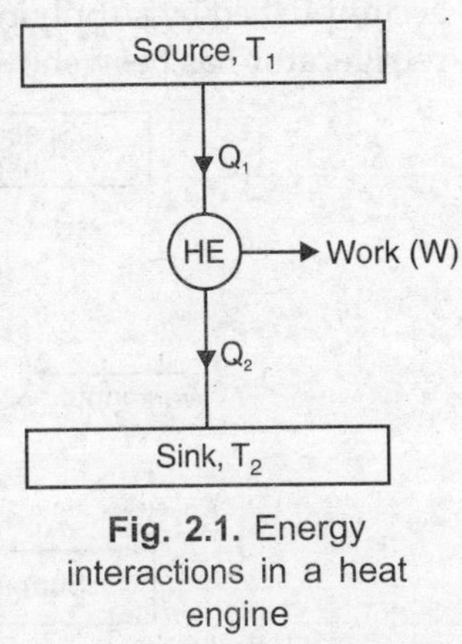

Fig. 2.1. Energy interactions in a heat engine

A heat engine is characterised by the following features:

- reception of heat Q_1 from a high temperature source at T_1
- partial conversion of heat received to mechanical work W
- rejection of remaining heat Q_2 to a low temperature sink at temperature T_2
- cyclic/continuous operation and
- working substance flowing through the engine.

The performance of any machine is expressed as the ratio of 'what we want ' to 'what we have to pay for`. In the context of an engine, work is obtained at the expense of heat input. Accordingly, the performance of a heat engine is given by net work output to the entire amount of heat supplied to the working medium, and this ratio is called ***thermal efficiency***, η_{th} (Thermal efficiency is a measure of the degree of useful utilization of heat received in a heat engine).

$$\eta_{th} = \frac{\text{net work output}}{\text{total heat supplied}}$$

Application of the principle of energy conservation (First law) to the heat engine, which undergoes a cycle gives : $W = Q_1 - Q_2$

$$\therefore \quad \eta_{th} = \frac{Q_1 - Q_2}{Q_1} = 1 - \frac{Q_2}{Q_1} \qquad ...(2.1)$$

Obviously, thermal efficiency of a heat engine operating between two thermal reservoirs is always less than unity. To increase the thermal efficiency, it is necessary to reduce Q_2 (heat rejected) with Q_1 (heat supplied) remaining constant. Thermal efficiency could be equal to unity if $Q_1 \to \infty$ and $Q_2 = 0$ which, however, can not be realized in practice.

Refrigerators and heat pumps are ***reversed*** heat engines. The adjective reversed means operating backwards. The direction of heat and work interactions are opposite to that of a heat engine, *i.e.*, work input and heat output. These machines (refrigerators and heat pumps) are used to remove heat from a body at low temperature level and then transfer this heat to another body at high potential of temperature. When the main purpose of the machine is to remove heat from the cooled space, it is called a ***refrigerator***. A refrigerator operates between the temperature of surroundings and a temperature below that of the surroundings. Refrigerators are essentially used to preserve food items and drugs at low temperature.

The term ***heat pump*** is applied to a machine whose objective is to heat a medium which may already be warmer than its surroundings. A heat pump thus operates between the temperature of the surroundings and a temperature above that of the surroundings. Heat pumps are generally used to keep the rooms warm in winter.

Source and sink contain infinite energy; source always considered to be at a temperature higher than that of sink. However, they are relative terms. Atmosphere can be used as source or sink. In the working of a refrigerator, the surroundings become a source compared with space being cooled. For a heat pump, the surroundings become a sink compared with the room being heated.

The transfer of heat against a reverse temperature gradient in a refrigerator and heat pump is accomplished by supplying energy to the machine. A schematic representation of heat pump and a refrigerator has been shown in Fig. 2.2.

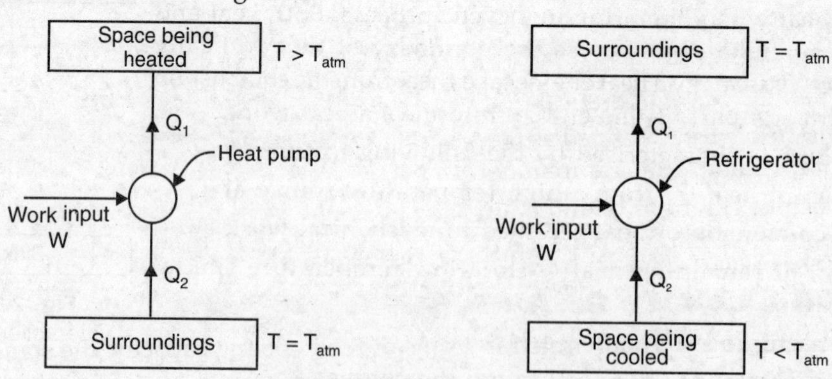

Fig. 2.2. Functional difference between a heat pump and a refrigerator

In the context of refrigerators and heat pumps, the performance is expressed in terms of ***coefficient of performance*** *(COP)* which represents the ratio of desired effect to work input

$$COP = \frac{\text{desired effect}}{\text{work input}}$$

In a ***refrigerator,*** the desired effect is the amount of heat Q_2 extracted from the space being cooled, *i.e.,* the space at low temperature.

$$(COP)_{ref} = \frac{\text{heat extracted at low temperature}}{\text{work input}} = \frac{Q_2}{W}$$

From the principle of energy conservation : $W = Q_1 - Q_2$

$$\therefore \quad (COP)_{ref} = \frac{Q_2}{Q_1 - Q_2}$$

Depending upon the type of refrigeration system, the value of $(COP)_{ref}$ may be less than one or greater than one. For most of the refrigerating machines, the values of *COP* lie between 3 and 4, and the *COP* are greatest when temperature differences are least.

In a ***heat pump,*** the desired effect is the amount of heat Q_1 supplied to the space being heated.

$$(COP)_{hp} = \frac{\text{heat rejected at high temperature}}{\text{work input}} = \frac{Q_1}{Q_1 - Q_2}$$

$$= 1 + \frac{Q_2}{Q_1 - Q_2} = 1 + (COP)_{ref}$$

Thus the *COP* of a machine operating as a heat pump is higher than the *COP* of the same machine when operating as a refrigerator by unity. The *COP* of a heat pump is some times termed as *energy performance ratio.*

Note : In the context of a refrigerator system :

1. The amount of heat extracted from the body at low temperature *i.e.,* the space being cooled is called *refrigerating effect* and is designated by *N.*
2. The *COP* of a refrigerator based on the theoretical values of refrigerating effect and work input is termed as *theoretical COP.* The theoretical refrigerating effect and work input are calculated by applying the laws of thermodynamics to the refrigeration cycle.
3. The *COP* of a refrigerator based on actual values of refrigerating effect and work input is termed as *actual COP.* The actual refrigerating effect and work input are obtained during test run on a refrigerating plant.
4. The ratio of actual *COP* to theoretical *COP* is known as *relative COP.*

$$\text{Relative } COP = \frac{\text{actual } COP}{\text{theroretical } COP}$$

5. ***Refrigeration efficiency*** is defined as the ratio of *COP* of a cycle to the *COP* of a Carnot cycle operating in the same temperature range.
6. A single machine can fulfil both the functions of cooling and heating simultaneously. For example, cool a food storage space as a refrigerator and heat a water system as a heat pump.

2.3. RATING OR CAPACITY OF A REFRIGERATING UNIT

The refrigerating machines are usually rated in terms of their cooling capacity; the standard unit being ***ton of refrigeration.***

One ton of refrigeration is defined as the refrigerating effect that freezes one ton (2000 pound mass) of liquid water during a period of 24 hours. The water is to be liquid at 0°C before and ice at 0°C after the process.

$$2000 \text{ pound mass} = \frac{2000}{2205} = 907 \text{ kg}$$

Enthalpy of fusion of water at 0°C = 333.43 kJ/kg

$$\therefore \quad 1 \text{ ton of refrigeration} = \frac{907 \times 333.43}{24 \times 60}$$

$$= 210 \text{ kJ/min} = 3.5 \text{ kJ/s} = 3.5 \text{ kW}$$

A ton of refrigeration is thus not a unit of mass but a measure of the rate of cooling. The refrigerating capacity decides the mass flow rate of a given working substance (refrigerant) working under specified conditions.

$$\text{Mass flow rate of refrigerant} = \frac{\text{refrigeration capacity}}{\text{refrigerating effect per unit mass}}$$

Quite often, power needed to produce a refrigeration effect equivalent of 1 ton of refrigeration is used as a measure to calculate the cost of operation or motor size of the refrigerating unit. Then

$$\text{kW per ton of refrigeration} = \frac{3.5}{COP}$$

Further, the ratio of heat removal rating (kJ/hr) of a refrigeration system to energy input (kWhr) of the machine is called energy efficiency ratio (EER).

A refrigerating system needs energy in the form of mechanical, electrical or thermal.

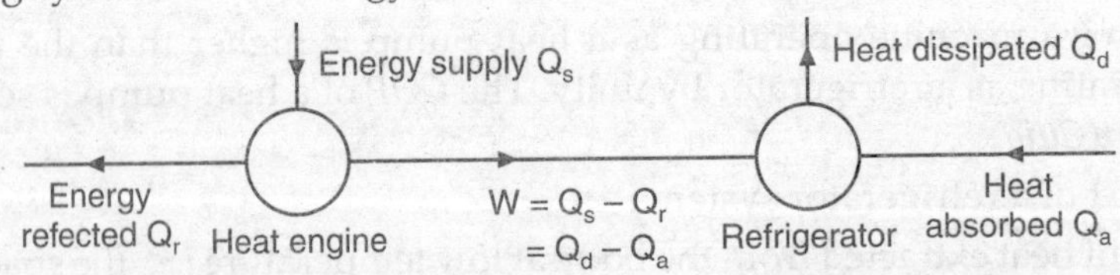

Fig. 2.3. Generalised representation of a refrigeration system

With respect to the generalised representation of a refrigeration system depicted in Fig. 2.3, a part of the energy supplied to the heat engine is rejected to surroundings and the remainder ($Q_s - Q_r$) is used to operate the refrigerator. The refrigerator absorbs heat from the confined space and produces cooling effect.

Mostly an electrical motor is used to operate the refrigerating device. The generator that runs the motor gets its motive mechanical power that results from :

(*i*) burning of fuel in the boiler furnace and steam is produced

(*ii*) steam expands in the turbine and mechanical power is produced

(*iii*) mechanical power drives the generator and the electrical power so produced is transmitted from the main station to sub-station through transmission systems to the place of work.

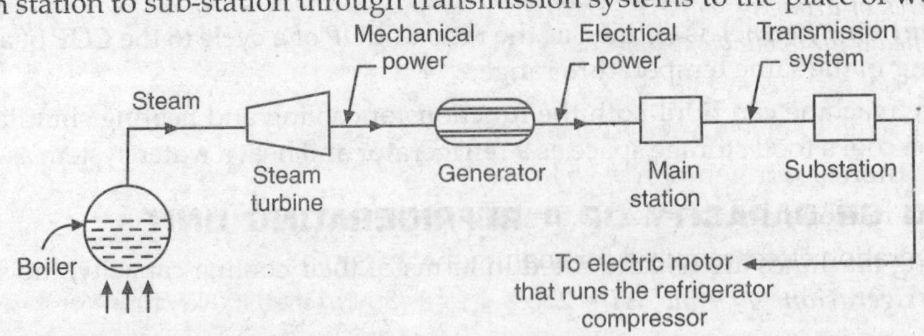

Fig. 2.4. Energy supply to refrigerator compressor from steam power plant

Apparently, the following efficiencies get involved in energy supply to the refrigerator compressor from the steam power plant.

(*i*) Thermal efficiency $\eta_{th} = \dfrac{\text{turbine output}}{\text{energy supply in fuel}}$

The thermal efficiency of a power plant is generally expressed as

$$\eta_{th} = \eta_{com}\, \eta_b\, \eta_t$$

where η_{com} is the combustion efficiency, η_b is the boiler efficiency and η_t is the turbine thermal efficiency. The practical values of these efficiencies are :

$$\eta_{com} = 0.96 - 0.98\,;\quad \eta_b = 00.84 - 0.86\,;\quad \eta_t = 0.38 - 0.40$$

(*ii*) Generator efficiency $\eta_g = \dfrac{\text{generator output}}{\text{input from turbine}} \simeq 0.38 - 0.40$

(*iii*) Transmission efficiency $\eta_{tr} = \dfrac{\text{energy transmitted}}{\text{input from generator}} \simeq 0.96 - 0.98$

(*iv*) Motor efficiency $\eta_m = \dfrac{\text{energy output from motor}}{\text{energy input to motor}} \simeq 0.8 - 0.82$

(*v*) Compressor efficiency $\eta_c = \dfrac{\text{energy used by compressor}}{\text{energy input to compressor}} \simeq 0.75 - 0.77$

Taking these efficiencies into account, the actual coefficient of performance of a refrigerating unit can be written as

$$(COP)_{actual} = \eta_{com}\, \eta_b\, \eta_t\, \eta_g\, \eta_{tr}\, \eta_m\, \eta_c \times (COP)_{theoretical}$$

and $$(COP)_{relative} = \frac{(COP)_{actual}}{(COP)_{theoretical}} = \eta_{com}\, \eta_b\, \eta_t\, \eta_g\, \eta_{tr}\, \eta_m\, \eta_c$$

Using the practical values of the efficiencies, the above identity can be written as

$$(COP)_{relative} = 0.96 \times 0.85 \times 0.38 \times 0.96 \times 0.9 \times 0.8 \times 0.75\, (COP)_{theoretical}$$
$$= 0.161\, (COP)_{theoretical}$$

The above result shows that even when all the units and devices of a refrigeration system operate efficiently, the relative *COP* is only one-sixth of the theoretical *COP*.

EXAMPLE 2.1

(i) The capacity of a refrigeration system is specified to be 12 tons. What is then the cooling rate of the machine ?

(ii) 250 litres of drinking water is required per hour at 10°C. Would the use of 1.5 ton refrigerating system be justified if the available water is at 30°C ?

Specific heat of water = 4.18 kJ/kg K

Enthalpy of solidification of water from and at 0°C = 335 kJ/kg

Solution : (*i*) 1 ton of refrigeration ≡ 3.5 kJ/s

∴ Cooling rate of machine = 12 × 3.5 = 42 kJ/s

(*ii*) Refrigeration effect required for cooling the water

$$= mc_p\, \Delta T = 250 \times 4.18 \times (30\text{-}10) = 20900 \text{ kJ/hr}$$

1 ton of refrigeration ≡ 3.5 kJ/s = 12600 kJ/hr

$$\therefore \quad \text{Tonnage required} = \frac{20900}{12600} = 1.658$$

As such, the use of 1.5 ton machine will not serve the purpose.

EXAMPLE 2.2

One ton of fruits are stored in a cold storage. The cold storage is maintained at – 5°C and the fruits get cooled to storage temperature in 12 hours. Find the refrigerating capacity of the plant. Assume that fruits freeze at –5°C and take latent heat of freezing as 105 kJ/kg and specific heat of fruit as 1.256 kJ/kg K.

Solution: For freezing the fruits, the heat to be removed by the plant is

$$Q = m\,[c_p \Delta T + L]$$

$$= 1000[\,1.256\{15 - (-5)\} + 105] = 130120 \text{ kJ in 12 hours}$$

$$= \frac{130120}{12 \times 3600} = 3.012 \text{ kJ/s}$$

1 ton of refrigeration = 3.5 kJ/s

$$\therefore \text{ Tonnage required, } \textit{i.e.}, \text{capacity of the plant is} = \frac{3.012}{3.5} = \mathbf{0.861}$$

EXAMPLE 2.3

An ice plant has been setup for producing 2000 kg of ice per hour at –10°C from water available at room temperature of 25°C. Determine the

(i) refrigerating effect and the tonnage required

(ii) power rating of the compressor if coefficient of performance of the plant is 3.2 and the overall electro-mechanical efficiency is 85 percent.

Take specific heat of water as 4.18 kJ/kg, enthalpy of solidication of ice as 335 kJ/kg and specific heat of ice below 0°C as 2.1 kJ/kg.

Solution: (*i*) Refrigeration effect

= removal of heat from water at 25°C to convert into ice at –10°C

$$= m\,[c_{pw}\,\Delta T_w + L + c_{pi}\,\Delta T_i]$$

$$= 2000\,[4.18\,(25 - 0) + 335 + 2.1\{0 - (-10)\}]$$

$$= 2000\,[\,104.5 + 335 + 21] = 921000 \text{ kJ/hr} = 255.8 \text{ kJ/s}$$

$$\therefore \text{ Tonnage required} = \frac{255.8}{3.5} = 73.09 \qquad (\because 1 \text{ ton} = 3.5 \text{ kJ/s})$$

$$(ii)\ COP = \frac{\text{refrigeration effect}}{\text{work input}}\,; \quad 3.2 = \frac{255.8}{W}$$

$$\therefore \quad W = \frac{255.8}{3.2} = 79.93 \text{ kJ/s} = 79.93 \text{ kW}$$

$$\therefore \quad \text{Power rating of the compressor motor} = \frac{W}{\eta} = \frac{79.93}{0.85} = \mathbf{94.03} \text{ kW}$$

EXAMPLE 2.4

An ice making plant produces 1000 kg of ice per day at – 10°C from water available at 25°C. The refrigerating system consumes 2.5 kW of power which is supplied from a thermal power plant. If the thermal plant has a specific coal consumption of 0.21 kg/s per MW, make calculations for

(i) coefficient of performance based on power and heat from coal

(ii) ice production per kg of kg

The heating value of coal is given to be 15500 kJ/kg.

Solution: Refrigerating effect per kg of ice

= removal of heat from water at 25°C to convert into ice at – 10°C

$$= c_{pw}\,\Delta T_w + L + c_{pi}\,\Delta t_i$$

$$= 4.18\,(25 - 0) + 335 + 2.1\,\{0 - (-10)\}$$

$$= 104.5 + 335 + 21 = 460.5 \text{ kJ/kg}$$

$$= 460.5 \times \frac{1000}{24 \times 3600} = 5.33 \text{ kJ/s}$$

Now, $$COP = \frac{\text{refrigerating effect}}{\text{power input}} = \frac{5.33}{2.5} = 2.13$$

Heat input from coal = power consumption × sp. coal consumption × heating value of coal

$$= \frac{2.5}{1000} \times 0.21 \times 15000 = 8.137 \text{ kJ/s}$$

Then, the COP based on coal used $= \dfrac{5.33}{8.137} = 0.655$

(*ii*) The refrigeration effect per kg of coal consumption

= COP based on coal used × heating value of coal

$$= 0.655 \times 15500 = 10152.5 \text{ kJ/kg of coal}$$

Then, ice production per kg of coal $= \dfrac{10152.5}{460.5} = 22.05$ kg/kg of coal

2.4. METHODS OF REFRIGERATION

Refrigeration is the technique of producing cooling effect by abstraction/withdrawl of heat so that temperature below that of surroundings is produced in a substance or within a space. The desired cooling effect can be produced by :

1. Evaporation

When a liquid evaporates, it absorbs heat from the surroundings equivalent to its latent heat of vaporisation, and that results in lowering the temperature of surroundings. For example, we feel cooling effect when there is evaporation of a drop of spirit placed on the palm of hand. Likewise, the evaporation of moisture from the skin of a human body helps to keep it cool. There is a common practice to cool the water for drinking purposes by keeping it in the porous earthen pots. The water evaporates through the pores and that produces cooling effect. The army people keep small water containers made of metal and covered with water soaked namada; the walls of the metallic container get cooled and that cools the water kept inside it. In the refrigeration literature, there is mention of an experiment where a pump was used to create partial vacuum over a container of ethyl ether. The liquid then boiled by absorbing heat from the surrounding air. The cooling effect so created even produced a small amount of ice.

The principle of evaporative refrigeration is employed in desert (room) coolers. The dry atmospheric air is made to pass through water soaked packings. When this water evaporates, it takes heat from the air causing it to cool.

With reference of Fig. 2.5, a volatile liquid (liquid nitrogen, liquid carbon dixoide) contained in a flask evaporates and gets converted into gas. For evaporation, it absorbs heat from the chamber and cooling effect is produced. The chamber is insulated to restrict the infiltration of heat from outside. The liquid N_2 and CO_2 are non-toxic and as such the liquid gas refrigeration finds application for keeping the perishable food articles cool when being transported.

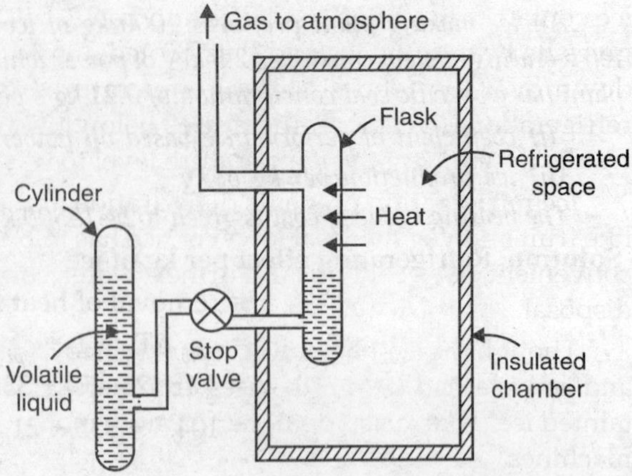

Fig. 2.5. Liquid gas refrigeration

2. Dissolution of salts in water

When certain salts are dissolved in water, they absorb heat and lower the temperature of water and create a sort of refrigeration bath for cooling substance. Sodium chloride lowers the water temperature upto – 20°C while calcium chloride upto – 50°C. The salt can be recovered by evaporation of water from the solution.

The method of producing cooling effect by dissolution of salt in water could not become feasible for commercial pursoses because,

(*i*) the refrigeration effect produced is quite small

(*ii*) the process of regaining salt is cumbersome.

There has been a practice in France to produce cold drinks and liqueurs (a strong alcoholic drink with a sweet taste) by spinning long necked bottles in water with dissolved saltpeter.

3. Ice refrigeration (change of phase)

The use of ice to refrigerate and thus preserve food goes back to the prehistoric times and the ancient cultures of Chinese, Greeks, Romans and Persians. Ice and snow were stored in caves or dugouts lined with straw or other insulating materials. This practice worked well down through the centuries, with ice houses remaining in use. Greater work was done on developing better insulation products for long distance shipment of ice and the ice harvesting became a big business.

The natural ice or artificially produced ice is brought into contact with the substance to be cooled. The ice melts and the heat required for melting of ice is supplied by the substance being cooled. The cooling effect produced by ice is

$$Q = \dot{m} \times h_{sf}$$

where $\dot{m}$ and h_{sf} are the rate of fusion of ice and enthalpy of fusion respectively. At normal atmospheric pressure of 1.01325 bar (1 atm), h_{sf} equals 335 kJ/kg.

The ice refrigerator consists of a cabinet which is completely insulated. The ice is kept in a container at the top, and a number of shelves are provided in the space below the ice container for storing the food stuff. When air comes in contact with ice, it becomes cool, dense and flows down over the shelves. It absorbs heat from the food stuff which gets

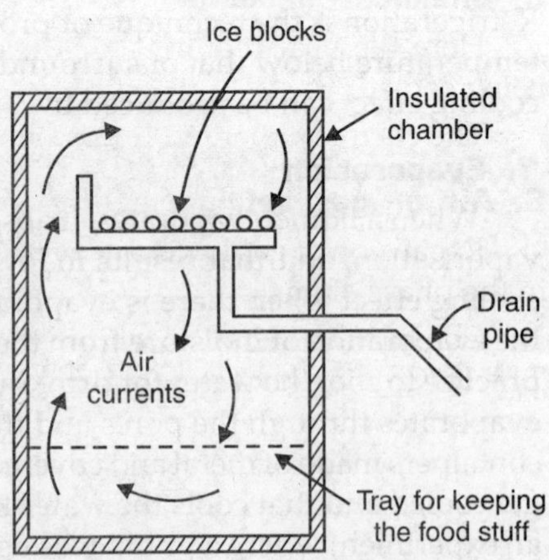

Fig. 2.6. Ice refrigeration

cooled. On absorption of heat, the air becomes warm. The warm air expands and returns back to the ice container from bottom, sides and back of the cabinet. When this warm air flows past the ice, it gives its heat to the ice and gets cooled. On cooling, the air becomes dense and once again flows down over the food shelves. Temperatures in the range of 5 to 10° can be obtained with ice refrigeration. In case, temperatures below this range are required, salt is mixed with ice and that results in reducing the temperature level to 0°C.

Ice refrigeration prevents dehydration and preserves the fresh appearance of eatable products like fruit and vegetables. However, ice has to be fed to the refrigerator of and on and that is neither convenient nor economical. Further, water coming out of the refrigerator poses a problem in its disposal.

Though the ice-harvesting industry had grown immensely by the turn of 20th century, pollution and sewage had begun to creep into natural ice and eventually breweries began to complain of tainted ice. This raised demand for more modern and consumer ready refrigeration and ice making machines.

4. Dry ice refrigeration (sublimation)

Solid carbon dioxide (called dry ice) has a peculiar characteristics that it changes from solid state to vapour state without passing through liquid state. During change of phase, it absorbs heat equivalent to its latent heat of vaporisation and produces cooling effect. This process occurs when the solid is maintained below triple point. Then,

$$Q = m\, h_{sv}$$

where h_{sv} is the enthalpy of evaporation.

At one atmospheric pressures, solid CO_2 produces 573 kJ/kg of refrigeration maintaining a temperature of – 78.5°C. Dry ice is used to preserve food stuff during transportation. The slabs of ice are usually packed on either side or on top of food packages in cartons. When dry ice evaporates, it absorbs heat from the food stuff and preserves it in frozen state.

The refrigeration methods (1) to (4) as mentioned above are called the ***natural methods.*** These methods are non-cyclic and the temperatures attainable are limited. Further, there is continuous consumption of the refrigerating substance and that necessitates replenishment. However, these methods are sometimes convenient forms of cooling where small refrigeration is required such as in the laboratory and workshop.

5. Chemical methods

Here the heat required for the completion of chemical reaction is taken from the substance being cooled.

The chemical method for producing cooling effect cannot be followed on a commercial scale.

6. Air or gas refrigeration

Expansion of gas lowers its pressure and temperature. This cooling effect results without change in the phase of gas.

Consider air initially compressed isentropically from atmospheric conditions (p_1 = 1 atm and T_1 = 15°C) to 5 atm. The temperature after compression then would be

$$T_2 = T_1\left(\frac{p_2}{p_1}\right)^{\frac{\gamma-1}{\gamma}} = 288\left(\frac{5}{1}\right)^{\frac{1.4-1}{1.4}} = 456.3 \text{ K}$$

The compressed air may be next cooled to initial (presume) temperature of 15°C in a heat exchanger without any loss of pressure. Then at state point 3, T_3 = 288 K and p_3 = 5 atm. Subsequently the cooled high pressure can be expanded in a suitable device to original pressure of 1 atm (state point 4). Then temperature after expansion will be

$$T_4 = T_3\left(\frac{p_4}{p_3}\right)^{\frac{\gamma-1}{\gamma}} = 288\left(\frac{1}{5}\right)^{\frac{1.4-1}{1.4}} = 181.8\text{ K} = -91.2°\text{C}$$

The different air refrigeration systems use this thermodynamic principle for producing low temperatures.

7. Throttling process

Throttling is the expansion of fluid from high pressure to low pressure. This process occurs when fluid passes through an obstruction (partially opened valve or a small orifice) placed in the fluid flow passage.

Fig. 2.7. shows the schematics of porous plug experiment performed by Joule and Thomson in 1852. A stream of incompressible fluid (gas) is made to pass steadily through a porous plug placed in an insulated and horizontal pipe. The upstream conditions of pressure p_i and temperature T_i are held constant and the corresponding values at exit are measured. The friction of the narrow passage causes the pressure to drop and accordingly the exit pressure p_e is less than the intake pressure p_i.

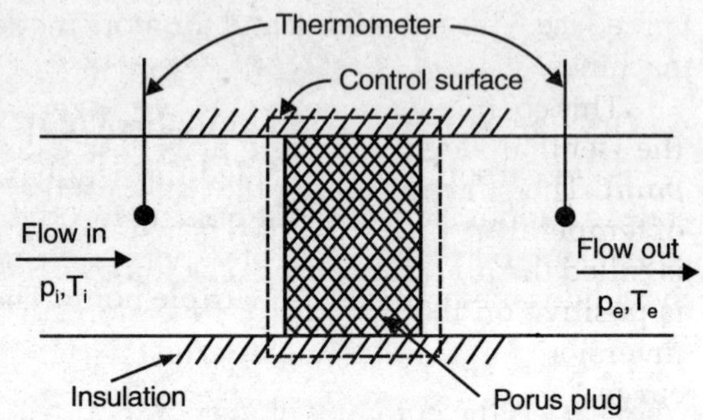

Fig. 2.7. Schematic of porous plug apparatus

A throttling process is characterised by the following features :

– no shaft work is involved
– no heat interaction as the pipe is thermally insulated
– no change in potential energy ($z_1 = z_2$) as the pipe is placed horizontally
– negligible changes in kinetic energy.

With these stipulations the steady flow energy equation,

$$h_1 + \frac{V_1^2}{2} + gz_1 + q = h_2 + \frac{V_2^2}{2} + gz_2 + w_s$$

transforms to

$h_1 = h_2$ *i.e.*, enthalpy of fluid remains constant during throttling.

Thus the throttling expansion process is an ***isenthalpic process***. If the fluid undergoing throttling behaves as an ideal gas for which $h = c_pT$, we get

$$c_pT_1 = c_pT_2\ ;\ T_1 = T_2$$

Again for a perfect gas, internal energy is a function of temperature alone and equality of temperature implies that $u_1 = u_2$. Apparently a throttling process takes place at constant enthalpy, constant temperature and constant internal energy.

Throttling is an irreversible process and involves degradation of energy and its dissipation in turbulence.

Joule- Thomson Coefficient, Inversion Point and Inversion Curve

For real gases, enthalpy is a function of both temperature and pressure. As such even though enthalpy remains constant during throttling, the temperature need not remain the same. Experimental test-runs can be conducted by keeping upstream conditions constant but with different down stream pressures. This is achieved by having porous plugs of different sizes. The exit temperature of the fluid at different exit pressures is measured. Since the upstream pressure and temperature conditions are kept constant, the enthalpy of the fluid for all measured conditions of exit pressure and temperature would be constant. The results are plotted as a constant enthalpy (isenthalpy) curve on $T-p$ diagram. Several enthalpy curves can be obtained by repeating the experiments with several inlet conditions.

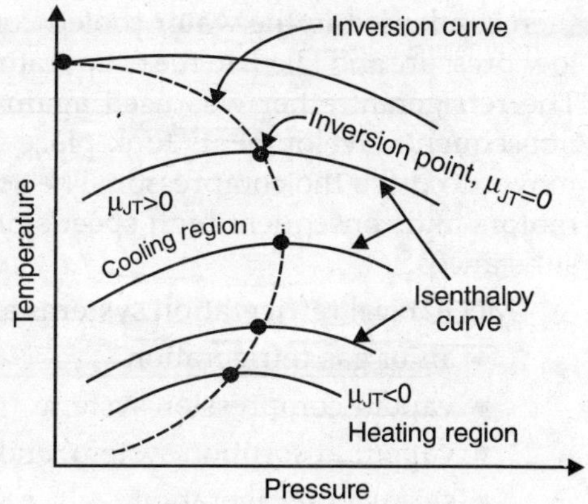

Fig. 2.8. Isenthalpy and inversion curves for a real gas

The slope of an isenthalpic curve is called the Joule-Thomson coefficient, μ_{JT}. That is

$$\mu_{JT} = \left(\frac{\partial T}{\partial p}\right)_{h=\text{constant}}$$

This coefficient may be +ve, –ve or zero. The point on the isenthalpic curve where $\mu_{JT} = 0$ is called the ***inversion point***. Thus the inversion point denotes the maximum value of temperature on $T-p$ plot. The locus of all inversion points is called the ***inversion curve***. The Joule - Thomson coefficient is positive on the left side of inversion curve, is zero at the inversion point, and is negative on the right side of inversion curve.

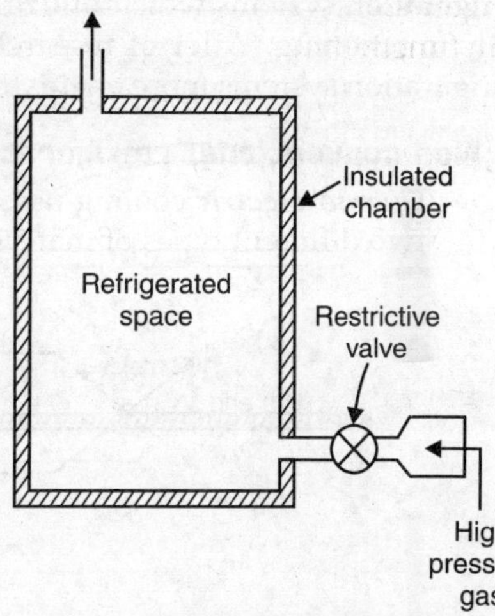

Fig. 2.9. Refrigeration by throttling of gas

Throttling is always accompanied by pressure drop, *i.e.*, Δp is –ve. That leads to drop in temperature when μ_{JT} is +ve. Accordingly when a real gas is throttled at the condition such that its state lies to the left of inversion curve, the gas will get cooled. Thus the region of $\mu_{JT} > 0$ represents the region of cooling. Likewise, when μ_{JT} is –ve, the temperature change is +ve and therefore, the throttling of real gas would produce heating effect.

A knowledge of the inversion temperatures and inversion curves of real gases is of considerable importance in the design of refrigeration and liquefaction equipment. The use of positive values of Joule-Thomson coefficient is made in the liquefication of gases such as air, nitrogen and oxygen.

Refer Fig. 2.9 which shows a simplified arrangement of refrigeration by throtting of a gas. The throtting process occurs when the gas at high pressure with its temperature below its critical temperature is made to pass through a valve with restricted opening. Upon throttling, the temperature of gas reduces and cooling effect is produced.

8. Mechanical refrigeration

The natural and chemical methods have been successfully replaced by mechanical or heat energy refrigeration techniques. In these methods, the heat is abstracted from the substance or space (which is to be cooled) is pumped to a system (which is at high temperature level) by taking energy from an external source as input to refrigerating machine. The refrigeration system consists of a cycle of processes with the same quantity of working fluid (refrigerant) in continuous circulation.

The first commercial hand operated refrigeration system was developed in UK by Perkins. The system consisted of a hand operated compressor, a water cooled condenser, a throttle valve and an evaporator. The working substance (ether) was compressed in the hand operated compressor and then condensed in the water cooled condensing unit. Thereafter, the liquid ether was throttled to low pressure and taken to the evaporator where heat was absorbed and cooling effect was produced. The refrigerant ether was used again and again in the cyclic process with negligible wastage. Subsequent developments took place in United States where steam engine was used as a prime mover to drive the compressor. The scope got further widened with the development of electric motors and consequent high speeds of the compressor. Ether too was replaced by new working substances.

Mechanical refrigeration systems are broadly classified into

- air or gas refrigeration
- vapour compression system
- vapour absorption system, and
- steam jet refrigeration.

Energy needed for the mechanical systems is essentially in the form of mechanical, electrical or thermal. Due to world energy crisis, concerted efforts are being made by various agencies to develop refrigeration systems which utilize waste heat, solar energy, wind energy and bio-energy etc for their functioning. A lot of research is being done to devise ways and means which make the refrigeration systems more energy efficient.

9. Non-conventional refrigeration systems

- ***Thermo-electric*** cooling uses the Peltier effect to create a heat flux between the junction of two different types of materials.

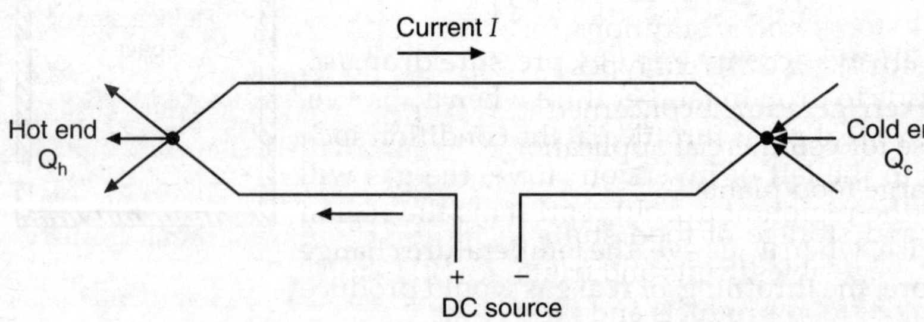

Fig. 2.10. Thermo-electric cooling : Peltier effect

When a direct current is made to pass through electrical junctions between unlike metals, heating or cooling effect is produced depending on the direction of flow of current. One junction gets heated and the other gets cooled. The cold junction is located inside the space desired to be cooled and the warmed junction lies outside in the surroundings. The hot and cold junctions get reversed when a change occurs in the direction of flow of current. Apparently the same setup can be used for heating and cooling of an insulated space.

With respect to Fig. 2.10, the energy balance of the system is,

$$Q_h - Q_c = EI$$

where E is the emf applied, I is the current, Q_h and Q_c are the actual amount of heat flows at the hot and cold junctions respectively.

This effect is commonly used in camping and portable coolers, and for cooling electronic components and small instruments.

- ***Magnetic refrigeration*** (adiabatic demagnetisation) is a cooling technology based on magnet caloric effect, an intrinsic property of magnetic solids. A strong magnetic field is applied to the refrigerant, forcing its various magnetic dipoles to align and putting these degrees of freedom of the refrigerant into a state of lowered entropy. A heat sink then absorbs the heat released by the refrigerant due to its loss of entropy. Thermal contact with the heat sink is then broken so that the system is insulated and the magnetic field is switched off. This increases the heat capacity of the refrigerant, thus decreasing its temperature below the temperature of heat sink.

The refrigerant is often a paramagnetic salt such as cerium magnesium nitrate, and the active magnetic dipoles are those of electron shells of the paramagnetic atoms.

- *Thermo acoustic* refrigeration which uses sound waves in a pressurised gas to drive heat exchange.
- *Vortex tube* that operates with compressed air and is used for spot cooling. A high pressure gas is allowed to expand through a nozzle fitted tangentially to a pipe. This causes simultaneous discharge of cool air at the core and hot air at the periphery.

The mechanical refrigeration and the non-conventional refrigeration systems have been dealt with in details at appropriate places.

2.5. APPLICATIONS OF REFRIGERATION

Refrigeration has played an important role in the growth and attainment of the present-dayu standard of living. Its various applications can be essentially grouped into the following categories :

1. Domestic refrigeration deals with providing a low temperature place for preserving eatables, drinks and medicines. The use of refrigerators in our kitchens for the storage of fruits and vegetables has allowed us to add fresh salads to our diets year round, and to store fish and meats for long periods.

2. Commercial refrigeration is related to refrigeration fixtures of the type used by restaurants and hotels, retail stores and institutions for storing, displaying, processing and dispensing of perishable commodities of different types.

3. Industrial refrigeration is concerned with systems which are much larger in size and cooling capacity that those for commercial applications. Typical industrial applications are :

- Ice and large food plants
- Chilling and storage of food stuffs including beverages, meat, poultry products, dairy products, vegetables, fruits and fruit juices etc
- Processing of food products and farm crops
- Processing of textiles, printing work and photographic materials
- Oil refineries and synthetic rubber manufacturing. In oil refineries, chemical manufacturing and petrochemical plants, refrigeration is used to maintain processes at their required low temperatures.

Refrigeration is also used to liquefy gases like oxygen, nitrogen, propane and methane. Further, in compressed air purification, it is used to condense water vapour from compressed air to reduce its moisture air content.

4. Transport refrigeration applies to refrigerated railway cars and trucks for local delivery and long distance transport of temperature sensitive food stuffs and other materials.

5. Air-conditioning refers to the simultaneous control of temperature, humidity, cleanliness and air motion. The conditioning of a space done to provide comfortable and healthy conditions for the occupants is called comfort air conditioning. Residences, offices, theatres and hospitals are the spaces air-conditioned for this purposes.

There are some manufacturing processes which need to be done under controlled environmental conditions. Even some of the sophisticated and precision instruments need controlled conditions for their effective working and upkeep. The conditioning done for this purpose is called industrial air-conditioning and this is concerned with production of environment suitable for :

- computer centres
- pharmaceutical units
- printing works and photographic products
- precision devices and production shop laboratories etc.

The applications cited above clearly indicate that refrigeration and air-conditioning which was considered luxury in the society a few decades ago has become the necessity of the present society and a tool for higher productivity.

REVIEW QUESTIONS

A. Conceptual and conventional questions.

1. What is a reversed heat engine? Point out the purpose for which it is used.
2. State the difference between a refrigerator and a heat pump. How do these machines satisfy the second law of thermodynamics?
3. Define the following terms:
   (*i*) refrigeration (*ii*) refrigerating effect (*iii*) ton of refrigeration
4. (*a*) What is meant by refrigerating effect?
   (*b*) Is it true that the ton unit of refrigeration is a unit of mass?
   (*c*) A refrigeration system is stated to be of 10 tons capacity. What would be the cooling rate of the machine?
5. Consider the following heat infiltrations/releases through a building which is required to be kept cool and maintained at a temperature of 25°C.
   (*i*) Heat transfer through the structure : 40000 kJ/hr
   (*ii*) Heat release from the persons sitting insider : 25000 kJ/hr
   (*iii*) Release of heat energy from the appliances operating inside : 5000 kJ/hr
   What tonnage of refrigeration is required for the given set of conditions? **[Ans. 5.55]**
6. The latent heat of fusion of water is 335 kJ/kg. What quantity of heat energy would be given up by 5 kg of water at 0°C when it freezes into ince at 0°C? **[Ans. 1675 kJ]**
7. A refrigerating machine consumes 1.25 kW of power when producing 25 kg/hr of ice at 0°C from water available at 30°C. Determine the refrigerating effect, tonnage and coefficient of performance of the machine. Take
   specific heat of water = 4.18 kJ/kg K
   Enthaly of solidification of water from and at 0°C = 335 kJ/kg. **[Ans. 11510 kJ/hr, 0.913 ton, 2.558]**
8. Setup an expression for the relative *COP* of a refrigeration system in terms of ideal COP and various efficiencies that get involved when voltage supply to the motor driving the refrigerator compressor is from a steam power plant.
9. What is meant by *COP*? What value of *COP* is desirable, large or small and why?
10. Set up a relation for the *COP* of a heat-pump and that of a refrigeration. Proceeed to show that
$$(COP)_{\text{heat pump}} = 1 + (COP)_{\text{refrigeration}}$$
11. State and briefly explain the various techniques of producing cooling.
12. Name and explain the commonly used unit of refrigeration.
13. Define refrigeration. How ice and dry ice are used for the purpose of refrigeration? Is it possible to get the temperature below zero degree with the help of ice? If yes, how?
14. Explain the statement : Refrigeration can be produced either by expansion of gas or by throttling of gas.
15. Mention the various applications of refrigeration.

16. Comment on the validity of following statements:
    (*i*) *COP* of heat pump is always greater than unity.
    (*ii*) One ton of refrigeration plant produces one ton of ice.

B. Fill in the blanks with appropriate word/words

1. The amount of cooling produced by a system is called ....
2. 1 ton of refrigeration imples refrigeration effect equivalent to .... kJ/min or .... kW
3. A scientific equipment of 3 kW rating will need a refrigeration capacity of .... tonnage
4. The performance of a refrigerator and heat pump is expressed in terms of ....
5. The ratio of actual *COP* to theoretical *COP* is known as ....
6. .... ratio is defined as ratio of kJ/hr rating of a refrigeration system to energy input (kW hr) of the machine.
7. In .... cooling, which of the junctions or ends will become cold or hot depends upon the direction of flow of current.
8. Literature suggests that cooling can be produced by the adiabatic demagnetization of a .... salt.
9. Temperatures down to zero can be obtained with ice and .... mixture.
10. .... of moisture from the skin surface of a person helps him to keep cool.
11. Use of .... values of Joule-Thomson coefficient is made in the liquefication of gases.

Answers

1. refrigerating effect 2. 210 kJ/min, 3.5 kW 3. one 4. *COP* 5. relative *COP* 6. energy efficiency 7. thermo-electric 8. paramagnetic 9. salt 10. evaporation 11. positive

C. Multiple choice questions.

1. A refrigeration system
   (*a*) removes heat from a system at low temperature and transfers the same to a system at high temperature
   (*b*) delivers less heat to the system at high temperature than it extracts from the system at low temperature
   (*c*) transfers heat from a high temperature source to a low temperature sink
   (*d*) violates second law of thermodynamics
2. The capacity of refrigerating machine is expressed as
   (*a*) inside volume of cabinet
   (*b*) lowest temperature attained
   (*c*) gross weight of machine in tons
   (*d*) rate of abstraction of heat from the space being cooled
3. One ton refrigeration machine implies that the machine
   (*a*) has a gross weight of 1 ton
   (*b*) consumes one ton of refrigerant in 24 hours
   (*c*) has the capacity to convert one ton of water into ice
   (*d*) is capable of extracting heat that would melt one ton of ice at 0°C into water at 0°C in a period of 24 hours
4. For the same operating temperature limits, the coefficient of performance (COP) of a heat pump equals
   (*a*) COP of refrigerator
   (*b*) 1 + (COP of refrigerator)
   (*c*) (COP of refrigerator) – 1
   (*d*) $\dfrac{1}{\text{(COP of refrigerator)}}$

5. The cooling effect in refrigeration is obtained by
   1. mechanical refrigeration technique
   2. passing direct current through junction of two dissimilar metals
   3. sublimation of carbon dioxide
   4. throttling of a real gas

   Which of the above statements are correct?

   (*a*) 1 and 2 (*b*) 1, 3 and 4 (*c*) 1, 2 and 4 (*d*) 1, 2, 3 and 4
6. Apart form human comfort, air-conditioning finds application in
   1. automatic telephone exchange
   2. computer centre
   3. measurement room in production industry
   4. broadcasting studio

   Mark the correct statements:

   (*a*) 1 and 2 (*b*) 2 and 3 (*c*) 1, 2 and 3 (*d*) 1, 2, 3 and 4
7. The throttling of certain gases may be used for getting the refrigeration effect. The value of Joule-Thomson coefficient μ for such a gas has to be

   (*a*) $= 0$ (*b*) $= 1$ (*c*) < 0 (*d*) > 1
8. A refrigerator having a coefficient of performance of 4 is made to work as a heat pump with in the same temperature limits. If the heat pump consumes 1 kW, the heating effect will be

   (*a*) 1 kW (*b*) 3 kW (*c*) 5 kW (*d*) 7 kW

Answers

1. (*a*) 2. (*d*) 3. (*d*) 4. (*b*) 5. (*d*) 6. (*d*) 7. (*d*) 8. (*c*)

Chapter – 3

Air Refrigeration Cycles

Learning objectives : A study of the subject matter presented in this chapter will enable the students/readers to explain and analyse

- *reversed Carnot cycle*
- *schematics of Bell-Coleman refrigerating system*
- *ideal and actual Bell-Coleman cycle*
- *open and closed Bell-Coleman refrigerating systems*

3.1. REVERSED CARNOT CYCLE

Refrigerators and heat pumps are reversed heat engines. They operate backwards and have heat and work interactions opposite to that of a heat engine. Heat at low temperature is removed and rejected at higher temperature and the task is accomplished by supplying work to the system.

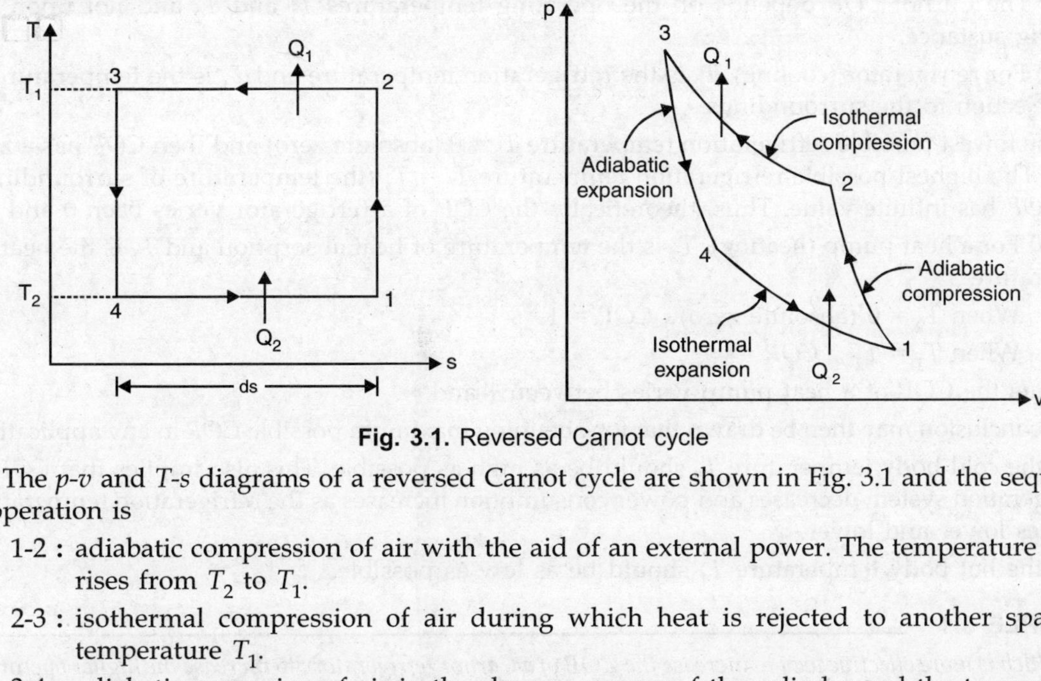

Fig. 3.1. Reversed Carnot cycle

The *p-v* and *T-s* diagrams of a reversed Carnot cycle are shown in Fig. 3.1 and the sequence of operation is

1-2 : adiabatic compression of air with the aid of an external power. The temperature of air rises from T_2 to T_1.

2-3 : isothermal compression of air during which heat is rejected to another space at temperature T_1.

3-4 : adiabatic expansion of air in the clearance space of the cylinder and the temperature of air falls from T_1 to T_2.

4-1 : isothermal expansion at which heat is absorbed at low temperature T_2 from the space being cooled.

Heat abstracted $Q_2 = T_2\, ds$

Heat rejected $Q_1 = T_1\, ds$

Work input W = heat rejected – heat abstracted = $(T_1 - T_2)\, ds$

$$\therefore \quad (COP)_{ref} = \frac{\text{heat abstracted at low temperature}}{\text{work input}}$$

$$= \frac{T_2\, ds}{(T_1 - T_2)\, ds} = \frac{T_2}{T_1 - T_2} = \frac{1}{\dfrac{T_1}{T_2} - 1} \quad \text{...(3.1)}$$

$$(COP)_{heat\ pump} = \frac{\text{heat rejected at high temperature}}{\text{work input}}$$

$$= \frac{T_1\, ds}{(T_1 - T_2)\, ds} = \frac{T_1}{T_1 - T_2} = \frac{1}{1 - \dfrac{T_2}{T_1}} \quad \text{...(3.2)}$$

Although theoretically more efficient, Carnot heat pumps and refrigerators like Carnot heat engines have not been used in practice. This is because the machine has to run at high speed during the portions of the cycle when adiabatic processes are being performed and run very slowly during other portions when isothermal processes are being carried out. The variation of speed during different processes constituting the cycle is not possible.

An examination of expressions 3.1 and 3.2 shows that

(*i*) The Carnot *COP* depends on the operating temperatures T_1 and T_2 and not upon the working sustance.

(*ii*) For refrigerator (cooling), T_2 is the refrigeration temperature and T_1 is the temperature of heat rejection to the surroundings

The lowest possible refrigeration temperature $T_2 = 0$ (absolute zero) and then *COP* has a zero value. The highest possible refrigeration temperature $T_2 = T_1$ (the temperature of surroundings) and *COP* has infinite value. Thus, theoretically the *COP* of a refrigerator varies been 0 and ∞.

(*iii*) For a heat pump (heating), T_2 is the temperature of heat absorption and T_1 is the heating temperature

When $T_2 = 0$ (absolute zero) : $COP = 1$

When $T_1 = T_2$; $COP = \infty$

Thus the COP of a heat pump varies between 1 and ∞.

A conclusion may then be drawn that for obtaining maximum possible *COP* in any application,

- the cold body temperature T_2 should be as high as possible. This also implies that *COP* of a refrigeration system decreases and power consumption increases as the refrigeration temperature becomes lower and lower.
- the hot body temperature T_1 should be as low as possible.

EXAMPLE 3.1

Which is more effective way to increase the COP of a Carnot refrigerator : to increase the higher operating temperature T_1 while the lower operating temperature T_2 is held constant or to decrease the temperature T_2 by the same amount while temperature T_1 is held constant?

Solution: Let $(COP)_1$ be the coefficient of performance of the refrigerator when the high temperature T_1 is increased by dT while the low temperature T_2 is held constant, and $(COP)_2$ be the coefficient of performance when the low temperature T_2 is decreased by dT while the high temperature T_1 is held constant. Then

$$(COP)_1 = \frac{T_2}{T_1 - T_2} = \frac{T_2}{(T_1 + dT) - T_2} \quad ...(i)$$

$$(COP)_2 = \frac{T_2}{T_1 - T_2} = \frac{T_2 - dT}{T_1 - (T_2 - dT)} = \frac{T_2 - dT}{(T_1 + dT) - T_2} \quad ...(ii)$$

It is to be noted that

- denominators for expressions (*i*) and (*ii*) are same
- numerator T_2 in expression (*i*) is more than the numerator $(T_2 - dT)$ in expression (*ii*)

Apparently then : $(COP)_1 > (COP)_2$. That is an increase in the high temperature (keeping the low temperature constant) is more effective in increasing the coefficient of performance of a Carnot refrigerator.

EXAMPLE 3.2

A refrigerating machine of 1 ton rating works on reversed Carnot cycle. It requires 1.5 kW to maintain low evaporator temperature of 250 K. Calculate coefficient of performance of the machine when working as a (i) refrigerator (ii) heat pump. Also determine the temperature at which heat is rejected.

Solution: 1 ton of refrigeration = 210 kJ/min = 3.5 kJ/s = 3.5 kW

$$(i)\ (COP)_{refrigerator} = \frac{\text{refrigerating effect}}{\text{work input}} = \frac{3.5}{1.5} = \mathbf{2.33}$$

$$(ii)\ (COP)_{heat\ pump} = 1 + (COP)_{refrigerator} = 1 + 2.33 = \mathbf{3.33}$$

(*b*) In terms of operating temperatures, the *COP* of a refrigerator is given as

$$COP = \frac{T_2}{T_1 - T_2};\quad 2.33 = \frac{250}{T_1 - 250}$$

$$\text{or} \quad T_1 = \frac{250 + 250 \times 2.33}{2.33} = \mathbf{357.3\ K}$$

EXAMPLE 3.3

*(**a**) A Carnot refrigerator extracts 500 kJ of heat per minute from a cold room which is maintained at –10°C. The discharge of the refrigerant is to the atmosphere which is maintained at 35°C. Determine the ideal power required to run the refrigerator.*

*(**b**) A reversed Carnot cycle is used for heating and cooling purposes. If the work supplied is 9.5 kW and COP is 3.6 for cooling, calculate*

(i) ratio of highest to lowest temperature, $Q_2 \times \frac{T_1}{T_2}$

(ii) refrigerating effect in tonnes of refrigeration

(iii) COP for heating

Solution: (***a***) Refer Fig. 3.2. The temperatures of the operating cycle are :

$$T_1 = 35°C = 308\ K \quad \text{and} \quad T_2 = -10°C = 263\ K$$

For a Carnot refrigerator that operates on reversed Carnot cycle,

$$\frac{T_2}{T_1 - T_2} = \frac{Q_2}{Q_1 - Q_2}$$

which gives :

$$\frac{Q_2}{T_2} = \frac{Q_1}{T_1} \quad ; Q_1 = Q_2 \times \frac{T_1}{T_2}$$

$$= 500 \times \frac{308}{263} = 585.5 \text{ kJ/min}$$

$$\text{Work input} = Q_1 - Q_2 = 585.5 - 500$$

$$= 85.5 \text{ kJ/min} = 1.425 \text{ kJ/s}$$

$$= \mathbf{1.425 \text{ kW}}$$

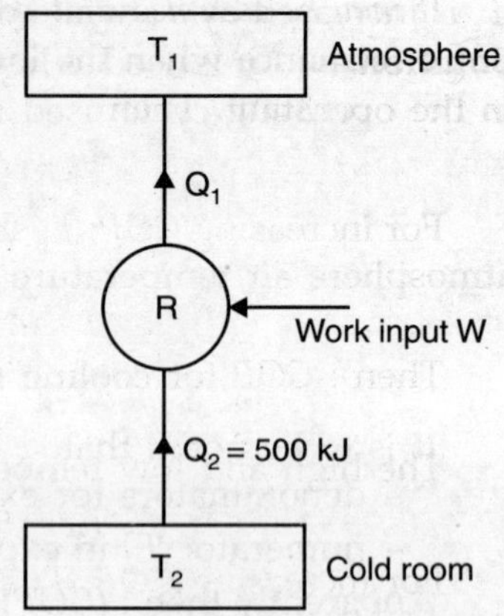

Fig. 3.2.

(*b*) The COP for cooling (refrigerator) is given by

$$(COP)_{\text{cooling}} = \frac{T_2}{T_1 - T_2}$$

where T_1 and T_2 are the highest and lowest temperatures in the cycle

$$\therefore \quad 3.6 = \frac{T_2}{T_1 - T_2} = \frac{1}{\frac{T_1}{T_2} - 1}$$

That gives : $\frac{T_1}{T_2} = \frac{1}{3.6} + 1 = \mathbf{1.278}$

(*ii*) $COP = \frac{\text{refrigerating effect}}{\text{work input}}$; $3.6 = \frac{\text{refrigerating effect}}{9.5}$

$\therefore$ Refrigerating effect = 3.6 × 9.5 = 34.2 kW

Since 1 tonne of refrigeration is equivalent to 3.5 kW, the refrigerant effect in tonnes is

$$= \frac{34.2}{3.5} = \mathbf{9.77 \text{ tonnes}}$$

(*iii*) The COP for heating (heat pump) is

$$(COP)_{\text{heating}} = \frac{T_1}{T_1 - T_2} = \frac{T_1/T_2}{T_1/T_2 - 1}$$

$$= \frac{1.278}{1.278 - 1} = 4.6$$

Alternatively : When working between the same temperature limits

$$(COP)_{\text{heat pump}} = 1 + (COP)_{\text{refrigerator}}$$

$$= 1 + 3.6 = 4.6$$

EXAMPLE 3.4

A reversed Carnot cycle is used for cooling in summer maintained at 5°C when the ambient temperature is 35°C. The same cycle is used for heating in winter at 50°C when the ambient temperature is 15°C. Assuming a temperature difference of 5°C in the heat exchanger that exchanges heat with the surroundings, compute the performance index both for the heating and cooling cycles.

(b) How the Carnot COP will be affected if the water at 28°C is used as cooling medium with 3°C temperature difference for summer air conditioning.

(c) What will be the power consumption per tonne of refrigeration in each case ?

It may be assumed that the air and water temperatures remain constant during heat transfer.

Solution: (*a*) Air is used as heat transfer medium. The high (suffix 1) and low (suffix 2) temperatures in the operating cycle used for cooling (refrigeration) are

$$T_1 = (35 + 5)°C = 313 \text{ K} \quad \text{and} \quad T_2 = 5°C = 278 \text{ K}$$

For increasing *COP*, T_1 should be kept as low as possible. However, it cannot be less than the atmosphere air temperature and accordingly T_1 has been taken as (35 + 5) = 40°C = 313 K.

$$\text{Then : } COP \text{ for cooling (refrigerator)} = \frac{T_2}{T_1 - T_2} = \frac{278}{313 - 278} = 7.94$$

The high and low temperatures in the operating cycle used for heating (heat pump) are

$$T_1 = 50°C = 323 \text{ K} \quad \text{and} \quad T_2 = (15 - 5)°C = 283 \text{ K}$$

For increasing *COP*, T_2 should be kept high but it cannot be increased above 15°C which is the required temperatures. According T_2 has been taken as (15 – 5) = 10°C = 283 K

$$\text{Then : } COP \text{ for heating (heat pump)} = \frac{T_1}{T_1 - T_2} = \frac{323}{323 - 283} = 8.07$$

(*b*) Water is used as heat transfer medium

$$T_1 = (28 + 3)°C = 304 \text{ K} \quad \text{and} \quad T_2 = 5°C = 278 \text{ K}$$

$$COP \text{ for cooling (refrigeration)} = \frac{T_2}{T_1 - T_2} = \frac{278}{304 - 278} = 10.69$$

(*c*) 1 tonne of refrigeration = 3.5 kW

$$COP = \frac{\text{heat absorbed (refrigeration effect)}}{\text{work input}} = \frac{N}{W}$$

∴ When air is used as working medium

$$W = \frac{N}{COP} = \frac{3.5}{7.94} = \mathbf{0.44} \text{ kW}$$

When water is used as working medium

$$W = \frac{N}{COP} = \frac{3.5}{10.69} = \mathbf{0.327} \text{ kW}$$

EXAMPLE 3.5

A refrigerating machine working between temperature limits of – 13°C and 37°C has 90% relative coefficient of performance. If it consumes 4.8 kW power, what is its tons of refrigeration capacity. For same TR capacity, how much power will be consumed by a Carnot refrigerator? Also for the same power consumption, determine TR capacity of a Carnot refrigerator operating between the same temperature limits.

(UPTU)

Solution: (*a*) The relative *COP* of a refrigerator is prescribed as

$$\text{Relative } COP = \frac{\text{actual } COP}{\text{theoretical } COP}$$

The theoretical *COP* here refers to the *COP* if the machine works on Carnot cycle for which

$$(COP)_{\text{Carnot}} = \frac{T_2}{T_1 - T_2} = \frac{260}{310 - 260} = 5.2$$

$\therefore$ Actual $COP = 5.2 \times 0.9 = 4.68$

Now : Actual $COP = \frac{R_n}{W}$ where R_n is the net refrigerating effect and W is the power consumed by the machine during the same time interval.

$$\therefore \quad R_n = 4.8 \times 4.68 = 22.464 \text{ kW}$$

Since 1 tonne of refrigeration is equivalent to 3.5 kW,

$$R_n = \frac{22.464}{3.5} = \mathbf{6.418} \text{ tonne}$$

(*b*) $R_n = 22.464$ kW ; $(COP)_{\text{Carnot}} = 5.12$

$$\therefore \quad \text{Power consumption} = \frac{R_n}{(COP)_{\text{Carnot}}} = \frac{22.464}{5.12} = \mathbf{4.387} \text{ kW}$$

(*c*) Power consumption = 4.8 kW ; $(COP)_{\text{Carnot}} = 5.12$

$$\therefore \quad \text{Refrigeration effect} = 4.8 \times 5.12$$

$$= 24.576 \text{ kW} \quad \text{or} \quad 7.022 \text{ tonne}$$

EXAMPLE 3.6

A refrigerator system operating on a reversed Carnot cycle produces 400 kg/hr of ice at – 5°C from water at 30°C. Make calculations for

(a) the power required to drive the machine,

(b) the heat rejected from the system.

Take latent heat of freezing = 335 kJ/kg and specific heat of ice = 2.1 kJ/kgK

Solution: $T_1 = 25°C = 298$ K and $T_2 = -5°C = 268$ K

In terms of operating temperatures, the COP of the machine is given as

$$COP = \frac{T_2}{T_1 - T_2} = \frac{268}{298 - 268} = 8.93$$

Heat absorbed or net refrigerating effect

$$= (5 \times 2.1 + 335 + 4.186 \times 30) \times 400$$

$$= 188432 \text{ kJ/hr} = 52.34 \text{ kJ/s}$$

Now, $$COP = \frac{\text{refrigerating effect}}{\text{work input}}\ ; \quad 8.93 = \frac{52.34}{W}$$

$$\therefore \quad \text{Work required to run the machine, } W = \frac{52.34}{8.93} = 5.86 \text{ kJ/s} = \mathbf{5.86 \text{ kW}}$$

(*b*) From the principle of energy conservation (1st law)

heat rejected = heat absorbed + work input

$$= 52.34 + 5.86 = \mathbf{58.2 \text{ kJ/s}}$$

EXAMPLE 3.7

A cold storage is to be maintained at – 5°C while the surroundings are at 35°C. The heat leakage from the surroundings into the cold storage is estimated to be 29 kJ/s and the actual coefficient of performance of the refrigeration plant is one-third of an ideal plant working between the same temperature limits. Find the power required in kW to drive the plant.

Solution : Refer Fig. 3.3 for the schematic diagram for the given system.

If the cold storage is to be maintained at a constant temperature of 268 K, it must reject heat at the same rate at which heat is leaking into it from the surroundings.

That is : $Q_2 = 29$ kJ/s

$$COP \text{ for ideal refrigerator} = \frac{T_2}{T_1 - T_2} = \frac{268}{308 - 268} = 6.7$$

$$\text{Actual } COP = \frac{1}{3} \times 6.7 = 2.233$$

Also : $$COP = \frac{\text{refrigerating effect}}{\text{work input}} = \frac{Q_2}{W}$$

$$\therefore \text{ Work input required} = \frac{Q_2}{COP} = \frac{29}{2.233}$$

$$= 12.987 \text{ kJ/s} = \mathbf{12.987 \text{ kW}}$$

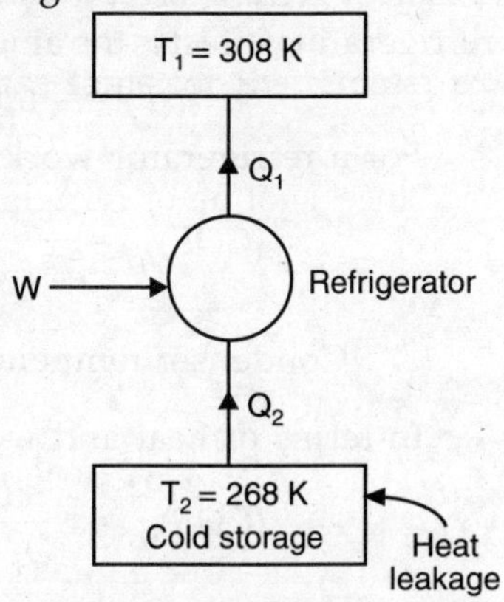

Fig. 3.3

EXAMPLE 3.8

A refrigerator working on Carnot cycle has a coefficient of performance of 5.0 when it extracts 8350 kJ/min from a space to be cooled. Determine (i) the heat rejected to the surroundings and (ii) the power required to run the compressor.

If the temperature maintained in the refrigerator is 253°C, determine the temperature at which heat is rejected.

Solution : Refer Fig. 3.4 for the schematic arrangement of the system. For a refrigerator,

$$COP = \frac{\text{heat absorbed from the system being cooled}}{\text{work input}}$$

$\therefore$ Work input to compressor

$$= \frac{8350}{5} = 1670 \text{ kJ/min} = 27.83 \text{ kJ/s}$$

Thus the compressor requires 27.83 kW power for its operation.

(*ii*) From the principle of energy conservation (1st law),

$$Q_1 = Q_2 + W = 8350 + 1670 = 10020 \text{ kJ/min}$$

Thus 10020 kJ/min of heat is rejected to the surroundings.

(*iii*) For a refrigerator operating on Carnot cycle, the *COP* is given by

$$COP = \frac{T_2}{T_1 - T_2} \;;\; 5 = \frac{253 + 273}{T_1 - (253 + 273)}$$

Solution gives : T_1 = **631.2 K** or **358.2°C**

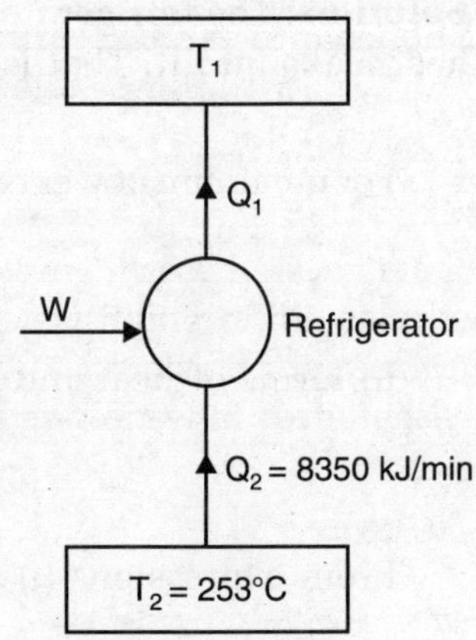

Fig. 3.4

EXAMPLE 3.9

A refrigerator operates on a reversed Carnot cycle whose coefficient of performance is 5.0. The evaporator is maintained at a temperature of – 6°C and the power required to run the refrigerator is 3.5 kW. Determine the refrigerating effect and the condenser temperature of the refrigerator.

Solution : Refer Fig 3.5.

In a refrigerator system, heat is removed from a body at low temperature (evaporator) and this heat is transferred to another body (condenser) at a high potential of temperature. Further, refrigerating effect is the amount of heat extracted from the body at low temperature (evaporator).

$$T_2 = -6°C = 273 + (-6) = 267 \text{ K}$$

For a refrigerator working on reversed Carnot cycle,

$$(COP)_{ref} = \frac{T_2}{T_1 - T_2}; \quad 5 = \frac{267}{T_1 - 267}$$

$\therefore$ Condenser temperature, $T_1 = 267 + \frac{267}{5} = \mathbf{320.4\ K}$

In terms of heat and work interactions,

$$(COP)_{ref} = \frac{Q_2}{Q_1 - Q_2} = \frac{Q_2}{W}$$

$\therefore$ Heat extracted from the space being cooled (refrigeration effect),

$$Q_2 = W \times (COP)_{ref}$$
$$= 3.5 \times 5 = \mathbf{17.5\ kW}$$

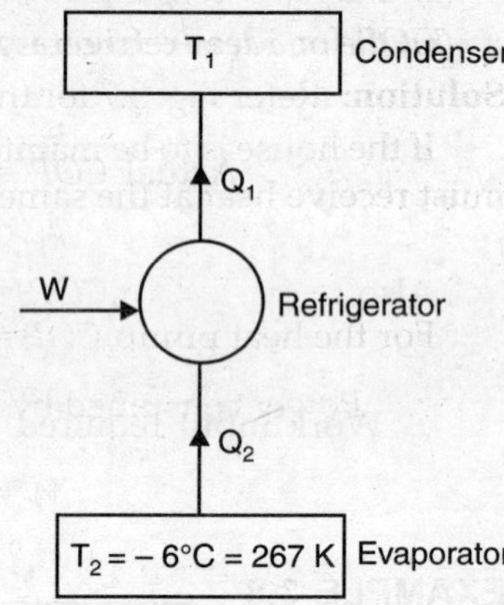

Fig. 3.5

EXAMPLE 3.10

A domestic food refrigerator maintains a temperature of –10°C whilst the ambient air temperature is – 30°C. The heat leakage into the freezer is estimated to be at the continuous rate of 2 kJ/s. Determine the least power needed to pump out this heat continuously.

Solution : The refrigerator is required to remove heat from the freezer at the same rate at which heat leaks into it. That is

$$Q_2 = 2 \text{ kJ/s}$$

For minimum power requirement, the refrigerator operates on reversed Carnot cycle. Accordingly

$$COP = \frac{T_2}{T_1 - T_2} \quad ...(i)$$

In terms of heat and work interactions

$$COP = \frac{Q_2}{W} = \frac{Q_2}{Q_1 - Q_2} \quad ...(ii)$$

From expressions (*i*) and (*ii*)

$$\frac{T_1 - T_2}{T_2} = \frac{Q_1 - Q_2}{Q_2} \text{ or } \frac{T_1}{T_2} = \frac{Q_1}{Q_2}$$

$$\therefore \quad Q_1 = \frac{T_1}{T_2} \times Q_2 = \frac{303}{263} \times 2 = 2.304 \text{ kJ/s}$$

and $\quad W = Q_1 - Q_2 = 2.304 - 2 = \mathbf{0.304}$ kJ/s

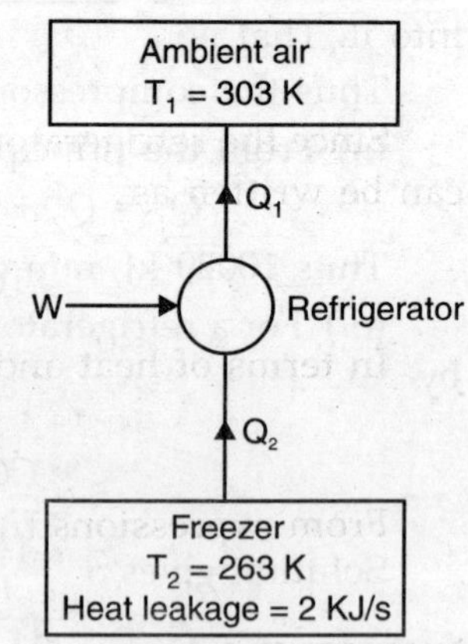

Fig. 3.6

Hence the least power required to pump the heat continuously is 0.304 kW

EXAMPLE 3.11

A heat pump is used to meet the heating requirements of a house and maintain it at 300 K. On a day when the outdoor air temperature drops to 275 K, it is estimated that the house loses heat at the rate of 75 MJ/hr. If the coefficient of performance of the heat pump under these conditions is 2.5, determine:

(a) the power consumed by the heat pump, and

(b) the rate at which heat is absorbed from the cold outside air.

Solution: Refer Fig. 3.7 for the schematic diagram for the given system.

If the house is to be maintained at a constant temperature of 300 K, it must receive heat at the same rate at which it is losing heat. That is

$$Q_1 = 75 \text{ MJ/hr}$$

For the heat pump; $COP = \dfrac{Q_1}{W}$

∴ Power consumed by the pump,

$$W = \frac{75}{2.5} = 30 \text{ MJ/hr}$$

$$W = \frac{30 \times 10^3}{3600} = 8.33 \text{ kJ/s} = 8.33 \text{ kW}$$

(*b*) From the principle of energy conservation; $Q_1 = W + Q_2$

∴ Rate at which heat is absorbed from the cold outside air is

$$Q_2 = Q_1 - W = 75 - 30 = \mathbf{45\ MJ/hr}$$

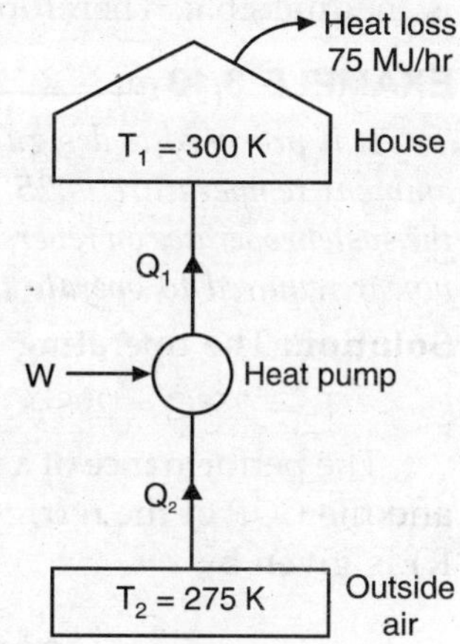

Fig. 3.7

EXAMPLE 3.12

A refrigerator operating on a reversed Carnot cycle consumes 150 W power in summer when the ambient atmosphere is at 310 K. The heat leakage into the refrigerator through the doors is estimated to be at the continuous rate of 15 W per degree temperature difference between the ambient air and the cold space of the refrigerator. For continuous operation of the refrigerator, determine the temperature at which the cold space is maintained.

Solution : The refrigerator is requried to remove heat from the cold space at which heat leaks into it. That is

$$Q_2 = 15(T_1 - T_2)$$

Since the refrigerator operates on the reversed Carnot cycle, its coefficient of performances can be written as

$$COP = \frac{T_2}{T_1 - T_2} \quad \text{...}(i)$$

In terms of heat and work interactions

$$COP = \frac{Q_2}{W} \quad \text{...}(ii)$$

From expressions (*i*) and (*ii*),

$$\frac{T_2}{T_1 - T_2} = \frac{15(T_1 - T_2)}{150}; \quad \frac{T_2}{310 - T_2} = \frac{310 - T_2}{10}$$

or $\quad (310 - T_2)^2 = 10\,T_2$

or $\quad 96100 + T_2^2 - 620\,T_2 = 10\,T_2$

or $\quad T_2^2 - 630\,T_2 + 96100 = 0$

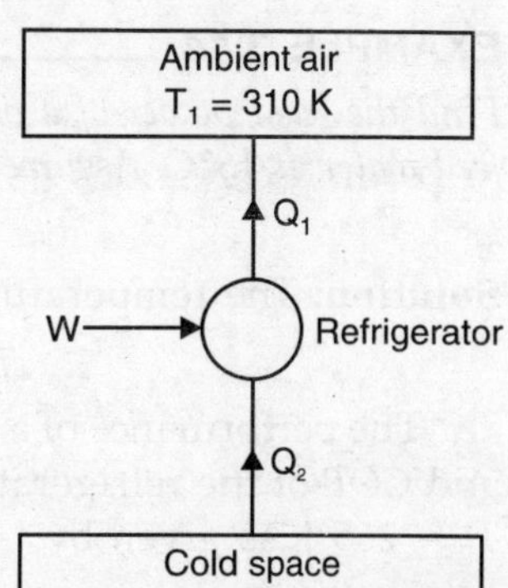

Fig. 3.8

$$T_2 = \frac{630 \pm \sqrt{396900 - 384400}}{2}$$

$$= \frac{630 \pm 111.80}{2} = 370.9 \text{ K or } 259.1 \text{ K}$$

The temperature $T_2 = 370.9$ K is greater than the temperature of ambient air and as such is inadmissible. Therefore the cold space has to be maintained at a temperature of **259.1 K.**

EXAMPLE 3.13

It is proposed to design a refrigeration plant for a food store which is to be maintained at –5 °C. The ambient temperature is 25 °C and the estimated heat transfer from the store is at the rate of 5 kW. If the system operates on reversed Carnot heat engine cycle, determine the performance index and the minimum power required to operate the refrigeration plant.

Solution: The operating temperatures of the system are

$T_2 = 25°C = 298$ K and $T_2 = -5°C = 268$ K

The performance of a refrigeration system is expressed in terms of coefficient of performance, and the *COP* of the refrigerator working between two thermal reservoirs T_1 (298 K) and T_2 (268 K) is given by

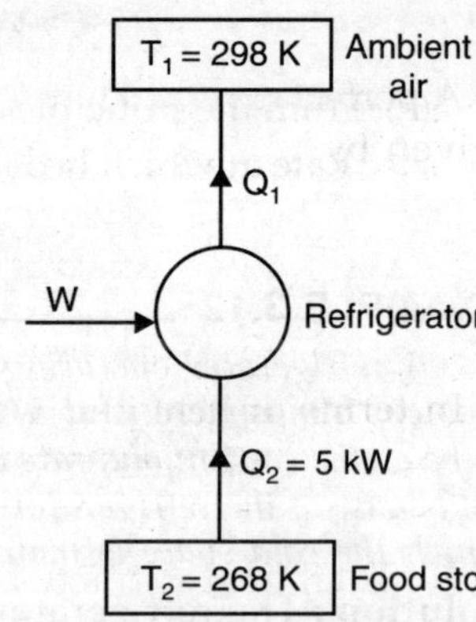

Fig. 3.9

$$(COP)_{ref} = \frac{T_2}{T_1 - T_2} = \frac{268}{298 - 268} = 8.93$$

In terms of heat interactions,

$$(COP)_{ref} = \frac{Q_2}{Q_1 - Q_2}; \quad 8.93 = \frac{5}{Q_1 - 5}$$

or $8.93\, Q_1 - 44.65 = 5$; $Q_1 = \dfrac{49.65}{8.93} = 5.56$ kW

From principle of energy conservation (1st law): $Q_1 = W + Q_2$

∴ Power required to operate the refrigerator,

$$W = Q_1 - Q_2 = 5.56 - 5 = \mathbf{0.56\ kW}$$

A reversible refrigerator requires the minimum power to operate it. Therefore, the minimum power required is 0.56 kW.

EXAMPLE 3.14

Find the least power of a perfect reversed heat engine that makes 400 kg of ice per hour at –8°C from feed water at 18°C. Assume specific heat of ice as 2.09 kJ/kg K and latent heat of solidification 334 kJ/kg.

(JNTU)

Solution: The temperature limits for the reversed heat engine are :

$T_1 = 18°C = 291$ K and $T_2 = -8°C = 265$ K

The performance of a refrigerating system is expressed in terms of coefficient of performance and *COP* of the refrigeration working reversibly between temperature limits of $T_1 = 291$ K and $T_2 = 265$ K is given by

$$COP = \frac{T_2}{T_1 - T_2} = \frac{265}{291 - 265} = 10.19$$

Heat removal from one kg of water at 18°C so that it becomes ice at –8°C is

$$= m[c_{pw}\ \Delta T_w + L + c_{pi}\ \Delta T_i]$$

$$= 1\ [4.186\ (18 - 0) + 334 + 2.09\ \{0 - (-8)\}] = 426.0\ \text{kJ/kg}$$

In terms of refrigerating effect and work input, the COP of the refrigerating machine is

$$COP = \frac{\text{refrigerating effect}}{\text{work input}}\ ;\ 10.19 = \frac{426.07 \times 400}{W}$$

$$\therefore \quad W = \frac{426.07 \times 400}{10.19} = 16725\ \text{kJ/hr} = 4.65\ \text{kJ/s} = 4.6\ \text{kW}$$

A reversible refrigerator requires the minimum power to operate it. Therefore the minimum power required is 4.6 kW.

EXAMPLE 3.15

A perfect reversed heat engine is used for making ice at– 5°C from water available at 25°C. The temperature of freezing mixture is – 10°C . Calculate the quantity of ice formed per kWh.

For ice : specific heat = 2.1 kJ/kgK and latent heat = 335 kJ/kg

Solution : The temperature limits for the reversed heat engine are:

$$T_1 = 25°\text{C} = 298\ \text{K and}\ T_2 = -10°\text{C} = 263\ \text{K}$$

A perfect reversed heat engine works on Carnot cycle for which *COP* in terms of temperatures is given by

$$COP = \frac{T_2}{T_1 - T_2} = \frac{263}{298 - 263} = 7.514$$

$$\text{Work input} = 1\ \text{kWh} = 3600\ \text{kJ}$$

In terms of heat and work interactions,

$$COP = \frac{Q_2}{Q_1 - Q_2} = \frac{Q_2}{W}$$

∴ Total heat extracted from ice,

$$Q_2 = W \times COP = 3600 \times 7.514 = 27050\ \text{kJ/hr}$$

Removal of heat from 1 kg of water at 25°C so that it becomes ice at – 5°C is given by

$$= m\ c_{pw}\ \Delta T + mL + m\ c_{pi}\ \Delta T$$

$$= 1 \times 4.186\ (25 - 0) + 1 \times 335 + 1 \times 2.1\ [0 - (-5)]$$

$$= 104.65 + 335 + 10.5 = 450.15\ \text{kJ}$$

$$\therefore \quad \text{Ice formed per kWh} = \frac{27050}{450.15} = \mathbf{60.09\ kg}$$

EXAMPLE 3.16

A cold storage plant is to be built to store 25 tons of fish. The fish is supplied at 30°C and it is required to be stored at – 10°C. If the required cooling effect is achieved within 10 hours, make calculations for :

(a) capacity of the refrigerating plant,

(b) coefficient of performance of Carnot cycle operating between the given temperature limits, and

(c) power required to run the compressor if actual COP is 40 percent of the Carnot COP.

The pertinent data related to first is

Specific heat of fish below freezing point = 1.26 kJ/kgK

Specific heat of fish above freezing point = 2.98 kJ/kgK

Freezing point of fish = – 3°C and
Latent heat of fish = 230 kJ/kg

Solution : Heat removal from each kg of fish in 10 hours

= heat removed from fish above freezing point + latent heat + heat removed from fish below freezing point

$$= 2.93\,[30 - (-3)] + 230 + 1.26\,[-3 - (-10)]$$

$$= 96.69 + 230 + 8.82 = 335.51 \text{ kJ/kg}$$

Total heat removed per second by the refrigerating plant

$$= \frac{335.51 \times (25 \times 1000)}{(10 \times 3600)} = 233 \text{ kJ/s}$$

(*a*) 1 ton of refrigeration = 3.5 kJ/s

$\therefore$ Capacity of the refrigerating plant $= \frac{233}{3.5}$ = **66.57 ton**

(*b*) Carnot $COP = \frac{T_2}{T_1 - T_2} = \frac{-10+273}{(30+273)-(-10+273)}$ = **6.575**

(*c*) Actual COP = 0.4 × Carnot COP = 0.4 × 6.575 = 2.63

In terms of refrigerating effect and work input, the actual *COP* is given as

$$\text{Actual } COP = \frac{\text{refrigerating effect}}{\text{work input}}$$

$$\therefore \quad \text{Work input} = \frac{233}{2.63} = 88.59 \text{ kJ/s}$$

Hence, power required to run the plant = **88.59 kW**

EXAMPLE 3.17

Two Carnot refrigerators A and B operate in series. The refrigerator A absorbs energy at the rate of 1 kJ/s from a body at 300 K and rejects energy as heat to a body at temperature T. The refrigerator B absorbs the same quantity of energy which is rejected by the refrigerator A from the body at T and rejects energy as heat to a body at 1000 K. If both the refrigerators have the same COP, calculate

(i) the temperature of the body,

(ii) the COP of the refrigerator, and

(iii) the rate at which energy is rejected as heat to the body at 1000 K.

Solution : Refer Fig. 3.10 for the schematics of the given refrigerators operating in series.

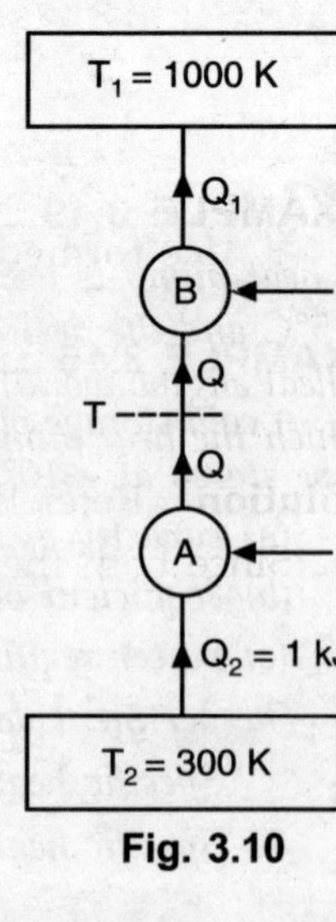

Fig. 3.10

Both the refrigerators have the some *COP*. Accordingly

$$\frac{T_2}{T - T_2} = \frac{T}{T_1 - T}$$

or $$\frac{300}{T - 300} = \frac{T}{1000 - T}$$

or $$300000 - 300\,T = T^2 - 300\,T$$

$\therefore$ Temperature of body,

$$T = \sqrt{300000} = \mathbf{547.72\ K}$$

(*ii*) *COP* of refrigerator A = $\dfrac{T_2}{T-T_2} = \dfrac{300}{547.72-300}$ = **1.21**

The *COP* of refrigerator B will also be equal to 1.21

(*iii*) For refrigerator A,

$$W_a = \frac{Q_2}{COP} = \frac{1}{1.21} = 0.826 \text{ kJ/s}$$

Then from the principle of energy conservation (1st law)

$$Q = Q_2 + W_a = 1 + 0.826 = 1.826 \text{ kJ/s}$$

For refrigerator B,

$$W_b = \frac{Q}{COP} = \frac{1.826}{1.21} = \mathbf{1.509\ kJ/s}$$

Then energy rejected as heat to the body at 1000 K

$$Q_1 = Q + W_b = 1.8267 + 1.509 = 3.3351 \text{ kJ/s.}$$

EXAMPLE 3.18

With reference to the adjoining figure, E is the heat engine with efficiency 0.4 and R is a refrigerator. If $Q_2 + Q_4 = 3Q$, compute the coefficient of performance for the refrigerator. *(GATE)*

Solution: (*a*) For the heat engine *E*,

$$\eta = \frac{\text{work done}}{\text{heat input}} = \frac{W}{Q_1} = \frac{Q_1-Q_2}{Q_1}$$

or $0.4 = \dfrac{Q_1-Q_2}{Q_1}$ This gives : $0.6\,Q_1 = Q_2$

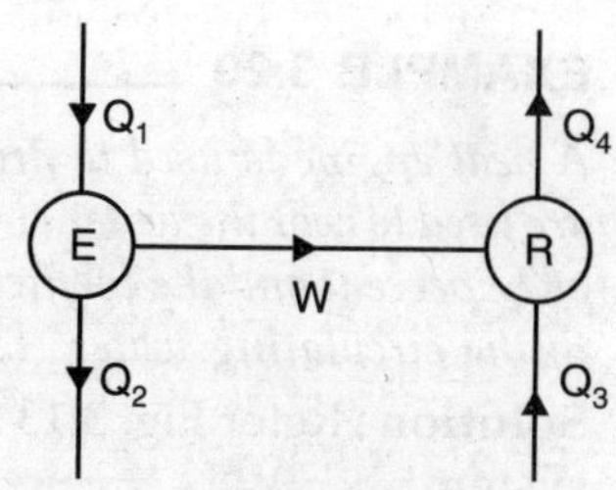

Fig. 3.11

For the refrigerator *R*,

$$W + Q_3 = Q_4$$

or $(Q_1 - Q_2) + Q_3 = Q_4$

or $Q_1 + Q_3 = Q_2 + Q_4 = 3Q_1$ (given) $\quad \therefore \quad 2\,Q_1 = Q_3$

The *COP* for the refrigerator is

$$COP = \frac{Q_3}{W} = \frac{Q_3}{Q_1-Q_2} = \frac{2Q_1}{Q_1-0.6Q_1} = 5.0$$

EXAMPLE 3.19

A heat pump is used for heating the interior of a house in a cold climate. The ambient temperature is – 5°C and the desired interior temperature is 25°C. The compressor of the heat pump is to be driven by a heat engine working between 1000°C and 25°C. Treating both cycles as reversible, calculate the ratio in which the heat pump and the heat engine share the load.

Solution : Refer Fig. 3.12 for the arrangement of the composite system.

Since both the engine *E* and heat pump *HP* operate on reversible cycles, we have

$$\frac{Q_1}{Q_2} = \frac{T_1}{T_2} = \frac{1273}{298} \quad \text{and} \quad \frac{Q_3}{Q_4} = \frac{T_3}{T_4} = \frac{268}{298}$$

The heat engine drives the pump and accordingly

$$W = (Q_1 - Q_2) = (Q_4 - Q_3)$$

or $$1 - \frac{Q_2}{Q_1} = \frac{Q_4}{Q_1} - \frac{Q_3}{Q_1}$$

$$1 - \frac{Q_2}{Q_1} = \frac{Q_4}{Q_1} - \frac{Q_3}{Q_4} \times \frac{Q_4}{Q_1}$$

$$= \frac{Q_4}{Q_1}\left(1 - \frac{Q_3}{Q_4}\right)$$

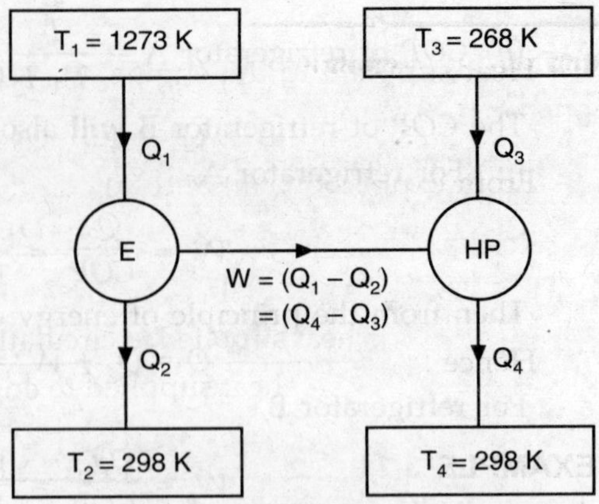

Fig. 3.12

Substituting the appropriate values, we get

$$1 - \frac{298}{1273} = \frac{Q_4}{Q_1}\left(1 - \frac{268}{298}\right)$$

or $$\frac{1273 - 298}{1273} = \frac{Q_4}{Q_1}\left(\frac{298 - 268}{298}\right)$$

That gives : $$\frac{Q_4}{Q_1} = \frac{1273 - 298}{1273} \times \frac{298}{298 - 268} = \mathbf{7.608}$$

EXAMPLE 3.20

A heat engine is used to drive a heat pump. The heat transfers from the heat engine and the heat pump are used to heat the water circulating though the radiators of a building. The engine has a thermal efficiency of 27 percent and the coefficient of performance for the heat pump is 4. Determine the ratio of heat transfer to the circulating water to the heat supplied to the heat engine.

Solution : Refer Fig. 3.13 for the arrangement of the composite system.

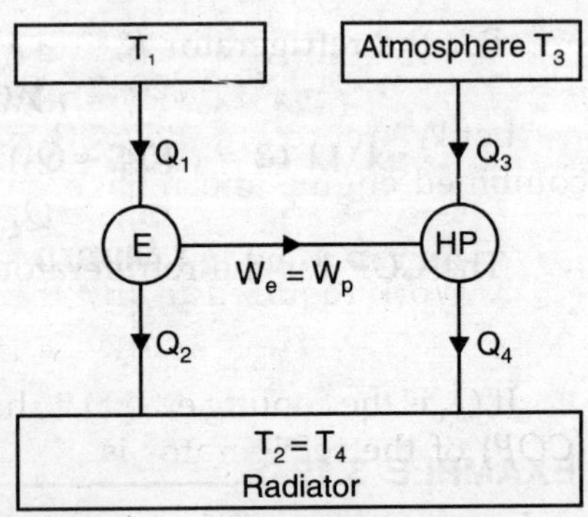

Fig. 3.13

For the engine, $$\eta = \frac{W_e}{Q_1} = \frac{W_e}{Q_2 + W_e} = 0.27$$

For the pump, $$COP = \frac{Q_4}{W_p} = \frac{Q_4}{W_e} = 4$$

From the above identities

$$0.27 = \frac{\frac{Q_4}{4}}{Q_2 + \left(\frac{Q_4}{4}\right)}$$

or $$0.27 = \frac{Q_4}{4Q_2 + Q_4}$$ or $$1.08\, Q_2 + 0.27\, Q_4 = Q_4$$

or $$1.08\, Q_2 = 0.73\, Q_4$$ or $$\frac{Q_4}{Q_2} = \frac{1.08}{0.73} = 1.48$$

$$\therefore \quad \frac{Q_2 + Q_4}{Q_2} = 2.48 \qquad \ldots(a)$$

Heat supplied to the engine, $Q_1 = Q_2 + W_e = Q_2 + \frac{Q_4}{4} = Q_2 + \frac{1.48Q_2}{4}$

$$\therefore \quad Q_1 = 1.37\ Q_2 \qquad ...(b)$$

From expressions (*a*) and (*b*)

$$\frac{Q_2+Q_4}{Q_1} = \frac{Q_2+Q_4}{Q_2} \times \frac{Q_2}{Q_1} = \frac{Q_2+Q_4}{Q_2} \times \frac{Q_2}{1.37\ Q_2} = \frac{2.48}{1.37} = 1.81$$

Hence : $\frac{\text{heat supplied to circulating water}}{\text{heat supplied to engine}} = \mathbf{1.81}$

EXAMPLE 3.21

A reversible heat engine operates between 875 K and 310 K and drives a reversible refrigerator operating between 310 K and 255 K. The engine receives 2000 kJ of heat and the net work output from the arrangement equals 350 kJ. Make calculations for the cooling effect.

Solution : Refer Fig. 3.14 for the arrangement of the composite system.

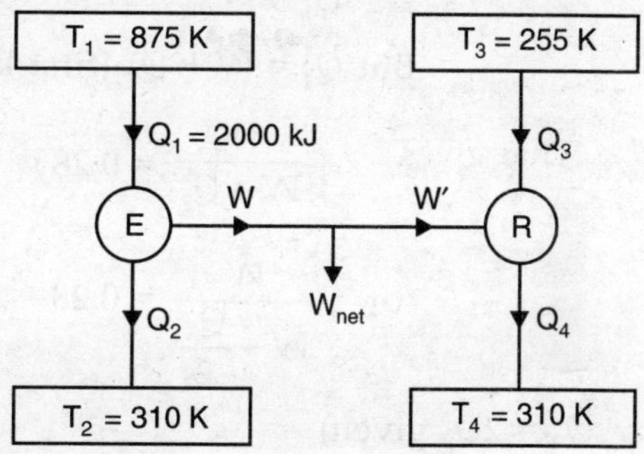

Fig. 3.14

If W is the work performed by the engine, then

$$\eta = \frac{W}{Q_1} = \frac{W}{2000}$$

Since the engine is reversible and it operates between temperatures T_1 and T_2, the efficiency is also given by

$$\eta = \frac{T_1 - T_2}{T_1} = \frac{875 - 310}{875} = 0.6457$$

$$\therefore \quad \frac{W}{2000} = 0.6457; \quad W = 1291.4 \text{ kJ}$$

Let W' represent the work needed to run the refrigerator. Then the net work output of the combined engine and refrigerator is

$$W_{net} = W - W'; \quad W' = W - W_{net}$$

∴ Work required to run the refrigerator,

$$W' = 1291.4 - 350 = 941.4 \text{ kJ}$$

If Q_3 is the cooling effect *i.e.*, heat extracted by the refrigerator, then coefficient of performance (*COP*) of the refrigerator is

$$COP = \frac{Q_3}{W'} = \frac{Q_3}{941.4}$$

Since the refrigerator is reversible and it operates between temperatures T_3 and T_4, its *COP* is also given by

$$COP = \frac{T_3}{T_4 - T_3} = \frac{255}{310 - 255} = 4.636$$

$$\therefore \quad \frac{Q_3}{941.4} = 4.636; \quad Q_3 = \mathbf{4364.3} \text{ kJ}$$

The refrigerator thus has a cooling effect equal to 4364.3 kJ

EXAMPLE 3.22

A heat engine operating between two reservoirs at 1000 K and 300 K is used to drive a heat pump which extracts heat from the reservoir at 300 K, at a rate twice that at which the engine rejects heat to it. If the efficiency of the engine is 40% of the maximum possible and the coefficient of performance of the heat pump is 50% of the maximum possible, make calculations for the temperature of the reservoir to which the heat pump rejects heat. Also workout the rate of heat rejection from the heat pump if the rate of supply of heat to the engine is 50 kW.

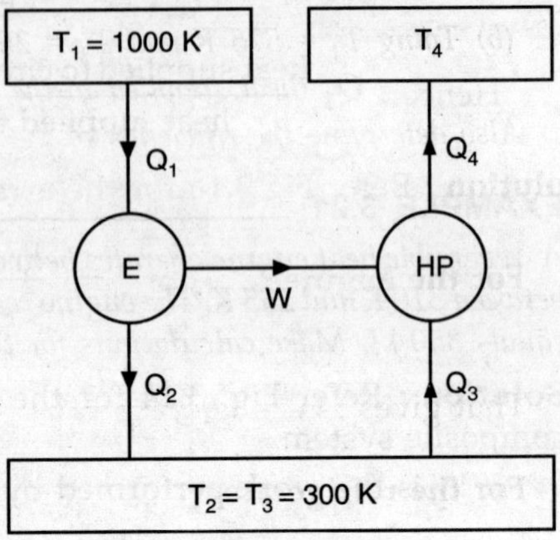

Fig. 3.15

Solution : Refer Fig 3.15.

For engine : $\eta_{max} = \frac{T_1 - T_2}{T_1} = \frac{1000 - 300}{1000} = 0.7$

$\eta_{actual} = 0.4 \times 0.7 = 0.28$

$\therefore \quad \frac{W}{Q_1} = 0.28$

But $Q_1 = W + Q_2$ (first law)

$\therefore \quad \frac{W}{W + Q_2} = 0.28$

or $\quad \frac{W}{W + \frac{Q_3}{2}} = 0.28$

($\because Q_3 = 2Q_2$ given)

That gives : $Q_3 = 5.14\ W$

For heat pump : $(COP)_{max} = \frac{T_4}{T_4 - T_3} = \frac{T_4}{T_4 - 300}$

$(COP)_{actual} = 0.5 \times \frac{T_4}{T_4 - 300}$

$\therefore \quad \frac{Q_4}{W} = 0.5 \times \frac{T_4}{T_4 - 300}$

But, $\quad Q_4 = Q_3 + W$ (1st law) $= 5.14\ W + W = 6.14\ W$

$\therefore \quad \frac{6.14\ W}{W} = 0.5 \times \frac{T_4}{T_4 - 300}\ ; \ T_4 = \mathbf{326.6\ K}$

Therefore, the temperature of the reservoir receiving heat from the heat pump is 326.6 K.

(*b*) $Q_1 = 50$ kW ; $\quad \frac{W}{Q_1} = 0.28\ ; W = 0.28 \times 50 = 14$ kW

$Q_2 = 50 - 14 = 36$ kW ; $\quad Q_3 = 2Q_2 = 36 \times 2 = 72$ kW

$\therefore$ Rate of heat rejection from the heat pump,

$$Q_4 = W + Q_3 = 14 + 72 = \mathbf{86\ kW}$$

EXAMPLE 3.23

A Carnot engine receives heat from the source at temperature T_1 and rejects heat to the sink at temeprature T_2. This engine drives a refrigerator which absorbs heat at T_3 and rejects at T_2.

(a) Determine the ratio $\frac{Q_3}{Q_1}$ where Q_3 is the heat absorbed at T_3 and Q_1 is the heat supplied to the engine at T_1.

(b) Taing T_1 = 575 K and T_3 = 250 K, determine temperature T_2 so that

Q_1 (heat supplied to the engine) = Q_3 (heat abosrbed by the refrigerator)

Also determine the efficiency of the engine and the coefficient of performance for the refrigerator.

Solution : Refer Fig 3.16 for the arrangement

For the engine *E*,
$$\eta_{eng} = \frac{W}{Q_1} = \frac{Q_1 - Q_2}{Q_1} = \frac{T_1 - T_2}{T_1}$$

That gives : $W = Q_1\left[\frac{T_1 - T_2}{T_1}\right]$

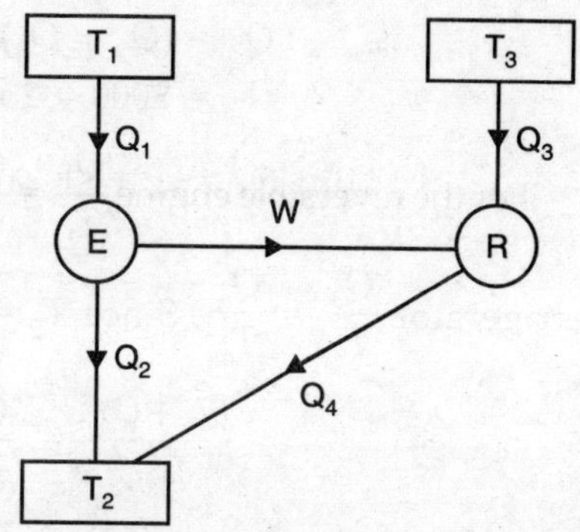

Fig. 3.16

For the refrigerator

$$(COP)_{ref} = \frac{Q_3}{W} = \frac{Q_3}{Q_4 - Q_3} = \frac{T_3}{T_2 - T_3}$$

That gives : $W = Q_3\left[\frac{T_2 - T_3}{T_3}\right]$

From the above two identities

$$Q_1\left[\frac{T_1 - T_2}{T_1}\right] = Q_3\left[\frac{T_2 - T_3}{T_3}\right] \quad \text{or} \quad \frac{Q_3}{Q_1} = \frac{T_1 - T_2}{T_1} \times \frac{T_3}{T_2 - T_3}$$

$$= \frac{T_3}{T_1}\left[\frac{T_1 - T_2}{T_2 - T_3}\right]$$

(*b*) T_1 = 575 K and T_3 = 250 K. Since $Q_1 = Q_3$ (given), we have

$$\frac{T_3}{T_1}\left[\frac{T_1 - T_2}{T_2 - T_3}\right] = 1 \quad \therefore \quad \frac{250}{575}\left[\frac{575 - T_2}{T_2 - 250}\right] = 1$$

or $$250 \times 575 - 250\,T_2 = 575\,T_2 - 575 \times 250$$

$$\therefore \quad T_2 = \frac{2(250 \times 575)}{575 + 250} = \mathbf{348.5\ K}$$

$$\eta_{engine} = \frac{T_1 - T_2}{T_1} = \frac{575 - 348.5}{575} = 0.394 \quad \text{or} \quad \mathbf{39.4\ \%}$$

$$(COP)_{ref} = \frac{T_3}{T_2 - T_3} = \frac{250}{348.5 - 250} = \mathbf{2.54}$$

EXAMPLE 3.24

A reversible heat engine absorbs 2500 kJ/cycle of heat from a constant temperature heat source at 2000 K, and rejects some energy as waste heat to a reservoir X. The work output from the engine is used to drive a reversible refrigerator ; its source temperature being 300 K. The heat outflow from the refrigerator is also taken to the reservoir X. If the total heat flow into the reservoir is 3000 kJ/cycle, make calculations for the temperature of reservoir X.

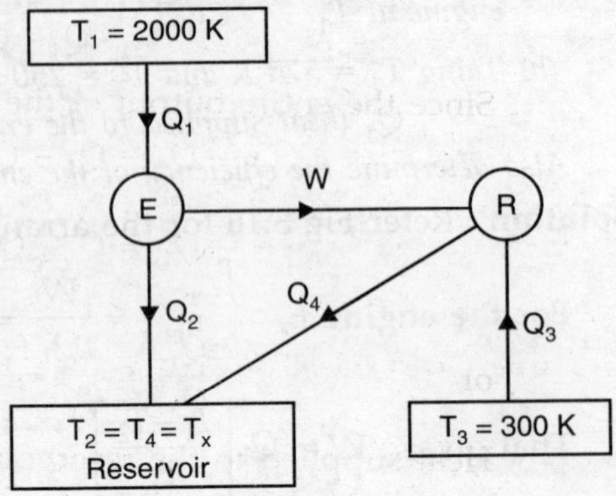

Fig. 3.17

Solution : Refer Fig 3.17

$$Q_2 + Q_4 = 3000 \text{ kJ/cycle}$$

The work developed by the engine equals the work required to run the refrigeration. That is

$$Q_1 - Q_2 = W = Q_4 - Q_3$$

$$\therefore \quad Q_3 = (Q_2 + Q_4) - Q_1$$

$$= 3000 - 2500 = 500 \text{ kJ/cycle}$$

For the reversible engine $\frac{Q_1}{T_1} = \frac{Q_2}{T_2}$ and for the reversible refrigerator $\frac{Q_3}{T_3} = \frac{Q_4}{T_4}$. Since $T_2 = T_4 = T_x$ (say), the summation of these expressions gives

$$\frac{Q_2 + Q_4}{T_x} = \frac{Q_1}{T_1} + \frac{Q_3}{T_3}$$

or $$\frac{3000}{T_x} = \frac{2500}{2000} + \frac{500}{300} = 2.917$$

$\therefore$ Temperature of the common reservoir X,

$$T_x = \frac{3000}{2.917} = \mathbf{1028.5 \ K}$$

EXAMPLE 3.25

A reversible heat engine operates within the higher and lower temperature limits of 1400 K and 400 K respectively. The entire output from this engine is utilized to operate a heat pump. The pump works on reversed Carnot cycle, extracts heat from a reservoir at 300 K and delivers it to the reservoir at 400 K. If 100 kJ/s of net heat is supplied to the reservoir at 400 K; calculate the heat supplied to the engine by the reservoir at 1400 K.

Solution : Refer Fig 3.18 for schematic arrangement of the composite system.

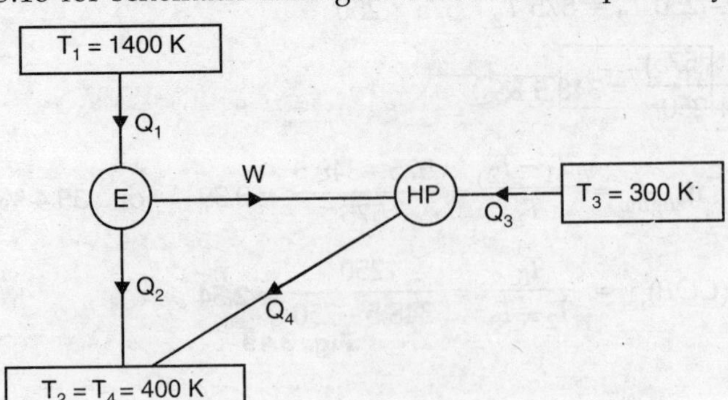

Fig. 3.18

For the reversible engine :

$$\eta = \frac{Q_1 - Q_2}{Q_1} = \frac{T_1 - T_2}{T_1}\ ;\ \frac{Q_1}{Q_2} = \frac{T_1}{T_2}$$

For the reversible pump :

$$COP = \frac{Q_4}{Q_4 - Q_3} = \frac{T_4}{T_4 - T_3}\ ;\ \frac{Q_3}{Q_4} = \frac{T_3}{T_4}$$

Since the entire output of the heat engine is utilised to drive the pump,

$$Q_1 - Q_2 = Q_4 - Q_3$$

$$Q_1 - \frac{T_2}{T_1}Q_1 = Q_4 - Q_4\frac{T_3}{T_4}$$

or

$$\frac{Q_1}{Q_4} = \frac{T_1}{T_4}\left(\frac{T_4 - T_3}{T_1 - T_2}\right) = \frac{1400}{400}\left(\frac{400 - 300}{1400 - 400}\right) = 0.35$$

Heat supplied to the reservoir at 400 K,

$$100 = Q_2 + Q_4 = Q_1\frac{T_2}{T_1} + \frac{Q_1}{0.35} = Q_1 \times \frac{400}{1400} + \frac{Q_1}{0.35} = 3.4284\ Q_1$$

∴ Energy taken from the reservoir at 1400 K,

$$Q_1 = \frac{100}{3.4284} = \mathbf{29.168\ kJ/s}$$

EXAMPLE 3.26

A Carnot heat engine operates between 825 K and 550 K constant temperature thermal reservoirs. Half of the work output from this engine drives a generator and the other half is used to run a reversed Carnot engine that abstracts heat from the 550 K temperature reservoir and rejects that to a thermal reservoir at 715 K. Determine the heat rejected to the reservoir by the reversed engine as a percentage of heat supplied from the 825 K reservoir to the heat engine. Proceed to calculate the heat received per hour if the generator output is 400 kW.

Solution : Refer Fig 3.19 for the schematic arrangement of the composite setup.

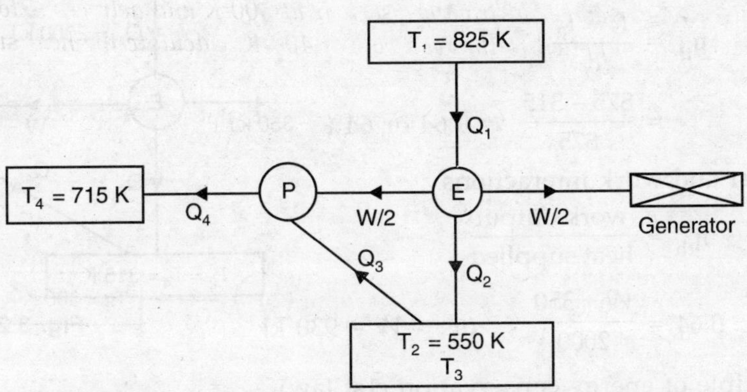

Fig. 3.19

The Carnot heat engine is a reversible engine and its efficiency is given by

$$\eta = \frac{T_1 - T_2}{T_1} = \frac{825 - 550}{825} = 0.333$$

If Q_1 and Q_2 are the heat interactions for the Carnot engine, then

$$\eta = \frac{Q_1 - Q_2}{Q_1} = \frac{W}{Q_1}$$

∴ Work output from the Carnot engine, $W = 0.333\, Q_1$

Thus, work input to the reversed Carnot engine $= \frac{W}{2} = 0.1665 Q_1$

$$COP \text{ of the reversed heat engine} = \frac{T_4}{T_4 - T_3} = \frac{715}{715 - 550} = 4.333$$

In terms of heat and work interactions,

$$COP = \frac{Q_4}{Q_4 - Q_3} = \frac{Q_4}{W_p}$$

where W_p is the work input to the reversed engine.

or $$4.333 = \frac{Q_4}{0.1165\, Q_1}\ ; \quad \frac{Q_4}{Q_1} = 0.7215 = \mathbf{72.15\%}$$

(*b*) Power output from the generator = 400 kW

∴ Total power developed by the Carnot engine = 800 kW

Thus, $0.333\, Q_1 = 800$; $Q_1 =$ **2402.4 kW**

EXAMPLE 3.27

A reversible heat engine working between two thermal reservoirs at 875 K and 315 K drives a reversible refrigerator which operates between the same 315 K reservoir and a reservoir at 260 K. The engine is supplied 2000 kJ of heat and the net work output from the composite system is 350 kJ. Make calculations for the heat transfer to the refrigerator and the net heat interaction with the reservoir at 315 K temperature.

Solution : Refer Fig 3.20.

For the reversible heat engine :

$$\eta_{th} = \frac{T_1 - T_2}{T_1} = \frac{875 - 315}{875} = 0.64 \text{ or } 64\%$$

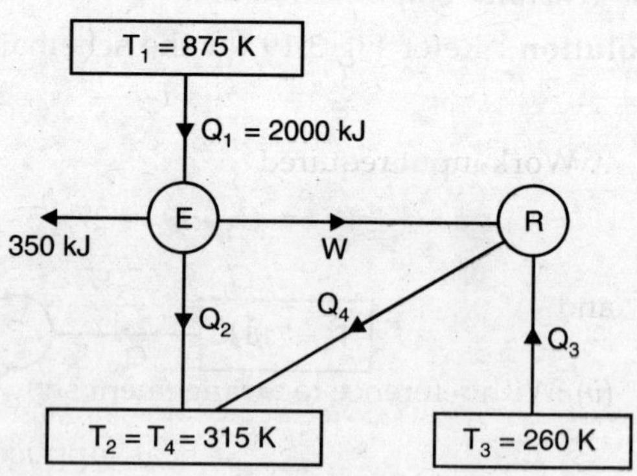

Fig. 3.20

In terms of heat and work interactions

$$\eta_{th} = \frac{\text{work output}}{\text{heat supplied}}$$

$$0.64 = \frac{W + 350}{2000} \quad \text{or} \quad W = 930 \text{ kJ}$$

From the principle of energy conservation (1st law)

$$Q_2 = 2000 - 350 - 930 = 720 \text{ kJ}$$

For the reversible refrigerator :

$$COP = \frac{T_3}{T_3 - T_4} = \frac{260}{315 - 260} = 4.727$$

In terms of heat and work interactions

$$COP = \frac{Q_3}{W}\ ;\ 4.727 = \frac{Q_3}{930} \quad \text{or} \quad Q_3 = \mathbf{4396\ kJ}$$

From the principle of energy conservation (1st law)

$$Q_4 = Q_3 + W = 4396 + 930 = 5326 \text{ kJ}$$

∴ Net heat transfer to the refrigerator at 315 K,

$$= Q_2 + Q_4 = 720 + 5326 = \mathbf{6046\ kJ}$$

EXAMPLE 3.28

A reversible heat pump is used to maintain a temperature of 0°C in a refrigerator when it rejects heat to the surroundings at 25°C. If the heat removal rate from the refrigerator is 25 kJ/s, determine coefficient of performance (COP) of the machine and the work input required.

(b) If the required input to run the pump is developed by a reversible engine which receives heat at 400°C and rejects heat to the atmosphere, then determine the overall COP of the system.

Solution: With reference to arrangement shown in Fig. 3.21,

$$COP = \frac{Q_1}{W} = \frac{Q_1}{Q_1 - Q_2}$$

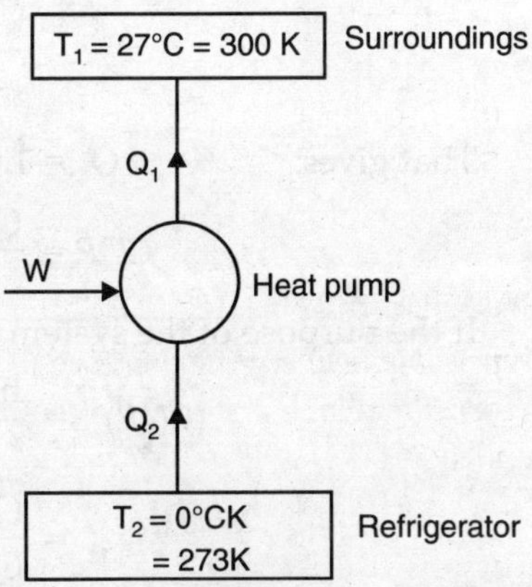

Fig. 3.21

Also $$COP = \frac{T_1}{T_1 - T_2}$$

Thus: $$\frac{Q_1}{Q_1 - Q_2} = \frac{T_1}{T_1 - T_2} \text{ and this identity gives}$$

$$\frac{Q_1}{T_1} = \frac{Q_2}{T_2}$$

$$Q_1 = Q_2 \times \frac{T_1}{T_2} = 25 \times \frac{300}{273} = 27.47 \text{ kJ/s}$$

∴ Work input required

$$W = Q_1 - Q_2 = 27.47 - 25 = 2.47 \text{ kJ/s or } 2.47 \text{ kW}$$

and $$COP = \frac{Q_1}{W} = \frac{27.47}{2.47} = \mathbf{11.12}$$

(b) With reference to arrangement shown in Fig. 3.22,

$$\text{Overall } COP = \frac{\text{heat supplied to surroundings}}{\text{heat supplied from source}} = \frac{Q_1 + Q_4}{Q_3}$$

For the heat pump: $Q_2 = 25$ kJ/s and $Q_1 = 27.47$ kJ/s

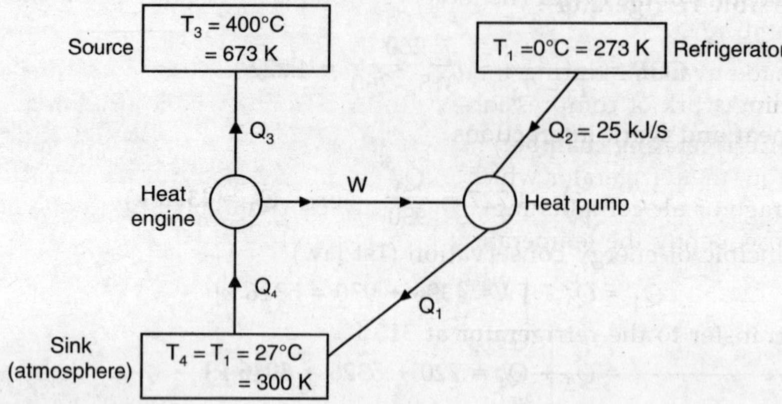

Fig. 3.22

For the reversible engine:

$$\frac{Q_3 - Q_4}{Q_3} = \frac{T_3 - T_4}{T_3}\ ; \quad \frac{Q_3}{T_3} = \frac{Q_4}{T_4}$$

$$\text{or} \qquad \frac{Q_4 + W}{T_3} = \frac{Q_4}{T_4}$$

$$\text{or} \qquad \frac{Q_4 + 2.47}{674} = \frac{Q_4}{300}$$

That gives: $\quad Q_4 = 1.986 \text{ kJ/s} \;$ and $\; Q_3 = 1.986 + 2.47 = 4.454 \text{ kJ/s}$

$$\therefore \qquad COP = \frac{Q_1 + Q_4}{Q_3} = \frac{27.47 + 1.986}{4.456} = 6.61$$

If the purpose of the system is to maintain the refrigerator at the prescribed temperature,

$$(COP)_{ref} = \frac{\text{heat removed from the refrigerator}}{\text{work input}}$$

$$= \frac{Q_2}{Q_3} = \frac{25}{4.456} = \mathbf{5.61}$$

3.2. BELL-COLEMAN CYCLE

3.2.1. Ideal Cycle

Fig. 3.23 shows the schematic arrangement *p-v* and *T-s* plots of the Bell-Coleman air refrigerator working on the reversed Brayton heat engine cycle.

The machine essentially consists of a refrigerating (cold) chamber, a compressor, an air cooler and an expander. The four stages of operation are :

Process 1-2 (Compressor) : The warm air leaving the refrigerating chamber at state point 1 enters the compressor where it is compressed isentropically. During compression, there occurs an increase both in temperature and pressure.

Process 2-3 (Air Cooler) : The hot compressed air leaving the compressor passes through pipes which are immersed in circulating water. The heat is rejected to water at constant pressure $p_2 = p_3$ and that cools the air to temperature T_3.

Process 3-4 (Expander) : The air at high pressure ($p_3 = p_2$) and moderate temperature T_3 is taken to expansion cylinder where it expands isentropically to lower pressure ($p_4 = p_1$) and lowest temperature T_4 existing in the cycle. The temperature T_4 is sufficiently below the temperature of refrigerating chamber.

The work made available during the expansion process assists in running the compressor unit. The remaining work of compression is supplied from an external source.

Process 4-1 (Refrigerating chamber) : The cold low pressure air corresponding to state point 4 absorbs heat from the refrigerator which may be brine solution (for making ice) or space to be cooled (cold storage or air conditioning). This heat absorption takes place at constant pressure ($p_4 = p_1$) and consequently the temperature increases from T_4 to T_1.

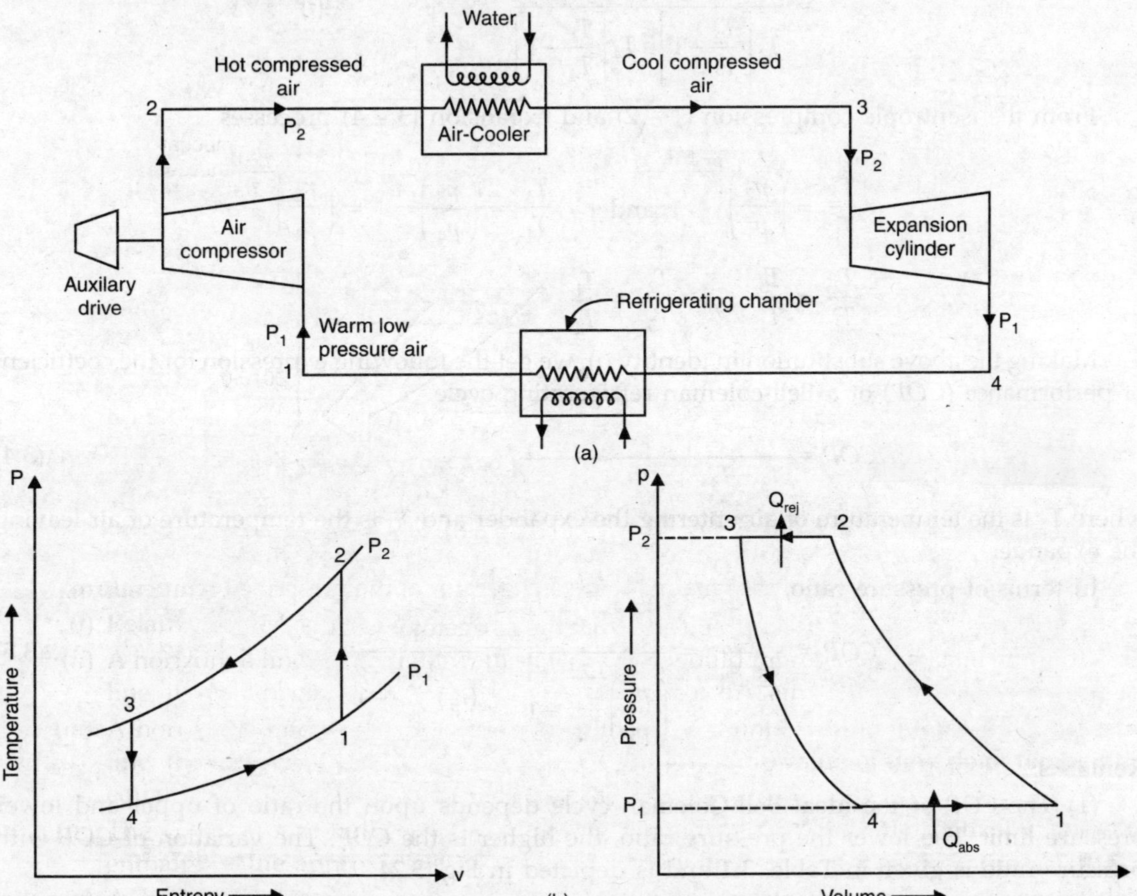

Fig. 3.23. Bell Coleman cycle

The air next enters the compressor and the cycle is repeated.

For unit mass flow of air,

Heat absorbed from the cold chamber, *i.e.*, refrigerating effect

$$Q_{abs} = c_p\,(T_1 - T_4)$$

Heat rejected to circulating water in the cooler

$$Q_{rej} = c_p\,(T_2 - T_3)$$

Work input in driving the compressor = heat rejected – heat absorbed

$$= c_p\,(T_2 - T_3) - c_p\,(T_1 - T_4)$$

$$\text{Coefficient of performance } (COP) = \frac{\text{refrigerating effect}}{\text{work input}}$$

$$= \frac{c_p(T_1 - T_4)}{c_p(T_2 - T_3) - c_p(T_1 - T_4)} = \frac{T_1 - T_4}{(T_2 - T_3) - (T_1 - T_4)} \quad ...(3.3)$$

$$= \frac{T_4\left[\frac{T_1}{T_4} - 1\right]}{T_3\left[\frac{T_2}{T_3} - 1\right] - T_4\left[\frac{T_1}{T_4} - 1\right]} \quad ...(i)$$

From the isentropic compression (1 - 2) and expansion (3 - 4) processes

$$\frac{T_2}{T_1} = \left(\frac{p_2}{p_1}\right)^{\frac{\gamma-1}{\gamma}} \quad \text{and} \quad \frac{T_3}{T_4} = \left(\frac{p_3}{p_4}\right)^{\frac{\gamma-1}{\gamma}} = \left(\frac{p_2}{p_1}\right)^{\frac{\gamma-1}{\gamma}}$$

$$\therefore \quad \frac{T_2}{T_1} = \frac{T_3}{T_4} \quad \text{or} \quad \frac{T_2}{T_3} = \frac{T_1}{T_4}$$

Making the above substitution in identity (*i*), we get the following expression for the coefficient of performance (*COP*) of a Bell-coleman refrigerating cycle.

$$COP = \frac{T_4}{T_3 - T_4} \quad ...(3.4)$$

where T_3 is the temperature of air entering the expander and T_4 is the temperature of air leaving the expander.

In terms of pressure ratio,

$$COP = \frac{1}{\frac{T_3}{T_4} - 1} = \frac{1}{\left(\frac{p_2}{p_1}\right)^{\frac{\gamma-1}{\gamma}} - 1} = \frac{1}{(r_p)^{\frac{\gamma-1}{\gamma}} - 1} \quad ...(3.5)$$

Remarks:

(1) The *COP* of the ideal Bell-Coleman cycle depends upon the ratio of upper and lower pressure limit. The lower the pressure ratio, the higher is the *COP*. The variation of *COP* with pressure ratio is given in Table. 3.1 and is depicted in Fig. 3.24.

Table 3.1

r_p	1	2	3	4	5	6
COP	∞	4.56	2.71	2.05	1.72	1.5

At a pressure ratio equal to unity, the *COP* is infinity but the refrigerating effect is zero. The effect of increasing pressure ratio is to increase the refrigerating effect. The optimum pressure is arrived at by striking a compromise between recurring cost of power consumption (governed by *COP*) and the capital cost of equipment (governed by refrigerating effect and hence capacity).

(2) The Bell-Coleman refrigerating systems had been built and operated extensively on ships for carrying frozen meat because the refrigerant used (air) is non-toxic, non-poisonous, easily available and non-expensive. The system still finds application in aircraft air-conditioning because of safety.

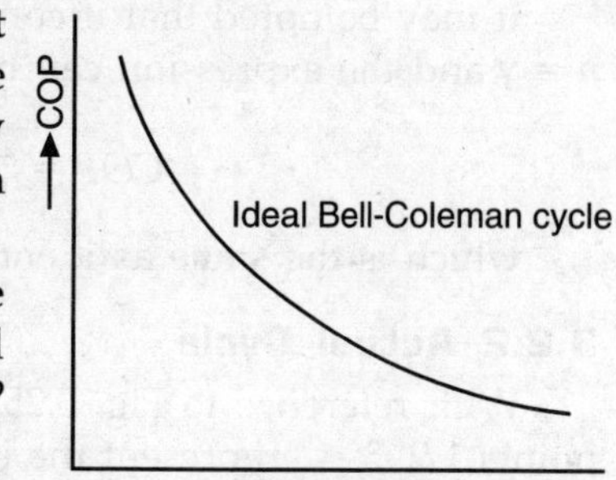

Fig. 3.24.

Due to low specific heat and high specific volume, a large amount of air has to be circulated for a given cooling load (refrigerating effect). The unit has a cumbersome design, low *COP* and high power consumption per ton of refrigeration.

(3) In modern air refrigeration cycles, the large displacement volumes can be handled best by the rotary type compressor and expander, as these are more compact for a given flow than the reciprocating machines.

(4) Quite often, the compression and expansion processes take place polytropically where pressure and volume are related as pv^n = constant. In that case

Work input during compression process 1-2

$$W_c = \frac{n}{n-1}(p_2 v_2 - p_1 v_1) = \frac{n}{n-1} R(T_2 - T_1) \qquad (\because pv = RT)$$

Work output during expansion process 3-4,

$$W_e = \frac{n}{n-1}(p_3 v_3 - p_4 v_4) = \frac{n}{n-1} R(T_3 - T_4)$$

Net work supplied during the cycle,

$$= W_c - W_e = \frac{n}{n-1} R\,[(T_2 - T_1) - (T_3 - T_4)]$$

Refrigerating effect = heat absorbed during constant pressure process 4-1,

$$= c_p\,(T_1 - T_4)$$

$$\therefore \quad COP = \frac{\text{heat absorbed}}{\text{work input}} = \frac{c_p(T_1 - T_4)}{\frac{n}{n-1} R\,[(T_2 - T_1) - (T_3 - T_4)]}$$

Substituting $R = c_p - c_v = c_v(\gamma - 1)$ in the above expression, we obtain,

$$COP = \frac{c_p(T_1 - T_4)}{\frac{n}{n-1} c_v(\gamma - 1)\,[(T_2 - T_1) - (T_3 - T_4)]}$$

$$= \frac{T_1 - T_4}{\frac{n}{n-1} \times \frac{\gamma - 1}{\gamma}\,[(T_2 - T_1) - (T_3 - T_4)]} \qquad \ldots(3.6)$$

An examination of the equations 3.3 and 3.6 does show that *COP* with polytropic process gets reduced by the factor $\left(\frac{n}{n-1} \times \frac{\gamma - 1}{\gamma}\right)$

The values of temperature T_2 and T_4 are calculated from the identities,

$$\frac{T_2}{T_1} = \left(\frac{p_2}{p_1}\right)^{\frac{n-1}{n}} \quad \text{and} \quad \frac{T_3}{T_4} = \left(\frac{p_3}{p_4}\right)^{\frac{n-1}{n}} = \left(\frac{p_2}{p_1}\right)^{\frac{n-1}{n}}$$

It may be noted that if compression and expansion processes processed isentropically, then $n = \gamma$ and the expression can be rewritten as

$$COP = \frac{T_1 - T_4}{(T_2 - T_1) - (T_3 - T_4)} = \frac{T_1 - T_4}{(T_2 - T_3) - (T_1 - T_4)}$$

which is the same as identity (*i*).

3.2.2 Actual Cycle

With reference to Fig. 3.25. the state points 1-2′-3′-4″ represent the ideal cycle whereas the actual cycle is represented by state points 1-2-3-4.

The actual cycle differs from the ideal (theoretical) cycle and the deviations are:

(*i*) Due to losses caused by internal friction and turbulence, the compression process is non-isentropic and tends to become irreversible. The temperature T_2 at the end of compression is higher than T_2' in an ideal case, and there is an increase in entropy.

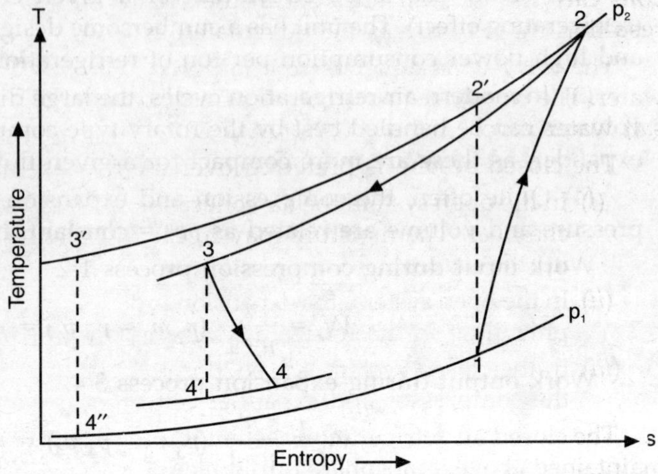

Fig. 3.25

Work input to compressor $W_c = c_p\,(T_2 - T_1)$

Efficiency of compression process (compressor) is

$$\eta_c = \frac{\text{isentropic temperature rise}}{\text{actual temperature rise}} = \frac{T_2' - T_1}{T_2 - T_1}$$

(*ii*) The friction causes drop in pressure in the air cooler and the state point at exit from the cooler is 3 rather than 3′. Apparently, the heat rejection does not take place at constant pressure.

(*iii*) Due to friction and turbulence, the expansion process in the turbine is non-isentropic. It follows the path 3-4 rather than 3-4′ resulting into an increase in temperature and entropy.

Work output from turbine

$$W_t = c_p\,(T_3 - T_4)$$

Efficiency of expansion process (turbine) is

$$\eta_t = \frac{\text{actual temperature drop}}{\text{isentropic temperature drop}} = \frac{T_3 - T_4}{T_3 - T_4'}$$

(*iv*) The friction causes drop in pressure in the brine tank and consequently the absorption of heat from the stuff or space being cooled does not take place at pressure p_1; it occurs at a pressure which is higher than p_1.

The performance and efficiency of the refrigerating plant is adversely affected due to non-isentropic compression and expansion processes, and pressure drops in the after cooler and cooling chamber.

The friction causes drop in pressure both in the air cooler and refrigerating chamber. Apparently the rejection of heat in the air cooler, and absorption of heat in the refrigerating chamber do not take place at constant pressure.

Obviously the ideal conditions do not prevail in a practical system. Due to pressure drop and frictional losses, the actual cycle is greatly distorted as shown in Fig. 3.21.

3.2.3. Open and Closed Systems

The Bell-Coleman refrigerator may work on the open system or closed system. In ***open system,*** the air comes in direct contact with the refrigerating chamber or the substances being cooled. In ***closed system,*** the air is passed through coils placed in the chamber. Quite often, the coils carrying cold air are kept in the brine tank. The air extracts heat from the brine and the resulting chilled brine is then circulated around the cold chamber.

The brine is an aqueous solution of sodium chloride (NaCl) or calcium chloride ($CaCl_2$) in water. The brine does not generally freeze at low temperature, and its freezing point depends on salt water ratio.

The closed system is preferred over the open system on account of the following reasons :

(*i*) In the open system the air collects moisture from the stuff being cooled in the refrigerating chamber. The moisture may freeze during expansion and the snow deposits may choke up the valves.

(*ii*) In the open system, the expansion process can be continued only upto atmospheric pressure prevailing in the refrigerated chamber. No such restriction exists in the closed system.

(*iii*) In the closed system, the suction to the compressor may be at high pressure. The size of the compressor and expander can be kept reasonably small by using dense air.

The closed air refrigeration system is often called *dense air system* because here the pressure is maintained above atmosphere through out.

EXAMPLE 3.29

Dense air is used as refrigerant in reverse Brayton or Bell-Coleman or Joule cycle. Represent the cycle on P-V and T-S diagrams and set up the expression for COP in terms of pressure ratio. For one such cycle the temperatures at the end of heat absorption and heat refection are 0°C and 30°C respectively, the pressure ratio is 4 and pressure in the cooler is 4 bar. Determine the temperatures at all the state points and volume flow rates at inlet to compressor and at exit of turbine. Assume that the machine has a cooling capacity equivalent to 1 ton of refrigeration. ***(Engg. Services)***

Solution: Refer to figure given below for the *p-v* and *T-s* plots of the reversed Brayton/Bell-Coleman cycle.

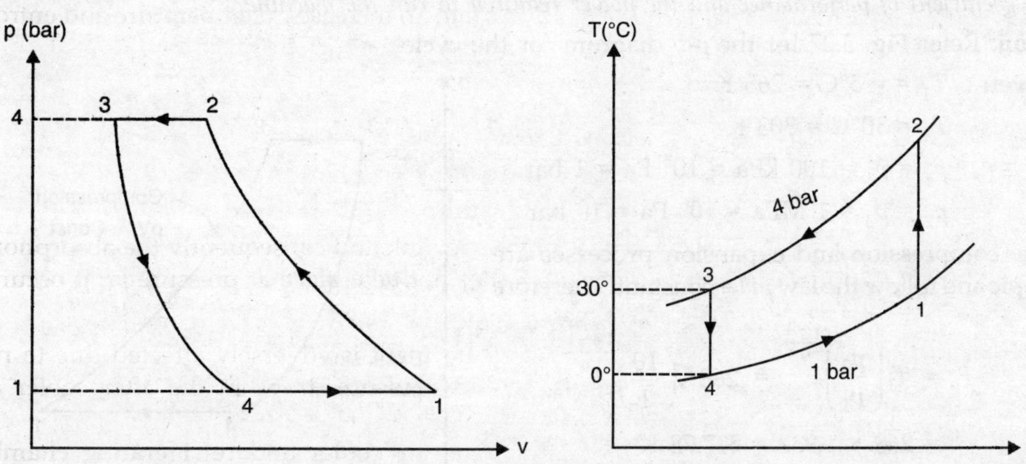

Fig. 3.26.

$$p_1 = p_4 = 1 \text{ bar} \quad \text{and} \quad p_2 = p_3 = 4 \text{ bar}$$
$$T_1 = 0°\text{C} = 273 \text{ K} \quad \text{and} \quad T_3 = 30°\text{C} = 303 \text{ K}$$

$$T_2 = T_1\left(\frac{P_2}{P_1}\right)^{\frac{\gamma-1}{\gamma}} = 273\times(4)^{\frac{1.4-1}{1.4}} = 405.68 \text{ K}$$

$$T_4 = \frac{T_3}{\left(\frac{P_3}{P_4}\right)^{\frac{\gamma-1}{\gamma}}} = \frac{303}{(4)^{\frac{1.4-1}{1.4}}} = 203.85 \text{ K}$$

1 ton of refrigeration capacity is equivalent to 3.5 kJ/s of heat absorption. Then

$$m_a\, c_p\, (T_1 - T_4) = 3.5$$

$$\therefore \text{ Volume flow rate of air, } m_a = \frac{3.5}{1.005\times(273-203.85)} = 0.0504 \text{ kg/s}$$

Volume flow rate at inlet to compressor,

$$V_1 = \frac{mRT_1}{p_1} = \frac{0.0504\times2.87\times273}{1\times10^5} = \mathbf{0.0395\ m^3/s}$$

Volume flow rate at exit of turbine

$$V_4 = \frac{mRT_4}{b_u} = \frac{0.0504\times287\times203.85}{1\times105} = 0.0295 \text{ m}^3/\text{s}$$

EXAMPLE 3.30

An air refrigeration open system operating between 1 MPa and 100 kPa is required to produce a cooling effect of 2000 kJ/min. The temperature of air leaving the cold chamber is – 5°C and at leaving the cooler is 30°C. Neglecting losses and clearance in the compressor and expander, determine :

(a) mass of air circulated per hour,

(b) compressor work, expander work and the cycle work,

(c) coefficient of performance and the power required to run the machine.

Solution: Refer Fig. 3.27 for the *p-v* diagram for the cycle.

Given : $T_1 = -\,5°\text{C} = 268 \text{ K}$

$T_3 = 30°\text{C} = 303 \text{ K}$

$p_4 = p_1 = 100 \text{ kPa} = 10^5 \text{ Pa} = 1 \text{ bar}$

$p_2 = p_3 = 1 \text{ MPa} = 10^6 \text{ Pa} = 10 \text{ bar}$

The compression and expansion processes are isentropic and follow the law pv^γ = constant. Therefore

$$T_2 = T_1\left(\frac{p_2}{p_1}\right)^{\frac{\gamma-1}{\gamma}} = 268\times\left(\frac{10}{1}\right)^{\frac{1.4-1}{1.4}}$$

$$= 268 \times 1.932 = 517.78 \text{ K}$$

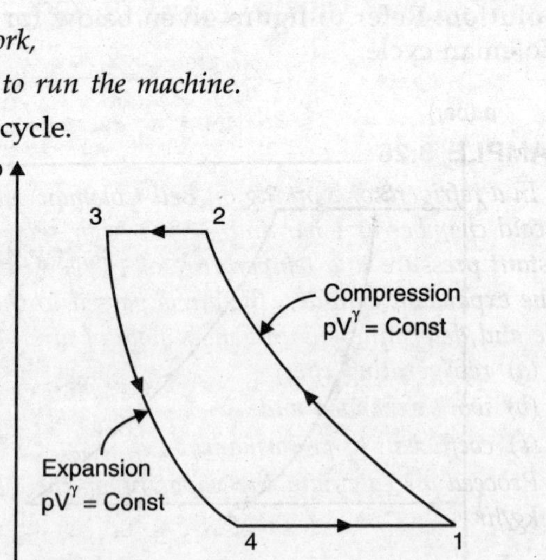

Fig. 3.27.

and $$T_4 = T_3\left(\frac{p_4}{p_3}\right)^{\frac{\gamma-1}{\gamma}} = 303\times\left(\frac{1}{10}\right)^{\frac{1.4-1}{1.4}} = \frac{303}{1.932} = 156.83 \text{ K}$$

(*a*) Refrigerating effect per kg of air flow

$$= c_p\,(T_1 - T_4) = 1.005\,(268 - 156.83) = 111.73 \text{ kJ/kg}$$

If m is the mass of air circulated per minute, then

$$111.73 \times m = 2000;\quad m = \frac{2000}{111.73} = \mathbf{17.9 \text{ kg/min}}$$

(*b*) Compression work $W_{comp} = \frac{\gamma}{\gamma-1}\,mR\,(T_2 - T_1)$

$$= \frac{1.4}{1.4-1} \times 17.9 \times 287\,(517.78 - 268)$$

$$= 4491182 \text{ J/min} = \mathbf{4491.18 \text{ kJ/min}}$$

Expander work $W_{exp} = \frac{\gamma}{\gamma-1}\,mR\,(T_3 - T_4)$

$$= \frac{1.4}{1.4-1} \times 17.9 \times 287\,(303 - 156.83)$$

$$= 2628217 \text{ J/min} = \mathbf{2628.22 \text{ kJ/min}}$$

Cycle work $W_{cycle} = W_{comp} - W_{exp}$

$$= 4491.18 - 2628.22 = \mathbf{1862.96 \text{ kJ/min}}$$

(*c*) Coefficient of performance, $COP = \frac{\text{refrigeration effect}}{\text{network expended}} = \frac{2000}{1862.96} = \mathbf{1.0736}$

Power required P = cycle work, *i.e.*, network expended per second

$$= \frac{1862.96}{60} = 31.05 \text{ kJ/s} = \mathbf{31.05 \text{ kW}}$$

EXAMPLE 3.28

In a refrigerator working on Bell-Coleman cycle, air is drawn into the cylinder of the compressor from the cold chamber at 1 bar and – 3°C. After reversible adiabatic compression to 5 bar, the air is cooled at constant pressure to a temperature of 17°C. After subsequent polytropic expansion ($pv^{1.25}$ = C) to 1 bar in the expansion cylinder, the air is passed to the cold chamber. Sketch the p-v and T-s diagrams of the cycle and determine for unit mass flow of air :

(a) refrigerating effect

(b) work expended and

(c) coefficient of performance

Proceed to calculate the capacity of the plant in tons of refrigeration for a mass flow rate of 100 kg/hr.

Solution : Refer Fig. 3.28 for the *p-v* and *T-s* diagrams for the cycle.

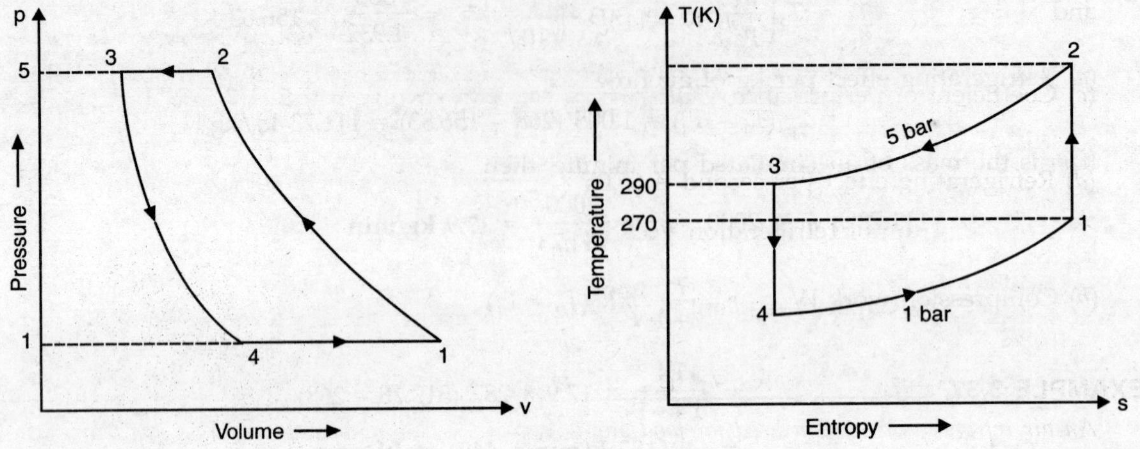

Fig. 3.28

Given: $p_4 = p_1 = 1$ bar ; $p_2 = p_3 = 5$ bar

$T_1 = -3 + 273 = 270$ K and $T_3 = 17 + 273 = 290$ K

The compression process 1-2 is isentropic and follows the law pv^γ = constant and therefore

$$T_2 = T_1\left(\frac{p_2}{p_1}\right)^{\frac{\gamma-1}{\gamma}} = 270\times\left(\frac{5}{1}\right)^{\frac{1.4-1}{1.4}} = 270 \times 1.584 = 427.69 \text{ K}$$

The expansion process 3-4 is polytropic and follows the law pv^n = constant and therefore

$$T_4 = T_3\left(\frac{p_4}{p_3}\right)^{\frac{n-1}{n}} = 290\times\left(\frac{1}{5}\right)^{\frac{1.25-1}{1.25}} = \frac{290}{1.379} = 210.25 \text{ K}$$

(*a*) Refrigerating effect, *i.e.,* heat extracted from the cold chamber

$$= c_p\,(T_1 - T_4)$$

$$= 1.005\,(270 - 210.25) = \mathbf{60.05\ kJ/kg}$$

(*b*) Work done during isentropic compression process 1-2

$$W_{comp} = \frac{\gamma}{\gamma-1}\,R\,(T_2 - T_1)$$

$$= \frac{1.4}{1.4-1} \times 287\,(427.68 - 270)$$

$$= 158390 \text{ J/kg} = 158.39 \text{ kJ/kg}$$

Work done during polytropic expansion process 3-4

$$W_{exp} = \frac{n}{n-1}\,R\,(T_3 - T_4)$$

$$= \frac{1.25}{1.25-1} \times 287\,(290 - 210.25)$$

$$= 114440 \text{ J/kg} = 114.44 \text{ kJ/kg}$$

$\therefore$ Net work expended $= W_{comp} - W_{exp}$

$= 158.39 - 114.44 = \mathbf{43.95\ kJ/kg}$

(c) Coefficient of performance, $COP = \dfrac{\text{refrigerating effect}}{\text{network expended}} = \dfrac{60.05}{43.95} = \mathbf{1.366}$

(d) Refrigerating effect per second $= 60.05 \times \dfrac{100}{3600} = 1.668$ kJ/s

1 ton of refrigeration = 3.5 kJ/s

$\therefore$ Capacity of plant $= \dfrac{1.668}{3.5} = \mathbf{0.476\ tons}$

EXAMPLE 3.32

An air refrigeration system used for food storage provides 25 tons of refrigeration. The temperature of air entering the compressor is 7°C and temperature at exit from the cooler is 27°C. The quantity of air circulated in the cooler is 3000 kg/hr. Both the compression and expansion follow the polytropic law $pv^{1.3}$ = constant. Compute

(a) COP of the cycle

(b) power required by the compressor per ton of refrigeration.

Solution: Refer Fig. 3.29 for *p-v* and *T-s* plot of the operating cycle

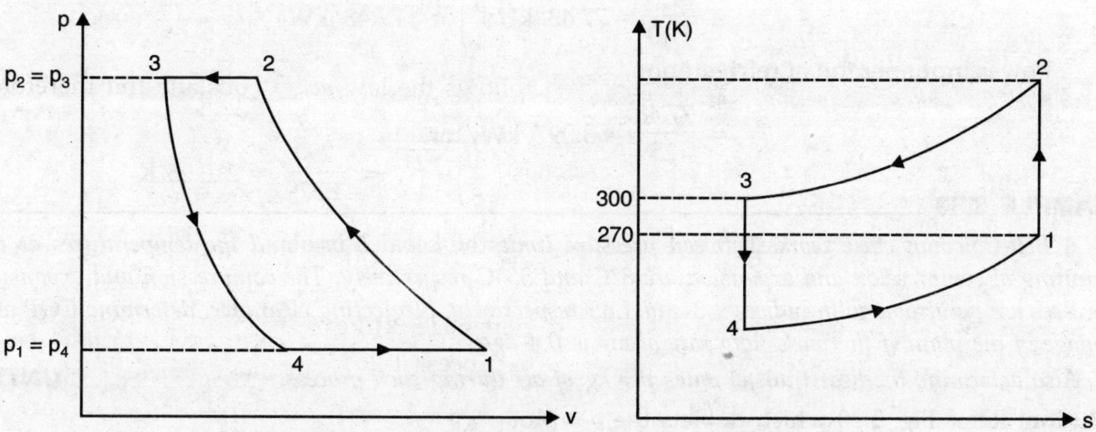

Fig. 3.29

Given : T_1 = 7°C = 280 K and T_3 = 27°C = 300 K

Refrigeration capacity = 25 ton = 25 × 3.5 = 87.5 kJ/s

If m_a is the mass flow rate of air, then

$$m_a\, c_p\, (T_1 - T_4) = 87.5 \quad ; \quad \frac{3000}{3600} \times 1 \times (280 - T_4) = 87.5$$

That gives : $T_4 = 280 - \dfrac{3600 \times 87.5}{3000} = 175$ K

The compression and expansion are stated to follow the polytropic law. Therefore

$$\frac{T_2}{T_1} = \left(\frac{p_2}{p_1}\right)^{\frac{n-1}{n}} \quad \text{and} \quad \frac{T_3}{T_4} = \left(\frac{p_3}{p_4}\right)^{\frac{n-1}{n}} = \left(\frac{p_2}{p_1}\right)^{\frac{n-1}{n}}$$

That gives : $\dfrac{T_2}{T_1} = \dfrac{T_3}{T_4}$ and so $T_2 = \dfrac{T_2 \times T_3}{T_4}$

$$= \frac{280 \times 300}{175} = 480 \text{ K}$$

(*a*) When compression and expansion processes take place polytropically, COP is given by

$$COP = \frac{T_1 - T_4}{\frac{n}{n-1} \times \frac{\gamma - 1}{\gamma}[(T_2 - T_3) - (T_1 - T_4)]}$$

$$= \frac{280 - 175}{\frac{1.3}{13.-1} \times \frac{1.4-1}{1.4}[(480 - 300) - (280 - 175)]} = \mathbf{1.13}$$

(*b*) Heat absorbed (refrigeration effect) during process 1-4

$$= m_a \, c_p \, (T_1 - T_4)$$

$$= \frac{3000}{3600} \times 1 \times (280 - 175) = 87.5 \text{ kJ/s}$$

$$\text{Work input to compressor} = \frac{\text{heat absorbed}}{COP}$$

$$= \frac{87.5}{1.13} = 77.43 \text{ kJ/s} \quad \text{or} \quad 77.43 \text{ kW}$$

∴ Power input per ton of refrigeration,

$$= \frac{77.43}{25} = 3.097 \text{ kW/ton}$$

EXAMPLE 3.33

A Bell-Coleman cycle works between pressure limits of 1 and 6 bar, and the temperatures at the beginning of compression and expansion are 8°C and 35°C respectively. The compression and expansion processes are polytropic with indices 1.3 and 1.35 respectively. Neglecting clearance, determine COP and tonnage of the plant if the mass flow rate of air is 0.4 kg/s.

Also determine the heat transfer rates per kg of air during each process. **(JNTU)**

Solution: Refer Fig. 3.30 which depicts the *p-v* plot for a refrigeration system operating on Bell-Coleman cycle.

T_1 = 8°C = 281 K and T_3 = 35°C = 308 K

Work supplied to compressor (area 1–2–5–6)

$$W_c = \frac{n_c}{n_c - 1}(p_2 v_2 - p_1 v_1)$$

$$= \frac{n_c}{n_c - 1} R(T_2 - T_1)$$

$$= \frac{n_c}{n_c - 1} R T_1 \left[\left(\frac{p_2}{p_1}\right)^{\frac{n_c - 1}{n_c}} - 1\right]$$

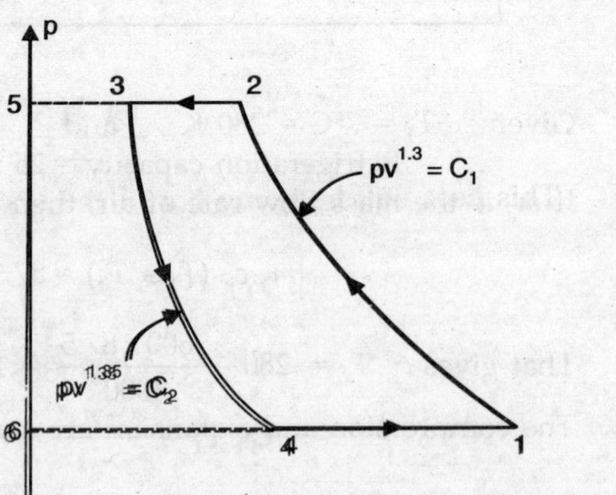

Fig. 3.30

$$= \frac{1.3}{1.3-1} \times 287 \times 281 \times \left[(6)^{\frac{0.3}{1.3}} - 1\right] = 178984 \text{ J/kg}$$

Work developed by expander (area 3-5-6-4)

$$W_e = \frac{n_e}{n_e - 1}(p_3 v_3 - p_4 v_4)$$

$$= \frac{n_e}{n_e - 1} R(T_3 - T_4) = \frac{n_e}{n_e - 1} R T_3 \left[1 - \left(\frac{p_3}{p_4}\right)^{\frac{n_e - 1}{n_e}}\right]$$

$$= \frac{1.35}{1.35-1} \times 287 \times 308 \times \left[1 - \left(\frac{1}{6}\right)^{\frac{0.35}{1.35}}\right] = 126666 \text{ J/kg}$$

Then net work supplied per kg of air circulated in the system

$$W_n = 178984 - 126666 = 52318 \text{ J/kj of air} \simeq 52.32 \text{ kJ/kg of air}$$

(*b*) From the expansion process 3-4,

$$T_4 = T_3 \times \left(\frac{p_4}{p_3}\right)^{\frac{n_e - 1}{n_e}} = 308 \left(\frac{1}{6}\right)^{\frac{0.35}{1.35}} = 195.58 \text{ K}$$

Refrigerating effect $= c_p\ (T_1 - T_4) = 1.005\ (281 - 193.58) = 87.86$ kJ/g of air

$$COP = \frac{\text{refrigerating effect}}{\text{work input}} = \frac{87.86}{52.32} = \mathbf{1.68}$$

For a mass flow rate of 0.4 kg/s,

Refrigeration effect = 0.4 × 87.86 = 35.144 kJ/s

Since 1 tonnage is equivalent to 3.5 kJ/s, the tonnage of the unit is

$$= \frac{35.144}{3.5} = \mathbf{10.04}$$

(*c*) Heat transfer rate during the different process constituting the cycle are

$$Q_{1-2} = \frac{\gamma - n_c}{\gamma - 1} \times \text{work done for process 1–2}$$

$$= \frac{\gamma - n_c}{\gamma - 1} \times \frac{p_2 v_2 - p_1 v_1}{n - 1} = \frac{\gamma - n}{\gamma - 1} \times \frac{R T_1}{(n-1)} \left[\left(\frac{p_2}{p_1}\right)^{\frac{n_c - 1}{n_c}} - 1\right]$$

$$= \frac{1.4 - 1.3}{1.4} \times \left(\frac{287 \times 281}{1.3 - 1}\right) \times \left[(6)^{\frac{1.3-1}{1.3}} - 1\right] = 34420 \text{ J} \simeq 34.4 \text{ kJ}$$

This heat will be going out of the system and will be taken negative

$$Q_{2-3} = c_p\ (T_3 - T_2) \quad \text{where } T_2 = T_1 \left(\frac{p_2}{p_1}\right)^{\frac{n-1}{n}} = 281 \times (6)^{\frac{0.3}{1.3}} = 424.92 \text{ K}$$

$$= 1.005\ (308 - 424.92) = -\ 117.50 \text{ kJ}$$

$$Q_{3-4} = \frac{\gamma - n_e}{\gamma - 1} \times \text{work output from process 3–4}$$

$$= \frac{\gamma - n_e}{\gamma - 1} \times \frac{p_3 v_3 - p_4 v_4}{n - 1} = \frac{\gamma - n}{\gamma - 1} \times \frac{R T_3}{n - 1} \left[1 - \left(\frac{p_4}{p_3}\right)^{\frac{n_e - 1}{n_e}}\right]$$

$$= \frac{1.4-1.35}{1.4-1} \times \left(\frac{287 \times 308}{1.35}\right) \times \left[1 - \left(\frac{1}{6}\right)^{\frac{1.35-1}{0.35}}\right] = 11728 \text{ J} = 117.3 \text{ kJ}$$

$$Q_{4-1} = c_p (T_1 - T_4) = 1.005 (281 - 193.58) = 87.86 \text{ kJ}$$

Check : Applying 1st law of thermodynamics to a cyclic process

$$\oint \delta W = \oint \delta Q = Q_{1-2} + Q_{2-3} + Q_{3-4} + Q_{4-1}$$

$$= -34.42 + (-117.50) + 11.73 + 87.86 = -52.33 \text{ kJ}$$

This is the same as calculated above

EXAMPLE 3.34.

A Bell-Coleman refrigeration system is to produce 10 tons of refrigeration. The cooler and the refrigerator pressures are 4.2 bars and 1.4 bars respectively. Air is cooled in the cooler to 45°C and the temperature of air at inlet to the compressor is – 20°C. For an ideal cycle, calculate:

(a) COP

(b) mass of air circulated per minute

(c) theoretical piston displacement of compressor, and

(d) power required per ton of refrigeration

Proceed to find the cylinder dimensions if the compressor is single-acting, single-cylinder with l/d ratio of 1.2 and runs at 600 rev/min. ***(MRIU, RTU, Engg. Services)***

Solution: Refer Fig. 3.31 for the *p-v* plot of the operating Bell-Coleman cycle.

$$p_1 = p_4 = 1.4 \text{ bar}$$
$$p_2 = p_3 = 4.2 \text{ bar}$$
$$T_1 = -20°\text{C} = 253 \text{ K}$$
$$T_3 = 45°\text{C} = 318 \text{ K}$$

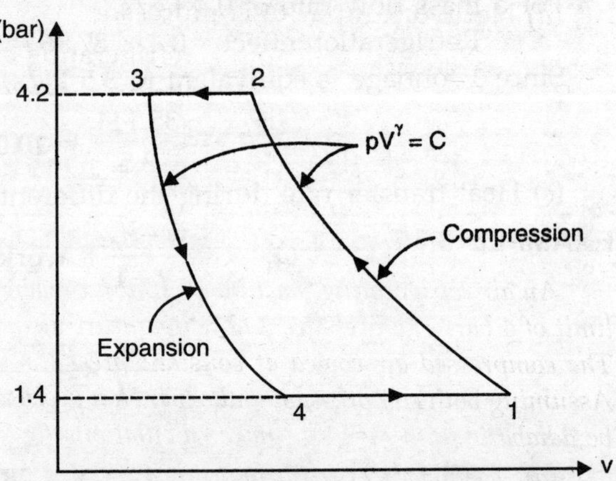

Fig. 3.31

The compression and expansion processes are isentropic and follow the law pv^γ = constant. Therefore

$$T_2 = T_1 \left(\frac{p_2}{p_1}\right)^{\frac{\gamma-1}{\gamma}} = 253 \times \left(\frac{4.2}{1.4}\right)^{\frac{1.4-1}{1.4}}$$

$$= 253 \times 1.369 = 346.3 \text{ K}$$

$$T_4 = T_3 \left(\frac{p_4}{p_3}\right)^{\frac{\gamma-1}{\gamma}} = 318 \times \left(\frac{1.4}{4.2}\right)^{\frac{1.4-1}{1.4}}$$

$$= \frac{318}{1.369} = 232.3 \text{ K}$$

(*a*) Since both the compression and expansion processes are isentropic,

$$COP = \frac{T_4}{T_3 - T_4} = \frac{232.3}{318 - 232.3} = 2.70$$

Alternatively: $$COP = \frac{1}{(r_p)^{\frac{\gamma-1}{\gamma}} - 1} = \frac{1}{\left(\frac{4.2}{1.4}\right)^{\frac{1.4-1}{1.4}} - 1} = \frac{1}{1.37 - 1} = 2.70$$

(*b*) If m_a is the mass of air circulated per minute, then

Refrigerating effect (heat absorbed) $= m_a c_{pa} (T_1 - T_4)$

$$= m_a \times 1.005\ (253 - 232.3)$$

$$= 20.8\ m_a \text{ kJ/min}$$

The absorbing capacity is stated to be 10 tons and one tonne is equivalent to 210 kJ/min. Therefore

$$20.80\ m_a = 10 \times 210\ ;\ m_a = \frac{10 \times 20}{20.80} = 100.96 \text{ kg/min}$$

(*c*) Theoretical piston displacement of compressor is given by the volume corresponding to state point 1.

$$V_1 = \frac{mRT_1}{p_1} = \frac{100.96 \times 287 \times 253}{1.4 \times 10^5} = 52.36 \text{ m}^3\text{/min}$$

$$\text{Swept volume per stroke} = \frac{52.36}{600} = 0.0873 \text{ m}^3 = 87266 \text{ cm}^3$$

If d is the diameter of the compressor cylinder and l is the stroke length, then

$$\frac{\pi}{4} d^2 l = 87266\ ;\ \frac{\pi}{4} d^2 \times 1.2d = 87266$$

$$d^3 = \frac{87266 \times 4}{\pi \times 1.2} = 92639\ ;\ d = (92639)^{1/3} = 45.08 \text{ cm}$$

$$l = 1.2\ d = 1.2 \times 45.08 = 54.03 \text{ cm}$$

(*d*) Net work input to compressor

$$= \frac{\text{refrigerating effect per ton of refrigeration}}{\text{coefficient of performance}}$$

$$= \frac{3.5}{2.70} = \mathbf{1.296} \text{ kW/ton of refrigeration}$$

EXAMPLE 3.35

An air refrigerating machine of 10 ton capacity operating on Bell-Coleman cycle has an upper pressure limit of 4 bar. The pressure and temperature at the start of compression are 1 bar and – 20°C respectively. The compressed air cooled at constant pressure to a temperature of 40°C enters the expansion cylinder. Assuming both compression and expansion processes to be adiabatic with γ = 1.4, make calculations for :

(a) coefficient of performance,
(b) quantity of air in circulation per minute,
(c) piston displacement of compressor and expander,
(d) bore of compressor and expansion cylinders,
(e) power required to drive the unit.

The unit turns 200 revolutions per minute, is double acting and the stroke length is 200 mm.

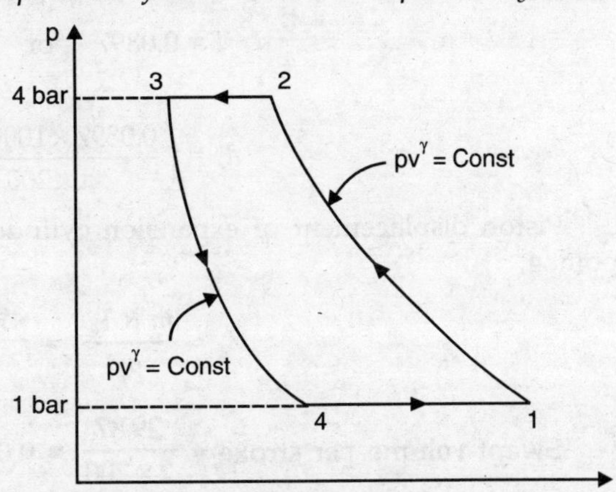

Fig. 3.32.

Solution : Refer Fig. 3.32 for the *p-v* plot of the operating cycle.

Given : $p_4 = p_1 = 1$ bar; $p_2 = p_3 = 4$ bar

$T_1 = -\ 20°\text{C} = 253$ K

and $T_3 = 40°\text{C} = 313$ K

The compression and expansion processes are isentropic and follow the law pv^γ = constant. Therefore

$$T_2 = T_1\left(\frac{p_2}{p_1}\right)^{\frac{\gamma-1}{\gamma}} = 253\times\left(\frac{4}{1}\right)^{\frac{1.4-1}{1.4}} = 253 \times 1.486 = 376 \text{ K}$$

$$T_4 = T_3\left(\frac{p_4}{p_3}\right)^{\frac{\gamma-1}{\gamma}} = 313\times\left(\frac{1}{4}\right)^{\frac{1.4-1}{1.4}} = \frac{313}{1.486} = 210.7 \text{ K}$$

(*a*) Since both the compression and expansion processes are isentropic

$$COP = \frac{T_4}{T_3 - T_4} = \frac{210.7}{313 - 210.7} = \mathbf{2.06}$$

(*b*) Refrigerating effect, *i.e.*, heat extracted from the cold chamber per kg of air

$$= c_p\ (T_1 - T_4) = 1.005\ (253 - 210.7) = 42.51 \text{ kJ/kg}$$

If m is the mass of air circulated per minute, then

Refrigerating effect = 42.51 m kJ/min

The absorbing capacity of the system is stated to be 10 ton or 10 × 210 kJ/min and therefore

$$42.51\ m = 10 \times 210$$

or $$m = \frac{10\times210}{42.51} = 49.4 \text{ kg/min}$$

(*c*) and (*d*) Piston displacement of compressor is given by the volume corresponding to state point 1.

$$V_1 = \frac{m\,R\,T_1}{p_1} = \frac{49.4\times287\times253}{1\times10^5} = \mathbf{35.87\ m^3/min}$$

$$\text{Swept volume per stroke} = \frac{35.87}{2\times200} = 0.0897 \text{ m}^3$$

The factor 2 accounts for the fact that the machine is double acting.

If d_c is the diameter of the compressor cylinder and l is the stroke length, then

$$\frac{\pi}{4}d_c^2\, l = 0.0897 \quad \text{or} \quad \frac{\pi}{4}d_c^2 \times \frac{200}{1000} = 0.0897$$

$$\therefore \quad d_c = \left(\frac{0.0897\times1000\times4}{\pi\times200}\right)^{\frac{1}{2}} = \mathbf{0.756\ m}$$

Piston displacement of expansion cylinder is given by the volume corresponding to state point 4.

$$V_4 = \frac{m\,R\,T_4}{p_4} = \frac{49.4\times287\times210.7}{1\times10^5} = \mathbf{29.87\ m^3/min}$$

$$\text{Swept volume per stroke} = \frac{29.87}{2\times200} = 0.0747 \text{ m}^3$$

If d_e is the diameter of the expansion cylinder, then

$$\frac{\pi}{4} d_e^2\, l = 0.0747 \quad \text{or} \quad \frac{\pi}{4} d_e^2 \times \frac{200}{1000} = 0.0747$$

$$\therefore \quad d_e = \left(\frac{0.0747 \times 1000 \times 4}{\pi \times 200}\right)^{\frac{1}{2}} = \mathbf{0.69\ m}$$

(e) Work done $= \dfrac{\text{refrigerating effect}}{\text{coefficient of performance}}$

$$= \frac{2100}{2.06} = 1019.42 \text{ kJ/min} = 16.99 \text{ kJ/s}$$

∴ Power required to drive the unit = **16.99 kW**

EXAMPLE 3.36

A 10 ton refrigerating air machine operating on the Bell-Coleman cycle draws air from the cold chamber at 1 bar and 10°C. After compression to 5 bar, the air is cooled at constant pressure and delivered to the expansion cylinder at 25°C where it expands to the initial pressure. Assuming that both the compression and expansion processes follow the polytropic law $pv^{1.3}$ = constant, make calculations for the following parameters :

(a) coefficient of performance for the refrigerating cycle,

(b) amount of air circulated per minute if the actual COP is 50 percent of the theoretical COP, and

(c) stroke and bore of the compressor presuming that the compressor turns 500 revolutions per minute, has volumetric efficiency of 85 percent and the stroke equals 1.5 times the bore.

Treat air as perfect gas with $\gamma = 1.4$ and $c_p = 1.005$ kJ/kgK

Solution : Refer Fig. 3.33 for the *p-v* plot of the operating cycle

$$p_1 = p_4 = 1 \text{ bar} \quad ; \quad p_2 = p_3 = 5 \text{ bar}$$

$$T_1 = 10°\text{C} = 283 \text{ K} \quad ; \quad T_3 = 25°\text{C} = 298 \text{ K}$$

For the compression4 process,

$$\frac{T_2}{T_1} = \left(\frac{p_2}{p_1}\right)^{\frac{n-1}{n}} = \left(\frac{5}{1}\right)^{\frac{1.3-1}{1.3}} = 1.45$$

$$\therefore \quad T_2 = 1.45 \times T_1 = 1.45 \times 283 = 410.24 \text{ K}$$

For the expansion process 3-4,

$$\frac{T_3}{T_4} = \left(\frac{p_3}{p_4}\right)^{\frac{n-1}{n}} = \left(\frac{5}{1}\right)^{\frac{1.3-1}{1.3}} = 1.45$$

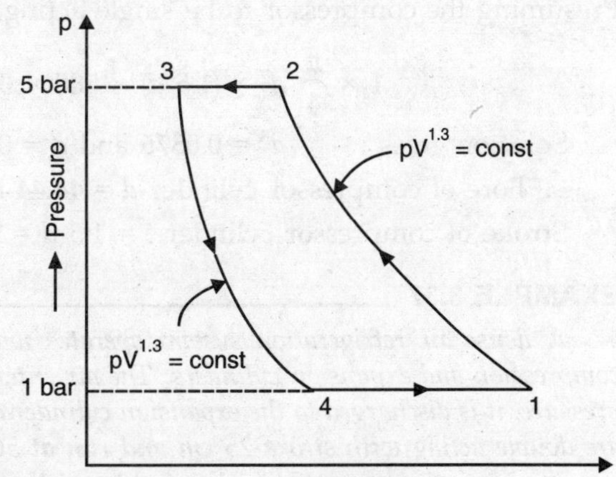

Fig. 3.33.

$$\therefore \quad T_4 = \frac{T_3}{1.45} = \frac{298}{1.45} = 205.57 \text{ K}$$

***(a)* Refrigerating effect (heat absorbed) per kg of air.**

$$= c_p(T_1 - T_4) = 1.005\,(283 - 205.57) = 77.82 \text{ kJ/kg}$$

Work input = compressor work – expander work

$$= \frac{n}{n-1} R\,(T_2 - T_1) - \frac{n}{n-1} R\,(T_3 - T_4)$$

$$= \frac{n}{n-1} R\,[(T_2 - T_1) - (T_3 - T_4)]$$

$$= \frac{1.3}{1.3-1} \times 0.287\,[(410.24 - 283) - (298 - 205.57)] = 43.20 \text{ kJ/kg}$$

$$\text{Theoretical } COP = \frac{\text{refrigerating effect}}{\text{work input}} = \frac{77.82}{43.20} = 1.8$$

(*b*) Actual refrigerating effect per kg of air

$$= 77.82 \times 0.5 = 38.91 \text{ kJ/kg}$$

If m is the mass of air circulated per minute, then refrigerating effect = 38.91 m kJ/min. The absorbing capacity of the system is stated to be 10 ton or 10 × 20 kJ/min and therefore

$$38.91\,m = 10 \times 210\ ; \quad m = \frac{10 \times 210}{38.91} = \mathbf{53.97\ kg/min}$$

(*c*) Piston displacement of the compressor is given by the volume corresponding to state point 1.

$$V_1 = \frac{mRT_1}{p_1} = \frac{53.97 \times 287 \times 283}{1 \times 10^5} = 43.83 \text{ m}^3\text{/min}$$

The volume V_1 is also given by

$$V_1 = n\left(\frac{\pi}{4} d^2 l\right) N \times \eta_v$$

The parameter n = 1 for single acting and n = 2 for double acting compression cylinder. Presuming the compressor to be single acting,

$$1 \times \frac{\pi}{4} d^2 \times (1.5\,d) \times 500 \times 0.85 = 43.83$$

Solution gives : $d^3 = 0.0876$ and d = 0.444 m = 44.44 cm

∴ Bore of compressor cylinder d = **44.44 cm**

Stroke of compressor cylinder l = 1.5 d = 1.5 × 44.44 = **66.66 cm.**

EXAMPLE 3.37

A dense air refrigeration system operates with pressure limits of 4 bar and 15 bar both for the compression and expansion cylinders. The air enters the compressor at 4°C and after cooling at constant pressure, it is discharged to the expansion cylinder at 20°C. Both the compression and expansion cylinders are double acting with stroke 25 cm and run at 500 rev/min. Calculate:

(a) bore of compression and expansion cylinders when the volumetric efficiency is 90%.

(b) power input to electric motor when the mechanical efficiency is 85% and

(c) ice formed from and at 0°C per day of 24 hours if the refrigerator has an absorbing capacity of 2100 kJ/min.

Assume compression and expansion to be isentropic and consider air to be perfect gas with γ = 1.4 and c_p = 1.005 kJ/kgK.

Solution : Refer Fig. 3.34 for the p-v plot of the operating cycle :

Given : $p_1 = p_4 = 4$ bar $\quad p_2 = p_3 = 16$ bar

$T_1 = 4°C = 277$ K $\quad T_3 = 20°C = 293$ K

From the isentropic compression process 1-2,

$$T_2 = T_1\left(\frac{p_2}{p_1}\right)^{\frac{\gamma-1}{\gamma}} = 277\times\left(\frac{16}{4}\right)^{\frac{1.4-1}{1.4}}$$

$$= 411.78 \text{ K}$$

From the isentropic expansion process 3-4,

$$T_4 = T_3\left(\frac{p_4}{p_3}\right)^{\frac{\gamma-1}{\gamma}} = 293\times\left(\frac{4}{16}\right)^{\frac{1.4-1}{4}}$$

$$= 197.17 \text{ K}$$

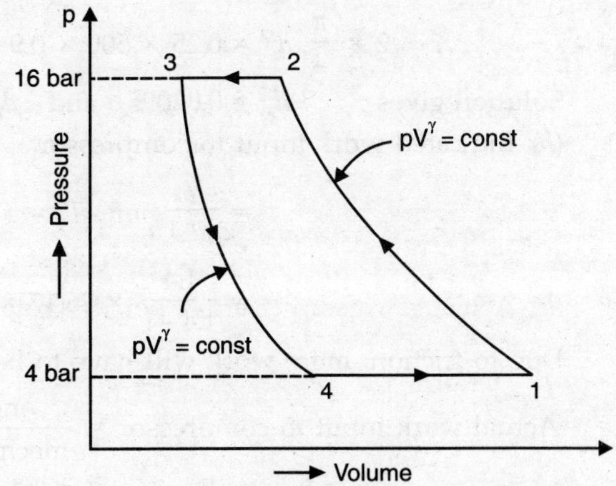

Fig. 3.34

If m is the mass of air circulated per minute, then

Refrigerating effect (heat absorbed) $= m\, c_p\, (T_1 - T_4)$

$= m \times 1.005\,(277 - 197.17) = 80.23\, m$ kJ/min

The absorbing capacity of the system is stated to be 2100 kJ/min and therefore

$$80.23\, m = 2100\ ; \quad m = \frac{2100}{80.23} = 26.17 \text{ kg/min}$$

(*a*) Piston displacement of the compressor is given by the volume corresponding to state point 1.

$$V_1 = \frac{mRT_1}{p_1} = \frac{26.17\times287\times277}{4\times10^5} = 5.20 \text{ m}^3/\text{min}$$

The volume V_1 is also given by

$$V_1 = 2\left(\frac{\pi}{4}d_c^2\, l\right) N \times \eta_v$$

The factor 2 accounts for the fact that the machine is double acting and d_c is the diameter of the compressor cylinder. Therefore

$$2 \times \frac{\pi}{4}\, d_c^{\,2} \times 0.25 \times 500 \times 0.9 = 5.20$$

Solution gives : $\quad d_c^{\,2} = 0.02944 \quad$ and $\quad d_c = 0.1715$ m $=$ **17.16 cm**

Piston displacement of the expander is given by the volume corresponding to state point 4.

$$V_4 = \frac{mRT_4}{p_4} = \frac{26.17\times287\times197.17}{4\times10^5} = 3.70 \text{ m}^3/\text{min}$$

The volume V_4 is also given by

$$V_4 = 2\left(\frac{\pi}{4}d_e^2\, l\right) N \times \eta_v\ ; \quad d_e \text{ is diameter of expansion cylinder}$$

$$\therefore \quad 2 \times \frac{\pi}{4} d_e^2 \times 0.25 \times 500 \times 0.9 = 3.70$$

Solution gives : $d_e^2 = 0.02095$ and $d_e = 0.1447$ m = **14.47 cm**

(*b*) Indicated work input to compressor

$$= \frac{\gamma}{\gamma - 1} mR\,(T_2 - T_1)$$

$$= \frac{1.4}{1.4-1} \times 26.17 \times 0.287\,(411.78 - 277) = 3543.06 \text{ kJ/min}$$

Due to friction, more work will have to be supplied and that is given by

$$\text{Actual work input to compressor} = \frac{\text{indicated work}}{\text{mechanical efficiency}}$$

$$= \frac{3543.06}{0.85} = 4168.30 \text{ kJ/min}$$

Indicated work obtained from the expansion cylinder

$$= \frac{\gamma}{\gamma - 1} mR\,(T_3 - T_4)$$

$$= \frac{1.4}{1.4-1} \times 26.17 \times 0.287\,(293 - 197.17) = 2519.16 \text{ kJ/min}$$

Due to friction, a portion of this work will be lost and work available is given by

Actual work available from the expansion cylinder

= indicated work × mechanical efficiency

= 2519.16 × 0.85 = 2141.29 kJ/min

∴ Net work required = 4168.30 – 2141.29 = 2027.01 kJ/min

$$= \frac{2027.01}{60} = 33.78 \text{ kJ/s} = \mathbf{33.78\ kW}$$

(*c*) Total heat absorbed by the refrigerator per day of 24 hours

$= 2100 \times 60 \times 24 = 3024 \times 10^3$ kJ

Taking latent heat of fusion of ice as 335 kJ/kg

$$\text{Ice formed} = \frac{3024 \times 10^3}{335} = 9027 \text{ kg} = \mathbf{9.027\ ton}$$

EXAMPLE 3.38

A dense air refrigeration machine operating on Bell-Coleman cycle operates within pressure limits of 3.4 bar and 17 bar. The air temperature after the cooler and after refrigeration are stated to be 15°C and 6°C respectively. For a referigeration capacity of 6 tonnes, calculate :

(*i*) *temperature after compression and expansion*

(*ii*) *air circulation per minute*

(*iii*) *work of compressor and expander*

(*iv*) *theoretical COP*

(*v*) *rate of water circulation in the cooler if the temperature rise is limited to 30°C.*

(Engg. services)

Solution: Refer Fig. 3.35 for the *p-v* plot of the operating cycle.

$p_1 = p_4 = 3.4$ bar ; $p_2 = p_3 = 17$ bar

$T_1 = 6°C = 279$ K ; $T_3 = 15°C = 288$ K

The compression and expansion processes are isentropic and follow the law pv^{γ} = constant. Therefore

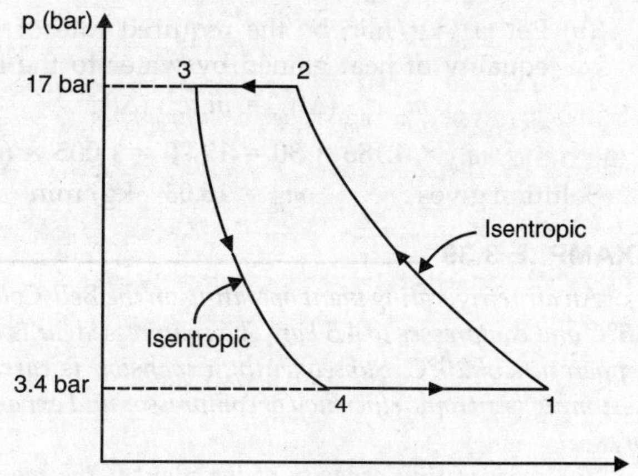

Fig. 3.35.

$$T_2 = T_1\left(\frac{p_2}{p_v}\right)^{\frac{\gamma-1}{\gamma}} = 279\times\left(\frac{17}{3.4}\right)^{\frac{1.4-1}{1.4}}$$

$$= 279 \times 1.597 = 445.7 \text{ K}$$

$$T_4 = T_3\left(\frac{p_4}{p_3}\right)^{\frac{\gamma-1}{\gamma}} = 288\times\left(\frac{3.4}{17}\right)^{\frac{1.4-1}{1.4}} = \frac{288}{1.597}$$

$$= 180.3 \text{ K}$$

(*i*) Temperature after compression, $T_2 = 445.7$ K

Temperature after expansion, $T_4 = 180.3$ K

(*ii*) If m is the mass of air circulated perminute, then

Refrigerating effect (heat absorbed) $= mc_p\,(T_1 - T_4)$

$= m \times 1.005\,(279 - 180.3) = 99.2$ m kJ/min

The absorbing capacity is stated to be 6 tonnes and one tonne is equivalent to 210 kJ/min. Therefore

$$99.2\, m = 210 \times 6\ ; \quad m = \frac{210\times 6}{99.2} = 12.70 \text{ kg/min}$$

(*iii*) Indicated work input to compressor

$$= \frac{\gamma}{\gamma-1} mR(T_2 - T_1)$$

$$= \frac{1.4}{1.4-1}\times\left(\frac{12.71}{60}\right)\times 287(445.7 - 279) = 354715 \text{ J/s}$$

$$= 35.47 \text{ kJ/s} = 35.47 \text{ kW}$$

Indicated work obtained from the expansion cylinder

$$= \frac{\gamma}{\gamma-1} mR(T_3 - T_4)$$

$$= \frac{1.4}{1.4-1}\times\frac{12.70}{60}\times 287(288 - 180.3) = 32880 \text{ J/s}$$

$$\simeq 22.9 \text{ kJ/s} = 22.9 \text{ kW}$$

(*iv*) Cylce work W = compressor work – expander work

$= 35.47 - 22.9 = 12.57$ kJ/s

$$\text{Refrigerating effect} = \frac{210\times 6}{60} = 21 \text{ kJ/s}$$

$$\therefore \text{ Coefficient of performance} = \frac{\text{refrigerating effect}}{\text{cycle work}} = \frac{21}{12.57} = 1.67$$

(*v*) Let m_w kg/min be the required rate of water circulation in the cooler. Then from the equality of heat gained by water to the heat lost by air gives

$$m_w \, c_{p_w} \, (\Delta t)_w = m_a \, c_{p_a} \, (\Delta t)_a$$

$$m_w \times 4.186 \times 30 = 12.70 \times 1.005 \times (445.7 - 288)$$

Solution gives : $m_w = \mathbf{16.03}$ kg/min

EXAMPLE 3.39

An air refrigerating plant operating on the Bell-Coleman cycle draws air from a cold chamber at 1 bar and – 5°C and compresses to 4.5 bar. The compressed air is cooled in the heat exchanger at constant pressure to a temperature of 20°C. Subsequently, expansion is carried out in the expansion cylinder to 1 bar pressure. Assuming isentropic efficiency of compressor and expander as 85% and 90% respectively, make calculations for :

(a) refrigerating capacity of the plant if the mass flow rate of air is 25 kg/min

(b) power input to compressor, and

(c) coefficient of performance of the system

Solution : Refer Fig. 3.36 for the *T-s* diagram. The subscripts denote the state points in the cycle.

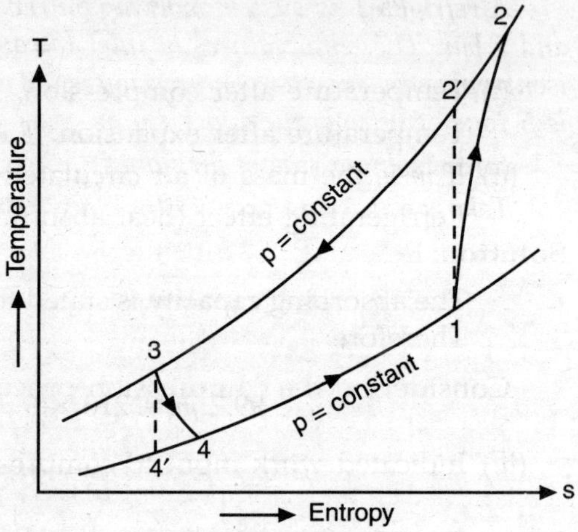

Fig. 3.36.

$$p_1 = p_4 = 1 \text{ bar} \quad \text{and} \quad T_1 = 268 \text{ K}$$

$$p_2 = p_3 = 4.5 \text{ bar} \quad \text{and} \quad T_3 = 293 \text{ K}$$

$$\eta_c = 0.8 \quad \text{and} \quad \eta_t = 0.9$$

Considering the compression process 1-2′,

$$T_2' = T_1 \left(\frac{p_2}{p_1}\right)^{\frac{\gamma-1}{\gamma}} = 268 \left(\frac{4.5}{1}\right)^{\frac{1.4-1}{1.4}} = 411.92 \text{ K}$$

$$\eta_c = \frac{T_2' - T_1}{T_2 - T_1} \; ; \quad 0.85 \; \frac{411.92 - 268}{T_2 - 268}$$

or $$T_2 = \frac{411.92 - 268}{0.85} + 268 = 437.32 \text{ K}$$

Considering the expansion process 3-4′

$$T_4' = T_3 \times \left(\frac{p_4}{p_3}\right)^{\frac{\gamma-1}{\gamma}} = 293 \times \left(\frac{1}{4.5}\right)^{\frac{1.4-1}{1.4}} = 196.63 \text{ K}$$

$$\eta_t = \frac{T_3 - T_4}{T_3 - T_4'} \; ; \quad 0.9 = \frac{293 - T_4}{293 - 190.63}$$

or $$T_4 = 293 - 0.9\,(293 - 190.63) = 200.87 \text{ K}$$

Work of compression $= c_p \, (T_2 - T_1)$

Work of expansion $= c_p \, (T_3 - T_4)$

∴ Net work input to compressor,

$$= c_p \, [(T_2 - T_1) - (T_3 - T_4)]$$

$$= 1.005 \, [(437.32 - 268) - (293 - 200.87)] = 77.57 \text{ kJ/kg}$$

Refrigerating effect $= c_p \, (T_1 - T_4) = 1.005 \, (268 - 200.87) = 67.46$ kJ/kg (*a*) Total refrigerating effect = 25 × 67.46 = 1686.5 kJ/min

1 ton of refrigeration = 210 kJ/min

$$\therefore \quad \text{Capacity of plant} = \frac{1686.5}{210} = \mathbf{8.03\ ton}$$

(*b*) Total net work supplied to compressor,

$$= 25 \times 77.57 = 1939.25 \text{ kJ/min}$$
$$= 32.32 \text{ kJ/s} = \mathbf{32.32\ kW}$$

(*c*) $$COP = \frac{\text{refrigerating effect}}{\text{work input}} = \frac{67.46}{77.57} = \mathbf{0.87}$$

The system is highly inefficient as *COP* is less than unity.

EXAMPLE 3.40.

A refrigerating system working on reversed Brayton cycle operates between pressure limits of 1 bar and 5 bar. The temperatures at inlet to compressor and expander are 30°C and 40°C respectively. The isentropic efficiency for compression is 0.8 and the same for expansion is 0.9. If the capacity of the system is 5 tons of refrigeration, determine

(a) mass flow side of air, and *(b) power input to compressor.*

Take $\gamma = 1.4$ and $c_p = 1.004$ kJ/kg K **(UPTU)**

Solution: Refer Fig. 3.37 for the *T-s* diagram. The subscripts denote the state points in the cycle.

$$p_1 = p_4 = 1 \text{ bar and } T_1 = 30°C = 303 \text{ K}$$
$$p_2 = p_3 = 5 \text{ bar and } T_3 = 40°C = 313 \text{ K}$$

Considering the compression process 1–2′,

$$T_2' = T_1\left(\frac{p_2}{p_1}\right)^{\frac{\gamma-1}{\gamma}} = 303 \times \left(\frac{5}{1}\right)^{\frac{1.4-1}{1.4}}$$
$$= 480.12 \text{ K}$$

$$\eta_c = \frac{T_2' - T_1}{T_2 - T_1};\ 0.8 = \frac{480.12 - 303}{T_2 - 303}$$

or $$T_2 = \frac{480.12 - 303}{0.8} + 303 = 524.4 \text{ K}$$

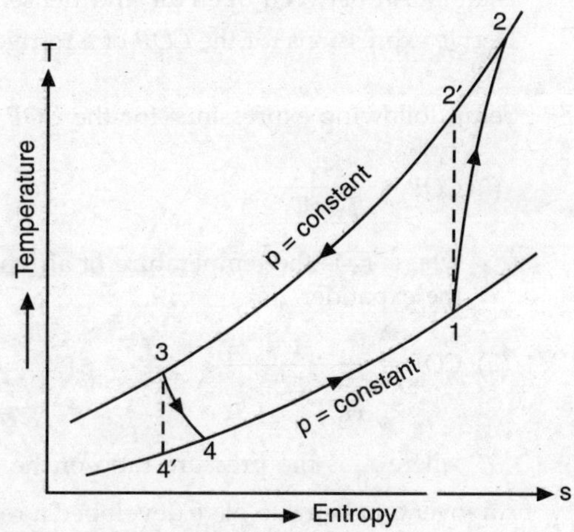

Fig. 3.37.

Considering the expansion process 3–4′,

$$T_4' = T_3\left(\frac{p_4}{p_3}\right)^{\frac{\gamma-1}{\gamma}} = 313 \times \left(\frac{1}{5}\right)^{\frac{1.4-1}{1.4}} = 197.56$$

$$\eta_e = \frac{T_3 - T_4}{T_3 - T_4'};\ 0.9 = \frac{313 - T_4}{313 - 197.5}$$

or $$T_4 = 313 - 0.9\ (313 - 197.5) = 209.05 \text{ K}$$

The capacity of the plant is stated to be 5 tons of refrigeration, and 1 ton of refrigeration = 210 kJ/min

∴ Total refrigerating effect = 5 × 210 = 10.50 kJ/min

(*a*) Let m_a be the mass flow rate of air in kg/min. Then

$$1050 = m_a\, c_{pa}\,(T_1 - T_4) = m_a \times 1.005\ (303 - 209.05)$$

That gives: $m_a = \dfrac{1050}{1.004(303-209.05)}$ = **11.12 kg/min**

(*b*) Work of compression = $m\ c_{pa}\ (T_2 - T_1)$

Work of expansion = $m\ c_{pa}\ (T_3 - T_4)]$

∴ Net work input to compressor

$= m\ c_{pa}\ [(T_2 - T_1) - (T_3 - 209.5)]$

$= 11.12 \times 1.004\ [(524.4 - 303) - (313 - 209.05)]$

= 1312.57 kJ/min = 21.87 kJ/s = 21.87 kW

REVIEW QUESTIONS

A. Conceptual and conventional questions.

1. State the limitations due to which it has not been possible to build a Carnot refrigeration system.
2. Carnot *COP* cannot be achieved in practice. Then what is the utility of its study?
3. A Carnot heat pump operates between higher and lower temperature limits of T_1 and T_2 respectively. Which is the more effective way to increase its coefficient of performance: increase the temperature T_1 while temperature T_2 is held constant or to decrease T_2 by the same amount while T_1 is held constant?
4. How is a reversed Brayton cycle used for refrigeration?
5. Sketch the schematics of a refrigeration system operating on Bell-Coleman cycle and explain its working. Represent the processes involved on *p-v* and *T-s* plots.
6. Distinguish between open air and dense air refrigeration cycles.
7. Derive expressions for the *COP* of a refrigerator and heat pump if both are working on reversed Carnot cycle.
8. Setup following expressions for the *COP* of a Bell-Coleman refrigeration cycle:

   (*i*) $COP = \dfrac{T_4}{T_3 - T_4}$

   where T_3 is the temperature of air entering the expander and T_4 is the temperature of air leaving the expander.

   (*ii*) $COP = \dfrac{1}{(r_p)^{\frac{\gamma-1}{\gamma}} - 1}$

   where r_p is the pressure ratio of the system.
9. An inventor claims to have developed a refrigerating machine which operates between – 20°C and 30°C and consumes 1 kW of power. The machine gives a refrigerating effect of 21.6 MJ/hr. Comment on the claim of the inventor. **[Ans.** False]
10. A water cooler is supplied water at 25°C at a rate of 0.75 litres per minute. The water gets cooled to 8°C and in the process, the unit consumes 200 watts power. Compare the *COP* of this refrigeration plant with that of the ideal refrigeration cycle for a similar situation.
11. For certain refrigerator working on reversed Carnot cycle, the *COP* is 5. It extracts 8350 kJ/min from a chamber which is to be maintained at a temperature lower than that of surroundings. Determine the heat rejected to the surroundings and the power required to run the compressor.

    If a heat pump is to work on this very cycle consuming the same power and extracting the same amount of heat, work out the reciprocal thermal efficiency of the heat pump.

    [Ans. 10020 kJ/min, 27.84 kW, 6]

12. Two Carnot refrigerators A and B operate in series. The refrigerator A absorbs energy at the rate of 1 kJ/s from a system at 300 K and rejects energy as heat to a system at *T*. The refrigerator B absorbs the same quantity of energy which is rejected by the refrigerator A to the system at A, and rejects energy as heat to a system at 1000 K. If both the refrigerators have the same *COP*, calculate
   (*a*) the temperature *T* of the intermediate system,
   (*b*) the *COP* of the refrigerators, and
   (*c*) the rate at which energy is rejected as heat to the system at 1000 K.
   [**Ans.** 547.7 K, 1.21, 3.33 kJ/s]
13. A Carnot refrigerator operates between temperatures – 10°C and 30°C. Determine the coefficient of performance. Subsequently it is desired to change the temperatures so as to make the coefficient of performance exactly equal to 7.0. The amount of increase in the higher temperature is equal to amount of decrease in the lower temperature. Workout the new temperatures. [**Ans.** 301.87 K, 264.13 K]
14. Two Carnot refrigerators A and B are arranged in series. Workout the following expression for COP of the composite system
$$COP = \frac{(COP)_A \times (COP)_B}{1 + (COP)_A + (COP)_B}$$
   where $(COP)_A$ is the coefficient of performance of refrigerator A
   and $(COP)_B$ is the coefficient of performance of refrigerator B
15. A heat pump is used for heating the interior of a house in a cold climate. The ambient temperature is – 5°C and the desired interior temperature is 25°C. The compressor of the heat pump is to be driven by a heat engine working between 1000°C and 25°C. Treating both cycles as reversible, calculate the ratio in which the heat pump and the heat engine share the heating load.
16. A gas refrigeration system using air as refrigerant operates on an ideal reversed Brayton cycle of pressure ratio 3 and minimum pressure 1.4 bar. The temperatures at inlet to compressor and expander are stated to be –20°C and 50°C respectively. If the system is to maintain a load of 10 tons, find:
   (*a*) *COP* of the system,
   (*b*) mass flow rate of air in kg/s,
   (*c*) volume flow rate entering the system, and
   (*d*) heat rejected in the condenser.
17. An air refrigeration open system operating between 1 MPa and 100 kPa is required to produce a cooling effect of 2000 kJ/min. The temperature of air leaving the cold chamber is – 5°C and that at leaving the cooler is 30°C. Determine:
   (*a*) mass of air circulated per minute,
   (*b*) coefficient of performance and power in kW required to operate the system.
   [**Ans.** 17.91 kg/min, 1.076, 30.97 kW]
18. An air refrigeration system working on Bell-Coleman cycle draws air from the refrigerated space at 1 bar pressure and 10°C temperature. After isentropic compression to 5 bar, the air is cooled at 25°C at constant pressure in the cooler. Subsequently the cooled air expands isentropically in the expansion-cylinder to a pressure of 1 bar. Make calculations for the refrigeration effect per kg of air and COP of the machine.
   How these parameters would be affected if the compression and expansion processes are polytropic according to relations $pv^{1.35}$ = constant and $pv^{1.3}$ = constant respectively.
   [**Ans.** 95.7 kJ/kg, 1.71, 78.2 kJ/kg, 1.75]
19. An air refrigerator working on Bell-Coleman cycle takes air into the compressor at 1 bar and at – 4°C. The air after compression to 5 bar is led into the condenser where it is cooled at constant pressure to a temperature of 26°C. The air is then expanded to 1 bar pressure and discharged into the cooling chamber. If the isentropic efficiency of the compressor and expander is 80% and 85% respectively, determine:

(a) the mass of the air circulated per minute to provide 15 ton of refrigeration
(b) power required to run the compressor. It may be presumed that actual COP is 50 percent of the theoretical.

Treat air is perfect gas and take $\gamma = 1.4$ and $c_p = 1.005$ kJ/kg K.

20. (a) Discuss the advantages of dense air refrigeration system over an open air refrigeration system.
(b) A dense closed cycle air refrigeration system working between 4 atm and 16 atm extracts 1.25×10^5 kJof heat per hour. The air enters the compressor at 5°C and into the expander at 20°C. If the unit runs at 300 rev/min, determine:
(a) power required to run the unit,
(b) bore of the compressor, and
(c) refrigerating capacity in tons of ice at 0°C per day

Take the following:

Compressor and expander are double acting; stroke for compressor and expander is 30 cm; mechanical efficiency is 80% for compressor and 85% for expander; both the compression and expansion processes are isentropic.

B. Fill in the blanks with appropriate word/words

1. A reversed Carnot cycle consists of two isothermal and two .... processes.
2. When the low temperature is fixed, the COP of a refrigerating machine can be improved by keeping the higher body temperature as .... as possible.
3. If a Carnot refrigeration cycle is to have a COP of 5, the ratio of maximum to minimum temperature in the cycle should be ....
4. A Carnot engine has an efficiency of 80 percent. If the cycle is reversed in direction and made to operate as a refrigerator, its COP will be ....
5. A Bell-Coleman cycle is a reversed ....
6. The Bell-Coleman cycle uses .... as the working fluid.
7. The COP of an ideal Bell-Coleman cycle depends upon the ratio of the upper and lower .... limits.
8. In a closed or dense air refrigeration cycle, the operating pressure ratio can be reduced and that results in .... coefficient of performance.

Answers

1. isentropic 2. low 3. 1.2 4. 0.25 5. Brayton cycle 6. air 7. pressure 8. high

C. Multiple choice questions.

1. With reference to refrigerating system shown in Fig. 3.38
The coefficient of performance of the refrigerator is
(a) $\frac{Q_2}{W}$ (b) $\frac{Q_1}{W}$
(c) $\frac{Q_1}{Q_2}$ (d) $\frac{Q_1 - Q_2}{W}$

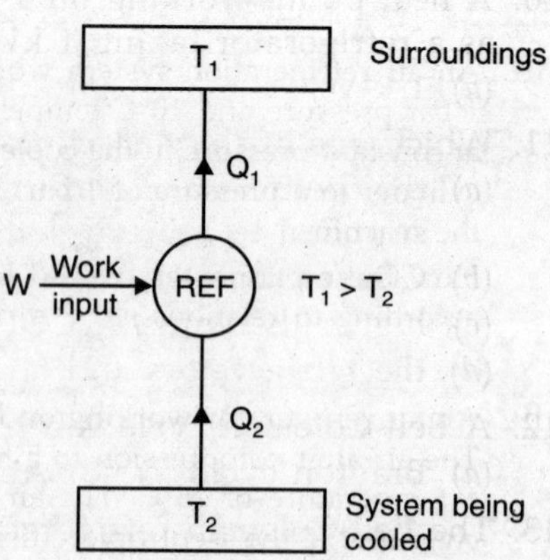

Fig. 3.38

2. The coefficient of performance of a refrigerator working on Carnot cycle is
(a) $\frac{T_1 - T_2}{T_1}$ (b) $\frac{T_1 - T_2}{T_2}$
(c) $\frac{T_2}{T_1 - T_2}$ (d) $\frac{T_1}{T_1 - T_2}$

where T_1 and T_2 are the highest and lowest operating

temperatures in the cycle.

3. The *COP* of a Carnot refrigeration cycle decreases on
   (*a*) decreasing the difference in operating temperatures
   (*b*) keeping the upper temperature constant and increasing the lower temperature
   (*c*) increasing the upper temperature and keeping the lower temperature constant
   (*d*) increasing the upper temperature and decreasing the lower temperature.
4. A Carnot refrigerating cycle used in house air-conditioning delivers 10 kW heat to the surroundings with work input at the rate of 10 kW of power. The coefficient of performance of this refrigerator would be
   (*a*) 1.5 (*b*) 1.67 (*c*) 2.5 (*d*) 0.6
5. A Carnot refrigerator rejects 3000 kJ of heat at 400 K while using 1000 kJ of work. The lowest operating temperature in the cycle should be
   (*a*) 15°C (*b*) 27°C (*c*) – 6°C (*d*) 0°C
6. A condenser of a refrigeration system rejects heat at a rate of 120 kW, while its compressor consumes a power of 30 kW. The coefficient of performance of the system would be
   (*a*) 2 (*b*) 3 (*c*) 4 (*d*) 5
7. The operating temperature of cold storage is 280 K and the heat leakage from the surroundings is 35 kW for the ambient temperature of 310 K. If the actual COP of the refrigeration plant is one-fourth of an ideal plant working between the same temperature limits, the power required to drive the plant would be.
   (*a*) 3.7 kW (*b*) 7.5 kW (*c*) 12 kW (*d*) 15 kW
8. Reference Fig. 3.39, E is a heat engine with efficiency 0.4 and R is a refrigerator Given that $Q_2 + Q_4 = 3\,Q_1$, the COP of the refrigerator is
   (*a*) 2.5 (*b*) 3.0
   (*c*) 4.0 (*d*) 5.0
9. A Carnot heat pump for domestic heating works between a cold system (the contents of refrigerator cabinet) at 0ºC and the water in the radiator system at 80ºC. The coefficient of performance of this heat pump would be about
   (*a*) 1.4 (*b*) 3.4 (*c*) 4.4
   (*d*) 6.2

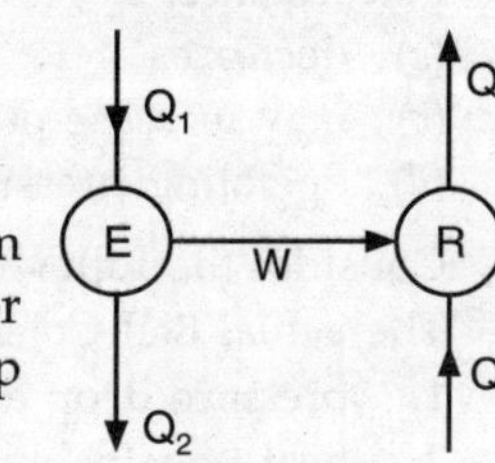

Fig. 3.39.

10. A heat pump working on a reversed Carnot cycle has a *COP* of 5. If it is made to work as a refrigerator taking 1 kW of work input, the refrigerating effect will be
    (*a*) 1 kW (*b*) 2 kW (*c*) 3 kW (*d*) 4 kW
11. Which one of the followings is a wrong statement?
    (*a*) the coefficient of performance (COP) refers to the ratio of desired effect to the energy supplied to achieve that effect
    (*b*) COP of most of the refrigerators and heat pumps is greater than one
    (*c*) higher the value of COP, more efficient the refrigerators and heat pumps would be
    (*d*) the refrigerators and heat pumps are reversible heat engines
12. A Bell-Coleman cycle is a reversed
    (*a*) Brayton cycle (*b*) Atkinson cycle (*c*) Ericsson cycle (*d*) Carnot cycle
13. The Bell-Coleman refrigeration cycle uses .... as the working fluid
    (*a*) hydrogen (*b*) carbon dioxide (*c*) air (*d*) any inert gas

14. Fig. 3.40 shows the P-V plot of Bell-Coleman cycle of refrigeration. When compression and expansion are both isentropic, the coefficient of performance of the machine is given by

(a) $\dfrac{T_c}{T_c - T_a}$ (b) $\dfrac{T_a}{T_d - T_a}$

(c) $\dfrac{T_c}{T_d - T_a}$ (d) $\dfrac{T_d}{T_d - T_b}$

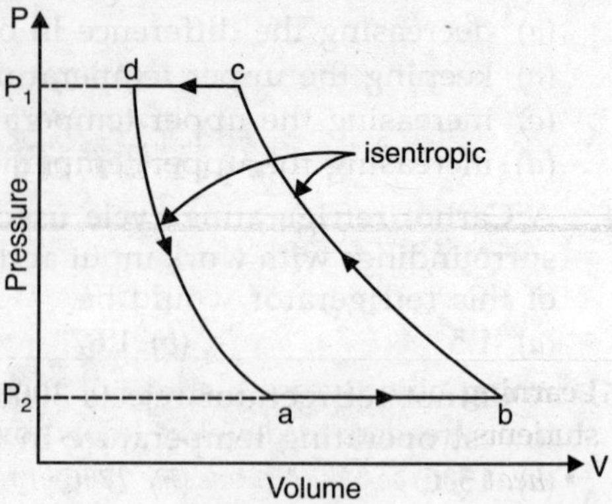

Fig. 3.40

15. Which aspect is not true in the context of a refrigerating machine working on Bell-Coleman cycle ?
(a) low coefficient of performance
(b) simplicity of design and operation
(c) high power consumption per ton of refrigeration
(d) large amount of air circulation per minute

16. All of the following statements about closed system of Bell-Coleman refrigeration cycle are correct, except
(a) air comes in direct contact with the space or substance being cooled
(b) less chances of choking of valves
(c) suction to the compression may be at high pressure
(d) size of the compressor and expander can be kept reasonably small

17. With an increase in the operating pressure difference in a Bell-Coleman cycle, the coefficient of performance
(a) decreases (b) increases
(c) may increase or decrease depending upon the pressure values
(d) operating pressure difference has no correlation with COP

18. Consider the following statements:
The actual Bell-Coleman cycle is distorted compared to the situation due to :
1. pressure drop as a result of fluid flow
2. heat transfer during compression and expansion
3. frictional losses due to blades

Which of these statements are correct?
(a) 1 and 2 (b) 2 and 3 (c) 1 and 3 (d) 1, 2 and 3

Answers

1. (a)	2. (d)	3. (d)	4. (a)	5. (b)	6. (c)	7. (d)	8. (d)
9. (c)	10. (d)	11. (d)	12. (d)	13. (c)	14. (b)	15. (b)	16. (a)
17. (a)	18. (d)						

□□□

Chapter – 4

Aircraft Refrigeration Systems

Learning objectives : A study of the subject matter presented in this chapter will enable the students/readers to develop knowhow of

- *heat sources that cause the temperature to rise and establish the need for aircraft refrigeration*
- *schematics and working of commonly used refrigeration systems for aircrafts*
- *analysis of aircraft refrigeration systems with respect to COP and power requirement*
- *comparison of aircraft refrigeration systems*
- *merits in the use of air as refrigerant the aircraft refrigeration systems*

4.1. NECESSITY OF AIRCRAFT REFRIGERATION

The higher we go, the cooler it becomes. Then why is the need of a refrigeration system in the aircraft when it is cruising at a high attitude. The answer to this question is reflected by the following aspects which cause the temperature to rise and establish the need for aircraft refrigeration.

1. Internal Sources

(*i*) Heat release from the occupants. Generally a human being releases about 300 kJ/hr of heat even while at rest. This value is likely to get enhanced due to nervousness which is quite common with some of the passengers.

(*ii*) Heat generated by control devices that operate on electrical energy. A medium size aeroplane needs about 10 to 15 kW of electric power for the operation of control devices installed in the plane.

(*iii*) Heat flow from the engine parts of the aircraft.

2. External Sources

(*i*) Heat transfer to the cabin from the solar radiation directly through the glass panes and through the airplane body.

(*ii*) Heat generated due to air resistance. There occurs an increase in the temperature of air due to skin friction when the fast moving aircraft passes through air. It is anticipated that an aircraft travelling at 1000 km/hr speed experiences a rise of about 50°C in temperature at its surface.

(*iii*) At high altitudes, the pressure is quite low and that would cause breathing problem to the passengers. Obviously, the air needs to be pressurized and that results into an increase in temperature.

(*iv*) Ramming (compression) of air. The ambient air is raised in pressure before it enters the cabin and the compression process results into an increase in temperature.

When the cumulative effect of the these factors is taken into consideration, the cabin temperature becomes as high as 40 to 50°C. Such a high temperature is quite uncomfortable against the comfortable value of 22°C and – 5°C required for preserving drinks and foods.

The cooling requirements per unit volume of the space are quite heavy due to compactness of all the equipments and engine. The refrigeration system required for an ordinary passenger plane is of 8 ton capacity and that for a jet fighter is of 20 ton capacity.

4.2. METHODS OF AIR REFRIGERATION SYSTEMS

The aircraft refrigeration systems essentially operate on the Bell-coleman cycle which became obsolete due to its low COP (low specific heat and poor conductivity of air), large power requirement per ton of refrigeration and bulk with use of reciprocating compressor and expander. While designing a refrigeration system for the aircraft, the weight of the refrigeration equipment is the major consideration. The weight has to be least and this factor overrides the effectiveness and operating cost of the system. With the advent of rotary compressors and turbines, the drawbacks of bulk and weight have been overcome to a large extent. Further, the air used as a refrigerant has not to be stored as it is cheaply and freely available from the surrounding atmosphere. Besides, air is non-flammable and its leakage is not a safety hazard. Again, space requirement of the air refrigeration system for the same capacity is less than any other type of refrigeration system.

When all these aspects are taken into consideration, the air cycle systems prove to be ideal for aircraft refrigeration.

Some of the commonly used air-refrigeration systems for aircrafts are :

(1) Basic or simple air-refrigeration system

(*a*) Without evaporating cooling (*b*) With evaporating cooling

(2) Boot strap air-refrigeration system

(*a*) Without evaporating cooling (*b*) With evaporating cooling

(3) Regenerative air-refrigeration system

(4) Reduced ambient air-refrigeration system

The salient aspects of these systems have been discussed briefly in the following paragraphs.

4.2.1 Basic or Simple Air-refrigeration System

(*a*) Without Evaporative Cooling :

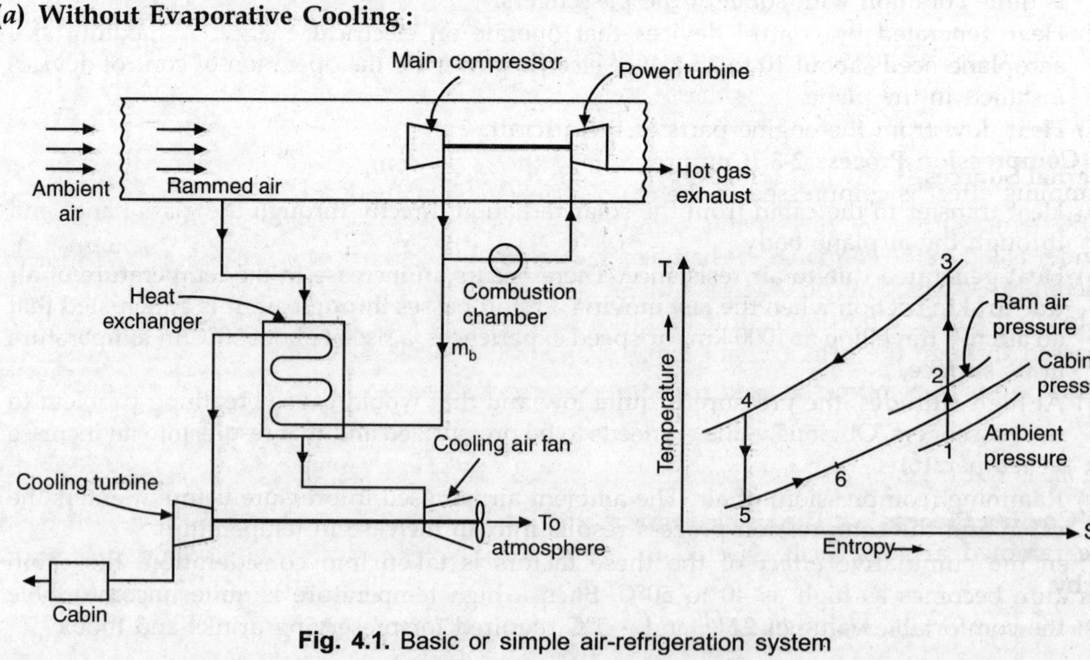

Fig. 4.1. Basic or simple air-refrigeration system

Fig. 4.1. shows the schematic arrangement of the various components of simple refrigeration system used for air-conditioning of the cabin and cockpit of the aircraft.

The various processes too have been shown on *T-s* plot and their sequence is as given below:

1. Ramming Process 1-2: The ambient air at a velocity equal to plane speed enters the inlet section which is designed as a diffuser. The air is slowed down and a part of the kinetic energy of the air stream is converted into pressure. The process is called rammed compression and the air so compressed is called rammed air.

Let C be the aircraft velocity or velocity of air relative to the aircraft in m/s. Then

$$\text{Kinetic energy of air} = \frac{C^2}{2} \text{ J/kg}$$

From the energy equation: $h_2 - h_1 = \dfrac{C^2}{2}$

or $$c_p (T_2 - T_1) = \frac{C^2}{2}$$

$$\therefore \quad T_2 = T_1 + \frac{C^2}{2c_p} \quad \text{or} \quad \frac{T_2}{T_1} = 1 + \frac{C^2}{2c_p T_1}$$

From the relation $c_p - c_v = R$ and $\dfrac{c_p}{c_v} = \gamma$, we have : $c_p = \dfrac{\gamma R}{\gamma - 1}$

Then

$$\frac{T_2}{T_1} = 1 + \frac{C^2 (\gamma - 1)}{2\gamma R T_1}$$

The parameter $\sqrt{\gamma R T_1}$ equals the sonic velocity *a*.

$$\therefore \quad \frac{T_2}{T_1} = 1 + \frac{\gamma - 1}{2} \frac{C^2}{a^2} = 1 + \frac{\gamma - 1}{2} M^2$$

where $M = \dfrac{C}{a}$ is the Mach number of flight.

2. Compression Process 2-3 (Compression in the main compressor): The ambient air after the ramming effect is compressed in the main compressor to about 4 to 5 bar pressure and 150 to 200°C temperature. A part of this compressed air is bled for refrigeration system and the remainder is led to the combustion chamber. The hot combustion products expand in the power turbine and the work thus obtained is utilized in running the compressor. Out of the total work input to compressor, a portion is utilized in compressing the ram air which is subsequently bled and passed through the heat exchanger for refrigeration system.

Work input to compressor used for refrigeration purpose,

$$= m_b \, c_p \, (T_3 - T_2)$$

where m_b is the mass of air bled from the main compressor for refrigeration system.

3. Cooling Process 3-4: The air bled for the refrigeration system is cooled in the heat exchanger by the rammed air. The heat rejected in the heat exchanger during the cooling process is given by

$$Q_{rej} = m_b \, c_p \, (T_3 - T_4)$$

4. Expansion Process 4-5 (Expansion in the cooling turbine): The cooled compressed air from the heat exchanger is taken to the cooling turbine where it expands. During expansion, there is fall both in pressure and temperature. The expansion is controlled and allowed to continue upto to cabin pressure. The pressure inside the cabin is higher than that of air outside at high altitude, and its limit is governed by the comfortable tolerance of the occupants.

The work made available during the expansion process is

$$W_{exp} = m_b\, c_p\, (T_4 - T_5)$$

The power output from the cooling turbine is utilized to drive the fan which draws the ram cooling air.

5. Refrigeration Process 5-6: The cooled low pressure air is made to pass through a water separator where traces of moisture associated with air are removed. The moisture free air enters the airplane cabin/cockpit, absorbs heat from the occupants and equipment, and produces the cooling effect.

The heat absorbed or refrigerating effect produced is

$$Q_{abs} = m_b\, c_p\, (T_6 - T_5)$$

where T_6 is the inside temperature of the cabin.

The different performance parameters for the system can then be worked out as :

$$COP = \frac{\text{refrigeration effect}}{\text{work input to compressor}}$$

$$= \frac{m_b\, c_p (T_6 - T_5)}{m_b\, c_p (T_3 - T_2)} = \frac{T_6 - T_5}{T_3 - T_2}$$

If the cooling system for the cabin is equivalent to a load of Q tons of refrigeration, then the mass of air required for refrigeration purposes is

$$m_b = \frac{\text{total cooling load}}{\text{refrigerating effect per kg of refrigerant}} = \frac{3.5\, Q}{c_p\, (T_6 - T_5)} \text{ kg/s}$$

Power required to take the cooling load in the cabin

$$= m_b\, c_p (T_3 - T_2) \text{ kJ/s or kW}$$

$$COP = \frac{\text{refrigerating effect}}{\text{power input}}$$

$$= \frac{3.5\, Q}{m_b\, c_p\, (T_3 - T_2)} = \frac{m_b\, c_p\, (T_6 - T_5)}{m_b\, c_p\, (T_3 - T_2)} = \frac{T_6 - T_5}{T_3 - T_2}$$

The analysis made above is valid under the following conditions :

(*i*) Both compression and expansion processes are isentropic

(*ii*) No pressure loss during heat exchange processes

(*iii*) Heat exchanger is 100 percent effective

(*b*) With Evaporative Cooling :

Fig. 4.2 shows the schematic arrangement of the various components of the basic air - refrigeration system with evaporative cooling. In addition to the elements of the basic system, this system has an evaporator which is installed between the heat exchanger and the cooling turbine. The evaporator provides an additional cooling effect through evaporation of a low boiling point liquid which may be alcohol, ammonia or even water.

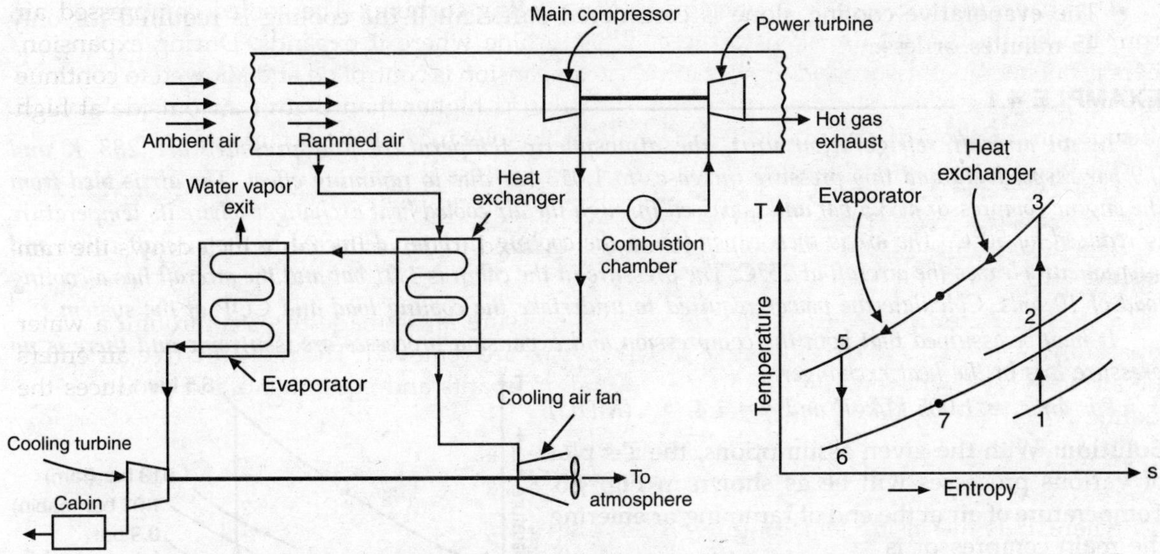

Fig. 4.2. Air-refrigeration system with evaporative cooling

The various processes involved in the evaporative cooling system are shown on *T-s* plot and their sequence is

1-2 : isentropic ramming action
2-3 : isentropic compression in the main compressor
3-4 : constant pressure cooling of bled air in the heat exchanger
4-5 : constant pressure cooling in the evaporator
The air at exit from the heat exchanger is passed through evaporation vessel containing a liquid of low boiling point. The liquid evaporates and that results in cooling of air.
5-6 : isentropic expansion of cooled compressed air in the cooling turbine upto the cabin pressure
6-7 : heating of air upto the temperature desired in the cabin

The warm air from the cabin finally exits to the atmosphere. Apparently the system follows an open cycle starting from and exhausting to the atmosphere.

- If the refrigerant plant is to handle a cabin load equivalent to Q tons of refrigeration, then mass of air to be bled for refrigeration purposes

$$m_b = \frac{3.5\,Q}{m_b\,c_p\,(T_7 - T_6)}\ \text{kg/s}$$

- Power required for the refrigeration system,

$$P = m_b\,c_p\,(T_3 - T_2)\ \text{kW}$$

- Coefficient of performance of the refrigeration system,

$$COP = \frac{3.5\,Q}{m_b\,c_p\,(T_3 - T_2)} = \frac{T_7 - T_6}{T_3 - T_2}$$

- The mass of evaporate required to be carried by the aircraft is given by

$$m_e = \frac{Q_e \times t}{h_{fg}}$$

where Q_e is the heat to be removed in evaporation (kJ/hr), t is the flight time in hours and h_{fg} is the latent heat of evaporation in kJ/kg.

- The evaporative cooling alone is considered sufficient if the cooling is required for only 45 minutes or less.

EXAMPLE 4.1

In an aircraft refrigeration unit, the atmospheric temperature and pressure are 283 K and 0.9 bar respectively and this pressure increases to 1.013 bar due to ramming effect. The air is bled from the engine compressor at 3.5 bar and is passed through the air cooled heat exchanger where its temperature is reduced by 50°C. The air is then expanded in the cooling turbine, delivered to the aircraft cabin and subsequently leaves the aircraft at 25°C. The pressure in the cabin is 1.01 bar and the aircraft has a cooling load of 10 tons. Calculate the power required to undertake the cooling load and COP of the system.

It may be assumed that both the compression and expansion processes are isentropic and there is no pressure loss in the heat exchanger.

For air c_p = 1.005 kJ/kgK and γ = 1.4. (MDU)

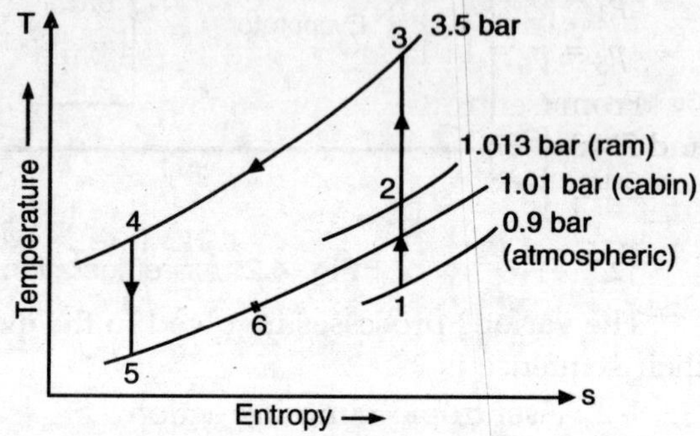

Fig. 4.3.

Solution: With the given assumptions, the *T-s* plot of various processes will be as shown in Fig. 4.3. Temperature of air at the end of ramming or entering the main compressor is

$$T_2 = T_1\left(\frac{p_2}{p_1}\right)^{\frac{\gamma-1}{\gamma}} = 283\times\left(\frac{1.013}{0.9}\right)^{\frac{1.4-1}{1.4}}$$

$$= 292.6 \text{ K}$$

Temperature of air leaving the main compressor is

$$T_3 = T_2\left(\frac{p_3}{p_2}\right)^{\frac{\gamma-1}{\gamma}} = 292.6\left(\frac{3.5}{1.013}\right)^{\frac{1.4-1}{1.4}} = 417 \text{ K} = 144°\text{C}$$

Due to 50°C reduction in temperature in the heat exchanger, the temperature of air at exit from the heat exchanger and at inlet to cooling turbine will be

$$T_4 = 144 - 50 = 94°\text{C} = 367 \text{ K}$$

With isentropic expansion upto cabin pressure of 1.03 bar in the cooling turbine

$$T_5 = T_4\left(\frac{p_5}{p_4}\right)^{\frac{\gamma-1}{\gamma}} = 367\left(\frac{1.01}{3.5}\right)^{\frac{1.4-1}{1.4}} = 257 \text{ K}$$

(*a*) Mass of bled air required for refrigeration purposes

$$m_b = \frac{\text{total cooling load (refrigeration effect)}}{\text{refrigerating effect per kg of air}}$$

$$= \frac{\text{total cooling load}}{c_p(T_6 - T_5)} = \frac{10\times3.5}{1.005(298-257)} = 0.8536 \text{ kg/s}$$

Power required to meet the cooling load in the cabin,

$$P = m_b\, c_p\, (T_3 - T_2) = 0.8536 \times 1.005\,(417 - 294.6)$$

$$= 106.72 \text{ kJ/s} = \mathbf{106.72\ kW}$$

$$(b)\ COP = \frac{\text{refrigerating effect}}{\text{power input}} = \frac{10\times3.5}{106.72} = \mathbf{0.328}$$

EXAMPLE 4.2

A simple air cooled system is used for an aeroplane having a load of 9 tonnes. The atmospheric pressure and temperature are 0.9 bar and 10°C respectively. During ramming pressure increases to 1.013 bar, and in the heat exchanger the temperature of air is reduced by 55°C. Measurements indicate that pressure in the cabin is 1.01 bar and the temperature of air leaving the cabin is 25°C. Determine :

(a) Power required to take the load of cooling in the cabin

(b) COP of the system ***(MRIU)***

Solution : Refer Fig. 4.4. for *T-s* plot of the operating cycle.

The pressure and temperature values at salient points as per given data are :

$T_1 = 10°C = 283\text{ K}$; $T_6 = 25°C = 298\text{ K}$

$p_1 = 0.9\text{ bar}$; $p_2 = 1.013\text{ bar}$

$p_5 = p_6 = 1.01\text{ bar}$; $p_3 = p_4 = 4\text{ bar}$

From isentropic compression process, 1 – 2 and 2 – 3

$$T_2 = T_1\left(\frac{p_2}{p_1}\right)^{\frac{\gamma-1}{\gamma}} = 283\times\left(\frac{1.013}{0.9}\right)^{\frac{1.4-1}{1.4}} = 292.5\text{ K}$$

$$T_3 = T_2\left(\frac{p_3}{p_2}\right)^{\frac{\gamma-1}{\gamma}} = 292.5\times\left(\frac{4}{1.013}\right)^{\frac{1.4-1}{1.4}} = 433\text{ K} = 160\text{ °C}$$

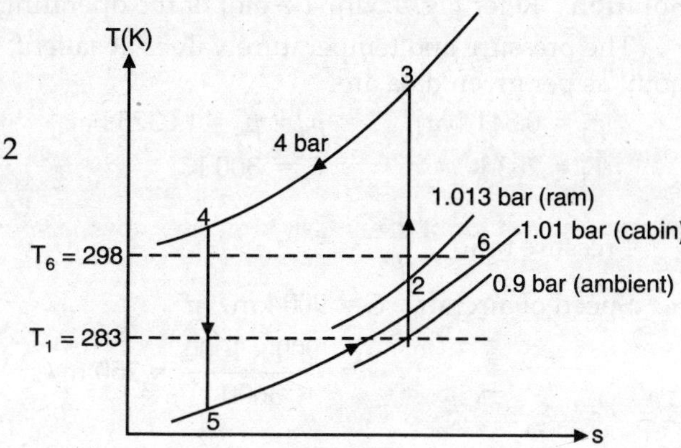

Fig. 4.4.

As a result of 55°C temperature drop in the heat exchanger, temperature of air leaving the heat exhanger is,

$$T_4 = (160 - 55) + 273 = 378\text{ K}$$

From isentropic expansion process 4 –5 ,

$$T_5 = T_4\left(\frac{p_5}{p_4}\right)^{\frac{\gamma-1}{\gamma}} = 378\times\left(\frac{1.01}{4}\right)^{\frac{1.4-1}{1.4}} = 255\text{ K}$$

(*a*) Since 1 tonne of refrigerating is equivalent to 3.5 kJ/s of cooling,

total refrigeration (cooling) effect = 9 × 3.5 = 31.5 kJ/s

Let m_a be the mass of air required for refrigeration purpose. Then,

$$m_a\, c_{p_a}(T_6 - T_5) = 31.5 \quad ; \quad m_a \times 1.005\,(298 - 255)$$

This gives : $$m_a = \frac{31.5}{1.005\,(298 - 255)} = 0.729\text{ kg/s}$$

Work input to compressor = $m_a\, c_{p_a}(T_3 - T_2)$

= 0.729 × 1.005 (433 – 292.5) = 102.93 kJ/s = **102.93** kW

(*b*) *COP* of the system

$$= \frac{\text{refrigerating effect}}{\text{work input}} = \frac{31.5}{102.93} = \mathbf{0.306}$$

EXAMPLE 4.3

An aircraft flying at an altitude of 8000 m, where the ambient air is at 0.341 bar pressure and 263 K temperature, has a speed of 900 km/hr. The pressure ratio of the air compressor is 5.

The cabin pressure is 1.013 bar and the temperature is 300 K. Determine (a) power requirement of the aircraft for pressurization (excluding the ram work), (b) additional power requirement for refrigeration and refrigerating capacity. Take mass flow rate of air at 1 kg/s. **(AFU)**

Solution : Refer Fig. 4.5 for *T-s* plot of the operating cycle.

The pressure and temperature values at salient point as per given data are:

$p_1 = 0.341$ bar, $\quad p_5 = p_6 = 1.0325$ bar

$T_1 = 263$ K, $\quad T_6 = 300$ K ;

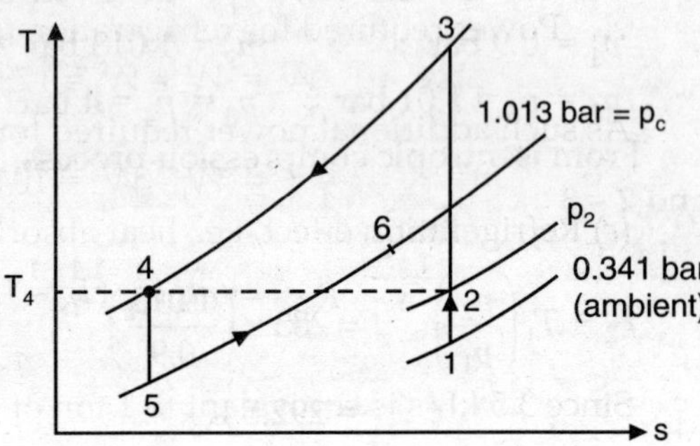

Fig. 4.5.

Pressure ratio, $\dfrac{p_3}{p_2} = 5$

Speed of aircraft, $C = 900$ km/hr

$$= \frac{900 \times 1000}{3600} = 250 \text{ m/s}$$

$$T_2 = T_1 + \frac{C^2}{2000\, C_p} = 263 + \frac{250}{2000 \times 1.005} = 294 \text{ K}$$

From isentropic process 1 – 2,

$$p_2 = p_1 \left(\frac{T_2}{T_1}\right)^{\frac{\gamma}{\gamma-1}} = 0.341 \times \left(\frac{294}{263}\right)^{\frac{1.4}{1.4-1}} = 0.504 \text{ bar}$$

The compression ratio for the air compressor is stated to be 5, and therefore

$$p_3 = 5\, p_2 = 5 \times 0.504 = 2.52 \text{ bar}$$

From isentropic compression process 2 – 3,

$$T_3 = T_2 \left(\frac{p_3}{p_2}\right)^{\frac{\gamma-1}{\gamma}} = 294 \times (5)^{\frac{1.4-1}{1.4}} = 465.5 \text{ K}$$

From isentropic expension process 4 – 5,

$$T_5 = T_4 \left(\frac{p_5}{p_4}\right)^{\frac{\gamma-1}{\gamma}} = 294 \times \left(\frac{1.013}{2.52}\right)^{\frac{1.4-1}{1.4}} = 226.3 \text{ K}$$

(*a*) Power required for pressurization (excluding ram work)

$$W_p = m_a\, c_{p_a} (T_c - T_2) = m_a\, c_{p_a} T_2 \left[\left(\frac{p_c}{p_2}\right)^{\frac{\gamma-1}{\gamma}} - 1\right]$$

$$= 1 \times 1.005 \times 294 \left[\left(\frac{1.013}{0.504} \right)^{\frac{1.4-1}{1.4}} - 1 \right] = 65 \text{ kJ/s} \quad \text{or} \quad \mathbf{65\ kW}$$

(*b*) Additional power required for refrigeration

$$W_c = m_a c_{p_a} (T_3 - T_2)$$

$$= 1 \times 1.005 \times (465.5 - 294) = 172.5 \text{ kJ/s} \quad \text{or} \quad 172.5 \text{ kW}$$

Work available from expansion process 4 – 5,

$$W_t = m_a c_{p_a} (T_4 - T_5)$$

$$= 1 \times 1.005\ (294 - 226.3) = 68 \text{ kJ/s} = 68 \text{ kW}$$

∴ Power required for refrigeration (excluding ram work)

$$W_r = W_c - W_t = 172.5 - 68 = 104.5 \text{ kW}$$

As such additional power required for refrigeration

$$= W_r - W_p = 104.5 - 65 = 39.5 \text{ kW}$$

(*c*) Refrigeration effect, i.e., heat absorbed

$$R_{eff} = m_a c_{p_a} (T_6 - T_5) = 1 \times 1.005\ (300 - 226.3)$$

$$= 74.07 \text{ kJ/s}$$

Since 3.5 kJ/s is equivalent to 1 ton of refrigeration.

$$\text{Refrigeration capacity} = \frac{74.07}{3.5} = \mathbf{21.16} \text{ tonne}$$

EXAMPLE 4.4

An air-conditioning system unit of a pressurized aircraft receives its air from the jet engine compressor at a pressure of 1.25 bar. The ambient pressure and temperature are 0.2 bar and 237 K respectively. The air-conditioning unit consists of a free-wheeling compression and a turbine mounted on the same shaft. The work produced by the turbine is sufficient to drive the compressor. The compressed air leaving the compressor is cooled in the cooler at constant pressure and subsequently expanded in the turbine to the cabin pressure of 1.0 bar and temperature of 280 K. Calculate (a) the compressor discharge pressure and (b) the cooler exit temperature.

Solution: Refer Fig. 4.6 for the schematic arrangement of the system and *T-s* plot of the processes involved.

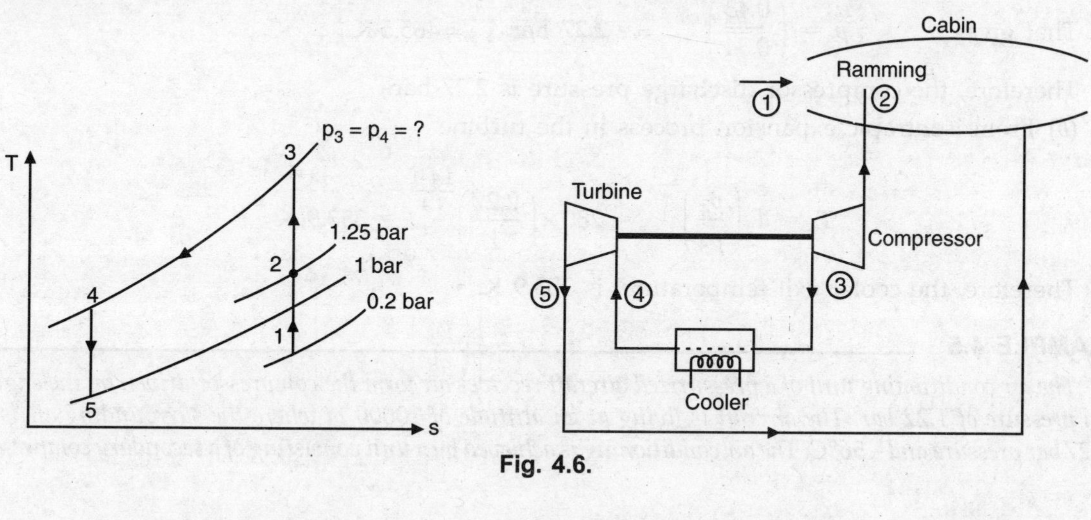

Fig. 4.6.

The solution is being worked out by assuming that both the compression and expansion processes are isentropic.

For the isentropic ramming process 1 – 2,

$$T_2 = T_1\left(\frac{p_2}{p_1}\right)^{\frac{\gamma-1}{\gamma}} = 237\times\left(\frac{1.25}{0.2}\right)^{\frac{1.4-1}{1.4}} = 400.5 \text{ K}$$

Since the work produced by the turbine is sufficient to drive the compressor, we have

work input to compressor = work output from turbine

$$m\, c_p\, (T_3 - T_2) = m\, c_p\, (T_4 - T_5)$$

or $$T_3 - T_2 = T_4 - T_5$$

or $$T_2\left[\frac{T_3}{T_2}-1\right] = T_5\left[\frac{T_4}{T_5}-1\right]$$

or $$\frac{T_2}{T_5}\left[\left(\frac{p_3}{p_2}\right)^{\frac{\gamma-1}{\gamma}}-1\right] = \left[\left(\frac{p_4}{p_5}\right)^{\frac{\gamma-1}{\gamma}}-1\right]$$

Substituting the appropriates values and nothing that $p_4 = p_3$ and $T_5 = 280$ K, we get

$$\frac{400.5}{280}\left[\left(\frac{p_3}{1.25}\right)^{\frac{1.4-1}{1.4}}-1\right] = \left[\left(\frac{p_3}{1}\right)^{\frac{1.4-1}{1.4}}-1\right]$$

or $$1.43\left[\left(\frac{p_3}{1.25}\right)^{0.286}-1\right] = (p_3)^{0.286} - 1$$

or $$\frac{1.43\times(p_3)^{0.286}}{(1.25)^{0.286}} - 1.43 = (p_3)^{0.286} - 1$$

or $$1.34\ (p_3)^{0.286} = (p_3)^{0.286} + 0.43$$

That gives : $$p_3 = \left(\frac{0.43}{0.34}\right)^{1/0.286} = \mathbf{2.27\ bar}$$

Therefore, the compressor discharge pressure is 2.27 bar.

(*b*) From isentropic expansion process in the turbine

$$T_4 = T_5\left(\frac{p_4}{p_5}\right)^{\frac{\gamma-1}{\gamma}} = 280\times\left(\frac{2.27}{1}\right)^{\frac{1.4-1}{1.4}} = \mathbf{353.9\ K}$$

Therefore, the cooler exit temperature, is 353.9 K.

EXAMPLE 4.5

The air conditioning unit of a pressurized aircraft receives air form the compressor driven by the engine at a pressure of 1.22 bar. The aircraft is flying at an altitude of 10000 m where the surrounding air is at 0.227 bar pressure and –56°C. The air conditioning is achieved by a unit consisting of a secondary compressor and a turbine mounted on the same shaft. The air leaves the turbine at 1 bar pressure and 7°C temperature. Assuming all processes to be reversible, determine pressure after the secondary compression and temperature of air at exit from the cooler.

Solution : Refer Fig. 4.7 for *T-s* plot of the operating cycle.

The compression process 1 – 2 is executed in the jet engine compressor and temperature at exit from the jet engine compressor is,

$$T_2 = T_1\left(\frac{p_2}{p_1}\right)^{\frac{\gamma-1}{\gamma}}$$

$$= 217 \times \left(\frac{1.22}{0.227}\right)^{\frac{1.4-1}{1.4}} = 351 \text{ K}$$

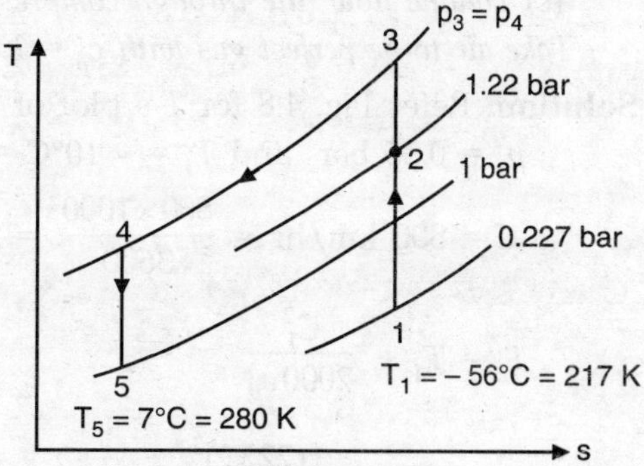

Fig. 4.7.

Work input to secondary compressor

$$W_{2-3} = c_{p_a}(T_3 - T_2) \quad \text{(taking unit mass of air)}$$

$$= c_{p_a} T_2\left[\left(\frac{p_3}{p_2}\right)^{\frac{\gamma-1}{\gamma}} - 1\right] = 1.005 \times 351\left[\left(\frac{p}{1.22}\right)^{\frac{1.4-1}{1}} - 1\right]$$

Work output from turbine during expansion process 4 – 5,

$$= c_p(T_4 - T_5) \quad \text{(taking unit mass of air)}$$

$$= c_{p_a} T_5\left[\left(\frac{p_4}{p_5}\right)^{\frac{\gamma-1}{\gamma}} - 1\right] = 1.005 \times 280\left[\left(\frac{p}{1}\right)^{\frac{1.4-1}{1}} - 1\right]$$

The work output from the turbine is utilized in running the secondary compressor. Therefore

$$1.005 \times 280\left[\left(\frac{p}{1}\right)^{0.286} - 1\right] = 1.005 \times 351\left[\left(\frac{p}{1.22}\right)^{0.286} - 1\right]$$

Solution gives : Pressure after secondary compression $p = 3.06$ bar

Temperature of air at exit form cooler

$$T_4 = T_5 \times \left(\frac{p_4}{p_5}\right)^{\frac{\gamma-1}{\gamma}} = 280 \times \left(\frac{3.06}{1}\right)^{\frac{1.4-1}{1.4}} = \mathbf{3.85\ K}$$

EXAMPLE 4.3

An aircraft moving with a speed of 800 km/hr uses simple air refrigeration system to take a load of 50 tons of refrigeration. The aircraft is flying at an altitude of 8000 m above sea level where the surrounding air pressure and temperature are 0.35 bar and – 10°C respectively. The main compressor has a pressure ratio of 4.5. The air leaving the compressor is cooled in the heat exchanger which has an effectivens of 90 percent. Finally the air is passed through the cooling turbine and is then supplied to the cabin at a pressure of 1.05 bar and 25°C. Assuming that isentropic efficiencies of compressor and turbine are 0.8 and 0.82 respectively, determine

(a) the pressure and temperature at all points of the cycle,
(b) the power required to take the load in the cabin and COP of the system,
(c) volume flow rate through compressor inlet and expander outlet.

Take air to be perfect gas with c_p = 1.005 kJ/kg K and γ = 1.4

Solution: Refer Fig. 4.8 for T-s plot of the operating cycle

$$p_1 = 0.35 \text{ bar and } T_1 = -10°C = 263 \text{ K}$$

$$C = 800 \text{ km/hr} = \frac{800 \times 1000}{3600} = 222.2 \text{ m/s}$$

$$T_2 = T_1 + \frac{C_1^2}{2000\, c_p}$$

$$= 263 + \frac{(222.2)^2}{2000 \times 1.005} = 287.56 \text{ K}$$

$$\frac{p_2}{p_1} = \left(\frac{T_2}{T_1}\right)^{\frac{\gamma}{\gamma-1}} = \left(\frac{287.56}{263}\right)^{\frac{1.4}{1.4-1}} = 1.367$$

$$\therefore\ p_2 = 1.367\, p_1 = 1.367 \times 0.35 = 0.478 \text{ bar}$$

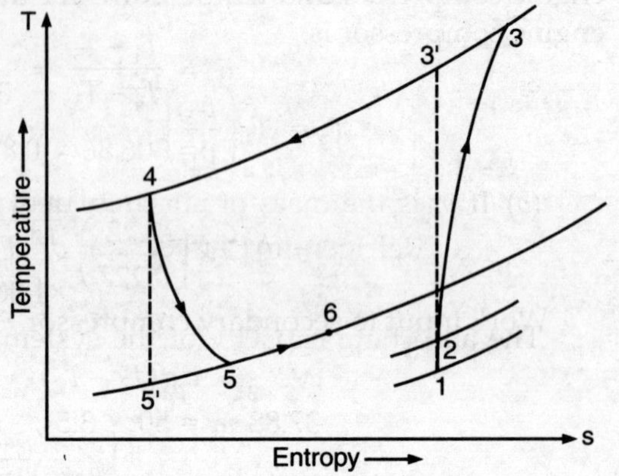

Fig. 4.8

Compression Process 2-3 :

The main compressor has a pressure ratio of 4.5 (given) and therefore pressure at exit from the compressor is

$$p_3 = 4.5\, p_2 = 4.5 \times 0.478 = 2.15 \text{ bar}$$

$$\frac{T_3'}{T_2} = \left(\frac{p_3}{p_2}\right)^{\frac{\gamma-1}{\gamma}} = \left(\frac{2.15}{0.478}\right)^{\frac{1.4-1}{1.4}} = 1.537$$

$$T_3' = 1.537\, T_2 = 1.537 \times 287.56 = 441.98 \text{ K}$$

$$\eta_c = \frac{T_3' - T_2}{T_3 - T_2} = \frac{441.98 - 287.56}{T_3 - 287.56}\ ; \quad \eta_c = 0.8 \text{ (Given)}$$

$$\therefore \quad T_3 = \frac{441.98 - 287.56}{0.8} + 287.56 = 480.58 \text{ K}$$

Constant pressure cooling in the heat exchanger 3-4 :

The effectiveness of the heat exchanger is prescribed as

$$\epsilon = \frac{T_3 - T_4}{T_3 - T_2}\ ; \quad \epsilon = 0.9 \text{ (Given)}$$

$$\therefore \quad T_4 = T_3 - \epsilon\,(T_3 - T_2) = 480.58 - 0.9\,(480.58 - 287.56)$$
$$= 480.58 - 173.72 = 306.86 \text{ K}$$

$$\text{and} \quad p_4 = p_3 = 2.15 \text{ bar}$$

Expansion process 4-5 :

$$p_5 = 1.05 \text{ bar (Given)}$$

$$\frac{T_4}{T_5'} = \left(\frac{p_4}{p_5}\right)^{\frac{\gamma-1}{\gamma}} = \left(\frac{2.15}{1.0.5}\right)^{\frac{1.4-1}{1.4}} = 1.227$$

$$T_5' = \frac{T_4}{1.227} = \frac{306.86}{1.227} = 250.09 \text{ K}$$

$$\eta_t = \frac{T_4 - T_5}{T_4 - T_5'} = \frac{306.86 - T_5}{306.86 - 250.09}\ ; \quad \eta_t = 0.82 \text{ (Given)}$$

$$\therefore \quad T_5 = 306.86 - 0.82\ (306.86 - 250.09) = 260.31 \text{ K}$$

(*b*) If m is the mass of air circulated per second, then

$$\text{Refrigerating effect} = m\, c_p\, (T_6 - T_5)$$
$$= m \times 1.005\ (298 - 260.31) = 37.88\, m$$

The absorbing capacity of the system is stated to be 50 ton or 50 × 3.5 kJ/s and, therefore

$$37.88\, m = 50 \times 3.5\ ; \quad m = \frac{50 \times 3.5}{37.88} = 4.62 \text{ kg/s}$$

$$\text{Work input to compressor} = m\, c_p\, (T_3 - T_2)$$
$$= 4.62 \times 1.005\ (480.58 - 287.56) = 896.2 \text{ kJ/s} = \mathbf{896.2\ kW}$$

$$COP = \frac{\text{refrigerating effect}}{\text{work input}} = \frac{50 \times 3.5}{896.2} = \mathbf{0.195}$$

(*c*) The volume flow rate through the compressor inlet is given by the volume corresponding to state point 2.

$$V_2 = \frac{mRT_2}{p_2} = \frac{4.62 \times 287 \times 287.56}{0.478 \times 10^5} = \mathbf{7.98\ m^3/s}$$

The volume flow rate through the expander outlet is given by the volume corresponding to state point 5.

$$V_5 = \frac{mRT_5}{p_5} = \frac{4.62 \times 287 \times 260.3}{1.05 \times 10^5} = \mathbf{3.29\ m^3/s}$$

EXAMPLE 4.7

The compreessed air from main compressor of an aircraft cooling system is bled off at 4.5 bar and 200°C. It is then passed through the heat exchanger in which the ram air is forced through for cooling purposes by the fan driven by cooling turbine.

The condition of air at inlet of cooling turbine is 4 bar and 30°C. The air expands in the cooling turbine upto 0.7 bar, the isentropic efficiency of cooling turbine is 30 kg/min. Show the process on T-s diagram and find

(a) actual exit temperature from the cooling turbine

(b) power delivered to the ram air blow fan

(c) tons of refrigeration if the cold air is tempered with by-passed warm air and delivered to the cabin cockpit area where it warms up to 25°C before exhausting out to waste.

Solution : The situation corresponds to simple air cooling system for which schematic arrangement of the different components is shown in Fig. 4.1. Refer Fig. 4.9 for various processes on *T-s* diagram.

The state point 1 refers to condition of air bled from the main compressor to the heat exchanger.

The state point 2 represent the condition of air leaving the heat exchanger and at entry to cooling turbine.

Isentropic expansion in the turbine gives,

$$T_3' = T_2\left(\frac{p_3}{p_2}\right)^{\frac{\gamma-1}{\gamma}} = 303\times\left(\frac{0.7}{4}\right)^{\frac{1.4-1}{1.4}}$$

$$= 184.15 \text{ K}$$

$$\eta_t = \frac{\text{actual temperature drop}}{\text{isontropic temperature drop}}$$

$$= \frac{T_2 - T_3}{T_2 - T_3'}$$

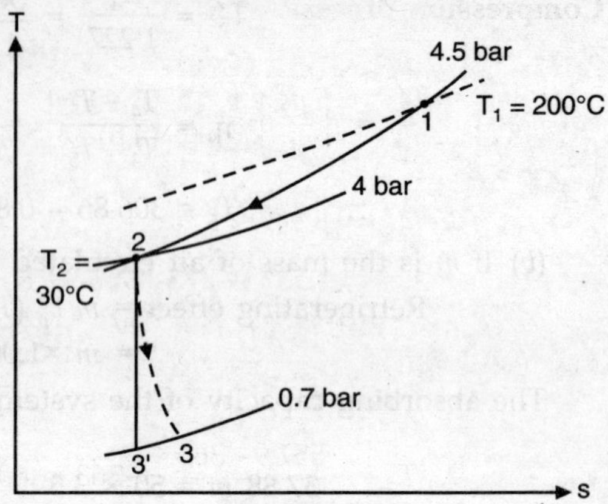

Fig. 4.9.

That gives: $T_3 = T_2 - \eta_t (T_2 - T_3')$

$$= 303 - 0.8\,(303 - 184.15) = 207.92 \text{ K}$$

(*b*) Power delivered to fan = work output form cooling turbine

$$= m_a c_p (T_2 - T_3) = \frac{30}{60} \times 1.005\,(303 - 207.192)$$

$$= 47.82 \text{ kJ/s} = 47.82 \text{ kW}$$

(*c*) Refrigerating effect $= m_a c_p (T_{\text{cabin}} - T_3)$

$$= \frac{30}{60} \times 1.005\ (298 - 207.92) = 45.26 \text{ kW}$$

Since 1 ton of refrigeration is equivalent to 3.5 kW,

$$\text{Refrigeration capacity} = \frac{45.26}{3.5} = \mathbf{12.93 \text{ tons}}$$

EXAMPLE 4.8

The following data refers to a simple aircraft refrigeration system:

Ram air condition : 0.84 bar and 90°C

Cockpit condition : 1.013 bar and 27°C

Cooling load : 8 tons

Pressure ratio and isentropic efficiency of compressor : 4.5 and 0.82

Temperature of air leaving the heat exchanger and entering the cooling turbine : 50°C

Isentropic efficiency of turbine : 0.83

Pressure drop in the heat exchanger = 0.55 bar

Pressure loss between the cooler turbine and cockpit : 0.25 bar

Treating air as perfect gas with c_p = 1.005 kJ/kg K and γ = 1.4, make calculation for the

(a) mass flow through cockpit

(b) power input to the compressor and COP of the system ,

(c) *mass of rammed air mixed with unit mass of cold air from the cooler turbine to get cockpit temperature condition.*

Solution: Refer Fig. 4.10 for *T-s* plot of the given cycle.

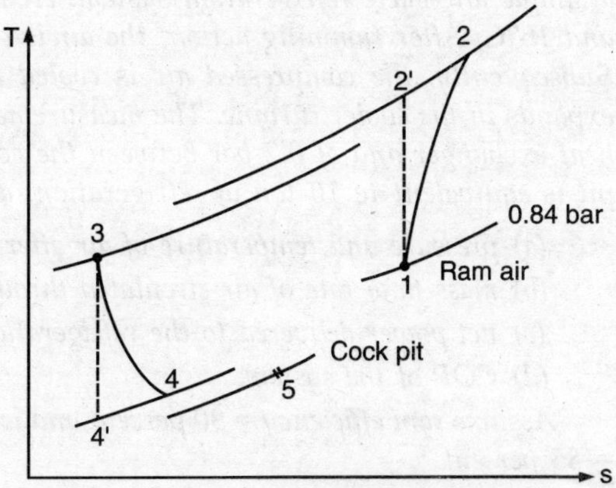

Fig. 4.10

Compression process:

$$\frac{T_2'}{T_1} = \left(\frac{p_2}{p_1}\right)^{\frac{\gamma-1}{\gamma}} = (4.5)^{\frac{1.4-1}{1.4}} = 1.537$$

$$T_2' = 1.537\ T_1$$

$$= 1.537 \times 373 = 557.9 \text{ K}$$

$$\eta_c = \frac{T_2'-T_1}{T_2-T_1};\ 0.82 = \frac{557.9-363}{T_2-363}$$

$$\therefore\ T_2 = \frac{557.9-363}{0.82} + 363 = 600.7 \text{ K}$$

Heat exchanger: $p_3 = p_2 - 0.55 = 4.5 \times 0.84 - 0.55 = 3.25$ bar

$T_3 = 273 + 50 = 323$ K

Expansion Process : $p_4' = p_4 = 1.013 + 0.25 = 1.263$ bar

$$T_4' = T_3\left(\frac{p_4'}{p_3}\right)^{\frac{\gamma-1}{\gamma}} = 323 \times \left(\frac{1.263}{3.25}\right)^{\frac{1.4-1}{1.4}} = 246.5 \text{ K}$$

$$\eta_t = \frac{T_3-T_4}{T_3-T_4'};\quad 0.83 = \frac{323-T_4}{323-246.56}$$

$$\therefore\quad T_4 = 323 - 0.823\ (323 - 246.56) = 259.55 \text{ K}$$

(*a*) Mass flow through the cockpit $= \dfrac{3.5\,Q}{c_p(T_5-T_4)}$

$$= \frac{3.5\times 8}{1.005(300-259.55)} = 0.689 \text{ kg/s}$$

(*b*) Power input to compressor $= m\,c_p\ (T_2 - T_1)$

$= 0.689 \times 1.005\ (600.7 - 363) = 164.59$ kJ/s $=$ **164.59 kW**

Coefficient of performance $COP = \dfrac{3.5\times 8}{164.59} =$ **0.17**

(*c*) Let *x* kg of ram air at 90°C be mixed with a unit mass of cold air from the cooler turbine at temperature $T_4 = 259.55 - 273 = -13.45$°C to get cockpit temperature of 27°C. Then for the mixing process.

$$(1 + x) \times 27 = 1 \times (-13.45) + (x \times 90)$$

or $$27 + 27x = -13.45 + 90x$$

Then : $x = \dfrac{13.45+27}{90-27} =$ **0.642** kg/kg of cold air

EXAMPLE 4.9

The cockpit of a jet plane flying at 1000 km/hr is to be maintained at 1 bar and 25°C by employing a simple air -cycle refrigeration system. At the altitude of flight, the ambient conditions are 0.85 bar and 15°C. After ramming action, the air is compressed in the main compressor with pressure ratio 4. Subsequently, the compressed air is cooled to 60°C temperature in a heat exchanger and thereafter expands in the cooler turbine. The measurements indicate that there is a pressure drop of 0.5 bar in the heat exchanger and of 0.2 bar between the cooler turbine and cock-pit. If the cooling load in the cock-pit is equivalent to 10 ton of refrigeration, determine :

(a) pressure and temperature of air after ramming,
(b) mass flow rate of air circulated through the cabin,
(c) net power delivered to the refrigeration system, and
(d) COP of the system.

Assume ram efficiency = 90 percent and isentropic efficiencies of main compressor and cooling turbine = 85 percent.

$$\text{For air } \gamma = 1.4 \quad \text{and} \quad c_p = 1 \text{ kJ/kg K}$$

Solution : Refer Fig. 4.11 for *T-s* plot of the working cycle.

Ramming Process 1-2 :

$$p_1 = 0.85 \text{ bar}$$

and $T_1 = 15°\text{C} = 288$ K

$$C = 1000 \text{ km/hr} = \frac{1000 \times 10^3}{3600} = 278 \text{ m/s}$$

$$T_2 = T_1 + \frac{C^2}{2000\,C_p} = 288 + \frac{(278)^2}{2000 \times 1} = 326.64 \text{ K}$$

where $T_2' = T_2$ = stagnation temperature of air at entry to main compressor

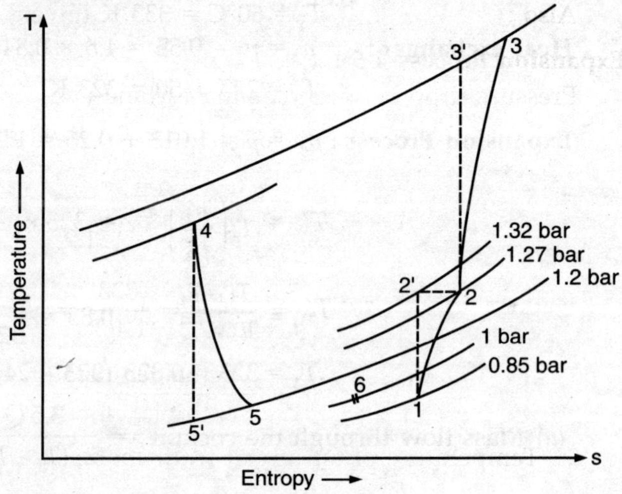

Fig. 4.11

$$\frac{p_2'}{p_1} = \left(\frac{T_2'}{T_1}\right)^{\frac{\gamma-1}{\gamma}} = \left(\frac{326.64}{288}\right)^{\frac{1.4}{1.4-1}} = 1.554$$

$\therefore$ p_2' (pressure after isentropic ramming) = 1.554 × 0.85 = 1.32 bar

$$\text{Ram efficiency} = \frac{\text{actual pressure rise}}{\text{isentropic pressure rise}} = \frac{p_2 - p_1}{p_2' - p_1}$$

$$\therefore \quad p_2 = p_1 + \eta_r\,(p_2' - p_1) = 0.85 + 0.9\,(1.32 - 0.85) = \mathbf{1.27 \text{ bar}}$$

Compression Process 2-3 :

The main compressor has a pressure ratio of 4 (given) and therefore pressure at exit from the compressor is

$$p_3 = 4\,p_2 = 4 \times 1.27 = 5.08 \text{ bar}$$

$$\frac{T_3'}{T_2} = \left(\frac{p_3}{p_2}\right)^{\frac{\gamma-1}{\gamma}} = \left(\frac{5.08}{1.27}\right)^{\frac{1.4-1}{1.4}} = 1.486$$

$$T_3' = 1.486\ T_2 = 1.486 \times 326.64 = 485.58 \text{ K}$$

$$\eta_c = \frac{\text{isentropic temperature rise}}{\text{actual temperature rise}} = \frac{T_3' - T_2}{T_3 - T_2}$$

∴ Temperature at exit from main compressor,

$$T_3 = \frac{T_3' - T_2}{\eta_c} + T_2 = \frac{485.58 - 326.64}{0.85} + 326.64 = 513.63 \text{ K}$$

Constant Pressure cooling in the heat exchanger 3-4 :

Pressure drop in heat exchanger = 0.5 bar.

∴ Pressure of air at entry to cooling turbine, $p_4 = p_3 - 0.5 = 5.08 - 0.5 = 4.58$ bar

Also : $T_4 = 60°\text{C} = 333$ K (given)

Expansion Process 4-5 :

Pressure drop between cooling turbine and cockpit = 0.2 bar

∴ $p_5 = p_6 + 0.2 = 1 + 0.2 = 1.2$ bar

$$\frac{T_4}{T_5'} = \left(\frac{p_4}{p_5}\right)^{\frac{\gamma-1}{\gamma}} = \left(\frac{4.58}{1.2}\right)^{\frac{1.4-1}{1.4}} = 1.467$$

$$T_5' = \frac{T_4}{1.467} = \frac{333}{1.467} = 227 \text{ K}$$

$$\eta_t = \frac{\text{actual temperature drop}}{\text{isentropic temperature drop}} = \frac{T_4 - T_5}{T_4 - T_5'}$$

∴ Temperature of air at exit from cooling turbine

$$T_5 = T_4 - \eta_t\,(T_4 - T_5')$$
$$= 333 - 0.85\,(333 - 227) = 242.9 \text{ K}$$

(*b*) If *m* is the mass of air circulated per second, then

Refrigerating effect $= m\,c_p\,(T_6 - T_5)$

$= m \times 1.005\,(298 - 242.9) = 55.37\,m$

The absorbing capacity of the system is stated to be 10 ton or 10 × 3.5 kJ/s and therefore,

$$55.37\,m = 10 \times 3.5\ ; \quad m = \frac{10 \times 3.5}{55.37} = 0.632 \text{ kg/s}$$

(*c*) Ram work (done by the engine in overcoming the drag of ram diffusion)

$$W_r = c_p\,(T_2 - T_1) = 1.005\,(326.64 - 288) = 38.88 \text{ kJ/kg}$$

Compressor work $W_c = c_p\,(T_3 - T_2) = 1.005\,(513.63 - 326.64) = 187.92$ kJ/kg

Expander work $W_e = c_p\,(T_4 - T_5) = 1.005\,(333 - 242.9) = 90.55$ kJ/kg

Network input $W = W_r + W_c - W_e = 38.83 + 187.92 - 90.55 = 136.2$ kJ/kg

∴ Net power delivered by the engine to the refrigeration unit

$$= 0.632 \times 136.2 = 86 \text{ kJ/s} = \mathbf{86\ kW}$$

(*d*) $COP = \dfrac{\text{refrigerating effect}}{\text{work input}} = \dfrac{10 \times 3.5}{86} = \mathbf{0.407}$

4.2.2 Boot-strap Air Refrigeration System

(*a*) Without Evaporative Cooling: Fig. 4.12 shows the schematic arrangement of the different components comprising the Boot-strap air refrigeration system (without evaporative cooling). The main components are : main compressor, two heat exchangers, secondary compressor and a cooling turbine.

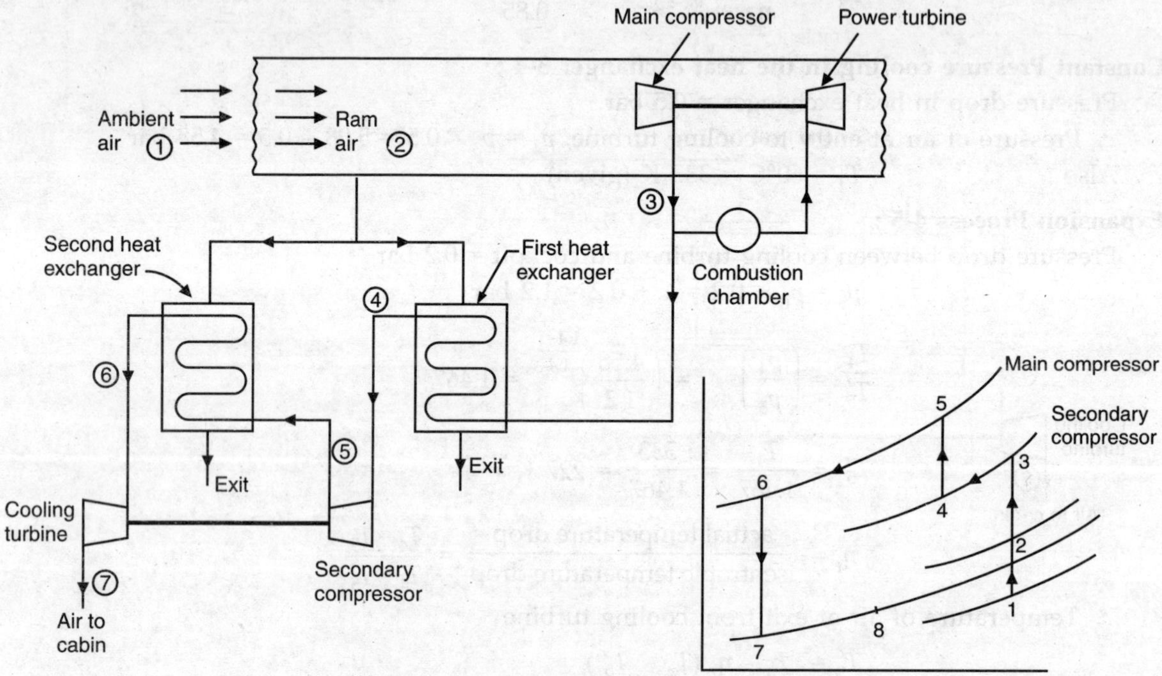

Fig. 4.12. Boot-strap air refrigeration system

The various processes involved in the system are shown on *T-s* plot and their sequence is :

1-2 : isentropic ramming of the ambient air.

2-3 : isentropic compression of rammed air in the main compressor. A part of compressed air goes to the combustion chamber, expands in the power turbine and the power so developed runs the main compressor. Finally this air expands through the nozzle for the propulsion of aircraft.

3-4 : cooling of air bled from the main compressor for refrigeration system at constant pressure in the first heat exchanger. The cooling medium is the stream of rammed air.

4-5 : isentropic compression of air (leaving the first heat exchanger) in the auxiliary or secondary compressor.

5-6 : cooling of the compressed air (leaving the secondary compressor) at constant pressure in the second heat exchanger by a stream of rammed air.

6-7 : isentropic expansion of cooled compressed air in the cooling turbine upto the cabin pressure. Power output from the cooling turbine is utilized to drive the auxiliary compressor.

7-8 : constant pressure heating of air upto the temperature designed in the cabin.

The warm air from the cabin is finally exhausted to atmosphere.

***(b)* With Evaporative Cooling:** Fig. 4.13 shows the schematic arrangement and *T-s* plot of the various processes involved in the Boot strap air-refrigeration system with evaporative cooling. In addition to the elements of basic Bootstrap system, this system has an evaporator which is installed between the second heat exchanger and the cooling turbine. The evaporator provides an additional cooling effect through evaporation of a low boiling point liquid.

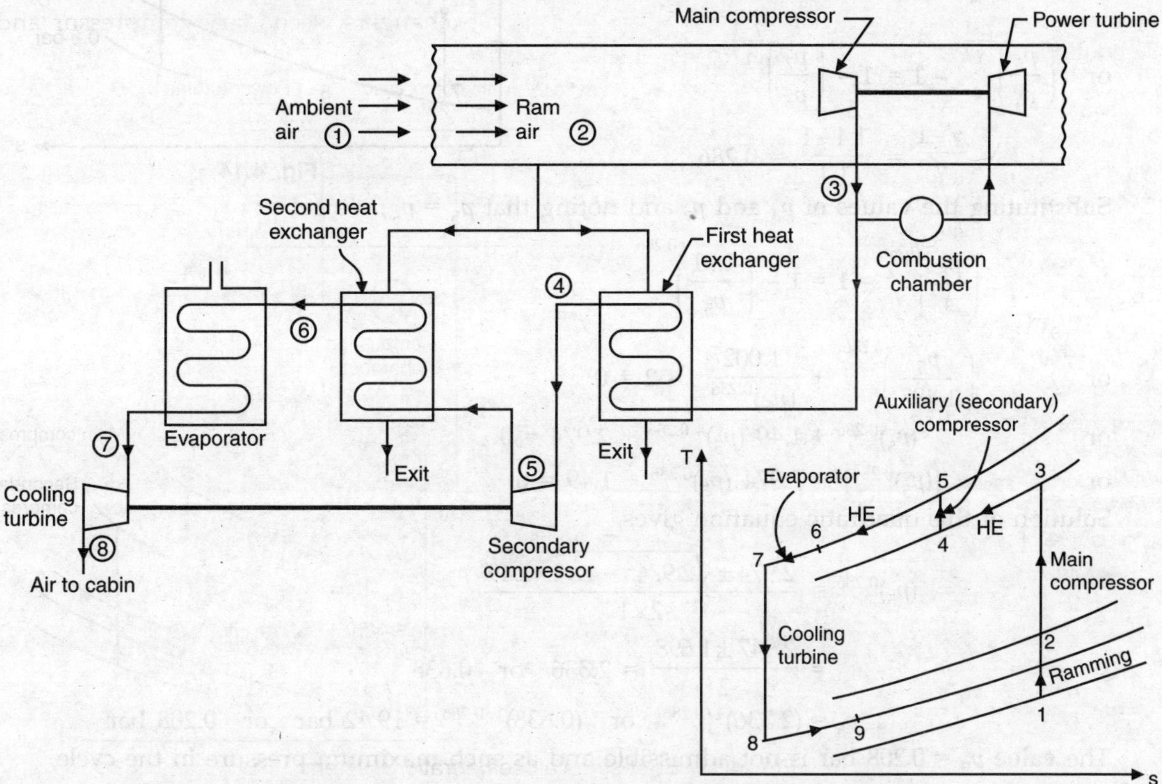

Fig. 4.13. Boot-strap air evaporative refrigeration system

As such the temperature of air at exit from the cooing turbine would be lower than that in a simple boot-strap system. Obviously then, the mass of air per ton of refrigeration will be less in boot-strap evaporative system.

EXAMPLE 4.10

An aircraft flying at an altitude of 2000 m employs a boot-strap air refrigeration system. The ambient condition is 0.8 bar and 0°C, and the ram air pressure and temperature are 1.05 bar and 17°C respectively. After isentropic compression to 4 bar, the air is cooled to 27°C using ram air. Subsequently the air is further compressed in auxiliary compressor which is driven by the cooling turbine. At this state, the air is again cooled to 27°C in another heat exchanger. Finally, the air expands to cabin pressure of 1.01 bar and 25°C temperature. Make calculations for the maximum pressure in the cycle, power input and COP of the system.

Solution: Refer Fig. 4.14 for *T-s* plot of the processes involved. The auxiliary compressor is driven by the cooing turbine. Then equating the work of compression (process 4-5) and expansion (process 6-7) we obtain,

$$c_p\,(T_5 - T_4)\;\; c_p\,(T_6 - T_7)$$

or $$T_4\left[\frac{T_5}{T_4}-1\right] = T_6\left[1-\frac{T_7}{T_6}\right]$$

Since $T_4 = T_6 = 27°\text{C}$, we can write

$$\frac{T_5}{T_4} - 1 = 1 - \frac{T_7}{T_6}$$

or $$\left(\frac{p_5}{p_4}\right)^{\frac{\gamma-1}{\gamma}} - 1 = 1 - \left(\frac{p_7}{p_6}\right)^{\frac{\gamma-1}{\gamma}}$$

$$\frac{\gamma-1}{\gamma} = \frac{1.4-1}{1.4} = 0.286$$

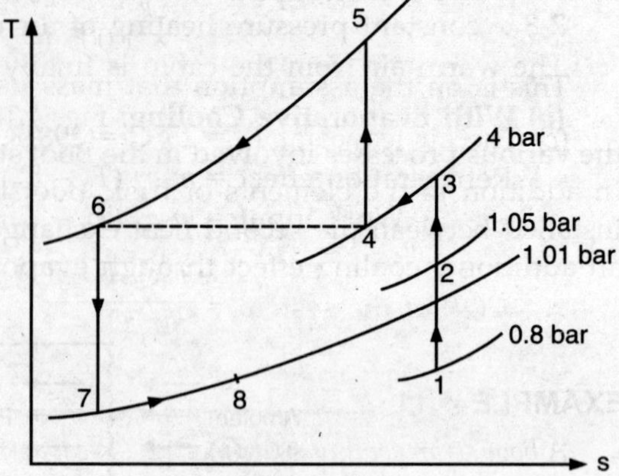

Fig. 4.14

Substituting the values of p_4 and p_7 and noting that $p_5 = p_6$, we get

$$\left(\frac{p_5}{4}\right)^{0.286} - 1 = 1 - \left(\frac{1.01}{p_5}\right)^{0.286}$$

or $$\left(\frac{p_5}{1.487}\right)^{0.286} + \frac{1.002}{(p_5)^{0.286}} - 2 = 0$$

or $$(p_5)^{0.286} + 1.49\,(p_5)^{-0.286} + 2.974 = 0$$

or $$[(p_5)^{0.286}]^2 - 2.974\,(p_5)^{0.286} + 1.49 = 0$$

Solution of this quadratic equation gives,

$$(p_5)^{0.286} = \frac{2.974 \pm \sqrt{2.974^2 - 4\times1\times1.49}}{2\times1}$$

$$= \frac{2.947 \pm 1.698}{2} = 2.336 \quad \text{or} \quad 0.638$$

$$\therefore \quad p_5 = (2.336)^{1/0.286} \quad \text{or} \quad (0.638)^{1/0.286} = 19.42 \text{ bar} \quad \text{or} \quad 0.208 \text{ bar}$$

The value $p_5 = 0.208$ bar is not admissible and as such maximum pressure in the cycle,

$$p_5 = \mathbf{19.42\ bar}$$

(*b*) From isentropic compression in main compressor,

$$T_3 = T_2\left(\frac{p_3}{p_2}\right)^{\frac{\gamma-1}{\gamma}} = (273 + 17) \times \left(\frac{4}{1.05}\right)^{0.286} = 425.13 \text{ K}$$

From isentropic expansion in cooling turbine,

$$\frac{T_6}{T_7} = \left(\frac{p_6}{p_7}\right)^{\frac{\gamma-1}{\gamma}} = \left(\frac{19.42}{1.01}\right)^{0.286} = 2.328$$

$$\therefore \quad T_7 = \frac{T_6}{2.328} = \frac{27+273}{2.328} = 128.86 \text{ K}$$

The auxiliary compressor is driven by the cooling turbine and as such power has to be supplied only to the main compressor.

$$\text{Work input} = m\,c_p\,(T_3 - T_2)$$
$$= 1 \times 1.005\,(425.13 - 290) = 135.8 \text{ kJ/s}$$

This is on the assumption that mass flow rate of air is 1 kg/s

(*c*) The air leaves at T_8 = 25°C = 298 K

$$\text{Refrigeration effect} = m\,c_p\,(T_8 - T_7)$$
$$\text{Work input} = m\,c_p\,(T_3 - T_2)$$

$$\therefore \quad COP \text{ of the system} = \frac{m\,c_p(T_8 - T_7)}{m\,c_p(T_3 - T_2)} = \frac{T_8 - T_7}{T_3 - T_2} = \frac{298 - 128.86}{425.13 - 290} = \mathbf{1.25}$$

EXAMPLE 4.11

A boot strap cooling system of 12 TR capacity is used in an aeroplane. The ambient air temperature and pressure are 20° C and 0.85 bar respectively. The presssure of air increases from 0.85 bar to 1 bar due to ramming action of air. The pressure of air discharged from the compressors is 3 bar. The pressure of air from auxiliary compressor is 4 bar. Isentropic efficiency of each of the compressors is 80% and that of the turbine is 85%. 50% of the enthalpy of air discharged from the main compressor is removed in the first heat exchanger and 30% of enthalpy of air discharged is removed in the second heat exchanger using ram air. Assuming ram action to be isentropic, the required cabin pressure is 0.9 bar and the temperature of air leaving the cabin as 20°C. Find:

(i) power required to operate the system and (ii) COP ***(UPTU)***

Solution : Refer Fig. 4.15 for the *T-s* diagram of the aircraft operating on boot strap cooling system

Ramming proces 1 – 2:

$$p_1 = 0.85 \text{ bar}, \quad T_1 = 20°\text{ C} = 293 \text{ K} \quad \text{and} \quad p_2 = 1 \text{ bar}$$

$$T_2 = T_1\left(\frac{p_2}{p_1}\right)^{\frac{\gamma-1}{\gamma}} = 213\left(\frac{1}{0.85}\right)^{\frac{1.4-1}{1.4}} = 306.94 \text{ K}$$

Compression process in main compressor 2 – 3

$$T_3' = T_2\left(\frac{p_3}{p_2}\right)^{\frac{\gamma-1}{\gamma}} = 306.94\left(\frac{3}{1}\right)^{\frac{1.4-1}{1.4}} = 420.25 \text{ K}$$

$$\eta_c = \frac{\text{isentropic temperature rise}}{\text{actual temperature rise}}$$

$$= \frac{T_3' - T_2}{T_3 - T_2}$$

$$\therefore \quad T_3 = T_2 + \frac{T_3' - T_2}{\eta_c}$$

$$= 306.94 + \frac{420.25 - 306.94}{0.80}$$

$$= 448.58 \text{ K} = 175.58° \text{ C}$$

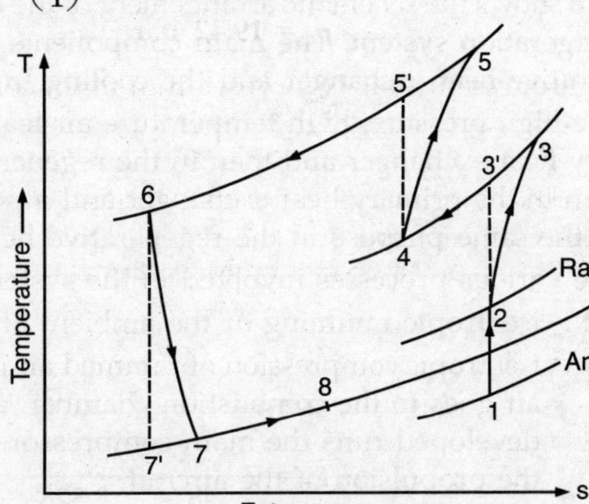

Fig. 4.15.

Since 30 percent of the enthalpy of air discharged from the auxiliary compressor is removed in the second heat exchanger, i.e., during constant cooling process 5-6, the temperature of air leaving the second heat exchanger is,

$$T_6 = (1 - 0.3) \times 126.47 = 87.53\ °C = 361.53\ K$$

Expansion process 6 – 7 :

$$T_7' = \frac{T_6}{\left(\frac{p_6}{p_7}\right)^{\frac{\gamma-1}{\gamma}}} = \frac{361.53}{\left(\frac{4}{0.9}\right)^{\frac{1.4-1}{1.4}}} = 235.98\ K$$

$$\eta_t = \frac{\text{actual temperature drop}}{\text{isentropic temperature rise}} = \frac{T_6 - T_7}{T_6 - T_7'}$$

$$\therefore \quad T_7 = T_6 - \eta_t\,(T_6 - T_7') = 361.53 - 0.85\,(361.53 - 235.98) = 254.81\ K$$

$$T_8 = \text{temperature of air leaving the cabin} = 20°\ C = 293\ K\ \text{(Given)}$$

(*a*) Since 1 ton of refrigeration is equivalent to 3.5 kJ/s of refrigerating effect,

Total refrigerating effect (heat absorbed from the cabin)

$$= 12 \times 3.5 = 42\ kJ/s$$

If m_b is the mass of air bled off the compressor and used for cooling the cabin, then

$$m_b\,c_p\,(T_8 - T_7) = 42$$

$$\therefore \quad m_b = \frac{42}{1.005\,(293 - 254.81)} = 1.094\ kg/s$$

Then power required to operate the system

$$P = m_b\,c_p\,(T_3 - T_2) = 1.094 \times 1.005\,(448.58 - 308.94)$$

$$= 155.73\ kJ/s = 155.73\ kW$$

(*b*) $\quad COP \text{ of the system} = \frac{\text{refrigerating effect}}{\text{power input}} = \frac{42}{155.73} = 0.27$

4.2.3 Regenerative Air-refrigeration System

Fig. 4.16 shows the schematic arrangement of the different components comprising the regenerative air-refrigeration system. The main components are: main compressor, primary heat exchanger, regenerative heat exchanger and the cooling turbine.

The high pressure, high temperature air leaving the main compresssor is first cooled in the primary heat exchanger and then in the regenerative heat exchanger. The ram air is the cooling medium in the primary heat exchanger and a portion of the cool air leaving the cooling turbine serves the same purpose in the regenerative heat exchanger.

The various processes involved in the system are shown on *T-s* plot and their sequence is :

1-2 : isentropic ramming of the ambient air.

2-3 : isentropic compression of rammed air in the main compressor. A part of the compressed air goes to the combustion chamber, expands in the power turbine and the power so developed runs the main compressor. Finally this air expands through the nozzle for the propulsion of the aircraft.

3-4 : cooling of air bled from the main compressor for the refrigeration system at constant pressure in the primary heat exchanger. The cooling medium is the stream of rammed air.

4-5 : cooling of air leaving the regenerative heat exchanger at constant pressure. This cooling occurs in the regenerator where the cooling medium is a portion of cool air delivered by the cooling turbine.

5-6 : isentropic expansion of cooled compressed air upto the cabin pressure. The work output from the cooling turbine is used to drive the exhaust fan which draws the rammed air through the heat exchanger.

6-7 : constant pressure heating of air upto the temperature desired in the cabin. The warm air from the cabin is finally exhausted to atmosphere.

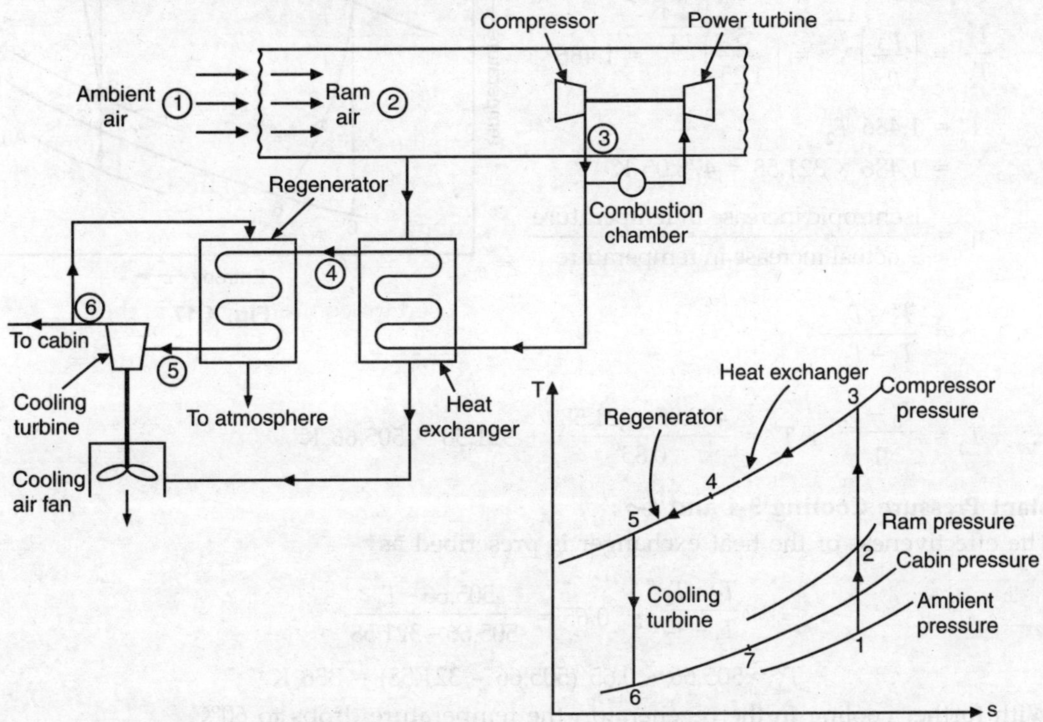

Fig. 4.16. Regenerative air-refrigeration system

EXAMPLE 4.12

In a regenerative type aircraft refrigeration system the ambient air at 15°C and 0.85 bar is rammed isentropically to 1.25 bar. Air leaves the main compressor at 5 bar and the bled off portion of air is cooled by the ram air in the heat exchanger whose effectiveness is 65 percent. Subsequent cooling of air to 60°C occurs in the regenerative heat exchanger where cooling medium is portion of air bled after expansion in the cooling turbine. Presuming that isentropic efficiencies of the compressor and turbine are 85 and 80 percent respectively and the system is desired to have 20 tons of cooling capacity, determine :

(a) the mass of air bled from cooling turbine to be used for regenerative cooling,

(b) the power required for cooling and pressurization of the cabin, and

(c) the COP of the system.

The cabin is maintained at 1.10 bar pressure, 25°C temperature and the temperature of air coming out of regenerative heat exchanger is 105°C.

Consider air to be perfect gas with $\gamma = 1.4$ and $c_p = 1.005$ kJ/kgK

Solution : Refer Fig. 4.17 for the *T-s* diagram of the given regenerative air refrigeration system for the aircraft.

Ramming Process 1-2 : $p_1 = 0.85$ bar, $T_1 = 15°C = 288$ K, $p_2 = 1.25$ bar

$$\frac{T_2}{T_1} = \left(\frac{p_2}{p_1}\right)^{\frac{\gamma-1}{\gamma}} = \left(\frac{1.25}{0.85}\right)^{\frac{1.4-1}{1.4}} = 1.117$$

$$\therefore \quad T_2 = 1.117\, T_1 = 1.117 \times 288 = 321.58 \text{ K}$$

Compression Process 2-3 :

$$\frac{T_3'}{T_2} = \left(\frac{p_3}{p_2}\right)^{\frac{\gamma-1}{\gamma}} = \left(\frac{5}{1.25}\right)^{\frac{1.4-1}{1.4}} = 1.486$$

$$T_3' = 1.486\, T_2$$

$$= 1.486 \times 321.58 = 478.05 \text{ K}$$

$$\eta_c = \frac{\text{isentropic increase in temperature}}{\text{actual increase in temperature}}$$

$$= \frac{T_3' - T_2}{T_3 - T_2}$$

$$\therefore \quad T_3 = \frac{T_3' - T_2}{\eta_c} + T_2 = \frac{478.05 - 321.58}{0.85} + 321.58 = 505.66 \text{ K}$$

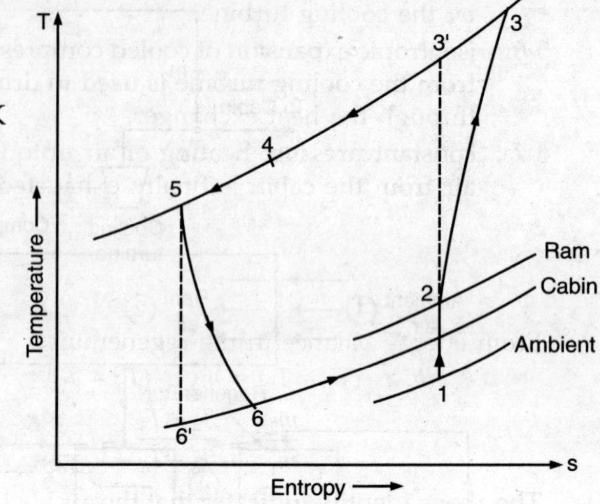

Fig. 4.17

Constant Pressure Cooling 3-4 and 4-5 :

The effectiveness of the heat exchanger is prescribed as

$$\in = \frac{T_3 - T_4}{T_3 - T_2}; \quad 0.65 = \frac{505.66 - T_4}{505.66 - 321.58}$$

$$\therefore \quad T_4 = 505.66 - 0.65\,(505.66 - 321.58) = 386 \text{ K}$$

With further cooling in the regenerator, the temperature drops to 60°C

$$\therefore \quad T_5 = 60°C = 333 \text{ K}$$

Expansion Process 5-6 :

$$p_6 = \text{cabin pressure} = 1.10 \text{ bar}$$

$$\frac{T_5}{T_6'} = \left(\frac{p_5}{p_6}\right)^{\frac{\gamma-1}{\gamma}} = \left(\frac{5}{1.1}\right)^{\frac{1.4-1}{1.4}} = 1.542$$

$$T_6' = \frac{T_5}{1.542} = \frac{333}{1.542} = 215.95 \text{ K}$$

$$\eta_t = \frac{\text{actual decrease in temperature}}{\text{isentropic decrease in temperature}} = \frac{T_5 - T_6}{T_5 - T_6'}$$

$$\therefore \quad T_6 = T_5 - \eta_t\,(T_5 - T_6') = 333 - 0.80\,(333 - 215.95) = 239.36 \text{ K}$$

$$T_7 = \text{temperature in the cabin} = 25°C = 298 \text{ K}$$

$$T_8 = \text{temperature of air at exit from regenerator} = 105°C = 378 \text{ K}$$

(*a*) Let m_c be the mass of air supplied to the cabin to provide 20 tons of cooling capacity. Then

$$m_c = \frac{3.5 \times 20}{c_p (T_7 - T_6)} = \frac{3.5 \times 20}{1.005 \times (298 - 239.36)} = 1.188 \text{ kg/s}$$

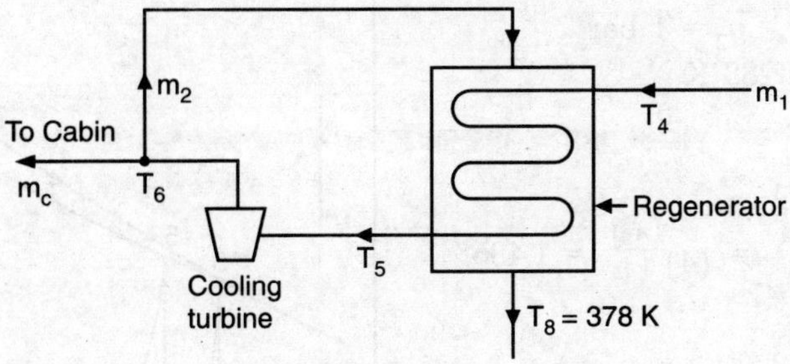

Fig. 4.18

From energy balance in the regenerator,

$$m_1 \, c_p \, (T_4 - T_5) = m_2 \, c_p \, (T_8 - T_6)$$

$$\frac{m_2}{m_1} = \frac{T_4 - T_5}{T_8 - T_6} = \frac{386 - 333}{378 - 239.36} = 0.382$$

The above identify indicates that the air bled from the cooling turbine for regenerative cooling is 38.2% of the total air bled from the main compressor.

Again from mass balance;

$$m_c = m_1 - m_2 = m_1 \left(1 - \frac{m_2}{m_1}\right) = m_1 \, (1 - 0.382)$$

$$\therefore \quad m_1 = \frac{m_c}{1 - 0.382} = \frac{1.188}{0.618} = 1.922 \text{ kg/s}$$

and $\quad m_2 = 0.382 \, m_1 = 0.382 \times 1.922 = \mathbf{0.734 \text{ kg/s}}$

(*b*) Power required for cooling and pressurization of the cabin

$$= m_1 \, c_p \, (T_3 - T_2) = 1.922 \times 1.005 \, (505.66 - 321.58)$$

$$= 355.57 \text{ kJ/s} = \mathbf{355.57 \text{ kW}}$$

(*c*) *COP* of the system $= \dfrac{\text{refrigerating effect}}{\text{work input}} = \dfrac{3.5 \times 20}{355.57} = \mathbf{0.197}$

EXAMPLE 4.13

A regenerative air refrigeration system for an airplane operates at an altitude where the ambient atmosphere has a pressure of 0.8 bar. Due to isentropic ramming action, the pressure and temperature attain values of 1 bar and 15°C respectively. The pressure ratio of the main compressor is 4 with isentropic efficiency of 0.8. The air bled off the main compressor is cooled by ram air in a heat exchanger which is 75 percent effective. The air from the heat exchanger is led to regenerator and subsequently to the cooling air turbine where expansion occurs with 0.8 isentropic efficiency. Some portion of air from the cooling turbine is led to the regenerator which reduces the temperature of air from the heat exchanger by 20°C. The cooling air gets heated to 27°C before being discharged to air. The cabin is pressurized to 1.01 bar and maintained at 27°C. Make calculations for :

(a) the amount of air bled for regeneration,

(b) power requirement, tonnage and COP of the system if 0.5 kg/s of air from the main compressor is used for air-conditioning.

Treat air as perfect gas with $\gamma = 1.4$ and $c_p = 1.005$ kJ/kgK

Solution : Refer Fig. 4.19 for *T-s* plot of the given regenerative air refrigeration system for the aircraft.

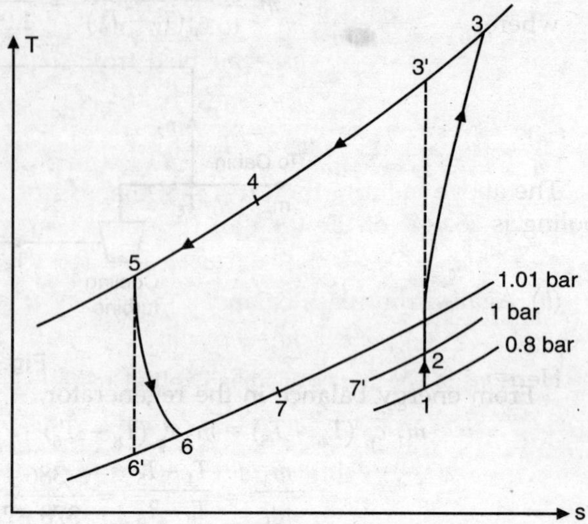

Fig. 4.19

Ramming process :

$p_1 = 0.8$ bar, $p_2 = 1$ bar

and $T_2 = 15°C = 288$ K

Compression Process :

$$\frac{T_3'}{T_2} = \left(\frac{p_3}{p_2}\right)^{\frac{\gamma-1}{\gamma}} = (4)^{\frac{1.4-1}{1.4}} = 1.486$$

$$T_3' = 1.486\ T_2 = 1.486 \times 288 = 428.13 \text{ K}$$

$$\eta_c = \frac{\text{isentropic temperature rise}}{\text{actual temperature rise}} = \frac{T_3' - T_2}{T_3 - T_2}$$

$$\therefore \quad T_3 = T_2 + \frac{T_3' - T_2}{\eta_c} = 288 + \frac{428.13 - 288}{0.8} = 463.16 \text{ K}$$

Constant Pressure Cooling 3-4 and 4-5 :

The effectiveness of the heat exchanger is prescribed as

$$\in = \frac{T_3 - T_4}{T_3 - T_2}; \quad 0.75 = \frac{463.16 - T_4}{463.16 - 288}$$

$$\therefore \quad T_4 = 463.16 - 0.75\ (463.16 - 288) = 331.79 \text{ K}$$

With further drop of 20°C in the regenerator,

$$T_5 = 331.79 - 20 = 311.79 \text{ K}$$

Expansion Process 5-6 :

p_6 = cabin pressure = 1.01 bar

$$\frac{T_5}{T_6'} = \left(\frac{p_5}{p_6}\right)^{\frac{\gamma-1}{\gamma}} = \left(\frac{4}{1.01}\right)^{\frac{1.4-1}{1.4}} = 1.482$$

$$T_6' = \frac{T_5}{1.482} = \frac{311.79}{1.482} = 210.38 \text{ K}$$

$$\eta_t = \frac{\text{actual fall in temperature}}{\text{isentropic fall in temperature}} = \frac{T_5 - T_6}{T_5 - T_6'}$$

$$\therefore \quad T_6 = T_5 - \eta_t\ (T_5 - T_6') = 311.79 - 0.8\ (311.79 - 210.38) = 230.66 \text{ K}$$

T_7 = temperature in the cabin = 27°C = 300 K

T_8 = temperature of air at exit from the regenerator = 27°C = 300 K (*a*) From energy balance in the regenerator,

$$m_1 c_p (T_4 - T_5) = m_2 c_p (T_8 - T_6)$$

where m_1 = total air bled from the main compressor

m_2 = air bled from the cooling turbine for regenerative cooling

or $$\frac{m_2}{m_1} = \frac{T_4 - T_5}{T_8 - T_6} = \frac{331.79 - 311.79}{300 - 230.66} = 0.2884$$

The above identity indicates that the air bled (m_2) from the cooling turbine for regenerative cooling is 28.84% of the total air (m_1) bled from the main compressor.

$$m_2 = 0.2884\, m_1 = 0.2884 \times 0.5 = 0.1442 \text{ kg/s}$$

(*b*) Again from mass balance :

$$m_c = m_1 - m_2$$

Here m_c is the mass of air supplied to the cabin to provide the necessary cooling effect

$$m_c = 0.5 - 0.1442 = 0.3558 \text{ kg/s}$$

$$\text{Power input} = m_1 c_p (T_3 - T_2) = 0.5 \times 1.005\,(463.16 - 288)$$

$$= 88.02 \text{ kJ/s} = \mathbf{88.02 \text{ kW}}$$

Refrigerating effect (heat absorbed)

$$= m_c c_p (T_7 - T_6)$$

$$= 0.3558 \times 1.005\,(300 - 230.66) = 24.79 \text{ kJ/s}$$

Since 1 ton of refrigeration is equivalent to 3.5 kJ/s, tonnage of the system is,

$$= \frac{27.79}{3.5} = \mathbf{7.08} \text{ ton}$$

$$COP = \frac{\text{refrigerating effect}}{\text{work input}} = \frac{24.79}{88.02} = \mathbf{0.2816}$$

EXAMPLE 4.14

A regenerative air refrigeration system for an airplane is designed to handle 25 tons of load. The pressure and temperature conditions of the ambient atmosphere are 20°C and 0.8 bar, and due to isentropic ramming action, the pressure is increased from 0.8 bar to 1.2 bar. The pressure ratio across the main compressor is 4 and 60 percent of the total heat of the air leaving the compressor is removed in the heat exchanger before passing the air to the cooling turbine. The temperature of rammed air which is used for cooling purpose is reduced to 40°C by mixing the air coming out from the cooling turbine. The mass of cooled air passing through the heat exchanger equals the mass of cooling air.

The pressure required in the cabin is 1 bar and the temperature of air leaving the cabin is not to exceed 25°C. Assuming isentropic efficiency of compressor and turbine as 85 percent and 80 percent respectively, make calculations for :

(a) ratio of the by-passed air to ram air used for cooling purpose,

(b) power required for cooling and pressurizing the cabin, and

(c) coefficient of performance of the system.

For air, take c_p = *1.005 kJ/kg K and* $\gamma = 1.4$

Solution : Refer Fig. 4.20 for the schematic arrangement and *T-s* plot for the cooling system adopted for the aircraft refrigeration.

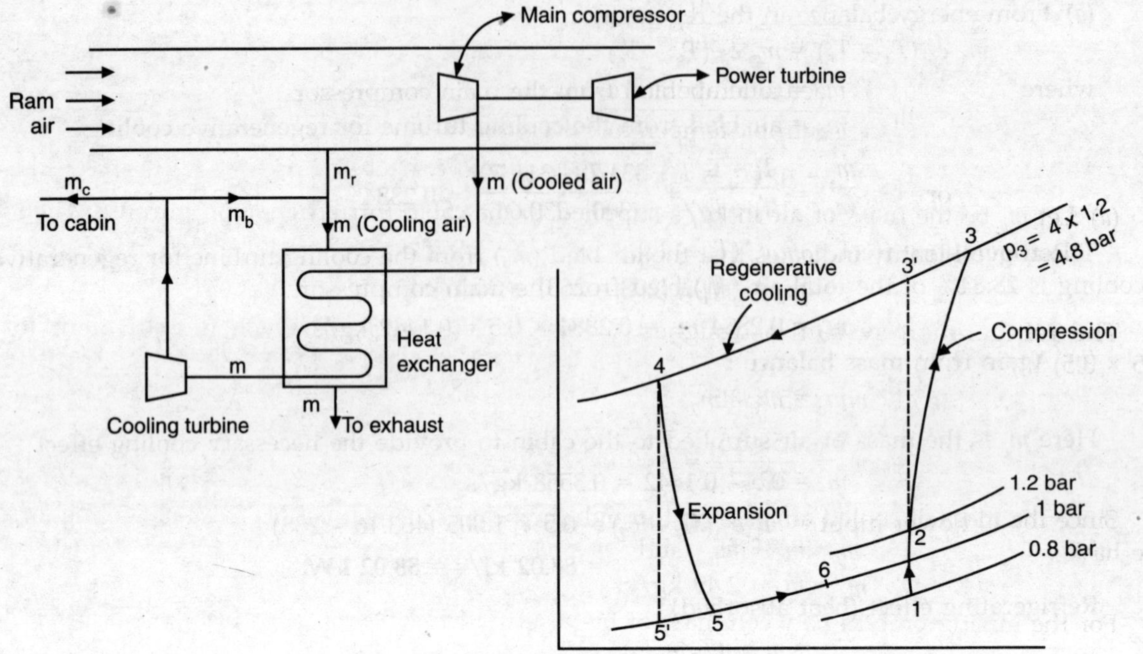

Fig. 4.20.

Ramming Process 1-2 :

$$p_1 = 0.8 \text{ bar}, \quad T_1 = 20°\text{C} = 293 \text{ K} \quad \text{and} \quad p_2 = 1.2 \text{ bar}$$

$$\frac{T_2}{T_1} = \left(\frac{p_2}{p_1}\right)^{\frac{\gamma-1}{\gamma}} = \left(\frac{1.2}{0.8}\right)^{\frac{1.4-1}{1.4}} = 1.123$$

$$\therefore \quad T_2 = 1.123\ T_1 = 1.123 \times 293 = 329 \text{ K} = 56°\text{C}$$

Compression Process 2-3 :

$$\frac{T_3'}{T_2} = \left(\frac{p_3}{p_2}\right)^{\frac{\gamma-1}{\gamma}} = (4)^{\frac{1.4-1}{1.4}} = 1.486$$

$$T_3' = 1.486\ T_2 = 1.486 \times 329 = 487.6 \text{ K}$$

$$\eta_c = \frac{\text{isentropic temperature rise}}{\text{actual temperature rise}} = \frac{T_3' - T_2}{T_3 - T_2}$$

$$\therefore \quad T_3 = T_2 + \frac{T_3' - T_2}{\eta_c} = 329 + \frac{487.6 - 329}{0.85} = 515.6 \text{ K} = 242.6°\text{C}$$

Since 60 percent of the enthalpy of air discharged from the main compressor is removed in the regenerator, the temperature of air leaving the regenerator and being passed to the cooling turbine is

$$T_4 = 242.6 \times 0.4 = 97°\text{C} = 370 \text{ K}$$

Expansion Process 4-5 :

$$\frac{T_5'}{T_4} = \left(\frac{p_5}{p_4}\right)^{\frac{\gamma-1}{\gamma}} = \left(\frac{1}{4.8}\right)^{\frac{1.4-1}{1.4}} = \frac{1}{1.566}$$

$$T_5' = \frac{T_4}{1.566} = \frac{370}{1.566} = 236.3 \text{ K}$$

$$\eta_t = \frac{\text{actual temperature decrease}}{\text{isentropic temperature decrease}} = \frac{T_4 - T_5}{T_4 - T_5'}$$

$$\therefore \quad T_5 = T_4 - \eta_t (T_4 - T_5') = 370.04 - 0.8 (370.04 - 226.3) = 263 \text{ K} = -10°\text{C}$$

(*a*) Let m_c be the mass of air in kg/s supplied to the cabin for refrigeration purpose. Then

$$\text{Refrigeration effect} = m_c c_p (T_6 - T_5)$$
$$= m_c \times 1.005 (298 - 263) = 35.17 m_c$$

The refrigeration system is designed to handle 25 tons load which is equivalent to (25 × 3.5) kJ/s

$$\therefore \quad 35.17 m_c = 25 \times 3.5$$

$$m_c = \frac{25 \times 3.5}{35.17} = 2.49 \text{ kg/s}$$

Since the mass of cooled air passing through the heat exchanger equals mass of cooling air, we have

$$m = m_r + m_b \quad \text{and} \quad m = m_b + m_c$$

$$\therefore \quad m_r = m_c = 2.49 \text{ kg/s}$$

For the mixing process between mass of by-passed air m_b and the mass of rammed air m_r,

enthalpy before mixing = enthalpy after mixing

$$m_b \times t_5 + m_r t_2 = m \times 40 = (m_b + m_r) \times 40$$

$$m_b \times (-10) + 2.49 \times 56 = 40 m_b + 2.49 \times 40$$

That gives: $$m_b = \frac{2.49 \times 56 - 2.49 \times 40}{40 + 10} = 0.797 \text{ kg/s}$$

$$\therefore \quad \frac{\text{mass of by-passed air, } m_b}{\text{mass of ram air, } m_r} = \frac{0.797}{2.49} = \mathbf{0.32}$$

(*b*) Mass of air bled from the compressor and also passing through the cooling turbine,

$$m = m_b + m_r = 0.797 + 2.49 = 3.287 \text{ kg/s}$$

Then power input to the system

$$= \text{compressor work} - \text{turbine work}$$
$$= m_c c_p (T_3 - T_2) - m_c c_p (T_4 - T_5)$$
$$= m_c c_p [(T_3 - T_2) - (T_4 - T_5)]$$
$$= 3.287 \times 1.005 [(515.6 - 329) - (370 - 263)]$$
$$= 262.95 \text{ kJ/s} = \mathbf{262.95 \text{ kW}}$$

(*c*) *COP* of the system $= \frac{\text{refrigerating effect}}{\text{work input}} = \frac{25 \times 3.5}{262.95} = \mathbf{0.332}$

4.2.4 Reduced Ambient Air Refrigeration System

Fig. 4.21 shows the schematic arrangement of the different components comprising the reduced air-refrigeration system. The main components are :

main compressor, heat exchanger and two cooling turbines

One of the cooling turbines is used to cool the rammed air which is then supplied as cooling medium in the heat exchanger.

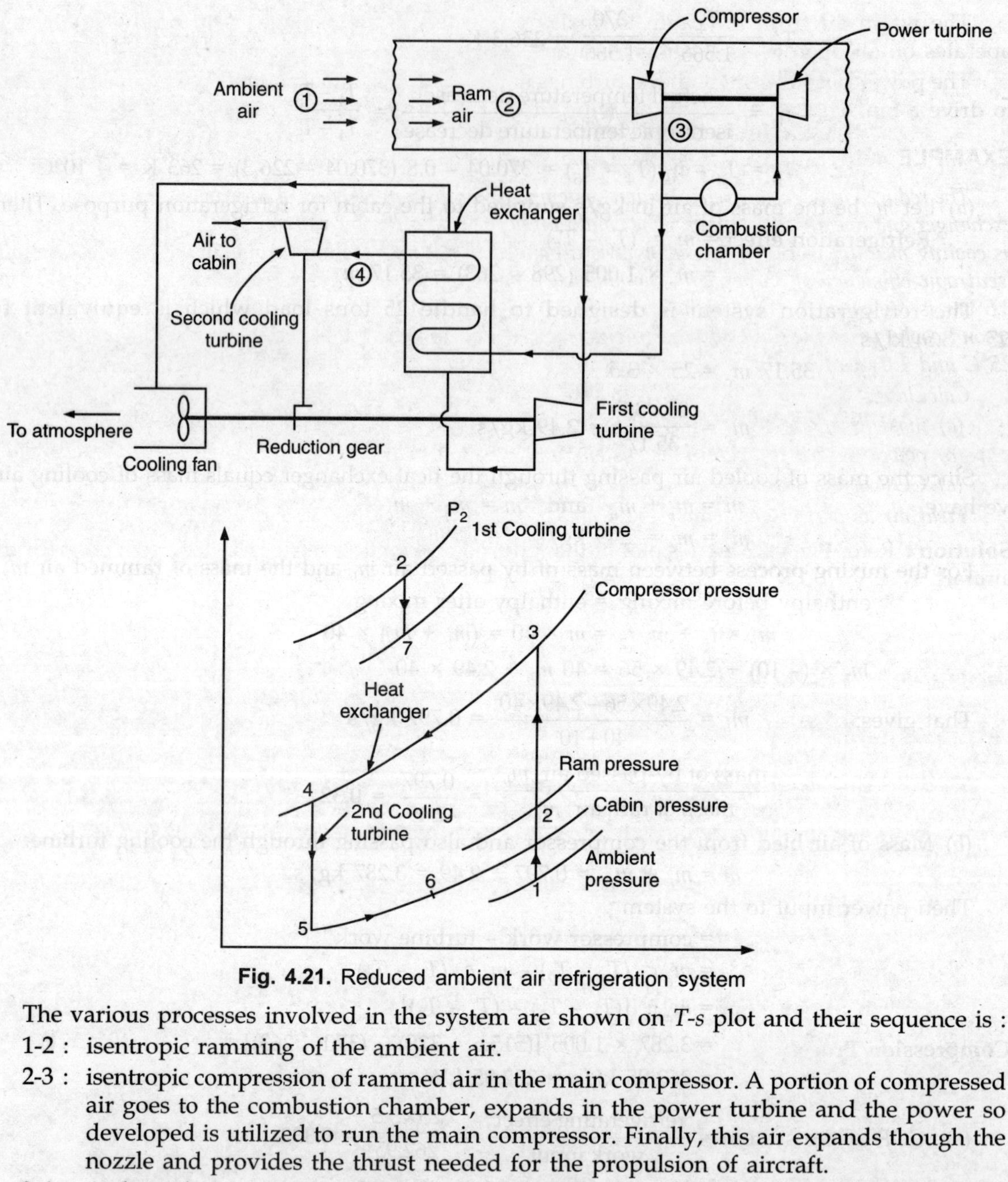

Fig. 4.21. Reduced ambient air refrigeration system

The various processes involved in the system are shown on *T-s* plot and their sequence is :

1-2 : isentropic ramming of the ambient air.

2-3 : isentropic compression of rammed air in the main compressor. A portion of compressed air goes to the combustion chamber, expands in the power turbine and the power so developed is utilized to run the main compressor. Finally, this air expands though the nozzle and provides the thrust needed for the propulsion of aircraft.

3-4 : cooling of compressed air bled for refrigeration purpose at constant pressure in the heat exchanger. The cooling medium is the stream of rammed air which has been cooled. With high speed aircrafts, the temperature of rammed air is quite high. Obviously then, its temperature has to be lowered before it is supplied as cooling medium in the heat exchanger. This is accomplished by expanding the ram air in the first cooling turbine.

4-5 : isentropic expansion of cooled compressed air in the second cooling turbine upto the cabin pressure.

5-6 : constant pressure heating of air upto the temperature desired in the air.

The warm air from the cabin is finally exhausted to atmosphere. Apparently, the system operates on an open cycle starting from and exhausting to the atmosphere.

The power output from both the turbines is supplied, through a suitable gearing arrangement, to drive a fan. The fan assists in drawing the rammed air through the heat exchanger.

EXAMPLE 4.15

The reduced ambient air refrigeration system used for aircraft consists of two cooling turbines, one heat exchanger and one air cooling fan. The condition of the ram air is 1.1 bar and 20°C. The rammed air used as cooling medium in the heat exchanger is expanded in the first cooling turbine to 0.8 bar pressure with isentropic efficiency of 85 percent. The air bled from the main compressor (isentropic efficiency = 0.8) at 3.5 bar is cooled in the heat exchanger with 82 percent effectiveness. After subsequent expansion in the second cooling turbine with 80 percent efficiency, the air is delivered to the cabin which is maintained at 25°C and 1.01 bar.

Calculate :

(a) mass flow rate of air supplied to the cabin to take a load of 20 tons of refrigeration,

(b) power needed to operate the system, and

(c) COP of the system.

Treat air as perfect gas with $\gamma = 1.4$ *and* $c_p = 1.005$ *kJ/kg K.*

Solution : Refer Fig. 4.22 for *T-s* plot of the reduced ambient air refrigeration system used for the aircraft.

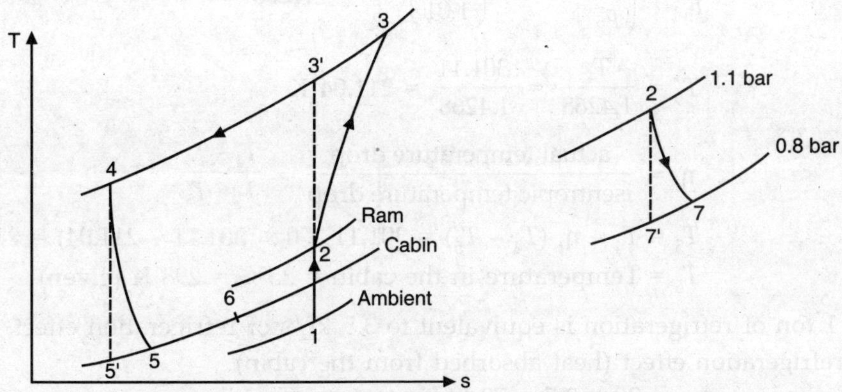

Fig. 4.22

$$p_2 = 1.15 \text{ bar}; \quad T_2 = 20°\text{C} = 293 \text{ K}$$

Compression Process 2-3 :

$$\frac{T_3'}{T_2} = \left(\frac{p_3}{p_2}\right)^{\frac{\gamma-1}{\gamma}} = \left(\frac{3.5}{1.1}\right)^{\frac{1.4-1}{1.4}} = 1.392$$

$$T_3' = 1.392\ T_2 = 1.392 \times 293 = 407.86 \text{ K}$$

$$\eta_c = \frac{\text{isentropic temperature rise}}{\text{actual temperature rise}} = \frac{T_3' - T_2}{T_3 - T_2}$$

$$\therefore \quad T_3 = T_2 + \frac{T_3' - T_2}{\eta_c} = 293 + \frac{407.86 - 293}{0.8} = 436.57 \text{ K}$$

Considering the expansion of ram air in the 1st cooling turbine,

$$\frac{T_2}{T_7'} = \left(\frac{p_2}{p_7}\right)^{\frac{1.4-1}{1.4}} = \left(\frac{1.1}{0.8}\right)^{\frac{1.4-1}{1.4}} = 1.095$$

$$T_7' = \frac{T_2}{1.095} = \frac{293}{1.095} = 267.56 \text{ K}$$

$$\eta_t = \frac{\text{actual temperature drop}}{\text{isentropic temperature drop}} = \frac{T_2 - T_7}{T_2 - T_7'}$$

$$\therefore \quad T_7 = T_2 - \eta_t (T_2 - T_7') = 293 - 0.85 (293 - 267.56) = 271.38 \text{ K}$$

Considering cooling process 3-4 in the heat exchanger,

$$\in = \frac{T_3 - T_4}{T_3 - T_7}; \quad 0.82 = \frac{436.57 - T_4}{436.57 - 271.38}$$

$$\therefore \quad T_4 = 436.57 - 0.82 (436.57 - 271.38) = 301.11 \text{ K}$$

Considering expansion process 4-5 in the second turbine,

$$\frac{T_4}{T_5'} = \left(\frac{p_4}{p_5}\right)^{\frac{1.4-1}{1.4}} = \left(\frac{3.5}{1.01}\right)^{0.286} = 1.4268$$

$$T_5' = \frac{T_4}{1.4268} = \frac{301.11}{1.4268} = 211.04 \text{ K}$$

$$\eta_t = \frac{\text{actual temperature drop}}{\text{isentropic temperature drop}} = \frac{T_4 - T_5}{T_4 - T_5'}$$

$$\therefore \quad T_5 = T_4 - \eta_t (T_4 - T_5') = 301.11 - 0.8 (301.11 - 211.04) = 229.05 \text{ K}$$

$$T_6 = \text{Temperature in the cabin} = 25°\text{C} = 298 \text{ K (given)}$$

(*a*) Since 1 ton of refrigeration is equivalent to 3.5 kJ/s of refrigeration effect

Total refrigeration effect (heat absorbed from the cabin)

$$= 20 \times 3.5 = 70 \text{ kJ/s}$$

If m_b is the mass of air bled off the main compressor and used for cooling the cabin, then

$$m_b \, c_p \, (T_6 - T_5) = 70$$

$$\therefore \quad m_b = \frac{70}{1.005(298 - 229.05)} = 1.01 \text{ kgs}$$

(*b*) Power required to operate the system

$$P = m_b \, c_p \, (T_3 - T_2) = 1.01 \times 1.005 \, (436.57 - 293)$$

$$= 145.73 \text{ kJ/s} = \mathbf{145.73 \text{ kW}}$$

(*c*) COP of the system $= \dfrac{\text{refrigerating effect}}{\text{work input}} = \dfrac{70}{145.73} = \mathbf{0.48}$

EXAMPLE 4.16

A reduced ambient system for an aircraft consists of two cooling turbines, one heat exchanger and one cooling fan. At the altitude of flight, the ambient conditions are 0.8 bar and 10°C, and the speed of aircraft corresponds to 1.22 Mach number. After ramming effect with 90 percent efficiency, a portion of the rammed air to be used for cooling is delivered to the cooling turbine where it expands to 0.8 bar pressure. After expansion, this air is delivered to the heat exchanger for cooling the air bled off from the main compressor at 6 bar. The cooling air from the heat exchanger is sucked by a fan and discharged to the atmosphere. The expansion of cooled compressed air is subsequently carried out in the second cooling turbine and discharged into the air cabin which is to be maintained at 1 bar and 20°C. The pressure loss between the second cooling turbine and cabin is 0.1 bar. If compression index for the main compressor is 1.5 and expansion index for both the cooling turbines is 1.35, make calculations for :

(a) mass flow rate of air supplied to the cabin to take a load of 12 tons of refrigeration,

(b) quantity of ram air passing through the heat exchanger,

(c) combined output of both cooling turbines driving the cooling fan with transmission efficiency of 70 percent, and

(d) COP of the refrigeration system considering only power input to compressor.

It may be pressumed that in the heat exchanger, the compressed air is to be cooled to 100°C and the temperature rise of ram air used for cooling is limited to 80 degree.

Treat air as perfect gas with

$$c_p = 1.005 \text{ kJ/kg and } \gamma = 1.4$$

Solution : Refer Fig. 4.23 for *T-s* plot of the various processes involved in the given reduced ambient refrigeration system.

Let p_2 = stagnation pressure of air entering the main compressor

p_2' = pressure of air at the end of isentropic ramming and

$T_2 = T_2'$ = stagnation temperature of air entering the main compresssor

$$T_2' = T_2 = T_1\left[1+\frac{\gamma-1}{2}M^2\right]$$

$$= (273 + 10)\left[1+\frac{1.4-1}{2}\times 1.25^2\right]$$

$$= 371.4 \text{ K}$$

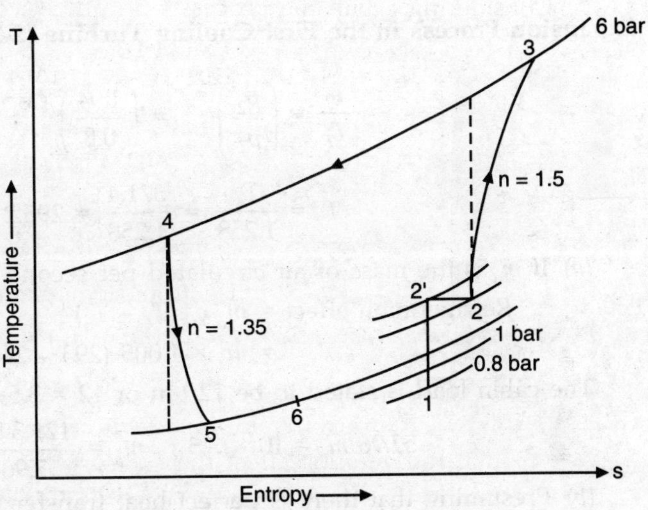

Fig. 4.23

For the isentropic ramming process 1-2′ :

$$\frac{p_2'}{p_1} = \left(\frac{T_2'}{T_1}\right)^{\frac{\gamma}{\gamma-1}} = \left(\frac{371.4}{283}\right)^{\frac{1.4}{1.4-1}} = 2.59$$

$$\therefore \quad p_2' = 2.59\ p_1 = 0.8 \times 2.59 = 2.07 \text{ bar}$$

$$\text{Ram efficiency } \eta_r = \frac{\text{actual pressure rise}}{\text{isentropic pressure rise}} = \frac{p_2 - p_1}{p_2' - p_1}$$

$$0.9 = \frac{p_2 - 0.8}{2.07 - 0.8} = \frac{p_2 - 0.8}{1.27}$$

$$\therefore \quad p_2 = 0.8 + 0.9 \times 1.27 = 1.94 \text{ bar}$$

Compression Process 2-3 :

$$T_3 = T_2\left(\frac{p_3}{p_2}\right)^{\frac{n-1}{n}} = 371.4\left(\frac{6}{1.94}\right)^{\frac{1.5-1}{1.5}} = 541.5 \text{ K}$$

Cooling Process 3-4 : The compressed air is cooled to 100°C. Therefore,

$$T_4 = 100 + 273 = 373 \text{ K}$$

Expansion Process in the Second Cooling Turbine 4-5 :

There is a pressure drop of 0.1 bar between the cabin and the second cooling turbine and therefore

$$p_5 = 1 + 0.1 = 1.1 \text{ bar}$$

$$\frac{T_4}{T_5} = \left(\frac{p_4}{p_5}\right)^{\frac{n-1}{n}} = \left(\frac{6}{1.1}\right)^{\frac{1.35-1}{1.35}} = 1.552$$

$$\therefore \quad T_5 = \frac{T_4}{1.552} = \frac{373}{1.552} = 240.3 \text{ K}$$

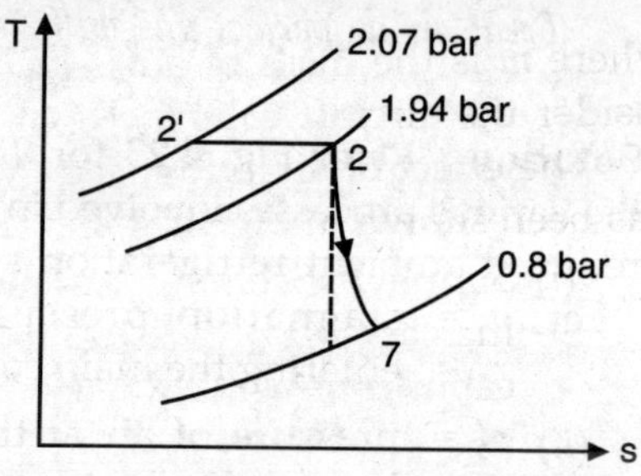

Expansion Process in the First Cooling Turbine (2-7) :

$$\frac{T_2}{T_7} = \left(\frac{p_2}{p_7}\right)^{\frac{n-1}{n}} = \left(\frac{1.94}{0.8}\right)^{\frac{1.35-1}{1.35}} = \mathbf{1.258}$$

$$\therefore \quad T_7 = \frac{T_2}{1.258} = \frac{371.4}{1.258} = 295.2 \text{ K}$$

(*a*) If m_a is the mass of air circulated per second for refrigeration purpose, then

$$\text{Refrigeration effect} = m_a\, c_p\, (T_6 - T_5)$$
$$= m_a \times 1.005\,(293 - 240.3) = 52.96\, m_a$$

The cabin load is stated to be 12 ton or 12 × 3.5 kJ/s and therefore

$$52.96\, m_a = 10 \times 3.5\,; \quad m_a = \frac{12 \times 3.5}{52.96} = \mathbf{0.79} \text{ kg/s}$$

(*b*) Presuming that there is perfect heat transfer in the heat exchanger,
heat gained by ram air = heat lost by the compressed air

$$m_r\,(T_8 - T_7) = m_a\,(T_3 - T_4)$$

$$\therefore \quad m_r = \frac{m_a(T_3 - T_4)}{T_8 - T_7} = \frac{0.79(541.5 - 373)}{80} = \mathbf{1.66\ kg/s}$$

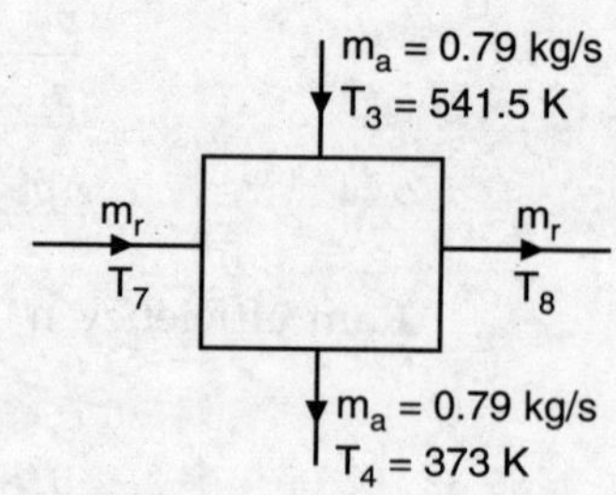

(*c*) Work output from the first cooling turbine,

$$W_{t_1} = m_r\, c_p\, (T_2 - T_7)$$
$$= 1.66 \times 1.005\,(371.4 - 295.2) = 127.2 \text{ kJ/s}$$

Work output from the second cooling turbine,

$$W_{t_2} = m_a\, c_p\, (T_4 - T_5)$$
$$= 0.79 \times 1.005\,(373 - 240.3) = 105.3 \text{ kJ/s}$$

Total power available from both the turbines,

$$W_t = W_{t_1} + W_{t_2} = 127.2 + 105.3 = 232.5 \text{ kJ/s} = 232.5 \text{ kW}$$

Since the work output from both the cooling turbines is used to drive the cooling fans, and the transmission efficiency is 70%.

Power used to drive the cooling fan = 232.5 × 0.7 = **162.75 kW**

(*d*) *COP* of the system $= \dfrac{\text{refrigerating effect}}{\text{work input to main compressor}}$

$$= \frac{m_a c_p (T_6 - T_5)}{m_a c_p (T_3 - T_2)} = \frac{T_6 - T_5}{T_3 - T_2} = \frac{293 - 240.3}{541.5 - 371.4} = \mathbf{0.31}$$

4.3. COMPARISON OF AIRCRAFT REFRIGERATION SYSTEMS

The comparison of different types of air refrigeration systems used for aircrafts is generally made on the basis of dry air rated temperature (DART). The DART is the temperature of air discharged from the expander if it contains no condensed moisture. The rating of the aircraft units is prescribed in terms of kg of air per unit volume at the design DART, and it equals

$$Q = m\, c_p\, (t_o - t_c)$$

where m is the mass of air is kg/s, t_o is the design DART and t_c is the temperature maintained insider the aircraft cabin.

The effect of aircraft speed (Mach number) and DART for different air refrigeration systems has been shown in Fig. 4.24 and the following useful information becomes evident from the plot.

(*i*) Simple air cooling system has low temperature for low speed of aircraft. As the speed increases (*i.e.*, Mach number rises), the discharge temperature becomes too high for effective cabin cooling. Obviously, the simple system is preferable at low speeds only.

(*ii*) The boot-strap system is also not very effective at higher Mach speeds. Moreover, the system requires the plane to be in flight so that ram air becomes available for cooling in heat exchangers. Alternatively, a part of the work output from the turbine can be used to drive a fan which would pull air over the secondary heat exchanger of the system. Essentially, this approach combines the features of simple and boot-strap system.

The effectiveness of the boot-strap system is greatly enhanced if evaporative cooling is used. As such boot-strap system with evaporative cooling is best suited for intermediate and high speed aircrafts.

(*iii*) For supersonic aircrafts and rockets, the reduced ambient system is the ideal choice because of its higher cooling capacity.

(*iv*) The regenerative system too has lower discharge temperature compared to simple cooling system, and accordingly is suitable for high speed aircrafts.

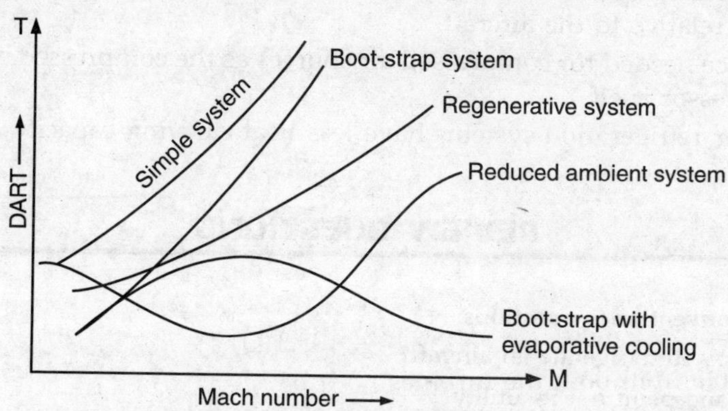

Fig. 4.24

From Fig. 4.24, it is also seen that the turbine discharge temperature of the air is variable which is not desirable. In order to maintain the constant temperature of supply air to the cabin, the aircrafts are generally equipped with the following three control systems :

- **Temperature control :** The signals from the temperature pick up and temperature selector are communicated to the actuator of a hot air modulating valve. The position of the valve then allows a portion of the hot air to by-pass the heat exchanger. This air then mixes with the cold air from the turbine discharge. The proportions from the two streams are so controlled that the mixture suits the cabin temperature for which the selector is set.
- **Air flow quantity control :** The control of air flow to the cabin is achieved by the throttle valve which is placed in the passage of air bled from the main compressor.
- **Water separation and anti-ice control :** The water separator is installed in the passage of cold air to the cabin, and serves to remove any moisture or water that may be present in the cold air.

The anti-ice valve is located between the turbine discharge and the water separation inlet, and apparently connects the hot air duct with the cold air duct. The anti-ice control system limits the DBT of air entering the water separator to a minimum. The thermostat senses the temperature of air entering the water separator and maintains the air at a temperature of about 2°C by positioning the anti-ice valve.

4.4. MERITS OF AIR CYCLE FOR AIRCRAFT REFRIGERATION

Towards the closure, it is worth while to list the merits in the use of air as a refrigerant in the aircraft refrigeration systems :

(*i*) air is non-toxic, non-flammable and hence safe
(*ii*) availability of air in mid-air and so no cost of refrigerant
(*iii*) leakage in small amounts can be tolerated vis-a-vis that of any costly refrigerant
(*iv*) system operates at low pressure and accordingly design and fabrication is with materials commonly available
(*v*) low maintenance due to absence of any complicated arrangement
(*vi*) comparative low weight per ton of refrigeration
(*vii*) chilled air is directly used for cooling and so no evaporator is used
(*viii*) efficient heat transfer due to direct mixing between the cold air and warm air in the cabin
(*ix*) cabin pressurization and air-conditioning can be carried out in one operation
(*x*) initial compression of ram air achieved through ram action of high kinetic energy of ambient air relative to the aircraft
(*xi*) no extra space needed for compressed air sources as the compressed air is taken from the main compressor itself

However, the air refrigeration systems have less heat carrying capacity and poor COP.

REVIEW QUESTIONS

A. Conceptual and conventional questions.

1. Why is it necessary to refrigerate an aircraft?
2. Define ramming and point out its utility.
3. What is meant by DART (dry air rated discharge temperature)?

4. List the advantages and disadvantages of using air as refrigerant.
5. Why pressurization of cabin is needed in case of aircraft refrigeration?
6. How aircraft refrigeration systems are classified?
7. Draw the schematic arrangement of the various components of the basic refrigeration system used for air-conditioning of an aircraft. Explain the working of the system with reference to T-s plot of the processes involved.
8. Explain, with a neat sketch, the operation of reduced ambient air refrigeration system.
9. Differentiate between simple air refrigeration system and boot-strap air refrigeration system.
10. State the function of an evaporative cooler in an aircraft refrigeration system.
11. Explain in detail the regenerative air refrigeration system. Derive an expression for its COP.
12. Compare the various cooling systems used for aircraft.
13. Draw T-s representation of the processes involved in a boot-strap air refrigeration system. Write formulas for mass flow rate, power input and COP of the system.
14. Discuss the factors affecting the performance of simple air refrigeration system used for aircrafts.
15. The COP of an air refrigeration system is very low, but still air refrigeration system is mot common to aircrafts. Why?
16. In an air cooling system of aircraft the compressed air bled off the main compressor is passed through the heat exchanger in which the rammed air is forced through for cooling purposes by the fan run by the cooling turbine. The condition of air at inlet to the cooling turbine is 4 bar and 305 K. The air expands in the cooling turbine to 0.72 bar with an isentropic efficiency of 80 percent. Presuming that mass flow rate of air in the cooling turbine is 25 kg/min, determine:
    (*a*) actual exit temperature from the cooling turbine,
    (*b*) power delivered to the ram air blow fan, and
    (*c*) capacity of the system. The cold air at exit from the turbine is mixed with by-passed warm air and delivered to the cabin cockpit area where it is warmed to 295 K before exhausting out.
    Treat air as perfect gas with $\gamma = 1.4$ and $c_p = 1.005$ kJ/kg K
    (**Ans.** – 62.58°C, 39.60 kW, 10.12 Ton)
17. An aircraft is flying at a speed of 1000 km/hr at a height where the surrounding air pressure and temperature are 0.35 bar and –15°C. The cockpit is to be maintained at 25°C and its cooling load is equivalent to 60 kW. The system operates on simple cycle with the following conditions:
    pressure ratio of jet compressor = 3
    pressure of air at exit from cooling turbine = 1.05 bar
    pressure in the cock pit = 1.01 bar
    pressure drop through the cooler coil = 0.1 bar
    Make calculations for the following parameters:
    (*a*) mass flow of air circulated for refrigeration in kg/min,
    (*b*) volume of air handled by compressor and expander in m^3/min,
    (*c*) net power delivered by the engine to the refrigeration unit, and
    (*d*) coefficient of performance of the system.
    Treat air as a perfect gas and take $\gamma = 1.4$ and $c_p = 1.005$ kJ/kg K
    (**Ans.** 100.62 kg/min, 150.2 m^3/min, 72.17 m^3/min, 193 kW, 0.31)
18. A bootstrap cooling system of 20 tons capacity is used in a airplane. The ambient air pressure and temperature are 0.85 bar and 5°C respectively. The pressure of air increases from 0.85 bar to 1.1 bar due to ramming action of air. The pressure of air bled off the main compressor is 3.5 bar and this air is further compressed in the auxiliary compressor to 4 bar. Isentropic efficiency of each compressor is 90 percent and that of cooling turbine is 80 percent, and both the heat exchangers are 60 percent effective. Assuming ramming action as isentropic, the required cabin pressure of 1 bar and temperature

of air leaving the cabin not to be more than 25°C, determine the power required to operate the system and its coefficient of performance. Treat air as perfect gas with γ = 1.4 and c_p = 1.005 kJ/kg K

(**Ans.** 156 kW, 0.45)

19. A regenerative aircooling system employed for an airplane is designed to take 30 tons of refrigeration load for the aircraft cabin. At the altitude of flight, the ambient air pressure and temperature are 0.8 bar and 7°C respectively. The ramming of ambient air increases its pressure from 0.8 bar to 1.2 bar. The air bled off the main compressor at 4.8 bar is cooled by the rammed air in a heat exchanger whose effectiveness is 0.6. The air from the heat exchanger is further cooled to 50°C in the regenerative heat exchanger with a portion of air bled after expansion in the cooling turbine. The air at exit from the regenerative heat exchanger is at 100°C. If the isentropic efficiencies of the compressor and cooling turbine are 90 percent and 80 percent respectively, make calculations for the following:
    (*a*) percentage of air bled from the cooling turbine to be used for regenerative cooling,
    (*b*) power required to maintain the cabin at required condition of 1 bar pressure and 25°C temperature,
    (c) coefficient of performance of the system.
    Treat air as perfect gas with γ = 1.4 and c_p = 1.005 kJ/kg K (**Ans.** 41%, 440 kW, 0.239)
20. The following data pertains to 100 kW power refrigeration system working on Bootstrap air refrigeration cycle:
    Ram air pressure and temperature: 1 bar and 288 K
    Pressures at exit from main and auxiliary compressor: 3.5 bar and 5.6 bar
    Heat exchanger effectiveness for both heat exchangers = 0.85
    Isentropic efficiency of main and auxiliary compressor = 0.83 and 0.75
    Turbine efficiency = 0.77
    If the cabin is to be maintained at 1 bar and 27°C, determine:
    (*a*) refrigeration capacity in ton,
    (*b*) *COP* of the system, and
    (*c*) percentage of turbine power used by auxiliary compressor. (**Ans.** 17.17 ton, 0.6, 66.25%)
21. A reduced ambient system for aircraft cooling comprises two cooling turbines, one intercooler and one blow out fan. The ram air driven turbine has ram air entry at 1.25 bar and 20°C and the exit at 0.9 bar. This exit air is delivered to a heat exchanger where it cools the air bled off from the main compressor at 5 bar. The compressed air is cooled at 55°C in the heat exchanger and the temperature rise in the heat exchanger for the ram air is limited to 35°C. The ram air from the heat exchanger is sucked by the fan and discharged to the atmosphere. The compressed cooled air expands in the second turbine to 1 bar and is delivered to the aircraft cabin from which it is exhausted at 25°C.
    The compression index for the main compressor is 1.45 and the expansion index for both the cooling turbines is 1.4. If the refrigeration capacity required is 15 tons, determine:
    (*a*) mass flow rate of cabin air,
    (*b*) output of both cooling turbines which drive the fan with 75% transmission efficiency,
    (*c*) *COP* of the system, and
    (*d*) capacity of heat exchanger. (**Ans.** 40.5 kg/min, 120 HP, 0.49, 4952 kJ/min)

B. Fill in the blanks with appropriate word/words

1. The refrigerator system required for an ordinary passenger plane is of about .... ton capacity.
2. The simple air cooling system is preferable for .... speed aircrafts.
3. There is/are .... heat exchanger/exchangers in a boot strap air cooling system.
4. In the .... air cycle, a part of air after cooling turbine is utilized to cool the air prior to its expansion in the turbine.
5. The comparison of different types of air refrigeration systems used for aircrafts is generally made as the basis of .... temperature.

Answers

1. eight 2. low 3. two 4. regenerative 5. dry air rated

C. Multiple choice questions

1. What is the prime consideration in the design of refrigeration system of an aircraft?
   (*a*) high coefficient of performance
   (*b*) low work input per ton of refrigeration
   (*c*) low weight of the refrigeration equipment
   (*d*) non-toxicity of the refrigerant
2. For aircraft cooling, the air refrigeration cycle is used because of
   (*a*) high coefficient of performance
   (*b*) low unit weight per ton of refrigeration
   (c) availability of cold air at high altitudes
   (*d*) high heat transfer rate
3. Identify the cooling system that is used to provide cooling effect in rocket and supersonic aircrafts.
   (*a*) simple air cooling (*b*) boot-strap air cooling
   (*c*) regenerative air cooling (*d*) reduced ambient air cooling
4. The cooling system used with a jet fighter has a capacity close to .... tons
   (*a*) 45 (*b*) 8 (*c*) 12 (*d*) 20
5. The evaporator unit of the boot strap evaporator cooling system is located between the
   (*a*) combustion chamber and the first heat exchanger
   (*b*) first heat exchanger and the secondary compressor
   (*c*) secondary compressor and the second heat exchanger
   (*d*) second heat exchanger and the cooling turbine
6. Which of the following aspects is true in the context of reduced ambient air-cooling system ?
   (*a*) one cooling turbine and one heat exchanger
   (*b*) one cooling turbine and two heat exchangers
   (*c*) two cooling turbines and one heat exchanger
   (*d*) two cooling turbines and two heat exchangers

Answers

1. (*c*) 2. (*b*) 3. (*d*) 4. (*d*) 5. (*d*) 6. (*d*)

Chapter – 5

Vapour Compression Refrigeration

Learning objectives : A study of the subject matter included in this chapter will enable the students/readers to

- *understand the concept of vapour compression refrigeration system*
- *explain the sequence of operation of different processes constituting the Carnot vapour compression cycle*
- *analyse the simple and actual vapour compression refrigeration cycle*
- *discuss the effect of different parameters on the performance of the vapour compression cycle*

5.1. CONCEPT OF VAPOUR COMPRESSION REFRIGERATION

Vapour compression refrigeration is the most important and practical form of refrigeration system for domestic and commercial utility. The working fluids (refrigerants) used are NH_3, CO_2, Freons etc., and these fluids have the characteristic ability to change alternately between the vapour and liquid phases without leaving the plant. The alternate condensation and evaporation takes place at temperature and pressure close to atmospheric conditions. During evaporation, the liquid refrigerant absorbs its latent heat from the space/product to be cooled and changes from the liquid to gaseous state. The vapour is next compressed (raised in temperature and pressure) and subsequently condensed. While condensing, the vapour liquefies by rejecting heat to an external system (typically available cooling water or cooling air). As such, this refrigeration system may be likened to a latent heat pump that pumps heat from the space being refrigerated to either water or air in the condenser. The system is called vapour compression as it is the compression of refrigerant that permits pumping of heat.

5.2. CARNOT VAPOUR-COMPRESSION REFRIGERATION CYCLE

Fig. 5.1 shows the schematic representation and *T-s* diagram of a Carnot vapour-compression refrigeration cycle.

The sequence of operation of different processes constituting the cycle is

1-2: The refrigerant as a mixture of liquid and vapour corresponding to state point 1 enters the compressor where isentropic compression takes place. The compression process increases the temperature of refrigerant from lower limit T_2 to the upper limit T_1. Work is supplied to the system and after compression, the vapour is wet or saturated but not superheated.

2-3: The refrigerant in the form of vapour enters the condenser at state 2 and heat is rejected at constant pressure and temperature. At exit from the condenser, the refrigerant becomes saturated liquid at state point 3.

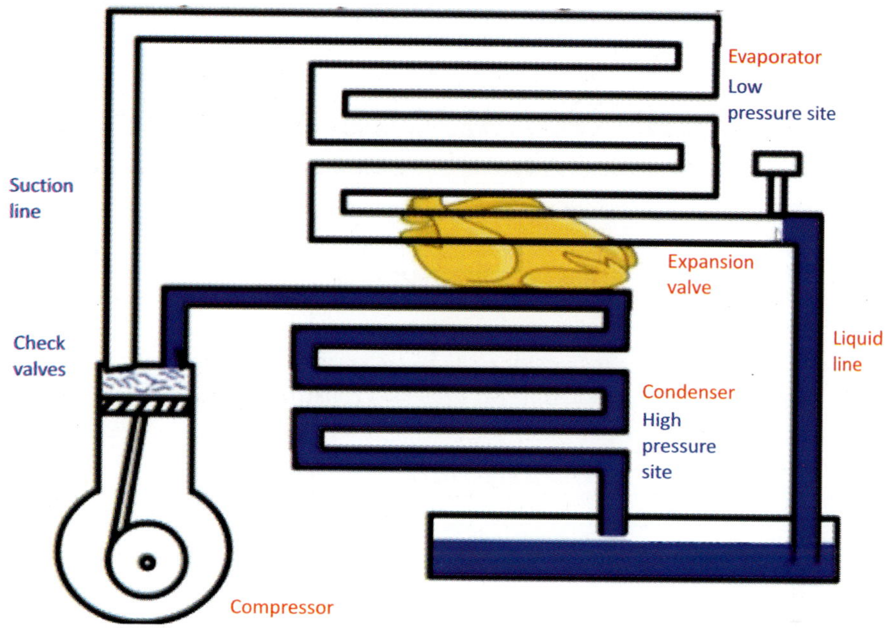

(a)

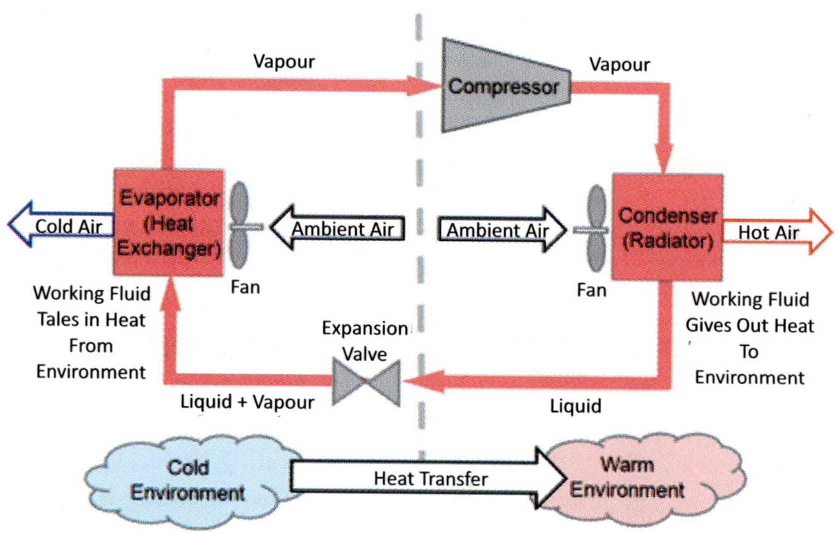

(b)

Vapour Compression Refrigeration Systems

3-4: The refrigerant at state point 3 enters the expansion cylinder, expands isentropically and its temperature drops to lower temperature T_2 at the end of the expansion process. Work is obtained during the expansion process.

4-1: The liquid refrigerant at point 4 enters the evaporator and extracts heat at constant pressure and temperature from the space or substance being cooled and thus produces refrigerating effect.

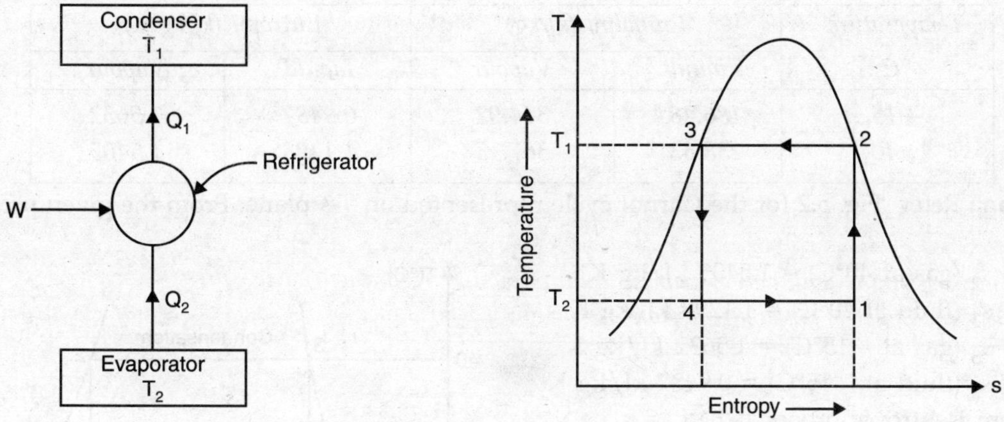

Fig. 5.1. Carnot vapour-compression refrigeration cycle

The cycle consists of two isothermals and two adiabatics and obviously is a reversible cycle. From the temperature-entropy plot,

Heat absorbed (refrigerating effect) during evaporation process (4 – 1)

$$Q_2 = T_2 \times ds$$

Heat rejected during condensation process (2 – 3)

$$Q_1 = T_1 \times ds$$

Work input W = heat rejected – heat absorbed = $(T_1 - T_2)\ ds$

$$\therefore \quad (COP)_{ref} = \frac{\text{heat absorbed at low temperature}}{\text{work input}} = \frac{T_2\ ds}{(T_1 - T_2)ds} = \frac{T_2}{T_1 - T_2}$$

$$(COP)_{heat\ pump} = \frac{\text{heat rejected at high temperature}}{\text{work input}} = \frac{T_1\ ds}{(T_1 - T_2)ds} = \frac{T_1}{T_1 - T_2}$$

It is to be noted that:

(*i*) The values of COP decrease with increased temperature difference. Also COP tends to infinity as T_2 approaches T_1. This aspect suggests that the temperature of refrigerant should not be at a lower temperature than is necessary.

(*ii*) A Carnot refrigeration cycle consists of all reversible processes. As such no refrigerator can have a COP greater than that of a Carnot refrigerator when operating between the same temperature limits. As such a Carnot refrigerator is considered as a standard of perfection against which the performance of a practical refrigerator can be compared.

(*iii*) A Carnot refrigerator has the practically difficulty due to isentropic expansion of saturated liquid to a low pressure. At low pressure, the liquid vaporises and the expander has to deal with a two phase mixture of relatively low quality.

EXAMPLE 5.1

A Carnot refrigerator has working temperatures of – 15°C and 40°C. If it operates with F-12 as working substance, calculate:

(a) the work of isentropic compression and expansion,

(b) refrigerating effect and heat rejected per kg of the refrigerant, and

(c) coefficient of performance of the cycle.

Use the following data in respect of properties of F-12

Temperature	*Enthalpy (kJ/kg)*		*Entropy (kJ/kg K)*	
°C	*Liquid*	*Vapour*	*Liquid*	*Vapour*
– 15	*186.28*	*344.92*	*0.9487*	*1.5632*
40	*238.53*	*367.15*	*1.1298*	*1.5405*

Solution : Refer Fig. 5.2 for the Carnot cycle represented on *T-s* plane. From the given property values:

s_2 (gas at 40°C) = 1.5405 kJ/kg K

s_3 (fluid at 40°C) = 1.1298 kJ/kg K

s_5 (gas at – 15°C) = 1.5632 kJ/kg K

s_6 (fluid at – 15°C) = 0.9487 kJ/kg K

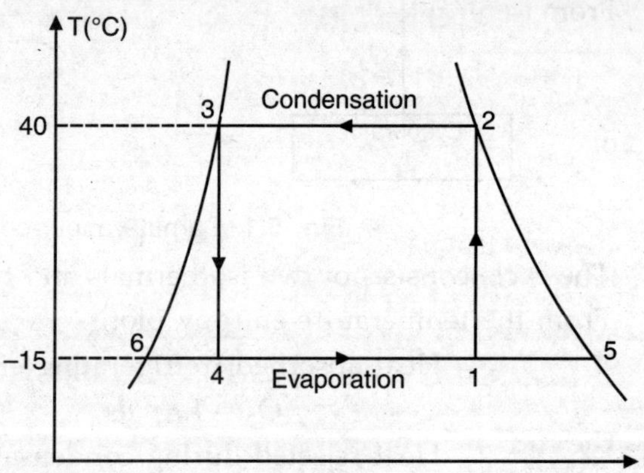

Fig. 5.2

From isentropic proces 1–2 : $s_1 = s_2$

$$s_{f1} + x_1(s_{g1} - s_{f1}) = s_2$$

or $$x_1 = \frac{s_2 - s_{f1}}{s_{g1} - s_{f1}} = \frac{s_2 - s_6}{s_5 - s_6}$$

or $$x_1 = \frac{1.5402 - 0.9487}{1.5632 - 0.9487} = 0.963$$

From isentropic process 3–4; $s_4 = s_3$

$$s_{f4} + x_4(s_{g4} - s_{f4}) = s_2$$

or $$x_4 = \frac{s_3 - s_{f4}}{s_{g4} - s_{f4}} = \frac{s_3 - s_6}{s_5 - s_6} = \frac{1.1298 - 0.9487}{1.5632 - 0.9487} = 0.296$$

The corresponding enthalpies, $h = h_f + x\,h_{fg}$, at these state points are :

$$h_1 = 186.28 + 0.963\ (344.92 - 186.28) = 339.05 \text{ kJ/kg}$$

$$h_4 = 186.28 + 0.295\ (344.92 - 186.28) = 233.08 \text{ kJ/kg}$$

(*a*) Work of compression $W_c = h_2 - h_1$ = 367.15 – 339.05 = **28.1** kJ/kg

Work of expansion $W_e = h_3 - h_4$ = 238.53 – 233.08 = **5.45** kJ/kg

(*b*) Refrigerating effect $N = h_1 - h_4$ = 339.05 – 233.08 = **105.97** kJ/kg

Heat rejected = $h_2 - h_3$ = 367.15 – 238.53 = **128.62** kJ/kg

(*c*) Net work done = work of compression – work of expansion

= 28.1 – 5.45 = 22.65 kJ/kg

$$COP = \frac{\text{refrigerant effect}}{\text{net work done}} = \frac{105.97}{22.65} = 4.68$$

EXAMPLE 5.2

A Carnot refrigeration system operates between temperature limits of 35°C (condenser) and – 30°C (evaporator) and uses refrigerant R-12 as the working fluid. Make calculations for:

(a) Work of comprission, refrigeration effect and coefficient of performance of the working cycle.

(b) What will be the power consumption and heat rejected per ton of refrigeration if the actual COP is 70 percent of the Carnot COP when working between the same temperature limits?

Consult the table of properties of R-12 for the relevant property values.

Solution : Refer Fig. 5.3 for the Carnot cycle representation on T-s plane, From the table of properties for R-12,

s_2 (gas at 35°C) = 0.684 kJ/kg K

s_3 (fluid at 35°C) = 0.256 kJ/kg K

s_5 (gas at –30°C) = 0.717 kJ/kg K

s_6 (fluid at –30°C) = 0.037 kJ/kg K

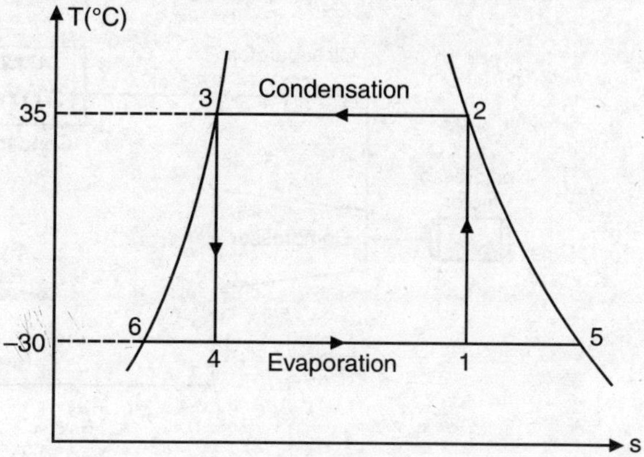

Fig. 5.3

From isentropic process 1 – 2 : $s_1 = s_2$

$$s_{f1} + x_1 (s_{g1} - s_{f1}) = s_2$$

or $$x_1 = \frac{s_2 - s_{f1}}{s_{g1} - s_{f1}} = \frac{s_2 - s_6}{s_5 - s_6} = \frac{0.684 - 0.037}{0.717 - 0.037}$$

$$= 0.95$$

From isentropic process 3 – 4 : $s_4 = s_3$

or $$x_4 = \frac{s_3 - s_{f4}}{s_{g4} - s_{f4}} = \frac{s_3 - s_6}{s_5 - s_6} = \frac{0.256 - 0.037}{0.717 - 0.037} = 0.32$$

The corresponding enthalpies, $h = h_f + x\,h_{fg} = h_f + x\,(h_g - h_f)$ are :

$$h_1 = 8.86 + 0.95\,(174.20 - 8.86) \simeq 165.93 \text{ kJ/kg}$$

$$h_4 = 8.86 + 0.32\,(174.20 - 8.86) = 61.77 \text{ kJ/kg}$$

(*a*) Work of compression $W_c = h_2 - h_1 = 201.45 - 165.93 = 35.52$ kJ/kg

Work of expansion $W_e = h_3 - h_4 = 69.56 - 61.77 = 7.79$ kJ/kg

Net work $W_n = W_c - W_e = 35.52 - 7.79 = 27.73$ kJ/kg

Refrigerating effect $N = h_1 - h_4 = 165.93 - 61.77 = 104.16$ kJ/kg

$$\text{Coefficient of performance } COP, = \frac{\text{refrigerating effect}}{\text{network}} = \frac{104.16}{27.73} = 3.75$$

Alternatively, the *COP* can also be obtained from the relation:

$$COP = \frac{T_l}{T_h - T_l} = \frac{(-30 + 273)}{(35 + 273) - (-30 + 273)} = 3.74$$

(*b*) Actual $COP = 0.7 \times 3.74 = 2.617$

$$\therefore \quad \text{Work done} = \frac{\text{refrigerating effect}}{COP} = \frac{210}{2.617} = \mathbf{80.24} \text{ kJ/min}$$

Heat rejected = work done + refrigerating effect

= 80.24 + 210 = **290.24** kJ/min

5.3. VAPOUR COMPRESSION REFRIGERATION CYCLE

Fig. 5.4 shows the schematic arrangement and *T-s* diagram of a simple vapour-compression refrigeration cycle.

The system comprises a number of flow processes carried out in separate components namely compressor, condenser, expansion valve or capillary tube and evaporator. The working medium is always a liquid gas, *i.e.*, CO_2, NH_3 and Freon etc. Liquid gas is a fluid which at normal temperature may exist either as a liquid or as a gas depending upon pressure. The refrigerant is not allowed to leave the system but is circulated again and again, alternately condensing and evaporating.

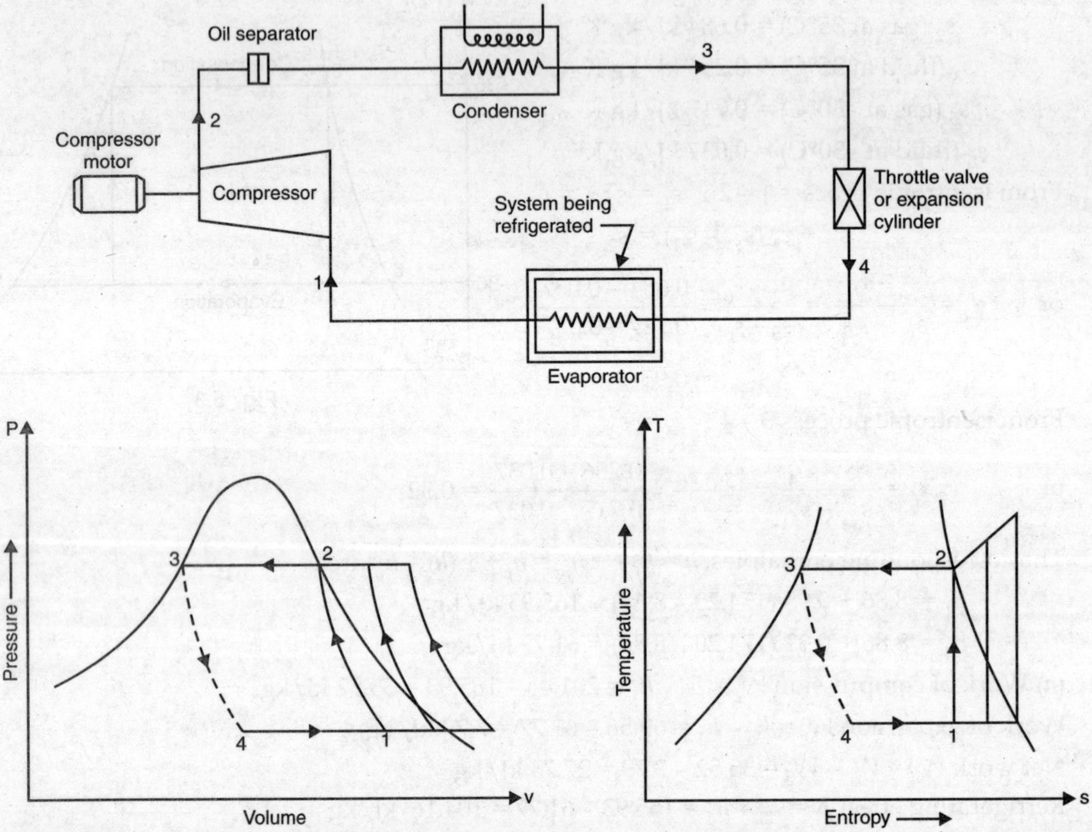

Fig. 5.4. Simple vapour compression cycle

The sequence of operation of the cycle is :

1. Reversible adiabatic compression (1-2) : The vapour at a low pressure p_1, temperature T_1 and preferably dry (state 1) is drawn from the evaporator during suction stroke of the compressor. During compression, the vapour is compressed isentropically to a higher pressure p_2 and temperature T_2. At compressor exit, the vapour may be wet, dry or superheated. In addition to raising the pressure of vapour, the compressor creates the pressure difference between the evaporator and condenser and that maintains a continuous flow of refrigerant through the system.

The compressor is of rotary type for domestic refrigerators and of reciprocating type for large plants and machines, *viz.*, ice plants and cold storages. For plants with capacity over 50 tons, centrifugal compressors are used and these are directly coupled with high speed gas turbines or electrical controls.

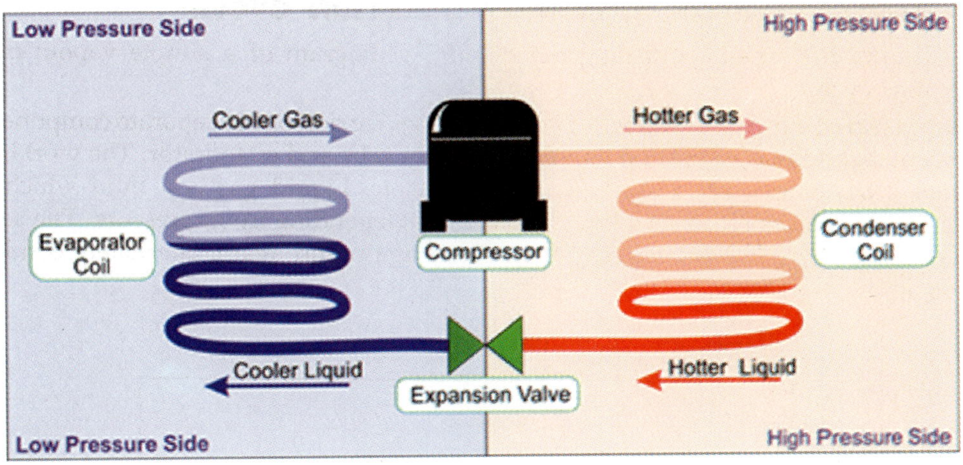

(a)

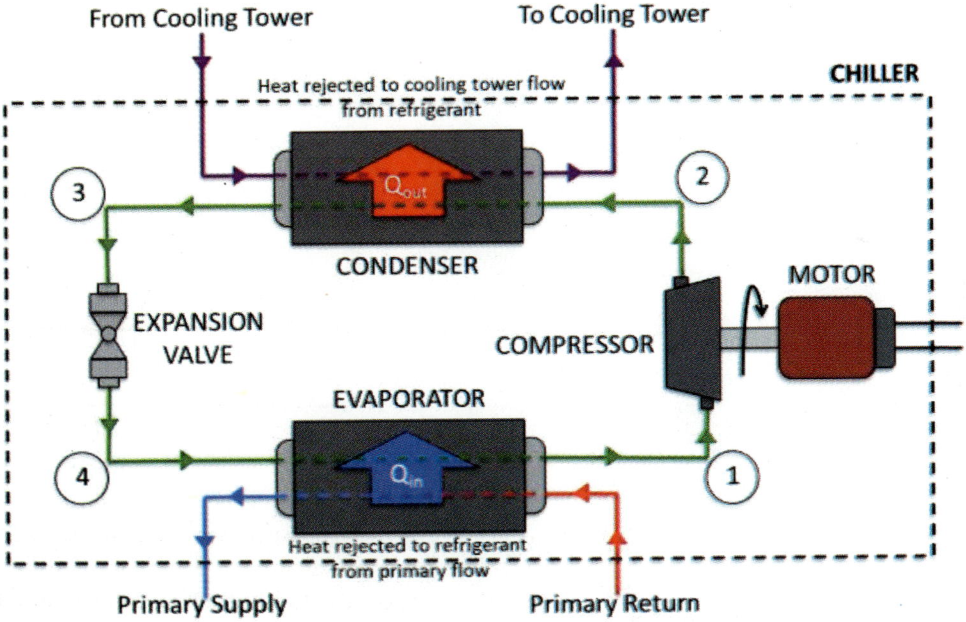

(b)

Refrigeration Cycle

2. Constant pressure condensation (2-3) : The vapour refrigerant at high temperature and pressure (state 2) coming from the compressor is delivered to the condenser where its condensation takes place at constant pressure. The condenser first absorbs the heat of superheat and the temperature of the vapour falls to saturation temperature. Subsequently the refrigerant loses its latent heat and is changed into a high pressure liquid (state 3).

For small machines (*i.e.* domestic refrigerators), the cooling medium is atmosphere and for larger units it may be a stream of water.

3. Throttling (3-4) : The expansion of the high pressure liquid refrigerant is done by the throttling process. During throttling pressure drops but the enthalpy remains constant. Due to drop in pressure, the liquid starts boiling and the required heat is provided by the refrigerant itself. Consequently the temperature falls and at throttle exit, the refrigerant is a mixture of liquid and vapour (very wet vapour of 10 to 20% quality). Throttling is an irreversible process and has been shown by dotted line on *T-s* plot.

The throttling device may be a capillary tube, a thermostatic expansion valve, an automatic expansion valve and float valve etc.

4. Constant pressure evaporation (4-1) : The wet vapour after throttling passes through evaporator coils placed in the medium (usually brine) which is to be cooled. Since the temperature of the vapour is below that of brine, it absorbs latent heat from the brine and gets evaporated at constant pressure. The vaporisation continues upto state point 1 and this stage of the refrigerant depends upon the heat absorbed. The refrigerant may remain wet, become dry saturated or get slightly superheated.

To complete the refrigerant cycle, the refrigerant vapour from the evaporator is routed back into the compressor.

The components of the refrigeration system are connected by pipes such as :

— Suction line that conveys the low pressure vapour from the evaporator to the suction inlet of the compressor.

— Discharge line (or hot gas line) that delivers the high pressure, high temperature refrigerant from the compressor to the condenser.

— Liquid line that carries the liquid refrigerant from the receiver tank to the expansion/throttle valve.

The receiver tank installed between the condenser and expansion device provides storage for the condensed liquid so that a constant supply of liquid is assured to the evaporator as per requirement.

5.3.1 COP from T-s Diagram

Refer Fig. 5.5, which depicts the *T-s* diagram for an ideal vapour compression cycle.

The following assumptions are made with reference to this diagram:

(*i*) compression is isentropic;

(*ii*) there is no undercooling of refrigerant, *i.e.,* the refrigerant after leaving the condenser is a saturated liquid;

(*iii*) there is no change of enthalpy during throttling process;

(*iv*) no leakage of refrigerant;

(*v*) no pressure loss due to friction and heat exchange in the pipelines.

Work input during compression

$$= \text{total heat at point 2} - \text{total heat at point 1}$$
$$= (\text{area 7-2-3-5-6}) - (\text{area 7-1-5-6})$$
$$W = \text{area 1-2-3-5}$$

Thus the compressor work is represented by the area 1-2-3-5.

Refrigerating effect (heat extracted by the refrigerant)

= heat at point 1 – heat at point 4

= (area 7-1-5-6) – area (8-4-5-6)

N = area 7-1-4-8

Thus the refrigerating effect is shown by the area 7-1-4-8. Knowing refrigerating effect N and work input to compressor W, the coefficient of performance can be determined.

$$COP = \frac{N}{W} = \frac{\text{area } 1\text{-}2\text{-}3\text{-}5}{\text{area } 7\text{-}1\text{-}4\text{-}8}$$

$$= \frac{h_2 - h_1}{h_1 - h_4}$$

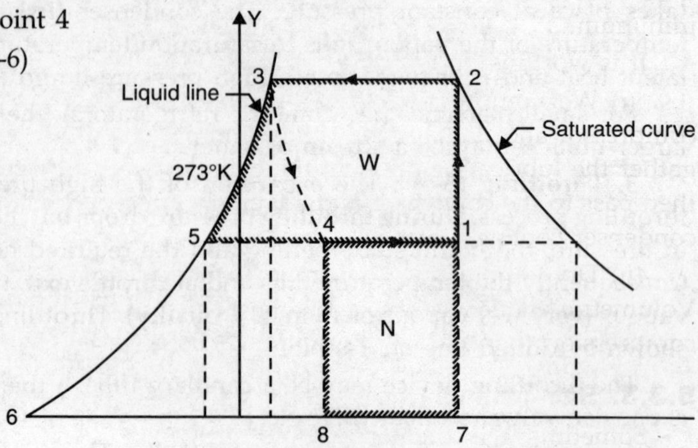

Fig. 5.5. COP from T-s diagram

5.3.2. Wet and Dry Compression

For maximum refrigerating effect, the abstraction of heat in the evaporator should be continued until the vapour is dry saturated or slightly superheated. If the vapour is perfectly dry and saturated at the beginning of compression, the compression is called ***dry compression.*** In dry compression, vapour throughout the compression process lies in the superheated region. In ***wet compression,*** the vapour is wet or just dry at the end of compression process. In wet compression, the vapour remains wet throughout the compression process. Fig. 5.6 shows *T-s* diagrams of wet and dry compression processes.

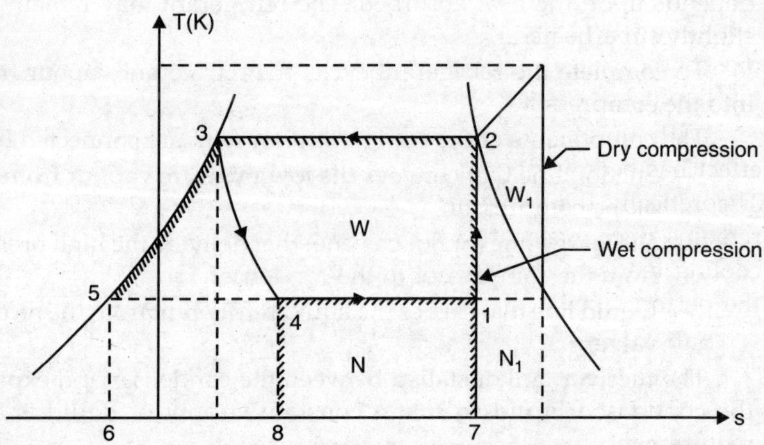

Fig. 5.6. Wet and dry compression

With wet compression process 1-2: $COP = \dfrac{N}{W}$

Let W_1 = additional work expended in compression process due to superheating, *i.e.*, due to dry compression

and N_1 = additional refrigerating effect as heat extracted from the medium being refrigerated with dry compression.

Then, *COP* with dry compression

$$= \frac{N + N_1}{W + W_1} = \frac{N}{W}\left[\frac{1 + \dfrac{N_1}{N}}{1 + \dfrac{W_1}{W}}\right]$$

The ratio of W_1 to W is greater than the ratio of N_1 to N. The coefficient of performance is thus lowered with dry compression and thus the amount of superheat needs to be kept to minimum.

It is worthwhile to point out that :

(*i*) Wet compression is not desirable as the liquid droplets in the refrigerant would enter the compressor and damage the valves and other moving parts. Further, the liquid droplets would gather the lubricating oil from the cylinder walls of a reciprocating compressor. The oil would then pass to the condenser and adversely affect the heat transfer rate from the refrigerant to the condenser cooling water.

(*ii*) Dry compression is safeguard to the compressor cylinder and is almost universally employed. Volumetric efficiency is also high if dry compression is used. During suction stroke with wet compression, there occurs evaporation of liquid particles and that reduces the volumetric efficiency.

5.3.3. Sub-cooling

Sometimes the condensed liquid refrigerant is sub-cooled (under-cooled) at constant pressure before entering the throttle valve. Sub-cooling means reducing the temperature of the liquid below saturation temperature corresponding to condenser pressure.

With reference to Fig. 5.7, the process 3-3′ represents under-cooling of the liquid refrigerant before throttling operation. For most of the refrigerants, under-cooling follows the liquid line.

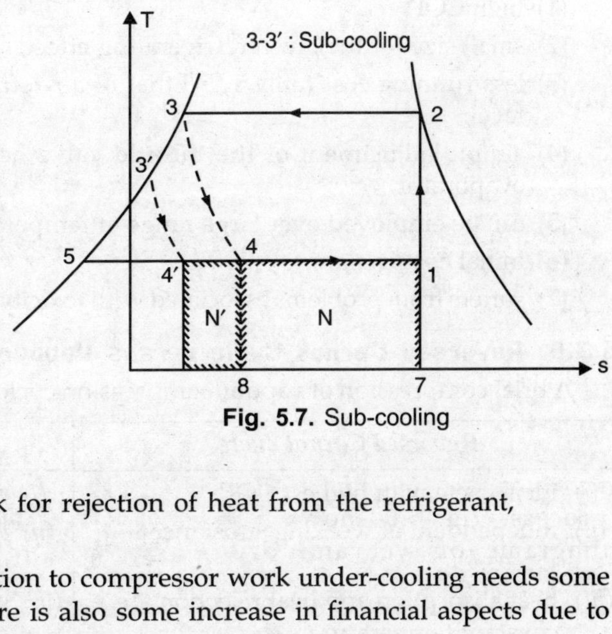

Fig. 5.7. Sub-cooling

With under-cooling, the refrigerating effect has been increased by the amount N'. Theoretically the work input to compressor remains unchanged and accordingly under-cooling tends to increase *COP* and improve the performance of the unit.

Sub-cooling is achieved by

(*i*) circulating greater quantity of cooling water in the condenser,

(*ii*) using low temperature water as sink for rejection of heat from the refrigerant,

(*iii*) incorporating a separate sub-cooler.

It is worthwhile to point out that in addition to compressor work under-cooling needs some extra energy to obtain the extra coolant. There is also some increase in financial aspects due to separate sub-cooler.

5.3.4. Throttle Valve *versus* Expansion Cylinder

The throttling process is highly irreversible and that makes the vapour compression cycle irreversible. Further, no work is obtained during throttling and the entire amount of work required for compression has to be provided by an external source. If the refrigerant after condensation process is expanded adiabatically in an expansion cylinder, the work is obtained and that can be fed to the compressing unit. That reduces the external mechanical work required in running the compressor. Also, the end state 4′ with expansion cylinder is at lower enthalpy when compared with enthalpy corresponding to end state 4 achieved with throttle valve. This aspect increases the refrigerating effect. The net result is that with expansion cylinder, the coefficient of performance of the machine can be improved.

The expansion cylinder system is, however, mechanically more complicated and involves greater first cost which is not justified for small gains in cooling capacity. Moreover, with throttle valve the flow of refrigerant can be controlled. The expansion cylinder has fixed cylinder volume which cannot accommodate varying flow. Modern vapour compression plants prefer throttle valve to simplify the design of installation and to improve the process of control.

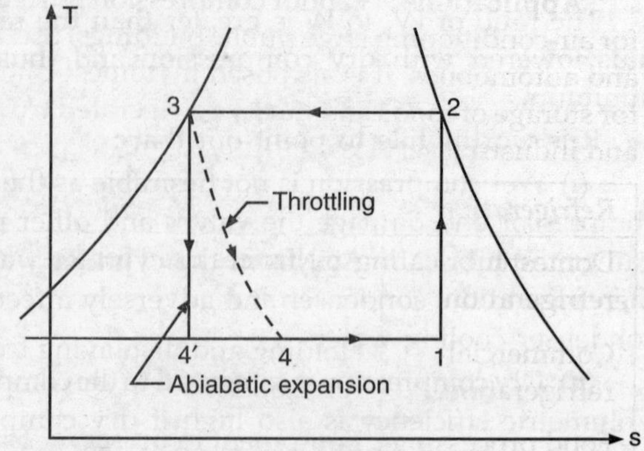

Fig. 5.8. Throttle valve versus expansion cylinder

5.3.5. Air-refrigeration System *versus* Vapour Compression System

Compared to the air-refrigeration system, the vapour-compression system has the following merits and demerits :

(1) high COP,

(2) small size for the given refrigerating effect,

(3) less running cost (only $1/5^{th}$ that of air-refrigeration system) when operated on ground level,

(4) simple adjustment of the throttle valve helps to achieve required temperature at the evaporator,

(5) can be employed over large range of temperatures,

(6) initial cost is high,

(7) suffers from problems associated with toxicity, inflammability and leakage of the refrigerant.

5.3.6. Reversed Carnot Cycle *versus* Vapour Compression Refrigeration

A brief compression of vapour compression cycle with the reverse Carnot cycle is given below:

Reversed Carnot cycle	*Vapour compression refrigeration*
(*i*) ideal cycle with highest COP	(*i*) practical cycle with lower COP
(*ii*) independent of working substance	(*ii*) performance depends on the type of vapour used
(*iii*) heat absorption and heat rejection are at constant temperature	(*iii*) heat absorption and heat rejection are at constant pressure as well as at constant temperature
(*iv*) involves an expansion process which is isentropic (reversible adiabatic)	(*iv*) expansion is by throttling which is irreversible and at constant enthalpy
(*v*) there is no superheating and no under-cooling of the refrigerant	(*v*) the refrigerant may be superheated in the compressor and undercooled in the condenser so as to improve the efficiency of the system
(*vi*) canot be practically implemented	(*vi*) most popular and widely used refrigeration system for commercial, industrial and domestic services

Applications : Vapour compression refrigeration has been and is the most widely used method for air-conditioning large public buildings, private residences, hotels, hospitals, theatres, restaurants and automobiles. It is also used in domestic and commercial refrigerators, large-scale warehouses for storage of foods and meats, refrigerated trucks and railroad cars, and a host of other commercial and industrial services.

Refrigeration	*Short description*	*Typical refrigerants used*
Domestic refrigeration	Appliances used for keeping food in dwelling units	R-600a, R-134a
Commercial refrigeration	Holding and displaying frozen and fresh food in retail outlets	R-134a, R-404A, R-507
Food processing and cold storage	Equipment to preserve, process and store food from its source to the wholesale distribution point	R-134a, R-407C, R-410A, R-507
Industrial refrigeration	Large equipment. typically 25 kW to 30 MW, used for chemical processing, cold storage, food proceasing and district heating and cooling	R-134a, R-404A, R-507, R-717
Transport refrigeration	Equipment to preserve and store goods, primarily foodstuffs, during transport by road, rail, air and sea	R-134a, R-404A, R-507
Electronic cooling	Low-temperature cooling of CMOS circuitry and other components in large computers and services	R-134a, R-404A, R-507.
Medical refrigeration		R-134a, R-404A, R-507
Cryogenic refrigeration		Ethylene, Helium

5.4. PRESSURE ENTHALPY (*p-h*) CHART

The solution to problems dealing with vapour compression cycle are easily and rapidly obtained by referring to *p-h* diagrams which have been prepared for different refrigerants by having the data from tables of refrigerant properties.

Fig. 5.9 shows a skelton *p-h* chart and its essential features are :

(*i*) The pressure is plotted on the axis of ordinate and enthalpy on the abscissa. Apparently the horizontal axis represents specific enthalpy in kJ/kg and vertical axis represents absolute pressure in bar. The enthalpy has a linear scale but to compress the chart, logarithmic scale is used for the pressure ordinate. Due to logarithmic scale equal pressure ratios are represented by equal vertical distance.

(*ii*) The plot has three regions which are separated from each other by the saturated liquid line AB and the saturated vapour line CD. There is

— a subcooled region to the let of bounding curve AB,

— a wet vapour region bounded by curves AB and CD and

— a region of superheated vapour to the right of curve CD

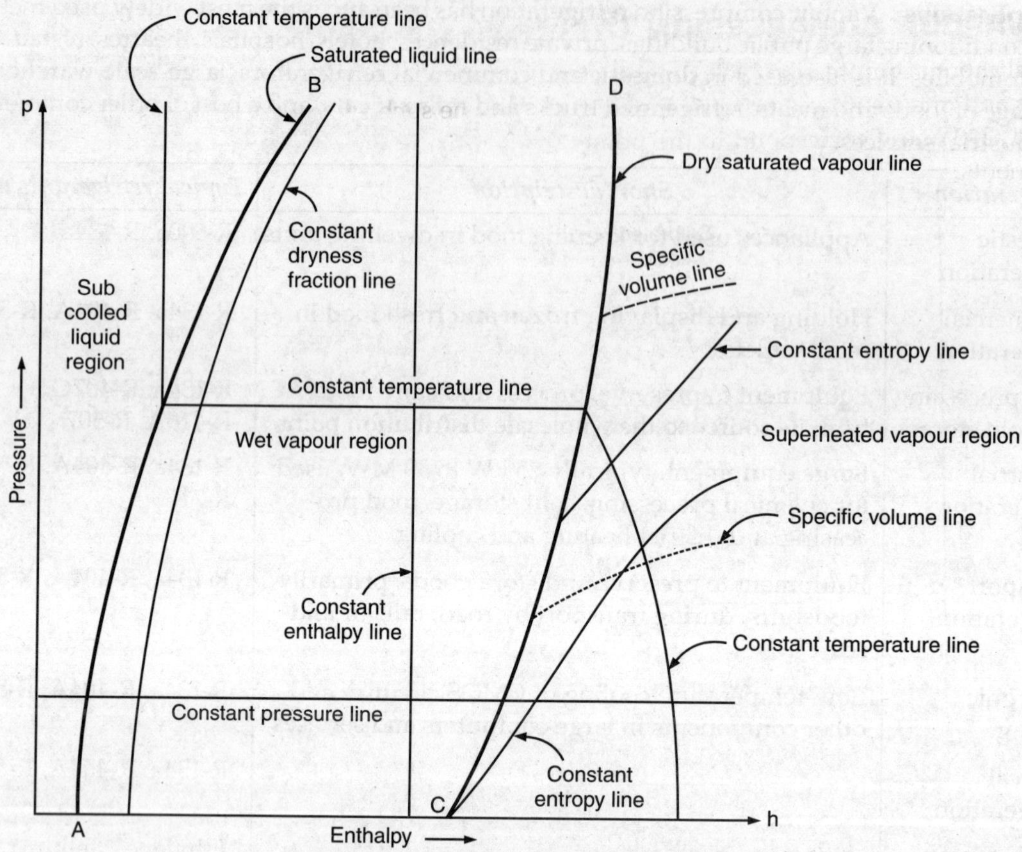

Fig. 5.9. Pressure-enthalpy (p-h) chart

(*iii*) At any point in the subcooled region, the refrigerant is in the liquid phase and its temperature is below the saturation temperature corresponding to its pressure. In the superheated region, the refrigerant is in the form of supherheated vapour and its temperature is above the saturation temperature.

The wet vapour zone is the two phase region ; here the refrigerant exists as a liquid-vapour mixture. This region represents a change in refrigerant from liquid to vapour phase.

(*iv*) The isotherms (constant temperature lines) are :

– almost vertical and parallel to lines of constant enthalpy lines in the wet region.
– parallel to and coincide with constant pressure lines in the wet-vapour zone
– curved with sharp fall toward the bottom in the superheated vapour region.

(*v*) The constant enthalpy lines extend diagonally and almost vertically across the supherheated vapour region.

(*vi*) The constant volume lines are curved, nearly horizontal and crossing the superheated region.

The *p-h* chart gives directly the changes in enthalpy and pressure during a process and accordingly is put to an extensive use in the thermodynamic analysis and investigation of vapour compression refrigerant cycles. The *p-h* charts are available for most common refrigerants.

Note: The property tables of *p-h* charts available have been prepared with different datums and as such the property values indicated there in are different. However, calculated values for different performance parameters workout to be same which ever property table or plot is used.

5.5. VAPOUR COMPRESSION CYCLE ON *p-h* CHART

An ideal vapour compression cycle consists of two constant pressure processes, one constant enthalpy process and one isentropic process. Fig. 5.10 shows one such typical cycle on *p-h* chart. The state points 1, 2, 3 and 4 correspond to the points marked in schematic diagram (Fig. 5.2) of the cycle components.

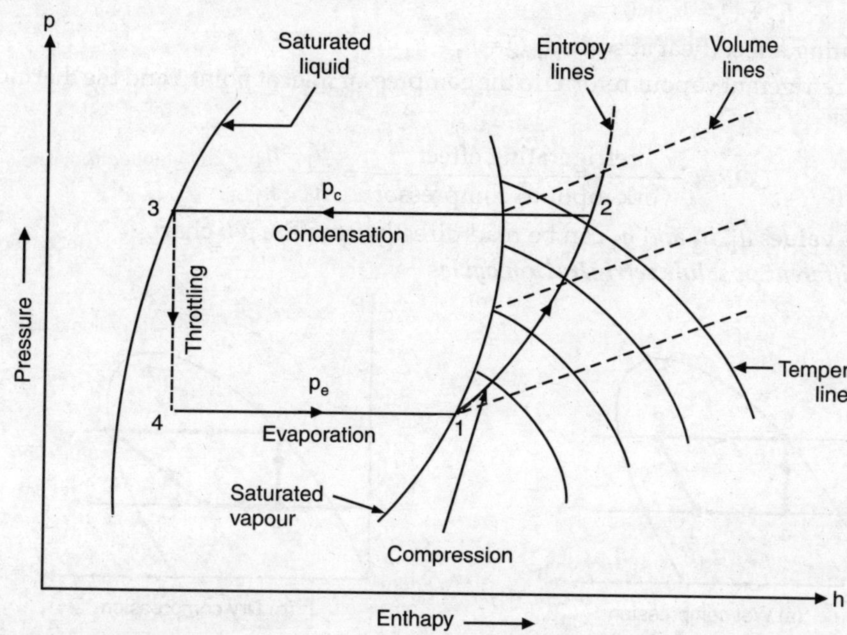

Fig. 5.10. Simple vapour compression cycle on p-h chart

The various processes are :

1-2 : The refrigerant is compressed adiabatically from low pressure p_1 (evaporator pressure) to higher pressure p_2 (condenser pressure). The refrigerant is assumed dry saturated at the start of compression (state 1) and it becomes superheated at exit (state 2) from the compressor.

Applying steady flow energy equation (SFEE),

$$h_1 + Q = h_2 + W; \quad Q = 0 \text{ as the compression process is adiabatic}$$

That gives : $\quad W = -(h_2 - h_1)$

$\therefore$ Work done on refrigerant $= h_2 - h_1$

2-3 : The supherheated vapour travels through the condenser which first cools and removes the superheat and then condenses the vapour into a liquid by abstracting additional heat at constant pressure and temperature.

From steady flow energy equation : $h_2 + Q_2 = h_3 + W$

As no work is done during condensation process, $W = 0$ and therefore

Heat rejected by the refrigerant $Q_2 = h_2 - h_3$

3-4 : Between state points 3 and 4, the liquid refrigerant goes through the expansion valve (also called throttle valve) where its pressure abruptly decreases. That results in a mixture of liquid and vapour at lower temperature and pressure as shown by state point 4.

From SFEE : $\quad h_3 + Q = h_4 + W$

During throttling, heat and work transfer are zero

$\therefore \quad h_3 = h_4$, *i.e.* enthalpy remains constant.

4-1 : After throttling, the liquid vapour mixture travels through evaporator coil and is completley vaporized by extracting heat from the space/product being refrigerated.

It is a constant pressure process and work transfer is zero. Therefore, from SFEE

$$h_4 + Q = h_1 + 0$$

$\therefore$ Refrigerating effect (heat absorbed) $Q = h_1 - h_4$

The resulting refrigerant vapour returns to the compressor inlet at point 1 and the thermodynamic cycle is completed

$$COP = \frac{\text{refrigerating effect}}{\text{work input to compressor}} = \frac{h_1 - h_4}{h_2 - h_1}$$

The enthalpy values h_1, h_2 and h_4 can be read directly from the *p-h* chart.

p-h plot of different possible refrigeration cycles

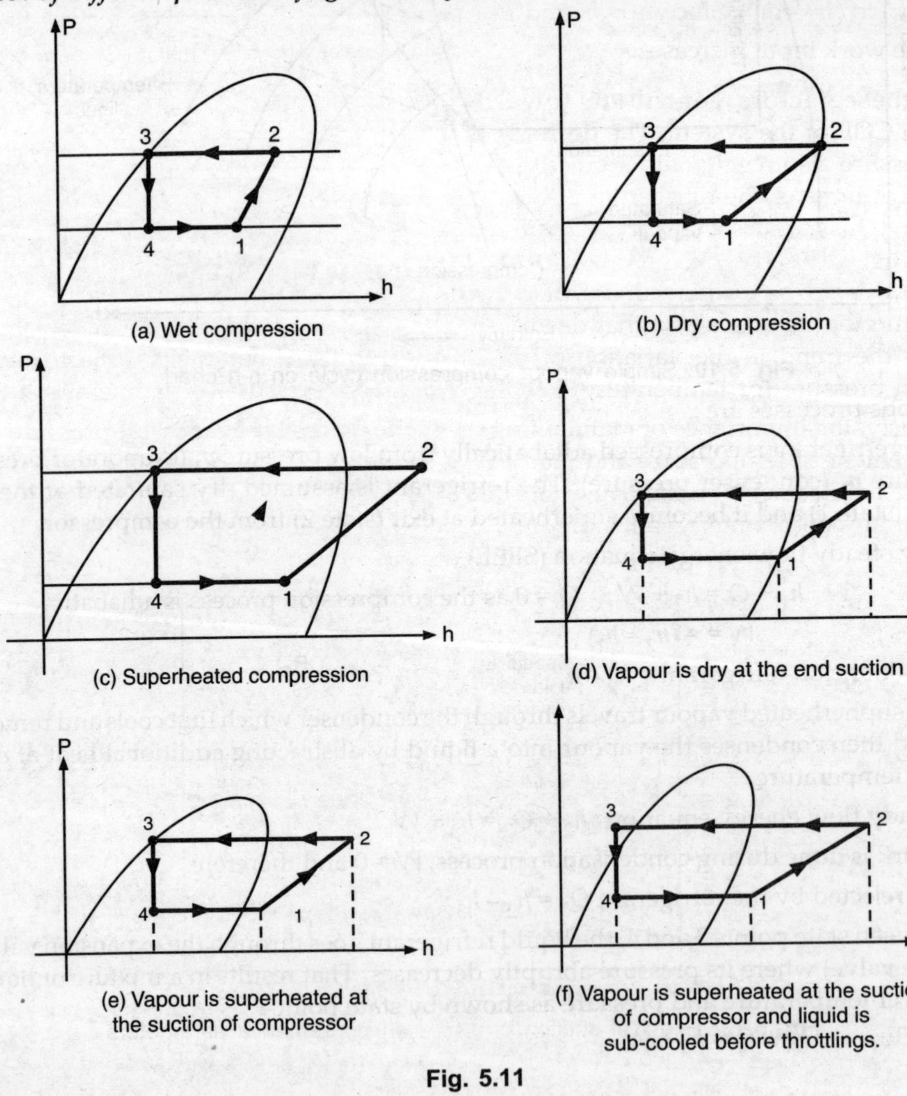

Fig. 5.11

Figure 5.11 shows the *p-h* representation of different possible refrigeration cycles.

Effect of different parameters on the performance of Simple Vapour Compression Cycles.

It will be now appropriate to use *p-h* chart for studying the effect of different parameters on the performance of a vapour compression refrigeration system.

- **Effect of evaporator or suction pressure :** Suction pressure, *i.e.*, the pressure in the evaporator is reduced by throttling in expansion valve, and the reduced suction pressure results in lowering of temperature in the evaporator.

From Fig. 5.10, it may be noted that due to decrease in suction pressure,

(*i*) The refrigerant effect changes from $(h_1 - h_4)$ to $(h_1' - h_4')$. Since $h_4 = h_4'$ and $h_1' < h_1$, the refrigerant effect decreases.

(*ii*) The work input to compressor changes from $(h_2 - h_1)$ to $(h_2' - h_1')$. Since $h_1' < h_1$ and $h_2' > h_2$, the work input increases.

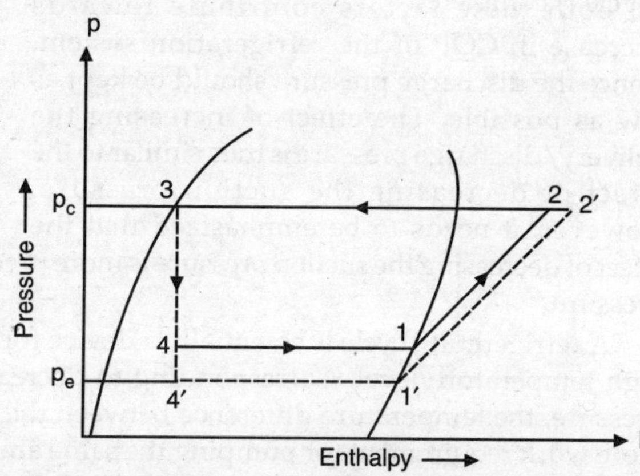

Fig. 5.12. Effect of decrease in suction pressure

Both these factors contribute towards decrease in COP of the system. The decrease in suction pressure also results into an increase in the specific volume of the refrigerant at inlet to compressor. This implies that for the required refrigerating capacity, large size of the compressor has to be employed. Further, the pressure ratio too increases and that decreases the volumetric efficiency of the compressor. Fig. 5.11 shows the trend in the variation of various performance parameters due to reduction in evaporation pressure (or temperature) keeping condenser pressure (or temperature) constant.

Obviously, the importance of maintaining the highest possible evaporator pressure consistent with requirement of temperature and load at the evaporator needs to be emphasized. Since the temperature requirements in the household refrigerators or ice plant are much lower than those required for air-conditioning plants, the COP of the air-conditioning plant is higher than that of other units.

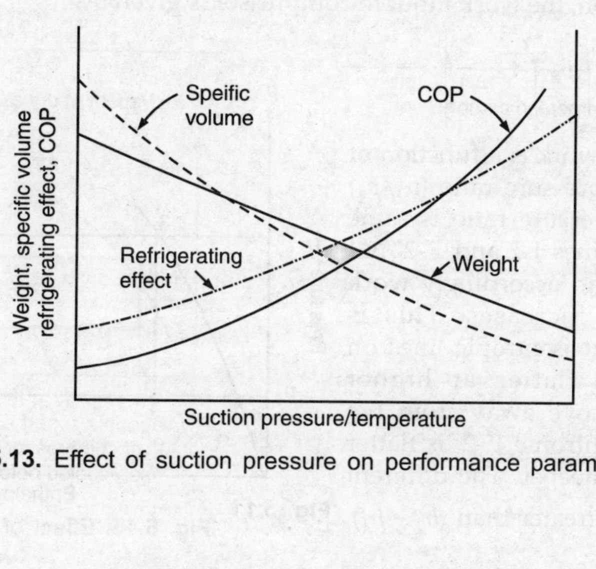

Fig. 5.13. Effect of suction pressure on performance parameters

- **Effect of delivery pressure :** From Fig. 5.14, it may be noted that due to increase in delivery (condenser) pressure,
  - (*i*) the refrigerating effect decreases from $(h_1 - h_4)$ to $(h_1 - h_4')$ because h_4' is greater than h_4.
  - (*ii*) the work input to compressor increases from $(h_2 - h_1)$ to $(h_2' - h_1)$ because h_2' is greater than h_2.

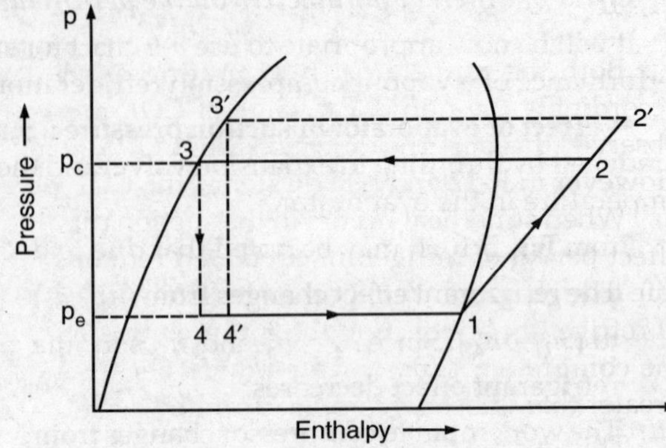

Fig. 5.14. Effect of increase in delivery pressure

Both these factors contribute towards decrease in COP of the refrigeration system. Hence the discharge pressure should be kept as low as possible. The effect of increasing the delivery/discharge pressure is thus similar to the effect of decreasing the suction pressure. However, it needs to be emphasized that the effect of decreasing the suction pressure is more predominant than the effect of increasing the delivery pressure.

A refrigerating cycle is essentially a device for pumping heat from a low temperature level to a high temperature level. Corresponding to decrease in suction pressure and increase in delivery pressure, the temperature difference between the two levels would increase. That would result in more work requirement for pumping the same amount of heat, and hence a decrease in the value of coefficient of performance of the system. As such, it is desirable that the pressure difference between the discharge pressure and suction pressure be kept as small as possible.

- **Effect of superheating:** In a simple saturated cycle, the refrigerant at exit from the evaporator or at suction to the compressor is dry saturated. In practice, the vaporising refrigerant continues to absorb heat and becomes superheated before it reaches the compressor suction. An examination of Fig. 5.13 shows that due to superheating
  - (*i*) the refrigerating effect, increases from $(h_1 - h_4)$ to $(h_1' - h_4)$. This is, however, on the assumption that superheating 1-1′ occurs within the evaporator, *i.e.*, the space being refrigerated.
  - (*ii*) the work input to compressor changes from $(h_2 - h_1)$ to $(h_2' - h_1')$.

For ideal compression, the work input to compressor is given by

$$W_c = \frac{\gamma}{\gamma - 1} RT_i \left[\left(\frac{p_2}{p_1} \right)^{\frac{\gamma-1}{\gamma}} - 1 \right]$$

i.e., the compressor work is a function of initial temperature T_i, pressure ratio (p_2/p_1) and exponent γ. The pressure ratio is same for both the isentropic lines 1-2 and 1′-2′ but temperature $T_1' > T_1$ and accordingly work input to compressor increases. This is precisely the reason that isentropic lines on *p-h* diagram become flatter at higher temperature as they move away from the saturated curve, *i.e.*, isentropic 1′-2′ is flatter while isentropic 1-2 is steeper. The different slopes make $(h_2' - h_1')$ greater than $(h_2 - h_1)$.

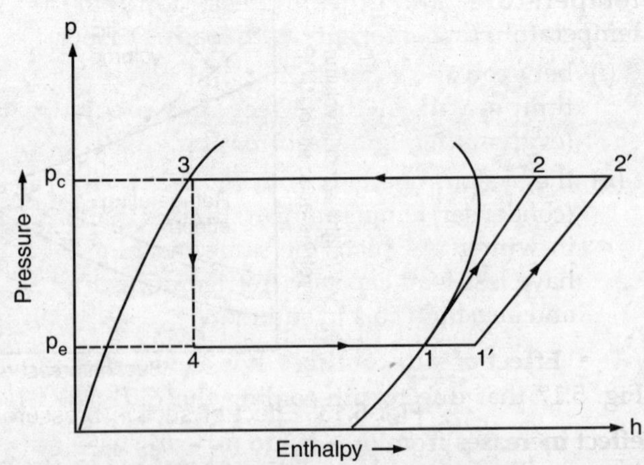

Fig. 5.15. Effect of supherheating

The COP of the cycle with superheating is given by

$$COP = \frac{h_1' - h_4}{h_2' - h_1'}$$

Both the numerator (refrigerating effect) and the denominator (work input) increase and accordingly the numerical value of COP may increase, decrease or remain the same. It has been observed that superheating increases COP when R-12 is the working fluid in the refrigeration system. However in R-12 and NH_3 systems, the COP decreases with superheating.

When superheat occurs in the suction pipe, *i.e.*, outside the refrigerated space, the refrigerating effect per kg of refrigerant circulated remains the same. But the system needs a large compressor due to high specific volume at suction. Also, there is increase in compressor work and temperature of refrigerant at exit from the compressor. Obviously with superheat outside the refrigerant space, the compressor capacity, the power per ton of refrigeration and the condenser capacity become greater than those for simple saturated cycle.

This degree of superheat can be reduced by locating the evaporator near the comperssor and insulating the pipe carrying low pressure gas.

- **Effect of suction temperature and condenser temperature :** It has been noted that COP and refrigeration capacity of a vapour compression system

(*i*) improve with increase in evaporator (suction) temperature

(*ii*) decrease with increase in condenser temperature and the nature of variation is as depicted in Fig. 5.16.

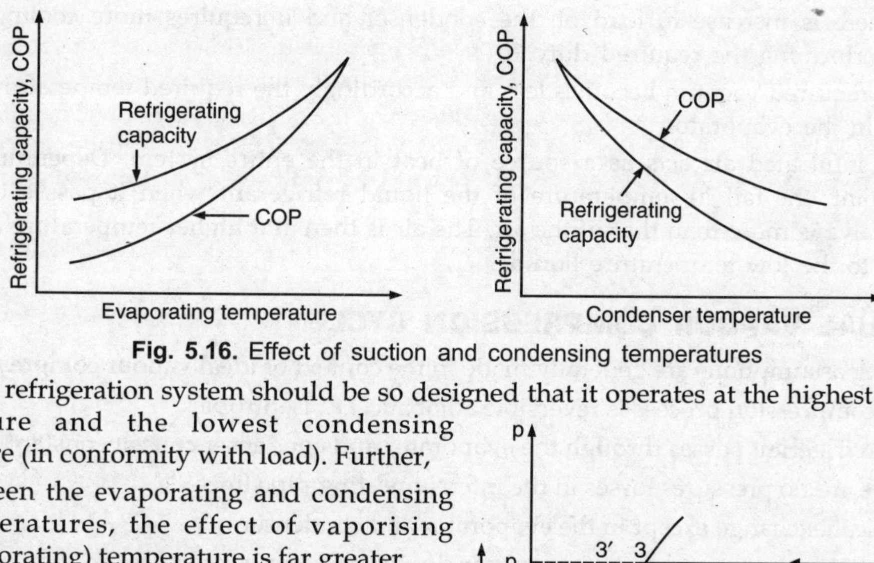

Fig. 5.16. Effect of suction and condensing temperatures

Thus a refrigeration system should be so designed that it operates at the highest evaporating temperature and the lowest condensing temperature (in conformity with load). Further,

(*i*) between the evaporating and condensing temperatures, the effect of vaporising (evaporating) temperature is far greater.

(*ii*) the system operates with higher discharge (condenser) temperature in summer than that in winter. As such, the same machine will have less load capacity (refrigerating effect) and also high cost in summer.

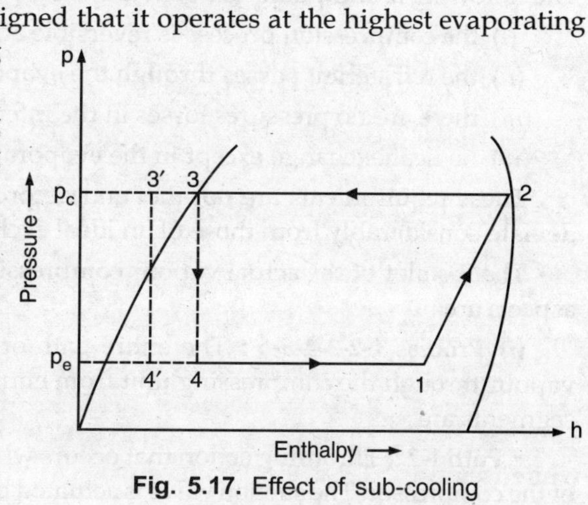

Fig. 5.17. Effect of sub-cooling

- **Effect of sub-cooling :** It is a evident from Fig. 5.17 that due to sub-cooling the refrigerant effect increases from $(h_1 - h_4)$ to $(h_1 - h_4')$ as h_4' is less than h_4. Since the work expended in running the compressor remains same at $(h_2 - h_1)$, subcooling results in an increase in coefficient of performance. However, it is being presumed that no energy has been spent in providing additional cooling to the refrigerant.

Further with subcooling,

– less mass flow of refrigerant per ton of refrigeration,

– reduced mass flow requires reduced piston displacement,

– power input per ton of refrigeration becomes less due to reduced mass flow rate.

The amount of undercooling is limited by the temperature and quantity of cooling water available, and the essential temperature difference required for the transfer of heat.

Effect of Air Leakage

There should not be any air leakage into the vapour compression refrigeration system as its presence adversely affects the performance of the system:

(*i*) The air does not take part in the refrigeration process but it consumes power when it is being compressed. As such the refrigerant effect decreases and work input becomes more per kg of refrigerant containing air. The net effect is reduction in the *COP* of the system.

(*ii*) The air comes in contact with the surfaces of the condenser and evaporator and that reduces the heat transfer coefficeint. Consequently

- there is decrease in the heat carried by the refrigerant in the evaporator.
- there is increase in load on the condenser and it requires more cooling water for performing the required duty.

(*iii*) The required vacuum becomes less and accordingly the required temperature cannot be maintained in the evaporator.

(*iv*) The infiltrated air acts as a source of heat in the entire system. Depending upon its inversion point, the fall in temperature of the liquid refrigerant when it passes through the expansion valve is more than that of the air. The air is then at a higher temperature and it gives out its heat to the low temperature liquid.

5.6. ACTUAL VAPOUR COMPRESSION CYCLE

The following assumptions are generally made in the context of ideal vapour compression cycle :

(*i*) the compression process is reversible adiabatic, *i.e.*, isentropic

(*ii*) the refrigerant passes through the evaporator and condenser coils at constant pressure

(*iii*) there are no pressures losses in the interconnecting pipe lines

(*iv*) no heat exchange except in the evaporator and condenser

These requirements are not met and accordingly the various processes in the practical cycle deviate considerably from those of an ideal cycle.

The *T-s* plot of the actual vapour compression cycle has been depicted in Fig. 5.18 and salient aspects are :

(*i*) Process 1-2-3-4-5-6 : The entire path or process represents the passage of the refrigerant vapour through the compressing unit from entry to exit. The changes during the different process segments are :

• **Path 1-2 :** Throttling action that occurs when the refrigerant passes through the suction valve of the compressor. The suction valve is actuated by pressures difference and accordingly the pressure inside the compressor has to be slightly less than the evaporator pressure. During throttling, there occurs slight increase in entropy and drop in temperature.

A similar situation prevails when the refrigerant leaves the compressor (path 5-6) through its exhaust valve.

- **Path 3-4 :** Compression process (assumed to be isentropic) is neither isentropic nor polytorpic. There occurs heat transfer at constant pressure indicated by paths 2-3 and 4-5.

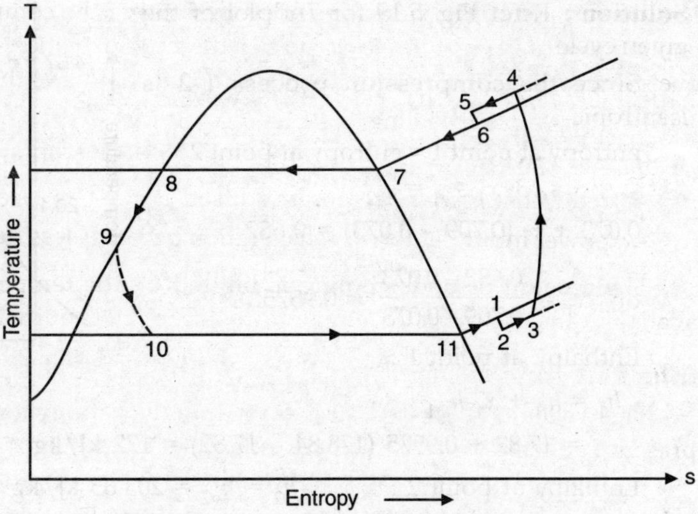

Fig. 5.18. Actual vapour compression cycle

(*ii*) Process 6-7-8-9 : The entire path represents flow of refrigerant through the condenser where the refrigerant loses heat to the cooling medium and it liquefies/condenses. The different heats lost are as :

"path 6-7 indicating removal of superheat, 7-8 the removal of latent heat and 8-9 the removal of heat of liquid, *i.e.*, subcooling"

- **Path 9-10 :** The path represents the passage of liquid refrigerant through expansion valve where the throttling process follows an irreversible adiabatic path.
- **Path 10-11-1 :** The evaporation process represents the passage of refrigerant through the evaporator and it is almost indentical to the constant pressure conditions assumed in the ideal cycle. During segment 10-11, there is absorption of latent heat of vaporization and during part 11-1 occurs the gain of superheat before the refrigerant vapour is to enter the compressor at state point 1.

The vapour enters the compressor and the cyclic is completed.

EXAMPLE 5.3

A standard vapour compression refrigerator using F-12 as the refrigerant operates between the condenser pressure of 10 bar and the evaporator pressure of 1.5 bar. The evaporator absorbs 75 kJ/min of energy as heat and the vapour is dry saturated at exit from the compressor.

Represent the cycle on T-s plane and calculate :

(a) flow rate of refrigerant (b) power consumed and (c) COP of the cycle.

Also calculate COP of the Carnot refrigerator operating between the same temperature limits.

The relevant properties of F-12 are given below :

Pressure bar	*Saturation temperature, °C*	*Enthalpy kJ/kg*		*Entropy kJ/kg K*	
		Liquid	*Vapour*	*Liquid*	*Vapour*
10	*41.7*	*76.8*	*203.65*		*0.682*
1.5	*– 20.1*	*17.82*	*178.84*	*0.073*	*0.709*

Solution : Refer Fig. 5.19 for *T-s* plot of the given cycle

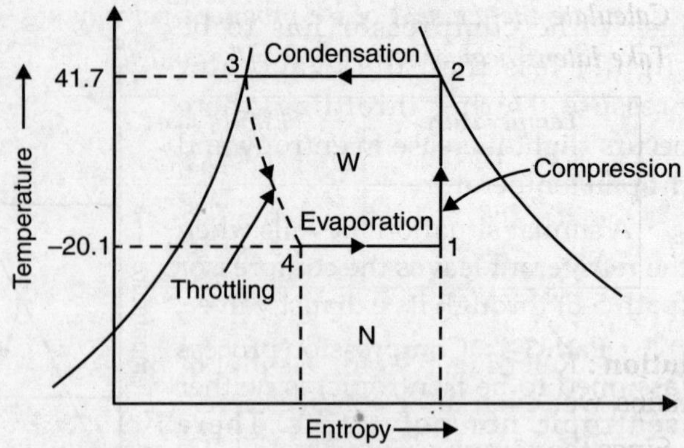

Fig. 5.19.

Since the compression process 1-2 is isentropic,

entropy at point 1 = entropy at point 2

$$s_{f1} + x_1 \, s_{fg1} = s_{g2}$$

$$0.073 + x_1 \, (0.709 - 0.073) = 0.682$$

or $$x_1 = \frac{0.682 - 0.073}{0.709 - 0.073} = 0.9575$$

Enthalpy at point 1 :

$$h_1 = h_{f1} + x_1 \, h_{fg1}$$
$$= 17.82 + 0.9575 \, (178.84 - 17.82) = 172 \text{ kJ/kg}$$

Enthalpy at point 2 : $h_2 = h_{g2} = 203.65$ kJ/kg

Enthalpy at point 3 : $h_3 = h_{f2} = 76.8$ kJ/kg

The process 3-4 is that of throttling, it takes place at constant enthalpy and accordingly

Enthalpy at point 4 : $h_4 = h_3 = 76.8$ kJ/kg

Refrigerating effect $N = h_1 - h_4 = 172 - 76.8 = 95.2$ kJ/kg

Work done $W = h_2 - h_1 = 203.65 - 172 = 31.65$ kJ/kg

(*a*) If *m* is the mass flow of refrigerant/min, then energy balance in the evaporator gives

$m \times 95.2 = 75$; $m =$ **0.788 kg/min**

(*b*) Power consumed = work done per second

$$= 31.65 \times \frac{0.788}{60} = 0.416 \text{ kJ/s} = \mathbf{0.416 \text{ kW}}$$

(*c*) Coefficient of performance of the cycle,

$$COP = \frac{\text{refrigerating effect}}{\text{work done}} = \frac{95.2}{31.65} = \mathbf{3.0}$$

The coefficient of performance of a refrigerator working on reversed Carnot cycle is given by

$$COP = \frac{T_2}{T_1 - T_2}$$

where T_2 = temperature in the evaporator = 273 + (– 20.1) = 252.9 K

T_1 = temperature in the condenser = 273 + 41.7 = 314.7 K

$$\therefore \quad COP = \frac{252.9}{314.7 - 252.9} = \mathbf{4.09}$$

EXAMPLE 5.4

A refrigerating plant running on CO_2 operates between saturation temperatures of 268 K (in the evaporator) and 298 K (in the condenser coils). The vapour has a dryness fraction of 0.6 at entry to the compressor, isentropic compression takes place in the compressor and there is no under-cooling of the working fluid in the condenser. The machine has a relative efficiency of 45% and 8.2 kg of CO_2 is circulated per minute.

Calculate the amount of ice produced per day at 0°C from water at 15°C.
Take latent heat of ice as 335 kJ/kg and use the following table for the properties of CO_2.

Temperature °C	*Liquid heat kJ/kg*	*Latent heat kJ/kg*	*Entropy of liquid kJ/kg K*
268	*– 7.53*	*245.8*	*– 0.04187*
298	*81.25*	*121.5*	*0.2513*

(WBUT)

Solution : Refer Fig. 5.20 for *T-s* plot of the given cycle. It has been presumed that the refrigerant remains wet even after compression.

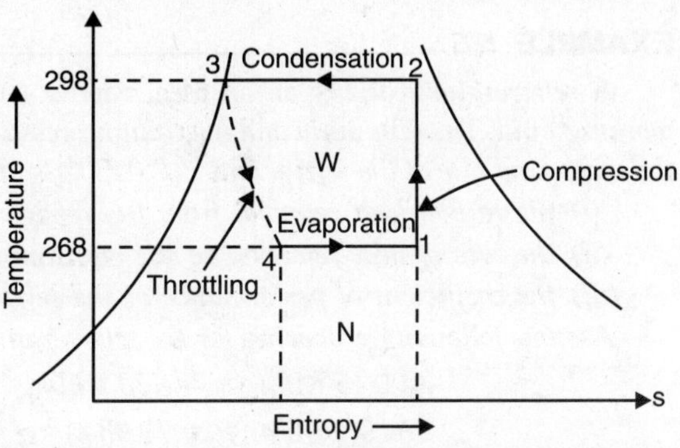

Fig. 5.20.

Since the compression process 1-2 is adiabatic, entropy at point 1 = entropy at point 2

$$s_{f1} + x_1 \frac{h_{fg1}}{T_1} = s_{f2} + x_2 \frac{h_{fg2}}{T_1}$$

or $$-0.04187 + 0.6 \times \frac{245.8}{268} = 0.2513 + x_2 \times \frac{121.5}{298}$$

or $$-0.04187 + 0.5503 = 0.2513 + 0.4077\, x_2$$

∴ Dryness fraction of vapour at exit from the compressor,

$$x_2 = \frac{-0.04187 + 0.5503 - 0.2513}{0.4077} = 0.631$$

Total heat at point 1 : $h_1 = h_{f1} + x_1\, h_{fg1} = -7.53 + 0.6 \times 245.8 = 139.95$ kJ/kg

Total heat at point 2 : $h_2 = h_{f2} + x_2\, h_{fg2} = 81.25 + 0.631 \times 121.5 = 157.92$ kJ/kg

Total heat at point 3 : h_3 = liquid heat corresponding to condenser temperature of 298 K = 81.25 kJ/kg

The expansion process 3-4 is of throttling during which enthalpy remains constant. That is

$$h_4 = h_3 = 81.25 \text{ kJ/kg}$$

Now :

$$\text{Work done } W = h_2 - h_1 = 157.92 - 139.95 = 17.97 \text{ kJ/kg}$$

$$\text{Refrigerating effect } N = h_1 - h_4 = 139.95 - 81.25 = 58.7 \text{ kJ/kg}$$

$$\text{Theoretical } COP = \frac{N}{W} = \frac{58.7}{17.97} = 3.266$$

The ratio of actual *COP* to the theoretical *COP* is known as relative coefficient of performance.

∴ Actual *COP* = 0.45 × 3.266 = 1.47

With 8.2 kg of CO_2 circulated per minute,

$$\text{Work done} = \frac{8.2}{60} \times 17.97 = 2.456 \text{ kJ/s}$$

∴ Actual refrigerating or cooling effect

$$= 2.456 \times 1.47 = 3.61 \text{ kJ/s}$$

Heat to be extracted during one day of 24 hours

$$= 24 \times 3600 \times 3.61 = 311932 \text{ kJ}$$

Heat extracted from 1 kg of water at 15°C for formation of 1 kg of ice at 0°C

$$= 1 \times 4.187 \times (15 - 0) + 335 = 397.80 \text{ kJ}$$

$$\therefore \text{Mass of ice formed} = \frac{311932}{397.80} = \mathbf{384\ kg}$$

EXAMPLE 5.5

A refrigerator working on an ideal vapour compression refrigeration cycle uses refrigerant-12 as the working fluid. The minimum and maximum pressures of the cycle are 0.15 MPa and 0.9 MPa respectively. If the mass flow rate of the refrigerant is 0.045 kg/s, determine:

(i) the rate of heat removal from the refrigerated space and power input to the compressor
(ii) the rate of heat rejection to the environment, and
(iii) the coefficient of performance of the refrigerator.

Assume following properties for the refrigerant :

At 0.15 MPa : h_g = 178.9 kJ/kg and s_g = 0.7089 kJ/kg K

At 0.9 MPa : h_f = 71.93 kJ/kg, h = 210.43 kJ/kg when s = 0.7089 kJ/kg K

Represent the cycle on T-s and p-h plots.

Solution : Refer Fig. 5.21 For *T-s* and *p-h* plots of the given refrigeration system :

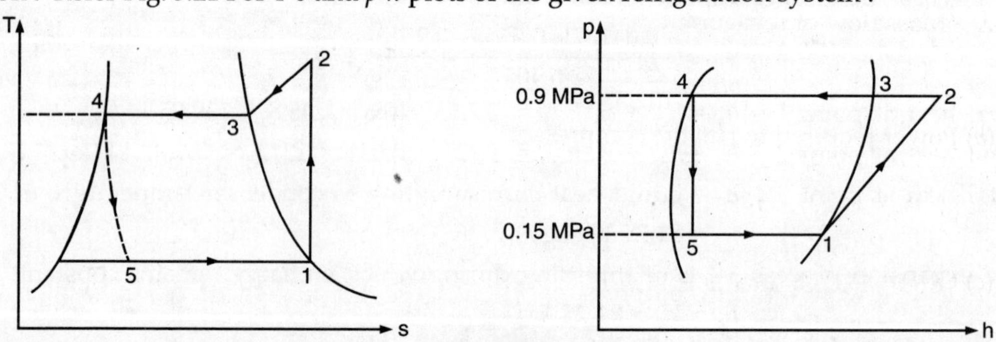

Fig. 5.21

At entry to compressor at 0.15 MPa pressure, the value of entropy is given for the gaseous state (h_g = 178.9 kJ/kg). This suggests that the refrigerant can be taken to be in the dry saturated state when it enters the compressing unit. Then from the given data :

$$h_1 = 178.9 \text{ kJ/kg}; \quad s_1 = 0.7089 \text{ kJ/kg K}$$

$$h_4 = h_5 = 71.93 \text{ kJ/kg} \quad \text{and} \quad h_2 = 210.43 \text{ kJ/kg}$$

(*i*) Rate of heat removal = $m_r \times (h_1 - h_5)$ = 0.045 (178.9 – 71.93) = 4.813 kJ/s

Power input to compressor = $m_r (h_2 - h_1)$ = 0.045 (210.43 – 178.9) = 1.419 kJ/s = **1.419 kW**

(*ii*) Rate of heat rejection to environment

$$= m_r (h_2 - h_4) = 0.045\ (210.43 - 71.93) = 6.232 \text{ kJ/s}$$

$$(iii)\ COP = \frac{\text{refrigerating effect (rate of heat removal)}}{\text{power input to compressor}} = \frac{4.813}{1.419} = \mathbf{3.392}$$

EXAMPLE 5.6

A R-12 vapour compression refrigeration system has a condensing temperature of 50° C. The refrigeration capacity is 7 tons. The liquid leaving the condenser is saturated liquid and compression is isentropic. Determine

(a) refrigerant flow rate (b) power required to run the compressor (c) COP of the system

Take enthalpy at the end of isentropic compression = 210 kJ/kg. Take following properties of R-12

Temperature (°C)	*Enthalpy (kJ/kg)*	
	Liquid	*Vapour*
50	84.868	206.298
0	36.022	187.397

(UPTU)

Solution : Refer Fig. 5.22 for *T-s* plot of the working cycle.

Given : $h_1 = 187.397$ kJ/kg and $h_2 = 210$ kJ/kg

Process 4 – 5 is of throttling during which enthalpy remains constant and therefore,

$$h_3 = h_{f4} = 84.868 \text{ kJ/kg}$$

Refrigerating capacity = 7 tons

$$= 7 \times 3.5 = 24.5 \text{ kJ/s}$$

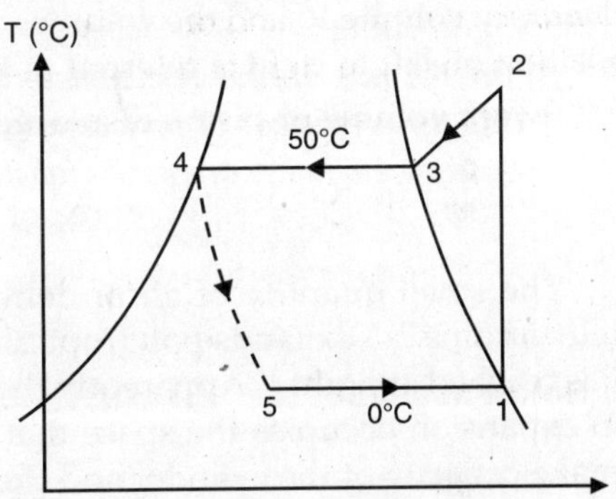

Fig. 5.22

(*a*) Refrigerating effect $= m_r (h_1 - h_5)$

$$= m_r (187.397 - 84.868)$$

$$= 102.53\, m_r$$

Equating it to the refrigeration capacity, we get

$$102.53\, m_r = 24.5$$

∴ Mass flow of refrigerant,

$$m_r = \frac{24.5}{102.53} = 0.239 \text{ kg/s}$$

(*b*) Power required to run the compressor,

$$= m_r (h_2 - h_1) = 0.239\,(210 - 187.397)$$

$$= 5.402 \text{ kJ/s} = 5.402 \text{ kW}$$

$$(c)\ COP = \frac{\text{refrigerating effect}}{\text{work input to compressor}} = \frac{h_1 - h_5}{h_2 - h_1}$$

$$= \frac{187.397 - 84.868}{210 - 187.397} = 4.54$$

Note: Quite often it is also required to determine the heat rejected in the condensing unit of the refrigeration. That is given by

$$= m_r (h_2 - h_4) = 0.239\,(210 - 84.868) = 29.90 \text{ kJ/s} = 29.90 \text{ kW}$$

5.7. VOLUMETRIC EFFICIENCY

A reciprocating compressor has a piston which reciprocates inside a cylinder. The piston derives its motion through a connecting rod and crank mounted in a crank case. The inlet and outlet valves are provided in the head of cylinder. When the piston is at its extreme left position, it does not flush with the cylinder head. A dead space remains between the cylinder head and the piston. The dead space is called clearance space. The clearance ensures mechanical freedom to the moving parts without much wear and tear and provides the necessary space for valve operation.

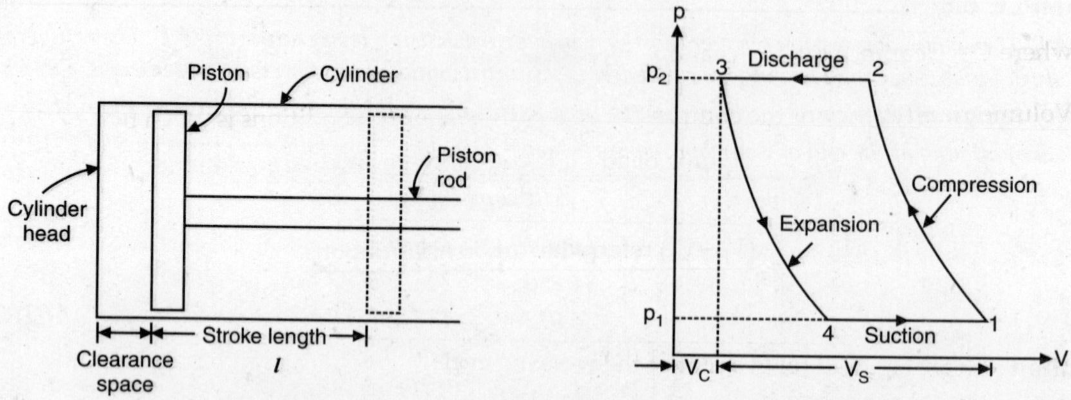

Fig. 5.23. Piston cylinder arrangement and pressure-volume variation

The volume occupied by the gas when the piston is at its extreme left position is known as clearance volume V_c and the volume swept by the piston as it moves in the cylinder from its extreme position on left to right is referred to as swept volume or piston displacement V_s.

Swept volume or piston displacement volume,

$$V_s = \text{area of the piston} \times \text{stroke length} = \frac{\pi}{4} d^2 l$$

The small quantity of air at delivery pressure p_2 remaining in the dead space at the end of delivery line 2-3 expands polytropically along 3-4. The expansion continues until suction pressure p_1 is reached at point 4. Apparently the quantity of compressed air remaining in the clearance space on expansion occupies the space that ought to be available for fresh air. This aspect reduces the intake capacity of the cylinder and the effective volume sucked in during suction equals $(V_1 - V_4)$.

The volumetric efficiency η_v of the reciprocating compressor is defined as the ratio of actual volume sucked in to the piston displacement or swept volume of the compressor. Thus

$$\eta_v = \frac{V_1 - V_4}{V_s} = \frac{V_s + V_c - V_4}{V_s} = 1 + \frac{V_c}{V_s} - \frac{V_4}{V_s} \quad \text{...(i)}$$

Considering polytropic expansion process 3-4, we have

$$p_3 V_3^n = p_4 V_4^n; \quad V_4 = V_3 \left(\frac{p_3}{p_4}\right)^{1/n}$$

But $\quad p_3 = p_2; \quad p_4 = p_1 \quad$ and $\quad V_3 = V_c$

$$\therefore \quad V_4 = V_c \left(\frac{p_2}{p_1}\right)^{1/n} \quad \text{...(ii)}$$

Substituting the value of V_4 as given by expression (*ii*) in expression (*i*), we get

$$\eta_v = 1 + \frac{V_c}{V_s} - \frac{V_c}{V_s}\left(\frac{p_2}{p_1}\right)^{1/n}$$

$$= 1 - \frac{V_c}{V_s}\left[\left(\frac{p_2}{p_1}\right)^{1/n} - 1\right] = 1 - C\left[(r_p)^{1/n} - 1\right] \quad \text{...(iii)}$$

where C = clearance ratio $\frac{V_c}{V_s}$ and r_p = pressure ratio $\frac{p_2}{p_1}$

Volumetric efficiency of the compressor referred to ambient conditions is given by

$$\eta_v = \frac{\text{volume of air sucked referred to ambient conditions}}{\text{swept volume}}$$

$$= \frac{(V_1 - V_4) \text{ referred to ambient conditions}}{V_s}$$

$$= \frac{(V_1 - V_4)\left(\frac{p_1 T_0}{p_0 T_1}\right)}{V_s} = \frac{p_1 T_0}{p_0 T_1} \frac{V_1 - V_4}{V_s}$$

The suffix 1 represents the suction and suffix 0 represents the ambient conditions. Thus

$$\eta_v \text{ (ambient suction condition)} = \frac{p_1 T_0}{p_0 T_1}\left[1 - C\{(r_p)^{1/n} - 1\}\right] \quad ...(iv)$$

When free air or ambient conditions are not given, they are assumed to be 1 bar and 15°C. Further, the expressions of volumetric efficiency as given by expressions (*iii*) and (*iv*) become identical when suction conditions are same as ambient conditions. The dimensions of the compression cylinder are determined by using the volumetric efficiency based on suction conditions.

The volumetric efficiency of a reciprocating compressor is influenced by the factors C, r_p and n. The volumetric efficiency decreases with increase of C, decreases with increase of r_p and increases with increase of expansion index n. The volumetric efficiency is further decreased by the following factors:

(*i*) Increase in temperature of fresh air when it comes in contact with cylinder walls. The mass of air actually present in the cylinder when compared to that at atmospheric conditions gets reduced in the ratio of absolute atmospheric temperature to that of air in the cylinder.

(*ii*) The fluid resistance through the intake pipe decreases the suction pressure and so the cylinder cannot be fully charged with air.

(*iii*) At very high speeds, the air does not get enough time to fill the cylinder fully during suction stroke.

(*iv*) Leakage in the system.

(*v*) Inertia effects in opening the inlet valve.

(*vi*) In actual operation, the inlet valve opens only when the high pressure air left in the dead space expands to a pressure below the atmospheric pressure. Some portion of the suction stroke may thus be wasted.

EXAMPLE 5.7

In a simple vapour compression cycle, the properties of the refrigerant R-12 at various sections are:

Compressor inlet : enthalpy = 183.2 kJ/kg and specific volume = 0.0797 m³/kg

Compressor discharge: enthalpy = 222.6 kJ/kg and specific volume = 0.00164 m³/kg

Condenser exit : enthalpy = 84.9 kJ/kg and specific volume = 0.00083 m³/kg

The piston displacement volume for the compressor is 1.5 litres per stroke and its volumetric efficiency is 80%. If the compressor turns 1600 rev/min, find the power rating of the compressor and the refrigerating effect.

(GATE)

Solution : In a simple vapour compression cycle, the refrigerant is taken to be dry saturated at entry to the compressor and there is no undercooling. Then the cycle on *p-h* plot depicts as shown in Fig. 5.24.

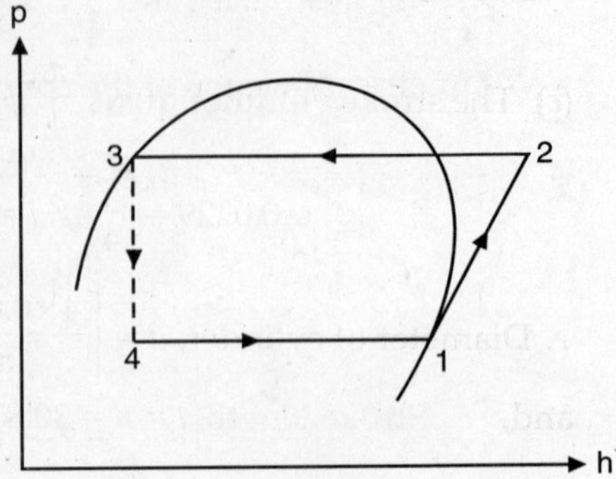

Fig. 5.24

Piston displacement = 1.5 litres per stroke

$= 1.5 \times 10^{-3}$ m³/stroke

The compressor turns 1600 rev/min and has a volumetric efficiency equal to 0.8, As such

Compressor discharge $= (1.5 \times 10^{-3}) \times 1600 \times 0.8$

$= 1.92$ m³/min

Mass flow rate of refrigerant,

$$m_r = \frac{\text{compressor discharge}}{\text{specific volume}} = \frac{1.92}{0.0767}$$

$= 25.03$ kg/min $= 0.417$ kg/s

(*a*) Power rating of compressor $= m_r (h_2 - h_1)$

$= 0.417 (222.6 - 183.2) = 16.43$ kJ/s $= 16.43$ kW

(*b*) Refrigerating effect $= m_r (h_1 - h_4)$

During throttling process, enthalpy of the refrigerant remains constant and so $h_4 = h_3$. Then

Refrigerating effect $= m_r (h_1 - h_3)$

$= 0.417(183.2 - 84.9) = 40.99$ kJ/s $\simeq$ **41 kW**

EXAMPLE 5.8

A vapour compression refrigerator using ammonia has pressure of 2.8 bar in the evaporator. The refrigerant at entry to the evaporator has a dryness fraction of 0.12 while at exit the dryness fraction is 0.92. A work of 0.172×10^6 Nm/kg is spent when ammonia is compressed. The compressor is single acting, turns 120 revolutions per minute and has a volumetric efficiency of 80 percent. For refrigerant flow rate of 5.5 kg/min, calculate : (a) COP (b) volume of refrigerant vapour entering the compressor per stroke (c) bore and stroke of the compressor if bore equals the stroke.

From ammonia tables, the latent heat at 2.8 bar is 1339.5 kJ/kg and specific volume is 0.435 m³/kg.

Solution : Enthalpy of refrigerant at entry to evaporator $= 0.12 \times 1339.5 = 160.74$ kJ/kg

Enthalpy of refrigerant at exit from evaporator $= 0.92 \times 1339.5 = 1232.34$ kJ/kg

Then, refrigerating effect (heat absorbed in the evaporator)

$= 1232.34 - 160.74 = 1071.6$ kJ/kg

Work spent during compression, $W = 0.172 \times 10^6$ Nm/kg $= 0.172 \times 10^6$ J/kg

$= 0.172 \times 10^3$ kJ/kg

$$\therefore \quad COP = \frac{\text{refrigerating effect}}{\text{work spent during compression}} = \frac{1071.6}{0.172 \times 10^3} = \mathbf{6.23}$$

(*b*) Volume of refrigerant at entry to compressor $= 0.435 \times 0.92 = 0.40$ m³/kg

Since the refrigerant flow rate is stated to be 5.5 kg/min,

Total volume of refrigerant entering the compressor

$V = 0.40 \times 5.5 = 2.2$ m³/min

Since the compressor is single acting, turns 120 revolutions per minute and has a volumetric efficiency of 80 percent,

Stroke volume of compressor = $\dfrac{2.2}{120 \times 0.8}$ = **0.0229 m³**

(c) The stroke volume equals $\dfrac{\pi}{4}\, d^2 l$ where d is the cylinder diameter and l is the stroke length.

$$\therefore \quad 0.0229 = \frac{\pi}{4} d^2 l = \frac{\pi}{4} d^3 \qquad (\because d = l)$$

$\therefore$ Diameter of cylinder, $d = \left[\dfrac{4 \times 0.0229}{\pi}\right]^{1/3}$ = 0.308 m = **30.8 cm**

and, Stroke length, $l = d$ = **30.8 m**

EXAMPLE 5.9

An ammonia ice plant operates between a condenser temperature of 35°C and an evaporator temperature of –15°C. It produces 10 tons of ice per day from water at 32°C to ice at –3°C. Assuming simple saturation cycle, determine:

(a) capacity of the refrigerant plant *(b) mass flow rate of refrigerant*

(c) compressor cylinder diameter and stroke if its volumetric efficiency is 70 %, rpm = 1250 and stroke to bore ratio is 1.3

(d) power of the compressor motor if the isentropic efficiency of the compressor = 0.8% and the mechanical efficiency = 0.94

(e) theoretical and actual COP

Given: Properties of NH_3 are

T °C	$v_f \times 10^3$ m³/kg	v_g m³/kg	h_f kJ/kg	h_g kj/kg	s_f kJ/kg K	s_g kJ/kg K
–15	1.52	0.508	131.3	1443.9	0.7426	5.8223
35	1.7	0.096	366.1	1488.6	1.566	5.2086

Average specific heat of ammonia vapour = 2.2 kJ/kg, Specific heat of water = 4.187 kJ/kg K
Specific heat of ice = 1.94 kJ/kg K, Latent heat of ice = 335 kJ/kg

(Engg. Services)

Solution : Refrigerating capacity,

= heat removed from the water at –32°C for its conversion into ice at –3°C

$$= m\left[c_{pw}\Delta t_w + L_i + c_{pi}\Delta t_i\right]$$

$$= \frac{10 \times 1000}{24 \times 3600}\left[4.187 \times (32 - 0) + 335 + 1.94\{0 - (-3)\}\right]$$

= 54.95 kJ/s = 54.95 kW

(b) With reference to T-s plot of the simple saturation cycle (refrigerant dry saturated at entry to compressor and no undercooling in the condenser),

$$h_1 = 1443.9 \text{ kJ/kg}; \quad s_1 = 5.8223 \text{ kJ/kg K}$$

$$h_5 = h_4 = h_f \text{ at } 35°\text{C} = 366.1 \text{ kJ/kg}$$

This is so as the enthalpy remains constant during the throttling process 4–5.

Refrigerating effect = $m_r\,(h_1 - h_5)$

$= m_r\,(1443.9 - 366.1) = 1077.8\,m_r$

Equating it to the refrigeration capacity, we get

Refrigerant mass flow $m_r = \dfrac{54.95}{1077.8} = 0.051$ kg/s

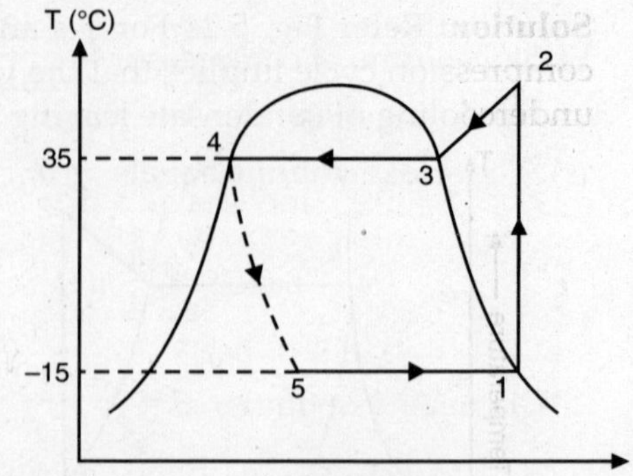

Fig. 5.25.

(*c*) Actual volume flow rate per stroke

$$= \frac{m_r \times v_1}{N} = \frac{0.051 \times 0.0508}{(1250/60)} = 1.243 \times 10^{-3}\ \text{m}^3$$

Volumetric efficiency of a compressor is defined as,

$$\eta_v = \frac{\text{actual air intake}}{\text{swept volume}}\ ;\quad 0.7 = \frac{1.243 \times 10^{-3}}{\frac{\pi}{4} d^2 l}$$

$$\therefore\quad \frac{\pi}{4} d^2 \times 1.3 d = \frac{1.243 \times 10^{-3}}{0.7}\ ;\quad d^3 = \frac{1.243 \times 10^{-3}}{0.7} \times \frac{4}{1.3\pi} = 1.74 \times 10^{-3}\ \text{m}^3$$

That gives: $d = 0.122$ m and $l = 1.3 \times 0.122 = 0.1586$ m

(*d*) For isentropic compression,

$$s_1 = s_2 = 5.8223\ \text{kJ/kg K} \quad \text{and} \quad s_2 = s_3 + c_p \log_e \frac{T_2}{T_3}$$

$$\therefore\quad 5.2086 + 2.2 \log_e \frac{T_2}{308} = 5.8223$$

$$\text{or}\quad \log_e \frac{T_2}{308} = \frac{5.8223 - 5.2086}{2.2} = 0.2789$$

That gives: $T_2 = 308 \times 1.321 = 406.9$ K

$$h_2 = h_3 + c_p\,(T_2 - T_3) = 1488.6 + 2.2(406.9 - 308) = 1706.18\ \text{kJ/kg}$$

Power consumption in the compressor,

$$= \frac{m_r (h_2 - h_1)}{\eta_{isen} \times \eta_m} = \frac{0.051\,(1706.18 - 1443.9)}{0.86 \times 0.94} = 16.56\ \text{kJ/s} = \mathbf{16.56}\ \text{kW}$$

EXAMPLE 5.10

A standard vapour compression cycle developing 50 kW of refrigeration using refrigerant 22 operates with a condensing temperature of 35°C and an evaporating temperature of –10°C. Calculate :

(a) the refrigerating effect in kJ/kg,

(b) the circulation rate of refrigerant in kg/s,

(c) the power required by the compressor in kW,

(d) the coefficient of performance,

(e) the volume flow rate measured at the compressor suction

(f) the power per kilowatt of refrigeration, and

(g) the compressor discharge temperature.

Solution: Refer Fig. 5.24 For *T-s* and *p-h* plots for the given refrigeration system. The standard compression cycle implies that the vapour is dry saturated at entry to compressor and there is no undercooling of condensate leaving the condenser.

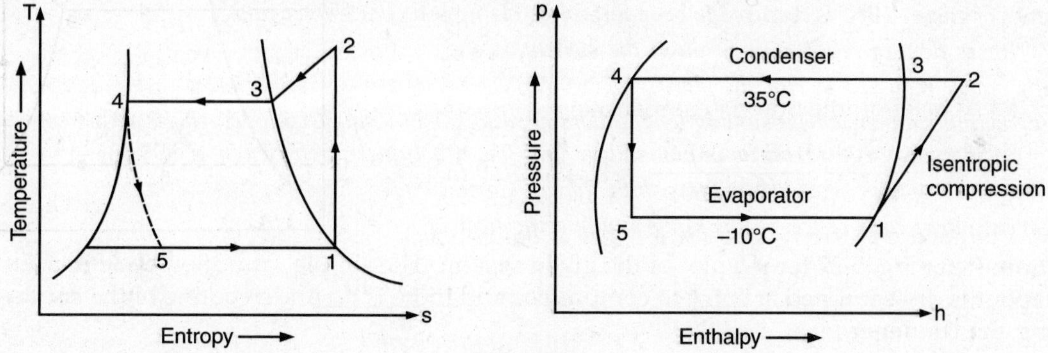

Fig. 5.26.

From the tables for refrigerant 22,

At $t_e = -10°C$: $p_e = 3.543$ bar and at $t_c = 35°C$: $p_c = 13.548$ bar

The cycle is then laid on the *p-h* chart for refrigerant 22 and the following values are read therefrom

$$h_1 = 401.5 \text{ kJ/kg} \; ; \quad h_2 = 435.5 \text{ kJ/kg}$$

$$h_5 = h_4 = 243 \text{ kJ/kg} \quad \text{and} \quad v_1 = 0.0654 \text{ m}^3\text{/kg} \; ;$$

The values of h_1, v_1 and h_4 can be checked and compared with the values given in the saturated property tables for refrigerant 22.

(*a*) Refrigerating effect = $h_1 - h_5 = 401.5 - 243 =$ **158.5** kJ/kg

(*b*) If m_r is the mass flow of refrigerant through the system, then

$$m_r (h_1 - h_2) = \text{total refrigerating effect}$$

$$m_r (401.5 - 243) = 50 \; ; \quad m_r = \frac{50}{401.5 - 243} = \mathbf{0.315} \text{ kg/s}$$

(*c*) Power required to run the compresor

= work input to compressor

$= m_r (h_2 - h_1) = 0.315 (435.5 - 401.5) = 10.71$ kJ/s = **10.71** kW

(*d*) Coefficient of performance of the system

$$COP = \frac{\text{refrigeration capacity}}{\text{work input to compressor}} = \frac{50}{10.71} = \mathbf{4.67}$$

(*e*) Volume flow rate = mass flow of refrigerant × specific volume

$= 0.315 \times 0.0654 = 0.0206$ m³/s

(*f*) Power per kW of refrigeration $= \dfrac{\text{power input to compressor}}{\text{refrigeration capacity}} = \dfrac{10.71}{50} = \mathbf{0.214}$ kW/kW

Essentially, it is reciprocal of the coefficient of performance

(*g*) The compressor discharge temperature is the temperature of superheated vapour at state point 2 which from the *p-h* plot is noted to be about 57°C

EXAMPLE 5.11

A simple saturation cycle using F-12 as refrigerant has been designed to produce 10 tons of ice per day from water at 35°C to ice at –5°C. For effective heat transfer at evaporator and condenser units, a temperature difference of atleast 10°C is required to be maintained. Using p-h chart, determine:

(a) mass flow of refrigerant through the system,

(b) power required to run the plant,

(c) cylinder dimensions assuming length/diamter ratio 1.0 for a single cylinder, single acting compressor if it turns 1200 revolutions per minute and has a volumetric efficiency of 90%.

How COP of this system compares with that of Carnot cycle?

Given: latent heat of ice = 335 kJ/kg and specific heat of ice = 1.92 kJ/kg K.

Solution: Refer Fig. 5.27 for *p-h* plot of the given system. The simple saturation cycle implies that the vapour is dry saturated at entry to compressor and there is no undercooling of the condensate leaving the condenser.

From the given condition,

$$T_c = 35 + 10 = 45°C \quad \text{and} \quad T_e = -5 + (-10) = -15°C$$

The cycle is then laid on the *p-h* plot for F-12 and enthalpy and specific volume as read from the plot are :

$h_1 = 345$ kJ/kg; $h_2 = 377$ kJ/kg

$h_5 = h_4 = 244$ kJ/kg and $v_1 = 0.09$ m^3/kg

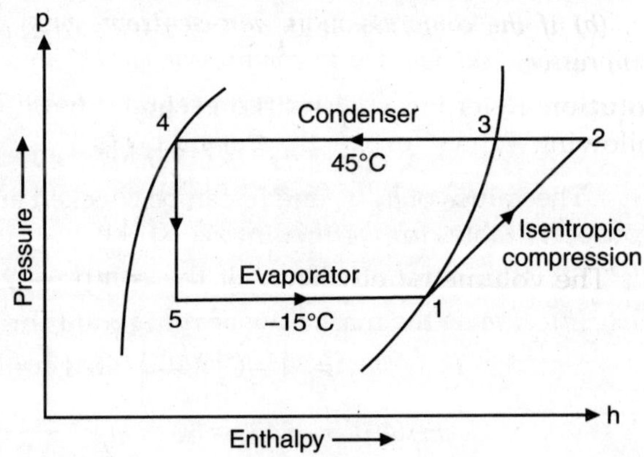

Fig. 5.27

(*a*) Heat to be removed from water for ice formation (per kg basis)

= sensible heat of water + latent heat of fusion + sensible heat of ice

$= 4.186 \times 35 + 335 + 1.92\,[0 - (-5)]$

$= 491.11$ kJ/kg

Refrigerating capacity of the evaporator,

$$= \frac{(10 \times 1000) \times 491.11}{24 \times 3600}$$

$= 56.84$ kJ/s $= 56.84$ kW (1 ton = 1000 kg)

If m_r is the mass flow of refrigerant through the system, then

$$m_r\,(h_1 - h_5) = 56.84\ ;\quad m_r = \frac{56.84}{345 - 244} = \mathbf{0.563}\ \text{kg/s}$$

(*b*) Power required to run the plant,

= work input to compressor

$= m_r\,(h_2 - h_1) = 0.563\,(377 - 345) = 18.02$ kJ/s = **18.02** kW

(*c*) Displacement volume = mass × specific volume

$= 0.563 \times 0.09 = 0.0507$ m^3/s $= 3.042$ m^3/min

This is equal to $\frac{\pi}{4} d^2 l\, N\, n\, \eta_v$ where *d* is the diameter, *l* is the stroke, *N* is rev/min, *n* is the number of cylinders of the compressor and η_v is the volumetric efficiency. As such

$$\frac{\pi}{4} d^2 \times d \times 1200 \times 1 \times 0.9 = 3.042\ ;\ d^3 = 3.588 \times 10^{-3}$$

∴ Diameter of cylinder, d = 0.153 m = **153** mm

and Length of cylinder, l = d = **153** mm

(*d*) Coefficient of performance of the system,

$$COP = \frac{\text{refrigeration capacity}}{\text{work input to compressor}} = \frac{56.84}{18.02} = \mathbf{3.15}$$

The coefficient of performance of a refrigerator working on reversed Carnot cycle is given by

$$COP = \frac{T_e}{T_c - T_e} = \frac{-15+273}{(45+273)-(-15+273)} = \mathbf{4.3}$$

EXAMPLE 5.12

The compressor unit of a simple vapour compression refrigeration unit operates within pressure limits of 2 bar (evaporator) and 12 bar (condenser). The compressor is single acting with 4-cylinders, 25 cm diameter × 30 cm stroke length, clearance factor 0.02 and it turns 360 rev/min. The vapours from the evaporator to the suction of compressor is dry saturated and there is no subcooling in the condenser. Taking ammonia as refrigerant, determine refrigeration capacity and power required to run the compressor if the compression process is isentropic with γ = 1.31.

(b) If the compression is non-isentropic with n = 1.2, determine power required to operatre the compressor, heat rejected to compressor jacket cooling water and cooling load on the condenser.

Solution: Refer Fig. 5.28 for representation of the given cycle on *p-h* plot and read there from the following values for the theoretical cycle 1-2-3-4

$$h_1 = 1439 \text{ kJ/kg} \; ; \quad h_2 = 1705 \text{ kJ/kg}$$

$$h_5 = h_4 = 347 \text{ kJ/kg} \; ; \quad v_1 = 0.595 \text{ m}^3\text{/kg}$$

The volumetric efficiency of the compressor is

$$= 1 - C\left[\left(\frac{p_c}{p_e}\right)^{1/\gamma} - 1\right] = 1 - 0.02\left[\left(\frac{12}{2}\right)^{1/1.31} - 1\right] = 0.941$$

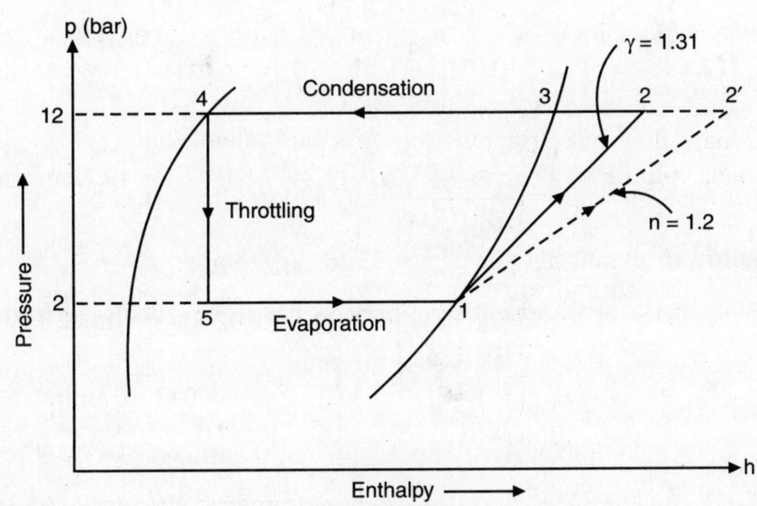

Fig. 5.28

Displacement volume of the compressor

$$= \text{number of cylinders} \times \frac{\pi}{4} d^2 l\, N\, \eta_v$$

$$= 4\left[\frac{\pi}{4}\times 0.25^2 \times 0.3\times 360\right]\times 0.941 = 19.94 \text{ m}^3/\text{min}$$

Mass flow of refrigerant through the system

$$m_r = \frac{\text{displacement volume}}{\text{specific volume}} = \frac{19.94}{0.595} = 33.51 \text{ kg/min}$$

∴ Refrigerating capacity of the system

$$= m_r\,(h_1 - h_5)$$

$$= 33.51\,(1439 - 347) = 36593 \text{ kJ/min} = \frac{36593}{210} = \mathbf{174.25} \text{ ton}$$

Power required to run the compressor

$$= m_r\,(h_2 - h_1)$$

$$= 33.51\,(1705 - 1439) = 8913.67 \text{ kJ/min} = \frac{8913.67}{60} = \mathbf{148.56} \text{ kJ/s or kW}$$

Note: The enthalpy increase $(h_2 - h_1)$ can also be calculated from the relation

$$h_2 - h_1 = \frac{\gamma}{\gamma-1} p_1 v_1 \left[\left(\frac{p_2}{p_1}\right)^{\frac{\gamma-1}{\gamma}} - 1\right]$$

$$= \frac{1.31}{1.31-1}\times 2\times 10^5 \times 0.595\left[\left(\frac{12}{2}\right)^{\frac{1.31-1}{1.31}} - 1\right]$$

$$= 265493 \text{ Nm/kg} = 265.49 \text{ kJ/kg}$$

(*b*) $$\eta_v = 1 - C\left[\left(\frac{p_c}{p_e}\right)^{1/n} - 1\right] = 1 - 0.02\left[\left(\frac{12}{2}\right)^{1/1.2} - 1\right] = 0.931$$

$$\text{Displacement volume} = 4\times\left[\frac{\pi}{4}\times 0.25^2 \times 0.3\times 360\right]\times 0.931 = 19.72 \text{ m}^3/\text{min}$$

$$\text{Mass flow of refrigerant } m_r = \frac{19.72}{0.595} = 33.16 \text{ kg/min}$$

The compression work for non-isentropic process has to be calculated from the relation,

$$W_c = \frac{n}{n-1} p_1 v_1 \left[\left(\frac{p_2}{p_1}\right)^{\frac{n-1}{n}} - 1\right]$$

$$= \frac{1.2}{1.2-1}\times 2\times 10^5 \times 0.595\left[\left(\frac{12}{2}\right)^{\frac{1.2-1}{1.2}} - 1\right]$$

$$= 248533 \text{ Nm/kg} = 248.53 \text{ kJ/kg}$$

Then, power required to run the compressor

$$= m_r \times W_c = 33.16 \times 248.53 = 8241.25 \text{ kJ/min}$$

$$= \frac{8241.25}{60} = \mathbf{137.35} \text{ kJ/s or kW}$$

The specific volume at point 2′ can be worked out from the relation $pv^{1.2}$ = constant. That is

$$v_2' = v_1\left(\frac{p_1}{p_2}\right)^{\frac{1}{1.2}} = 0.595\times\left(\frac{1}{6}\right)^{0.833} = 0.1337 \text{ m}^3/\text{kg}$$

Corresponding to specific volume 0.1337 m³/kg and pressure 12 bar, the enthalpy value as read from *p-h* plot is

$$h_2' = 1560 \text{ kJ/kg}$$

From steady flow energy equation (neglecting kinetic energy and potential energy changes),

$$h_1 + \text{work input} = h_2' + \text{heat rejected}$$

∴ Heat rejected to compressor jacket cooling water

$$= \text{work input} - (h_2' - h_1)$$
$$= 248.53 - (.560 - 1439) = 127.53 \text{ kJ/kg}$$
$$= 33.16 \times 127.53 = \mathbf{4229} \text{ kJ/min}$$

Cooling load on condenser

= heat removed in the condenser

$$= m_r\,(h_2' - h_3) = 33.16\,(1560 - 347) = \mathbf{40233} \text{ kJ/min}$$

EXAMPLE 5.13

An ideal vapour compression refrigerator using Freon-12 operates between temperature limits of –10°C and 40°C. The refrigerant leaves the condenser dry saturated and the rate of flow of refrigerant through the unit is 150 kg/hr. Calculate the refrigerant effect per kg of refrigerant, the coefficient of performance and the capacity of plant if:

(i) the refrigerant leaves the condenser dry saturated as stated,

(ii) the refrigerant is subcooled to 20°C before throttling. Assume that the enthalpy of refrigerant before throttling is approximately equal to the enthalpy of liquid refrigerant at undercooled temperature of 20°C and take it as h_f = 50.59 kJ/kg.

Give your comments on the results **(Engg. Services)**

Solution : Refer Fig. 5.30 for the *T-s* and *p-h* plots of the refrigeration system;

$$T_3 = 40\text{°C} = 313 \text{ K}$$

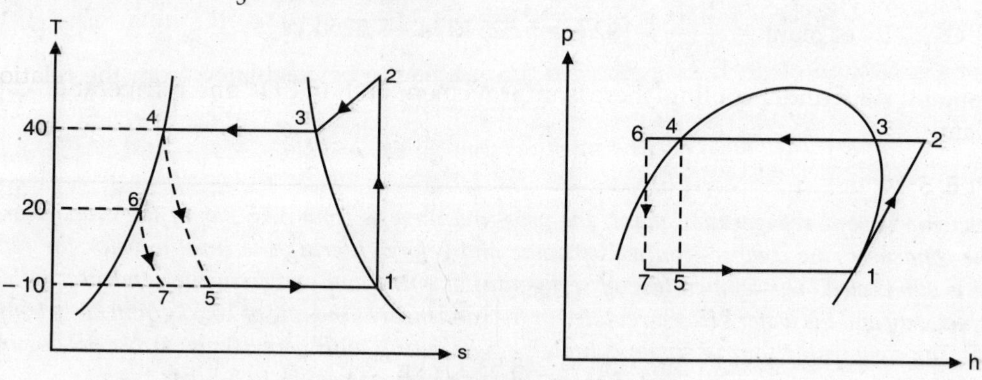

Fig. 5.29

From table of properties for Freon – 12,

$$h_1 = 183.19 \text{ kJ/kg} \quad ; \quad s_1 = 0.719 \text{ kJ/kg K} \qquad (\text{At } t = -10°\text{C})$$
$$h_3 = 203.20 \text{ kJ/kg} \quad ; \quad s_3 = 0.6825 \text{ kJ/kg K} \qquad (\text{At} = 40°\text{C})$$

Average specific heat = 0.609 kJ/kg K

Since the compression process is 1 – 2 is isentropic, $s_1 = s_2$ and $s_2 = s_3 + c_p \log_e \frac{T_2}{T_3}$

$$\therefore \quad 0.7019 = 0.6825 + 0.609 \log_e \frac{T_2}{313} \; ; \quad \text{Solution gives : } T_2 = 323.13 \text{ K}$$

Enthalpy at state point 2,

$$h_2 = h_3 + c_p (T_2 - T_3)$$
$$= 203.20 + 0.609 (323.13 - 313) = 209.38 \text{ kJ/kg}$$

h_4 = liquid enthalpy at 40°C as read from tables = 74.59 kJ/kg

The process 4-5 is of throttling during which enthalpy remains constant and therefore,

$$h_5 = h_4 = 74.59 \text{ kJ/kg}$$

Refrigerating effect $N = h_1 - h_5 = 183.19 - 74.59 = 108.6$ kJ/kg

Work spent during compression $W = h_2 - h_1 = 209.38 - 183.19 = 26.19$ kJ/kg

$$\therefore \text{ Coefficient of performance} = \frac{N}{W} = \frac{108.6}{26.19} = \mathbf{4.15}$$

Refrigerating capacity of the plant

$$= m_r \times \text{refrigerating effect per unit kg}$$
$$= \frac{150}{3600} \times 108.6 = 1.525 \text{ kJ/s} = \mathbf{4.525 \text{ kW}}$$

(*b*) With under cooling to state point 6; $h_6 = 50.59$ kJ/kg.

The enthalpy remains constant during the throttling process 6 – 7 and therefore,

$$h_7 = h_6 = 50.59 \text{ kJ/kg}$$

Refrigerating effect $N = h_1 - h_7 = 183.17 - 50.59 = 132.6$ kJ/kg

Compression work $W = h_2 - h_1 = 209.38 - 183.19 = 26.19$ kJ/kg

$$COP = \frac{N}{W} = \frac{132.6}{26.19} = 5.06$$

$$\text{Capacity of plant} = \frac{150}{3600} \times 132.6 = 5.525 \text{ kJ/s} = 5.525 \text{ kW}$$

Comment: With under cooling, there is improvement both in *COP* and refrigeration capacity of the plant.

EXAMPLE 5.14

In an ammonia vapour refrigeration plant, the pressure range is from 3.15 bar in the evaporator to 10.5 bar in the condenser, the compression is isentropic and before entering the throttle valve the refrigerant ammonia is sub-cooled. The temperature of refrigerant at entry and exit from the condenser is 50°C and 20°C respectively and the water being circulated in the condenser at the rate of 10.5 kg/min has a temperature rise 10°C. The compressor unit is single cylinder, single acting with bore 10 cm, stroke 15 cm and while running at 200 rev/min has an indicated mean effective pressure equal to 3.5 bar.

If the plant forms 50 kg/hr of ice at 0°C from water at 15°C, determine :
(a) COP (b) flow rate of refrigerant and (c) condition of vapour at entry to compressor.
Take latent heat of ice as 335 kJ/kg. The relevant properties of NH_3 are as given below :

Pressure bar	*Saturation temperature, K*	*Enthalpy kJ/kg*		*Specific heat kJ/kg K*	
		Liquid	*Vapour*	*Liquid*	*Vapour*
3.5	264	– 35.6	1264	–	–
10.5	304	134	1294	4.6	2.8

Solution : Refer Fig. 5.30 for *T-s* plot of cycle. Since the temperature of refrigerant at compressor exit (condenser inlet) T_2 = 273 + 50 = 323 K is higher than the saturation temperature of 304 K corresponding to compressor exit pressure of 10.5 bar, the refrigerant at compressor exit would be superheated.

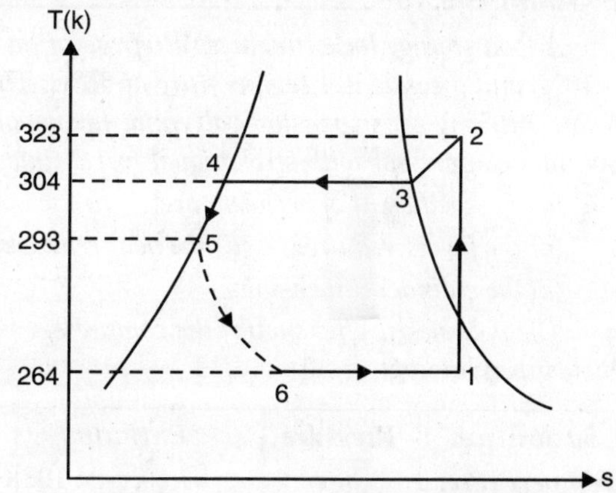

Fig. 5.30.

Compressor work = p_m *LAN*

$$= (3.5\times10^5)\times0.15\times\frac{\pi}{4}(0.1)^2\times200$$

$$= 82420 \text{ J/min} = \mathbf{82.42\ kJ/min}$$

Heat extracted for ice formation (refrigerating effect)

$$= (4.186 \times 15) + 335$$

$$= 397.8 \text{ kJ/kg of ice formed}$$

$$= 397.8 \times \frac{50}{60} = 331.5 \text{ kJ/min}$$

(*a*) Coefficient of performance, *COP*

$$= \frac{\text{refrigerating effect}}{\text{work input}} = \frac{331.5}{82.42} = \mathbf{4.02}$$

(*b*) Let *m* be the mass flow of refrigerant per minute. Then energy balance in the condenser gives :

heat lost by refrigerant = heat gained by condenser cooling water

$$m\,(h_2 - h_5) = m_w\, c_{pw}\, \Delta t$$

From the given data :

$$h_4 - h_5 = (c_p)_l \times (T_4 - T_5) = 4.6(304 - 293) = 50.6 \text{ kJ/kg}$$

$$h_2 - h_3 = (c_p)_v \times (T_2 - T_3) = 2.8(323 - 304) = 52.2 \text{ kJ/kg}$$

$$h_3 - h_4 = 1294 - 134 = 1160 \text{ kJ/kg}$$

Upon adding : $h_2 - h_5 = 50.6 + 53.2 + 1160 = 1263.8$ kJ/kg

∴ Mass flow rate of refrigerant $m = \dfrac{m_w c p_w \Delta t}{h_2 - h_5} = \dfrac{10.5\times4.186\times10}{1263.8}$ = **0.348 kg/min**

(*c*) Refrigerating effect $(h_1 - h_6)$ = 331.5 kJ/min = $\dfrac{331.5}{0.348}$ = 952.6 kJ/kg of refrigerant

During throttling operation 5-6, the enthalpy remains constant

$$\therefore \quad h_6 = h_5 = h_4 - (h_4 - h_5)$$
$$= 134 - 50.6 = 83.4 \text{ kJ/kg}$$

That gives : $h_1 = 952.6 + 83.4 = 1036$ kJ/kg

Further
$$h_1 = h_{f1} + x_1 h_{fg1}$$
$$= -35.6 + x_1 [1264 - (-35.6)] = -35.6 + 1229.6\, x_1$$

$$\therefore \quad 1036 = -35.6 + 1229.6\, x_1$$

That gives
$$x_1 = \frac{1036 + 35.6}{1229.6} = \mathbf{0.87}$$

Thus the refrigerant at entry to the compressor has a dryness fraction of 0.87.

EXAMPLE 5.15

A food storage locker requires a refrigeration system of 50 kW capacity at an evaporator temperature of – 10°C and a condenser temperature of 35°C. The refrigerant used is ammonia which is subcooled by 5°C before entering the expansion valve and the vapour is dry saturated before leaving the evaporator. The two cylinder compressor with stroke equal to 1.2 times the bore operates at 1000 rev/min. Determine :

(a) the coefficient of performance, (b) the mass of refrigerant to be circulated per minute,

(c) the power required, (d) the heat removed through condenser, and

(e) the cylinder dimensions.

The volumetric efficiency of the compressor is 80 percent. You may use the extract of ammonia property table given below :

Saturation temperature, °C	*Pressure bar*	*Enthalpy kJ/kg*		*Entropy kJ/kg K*		*Specific volume m³/kg*		*Specific heat kJ/kg K*	
		Liquid	*Vapour*	*Liquid*	*Vapour*	*Liquid*	*Vapour*	*Liquid*	*Vapour*
– 10	*2.91*	*154*	*1450*	*0.83*	*5.75*	—	*0.42*	—	*2.49*
35	*13.5*	*366*	*1489*	*1.56*	*5.21*	*1.7*	*0.096*	*4.56*	*2.90*

Solution: Refer Fig. 5.31, for *T-s* plot of the given cycle

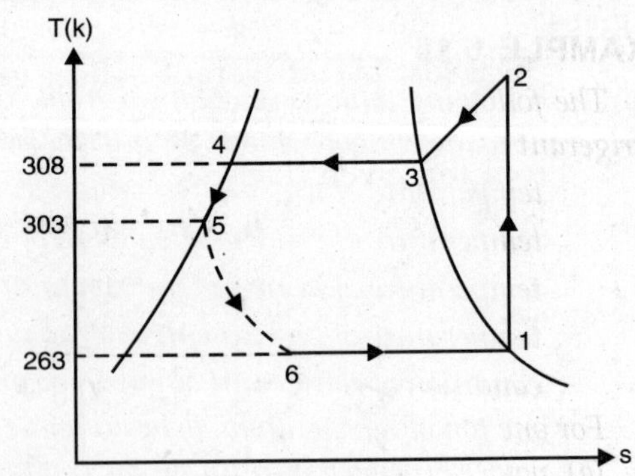

Fig. 5.31.

$$T_1 = T_6 = -10°\text{C} = 263 \text{ K}$$
$$T_3 = T_4 = 35°\text{C} = 308 \text{ K}$$
$$T_5 = 308 - 5 = 303 \text{ K}$$
$$h_1 = 1450 \text{ kJ/kg}$$
$$h_3 = 1489 \text{ kJ/kg}$$
$$h_4 = 366 \text{ kJ/kg}$$
$$h_5 (= h_6) = h_4 - c_{pl} \times (T_4 - T_5)$$
$$= 366 - 4.56 (308 - 303)$$
$$= 366 - 22.8 = 343.2 \text{ kJ/kg}$$

Since the compression process 1-2 is isentropic, we have

$$s_1 = s_2 \quad \text{or} \quad 5.75 = s_3 + c_p \log_e \left(\frac{T_2}{T_3}\right) = 5.21 + 2.9 \log_e \frac{T_2}{308}$$

$$\text{or } \log_e\left(\frac{T_2}{308}\right) = \frac{5.75 - 5.21}{2.9} = 0.1862 \quad \text{or} \quad \frac{T_2}{308} = e^{0.1862} = 1.2047$$

∴ Temperature at exit from the compressor, $T_2 = 308 \times 1.2047 = 371$ K

Now, $h_2 = h_3 + c_{pv} \times (T_2 - T_3) = 1489 + 2.90\ (371 - 308) = 1671.7$ kJ/kg

(*a*) Refrigerating effect, $N = h_1 - h_6 = 1450 - 343.2 = 1106.8$ kJ/kg

Work input, $W = h_2 - h_1 = 1671.7 - 1450 = 221.7$ kJ/kg

$$\text{Coefficient of performance, } COP = \frac{N}{W} = \frac{1106.8}{221.7} = \mathbf{4.99}$$

$$\text{(b) Mass of refrigerant } m = \frac{\text{total refrigeration effect}}{\text{refrigeration effect per kg of refrigerant}}$$

$$= \frac{50}{1106.8} = \mathbf{0.04517\ kg/s}$$

(*c*) Power required = 221.7 kJ/kg = 221.7 × 0.04517 = 10.01 kJ/s = **10.01 kW**

(*d*) Heat removed through condenser = $m\ (h_2 - h_5) = 0.04517\ (1671.7 - 343.2)$ = **60 kJ/s**

(*e*) Volume flow rate = m × specific volume of vapour at entry to compressor

= 0.04517 × 0.42 = 0.01897 m^3/s

$$\text{Compressor displacement} = \frac{0.01897}{0.8} = 0.0237\ \text{m}^3/\text{s} = 1.422\ \text{m}^3/\text{min}$$

This is equal to $\frac{\pi}{4} d^2\, l\, N\, n$ where d is the diameter, l is the stroke, N is rev/min and n is the number of cylinders of the compressor. As such

$$\frac{\pi}{4} d^2 \times 1.2d \times 1000 \times 2 = 1.422\,; \quad d^3 = 7.548 \times 10^{-4}$$

∴ Diameter of cylinder, d = 0.09127 m = **91.27 mm**

and Length of cylinder, l = 1.2 × 91.27 = **109.52 mm**

EXAMPLE 5.16

The following data were obtained from test on a 20 ton ice making refrigerating plant with NH_3 as refrigerant using chilled brine as an intermediate circuit.

temperature of water entering and leaving the condenser = 20°C and 27°C
temperature of condensation of NH_3 = 25°C
temperature of brine in the evaporator = – 15°C
temperature of NH_3 at entry to expansion valve = 20°C
condition of refrigerant at entry to compressor = dry saturated

For one ton of refrigeration, determine

(a) power expended in running the compressor,
(b) amount of cooling water in the condenser, and
(c) COP of the plant.

How does COP of this cycle compare with that of a reversed Carnot cycle operating between the same evaporator and condenser temperatures.

The relevant properties of NH_3 are given below :

Unit	*Saturation temperature*	*Enthalpy kJ/kg*		*Entropy kJ/kg K*		*Specific heat kJ/kg K*	
	°C	*Liquid*	*Vapour*	*Liquid*	*Vapour*	*Liquid*	*Vapour*
Evaporator	– 15	350	1663	3.82	9.01	4.39	2.30
Condenser	25	536	1703	4.59	8.51	4.65	2.80

Solution : Refer Fig. 5.32, for *T-s* plot of the given cycle.

$$T_1 = T_6 = -15°C = 258\text{ K}$$

$$T_3 = T_4 = 25°C = 298\text{ K}$$

$$T_5 = 20°C = 293\text{ K}$$

$$h_1 = 1663\text{ kJ/kg}$$

$$h_3 = 1703\text{ kJ/kg}$$

$$h_4 = 536\text{ kJ/kg}$$

$$h_5\ (= h_6) = h_4 - c_{pl} \times (T_4 - T_5)$$

$$= 536 - 4.65 \times (298 - 293)$$

$$= 512.75\text{ kJ/kg}$$

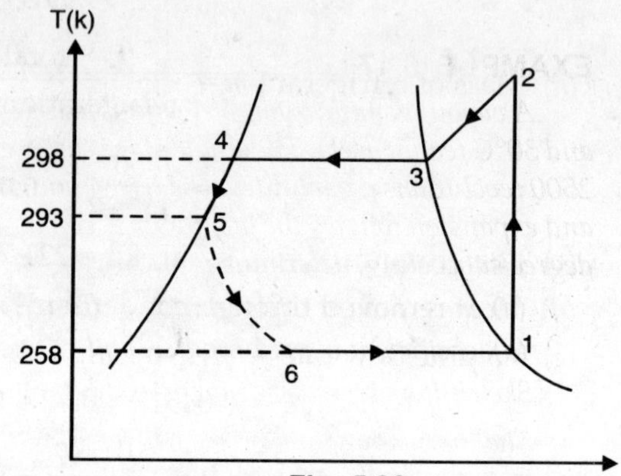

Fig. 5.32.

Since the compression process 1-2 is adiabatic

entropy at 1 = entropy at 2

$$9.01 = 8.51 + 2.80 \log_e \frac{T_2}{298}$$

or $\quad \log_e \dfrac{T_2}{298} = \dfrac{9.01-8.51}{2.80} = 0.17857 \quad$ or $\quad \dfrac{T_2}{2.98} = e^{0.17857} = 1.1955$

∴ Temperature at exit from the compressor, $T_2 = 298 \times 1.1955 = 356$ K

Now, $h_2 = h_3 + c_{pv}(T_2 - T_3) = 1703 + 2.8\,(356 - 298) = 1865.4$ kJ/kg

(*a*) Work done $W = h_2 - h_1 = 1865.4 - 1663 = 202.4$ kJ/kg

Refrigerating effect $N = h_1 - h_6 = 1663 - 512.75 = 1150.25$ kJ/kg

Coefficient of performance $COP = \dfrac{N}{W} = \dfrac{1150.25}{202.4} = \mathbf{5.68}$

(*b*) Let m_r be the refrigerant flow rate per ton of refrigeration. Then

$$m_r \times 1150.25 = 210\ ; \qquad m_r = 0.1826\text{ kg/min}$$

∴ Power input per ton of refrigeration

$$= 202.4 \times 0.1826 = 36.95\text{ kJ/min}$$

$$= \frac{36.95}{60} = 0.616\text{ kJ/s} = \mathbf{0.616\ kW}$$

(*c*) Energy balance in the condenser gives

heat lost by refrigerant = heat gained by condenser cooling water

$$\dot{m}_r (h_2 - h_5) = m_w\, c_p\, \Delta t$$

$$0.1826\ (1865.4 - 512.75) = m_w \times 4.186 \times (27 - 20)$$

∴ Mass flow rate of cooling water in the condenser

$$= \frac{0.1826(1865.4 - 512.75)}{4.186 \times (27 - 20)} = 8.43 \text{ kg/min}$$

Thus the amount of cooling water is 8.43 kg/min or 8.43 kg per ton of refrigeration.

(*d*) *COP* of reversed Carnot cycle

$$= \frac{T_e}{T_h - T_e} = \frac{258}{298 - 258} = \mathbf{6.45}$$

EXAMPLE 5.17

A vapour compression refrigerator uses R-134a and works in the temperature range of –10°C (evaporator) and 30°C (condenser). The single stage compressor has clearance ratio 0.03, swept volume 265 cm³, and turns 2500 revolutions per minutes. The refrigerant at suction in is dry saturated, the compression process is isentropic and expansion follows the law $pv^{1.12}$ = constant. Presuming that the refrigerant leaves the condenser with 5-degree subcooling, determine :

(a) mass flow of refrigerant, (b) refrigeration capacity of plant, (c) power input to compressor,
(d) coefficient of performance, and (e) condensor heat rejection.

Sketch the given cycle on T-s and p-h plots.

The thermodynamic properties for R-134a are given below :

t °C	p MPa	v_g m³/kg	h_f kJ/kg	h_g kJ/kg	s_f kJ/kg K	s_g kJ/kg K
– 10	0.202	0.0992	186.72	392.28	0.951	1.732
30	0.771	0.0267	241.79	415.07	1.144	1.715

Solution : Refer Fig. 5.33 For *T-s* and *p-h* plot the given system :

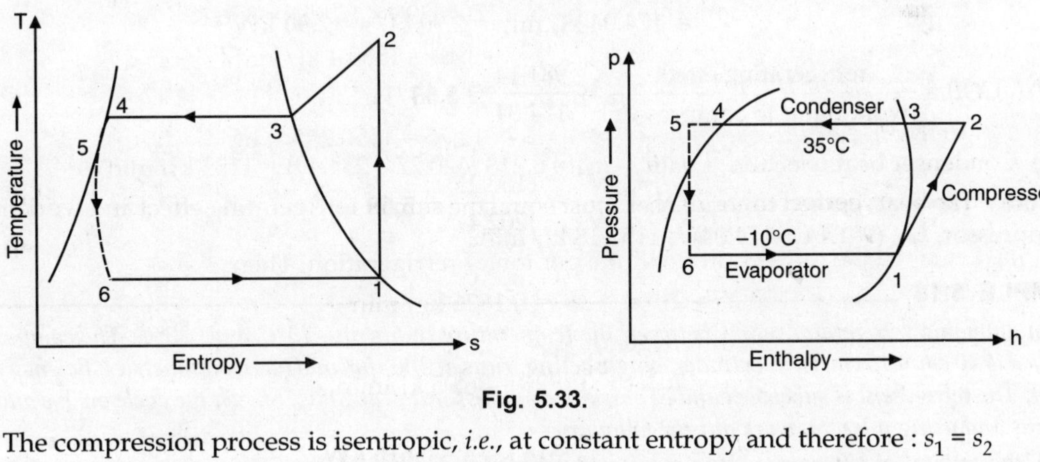

Fig. 5.33.

The compression process is isentropic, *i.e.*, at constant entropy and therefore : $s_1 = s_2$

$$1.732 = 1.715 + 1.15 \log_e \frac{T_2}{30 + 273}$$

$$\therefore \quad T_2 = 303 \times \text{antilog}\left(\frac{1.732 - 1.715}{1.15}\right) = 307.52 \text{ K}$$

$$h_2 = h_3 + c_p (T_2 - T_3) = 415.07 + 1.15 (307.52 - 303) = 420.27 \text{ kJ/kg}$$
$$h_5 = h_4 - c_{pl} (T_4 - T_5) = 241.79 - 1.46 (303 - 298) = 234.49 \text{ kJ/kg}$$
$$h_6 = h_5 = 234.49 \text{ kJ/kg}$$ ($\because$ Enthalpy remains constant during throttling process 5-6)

(a) Volumetric efficiency $\eta_v = 1 + C - C\left(\frac{p_2}{p_1}\right)^{1/4}$

$$= 1 + 0.03 - 0.03\left(\frac{0.771}{0.202}\right)^{1/1.12} = 0.931$$

$$\text{Mass of air circulated} = \frac{V_s \eta_v N n}{\text{specific volume}}$$

where V_s is the swept volume per cycle, N is speed of compressor and $n = 1$ for single acting compresssor.

$$\therefore \quad m = \frac{(265\times10^{-6})\times0.931\times2500\times1}{0.0992} = \mathbf{6.218 \text{ kg/min}}$$

(b) Refrigerating effect $= m(h_1 - h_6)$

$$= 6.218 (392.28 - 234.49) = 981.14 \text{ kJ/min}$$

1 ton of refrigeration = 3.5 kJ/s = 210 kJ/min

$$\therefore \text{ Refrigeration capacity of plant} = \frac{981.14}{210} = \mathbf{4.67 \text{ ton}}$$

(c) Work input to compressor $= m(h_2 - h_1)$

$$= 6.218 (420.27 - 392.28)$$
$$= 174.04 \text{ kJ/min} = 2.90 \text{ kJ/s} = \mathbf{2.90 \text{ kW}}$$

(d) $$COP = \frac{\text{refrigerating effect}}{\text{work input to compressor}} = \frac{981.14}{174.04} = \mathbf{5.63}$$

(e) Condenser heat rejection $= m(h_2 - h_5) = 6.218 (420.27 - 234.49) = 1155$ kJ/min

Note : The heat rejected to condenser must equal the sum of refrigerating effect and work input to compressor, *i.e.*, (981.14 + 174.04) = 1155.18 kJ/min.

EXAMPLE 5.18

An ammonia refrigerator works between the temperature limits of – 15°C and 30°C. The compressor employed is 20 cm × 15 cm, twin cylinder, single acting, runs at 300 rpm and has a volumetric efficiency of 80 percent. The refrigerant is subcooled and at entry to compressor it is at 21°C. Sketch the cycle on T-s and p-h diagrams and using p-h plot, work out the followings :

(i) the amount of refrigerant being circulated per minute,

(ii) the ton of refrigeration and COP of the machine.

It may be assumed that vapours leaving the evaporator are dry saturated.

Solution : Refer Fig. 5.35 for the *T-s* and *p-h* plots for the given refrigerating system :

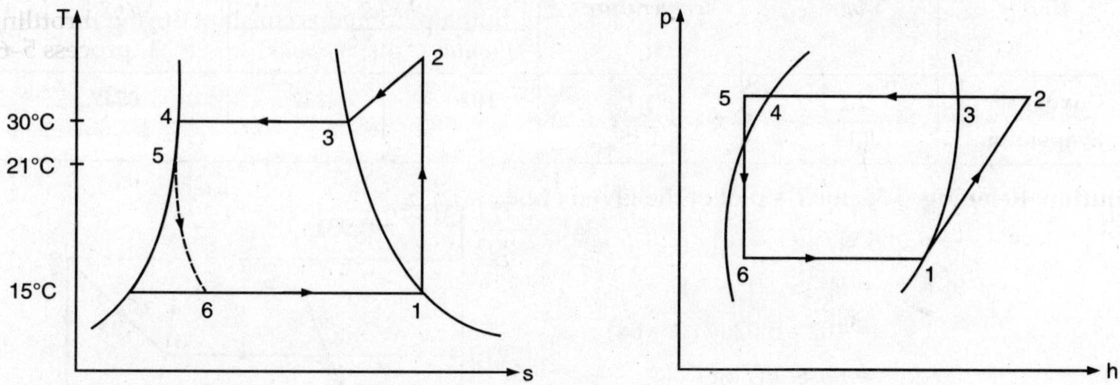

Fig. 5.34

The following property values can be read from the *p-h* chart when the given cycle is laid on it.

$$h_1 = 1665 \text{ kJ/kg}; \quad h_2 = 1900 \text{ kJ/kg}$$

$$h_6 = h_5 = 430 \text{ kJ/kg} \quad \text{and} \quad v_1 = 0.51 \text{ m}^3/\text{kg}$$

(*i*) Displacement volume $= \left(\frac{\pi}{4} d^2 l \times N\right) \times \eta_v \times$ number of cylinders

$$= \left[\frac{\pi}{4}(0.2)^2 \times 0.15 \times 300\right] \times 0.8 \times 2 = 2.261 \text{ m}^3 \text{ min}$$

Mass of vapour circulated $= \dfrac{2.261}{0.51} =$ **4.433 kg/min**

(*ii*) Refrigerating effect $= m_r(h_1 - h_6) = 4.433\ (1665 - 430) = 5474.75$ kJ/min

1 ton of refrigerant = 210 kJ/min

$\therefore$ Capacity of the refrigerator $= \dfrac{5474.75}{210} = 26.07$ ton

(*iii*) Work input to compressor $= m_r(h_2 - h_1)$

$= 4.433\ (1900 - 1665) = 1041.75$ kJ/min

$$COP = \frac{\text{refrigerating effect}}{\text{work input}} = \frac{5474.75}{1041.75} = \mathbf{5.255}$$

EXAMPLE 5.19

A heat pump using Freon as the working fluid is used to carry a part of the heating load of a factory. The heat pump raises the temperature of 6000 litres of circulating water from 35°C to 46°C. The unit operates between evaporator and condenser pressure limits of 0.14 bar and 2.1 bar. The vapour is in the superheated state at entry to the compressor and has an enthalpy 205 kJ/kg and entropy equal to 0.75 kJ/kg K. After adiabatic compression, the refrigerant Freon attains a temperature of 73.5°C. After condensation, refrigerant enters the throttle valve at 68°C and the vapour gets throttled to evaporator pressure of 0.14. Sketch the cycle on T-s plot and calculate the coefficient of performance and the power input required to run the unit.

Take specific heat of Freon liquid = 5.02 kJ/kg K and specific heat of superheated vapour at 2.1 bar = 1.3 kJ/kg K. Further the relevant properties of Freon are as given below :

Unit	Pressure bar	Saturation temperature °C	Enthalpy kJ/kg		Entropy of vapour kJ/kg K
			Liquid	Vapour	
Condenser	2.1	71.1	102	237	0747
Evaporator	0.14	- 1.15	—	—	

Solution: Refer Fig. 5.35, for *T-s* plot of the given cycle.

$$h_1 = 205 \text{ kJ/kg}$$

$$h_6 = h_5 = h_4 - c_{pl}(t_4 - t_5)$$

$$= 102 - 5.02\,(71.1 - 68)$$

$$= 86.44 \text{ kJ/kg}$$

$$h_2 = h_3 + c_{pv}(t_2 - t_3)$$

$$= 237 + 1.3\,(73.5 - 71.1)$$

$$= 240.12 \text{ kJ/kg}$$

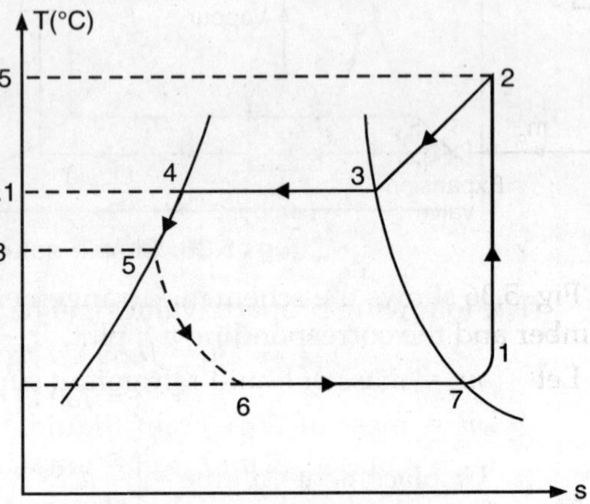

Fig. 5.35.

(*a*) Work done $W = h_2 - h_1 = 240.12 - 205$

$$= 35.12 \text{ kJ/kg}$$

Heat given by refrigerant to water in the condenser,

$$(h_2 - h_5) = 240.12 - 86.44 = 153.68 \text{ kJ/kg}$$

$$\therefore \quad COP \text{ of heat pump} = \frac{153.68}{35.12} = \mathbf{4.376}$$

(*b*) If m_r is the mass flow of refrigerant per second, then energy balance in the condenser gives :

heat given by refrigerant = heat taken by water

$$m_r \times 153.68 = \frac{6000}{3600} \times 4.186 \times (46 - 35)\ ;\ m_r = 0.4988 \text{ kg/s}$$

∴ Power input to compressor = 35.12 × 0.4988 = 17.518 kJ/s = **17.518 kW**

5.8. SIMPLE SATURATION CYCLE WITH FLASH CHAMBERS

When high pressure liquid refrigerant from the condenser passes through the expansion valve, it undergoes partial evaporation. The vapour thus formed is called ***flash*** and the process of partial evaporation of liquid is called ***flashing*** or ***flash process***. The flash vapour would not make much contribution towards producing cooling effect in the evaporator as liquid absorbs maximum heat. Thus it is better for the flash vapour to by-pass the evaporator and be directly supplied to the compressor suction. To accomplish this task, a flash chamber is introduced between the expansion valve and the evaporator. The flash chamber is essentially an insulated container which separates the liquid and vapour by centrifugal effect. The liquid refrigerant from the flash chamber is supplied to the evaporator and the vapour passes directly to the compressor. Obviously, there is decrease in the mass of refrigerant flowing through the evaporator and hence a reduction in its size.

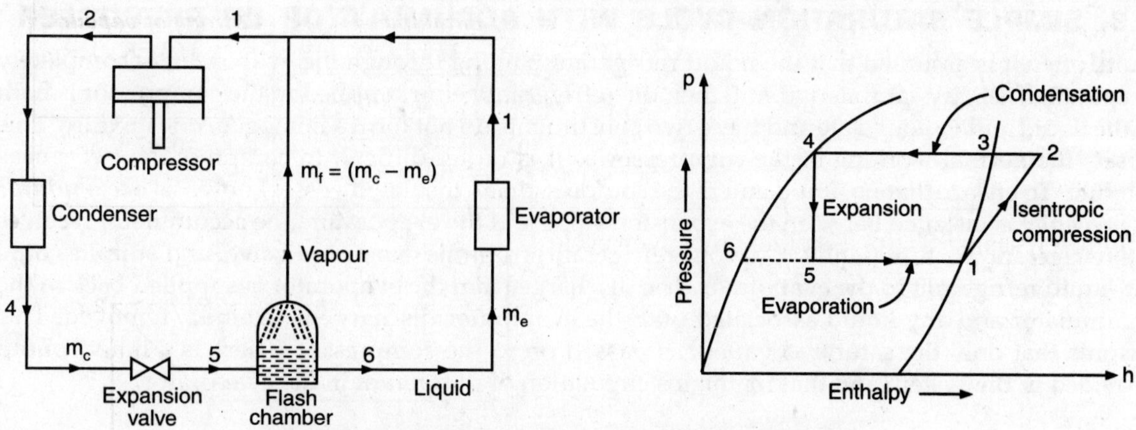

Fig. 5.36. Simple saturation cycle with flash chamber

Fig. 5.36 shows the schematic arrangement of the simple vapour compression cycle with flash chamber and the corresponding *p-h* plot.

Let m_e = mass of liquid refrigerant supplied to the evaporator

m_c = mass of refrigerant (liquid + vapour) which circulates through the compressor, condenser and expansion valve.

Then the mass of the refrigerant vapour which flows directly from the flash chamber to the compressor,

$$m_f = m_c - m_e$$

Since the flash chamber is insulated, it does not exchange any heat with the surroundings. Accordingly at the state of thermal equilibrium,

heat taken by the flash chamber = heat given out by the flash chamber

$$m_c h_5 = m_e h_6 + m_f h_1$$

$$= m_e h_6 + (m_c - m_e)\, h_1$$

or $$m_c\,(h_1 - h_5) = m_e\,(h_1 - h_6)$$

$\therefore$ $$m_e = m_c \left(\frac{h_1 - h_5}{h_1 - h_6}\right)$$

Heat absorbed in the evaporator, *i.e.* refrigerating effect

$$= m_e\,(h_1 - h_6)$$

$$= m_c \left(\frac{h_1 - h_5}{h_1 - h_6}\right) \times (h_1 - h_6) = m_c\,(h_1 - h_5)$$

Work input to compressor $= m_c\,(h_2 - h_1)$

$$\text{Coefficient of performance } COP = \frac{\text{refrigerant effect}}{\text{work input to compressor}} = \frac{m_c(h_1 - h_5)}{m_c(h_2 - h_1)} = \frac{h_2 - h_5}{h_2 - h_1}$$

From the expressions set up above, it is evident that the refrigerating effect, work input to compressor and *COP* when flash chamber is used are the same as that for a simple saturation cycle. However, there is reduction in the mass of refrigerant flowing through the evaporator and hence the reduction in the size of evaporator.

5.9. SIMPLE SATURATION CYCLE WITH ACCUMULATOR OR PRECOOLER

Quite often it is assumed that the liquid refrigerant passing through the evaporator is completely evaporated, *i.e.*, dry saturated. If not, then the refrigerant being supplied to the compressor is both in the liquid and vapour state and these two constituents do not form a homogeneous mixture. This upsets the normal working of the compressor as it is rather difficult to compress the two phase mixture. To ensure that no liquid enters the compressor, an insulated vessel known as ***accumulator*** or ***precooler*** is installed between the expansion valve and the evaporator. The accumulator receives the refrigerant mixture (liquid + vapour refrigerant) from the expansion valve, and supplies only the liquid refrigerant to the evaporator. The discharge from the evaporator is supplied back to the accumulator and any liquid associated with the evaporator discharge is retained/removed. This ensures that only the saturated vapour is passed on to the compressor. There is a liquid pump provided in the system and that maintains circulation of refrigerant in the evaporator.

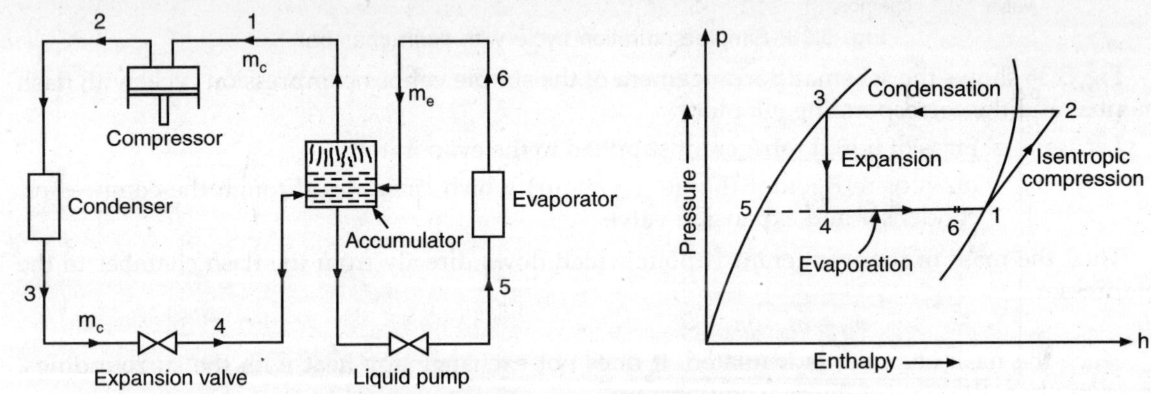

Fig. 5.37. Simple saturation cycle with accumulator

Fig. 5.37 shows the schematic arrangement of the simple vapour compression cycle with accumulator and the corresponding *p-h* plot.

Let m_e = mass of liquid refrigerant circlulating through the evaporator

m_c = mass of refrigerant which passes through the compressor, condenser and expansion valve.

Since the accumulator is insulated, it does not exchange any heat with the surroundings. Accordingly at the state of thermal equilibrium,

heat taken in by the accumulator = heat given out by the accumulator

$$m_c h_4 + m_e h_6 = m_c h_1 + m_e h_5$$

or $$m_e (h_6 - h_5) = m_c (h_1 - h_4)$$

$$\therefore \quad m_e = m_c \left(\frac{h_1 - h_4}{h_6 - h_5} \right)$$

Heat absorbed in the evaporator, *i.e.* refrigerating effect

$$= m_e (h_6 - h_5)$$

$$= m_c \left(\frac{h_1 - h_4}{h_6 - h_5} \right) \times (h_6 - h_5) = m_c (h_1 - h_4)$$

Work input to compressor = $m_c (h_2 - h_1)$

$$\text{Coefficient of performance } COP = \frac{\text{refrigerant effect}}{\text{work input to compressor}} = \frac{m_c(h_1 - h_4)}{m_c(h_2 - h_1)} = \frac{h_1 - h_4}{h_2 - h_1}$$

From the expressions set up above, it is evident that the refrigerating effect, work input to compressor and *COP* when accumulator is used are the same as that for a simple saturation cycle. The accumulator serves only to arrest the liquid refrigerant from flowing into the compressor and accordingly dry compression is always assured.

5.10. LIQUID SUCTION HEAT EXCHANGER

The liquid refrigerant leaves the condenser at a temperature higher than that at which the vapour refrigerant leaves the evaporator. This suggests the use of vapour refrigerant to sub-cool the liquid refrigerant by incorporating a heat exchanger in the circuit. Fig. 5.38 shows the schematic arrangement of such a system and the corresponding *p-h* plot.

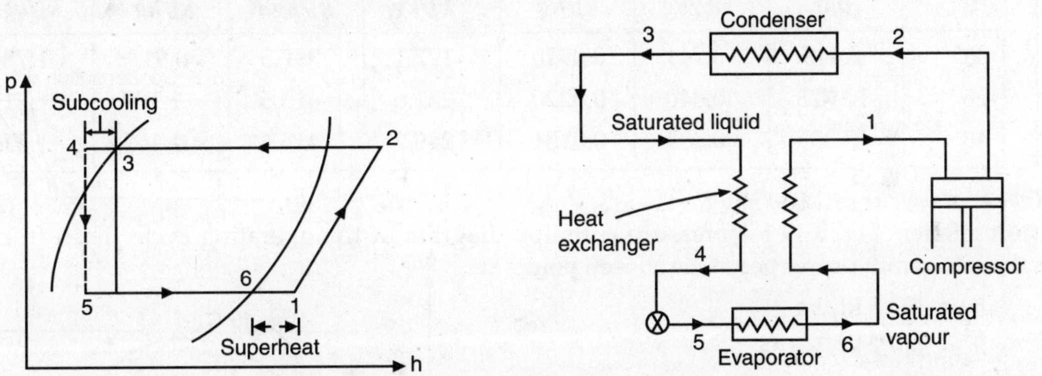

Fig. 5.38. Schematic of liquid heat exchanger

The saturated vapour after evaporator is passed in counter flow direction to the saturated liquid coming form the condenser. The heat exchange causes the liquid to get cooled further or sub-cooled, and consequently the saturated vapour gets superheated.

Assuming no heat loss from heat exchanger,

heat absorbed by vapour in getting superheated = heat rejected by liquid getting subcooled

$$m_v\, c_{pv}\, (T_1 - T_6) = m_l\, c_{pl}\, (T_3 - T_4)$$

The suffix v and l refer to vapour and liquid respectivley.

It is to be noted that mass of liquid and vapour is same ($m_v = m_l$), the specific heats for the liquid and vapour phases are different and as such the degree of supherheat ($T_1 - T_6$) and the degree of subcool ($T_3 - T_4$) would not be same. For the refrigerants used c_{pl} is the more than c_{pv} and therefore

$$(\Delta T) \text{ subcooled} = C\, (\Delta T) \text{ superheated}$$

where $C = \dfrac{c_{pv}}{c_{pi}}$ = 0.6 for *R*-12, 0.5 for *R*-22, and 0.55 for *R*-502

The system, referred to as liquid suction heat exchanger, serves to :

(*i*) sub-cool the liquid and that increases the refrigerating effect

(*ii*) superheat the suction vapour; no liquid droplets enter the compressor and that prevents any corrosive damage to it

(*iii*) increase the compressor work

Since there is increase both in refrigeraint effect and compressor work, the coefficient of performance may increase or decrease or remain same depending on the refrigerant used and the operating conditions. In general, the evaporator capacity can be increased by about 5% by superheating the suction gas and sub-cooling the refrigerant liquid. The suction volume per ton and kW power per ton have also been noted to reduce for R-12.

EXAMPLE 5.20

For a vapour compression system using R-22 as refrigerant, condensor outlet temperature is 40°C and evaporator inlet temperature is –20°C. In order to avoid flashing of refrigerant, a liquid suction vapour heat exchanger is provided where liquid is subcooled to 26°C. The refrigerant leaves the evaporator as saturated vapour. The compression process is isentropic. Find the power requirement and coefficient of performance if the system has a capacity of 10 kW at – 20°C.

The thermodynamic properties of R-22 are as given below:

t °C	p bar	v_f m³/kg	v_g kJ/kg	h_f kJ/kg	h_g kJ/kg K	s_f kJ/kg K	s_g kJ/kg K
–20	2.448	0.741	0.0928	177.1	397.5	0.9139	1.784
26	10.723	0.840	0.0220	231.6	413.5	1.109	1.717
40	13.335	0.884	0.0151	249.7	416.6	1.167	1.677

Take c_p of vapour as 1.03 kJ/kg K. **(PTU)**

Solution : Refer Fig. 5.39 for pressure enthalpy diagram of the operating cycle. Then from the table, the refrigerant properties at the chosen points are,

$h_7 = 397.5$ kJ/kg

$h_4 = 249.7$ kJ/kg

$h_5 = 231.6$ kJ/kg

$h_6 = h_5$ as enthalpy remains constant during throttling

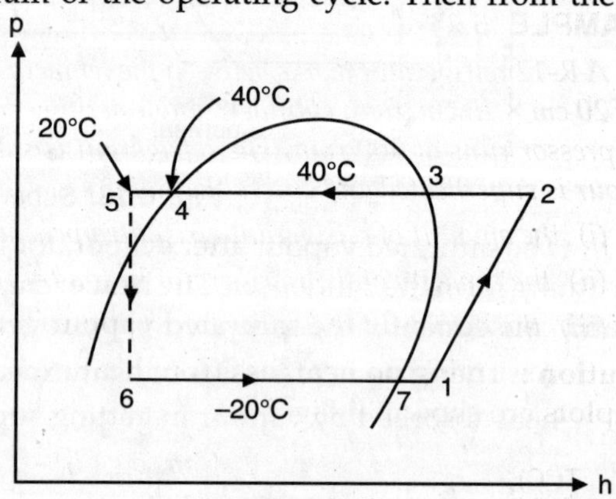

Fig. 5.39.

As per energy exchange in the heat exchanger,

$$h_4 - h_5 = h_1 - h_7$$

$$\therefore \quad h_1 = h_4 - h_5 + h_7 = 249.7 - 231.6 + 397.5 = 415.6 \text{ kJ/kg}$$

In terms of temperatures, the enthalpy difference during superheating is given as,

$$c_{pv}(t_2 - t_1) = h_1 - h_7 = 415.6 - 397.5 = 181.1$$

or, $$1.03\,[t_2 - (-20)] = 18.1$$

$$\therefore \quad t_2 = \frac{18.1}{1.03} - 20 = -2.43°\text{C} = 270.57 \text{ K}$$

During isentropic compression process, entropy remains constant, therefore

$$s_{g2} + c_{pv} \log_e \frac{T_2}{T_1} = s_{g3} + c_{pv} \log_e \frac{T_3}{T_4}$$

or $$1.784 + 1.03 \log_e \frac{270.57}{253} = 1.699 + 1.03 \log_e \frac{T_3}{313}$$

or $$1.853 = 1.699 + 1.03 \log_e \frac{T_3}{313}$$

That gives : $$T_3 = \text{anti-log}_e \left[\frac{1.853 - 1.699}{1.03}\right] \times 313 = 363.47 \text{ K}$$

Then : $$h_2 = h_3 + C_{pv}(T_3 - T_4)$$
$$= 416.6 + 1.03(363.47 - 313) = 468.5 \text{ kJ/kg}$$

Total refrigerant effect (heat absorbed in the evaporator) = 10 kW = 10 kJ/s

Presuming that superheating 7–1 occurs within the evaporator, *i.e.*, the space being refrigerated

$$m_r(h_1 - h_6) = 10$$

or Mass flow of refrigerant $$m_r = \frac{10}{h_1 - h_6} = \frac{10}{397.5 - 231.6} = 0.0603 \text{ kg/s}$$

Power required to operate the compressor

$$= m_r(h_2 - h_1) = 0.0603(468.5 - 415.6) = 3.19 \text{ kJ/s} = 3.19 \text{ kW}$$

$$COP = \frac{\text{refrigerating effect}}{\text{power input}} = \frac{10}{3.19} = 3.13$$

EXAMPLE 5.21

A R-12 refrigerator works between the temperature limits of – 10°C and + 30°C. The compressor employed is of 20 cm × 15 cm, twin cylinder, single acting compressor having a volumetric efficiency of 85 percent. The compressor runs at 500 rpm. The refrigerant is subcooled and it enters at 22°C in the expansion valve. The vapour is superheated and enters the compressor at – 2°C. Make calculations for the followings :

(i) the amount of refrigerant circulated per minute,

(ii) the capacity of the refrigerator in tons of refrigeration, and

(iii) the coefficient of performance of the system.

Solution : The various processes constituting the given refrigerating system as plotted on *T-s* and *p-h* plots are shown in Fig. 5.40.

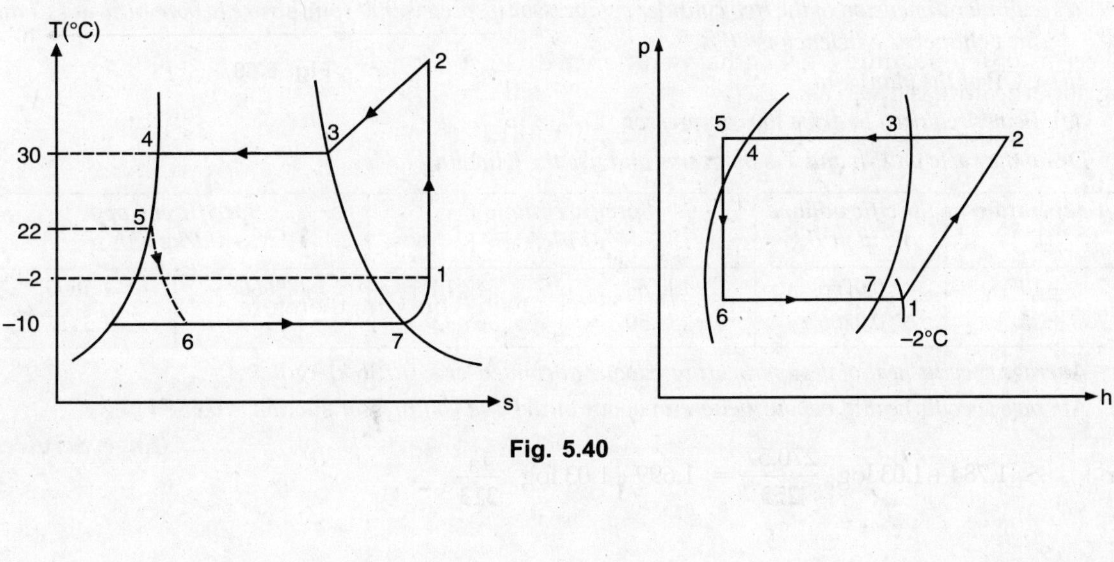

Fig. 5.40

The cycle can be plotted on the *p-h* chart and the enthalpy values read at the salient points.

$$h_1 = 350 \text{ kJ/kg}; \quad h_2 = 375 \text{ kJ/kg}$$

$$h_6 = h_5 = 220 \text{ kJ/kg} \quad \text{and} \quad v_1 = 0.08 \text{ m}^3/\text{kg}$$

(*i*) For the twin cylinder single acting reciprocating compressor, volume of refrigerant admitted per minute.

$$= 2 \times \frac{\pi}{4} d^2 l \times rpm \times \eta_v = 2 \times \left[\frac{\pi}{4}(0.2)^2 \times 0.15\right] \times 500 \times 0.55 = 4 \text{ m}^3/\text{min}$$

Mass of refrigerant circulated $m_r = \frac{4}{0.08} = \mathbf{50\ kg/min}$

Presuming that superheating 7–1 occurs within the evaporator, *i.e.*, the space being refrigerated.

(*ii*) Refrigerating effect per minute = $m_r(h_1 - h_6) = 50\ (350 - 220) = 6550$ kJ/min

1 ton of refrigeration ≡ 210 kJ/min

$\therefore$ Cooling capacity of the system = $\frac{6550}{210} = \mathbf{31.19\ ton}$

(*iii*) Work input $W = h_2 - h_1 = 375 - 350 = 25$ kJ/kg

Refrigerating effect $N = h_1 - h_6 = 350 - 220 = 130$ kJ/kg

$$COP = \frac{N}{W} = \frac{130}{25} = \mathbf{5.2}$$

EXAMPLE 5.22

An R-12 vapour compression plant producing 10 tonnes of refrigeration operates with condensing and evaporating temperature of 35°C and –10°C respectively. A suction line heat exchanger is used to subcool the saturated liquid leaving the condenser. Saturated vapour leaving the evaporator is superheated in the line heat exchanger to the extent that a discharge temperature of 60°C is obtained after isentropic compression. Calculate:

(a) subcooling achieved in the heat exchange

(b) refrigerant flow rate in kg/s

(c) cylinder dimension of the two cylinder compressor if speed is 900 rpm, stroke to bore ratio is 1 : 1 and the volumetric efficiency is 80%,

(d) COP of the plant and

(e) Power required to drive the compressor

Draw the cycle on P-h and T-s diagrams and use the following table.

Temperature °C	*Specific volume Vg in m³/kg*	*Specific enthalpy (kJ/kg)*		*Specific entropy (kJ/kg K)*	
– 10	0.0766	190.88	347.13	0.9660	1.5600
+ 35	0.0206	233.50	365.92	1.1139	1.5417

Average specific heat of desuperheating vapour at condenser = 0.796 kJ/kg K

Average specific heat of vapour between vapour outlet and compressor suction = 0.658 kJ/kg K

(Engg. Services)

Solution : Refer to Fig. 5.41 for *p-h* and *T-s* plots of the working cycle.

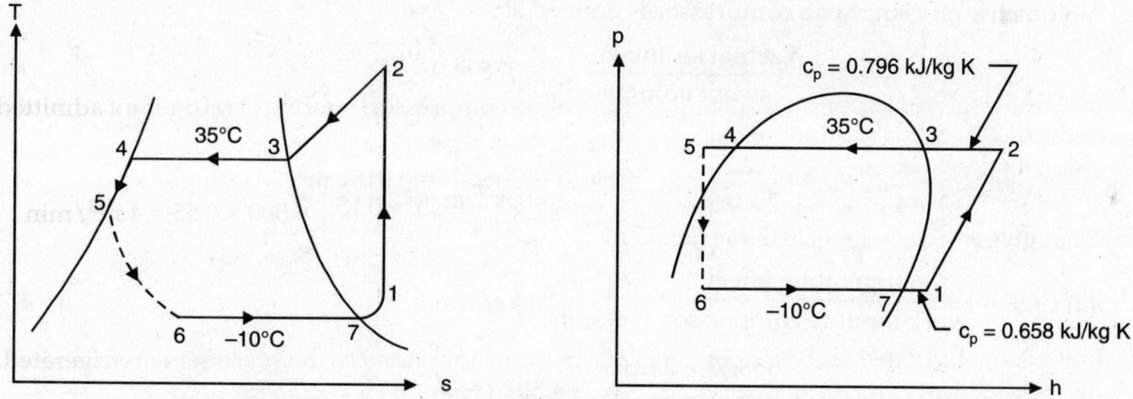

Fig. 5.41

The temperatures at the salient points of the cycle are :

$$T_7 = -10°C = 263 \text{ K}; \quad T_2 = 60°C = 333 \text{ K} \quad \text{and} \quad T_3 = T_4 = 35°C = 308 \text{ K}$$

$$\text{Entropy at state point 1} = s_7 + c_p \log_e \frac{T_1}{T_7} = 1.56 + 0.658 \log_e \frac{T_1}{263}$$

$$\text{Entropy at state point 2} = s_3 + c_p \log_e \frac{T_2}{T_3}$$

$$= 1.5419 + 0.796 \log_e \frac{333}{308} = 1.602 \text{ kJ/kgK}$$

For the isentropic compression process 1–2, the entropy remains constant.

$$\text{Therefore,} \quad 1.56 + 0.658 \log_e \frac{T_1}{263} = 1.602$$

$$\log_e \frac{T_1}{263} = \frac{1.602 - 1.56}{0.658} = 0.0638; \quad \frac{T_1}{263} = 1.0658$$

$$\therefore \quad T_1 = 1.0658 \times 263 = 280.3 \text{ K}$$

Now,

$$h_1 = h_1 + c_p \Delta T = 347.13 + 0.658\,(280.3 - 263) = 358.58 \text{ kJ/kg}$$

$$h_2 = h_2 + c_p \Delta T = 365.92 + 0.796\,(333 - 308) = 385.82 \text{ kJ/kg}$$

From energy balance in the regenerative heat exchanger,

$$h_4 - h_5 = h_1 - h_7 \quad ; \quad h_5 = h_4 - h_1 + h_7$$

$$= 233.50 - 358.58 + 347.13 = 222.05 \text{ kJ/kg}$$

(*a*) Subcooling achieved = $h_4 - h_5$ = 233.50 – 222.05 = 11.45 kJ/kg

(*b*) 1 ton of refrigeration is equivalent to absorption of 3.5 kJ/s of heat in the evaporator. Therefore,

$$m_r\,(h_1 - h_6) = 3.5 \times 10$$

For the throttling process 5-6 : $h_6 = h_5$. Then

$$\text{Refrigerant flow rate, } m_r = \frac{3.5 \times 10}{347.13 - 222.05} \simeq 0.28 \text{ kg/s}$$

(*c*) Applying Charles law for the process 7–1,

$$v_1 = v_7 \times \frac{T_1}{T_7} = 0.0766 \times \frac{280.3}{263} = 0.081 \text{ m}^3\text{/kg}$$

$\therefore$ Actual air intake per cylinder $V = m\,v_1 = 0.28 \times 0.081 = 0.0227\ \text{m}^3$

Volumetric efficiency of a compressor is defined as,

$$\eta_v = \frac{\text{actual air intake}}{\text{swept volume}} \quad ; \quad 0.8 = \frac{0.0227/2}{\frac{\pi}{4}d^2 l}$$

$$\therefore \quad \frac{\pi}{4}d^2 \times d = \frac{0.227}{2\times 0.8} \quad ; \quad d^3 = \frac{0.0227}{2\times 0.8}\times\frac{4}{\pi} = 0.0181\ \text{m}^3$$

That gives : $d = l = 0.2635\ \text{m} = 26.35\ \text{m}$

(*d*) $COP = \dfrac{\text{refrigerating effect}}{\text{work input to compressor}} = \dfrac{h_1 - h_6}{h_3 - h_1}$

$$= \frac{358.58 - 222.05}{385.82 - 358.58} = 4.59$$

(*e*) Power required to run the compressor $= m_r\,(h_2 - h_1)$

$= 0.28\,(385.82 - 358.58) = 7.63\ \text{kJ/s} = 7.63\ \text{kW}$

EXAMPLE 5.23

A single stage vapour refrigerating plant of 10 ton capacity operates between evaporator and condenser pressure limits of 2.35 bar and 7.45 bar respectively. The refrigerant F-12 is subcooled by 5°C before entering into the expansion valve and superheated by 6°C before leaving the evaporator. There occurs pressure drop of 0.2 bar in the suction valve and 0.1 bar in the discharge valve. Assuming isentropic compression, make calculations for the following parameters:

(a) power required to run the compressor, *(b) coefficient of performance of the plant,*
(c) load on the condenser, and *(d) piston displacement.*

Use p-h chart.

Solution: Refer Fig. 5.42 for *p-h* plot of the given system. Corresponding to the operating pressures,

evaporation temperature = – 8°C ; condender temperature = 25°C

Then : $t_2 = -8 + 6 = -2°\text{C}$ $\qquad t_7 = 30 - 5 = 25°\text{C}$

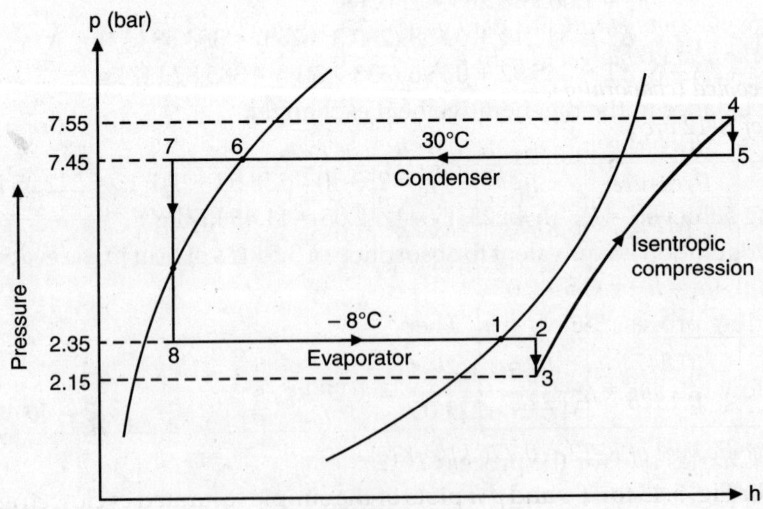

Fig. 5.42

Also, considering pressure drops in the suction and discharge valves, the pressures at state points 3 and 4 are :

$$p_4 = 7.45 + 0.1 = 7.55 \text{ bar}$$
$$p_3 = 2.35 - 0.2 = 2.15 \text{ bar}$$

The enthalpy values as read from the *p-h* plot are :

$$h_2 = h_3 = 355 \text{ kJ/kg} ; \qquad h_4 = h_5 = 377 \text{ kJ/kg}$$
$$h_7 = h_8 = 224 \text{ kJ/kg} ; \quad \text{and } v_3 = 0.088 \text{ m}^3\text{/kg}$$

(*a*) 1 ton of refrigeration ≡ 3.5 kJ/s

If m_r is the mass flow of refrigerant, then

$$m_r\,(h_2 - h_8) = 3.5\,Q$$
$$m_r\,(355 - 224) = 3.5 \times 10 ; \qquad m_r = 0.2672 \text{ kg/s}$$

Power required to run the compressor

$$= \text{work input to compressor}$$
$$= m_r\,(h_4 - h_3) = 0.2672\,(377 - 355) = 5.875 \text{ kJ/s} = \mathbf{5.878\ kW}$$

(*b*) Coefficient of performance of the system,

$$COP = \frac{\text{total refrigerating effect}}{\text{work input to compressor}}$$

$$= \frac{3.5 \times 10}{5.878} = \mathbf{5.95}$$

(*c*) Load on the condenser = heat removed in the condenser

$$= m_r\,(h_5 - h_7) = 0.2672\,(377 - 224) = \mathbf{40.88\ kJ/s}$$

(*d*) Piston displacement (theoretical) $= m_r\,v_3 = 0.2672 \times 0.088 = \mathbf{0.0235\ m^3/s}$

EXAMPLE 5.24

A refrigeration unit works between an evaporator temperature of –15°C and a condenser saturation temperature of 35°C with R-12 as refrigerant. Determine coefficient of performance if it operates on standard saturation cycle.

What will be the change in COP if a liquid vapour heat exchanger is installed in the system with the vapour leaving the heat exchanger at – 15°C ? Pressure that superheating occurs in connecting pipe outside the space being cooled (evaporator).

Properties of R-12 are :

Saturation temperature °C	Pressure MPa	h_f kJ/kg	h_g kJ/kg	s_g kJ/kg K	Degree 20 K		40 K	
					h kJ/kg	s kJ/kg K	h kJ/kg	s kJ/kg K
35	0.847	69.5	201.5	0.6839	216.4	0.731	231.0	07741
– 15	0.1826	22.3	181.0	0.7052	193.2	0.751	205.7	0.7942

Average specific heat of R-12 = 0.776 kJ/kgK.

Solution : Refer Fig. 5.43 for *T-s* and *p-h* plots of the simple saturated cycle (refrigerant dry saturated at compressor inlet and no undercooling).

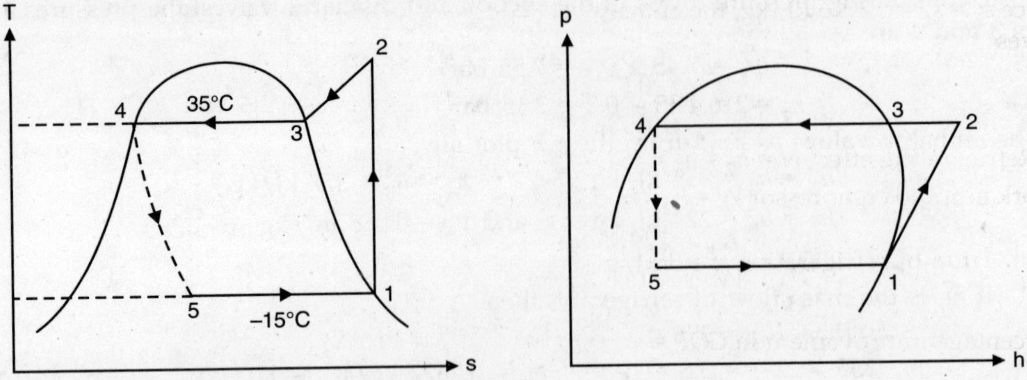

Fig. 5.43

Process 1-2 : The compression is isentropic, *i.e.*, at constant entropy. That is $s_2 = s_1$

$$s_3 + c_p \log_e \frac{T_2}{T_3} = 0.7052$$

$$0.6839 + 0.776 \log_e \frac{T_2}{308} = 0.0274$$

$$\log_e \frac{T_2}{308} = \frac{0.7052 - 0.6839}{0.776} = 0.0274$$

That gives : $\quad \dfrac{T_2}{308} = 1.0278 \quad$ and $\quad T_2 = 316.56$ K

$$\therefore \quad h_2 = h_3 + c_p\,(T_2 - T_3) = 201.5 + 0.776\,(36.56 - 308)$$

$$= 208.14 \text{ kJ/kg}$$

$$COP = \frac{\text{refrigerating effect}}{\text{work input to compressor}} = \frac{h_1 - h_5}{h_2 - h_1}$$

The expansion process 4 – 5 is of throttling during which enthalpy remains constant.

$$\therefore \quad COP = \frac{h_1 - h_4}{h_2 - h_1} = \frac{181.0 - 69.5}{208.14 - 181.0} = 4.11$$

(*b*) With superheating and subcooling

Degree of superheat at inlet state 1 to compressor,

$$= 15 - (-15) = 30°\text{C}$$

Then by interpolations,

$$h_1 = \frac{193.2 + 205.7}{2} = 199.45 \text{ kJ/kg}$$

$$s_1 = \frac{0751 + 0.7942}{2} = 0.7726 \text{ kJ/kg K}$$

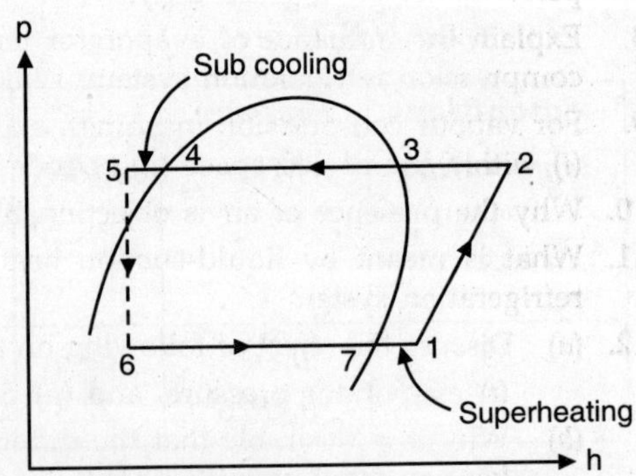

Fig. 5.44

With interchange of heat exchange in the liquid vapour heat exchanger,

$$h_1 - h_7 = h_4 - h_5$$

$$h_5 = h_4 - h_1 + h_7$$

$$= 69.5 - 199.45 + 181 = 51.05 \text{ kJ/kg}$$

Since $s_2 = s_1 = 0.7726$ kJ/kg, the enthalpy at state point 2 can be obtained by interpolation and that gives

$$h_2 = 216.4 + \frac{0.7726 + 0.731}{0.7741 + 0.731} \times (231 - 216.4) = 230.5 \text{ kJ/kg.}$$

Refrigerating effect $N = h_7 - h_6 = 181 - 51.05 = 129.95$ kJ/kg

Work input to compressor $W = h_2 - h_1 = 230.50 - 199.45 = 31.05$ kJ/kg

$$COP = \frac{N}{W} = \frac{129.95}{31.05} = 4.186$$

$$\text{Percentage improvement in } COP = \frac{4.186 - 4.11}{4.11} \times 100 = \mathbf{1.85\ \%}$$

REVIEW QUESTIONS

A. Conceptual and conventional questions.

1. (*a*) Explain the concept of vapour compression.
   (*b*) Compare air refrigeration with vapour compression cycle.
   (*c*) Compare the vapour compression cycle with Carnot cycle.
2. Describe, with a neat schematic diagram, the vapour compression refrigeration cycle. Represent the cycle on *T-s* and *p-h* plots.
3. Differentiate between wet and dry compression. Why is wet compression not employed through it is thermodynamically better ?
4. For a vapour compression machine, explain the effect of undercooling and superheating on coefficient of performance.
5. Sketch a practical vapour compression cycle and point out its salient features.
6. (*a*) State the functions of the following components of a simple vapour compression refrigerant system.
   (*i*) Compressor (*ii*) Condenser (*iii*) Expansion valve and (*iv*) Evaporator
   (*b*) Why in practice a throttle valve is used in vapour compression refrigerator rather than an expansion cylinder?
7. Write energy equations for each component of the simple vapour compression system of refrigeration. Obtain there from the expressions for mass of refrigerant in circulation, capacity and coefficient of performance for a given duty.
8. Explain the influence of evaporator temperature and condenser temperature on *COP* of the vapour compression refrigeration system. Which of the two temperatures has more influence on *COP* ?
9. For vapour compression machines, explain the effect of superheating on *COP* if superheating occurs (*i*) within the cooled space (*ii*) outside the cooled space.
10. Why the presence of air is objectionable in a vapour compression refrigeration system ?
11. What is meant by liquid-suction heat exchange ? Why is it incorporated in vapour compression refrigeration system ?
12. (*a*) Discuss the effect of following on the performance of a vapour compression refrigeration system:
    (*i*) evaporator pressure, and (*ii*) condenser pressure.
    (*b*) Why it is desirable that the difference between the discharge pressure and suction pressure be kept as small as possible?
13. (*a*) Differentiate between the function of a flash chamber and an accumulator in the context oftheir use in a vapour compression cycle.
    (*b*) Set up expressions to show that the use of flash chamber has no effect on the thermodynamics of the opening cycle.

14. Define and explain the terms clearance space, piston displacement and volumetric efficiency for a compressor.
15. Ammonia is the working fluid in an ideal vapour compression cycle that communicates thermally with cold region at –20°C and a warm region at 32°C. Determine the refrigerating effect per unit mass of refrigerant and COP for the following cases:
    (*a*) the system operates on reversed Carnot cycle
    (*b*) the system delivers dry saturated vapour to the condenser after isentropic compression, and there is no undercooling of the condensed liquid
    (*c*) the system delivers dry saturated vapour to the compressor where it is compressed isentropically and there is no sub-cooling of the condensed liquid
    (*d*) the system delivers dry saturated vapour to the compressor where it is compressed isentropically, and the liquid after condensation is sub-cooled by 10 K. Assume that undercooling takes place along the saturated liquid line.

    Take relevant data from table of properties of ammonia.

    (**Ans.** (a) 942.8 kJ/kg, 4.86 (*b*) 920 kJ/kg, 4.2 (c) 1087 kJ/kg, 3.9 (d) 1135 kJ/kg, 4.08)
16. A refrigerating machine operating on the vapour compression cycle operates between pressure limits of 2.5 bar in the evaporator and 9 bar in the condenser. Estimate the theoretical coefficient of performance if the vapour is in dry saturated condition at entry to the compressor, the compression is isentropic and there is no undercooling in the condenser.

    If the actual coefficient of performance is 0.65 of theoretical value, calculate the net cooling effect in tons of refrigeration. The refrigerant flow rate is 5 kg/min and the properties of refrigerant are:

Pressure bar	*Saturation temperature, °C*	*Enthalpy, kJ/kg*		*Entropy of saturated vapour, kJ/kgK*
		liquid	*vapour*	
2.5	– 7	412.5	570.3	4.76
9.0	36	456.4	585.3	4.74

(**Ans.** 5.37, 1.76 tons)

17. A refrigeration machine using R-12 as refrigerant operates between the pressures 2.5 bar and 9 bar. The compression is isentropic, there is no undercooling in the condenser and the vapour is in dry saturated condition at the beginning of the compression process. Represent the processes on *T-s* and *p-h* plots. Estimate the theoretical coefficient of performance. If the actual coefficient of performance is 0.65 of theoretical value and the refrigerant flow is 5 kg/min, calculate the net cooling produced per hour.

    Properties of the refrigerant are :

Pressure bar	*Saturation temperature K*	*Enthalpy kJ/kg*		*Entropy of saturated vapour kJ/kgK*
9.0	309	70.55	201.8	0.6836
2.5	266	29.62	184.5	0.7001

Take specific heat of supherheated vapour at 9 bar equal to 0.75 kJ/kg°C.

(**Ans.** 5.076, 34185 kJ/hr)

18. An ammonia vapour compression refrigerator has an effective swept volume of 0.3 m^3/min. Condensation and evaporation take place at 28.9°C and – 12.2°C respectively. There is no undercooling and the gas temperature after compression is 48.9°C. Taking c_p for the superheated vapour as 2.93 kJ/kg K, determine :
    (*i*) the dryness fraction of vapour as it enters the compressor,
    (*ii*) the rate of circulation of ammonia,

(*iii*) the rate of extraction of heat in kJ/min,

(*iv*) the heat rejected in the condenser per minute,

Use the following data for the properties of refrigerant ammonia :

Temperature	*Specific enthalpy kJ/kg*		*Specific entropy kJ/kgK*		*Specific volume* m^3/kg
	liquid	*latent*	*liquid*	*latent*	
28.9	320.2	1146.9	1.1947	3.7942	0.106
– 12.2	125.6	1306.0	0.5057	5.0018	0.406

19. A food storage locker requires 20 tons of refrigeration to keep the food stuff in better condition. This system operates between temperature limits of –8°C in the evaporator and 30°C in the condenser. The refrigerant F-12 used in the plant is subcooled to 25°C before entering the expansion valve and the vapour gets superheated to 6°C before leaving the evaporator coil. The compressor employed has two cylinders, is single-acting with stroke equal to 1.5 times the bore. The compressor turns 800 rev/min and compression is assumed as isentropic. Represent the process on *p-h* chart and workout the followings:
    (*i*) refrigerating effect and mass-flow rate of refrigerant circulated;
    (*ii*) theoretical power input and COP of the machine;
    (*iii*) heat rejected in the condenser per kg of refrigerant;
    (*iv*) theoretical piston displacement, bore and stroke of the compressor.
    (**Ans.** 128 kJ/kg, 32.8 kg/min; 12 kW, 5.82; 150 kJ/kg; d = 10.93 cm and l = 16.4 cm)

20. A refrigerating plant of 12 tons capacity uses F-12 as refrigerant. The evaporator temperature is –28°C and the condenser temperature is 26°C. The refrigerant is subcooled by 4°C before entering the throttle valve and is superheated by 5°C before entering the compressor where isentropic compression occurs. The refrigerating machine has six cylinder single acting compressor with stroke equal to 1.25 times the bore. The machine runs at 1000 rpm and has a volumetric efficiency of 87 percent. Using *p-h* plot, estimate
    (*a*) mass of refrigerant circulated per minute,
    (*b*) power required to run the machine and *COP* of the plant,
    (*c*) bore and stroke of the cylinder.
    Take $(c_p)_{liquid}$ = 0.96 kJ/kg K and $(c_p)_{vapour}$ = 0.615 kJ/kgK.
    (**Ans.** 22.15 kg/min, 12 kW, 3.48, 86.5 mm and 108.12 mm)

21. A twin cylinder single acting compressor of a F-12 refrigerant plant has 90 mm bore, 70 mm stroke and it works between pressure limits of 10.84 bar and 1.82 bar. The temperatures of the refrigerant at the entry and the exit of the condenser are 65°C and 32°C respectively and the rate of flow of cooling water is 13 kg/min with temperature rise of 8°C. The mean effective pressure in the compressor (from indicator diagram) is 3.5 bar. The plant produces 45 kg of ice per hour from water at 27°C (latent heat of ice 335 kJ/kg). Determine:
    (*a*) the theoretical and actual *COP* of the plant, and
    (*b*) mass flow of the refrigerant
    Draw up heat balance sheet for the plant on minute basis.
    The relevant properties of F-12 are :

Pressure bar	*Saturation temperature, °C*	*Specific enthalpy kJ/kgK*		*Specific entropy kJ/kgK*	
		liquid	*vapour*	*liquid*	*vapour*
10.84	45	243.66	368.82	1.1457	1.5391
1.82	– 15	186.28	344.92	0.9487	1.5632

The average specific heat for liquid is 1.256 kJ/kgK and for superheated vapour is 0.645 kJ/kgK.

22. A simple saturation cycle using ammonia as refrigerant works between temperature limits of –10°C and 40°C. If the refrigerant load on the system is 5 tons, determine power required to run the system and its coefficient of performance.

How these parameters would be affected if the temperature of refrigerant required in the evaporator is – 20°C ? Comment on the results. Use *p-h* chart for making the calculations.

(Ans. 7.1 kW, 4.46 ; 9.32 kW, 3.42)

B. Fill in the blanks with appropriate word/words

1. Compression, condensation, expansion and .... are the process that comprise a simple vapour compression cycle.
2. The compression process is called .... if the vapour gets superheated at the end of compression.
3. The employment of expansion cylinder in place of throttle valve .... the *COP* of a vapour compression refrigerator.
4. A decrease in the suction pressure results into a .... in the refrigerating capacity and *COP* of a vapour compression refrigeration system.
5. .... is the process of cooling the liquid refrigerant below the condensing temperature for a given pressure.
6. The changes in enthalpy and pressure required for analysing a thermodynamic process can be obtained directly from the .... chart.
7. Within the limits of application requirements, the refrigeration system needs to be designed to operate at the .... possible vaporising temperature and .... possible condensing temperature.
8. When high pressure liquid refrigerant from the condenser passes through the expansion valve, it undergoes partial evaporation and the vapour thus formed is called ....
9. For better COP of a refrigerator, the pressure range between evaporator and condenser should be ....
10. Increase in condenser pressure .... the refrigerating effect per kg of refrigerant

Answers

1. evaporation 2. dry 3. increases 4. decrease 5. subcooling 6. *p-h* 7. highest, lowest
8. flash 9. low 10. decreases

C. Multiple choice questions.

1. The vapour compression refrigeration cycle operates on reversed
   (*a*) Rankine cycle (*b*) Joule's cycle (*c*) Atkinson cycle (*d*) Ericsson cycle
2. The sequential order of different components of a vapour compression refrigeration system is
   (*a*) compressor, condenser, throttle valve, evaporator
   (*b*) compressor, throttle valve, evaporator, condenser
   (*c*) compressor, evaporator, throttle valve, condenser
   (*d*) compressor, throttle valve, condenser, evaporator
3. In a refrigeration system, expansion valve is incorporated between
   (*a*) evaporator and compressor (*b*) condenser and evaporator
   (*c*) compressor and condenser (*d*) superheater and subcooler
4. Which part of the vapour compression refrigeration cycle produces the refrigeration effect ?
   (*a*) condenser (*b*) throttle valve (*c*) evaporator (*d*) compressor
5. During a refrigeration cycle based on vapour compression system, the highest temperature of the refrigerant occurs when the vapour leaves the
   (*a*) condenser (*b*) throttle valve (*c*) evaporator (*d*) compressor

6. In the vapour compression refrigeration cycle, the refrigerant is generally in the form of fairly wet vapour at entry to the
(*a*) condenser (*b*) throttle valve (*c*) evaporator (*d*) compressor

7. During refrigeration cycle based on vapour compression system, the refrigerant exists as high pressure liquid at exit from the
(*a*) condenser (*b*) throttle valve (*c*) evaporator (*d*) compressor

8. In the vapour compression refrigeration system, the refrigerant before entering the expansion device or throttle valve is in the form of
(*a*) wet vapour at low pressure
(*b*) wet vapour at high pressure
(*c*) high pressure saturated liquid
(*d*) low pressure saturated liquid

9. After passing through the expansion or throttle valve, the condition of refrigerant in a vapour compression refrigeration cycle is
(*a*) high pressure saturated liquid
(*b*) very wet vapour
(*c*) fairly dry saturated vapour
(*d*) low pressure superheated vapour

10. The superheated vapour state of refrigerant in a vapour compression cycle occurs
(*a*) before entering the expansion valve
(*b*) after leaving the evaporator
(*c*) at exit from the compressor
(*d*) after passing through the condenser

11. During which component of vapour compression refrigeration system, the enthalpy remains constant ?
(*a*) condenser (*b*) throttle valve (*c*) evaporator (*d*) compressor

12. During refrigeration cycle based on vapour compression system, the heat is rejected by the refrigerant in
(*a*) condenser (*b*) expansion valve (*c*) evaporator (*d*) compressor

13. All of the following pipe connections in a vapour compression system require thermal insulation, except
(*a*) between compressor outlet and condenser
(*b*) between condenser outlet and expansion valve
(*c*) between expansion valve and evaporator
(*d*) between evaporator and compressor

14. The throttling operation in a refrigeration cycle is carried out in
(*a*) evaporator (*b*) discharge valve (*c*) capillary tube (*d*) expansion valve

15. A capillary tube is used in a small refrigerator to serve the purpose of
(*a*) thermostat (*b*) expansion valve (*c*) drier (*d*) evaporator

16. The pressure in a capillary tube decreases due to which of the followings :
(*i*) frictional resistance offered by the tube wall
(*ii*) acceleration of refrigerant in the tube
(*iii*) heat transfer from the tube
(*iv*) decrease in potential energy
Select your answer from the following code :
(*a*) 1 and 2 (*b*) 1, 2 and 3 (*c*) 1, 2 and 4 (*d*) 1 and 4

17. A capillary tube is employed in a refrigeration cycle to
(*a*) produce cooling effect
(*b*) convert refrigerant from liquid to vapour state
(*c*) control the temperature of refrigerated space
(*d*) control the flow of refrigerant

18. In a vapour compression refrigeration system, a throttle valve is used in place of expander because it
(*a*) it leads to significant cost reduction
(*b*) it considerably reduces the system work
(*c*) positive work in isentropic expansion of liquid is very small
(*d*) improves COP

19. The bank of tubes at the back of a domestic refrigerator of vapour compression type are
(*a*) condenser tubes (*b*) evaporator tubes (*c*) capillary tubes
(*d*) tubes carrying electric wires

20. Which of the followings is not a true statement in the context of subcooling of refrigerant before throttling in a vapour compression cycle ?
(*a*) cooling of refrigerant below saturation temperature corresponding to condenser pressure
(*b*) increase in refrigerant effect
(*c*) reduced power consumption
(*d*) improvement in coefficient of performance

21. The subcooling is a process of cooling the refrigerant in vapour compression refrigeration system before
(*a*) compression (*b*) condensation (*c*) throttling (*d*) evaporation

22. Which of the following aspects is not true in the context of dry compression in a vapour compression refrigeration cycle ?
(*a*) increase in refrigerating effect
(*b*) increase in power consumption
(*c*) improvement in coefficient of performance
(*d*) complete evaporation in the evaporator

23. Which is usually the costliest item in a refrigeration system ?
(*a*) condenser (*b*) capillary tube (*c*) compressor (*d*) evaporator

24. The location of freezer in the top portion of a refrigerator results in
(*a*) less noise in the system (*b*) quick cooling of entire chamber
(*c*) no overheating of the motor (*d*) no adverse effect on environment

25. Chose the false statement in the context of vapour compression refrigeration system :
(*a*) The efficiency of performance is the ratio of the refrigeration effect to the enthalpy increase in the compressor
(*b*) For obtaining high COP, the pressure range of the compressor should be high
(*c*) Dry compression permits high speeds to be used
(*d*) The pipe diameter of the compressor on the suction side is bigger than that on the delivery side.

26. Identify the purpose served by receiver fitted in a refrigeration unit :
(*a*) to reduce the work input to compressor
(*b*) to conserve heat
(*c*) to allow for variation of load
(*d*) to dampen the noise generated by moving parts of the compressor

27. Which of the following statements is not true in the context of vapour compression refrigeration cycle ?
    (*a*) the refrigerant occurs as high pressure saturated liquid between the condenser and expansion valve
    (*b*) the refrigerant is generally in the form of wet vapour in the evaporator
    (*c*) the refrigerant occurs as high pressure liquid in the condenser
    (*d*) the high temperature of refrigerant occurs when the vapour leaves the throttle valve

28. In a vapour compression refrigeration system, the high pressure liquid from the condenser/receiver is cooled below its saturation temperature to
    (*a*) reduce the net work per cycle
    (*b*) reduce the net refrigerant effect
    (*c*) increase the net refrigerant effect
    (*d*) reduce the pressure on high pressure side

29. In a VCR plant, the refrigerant leaves the compressor and evaporator with enthalpy of 205 kJ/kg and 177 kJ/kg respectively. Enthalpy of refrigerant leaving the condenser is 105 kJ/kg. If the mass flow of refrigerant is 0.2 kg/s, the refrigerant effect will be .... kW
    (*a*) 12.2 (*b*) 14.4 (*c*) 16.4 (*d*) 20.2

30. In an ideal vapour compression refrigeration cycle, the enthalpy values at salient points are as follows :

    At inlet to compressor = 1500 kJ/kg
    At outlet from compressor = 1800 kJ/kg
    At inlet to evaporator = 300 kJ/kg

    What is the *COP* of the cycle ?
    (*a*) 3 (*b*) 4 (*c*) 5 (*d*) 6

31. The capillary tube is not used in large capacity refrigeration system because
    (*a*) there can be leakage of gas
    (*b*) cost of capillary tube becomes high
    (*c*) capacity control is not possible
    (*d*) the pressure drop obtained is not adequate

32. The vapour compression refrigeration cycle is represented as shown in the adjoining figure with state 1 being the exit of evaporator. The co-ordinate system used in the figure is
    (*a*) *p-h* (*b*) *T-s*
    (*c*) *p-s* (*d*) *T-h*

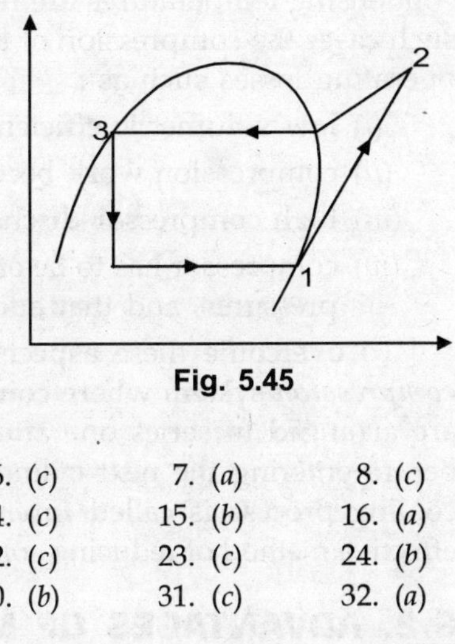

Fig. 5.45

Answers

1. (*a*)	2. (*a*)	3. (*b*)	4. (*c*)	5. (*d*)	6. (*c*)	7. (*a*)	8. (*c*)
9. (*b*)	10. (*c*)	11. (*b*)	12. (*a*)	13. (*a*)	14. (*c*)	15. (*b*)	16. (*a*)
17. (*d*)	18. (*c*)	19. (*a*)	20. (*c*)	21. (*c*)	22. (*c*)	23. (*c*)	24. (*b*)
25. (*b*)	26. (*c*)	27. (*d*)	28. (*c*)	29. (*b*)	30. (*b*)	31. (*c*)	32. (*a*)

Chapter – 6

Compound Vapour Compression Refrigeration Systems

Learning objectives : A study of the subject matter presented in this chapter will enable the students/readers to understand

- *multistage compression and its advantages*
- *usefulness of intercooler, subcooler and flash chamber in compound compression systems*
- *multi-evaporator systems with individual and multi-expansion valves*
- *working of cascade vapour compression and its advantages over multi-stage compression system*

6.1. MULTISTAGE COMPRESSION AND IT NECESSITY

The vapour compression refrigeration systems employing a single stage compression are essentially suitable for low delivery pressures. In situations where delivery pressure has to be large, the pressure ratio would become high. High pressure ratios (more than 4) occur when either high condensing temperatures are required and/or very low evaporator temperatures are desired. In such cases the compression of the vapour refrigerant in a single compression would entail heavy operating losses such as :

(*i*) low volumetric efficiency

(*ii*) compression work becomes more

(*iii*) high compressor discharge temperature which may damage the lubrication system

(*iv*) compressor has to be of robust construction so that the cylinder walls can withstand high pressures, and that adds to capital cost

To overcome these aspects, it is considered advantages to have ***compound*** or ***multistage compression*** system where compression is carried out in more than one cylinder. The cylinders are arranged in series one after the other. The gas after compression in the first cylinder and before entering the next cylinder is cooled at constant pressure by some suitable method; the cooling process is called ***inter-cooling***. Multistage compression with inter-stage cooling is an effective method of reducing compression work by working on isentropics close to saturation curve.

6.2. ADVANTAGES OF MULTISTAGE COMPRESSION

Compared to single stage compression, the multistage compression provides the following main advantages :

1. Improvement in volumetric efficiency of the compressor
2. Reduction in work input to compressor. This is achieved by cooling of compressed refrigerant vapour before its compression in the succeeding stage
3. Low operating temperatures and consequently there is little chance of lubrication problems

(*i*) Refrigerating effect (heat absorbed in the evaporator)

$$= m_1 (h_1 - h_6) = m_1 (h_1 - h_5)$$

If load on the evaporator is equivalent to Q tons of refrigeration and since one ton of refrigeration equals heat absorption rate of 3.5 kJ/s

$$m_1 (h_1 - h_5) = 3.5\ Q\ ;\quad m_1 = \frac{3.5Q}{h_1 - h_2}\ \text{kg/s}$$

(*ii*) Power required to drive the system,

= work input to low peressure compressor + work input to high pressure compressor

$$= m_1(h_2 - h_1) + m_2 (h_4 - h_3)$$

Coefficient of performance of the system,

$$COP = \frac{\text{refrigerating effect}}{\text{work input to both the compressors}}$$

$$= \frac{m_1(h_1 - h_5)}{m_1(h_2 - h_1) + m_2(h_4 - h_3)}$$

and load on the condenser

$$= m_2 (h_4 - h_5)$$

6.4. TWO-STAGE COMPRESSION WITH WATER INTERCOOLER AND LIQUID SUBCOOLER

Fig. 6.3 shows the schematic arrangement of the different components of a two-stage compression with water intercoolers and liquid subcooler, along with *p-h* diagram of the processes involved.

The main components of the system are : an evaporator, a low pressure compressor, a water intercooler, a high pressure compressor, a condenser, a liquid subcooler and an expansion valve.

Operation : The sequence of operation is

1-2 : Isentropic compression of saturated vapour (supplied from the evaporator) from suction pressure p_e to intermediate pressure p_i in the low pressure compressor. At exit from the compressor, the refrigerant vapour is in superheated condition (state point 2).

2-3 : Cooling or desuperheating of the superheated vapour leaving the low pressure compressor. This cooling is at constant pressure and is accomplished in the water intercooler where cooling medium (water) is from an external source. At exit from the water cooler, the refrigerant is still in the superheated condition corresponding to state plant 3.

3-4 : Isentropic compression of refrigerant vapours from intermediate (or intercooler) pressure p_i to the condenser pressure p_c in the high pressure compressor.

4-5 : Condensation of high pressure refrigerant in the condenser at constant pressure p_c. During condensation, the state of refrigerant changes from superheated vapour to saturated liquid.

5-6 : Sub-cooling of saturated liquid refrigerant by passing it through a liquid sub-cooler. This temperature reduction occurs at constant pressure.

6-7-1 : Expansion of liquid refrigerant from the subcooler in an expansion valve (6-7) and its subsequent evaporation in the evaporator to saturated vapour condition (process 7-1).

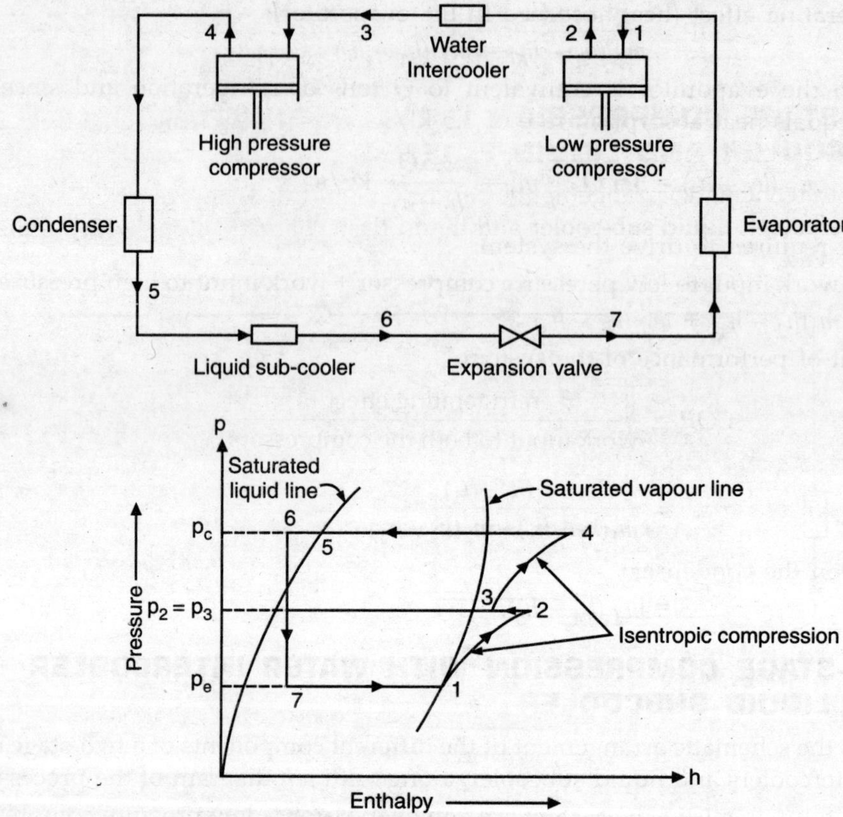

Fig. 6.3. Two-stage compression with water intercooler and liquid sub-cooler

Analysis :

Let m = mass flow rate of refrigerant through low pressure compressor or evaporator

Then

(*i*) Refrigerating effect (heat absorbed in the evaporator),

$$= m\,(h_1 - h_7)$$

$$= m\,(h_1 - h_6) \qquad (\because h_7 = h_6)$$

If load on the evaporator is equivalent to Q tons of refrigeration and since one ton of refrigeration equals heat absorption rate of 3.5 kJ/s

$$m(h_1 - h_6) = 3.5\,Q\;; \quad m = \frac{3.5\,W}{h_1 - h_6}\ \text{kg/s}$$

(*ii*) Power required to drive the system,

= work done in LP compressor + work done in HP compressor

$$= m(h_2 - h_1) + m(h_4 - h_3) = m[(h_2 - h_1) + (h_4 - h_3)]$$

This is because the mass flow rate of refrigerant passing through both the compressors is same.

(*iii*) Coefficient of performance of the system,

$$COP = \frac{\text{refrigerating effect}}{\text{work input to compressors}}$$

$$COP = \frac{m_1(h_1 - h_6)}{m[(h_2 - h_1) + (h_4 - h_3)]} = \frac{(h_1 - h_6)}{(h_2 - h_1) + (h_4 - h_3)}$$

6.5. TWO-STAGE COMPRESSION WITH WATER INTERCOOLER, LIQUID SUB-COOLER AND LIQUID FLASH CHAMBER

Fig. 6.4 shows the schematic arrangement of the different components of a two stage compression with water intercooler, liquid sub-cooler and liquid flash chamber, along with *p-h* diagram of the processes involved.

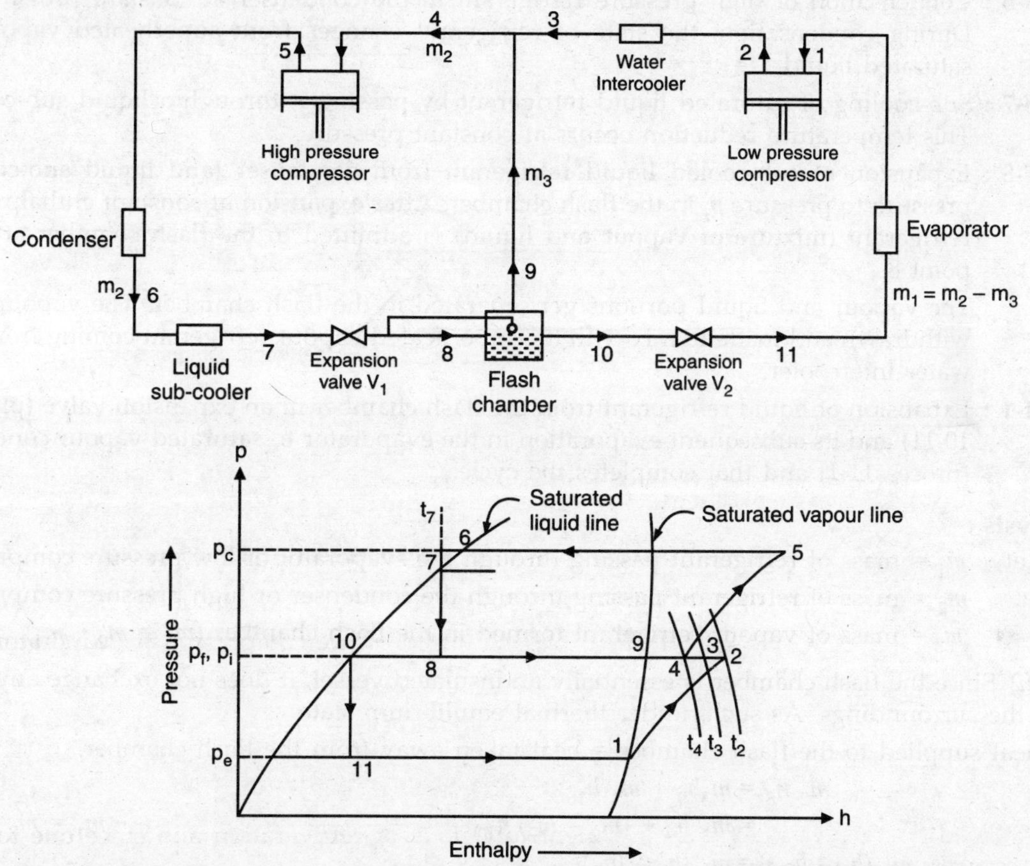

Fig. 6.4. Two-stage compression with water intercooler, liquid sub-cooler and liquid flash chamber

The main components of the system are : an evaporator, a low pressure compresses, a water intercooler, a high pressure compressor, a condenser, a liquid sub-cooler, a flash chamber and an expansion valve.

Operation : The sequence of operation is

1-2 : Isentropic compression of saturated vapour (supplied from the evaporator) from suction pressure p_e to intermediate pressure p_i that exists in the water intercooler and flash chamber. At exit from the compressor, the refrigerant vapour is in superheated condition (state point 2) at intermediate pressure p_i.

2-3 : Cooling or desuperheating of the superheated vapours leaving the low pressure compressor. This cooling is at constant pressure and is accomplished in the water intercooler where cooling medium (water) is supplied from an external source. At exit from the water cooler, the refrigerant is still in the superheated condition corresponding to state plant 3.

3-4 : Mixing of supherheated vapour refrigerant (leaving the water cooler at condition 3) with saturated vapour refrigerant (supplied by flash chamber at condition 9). The condition of the refrigerant after mixing corresponds to state point 4.

4-5 : Isentropic compression of refrigerant vapours from intermediate pressure p_i to the condenser pressure p_c in the high pressure compressor.

5-6 : Condensation of high pressure refrigerant in the condenser at constant pressure p_c. During condensation, the state of refrigerant changes from superheated vapour to saturated liquid.

6-7 : Sub-cooling of saturated liquid refrigerant by passing it through a liquid sub-cooler. This temperature reduction occurs at constant pressure.

7-8 : Expansion of sub-cooled liquid refrigerant from condenser (and liquid sub-cooler) pressure to pressure p_i in the flash chamber. After expansion at constant enthalpy, the refrigerant (mixture of vapour and liquid) is admitted to the flash chamber at state point 8.

The vapour and liquid portions get separated in the flash chamber. The vapours are withdrawn and made to mix with the superheated vapour refrigerant coming from the water intercooler.

10-11-1 : Expansion of liquid refrigerant from the flash chamber in an expansion valve (process 10-11) and its subsequent evaporation in the evaporator to saturated vapour condition (process11-1) and that completes the cycle.

Analysis :

Let m_1 = mass of refrigerant passing through the evaporator or low pressure compressor

m_2 = mass of refrigerant passing through the condenser or high pressure compressor

m_3 = mass of vapour refrigerant formed in the flash chamber ($m_3 = m_2 - m_1$)

(*ii*) Since the flash chamber is essentially an insulated vessel, it does not exchange any heat with the surroundings. As such, under thermal equilibrium state,

heat supplied to the flash chamber = heat taken away from the flash chamber.

$$m_2 h_8 = m_3 h_9 + m_1 h_{10}$$

$$= m_3 h_9 + (m_2 - m_3) h_{10} \qquad (\because m_1 = m_2 - m_3)$$

or $$m_2 (h_8 - h_{10}) = m_3 (h_9 - h_{10})$$

$$\therefore \quad m_3 = m_2 \frac{h_8 - h_{10}}{h_9 - h_{10}} = m_2 \frac{h_7 - h_{10}}{h_9 - h_{10}} \qquad (\because h_8 = h_7)$$

(*i*) Refrigerating effect (heat absorbed in the evaporator)

$$= m_1 (h_1 - h_{11})$$

$$= m_1 (h_1 - h_{10}) \qquad (\because h_{11} = h_{10})$$

If load on the evaporator is equivalent to Q tons of refrigeration and since one ton of refrigeration equals heat absorption rate of 3.5 kJ/s

$$m_1 (h_1 - h_{10}) = 3.5\, Q \; ; \quad m_1 = \frac{3.5\, Q}{h_1 - h_{10}} \text{ kg/s}$$

(*iii*) The enthalpy of the mixed refrigerant (represented by state point 4) can be determined by writing the energy equation for the mixing process which involves :

– vapour refrigerant from the water intercooler (state point 3)

– vapour refrigerant from the flash chamber (state point 9)

$$m_2 h_4 = m_3 h_9 + m_1 h_3$$
$$= (m_2 - m_1) h_9 + m_1 h_3$$

(*iv*) Work input to compressor,

$$= \text{work done in LP compressor} + \text{work done in HP compressor}$$
$$= m_1(h_2 - h_1) + m_2(h_5 - h_4)$$

$$COP \text{ of the system} = \frac{\text{refrigerating effect}}{\text{work input to compressors}}$$

$$= \frac{m_1(h_1 - h_{10})}{m_1(h_2 - h_1) + m_2(h_5 - h_4)}$$

(*vi*) The vapour refrigerant having mass m_1 enters the water intercooler where it cools from condition 2 to 3. Therefore,

$$\text{cooling capacity of the intercooler} = m_1(h_2 - h_3)$$

6.6. TWO-STAGE COMPRESSION WITH WATER INTERCOOLER, LIQUID SUB COOLER AND FLASH INTERCOOLER

Fig. 6.5 shows the schematic arrangement of the different components of a two-stage compression with water intercooler, liquid sub-cooler and flash intercooler, alongwith *p-h* diagram of the processes involved.

The main components of the system are : an evaporator, a low pressure compressor, a water intercooler, a flash chamber, a high pressure compressor, a condenser, a liquid sub-cooler and an expansion valve.

Operation : The sequence of operation is

1-2 : Isentropic compression of saturated vapour (supplied from the evaporator) from suction pressure p_e to intermediate pressure (flash intercooler pressure) p_f in the low pressure compressor. At exit from the compressor, the refrigerant vapour is in superheated condition (state point 2).

2-3 : Cooling or desuperheating of the superheated vapours leaving the low pressure compressor. This cooling is at constant pressure and is accomplished in the water intercooler where cooling medium (water) is supplied from an external source. At exit from the water cooler, the refrigerant is still in the superheated condition corresponding to state plant 3.

3-4 : Cooling of supherheated refrigerant vapours (leaving the water cooler) to saturated vapour state. This cooling occurs in the flash chamber and is due to evaporation of a part of the liquid refrigerant available in the chamber.

4-5 : Isentropic compression of saturated refrigerant vapours from intermediate (or flash intercooler) pressure p_f to the condenser pressure p_c in the high pressure compressor.

5-6 : Condensation of high pressure refrigerant in the condenser at constant pressure p_c. During condensation, the state of refrigerant changes from superheated vapour to saturated liquid.

6-7 : Sub-cooling of saturated liquid refrigerant by passing it through a liquid sub-cooler. This temperature reduction occurs at constant pressure.

7-8 : Expansion of sub-cooled liquid refrigerant from condenser (and liquid sub-cooler) pressure to pressure p_f in the flash chamber. The expanded refrigerant is a mixture of liquid and vapour refrigerant, and it enters the flash chamber at state point 8. The flash chamber acts as flash separator, *i.e.*, the vapours separate and are led at state point 4 to the high pressure compressor.

9-10-1 : Expansion of liquid refrigerant leaving the flash chamber at condition 9 in an expansion valve (process 9-10) and its subsequent evaporation in the evaporator to saturated vapour condition (process 10-1) and that completes the cycle.

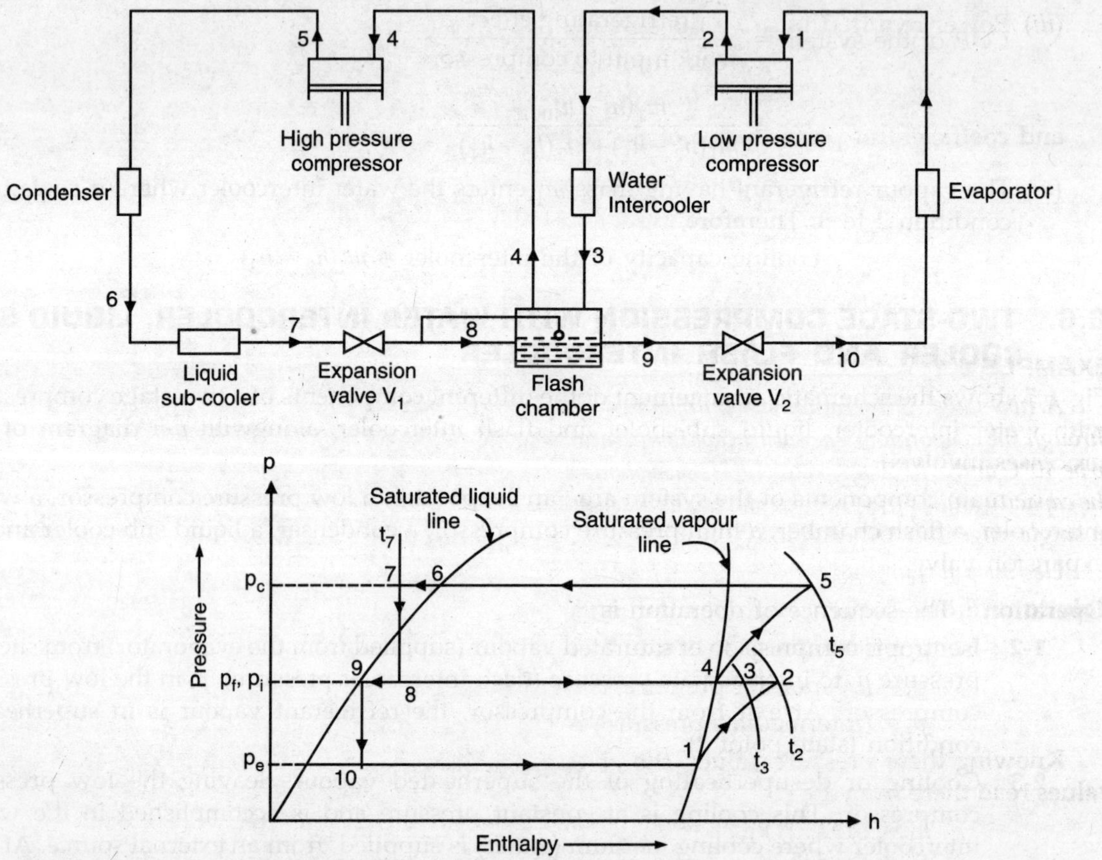

Fig. 6.5. Two-stage compression with water intercooler, liquid sub-cooler and flash intercooler

Analysis:

Let m_1 = mass of refrigerant passing through the evaporator or low pressure compressor

m_2 = mass of refrigerant passing through the condenser or high pressure compressor

(*i*) Refrigerant effect (heat absorbed in the evaporator)

$$= m_1(h_1 - h_{10})$$

$$= m_1(h_1 - h_9) \qquad (\because h_{10} = h_9)$$

If load on the evaorator is equivalent to Q tons of refrigeration and since one ton of refrigeration equals heat absorption rate of 3.5 kJ/s

$$m_1(h_1 - h_9) = 3.5\ Q\ ; \quad m_1 \frac{3.5\ Q}{h_1 - h_9}\ \text{kJ/s}$$

(ii) Since the flash chamber is essentially an insulated vessel, it does not exchange any heat with the surroundings. As such, under thermal equilibrium state,

heat supplied to the flash chamber = heat taken away from the flash chamber

$$m_2\ h_8 + m_1\ h_3 = m_2\ h_4 + m_1\ h_9$$

or
$$m_2\ (h_4 - h_8) = m_1\ (h_3 - h_9)$$

$$\therefore \quad m_2 = m_1 \frac{(h_3 - h_9)}{(h_4 - h_8)}$$

(iii) Power required to drive the system,

= work input to LP compressor + work input to high pressure

= $[m_1\ (h_2 - h_1) + m_2\ (h_5 - h_4)]$ kJ/s or kW

and coefficient of performance of the system,

$$COP = \frac{\text{refrigerating effect}}{\text{work input to compressor}}$$

$$= \frac{m_1(h_1 - h_9)}{m_1(h_2 - h_1) + m_2(h_5 - h_4)}$$

EXAMPLE 6.1

A two-stage refrigeration system operates with refrigerant ammonia flowing at the rate of 15 kg/min through the evaporator. The saturation temperatures in the condenser and evaporator units have been noted to be 40°C and – 15°C respectively. If the system has intercooling by liquid refrigerant at 4.25 bar, determine the capacity and COP of the system. How these parameters would be affected if the compression is carried out in single stage unit; the operating temperature limits remaining same ?

Use the p-h chart and properly tables for saturated ammonia refrigerant.

Solution : From property tables for ammonia

p_c = (condenser pressure) corresponding to 40°C = 15.57 bar

p_e = (evaporator pressure) corresponding to – 15°C = 2.37 bar

p_i = (intermediate pressure) = 4.25 bar (given)

Knowing these pressure values, the given cycle can be drawn on *p-h* chart and the relevant values read there from are :

h_1 = 1442 kJ/kg

h_2 = 1473 kJ/kg

h_3 = 1460 kJ/kg

h_4 = 1512 kJ/kg

$h_5 = h_6$ = 390 kJ/kg

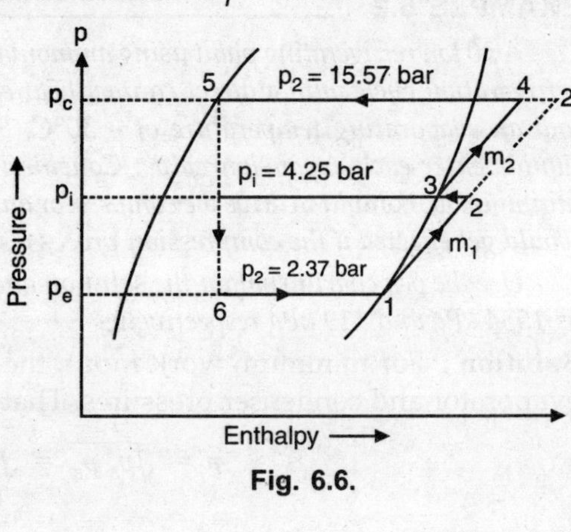

Fig. 6.6.

The values of h_1, h_3 and h_5 can be checked and compared with the values given in the saturated property tables.

Mass flow of refrigerant through the evaporator (or low pressure compressor)

$$m_1 = 15\ \text{kg/min} \quad \text{(given)}$$

Then : m_2 = (mass flow through high pressure compressor or condenser)

$$= \frac{m_1(h_2 - h_5)}{(h_3 - h_5)} = \frac{15(1473-390)}{(1460-390)} = 15.18 \text{ kg/min}$$

Refrigerating effect (heat absorbed in the evaporator)

$$m_1 = (h_1 - h_6) = 15(1442 - 390) = 15780 \text{ kJ/min}$$

Since 1 ton of refrigeration is equivalent to heat absorption rate of 210 kJ/min,

$$\text{Capacity of refrigeration system} = \frac{15780}{210} = \mathbf{75.14} \text{ ton}$$

Work input in driving both the compressors,

$$= m_1 (h_2 - h_1) + m_2(h_4 - h_3)$$

$$= 15 (1473 - 1442) + 15.18 (1512 - 1460) = 1254.36 \text{ kJ/min}$$

Coefficient of performance of the system,

$$COP = \frac{\text{refrigerating effect}}{\text{work input}} = \frac{15780}{1254.36} = \mathbf{12.58}$$

(*b*) When single stage compression is used

– compression of refrigerant follows the path 1-2′ and mass of refrigerant passing through the compressor equals the mass flow rate of 15 kg/min through the evaporator. The enthalpy of refrigerant of state point 2′ as read from *p-h* chart is 1528 kJ/min

– evaporator process 6-1 remains unchanged

Then :

Refrigerant effect = 15780 kJ/min and capacity = 75.14 ton as calculated above

Work input in driving the compressor

$$= m (h_2' - h_1) = 15(1528 - 1442) = 1290 \text{ kJ/min}$$

$$COP \text{ of the system} = \frac{15780}{1290} = \mathbf{12.23}$$

Comments : With liquid refrigerant cooling,

(*i*) The work requirement has been reduced from 1290 kJ/min to 1254.36 kJ/min

(*ii*) The COP of the system has increased from 12.23 to 12.58

EXAMPLE 6.2

A 30 ton refrigerating plant using ammonia as a refrigerant is to operate on a two-stage vapour compression refrigeration cycle with a direct contact heat exchanger. The system has a condensing temperature of 40°C and an evaporating temperature of – 30°C. Saturated vapour enters both the compressors and saturated liquids enter each expansion valve. Considering both stages of compression to be isentropic, estimate the minimum work input to drive the compressor and coefficient of performance of the system. How these parameters would get effected if the compression process is done in a single stage ?

Use the p-h chart to obtain the solution and take saturation pressures corresponding to 40°C and – 30°C as 1554 kPa and 119 kPa respectively.

Solution : For minimum work input, the intermediate pressures p_i is the geometric mean of the evaporator and condenser pressures. That is

$$p_i = \sqrt{p_c\, p_e} = \sqrt{1554 \times 119} = 430 \text{ kPa}$$

The schematic arrangement of the system and *p-h* plot of the processes involved in the given cycle is shown in Fig. 6.7 and the various values as read from *p-h* diagram for ammonia are :

$$h_1 = 1405 \text{ kJ/kg} \; ; \quad h_2 = 1575 \text{ kJ/kg} \; ; \quad h_3 = 1444 \text{ kJ/kg} \; ;$$
$$h_4 = 1630 \text{ kJ/kg} \; ; \quad h_5 = h_6 = 372 \text{ kJ/kg} \quad \text{and} \quad h_7 = h_8 = 182 \text{ kJ/kg}$$

Since 1 ton of refrigeration is equivalent to heat absorption rate of 3.5 kJ/s, mass of the refrigerant passing through the evaporator (or low pressure compressor) is

$$m_1 = \frac{3.5\,Q}{h_1 - h_8} = \frac{3.5 \times 30}{1405 - 182} = 0.0858 \text{ kg/s}$$

From energy balance of the direct contact heat exchanger,

$$m_1\, h_2 + m_2\, h_6 = m_2\, h_3 + m_1\, h_7$$

∴ Mass of refrigerant passing through the condenser (or high pressure compressor),

$$m_2 = m_1 \frac{(h_2 - h_7)}{(h_3 - h_6)} = 0.0858 \times \frac{1575 - 182}{1444 - 372} = 0.1115 \text{ kg/s}$$

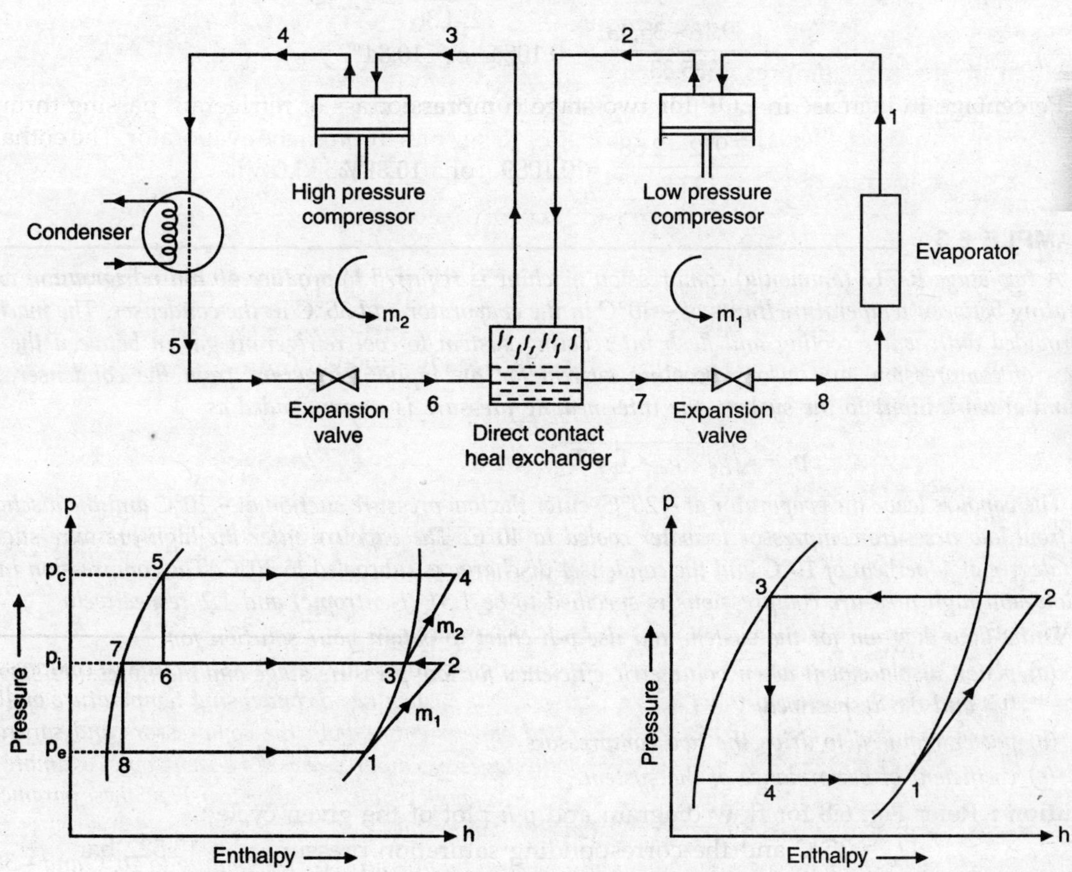

Fig. 6.7.

(*a*) Total work of compression $= m_1 (h_2 - h_1) + m_2 (h_4 - h_3)$

$$= 0.0858\ (1575 - 1405) + 0.1115\ (1630 - 1444)$$
$$= 14.59 + 20.74 = 35.33 \text{ kJ/s} = 35.33 \text{ kW}$$

Coefficient of performance of the system,

$$COP = \frac{\text{refrigerant effect}}{\text{work input}} = \frac{3.5 \times 30}{35.33} = \mathbf{2.972}$$

(*b*) When compression is carried out in single stage compression;

$$h_1 = 1405 \text{ kJ/kg} \; ; \; h_2 = 1805 \text{ kJ/kg} \; ; \; \text{ and } h_3 = h_4 = 372 \text{ kJ/kg}$$

Mass flow rate of refrigerant,

$$m = \frac{3.5Q}{h_1 - h_4} = \frac{3.5 \times 30}{1405 - 372} = 0.0979 \text{ kg/s}$$

Work input to compressor = $m(h_2 - h_1)$

$$W = 0.0979\ (1805 - 1405) = 39.16 \text{ kJ/s} = 39.16 \text{ kW}$$

$$\text{Coefficient of performance } COP = \frac{3.5Q}{W} = \frac{3.5 \times 30}{39.16} = \mathbf{2.68}$$

Thus, percentage increase in compression work for single stage compression

$$= \frac{39.16 - 35.33}{35.33} = 0.1084 \text{ or } 10.84\%$$

Percentage in increase in *COP* for two-stage compression,

$$= \frac{2.972 - 2.68}{2.68} = 0.1089 \text{ or } \mathbf{10.89\%}$$

EXAMPLE 6.3

A two-stage R-717 (ammonia) compression machine is required to produce 40 ton refrigeration while operating between temperature limits of – 30°C in the evaporator and 35°C in the condenser. The machine is provided with water cooling and flash intercooling system to cool refrigerant gas in between the two stages of compression, and also subcooling system for the liquid refrigerant from the condenser. For minimum work input to the system, the intermediate pressure is recommended as

$$p_i = \sqrt{p_c \times p_e \times T_c / T_e}$$

The vapours leave the evaporator at – 20°C, enter the low pressure suction at – 10°C and the discharge gas from low pressure compressor is water cooled to 40°C. The vapours enter the high pressure suction with degree of superheat of 10°C and the condenser discharge is subcooled to 30°C. The compression index for low and high pressure compressions is specified to be 1.31 (isentropic) and 1.2 respectively

Draw flow diagram for the system and use p-h chart to obtain your solution for

(a) piston displacement when volumetric efficiency for low pressure stage and high pressure stage is 0.7 and 0.8 respectively

(b) power required to drive the two compressors

(c) coefficient of performance of the system.

Solution : Refer Fig. 6.8 for flow diagram and *p-h* plot of the given cycle :

t_c = 35°C and the corresponding saturation pressure p_c = 13.522 bar

t_e = – 30°C and the corresponding saturation pressure p_e = 1.199 bar

$$\text{Then :} \quad p_i = \sqrt{13.522 \times 1.199 \times \frac{273 + 35}{273 - 30}} = 4.533 \text{ bar}$$

The values of enthalpy and specific volume as read at appropriate state points of the cycle are :

$$h_1 = 1468 \text{ kJ/kg}\ ;\quad h_2 = 1663 \text{ kJ/kg}\ ;\quad h_3 = 1562 \text{ kJ/kg}\ ;\quad h_4 = 1462 \text{ kJ/kg}$$

$$h_8 = h_9 = 340 \text{ kJkg}\ ;\quad h_{10} = h_{11} = 204 \text{ kJ/kg}\ ;\quad h_{12} = 1444 \text{ kJ/kg}$$

$$v_1 = 1.042 \text{ m}^3\text{/kg} \quad \text{and} \quad v_5 = 0.286 \text{ m}^3\text{/kg}$$

1 ton of refrigerant is equivalent to heat absorption of 3.5 kJ/s

∴ Mass flow of refrigerant in the low pressure circuit,

$$m_1 = \frac{3.5\,Q}{h_{12} - h_{11}} = \frac{3.5 \times 40}{1444 - 204} = 0.1129 \text{ kg/s}$$

From energy balance in the flash intercooler,

$$m_2\, h_9 + m_1\, h_3 = m_2\, h_4 + m_1\, h_{10}$$

where m_2 is the mass flow of refrigerant in the high pressure circuit

$$m_2 = \frac{m_1 (h_3 - h_{10})}{(h_4 - h_9)} = \frac{0.1129(1562 - 204)}{(1462 - 340)} = 0.1366 \text{ g/s}$$

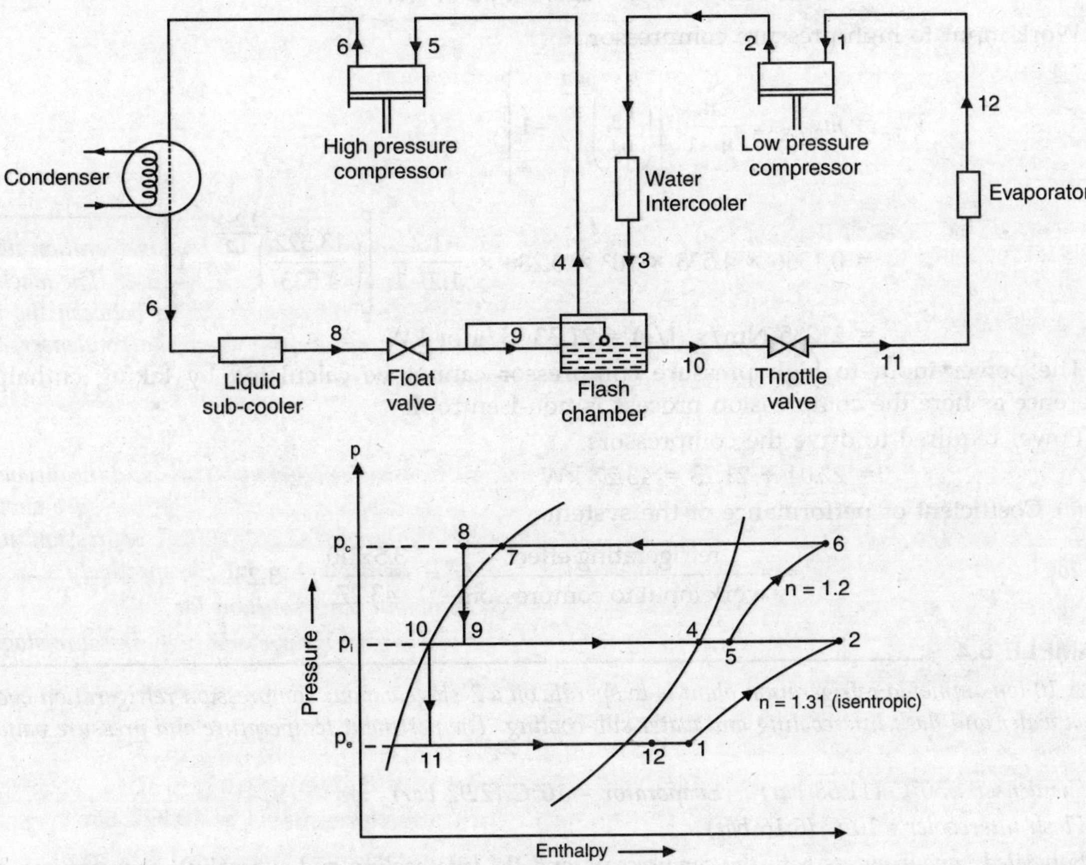

Fig. 6.8.

(*a*) Piston displacement of low pressure compressor

$$= \frac{m_1 v_1}{\eta_v} = \frac{0.1129 \times 1.042}{0.7} = 0.168 \text{ m}^3\text{/s}$$

Piston displacement of high pressure compressor,

$$= \frac{m_2 v_5}{\eta_v} = \frac{0.1366 \times 0.286}{0.8} = 0.0488 \text{ m}^3/\text{s}$$

(*b*) Work input to low pressure compressor

$$W_{lp} = m_1 (h_2 - h_1) = 0.1129 (1663 - 1468) = 22.01 \text{ kJ/s or kW}$$

Alternatively :

$$W_{lp} = m_1 p_1 v_1 \frac{\gamma}{\gamma - 1}\left[\left(\frac{p_i}{p_e}\right)^{\frac{\gamma-1}{\gamma}} - 1\right]$$

$$= 0.1129 \times 1.199 \times 10^5 \times 1.042 \times \frac{1.31}{1.31-1}\left[\left(\frac{4.533}{1.199}\right)^{\frac{1.31-1}{1.31}} - 1\right]$$

$$= 22041 \text{ Nm/s (J/s)} = 22.04 \text{ kJ/s or kW}$$

Work input to high pressure compressor,

$$W_{hp} = m_2 p_5 v_5 \frac{n}{n-1}\left[\left(\frac{p_6}{p_5}\right)^{\frac{n-1}{n}} - 1\right]$$

$$= 0.1366 \times 4.533 \times 10^5 \times 0.286 \times \frac{1.2}{1.2-1}\left[\left(\frac{13.522}{4.533}\right)^{\frac{1.2-1}{1.2}} - 1\right]$$

$$= 21235 \text{ Nm/s (J/s)} = 21.33 \text{ kJ/s or kW}$$

The power input to high pressure compressor cannot be calculated by taking enthalpy difference as here the compression process is non-isentropic

Power required to drive the compressors

$$= 22.04 + 21.23 = 43.27 \text{ kW}$$

(*c*) Coefficient of performance of the system,

$$= \frac{\text{refrigerating effect}}{\text{work input to compressors}} = \frac{3.5 \times 40}{43.27} = \mathbf{3.23}$$

EXAMPLE 6.4

A 10 ton ammonia refrigeration plant is to operate on a 2-stage vapour compression refrigeration cycle with a water and flash intercooling and water sub-cooling. The pertinent temperature and pressure values are :

Condenser : 30°C (11.68 bar) ; Evaporator – 10°C (2.92 bar)

Flash intercooler : 10°C (6.16 bar)

Saturated vapour enters both the compressors and the intercooling and subcooling temperatures are limited to 20°C. Determine :

(a) the total power required to drive the compressors

(b) the coefficient of performance of the system

(c) the piston displacement of each compressor if the volumetric efficiency is 80% for each machine.

Solution : Refer Fig. 6.9 for the schematic arrangement of the system and the corresponding *p-h* diagram of the processes involved in the operating cycle.

Since the temperatures of de-superheated vapour and subcooled refrigerant are limited to 20°C, we have

$$t_3 = t_7 = 20°C$$

From the data given, the cycle can be drawn on *p-h* chart for ammonia. The points 2 and 5 are located from the condition that compression processes 1-2 and 4-5 are isentropic, i.e., $s_1 = s_2$ and $s_4 = s_5$. Further, the points 3 and 7 are marked from the given condition that $t_3 = t_7 = 20°C$

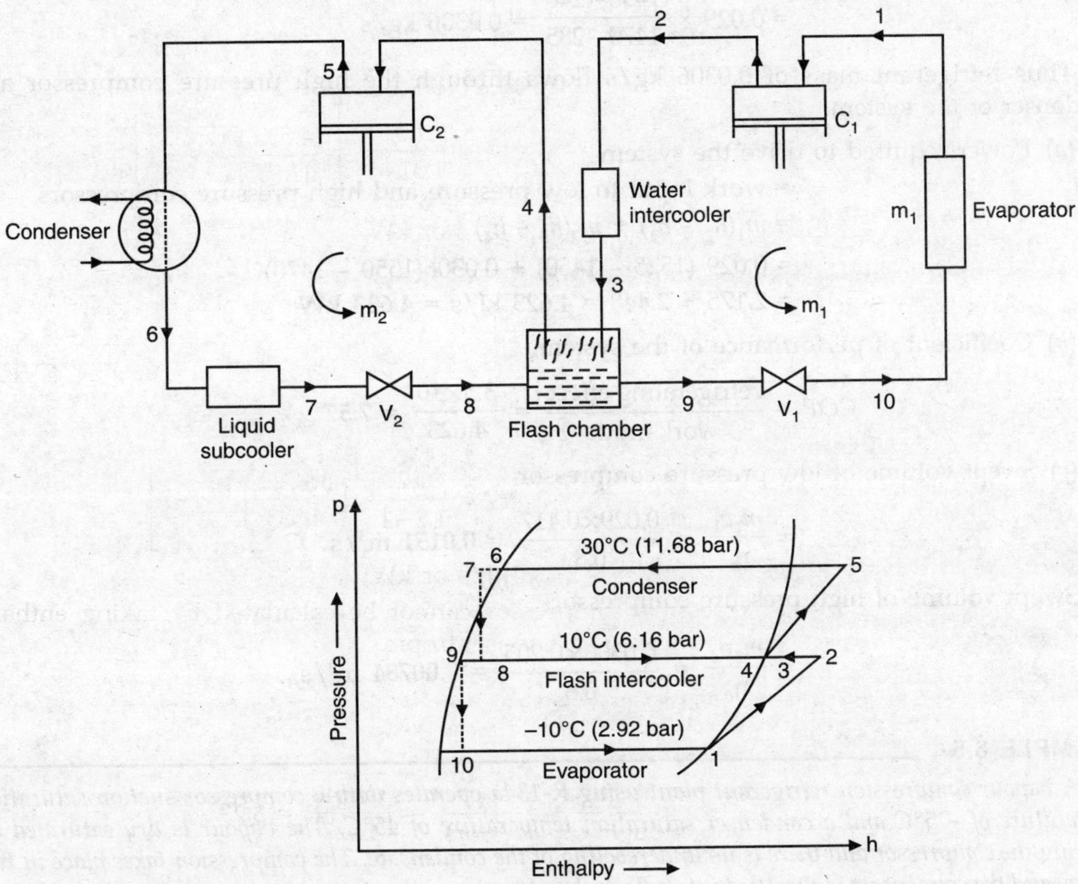

Fig. 6.9.

The relevant values as read from the *p-h* chart are :

$$h_1 = 1450 \text{ kJ/kg} \; ; \; v_1 = 0.417 \text{ m}^3\text{/kg}; \quad h_3 = 1498 \text{ kJ/kg}$$

$$h_4 = 1470 \text{ kJ/kg} \; ; \; v_4 = 0.205 \text{ m}^3\text{/kg}; \quad h_9 = h_{10} = 246 \text{ kJ/kg}$$

$$h_2 = 1525 \text{ kJ/kg} \; ; \; h_5 = 1550 \text{ kJ/kg}; \quad h_7 = h_8 = 285 \text{ kJ/kg}$$

The values of h_1, v_1, h_4, v_4, h_9 can be checked and compared with the values given in the saturated property tables.

Since 1 ton of refrigeration is equivalent to heat absorption rate of 3.5 kJ/s, the mass flow of refrigerant passing through the evaporator or low pressure compressor,

$$m_1 = \frac{3.5Q}{h_1 - h_{10}} = \frac{3.5 \times 10}{1450 - 246} = 0.029 \text{ kg/s}$$

Considering energy balance at flash intercooler,

$$m_1\, h_3 + m_2\, h_8 = m_1\, h_9 + m_2\, h_4$$

or

$$m_2 = m_1 \frac{(h_3 - h_9)}{(h_4 - h_8)}$$

$$= 0.029 \times \frac{1498 - 246}{1470 - 285} = 0.0306 \text{ kg/s}$$

Thus refrigerant mass of 0.0306 kg/s flows through the high pressure compressor and condenser of the system.

(*a*) Power required to drive the system

= work input to low pressure and high pressure compressors

$= m_1(h_2 - h_1) + m_2(h_5 - h_4)$

= 0.029 (1525 – 1450) + 0.0306(1550 – 1470)

= 2.175 + 2.448 = 4.623 kJ/s = **4.623 kW**

(*b*) Coefficient of performance of the system,

$$COP = \frac{\text{refrigerating effect}}{\text{work input}} = \frac{3.5 \times 10}{4.623} = \mathbf{7.57}$$

(*c*) Swept volume of low pressure compressor

$$= \frac{m_1 v_1}{\eta_v} = \frac{0.029 \times 0.417}{0.8} = \mathbf{0.0151} \text{ m}^3\text{/s}$$

Swept volume of high pressure compressor

$$= \frac{m_2 v_4}{\eta_v} = \frac{0.0306 \times 0.205}{0.8} = \mathbf{0.00784} \text{ m}^3\text{/s}$$

EXAMPLE 6.5

A vapour compression refrigerant plant using R-134a operates with a compressor suction saturation temperature of – 5°C and a condenser saturation temperature of 45°C. The vapour is dry saturated on entering the compressor and there is no undercooling of the condensate. The compression takes place in two stages, and the condensate is throttled into a flash chamber at an interstage saturation temperature of 15°C. The dry saturated vapour from the flash chamber is drawn off to mix with the refrigerant from LP compressor before entry to the HP compressor. The liquid from the flash chamber is throttled into the evaporator. Assuming isentropic compression in both compressors and neglecting all losses, calculate :

(a) the amount of vapours bled off at the flash chambers,

(b) the state of vapour at inlet to the second stage of compression,

(c) the refrigerating effect, work done and coefficient of performance of the system.

Make your calculations per unit mass of refrigerant in the condenser and take the properties for R-134a as given below :

Saturation					Superheat			
t	h_f	h_g	s_f	s_g	10 K		20 K	
°C	kJ/kg		kJ/kg K		h kJ/kg	s kJ/kg K	h kJ/kg	s kJ/kg K
– 5	93.46	291.77	0.976	1.715	302.18	1.754	312.05	1.788
15	120.06	303.38	1.071	1.707	314.86	1.746	325.59	1.782
45	162.93	319.54	1.210	1.703	332.87	1.744	345.44	1.780

Solution : Refer Fig. 6.10 for schematic arrangement of the given refrigeration system and *p-h* plot of the processes involved in the given cycle.

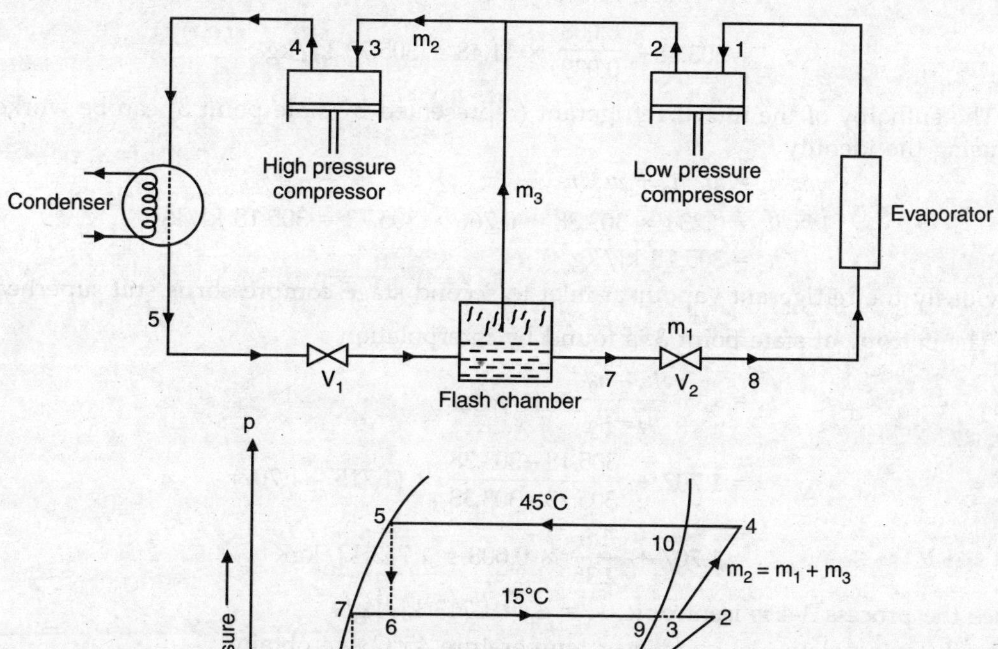

Fig. 6.10.

From the given data: $m_2 = 1$ kg

$$h_1 = 319.54 \text{ kJ/kg} \; ; \quad s_1 = s_2 = 1.703 \text{ kJ/kg K}$$
$$h_9 = 303.38 \text{ kJ/kg} \; ; \quad h_{10} = 319.54 \text{ kJ/kg K}$$
$$h_6 = h_5 = 162.93 \text{ kJ/kg} \; ; \quad h_8 = h_7 = 120.06 \text{ kJ/kg s}$$

(*a*) From heat balance equation for the flash chamber,

$$m_2 h_6 = m_3 h_9 + m_1 h_7$$
$$= m_3 h_9 + (m_2 - m_3) h_7$$

or $m_2(h_6 - h_7) = m_3(h_9 - h_7)$

∴ Mass of vapours bled off at the flash chamber,

$$m_3 = m_2 \frac{(h_6 - h_7)}{(h_9 - h_7)} = m_2 \frac{(h_5 - h_7)}{(h_9 - h_7)}$$

$$= 1 \times \frac{(162.93 - 120.06)}{(303.38 - 120.06)} = 0.234 \text{ kg per kg in the condenser}$$

and $m_1 = m_2 - m_3 = 1 - 0.234 = 0.766$ kg per kg in the condenser

Since the process 1-2 is isentropic, $s_2 = s_1 = 1.715$ kJ/kg K. Then by interpolation of given data at evaporator temperature – 5°C, we obtain,

$$h_2 = 303.38 + \frac{1.715 - 1.707}{1.746 - 1.707} \times (314.86 - 303.38)$$

$$= 303.38 + \frac{0.008}{0.039} \times 11.48 = 305.73 \text{ kJ/kg}$$

(*b*) The enthalpy of the mixed refrigerant (represented by state point 3) can be worked out by the using the identity

$$m_2\, h_3 = m_3\, h_9 + m_1\, h_2$$

$$1 \times h_3 = 0.234 \times 303.38 + 0.766 \times 305.73 = 305.18 \text{ kJ/kg}$$

$$\therefore \quad h_3 = 305.18 \text{ kJ/kg}$$

Obviously the refrigerant vapour at inlet to second stage compressor is still superheated.

(*c*) The entropy at state point 3 is found by interpolation

$$s_3 = s_9 + \frac{h_3 - h_9}{h_2 - h_9} \times (s_2 - s_9)$$

$$= 1.707 + \frac{305.18 - 303.38}{305.73 - 303.38} \times (1.715 - 1.707)$$

$$= 1.707 + \frac{1.8}{2.35} \times 0.008 = 1.713 \text{ kJ/kgK}$$

Since the process 3-4 is isentropic, $s_4 = s_3 = 1.713$ kJ/kgK

Then the interpolation at condenser temperature 45°C, we obtain

$$h_4 = 319.54 + \frac{1.713 - 1.703}{1.744 - 1.703} \times (332.87 - 319.54)$$

$$= 319.54 + \frac{0.01}{0.041} \times 13.33 = 322.79 \text{ kJ/kg}$$

Refrigerating effect $= m_1\,(h_1 - h_8)$
$= m_1\,(h_1 - h_7)$
$= 0.766\,(291.77 - 120.06) =$ **131.53** kJ/kg

Work input to compressors $= m_1(h_2 - h_1) + m_2\,(h_4 - h_3)$
$= 0.766\,(305.73 - 291.77) + 1\,(322.79 - 305.18)$
$= 10.69 + 17.61 =$ **28.3** kJ/kg

Coefficient of performance of the systems,

$$\text{COP} = \frac{\text{refrigerating effect}}{\text{work input to compressors}} = \frac{131.53}{28.3} = \mathbf{4.648}$$

EXAMPLE 6.6

The following data relates to a 9 tonnes refrigeration three-stage compression system :

Condenser pressure = 15 bar ; evaporative pressure = 2.5 bar

intermediate pressures = 5 bar and 10 bar

Assuming NH_3 as working substance, determine the power required to drive the system when it is provided with water cooling.

Compare the COP of this arrangement with that of simple saturation cycle working between the same overall pressure limits. **(MDU)**

Solution : Refer Fig. 6.11 which shows the *p-h* plot for the three stage compression with water intercooling

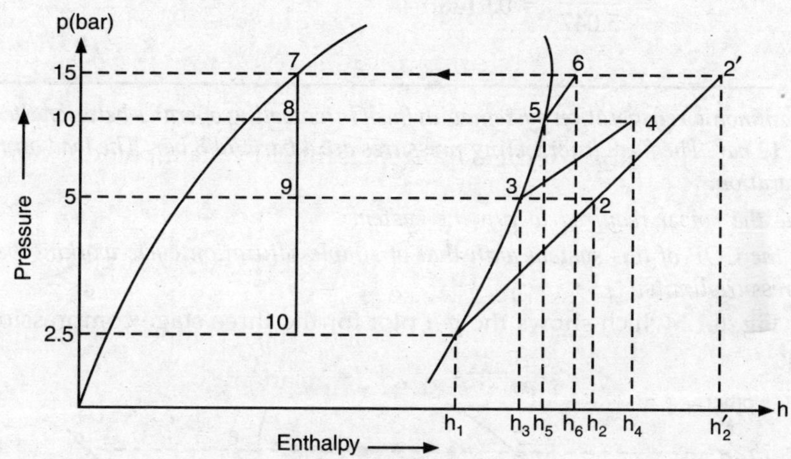

Fig. 6.11.

The enthalpy values of the refrigerant at the salient points indicated on the *p-h* plot are :

$h_1 = 1425$ kJ/kg ; $h_2 = 1515$ kJ/kg ; $h_3 = 1440$ kJ/kg

$h_4 = 1525$ kJ/kg ; $h_5 = 1460$ kJ/kg ; $h_6 = 1500$ kJ/kg

$h_7 = h_8 = h_9 = h_{10}$ 340 kJ/kg and $h_2' = 1660$ kJ/kg

(*a*) *Water intercooling :* 1 ton of refrigerant is equivalent to 3.5 kJ/s and then

$$m_r(h_1 - h_{10}) = 9 \times 3.5 \; ; \; m_r(1425 - 340) = 9 \times 3.5$$

∴ Mass of refrigerant to be circulated through the evaporator is

$$m_r = \frac{9 \times 3.5}{1425 - 340} = 0.02903 \text{ kg/s}$$

Work done in three compressors

$$= m_r[(h_2 - h_1) + (h_4 - h_3) + (h_6 - h_5)]$$

$$= 0.02903[(1515 - 1425) + (1525 - 1440) + (1500 - 1460)]$$

$$= 0.02903[90 + 85 + 40] = 6.241 \text{ kJ/s} = 6.241 \text{ kW}$$

(*b*) For a simple saturation cycle working between the same pressure limits of 2.5 bar and 15 bar, the superheated vapour leaves the compressor at point 2′, and the enthalpy at this point as read from *p-h* chart is 1660 kJ/kg. Then

Working input to compressor for simple saturation cycle

$$= m(h_2' - h_1) = 0.02903\ (1660 - 1425)$$

$$= 6.822\ \text{kJ/s} = 6.822\ \text{kW}$$

Then : COP of the given arrangement $= \dfrac{9\times3.5}{6.241} = 5.047$

COP for saturation cycle $= \dfrac{9\times3.5}{6.241} = 5.047$

∴ Ratio of COP's with water intercooling and simple saturation cycle is

$$= \frac{4.617}{5.047} = \mathbf{0.9148}$$

EXAMPLE 6.7

A three-stage ammonia refrigeration system with flash intercooling operates between the overall pressure limits of 2 bar and 12 bar. The flash intercooling pressures are 4 bar and 8 bar. The load on the evaporator is 10 tonnes of refrigeration.

(a) Determine the power required to run the system

(b) Compare the COP of this system with that of simple saturation cycle working between the same overall pressure limits.

Solution : Refer Fig. 6.12 which shows the *p-h* plot for the three stage compression system with flash intercooling

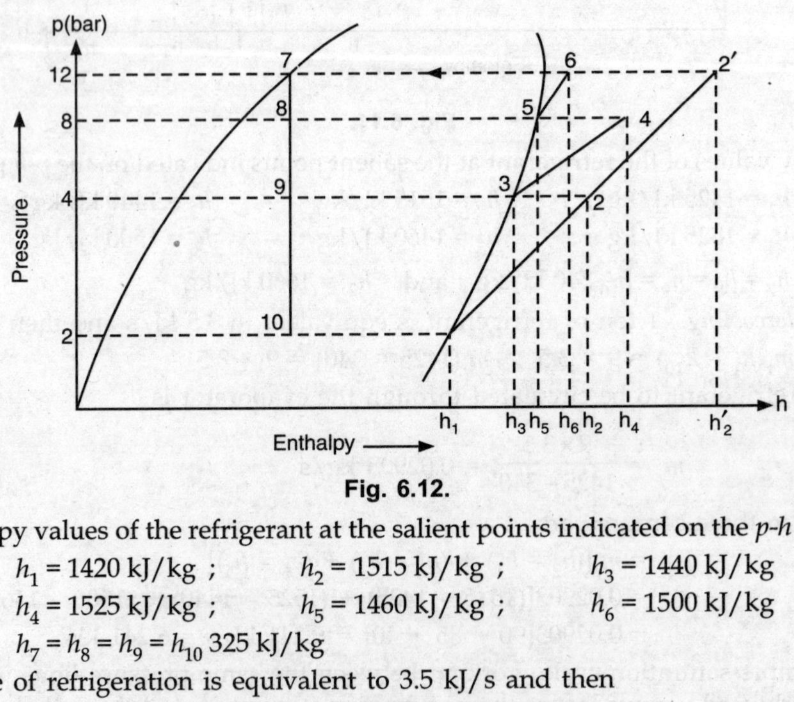

Fig. 6.12.

The enthalpy values of the refrigerant at the salient points indicated on the *p-h* plot are :

$h_1 = 1420$ kJ/kg ; $h_2 = 1515$ kJ/kg ; $h_3 = 1440$ kJ/kg

$h_4 = 1525$ kJ/kg ; $h_5 = 1460$ kJ/kg ; $h_6 = 1500$ kJ/kg

$h_7 = h_8 = h_9 = h_{10}$ 325 kJ/kg

(*a*) 1 tonne of refrigeration is equivalent to 3.5 kJ/s and then

$$m_1(h_1 - h_{10}) = 10 \times 3.5\ ;\quad m_1(1420 - 325) = 35$$

∴ Mass of refrigerant passing through the evaporator or low pressure compressor,

$$m_1 = \frac{35}{1420 - 325} = 0.03196\ \text{kg/s}$$

Mass m_2 of liquid refrigerant evaporated in the flash intercooler after first(low) pressure stage of compression

$$m_2 = \frac{m_1(h_2 - h_3)}{h_3 - h_9} = \frac{0.03196\,(1515 - 1440)}{1440 - 325} = 0.00215 \text{ kg/s}$$

∴ Mass of refrigerant passing through the intermediate pressure compressor

$$m_3 = m_1 + m_2 = 0.03196 + 0.00215 = 0.03411 \text{ kg/s}$$

Mass m_4 of liquid refrigerant evaporated in the flash chamber after the second stage of compression

$$m_4 = \frac{m_3(h_4 - h_5)}{h_5 - h_8} = \frac{0.03411(1525 - 1460)}{1460 - 325} = 0.001953 \text{ kg/s}$$

∴ Mass of refrigerant passing through the high pressure compressor

$$m_5 = m_3 + m_4 = 0.03411 + 0.001953 = 0.03606 \text{ kg/s}$$

Total work done in the three compressors

$$= W_{lp} + W_{ip} + W_{hp}$$

$$= m_1(h_2 + h_1) + m_3(h_4 + h_3) + m_5(h_6 + h_5)$$

$$= 0.03196\,(1515 - 1420) + 0.03411\,(1525 - 1440) + 0.03606\,(1500 - 1460)$$

$$= 3.0362 + 2.8993 + 1.4424 = 7.378 \text{ kJ/s} \rightarrow 7.378 \text{ kW}$$

(*b*) *COP* of the given system

$$= \frac{\text{refrigerating effect}}{\text{work input to compressors}} = \frac{10 \times 3.5}{7.378} = 4.774$$

For a simple saturation cycle working between the same pressure limits of 2 bar and 12 bar, the superheated vapour leaves the compressor at point 2′, and the enthalpy at this point as read from *p-h* chart is 1670 kJ/kg. Then

Work input to compressor for simple saturation cycle

$$= m_1(h_2' - h_1) = 0.03196\,(1670 - 1420)$$

$$= 7.99 \text{ kJ/s}$$

$$COP \text{ for saturation cycle} = \frac{10 \times 3.5}{7.99} = 4.380$$

Then percentage improvement in *COP* for three stage compression as compared to simple saturation cycle is

$$= \frac{4.744 - 4.380}{4.380} \times 100 = \mathbf{8.3\%}$$

6.7. MULTI-EVAPORATOR SYSTEM AT THE SAME TEMPERATURE WITH SINGLE COMPRESSOR AND EXPANSION VALVE

Fig. 6.13 shows the schematic arrangement of the different components of a single compressor with multiple evaporators maintained at the same temperature. The *p-h* diagram of the processes involved is also shown along with.

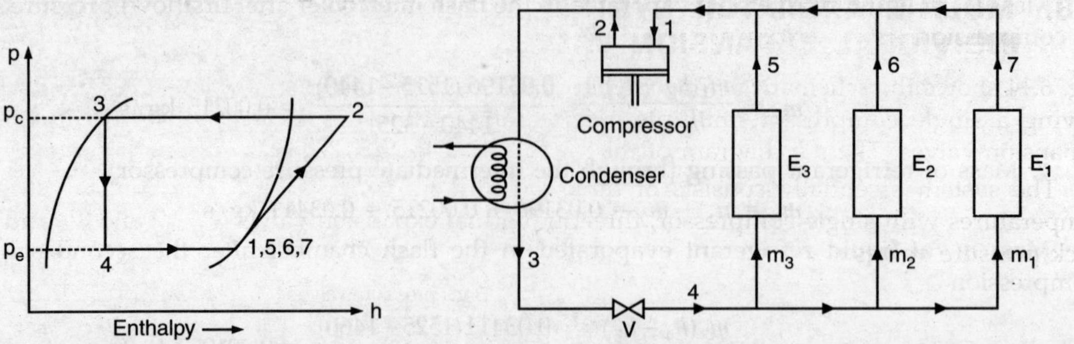

Fig. 6.13. Multiple evaporators at the same temperature with single compressor and expansion valve

Operation : The saturated vapour entering the compressor at state 1 is compressed isentropically (process 1-2) and thereafter the superheated vapour enters the condenser. During condensation (process 2-3), the refrigerant vapour rejects heat, both superheat and latent heat, to the stream of cooling water. The resultant saturated liquid is taken to expansion valve where it expands at constant enthalpy (process 3-4). The expanded refrigerant then enters the evaporators and undergoes evaporation at the same temperature. The pressures in the different evaporators are same and equal to that at entry to compressor. The flows leaving the evaporators at points 5, 6 and 7 are joined and the mixed flow of refrigerant vapour is passed onto the compressor inlet at point 1.

Analysis : Let

m_1 , Q_1 = mass of refrigerant flowing through evaporator E_1 and its load in tons of refrigeration

m_2 , Q_2 = mass of refrigerant flowing through evaporator E_2 and its load in tons of refrigeration

m_3 , Q_3 = mass of refrigerant flowing through evaporator E_3 and its load in tons of refrigeration

Refrigerating effect (heat absorbed) for evaporator $E_1 = m_1 (h_1 - h_4)$
Refrigerating effect (heat absorbed) for evaporator $E_2 = m_2 (h_1 - h_4)$
Refrigerating effect (heat absorbed) for evaporator $E_3 = m_3 (h_1 - h_4)$
Since 1 ton of refrigeration is equivalent to heat absorption rate of 3.5 kJ/s

$$m_1 = \frac{3.5Q_1}{h_1 - h_4} \; ; \; m_2 = \frac{3.5Q_2}{h_1 - h_4} \quad \text{and} \quad m_3 = \frac{3.5Q_3}{h_1 - h_4}$$

Power required to drive the system (work input to compressor)

$$= (m_1 + m_2 + m_2)\,(h_2 - h_1) \text{ kJ/s or kW}$$

and coefficient of performance of the system,

$$COP = \frac{\text{refrigerating effect}}{\text{work input to compressor}}$$

$$= \frac{(m_1 + m_2 + m_3)(h_1 - h_4)}{(m_1 + m_2 + m_3)(h_2 - h_1)} = \frac{h_1 - h_4}{h_2 - h_1}$$

This arrangement of refrigeration system is used when number of cooling loads kept in different compartments are to be maintained at the same temperature.

6.8. MULTI-EVAPORATOR SYSTEM WITH SINGLE COMPRESSION AND INDIVIDUAL EXPANSION VALVES

Fig. 6.14 shows the schematic arrangement of the different components of a refrigeration system having a single compressor, multiple evaporators at different temperatures with individual expansion valves. The *p-h* diagram of the various processes involved is also shown along side.

The system essentially consists of three evaporators E_1, E_2 and E_3 operating at different temperatures with single compressor, three individual expansion valves V_1, V_2 and V_3 and two back pressure valves.

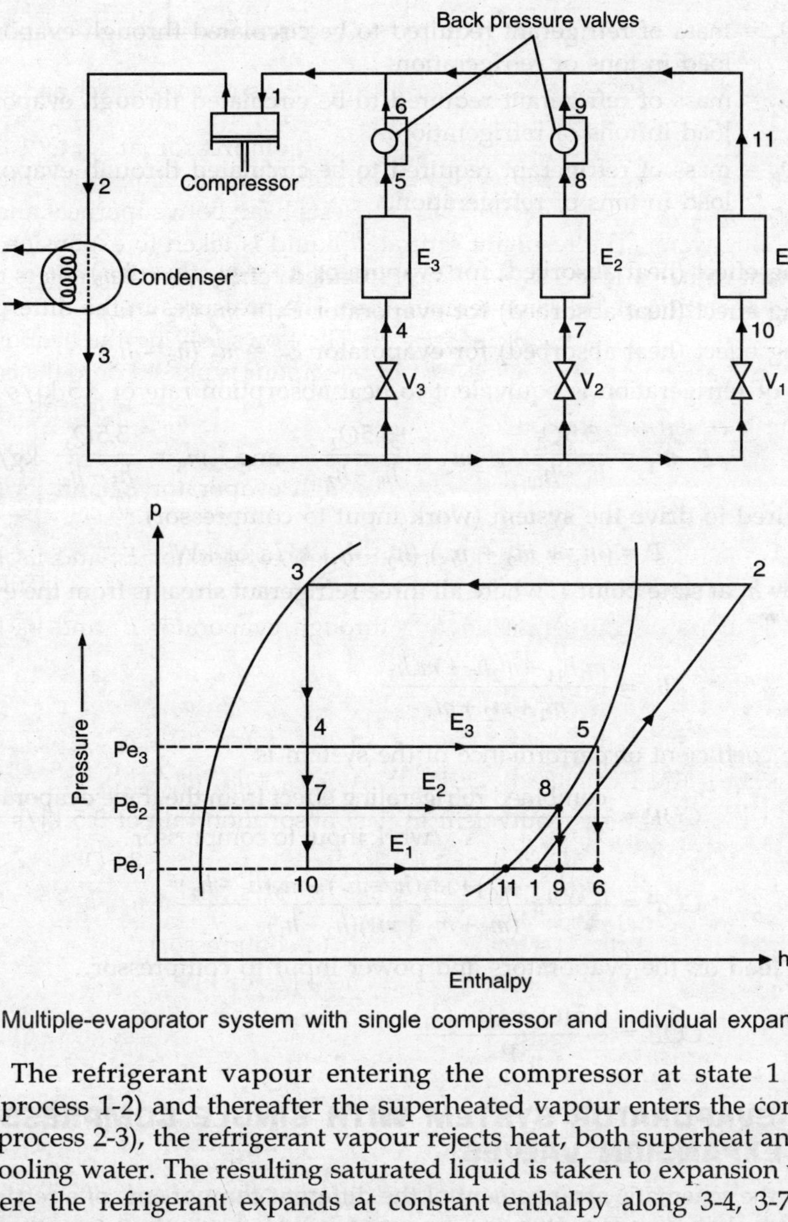

Fig. 6.14. Multiple-evaporator system with single compressor and individual expansion valves

Working : The refrigerant vapour entering the compressor at state 1 is compressed isentropically (process 1-2) and thereafter the superheated vapour enters the condenser. During condensation (process 2-3), the refrigerant vapour rejects heat, both superheat and latent heat, to the stream of cooling water. The resulting saturated liquid is taken to expansion valves arranged in parallel where the refrigerant expands at constant enthalpy along 3-4, 3-7 and 3-10. The expanded refrigerant enters the evaporators and produces the cooling affect at different temperatures. Further, it is to be noted that :

(*i*) The refrigerant leaving the evaporators E_3 at pressure Pe_3 expands further (process 5-6) in the back pressure valve to pressure Pe_1, the pressure in the evaporator E_1.

(*ii*) The refrigerant at exit from the evaporator E_2 at pressure Pe_2 expands further (process 8-9) in the back pressure valve to the pressure Pe_1 of the evaporator E_1.

(*iii*) The two streams leaving the back pressure valves at state points 6 and 9 are mixed with the refrigerant stream leaving the evaporator E_1 at point 11.

The mixture is then passed onto the suction of the compressor and the cycle is repeated.

Analysis : Let

m_1 , Q_1 = mass of refrigerant required to be circulated through evaporator E_1 and its load in tons of refrigeration.

m_2 , Q_2 = mass of refrigerant required to be circulated through evaporator E_2 and its load in tons of refrigeration.

m_3 , Q_3 = mass of refrigerant required to be circulated through evaporator E_3 and its load in tons of refrigeration.

Then :

Refrigerating effect (heat absorbed) for evaporator $E_1 = m_1\,(h_1 - h_{10})$

Refrigerating effect (heat absorbed) for evaporator $E_2 = m_2\,(h_8 - h_7)$

Refrigerating effect (heat absorbed) for evaporator $E_3 = m_3\,(h_5 - h_4)$

Since 1 ton of refrigeration is equivalent to heat absorption rate of 3.5 kJ/s

$$m_1 = \frac{3.5Q_1}{h_1 - h_{10}}\ ;\quad m_2 = \frac{3.5Q_2}{h_8 - h_7}\quad \text{and}\quad m_3 = \frac{3.5Q_3}{h_5 - h_4}\ \text{kg/s}$$

Power required to drive the system (work input to compressor)

$$P = (m_1 + m_2 + m_3)\,(h_2 - h_1)\ \text{kJ/s or kW}$$

The enthalpy h_1 at state point 1, where all three refrigerant streams from the evaporators mix, is given by

$$h_1 = \frac{m_1 h_{11} + m_2 h_8 + m_3 h_5}{m_1 + m_2 + m_3}$$

Further, the coefficient of performance of the system is

$$COP = \frac{\text{combined refrigerating effect from the three evaporators}}{\text{work input to compressor}}$$

$$COP = \frac{m_1(h_1 - h_{10}) + m_2(h_8 - h_7) + m_3(h_5 - h_4)}{(m_1 + m_2 + m_3)(h_2 - h_1)}$$

In terms of load on the evaporators and power input to compressor,

$$COP = \frac{3.5(Q_1 + Q_2 + Q_3)}{P}$$

6.9. MULTI-EVAPORATOR SYSTEM WITH SINGLE COMPRESSOR AND MULTI-EXPANSION VALVES

Fig. 6.15. shows the schematic arrangement of the different components of a refrigeration system having multiple evaporators at different temperatures with single compressor and multiple expansion valves. The *p-h* diagram of the various processes involved is also shown along side.

The system essentially consists of three evaporators E_1, E_2 and E_3 operating at different temperatures with single compressor, three expansion valves V_1, V_2 and V_3 and two back pressure valves.

Working : The refrigerant vapour entering the compressor at state 1 is compressed isentropically (process 1-2) and thereafter the superheated vapour enters the condenser. During condensation (process 2-3), the refrigerant vapour rejects heat, both superheat and latent heat, to the stream of cooling water.

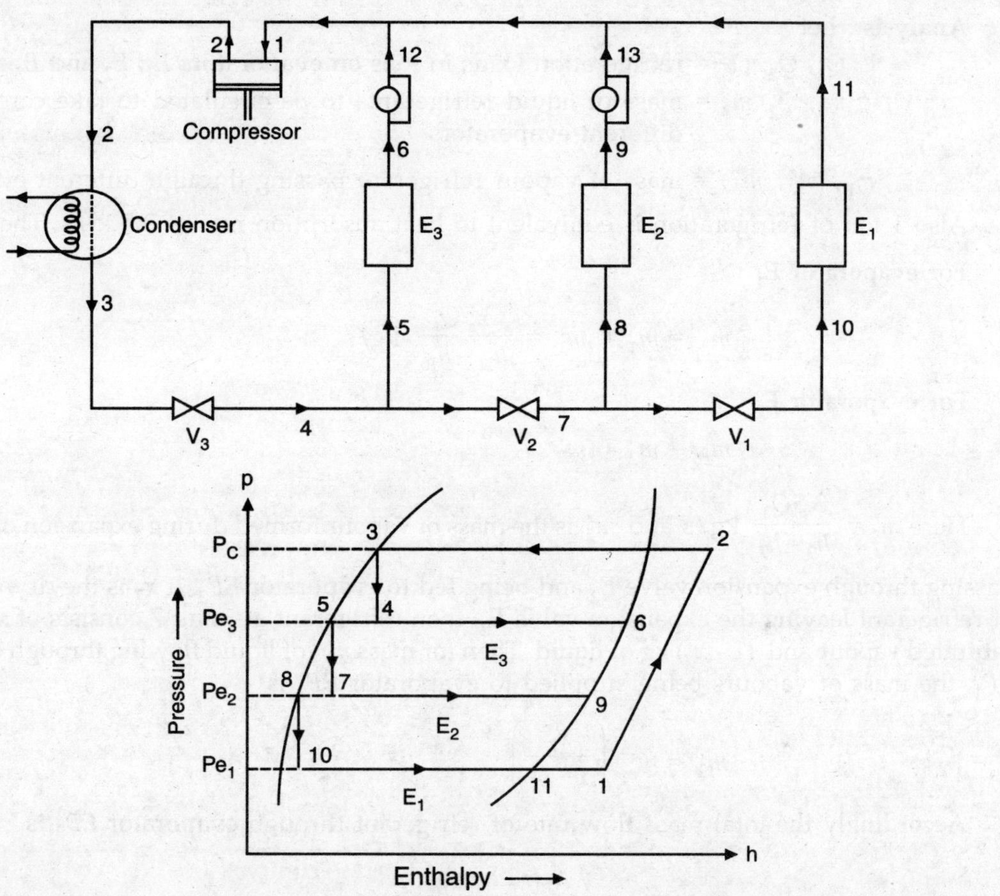

Fig. 6.15. Multiple evaporaotr systems with single compressor and multiple expansion valves

The resulting saturated liquid is taken to expansion valves V_3 where it expands isentropically (process 3-4) to pressure Pe_3 corresponding to evaporator E_3. The vapour and some of the liquid then flows to evaporator E_3, and the rest of refrigerant enters the expansion valve V_2. The refrigerant moving to the evaporator E_3 absorbs heat and produces the cooling affect at the required temperature. Further, it is to be noted that ;

(*i*) The refrigerant entering the expansion valve V_2 expands isentropically (process 5-7) and its pressure gets reduced from Pe_3 to Pe_2.

(*ii*) A part of the vapours and liquid refrigerant enters the evaporator E_2 and the remaining refrigerant is taken to expansion valve V_1 where it expands from pressure Pe_2 to pressure Pe_1, *i.e.*, the pressure for the evaporator E_1.

(*iii*) The evaporation (heat absorption) processes in the evaporators E_2 and E_1 are along the constant pressure paths 8-9 and 10-11 respectively. Due to heat absorption the required cooling effect is produced.

(*iv*) The exit pressures Pe_3 and Pe_2 from the evaporators E_3 and E_2 are brought to the level of pressure Pe_1

Finally, the refrigerant streams (state points 11, 12 and 13) from the different evaporators mix together and the mixture is passed on to the compressor inlet, and the cycle is repeated.

Analysis : Let

Q_1, Q_2, Q_3 = refrigeration loads in tons on evaporators E_1, E_2 and E_3 respectively

m_1, m_2, m_3 = mass of liquid refrigerants to be circulated to take care of load of different evaporators

m_1', m_2', m_3', = mass of vapour refrigerant passing through different evaporators

Also 1 ton of refrigeration is equivalent to heat absorption rate of 3.5 kJ/s. Then

For evaporator E_1

$$m_{e_1} = m_1 + m_1' = \frac{3.5Q_1}{h_{11} - h_{10}} \text{ kg/s} \qquad (m_1' = 0)$$

For evaporator E_2

$$m_{e_2} = m_2 + m_2'$$

Here $m_2 = \dfrac{3.5Q_2}{h_9 - h_7}$ kg/s and m_2' is the mass of vapour formed during expansion of refrigerant passing through expansion valve E_2 and being fed to evaporator EP_2. If x_7 is the dryness fraction of refrigerant leaving the expansion valve E_2, then refrigerant at point 7 consists of x_7 kg of dry saturated vapour and $(1 - x_7)$ kg of liquid. Then for mass m_{e_1} of liquid flowing through evaporation EP_1, the mass of vapours being supplied to evaporator EP_2 is

$$m_2' = m_{e_1}\left(\frac{x_7}{1-x_7}\right)$$

Accordingly the total mass flow rate of refrigerant through evaporator EP_2 is

$$m_{e_2} = m_2 + m_{e_1}\left(\frac{x_7}{1-x_7}\right)$$

For evaporator E_3

$$m_{e_3} = m_3 + m_3'$$

Here $m_3 = \dfrac{3.5\,Q_3}{h_6 - h_4}$ kg/s and m_3' is the mass of vapour formed during expansion of refrigerant passing through expansion valve E_2 and being fed to evaporator EP_3. If x_4 is the dryness fraction of refrigerant leaving the expansion valve E_3, then mass of vapour formed

$$m_3' = (m_{e_1} + m_{e_2})\left(\frac{x_4}{1-x_4}\right)$$

Accordingly, the total mass flow rate of refrigerant through evaporator EP_3 is

$$m_{e3} = m_3 + (m_{e1} + m_{e2})\left(\frac{x_4}{1-x_4}\right) \text{ kg/s}$$

- Power required to drive the system (work input to compressor)

$$P = (m_{e1} + m_{e2} + m_{e3})(h_2 - h_1) \text{ kJ/s or kW}$$

The enthalpy h_1 at state point 1, where all the three refrigerant streams from the evaporators mix, is given by

$$h_1 = \frac{m_{e_1} h_{11} + m_{e_2} h_{13} + m_{e_3} h_{12}}{m_{e_1} + m_{e_2} + m_{e_3}}$$

$$= \frac{m_{e_1} h_{11} + m_{e_2} h_9 + m_{e_3} h_6}{m_{e_1} + m_{e_2} + m_{e_3}} \qquad (\because h_{13} = h_9 \text{ and } h_{12} = h_6)$$

- Refrigerating effect of the system,

$$= m_1(h_{11} - h_{10}) + m_2(h_9 - h_7) + m_3(h_6 - h_4)$$

It should be noted that vapours formed during expansion while passing through expansion valves E_2 and E_3 do not take part in producing cooling effect.

- Coefficient of performance of the system,

$$COP = \frac{\text{combined refrigerating effect from the three evaporators}}{\text{work input to compressor}}$$

$$= \frac{m_1(h_{11} - h_{10}) + m_2(h_9 - h_7) + m_3(h_6 - h_4)}{(m_{e_1} + m_{e_2} + m_{e_3})(h_2 - h_1)}$$

In terms of load on the evaporators and power input to compressor,

$$COP = \frac{3.5(Q_1 + Q_2 + Q_3)}{P}$$

6.10. MULTI-EVAPORATOR SYSTEM AT DIFFERENT TEMPERATURES WITH MULTI-COMPRESSORS AND INDIVIDUAL EXPANSION VALVE

Fig. 6.16 shows the schematic arrangement of the different components of a refrigeration system having multi-evaporators at different temperatures with multi-stage compressors and individual expansion valves. The *p-h* diagram of the various processes involved is also shown along side.

The systems essentially consists of three compressors C_1, C_2, C_3 with three evaporators E_1, E_2, E_3 and their individual expansion valves V_1, V_2 and V_3.

Working : The saturated vapour enters the compressors C_1, C_2, C_3 at state points 1, 3, 5 and is compressed isentropically to superheated conditions designated 2, 4, 6. The superheated vapours at exit from all the three compressors are mixed and the mixed stream taken to condenser. During condensation, the refrigerant vapour rejects heat, both heat of superheat and latent heat, to a stream of cooling water. The resulting saturated liquid then expands in the different expansion valves at constant enthalpy processes 7-8, 7-9 and 7-10. The expanded refrigerant streams are then made to enter the evaporators. The vapour from each of the evaporators next enters the corresponding compressor and the cycle is repeated.

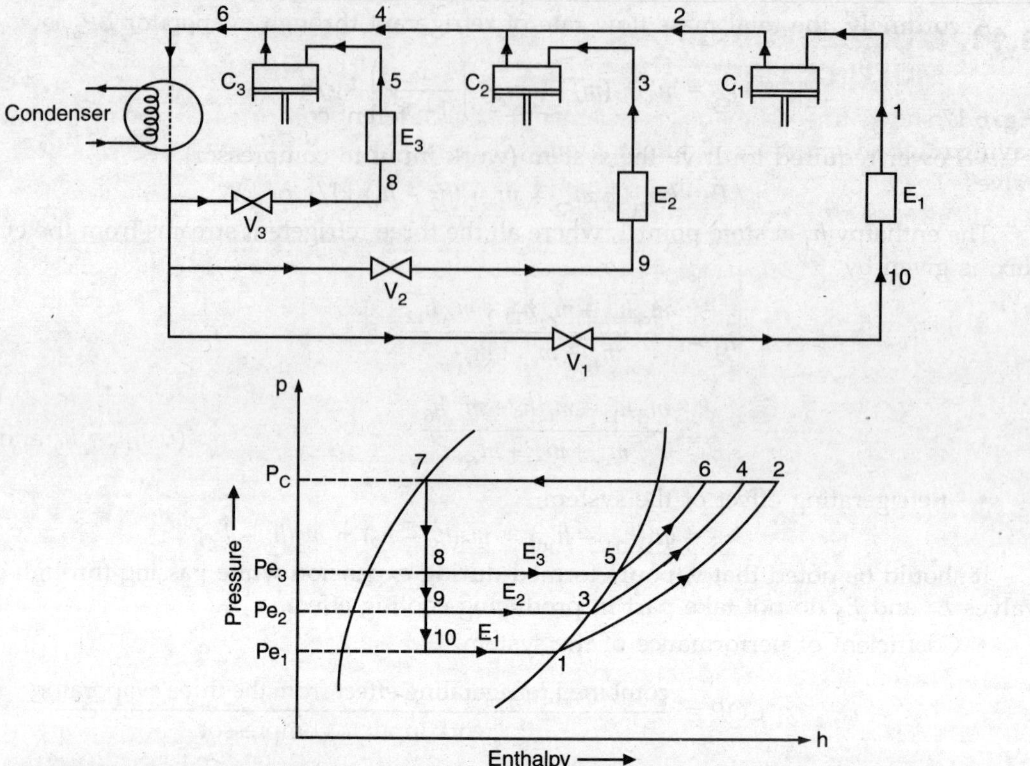

Fig. 6.16. Multi-evaporator system at different temperatures with multi-compressors and individual expansion valves

Analysis : Let

Q_1, Q_2, Q_3 = refrigeration loads in tons on evaporators E_1, E_2 and E_3 respectively

m_1, m_2, m_3 = mass of liquid refrigerants to be circulated to take care of load of different evaporators

Also 1 ton of refrigeration is equivalent to heat absorption rate of 3.5 kJ/s. Then

$$m_1 = \frac{3.5\,Q_1}{h_1 - h_{10}}, \quad m_2 = \frac{3.5\,Q_2}{h_3 - h_9} \quad \text{and} \quad m_3 = \frac{3.5\,Q_3}{h_5 - h_8}$$

- Refrigerating effect of the system,

$$= m_1(h_1 - h_{10}) + m_2(h_3 - h_9) + m_3(h_5 - h_8)$$

- Power required to drive the system (work input to compressors)

$$P = m_1(h_2 - h_1) + m_2(h_4 - h_3) + m_3(h_6 - h_5)$$

- Coefficient of performance of the system,

$$COP = \frac{\text{combined refrigerating effect from the three evaporators}}{\text{work input to compressors}}$$

$$= \frac{m_1(h_1 - h_{10}) + m_2(h_3 - h_9) + m_3(h_5 - h_8)}{m_1(h_2 - h_1) + m_2(h_4 - h_3) + m_3(h_6 - h_5)}$$

In terms of load on the evaporators and power input to compressors,

$$COP = \frac{3.5(Q_1 + Q_2 + Q_3)}{P}$$

6.11. MULTI-EVAPORATORS AT DIFFERENT TEMPERATURES WITH MULTI-COMPRESSORS AND MULTI-EXPANSION VALVES

Fig. 6.17 shows the schematic arrangement of the different components of a refrigeration system having multi-evaporators at different temperatures with multi-compressors and multi-expansion valves. The *p-h* diagram of the various processes involves is also shown along side.

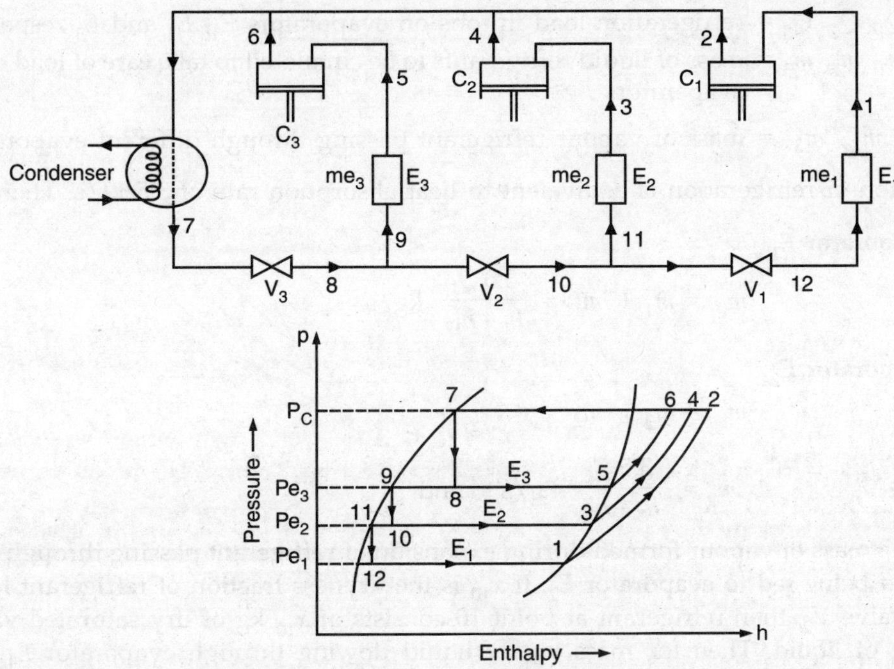

Fig. 6.17. Multi-evaporators at different temperatures with multi-compressors and individual expansion valve

The system essentially consists of three compressors C_1, C_2, C_3, three evaporators E_1, E_2, E_3 and their individual expansion valves V_1, V_2 and V_3.

Working : The saturated vapour enters the compressors C_1, C_2, C_3 at state points, 1, 3, 5 and is compressed isentropically to superheated conditions designated 2, 4, 6. The superheated vapours at exit from all the three compressors are mixed and the mixed stream is taken to condenser. During condensation, the refrigerant vapour loses heat, both heat of superheat and latent heat, to a stream of cooling water. The resulting saturated liquid then flows through the expansion valve V_3 and expands isentropically (process 7-8) and the pressure drops from P_C to Pe_3. The vapour and some of the liquid then flows to E_3, and the rest of refrigerant enters the expansion valve V_2. The refrigerant moving to evaporator E_3 absorbs heat and produces the cooling effect at the desired temperature. After evaporation, this refrigerant enters the compressor C_3. Further, it is to be noted that :

(*i*) The refrigerant entering the expansion valve V_2 expands isentropically (process 9-10) and its pressure gets reduced from Pe_3 to Pe_2. A part of the vapours and liquid refrigerant enters the evaporator and the remaining refrigerant is taken to expansion valve V_1 where it expands from pressure Pe_2 to pressure Pe_1.

(*ii*) The evaporation (heat absorption) processes in the evaporators E_2 and E_1 are along the constant pressure paths 10-3 and 12-1 respectively. Due to heat absorption the required cooling effect is produced.

(*iii*) The refrigerant vapours at exit from the evaporators, are taken to respective compressors for compression and the cycle is repeated.

Analysis : Let

Q_1, Q_2, Q_3 = refrigeration load in tons on evaporators E_1, E_2 and E_3 respectively

m_1, m_2, m_3 = mass of liquid refrigerants to be circulated to take care of load of different evaporators

m_1', m_2', m_3', = mass of vapour refrigerant passing through different evaporators

Also 1 ton of refrigeration is equivalent to heat absorption rate of 3.5 kJ/s. Then

For evaporator E_1

$$m_{e1} = m_1 + m_1' = \frac{3.5\,Q_1}{h_1 - h_{12}} \text{ kg/s} \qquad (m_1' = 0)$$

For evaporator E_2

$$m_{e2} = m_2 + m_2'$$

Here
$$m_2 = \frac{3.5\,Q_2}{h_3 - h_{10}} \text{ kg/s} \quad \text{and}$$

m_2' is the mass of vapour formed during expansion of refrigerant passing through expansion valve V_2 and being fed to evaporator E_2. If x_{10} is the dryness fraction of refrigerant leaving the expansion valve V_2, then refrigerant at point 10 consists of x_{10} kg of dry saturated vapour and $(1 - x_{10})$ kg of liquid. Then for mass m_{e1} of liquid flowing through evaporator E_1, the mass of vapours being supplied to evaporator E_2 is

$$m_2' = m_{e1}\left(\frac{x_{10}}{1 - x_{10}}\right) \text{ kg/s}$$

Accordingly the total mass flow rate of refrigerant through evaporator E_2 is

$$m_{e2} = m_2 + m_{e1}\left(\frac{x_{10}}{1 - x_{10}}\right)$$

For evaporator E_3

$$m_{e3} = m_3 + m_3'$$

Here
$$m_3 = \frac{3.5\,Q_3}{h_5 - h_8} \text{ kg/s and}$$

m_3' is the mass of vapour formed during expansion of refrigerant of mass $(m_{e1} + m_{e2})$ kg/s passing through expansion valve V_2 and being fed to evaporator E_3. If x_8 is the dryness fraction of refrigerant leaving the expansion valve V_3, then mass of vapour formed

$$m_3' = (m_{e1} + m_{e2})\left(\frac{x_8}{1 - x_8}\right) \text{ kg/s}$$

Accordingly, the total mass flow rate of refrigerant through evaporator E_3 is

$$m_{e3} = m_3 + (m_{e1} + m_{e2}) \left(\frac{x_8}{1 - x_8} \right) \text{ kg/s}$$

- Power required to drive the system (work input to compressors)

$$P = m_{e1}(h_2 - h_1) + m_{e2}(h_4 - h_3) + m_{e3}(h_6 - h_5)$$

- Refrigerating effect of the system,

$$= m_1(h_1 - h_{12}) + m_2(h_3 - h_{10}) + m_3(h_5 - h_9)$$

It should be noted that vapours formed during expansion while passing through expansion valves V_2 and V_3 do not take part in producing cooling effect.

- Coefficient of performance of the system,

$$COP = \frac{\text{combined refrigerating effect from the three evaporators}}{\text{work input to compressors}}$$

$$= \frac{m_1(h_1 - h_{12}) + m_2(h_3 - h_{10}) + m_3(h_5 - h_9)}{m_{e_1}(h_2 - h_1) + m_{e_2}(h_4 - h_3) + m_{e_3}(h_6 - h_5)}$$

In terms of load on the evaporators and power input to compressors,

$$COP = \frac{3.5(Q_1 + Q_2 + Q_3)}{P}$$

6.12. MULTI-EVAPORATORS SYSTEM AT DIFFERENT TEMPERATURES WITH MULTI-COMPRESSORS, INDIVIDUAL EXPANSION VALVES AND FLASH CHAMBER

Fig. 6.18 shows the schematic arrangement of the different components of a refrigeration system having multi-evaporators at different temperature with multi-compressors, individual expansion valve and flash chamber. The *p-h* diagram of the various pressures involved is also shown along side.

The systems essentially consists of three compressors C_1, C_2, C_3, three evaporators E_1, E_2, E_3, three expansion valves V_1, V_2 and V_3 and two flash chambers.

Working : The salient points of operation of the system are :

(*i*) The saturated vapour at exit from the evaporator E_1 enters the compressor C_1 where it is compressed isentropically (process 1-2) and its pressure increases from Pe_1 to Pe_2. The superheated vapour at state 2 enters the chamber F_1 where it is desuperheated by the liquid refrigerant extracted at point 9 from the refrigerant stream leaving the expansion valve V_2. There is mixing of cooled vapours leaving the flash chamber F_1 and the saturated vapours leaving the evaporators E_2.

(*ii*) The mixed stream enters the compressors C_2 where isentropic compression occurs (process 3-4) and pressure of the vapours increases form Pe_2 to Pe_3. The superheated vapours enter the flash chamber F_2 and are desuperheated by the liquid refrigerant extracted at point 8 from the refrigerant stream leaving the expansion valve V_3. The cooled vapours from the flash chamber enter the compressor C_3 along with saturated vapours leaving the evaporator E_3. The vapours undergo isentropic compression (process 5-6) in compressors C_3.

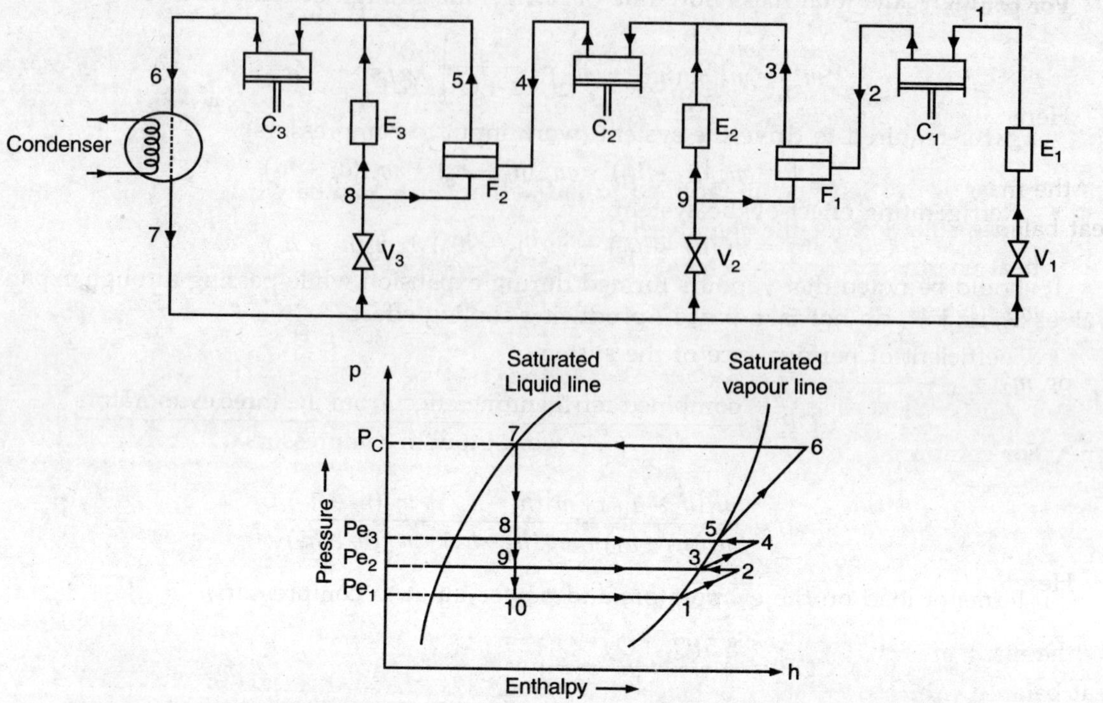

Fig. 6.18. Multi-evaporator system at different temperatures with multi-compressors individual expansion valves and flash chamber

(*iii*) The superheated vapours at exit from compressor C_3 are taken to condenser. During condensation, the refrigerant vapours lose, both heat of superheat and latent heat, to a stream of the cooling water.

(*iv*) The resulting saturated liquid is divided into three streams which enter the expansion valves V_1, V_2 and V_3. The refrigerant expands isentropically and vapour-liquid mixture is produced. A portion of the liquid refrigerant leaving the expansion valves V_3 and V_2 is extracted and sent to the flash chambers to desuperheat the refrigerant vapours leaving the compressors C_1 and C_2. The remaining refrigerant enters the evaporators E_3 and E_2 and provides the desired refrigeration effect. No flash chamber is provided for evaporator E_1 and the refrigerant directly enters it after expansion in valve V_1. The refrigerant vapour from evaporator E_1 then enters the compressor C_1 and the cycle is repeated.

Analysis : Let

Q_1, Q_2, Q_3 = refrigeration load in tons on evaporators E_1, E_2 and E_3 respectively

m_1, m_2, m_3 = mass of liquid refrigerants to be circulated to take care of load of different evaporators

m_{e1}, m_{e2}, m_{e3} = mass of refrigerant circulated through different compressors

m_2', m_3' = mass of refrigerant supplied to flash chambers F_1 and F_2

Also 1 ton of refrigeration is equivalent to heat absorption rate of 3.5 kJ/s. Then

• **For compressor C_1**

$$m_{c1} = m_1 = \frac{3.5\,Q_1}{h_1 - h_{10}} \text{ kg/s}$$

For compressor C_2

$$m_{c2} = m_{c1} + m_2 + m_2'$$

Here $$m_2 = \frac{3.5\,Q_2}{h_3 - h_9} \text{ kg/s} \quad \text{and}$$

the mass m_2' of refrigerant supplied to flash chamber F_1 can be worked out by writing the heat balance equation for the chamber. That is

heat energy entering the chamber = heat energy leaving the chamber

$$m_{c1}\,h_2 + m_2'\,h_9 = m_{c1}\,h_3 + m_2'\,h_3$$

or $$m_2' = \frac{m_{c_1}(h_2 - h_3)}{(h_3 - h_9)} \text{ kg/s}$$

- **For compressor C_3**

$$m_{c3} = m_{c2} + m_3 + m_3'$$

Here $$m_3 = \frac{3.5\,Q_3}{h_5 - h_8} \text{ kg/s} \quad \text{and}$$

the mass m_3' of refrigerant supplied to flash chamber F_2 can be determined by writing the heat balance equation for the chamber. That is

$$m_{c2}\,h_4 + m_3'\,h_8 = m_{c2}\,h_5 + m_3'\,h_5$$

or $$m_3' = \frac{m_{c_2}(h_4 - h_5)}{(h_5 - h_8)} \text{ kg/s}$$

- Power required to drive the system (work input to compressors)

$$P = m_{c1}(h_2 - h_1) + m_{c2}(h_4 - h_3) + m_{c3}(h_6 - h_5)$$

- Refrigerating effect of the system,

= heat absorbed in the evaporators

$$= m_1(h_1 - h_{10}) + m_2(h_3 - h_9) + m_3(h_5 - h_8)$$

- Coefficient of performance of the system,

$$COP = \frac{\text{combined refrigerating effect from the three evaporators}}{\text{work input to compressors}}$$

$$COP = \frac{m_1(h_1 - h_{10}) + m_2(h_3 - h_9) + m_3(h_5 - h_8)}{m_{c_1}(h_2 - h_1) + m_{c_2}(h_4 - h_3) + m_{c_3}(h_6 - h_5)}$$

In terms of load on the evaporators and power input to compressors,

$$COP = \frac{3.5(Q_1 + Q_2 + Q_3)}{P}$$

6.13. CASCADE VAPOUR COMPRESSION REFRIGERATION SYSTEM

Fig. 6.19. shows the schematic arrangement of the different components of a two stage cascade system along with *p-h* plot of processes involved.

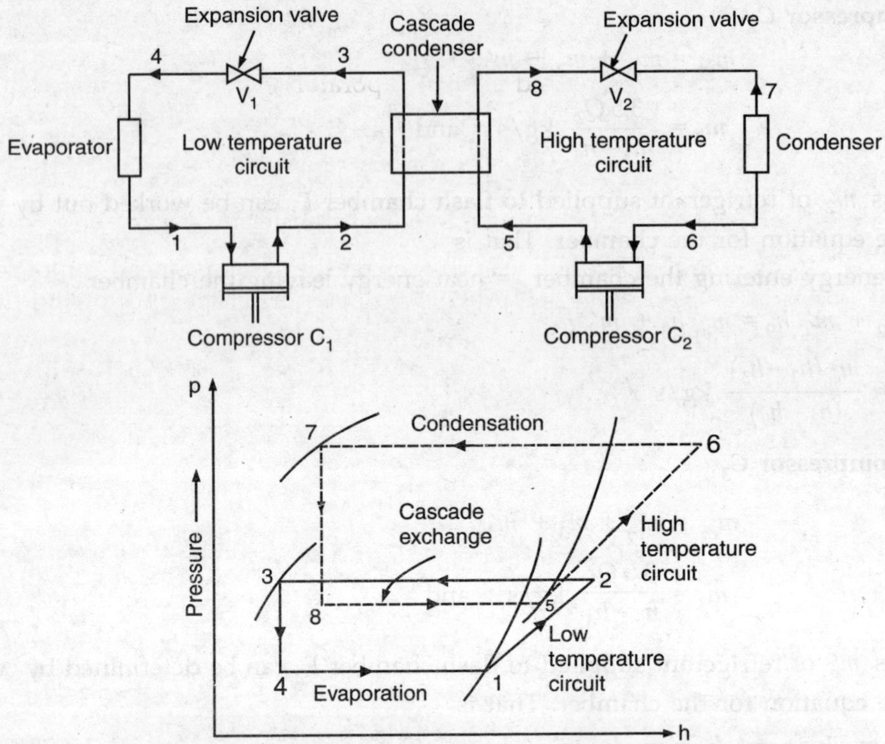

Fig. 6.19. Cascade vapour compression refrigeration system

The two stage cascade system essentially comprises two circuits and the two circuits use different refrigerants which have progressively lower boiling temperatures. The main components of the two circuits are :

- low and high temperature compressors
- evaporator, condenser and a cascade condenser
- expansion valves

Working : The salient points of operation of the system are :

(*i*) The cascade system uses two different refrigerants :

- low boiling temperature refrigerant such as R-13 for the low temperature circuit, and
- high boiling temperature refrigerant such as R-12 or R-22 for the high temperature circuit.

(*ii*) The saturated vapour at exit from the evaporators enters the compressor C_1 where these are compressed isentropically (process 1-2). The compressed vapours are made to pass through cascade condenser where the vapours lose both heat of supherheat and latent heat to the refrigerant stream coming from exapnsion valve V_2 of the high temperature circuit. The resulting saturated liquid enters the expansion valve V_1 and expands isentropically and vapour-liquid mixture is produced (process 3-4). The mixture is led to the evaporator where occurs evaporation (heat absorption) and the required cooling effect is produced. The refrigerant vapours at exit from the evaporator are taken to compressor for compression and the cycle is repeated.

(*iii*) The similar operations occur in the high temperature circuit with the difference that here the cascade condenser serves as an evaporator, it provides cooling to the compressed refrigerant vapour of the low temperature circuit.

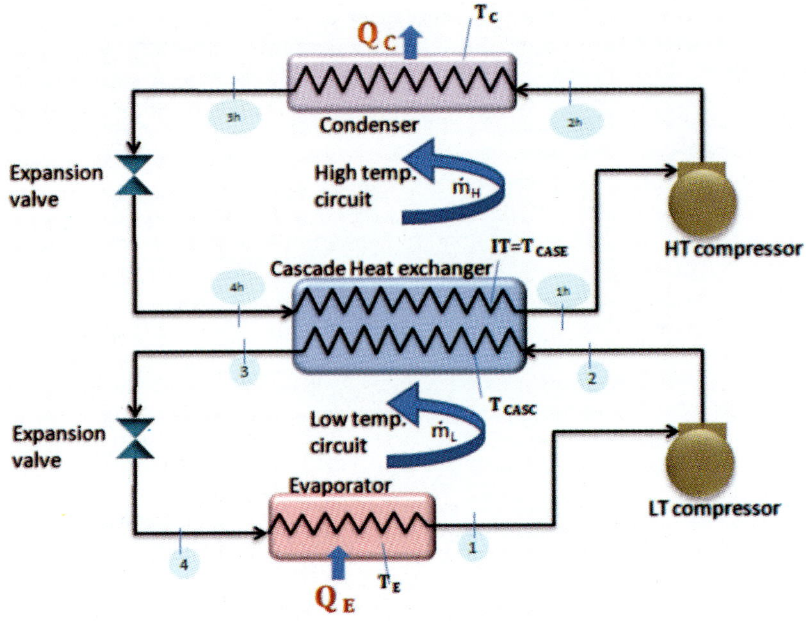

(a)

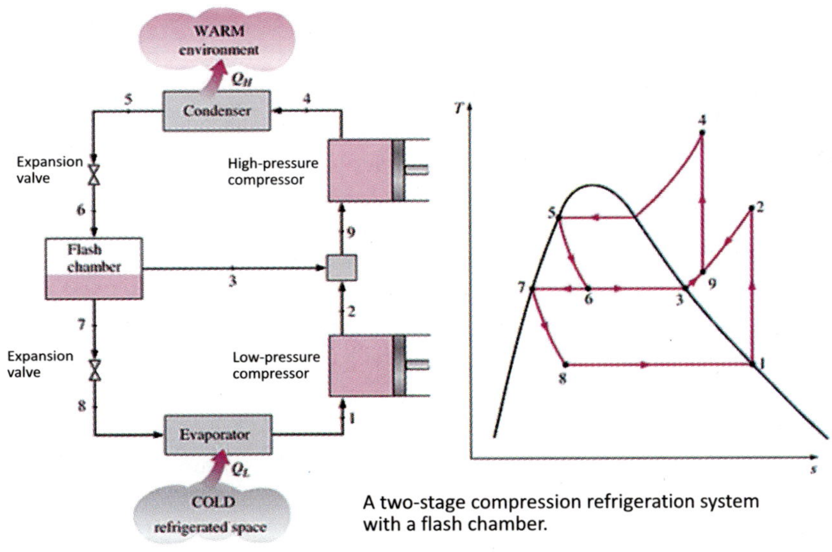

A two-stage compression refrigeration system with a flash chamber.

(b)

Cascade Refrigeration System

Thus in the cascade system, the cascade condenser serves as an evaporator for the high temperature circuit and as a condenser for the low temperature circuit. This also indicates that the useful refrigeration effect is produced in the evaporator of low temperature circuit.

(*iv*) The difference in low temperature cascade condenser temperature and high temperature cascade evaporator temperature is called ***temperature overlap***. If these temperatures are equal, then it is called ***intermediate temperature***.

Temperature overlap is necessary for effective heat transfer.

Analysis : Let Q be the refrigeration load in tons on the system. Since 1 ton of refrigeration is equivalent to heat absorption rate of 3.5 kJ/s, the mass of refrigerant flowing through low temperature cascade is

$$m_1 = \frac{3.5\,Q_1}{h_1 - h_4} \text{ kg/s}$$

The mass flow rate of refrigerant through the high temperature cascade can be determined by writing an expression of energy balance for the cascade condenser. That is

$$m_2\,(h_5 - h_8) = m_1(h_2 - h_3)$$

$$\therefore \quad m_2 = \frac{m_1(h_2 - h_3)}{(h_5 - h_8)} = \frac{m_1(h_2 - h_4)}{(h_5 - h_8)} \qquad (\because h_3 = h_4)$$

- Power required to drive the system (work input to compressors)
  $= [m_1(h_2 - h_1) + m_2(h_6 - h_5)]$ kJ/s or kW
- Refrigerating effect (heat absorbed in the evaporators of low temperatures circuit)
  $= m_1(h_1 - h_4)$ kJ/s
- Coefficient of performance of the system,

$$COP = \frac{\text{refrigerating effect}}{\text{work input to compressors}} = \frac{m_1(h_1 - h_4)}{m_1(h_2 - h_1) + m_2(h_6 - h_5)}$$

In terms of load on the evaporators and power input to compressors,

$$COP = \frac{3.5\,Q}{P}$$

The cascade system has proven its utility for low temperature refrigeration ; particularly for liquefaction of oxygen.

The cascade system scores over multi-stage compression system in the following aspects :

(*i*) The use of two or more different refrigerants makes each circuit comparatively simple.

(*ii*) The choice of refrigerants can be so made that each performs best within the required temperatures and pressure ranges.

(*iii*) Each circuit operates separately and as such the lubricant oil from one compressor cannot flow to the other compressor as generally happens in multi-stage systems.

(*iv*) The low boiling temperature refrigerant has extremely high pressure and that ensures small compressor displacement and provides better performance.

EXAMPLE 6.8

A single compressor using Freon-12 as refrigerant has three evaporators of capacity 8 ton, 15 ton and 25 ton. All the evaporators operate at the same temperature of – 18°C and the vapours leaving the evaporators are dry saturated. The condenser temperature is 40°C and the liquid refrigerant is subcooled in the sub-cooler section of condenser by 10°C. Assuming the compression to be isentropic, determine :

(a) the refrigerating effect in kJ/kg of flow rate,

(b) *the mass of refrigerating flowing through the evaporators,*

(c) *the theoretical power required to drive the compressor, and*

(d) *the coefficient of performance of the system.*

Solution : Refer Fig. 6.20 for the flow diagram and *p-h* representation of the system

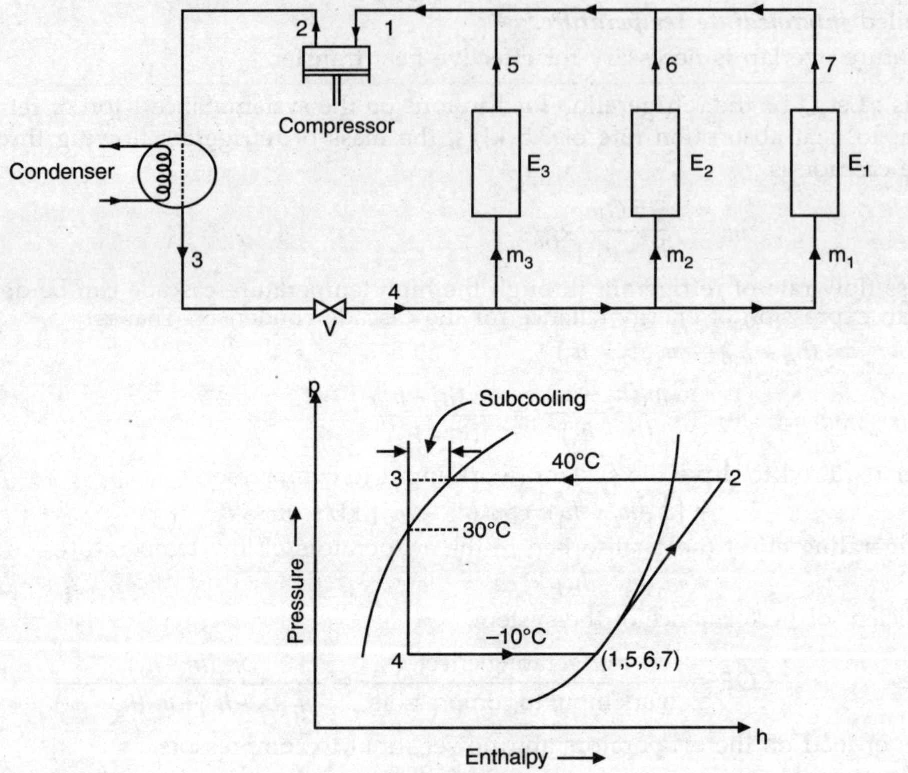

Fig. 6.20.

Due to subcooling, the temperature of liquid refrigerant at state point 3 is 40 – 10 = 10°C. The values of enthalpy as read from *p-h* diagram for R-12 are :

$$h_1 = h_5 = h_6 = h_7 = 347 \text{ kJ/kg} ;$$

$$h_2 = 374 \text{ kJ/kg} \quad \text{and} \quad h_3 = h_4 = 328 \text{ kJ/kg}$$

(*a*) Refrigerating effect = $h_1 - h_4$ = 347 – 228 = **119** kJ/kg

(*b*) Since 1 ton of refrigeration is equivalent to heat absorption rate of 3.5 kJ/s, the refrigerant mass flow rates through different evaporators are:

$$m_1 = \frac{3.5Q}{h_1 - h_4} = \frac{3.5\times 8}{119} = 0.235 \text{ kg/s}$$

$$m_2 = \frac{3.5\times 15}{119} = 0.441 \text{ kg/s} \quad \text{and} \quad m_3 = \frac{3.5\times 25}{119} = 0.735 \text{ kg/s}$$

Then the total mass flow of refrigerant through all the evaporators is

$$m = m_1 + m_2 + m_3 = 0.235 + 0.441 + 0.735 = 1.411 \text{ kg/s}$$

(*c*) Theoretical power required to drive the compressor

= work input to compressor

= $m(h_2 - h_1)$ = 1.411 (374 – 347) = **38.1** kJ/s or kW

(*d*) Coefficient of performance of the system

$$COP = \frac{\text{refrigerating effect}}{\text{work input to compressors}}$$

$$= \frac{m_1(h_1 - h_4)}{m(h_2 - h_1)} = \frac{h_1 - h_4}{h_2 - h_1} = \frac{347 - 228}{374 - 347} = \mathbf{4.41}$$

EXAMPLE 6.9

A refrigeration installation using Freon-12 as refrigerant has three evaporators of capacities 8 ton, 15 ton, 25 ton and operating at 10°C, 5°C and – 10°C respectively. The system uses individual expansion valves and individual compressors but one condenser operating at 40°C. All the evaporators discharge dry saturated refrigerant to the corresponding compressors, and the liquid refrigerant is subcooled in the subcooler section of condenser by 10°C. Assuming isentropic compression in each compressor, determine :

(*a*) *refrigerating effect and mass flow rate of refrigerant in each evaporator,*

(*b*) *power required to drive the compressors,*

(*c*) *load on the condenser, and*

(*d*) *COP of the system.*

Draw schematic arrangement of the system and use p-h chart to obtain the solution.

Solution : Refer Fig. 6.21 for the schematic arrangement and *p-h* plot of the given system. Due to subcooling, t_7 = 40 – 10 = 30°C.

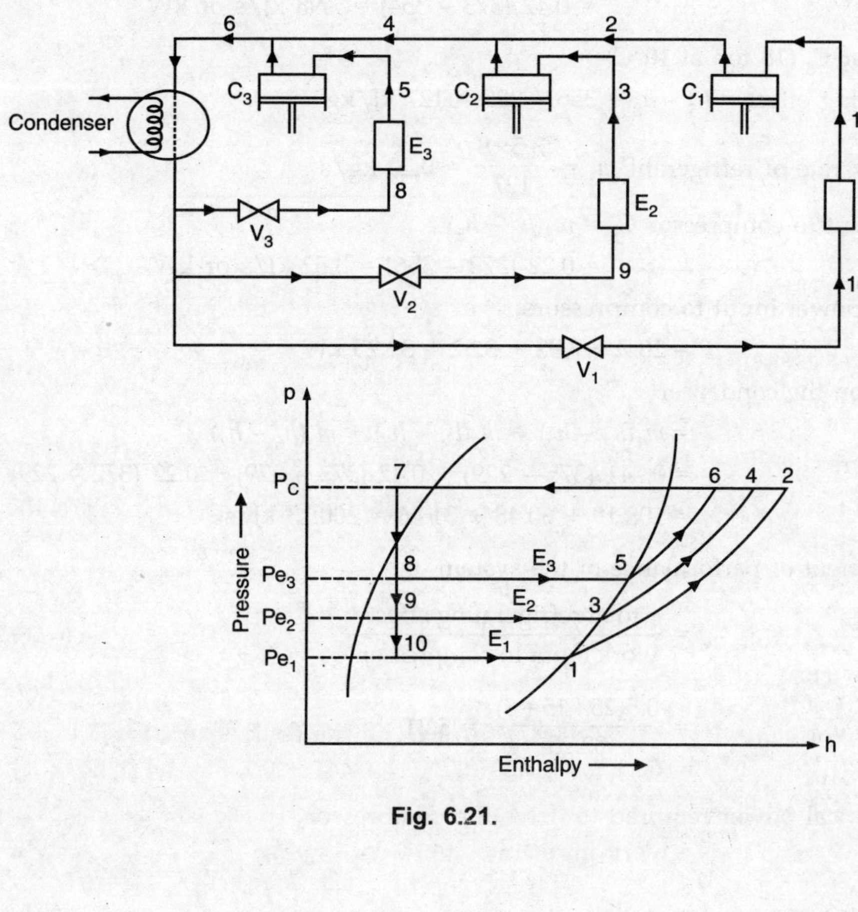

Fig. 6.21.

The enthalpy values as read from *p-h* plot for Freon-12 are :

$$h_1 = 347 \text{ kJ/kg} ; \quad h_3 = 354 \text{ kJ/kg} ; \quad h_5 = 356 \text{ kJ/kg} ;$$
$$h_2 = 375 \text{ kJ/kg} ; \quad h_4 = 373 \text{ kJ/kg} ; \quad h_6 = 372 \text{ kJ/kg} ;$$
$$h_7 = h_8 = h_9 = h_{10} = 229 \text{ kJ/kg}$$

Further, 1 ton of refrigerating is equivalent to heat absoption of 3.5 kJ/s

For Evaporator E_1 (25 ton at – 10°C)

Refrigerating effect $= h_1 - h_{10} = 347 - 229 = 118$ kJ/kg

Mass flow rate of refrigerant $m_1 = \dfrac{3.5\times25}{118} = 0.741$ kg/s

Power input to compressor $C_1 = m_1(h_2 - h_1)$

$= 0.741\,(375 - 347) = 20.75$ kJ/s or kW

For Evaporator E_2 (15 ton at 5°C)

Refrigerating effect $= h_3 - h_9 = 354 - 229 = 125$ kJ/kg

Mass flor rate of refrigerant $m_2 = \dfrac{3.5\times15}{125} = 0.42$ kg/s

Power input to compressor $C_2 = m_2(h_4 - h_3)$

$= 0.42\,(373 - 354) = 7.98$ kJ/s or kW

For Evaporator E_3 (18 ton at 10°C)

Refrigerating effect $= h_5 - h_8 = 356 - 229 = 127$ kJ/kg

Mass flow rate of refrigerant $m_3 = \dfrac{3.5\times8}{127} = 0.22$ kg/s

Power input to compressor $C_3 = m_3(h_6 - h_5)$

$= 0.22\,(372 - 356) = 3.52$ kJ/s or kW

(*b*) Total power input to compressors

$$P = 20.7 + 7.98 + 3.52 = 32.25 \text{ kW}$$

(*c*) Load on the condenser

$$= m_1(h_2 - h_7) + m_2(h_4 - h_7) + m_3(h_6 - h_7)$$
$$= 0.741\,(375 - 229) + 0.42\,(373 - 229) + 0.22\,(372 - 229)$$
$$= 108.19 + 60.48 + 31.46 = 200.13 \text{ kJ/s}$$

(*d*) Coefficient of performance of the system

$$= \frac{\text{total refrigerating effect}}{\text{work input to compressors}}$$

$$= \frac{3.5(25+15+8)}{32.25} = \mathbf{5.21}$$

EXAMPLE 6.10

An arrangement of the refrigeration system using F-12 as refrigerant is shown the figure given below :

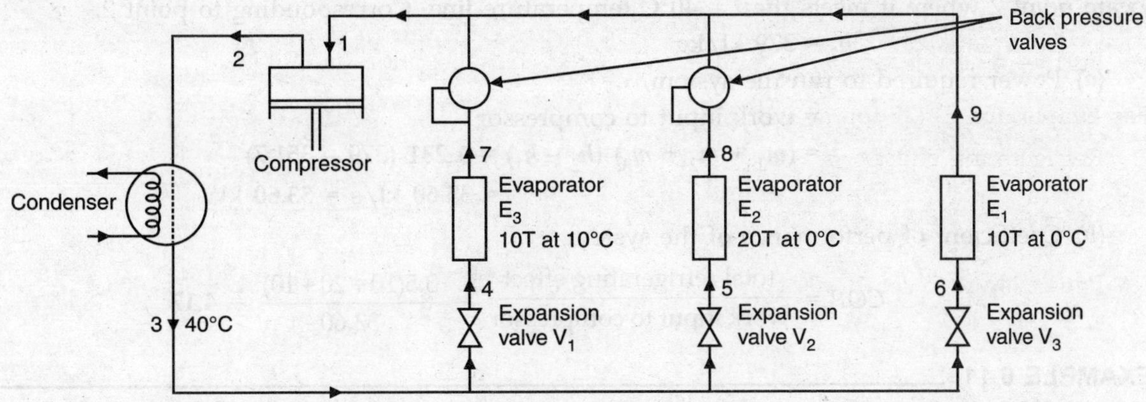

Using p-h chart, make calculations to workout the power required to run the system and its coefficient of performance.

Solution : Refer Fig. 6.22 for *p-h* plot of the given cycle. The enthalpy values as read from the plot with F-12 as refrigerant are :

$$h_3 = h_4 = h_5 = h_6 = 238 \text{ kJ/kg}$$
$$h_7 = 355 \text{ kJ/kg}$$
$$h_8 = 352 \text{ kJ/kg}$$
$$h_9 = 348 \text{ kJ/kg}$$

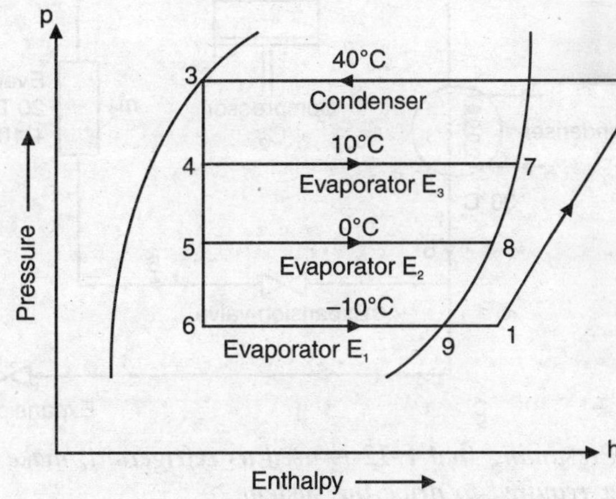

Fig. 6.22.

Since 1 ton of refrigeration is equivalent to heat abstraction of 3.5 kJ/s

Mass flow of refrigerant through evaporator E_1,

$$m_1 = \frac{3.5\,Q}{h_9 - h_6}$$
$$= \frac{3.5 \times 10}{348 - 238} = 0.318 \text{ kg/s}$$

Mass flow of refrigerant through evaporator E_2,

$$m_2 = \frac{3.5\,Q}{h_8 - h_5} = \frac{3.5 \times 20}{352 - 238} = 0.614 \text{ kg/s}$$

Mass flow of refrigerant through evaporator E_3,

$$m_3 = \frac{3.5\,Q}{h_7 - h_4} = \frac{3.5 \times 10}{355 - 238} = 0.299 \text{ kg/s}$$

The enthalpy of the refrigerant at state point 1 (suction to compressor) where all the three streams meet is calculated by using the relation,

$$h_1 = \frac{m_1 h_9 + m_2 h_8 + m_3 h_7}{m_1 + m_2 + m_3} = \frac{0.318 \times 348 + 0.614 \times 532 + 0.299 \times 355}{0.318 + 0.614 + 0.299}$$

$$= \frac{110.66 + 216.13 + 106.14}{1.231} = 351.7 \text{ kJ/kg}$$

This helps to mark point 1. Move therefrom along constant entropy line (isentropically) and locate point 2 where it meets the t = 40°C temperature line. Corresponding to point 2,

$$h_2 = 379 \text{ kJ/kg}$$

(*a*) Power required to run the system,

$$= \text{work input to compressor}$$

$$= (m_1 + m_2 + m_3)(h_2 - h_1) = 1.231\ (379 - 351.7)$$

$$= 33.60 \text{ kJ/s} = \mathbf{33.60 \text{ kW}}$$

(*b*) Coefficient of performance of the system,

$$COP = \frac{\text{total refrigerating effect}}{\text{work input to compressor}} = \frac{3.5(10 + 20 + 10)}{33.60} = \mathbf{4.17}$$

EXAMPLE 6.11

The schematic arrangement of a compound refrigeration plant used for multi-load system is depicted below :

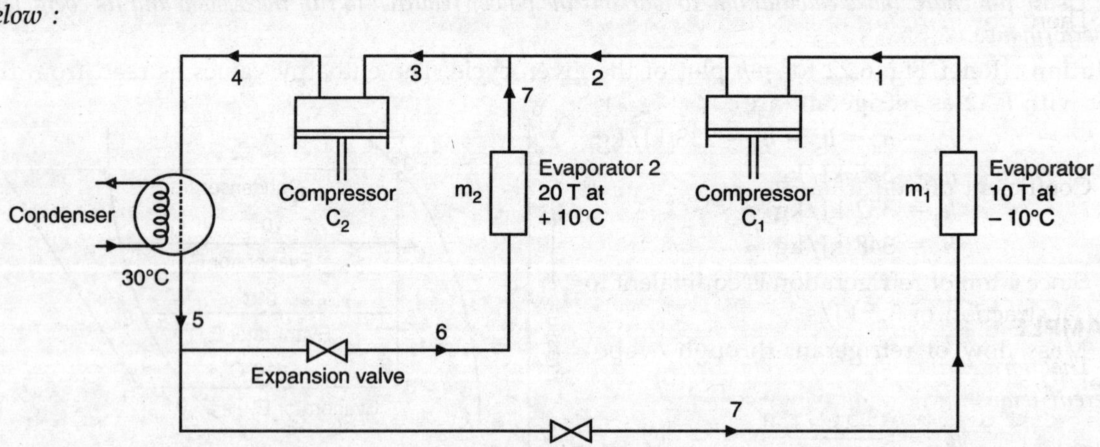

Presuming that F-12 is used as refrigerant, make calculations for the coefficient of performance and power required to drive the system.

Solution : Refer Fig. 6.23 for *p-h* plot of the processes involved in the given cycle.

From tables for F-12 refrigerant,

$$\text{at } 30°\text{C} : p_c = 7.45 \text{ bar}$$

$$\text{at } 10°\text{C} : p_i = 4.23 \text{ bar}$$

$$\text{at } -10°\text{C} : p_e = 2.19 \text{ bar}$$

The various enthalpy values as read from property tables and *p-h* plot are :

$$h_5 = h_6 = h_8 = 64.10 \text{ kJ/kg}$$

$$h_1 = 183.2 \text{ kJ/kg}$$

$$h_7 = 191.7 \text{ kJ/kg}$$

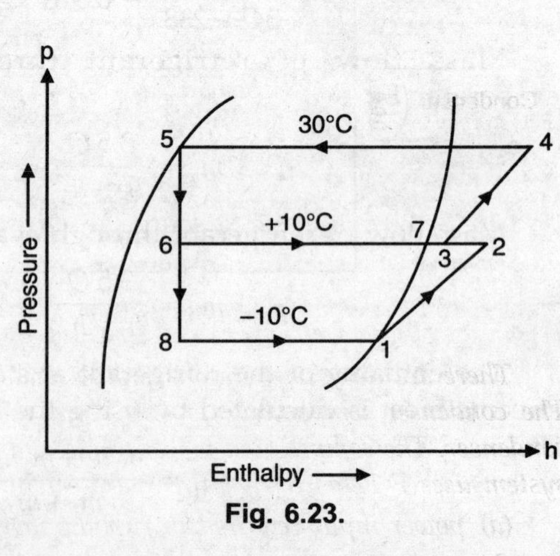

Fig. 6.23.

Moving along constant entropy (isentropic) line 1-2;

$$h_2 = 195 \text{ kJ/kg}$$

$$m_1 \text{ (mass flow through evaporator 1)} = \frac{3.5\,Q_1}{h_1 - h_8} = \frac{3.5 \times 10}{183.2 - 64.6} = 0.295 \text{ kg/s}$$

$$m_2 \text{ (mass flow through evaporator 2)} = \frac{3.5\,Q_2}{h_7 - h_6} = \frac{3.5 \times 20}{191.7 - 64.6} = 0.55 \text{ kg/s}$$

Considering mixing of refrigerant streams leaving the compressor C_1 and coming out of evaporator 2, we get

$$m_1\,h_2 + m_2\,h_7 = (m_1 + m_2)\,h_3$$

$$\therefore \qquad h_3 = \frac{m_1 h_2 + m_2 h_7}{m_1 + m_2} = \frac{0.295 \times 195 + 0.55 \times 191.7}{0.295 + 0.55} = 192.85 \text{ kJ/kg}$$

This helps to locate point 3 on the *p-h* plot. Then moving along isentropic line 3-4, point 4 is located and value of enthalpy at this state is read. That givers

$$h_4 = 205 \text{ kJ/kg}$$

Then, power required to drive the system (work input to compressors)

$$= m_1(h_2 - h_1) + (m_1 + m_2)(h_4 - h_3)$$

$$= 0.295\,(195 - 183.2) + (0.295 + 0.55)\,(205 - 192.85)$$

$$= 3.481 + 10.267 = 13.748 \text{ kJ/s} \simeq 13.75 \text{ kW}$$

Coefficient of performance of the system,

$$COP = \frac{\text{total refrigerating effect}}{\text{work input to compressors}} = \frac{3.5(10 + 20)}{13.748} = \mathbf{7.637}$$

EXAMPLE 6.12

The refrigeration system shown in the figure given below represents a multi-evaporation unit at different temperatures with single compressor and multi-expansion valves.

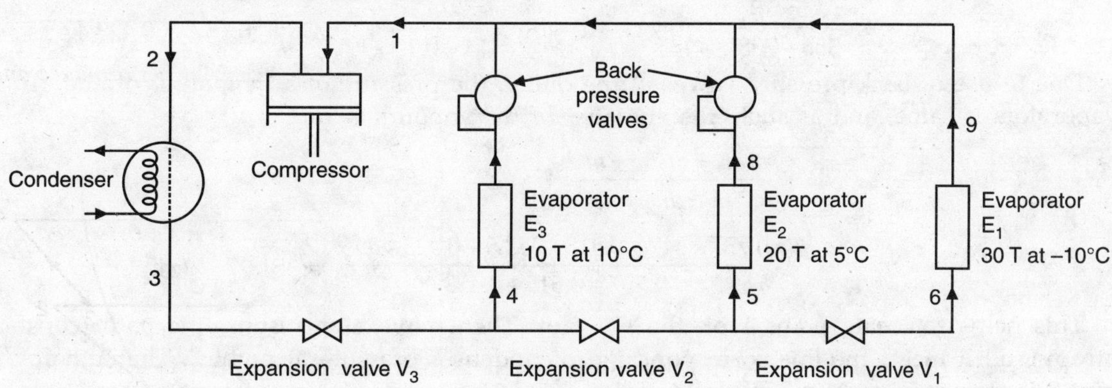

There are three evaporators having loads of 30 ton at – 10°C, 20 tons at 5°C and 10 tons at 10°C. The condenser temperature is 40°C and the liquid refrigerant is subcooled to 10°C after leaving the condenser. The compression is isentropic and the vapours leaving the evaporators are dry saturated. If the system uses F-12 as refrigerant, make calculations for

(a) power input required to run the machine, and

(b) coefficient of performance of the system.

Solution : Refer Fig. 6.24 for *p-h* plot of the given cycle. The enthalpy values as read from the plot with F-12 as refrigerant are :

$h_3 = h_4 = 208$ kJ/kg

$h_5 = h_{10} = 209$ kJ/kg

$h_6 = h_{11} = 204.5$ kJ/kg

$h_7 = 356$ kJ/kg

$h_8 = 353.5$ kJ/kg

$h_9 = 347$ kJ/kg

$x_4 = 0.13$ and $x_5 = 0.04$

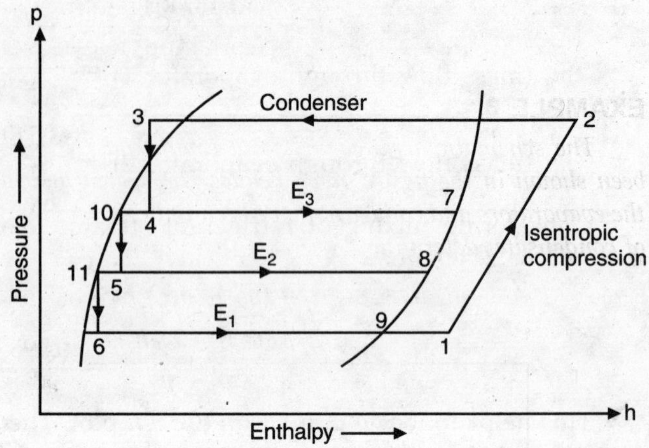

Fig. 6.24.

Since 1 ton of refrigeration is equivalent to heat abstraction of 3.5 kJ/s

Mass flow of refrigerant through evaporator E_1,

$$m_1 = \frac{3.5\,Q_1}{h_9 - h_6} = \frac{3.5 \times 30}{347 - 204.5} = 0.737 \text{ kg/s}$$

Mass flow of refrigerant through evaporator E_2,

$$m_2 = \frac{3.5\,Q_2}{h_8 - h_5} = \frac{x_5}{1 - x_5} \times m_1$$

where $\frac{x_5}{1-x_5}\, m_1$ is the mass of vapour carried in evaporation E_2

$$m_2 = \frac{3.5 \times 20}{353.5 - 209} + \frac{0.04}{1 - 0.04} = 0.484 + 0.031 = 0.515 \text{ kg/s}$$

Mass flow of refrigerant through evaporator E_3,

$$m_3 = \frac{3.5\,Q_3}{h_7 - h_4} = (m_1 + m_2) \times \frac{x_4}{1 - x_4}$$

$$= \frac{3.5 \times 10}{356 - 208} + (0.737 + 0.515) \times \frac{0.13}{1 - 0.13} = 0.236 + 0.187 = 0.423 \text{ kg/s}$$

Due to use of back pressure valves at the outlet, the pressure of all vapours coming out of evaporators is same, and as such, the enthalpy of the vapours at mix is

$$h_1 = \frac{m_1 h_9 + m_2 h_8 + m_3 h_7}{m_1 + m_2 + m_3}$$

$$= \frac{0.737 \times 347 + 0.515 \times 353.5 + 0.423 \times 356}{0.737 + 0.515 + 0.423} = 351.27 \text{ kJ/kg}$$

This helps to locate point 1 on the *p-h* plot. Then move along isentropic path (constant entropy), till it meets the line corresponding to condenser pressure at point 2. The enthalpy at point 2 is then read as $h_2 = 378$ kJ/kg

(*a*) Power required to run the machine,

$$= \text{work input to compressor}$$

$$= (m_1 + m_2 + m_3)\,(h_2 - h_1) = (0.737 + 0.515 + 0.423)\,(378 - 351.27)$$

$$= 44.77 \text{ kJ/s} = 44.77 \text{ kW}$$

(*b*) Coefficient of performance of the system,

$$COP = \frac{\text{total refrigerating effect}}{\text{work input to compressor}} = \frac{3.5(10+20+30)}{44.77} = \mathbf{4.69}$$

EXAMPLE 6.13

The schematic arrangement of a compound refrigeration system used for multi-load application has been shown in the figure given below. The system uses R-12 as a refrigerant and the pressure limits for the evaporators and condenser have been indicated on the diagram. Presuming that there is no undercooling of condensate, determine

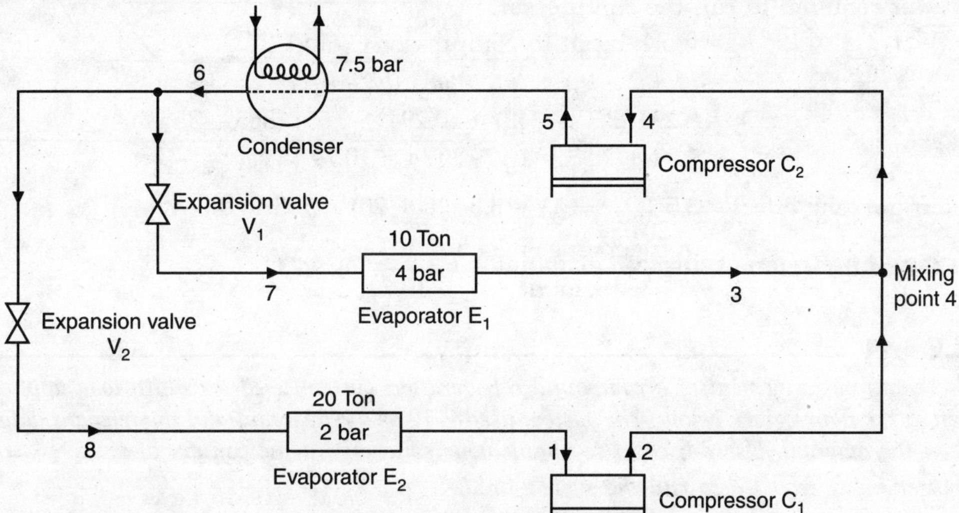

(*a*) *power input required to drive the compressors, and*

(*b*) *coefficient of performance of the combined system.*

Use p-h chart to obtain your solution.

Solution : Refer Fig. 6.25 for *p-h* plot of the given cycle. The enthalpy values as read from the plot with R-12 as a refrigerant are :

h_1 = 347 kJ/kg

h_2 = 357 kJ/kg

h_3 = 355 kJ/kg

$h_6 = h_7 = h_8$ = 228 kJ/kg

Since 1 ton of refrigeration is equivalent to heat abstraction of 3.5 kJ/s

Mass flow of refrigerant through evaporator E_1,

$$m_1 = \frac{3.5\,Q}{h_1 - h_8} = \frac{3.5 \times 10}{347 - 228} = 0.294 \text{ kg/s}$$

Mass of refrigerant passing through evaporator E_2,

$$m_2 = \frac{3.5\,Q}{h_3 - h_7} = \frac{3.5 \times 20}{355 - 228} = 0.55 \text{ kg/s}$$

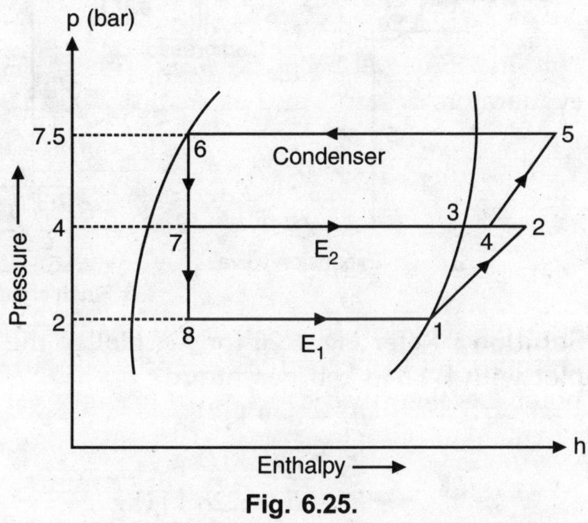

Fig. 6.25.

Enthalpy at point 4 is given by,

$$h_4 = \frac{m_1 h_2 + m_2 h_3}{m_1 + m_2}$$

$$= \frac{0.294 \times 357 + 0.55 \times 355}{0.294 + 0.55} = 355.7 \text{ kJ/kg}$$

After marking point 4 on the *p-h* plot, move along the constant entropy (isentropic line) from point 4 till it intersects the condenser pressure line p_c = 7.5 bar at point 5. Corresponding to point 5, h_5 = 365 kJ/kg.

(*a*) Power required to run the compressor,

= work input to compressor C_1 and C_2

$= m_1 (h_2 - h_1) + (m_1 + m_2)(h_5 - h_4)$

= 0.294 (357 – 347) + (0.294 + 0.55) (365 – 355.7)

= 2.94 + 7.85 = 10.79 kJ/s = 10.79 kW

(*b*) Refrigerating effect = 3.5 (Q_1 + Q_2) = 3.5 (10 + 20) = 105 kJ/s

$$COP \text{ of the system} = \frac{\text{refrigerating effect}}{\text{work input}} = \frac{105}{10.79} = \mathbf{9.73}$$

EXAMPLE 6.14

The schematic arrangement of a compound refrigeration system used for multi-load application has been shown in the figure given below. The system used F-12 as a refrigerant and the relevant data has been indicated on the diagram. Use p-h chart to obtain your solution in the context of :

(a) power input required to run the system, and

(b) coefficient of performance of the combined system.

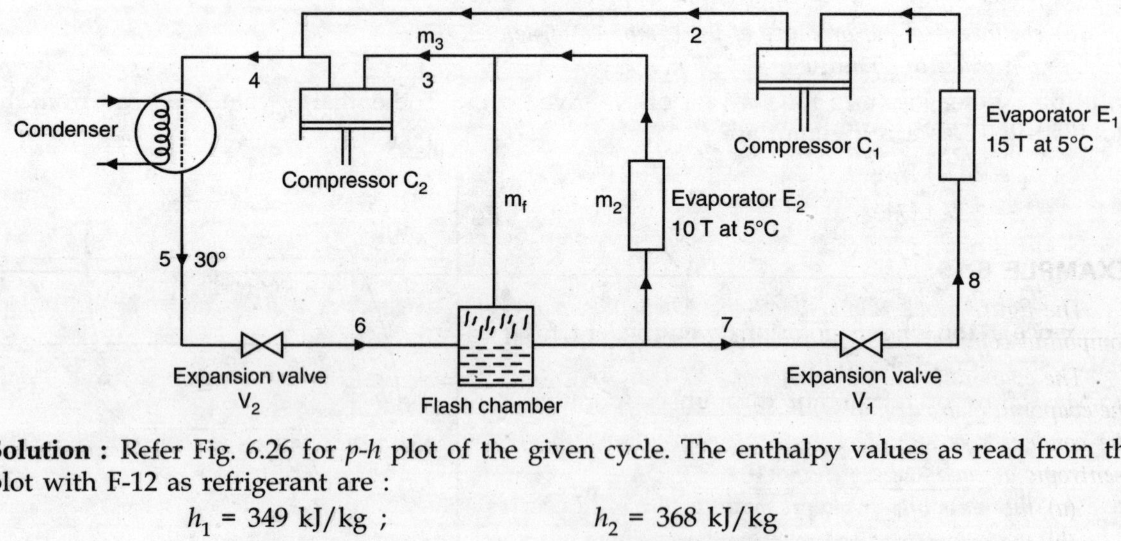

Solution : Refer Fig. 6.26 for *p-h* plot of the given cycle. The enthalpy values as read from the plot with F-12 as refrigerant are :

h_1 = 349 kJ/kg ; h_2 = 368 kJ/kg

h_3 = 355 kJ/kg ; h_4 = 366 kJ/kg

$h_5 = h_6$ = 228 kJ/kg ; $h_7 = h_8$ = 205 kJ/kg

Since 1 ton of refrigeration is equivalent to heat abstraction of 3.5 kJ/s

Mass of refrigerant passing through evaporator E_1, and

$$m_1 = \frac{3.5\,Q_1}{h_1 - h_8}$$

$$= \frac{3.5\times15}{349-205} = 0.3646 \text{ kg/s}$$

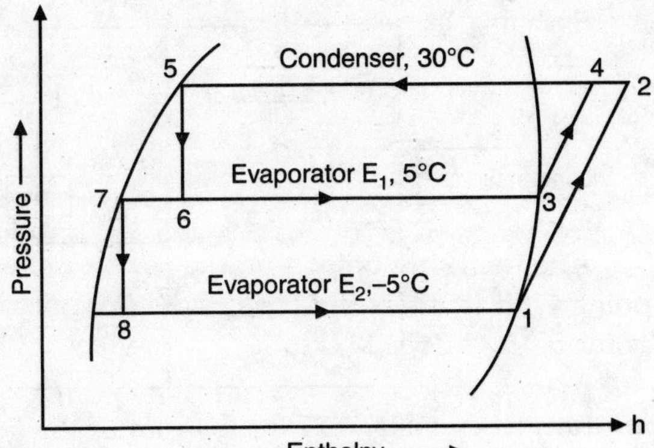

Fig. 6.26.

Mass of refrigerant passing through evaporator E_2,

$$m_2 = \frac{3.5\,Q_2}{h_3 - h_7}$$

$$= \frac{3.5\times10}{355-205} = 0.2333 \text{ kg/s}$$

The dryness fraction of vapour at state point 6 is 0.16 (from *p-h* plot) and as such the mass of vapours driven from the flash chamber is

$$m_f = (m_1 + m_2) \times \frac{x}{1-x}$$

$$= (0.3646 + 0.2333) \times \frac{0.16}{1-0.16} = 0.1139 \text{ kg/s}$$

∴ Mass of refrigerant being compressed in compressor C_2,

$$m_3 = m_2 + m_f = 0.2333 + 0.1139 = 0.3472 \text{ kg/s}$$

(*a*) Power required to run the system,

= work input to compressor C_1 and C_2

$= m_1 (h_2 - h_1) + m_3(h_4 - h_3)$

$= 0.3646\ (368 - 349) + 0.3472(366 - 355)$

= 10.746 kJ/s = **10.746 kW**

(*b*) Coefficient of performance of the system,

$$COP = \frac{\text{refrigerating effect}}{\text{work input}} = \frac{3.5(15+10)}{10.746} = \mathbf{8.14}$$

EXAMPLE 6.15

The figure given below shows the schematics of multiple evaporators at different temperatures with compound compression, individual expansion valves and flash intercoolers.

The evaporators are of capacitors 20 T at – 5°C, 30 T at 0°C and 10 T at 5°C. The vapours leaving the evaporators are dry and saturated, the condenser temperature is 40°C and the liquid refrigerant leaving the condenser is subcooled by 10°C. Presuming that F-12 is the refrigerant used and compression is isentropic at each stage, determine :

(a) the mass of refrigerant flowing through each evaporator,

(b) the power required to drive the compressing units, and

(c) the coefficient of performance of the system.

Use p-h chart for F-12 to workout your solution.

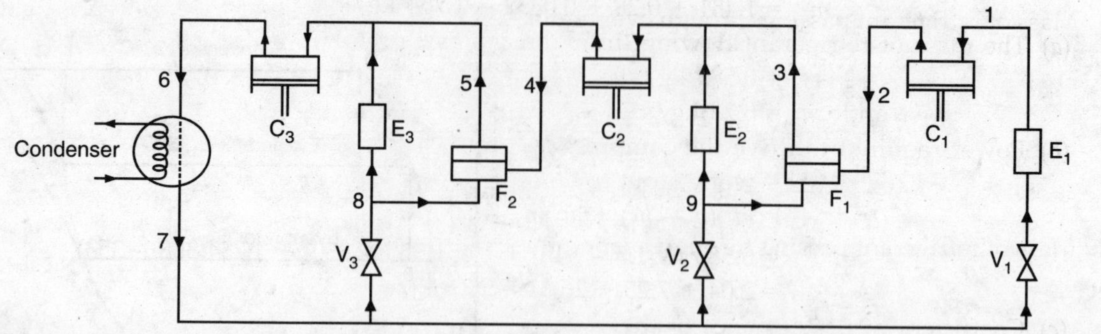

Solution : Refer Fig. 6.27 for *p-h* plot of the given cycle. From the given data $t_7 = 40 - 10 = 30°C$ The enthalpy values as read from the *p-h* plot are :

h_1 = 348 kJ/kg

h_2 = 351.5 kJ/kg

h_3 = 350 kJ/kg

h_4 = 355 kJ/kg

h_5 = 353 kJ/kg

h_6 = 368 kJ/kg

$h_7 = h_8 = h_9 = h_{10}$ = 228 kJ/kg

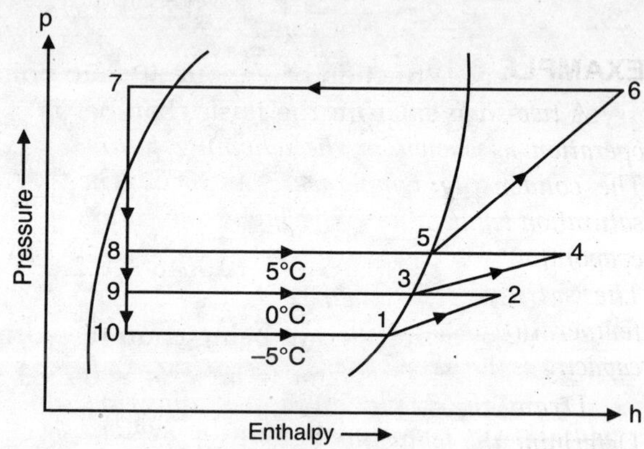

Fig. 6.27.

It is to be noted that there is only small difference between different enthalpy values and as such there is likely to be some difficulty in correctly reading from the *p-h* plot.

1 ton of refrigeration is equivalent to heat absorption of 3.5 kJ/s. Then

For compressor C_1

$$m_{c_1} = m_1 = \frac{3.5\,Q_1}{h_1 - h_{10}} = \frac{3.5 \times 20}{348 - 228} = 0.583 \text{ kg/s}$$

where m_1 is the mass of liquid refrigerant to take care of load of evaporator E_1

For compressor C_2

$$m_{c_2} = m_{c_1} + m_2 + m_2'$$

Here $$m_2 = \frac{3.5\,Q_2}{h_3 - h_9} = \frac{3.5 \times 30}{350 - 228} = 0.961 \text{ kg/s and}$$

m_2' (mass of refrigerant supplied to flash chamber F_1)

$$= \frac{m_{c_1}(h_2 - h_3)}{h_3 - h_9} = 0.583 \times \frac{351.5 - 350}{350 - 228} = 0.00717 \text{ kg/s}$$

$$\therefore \quad m_{c_2} = 0.583 + 0.861 + 0.00713 = 1.451 \text{ kg/s}$$

For compressor C_3

$$m_{c_3} = m_{c_2} + m_3 + m_3'$$

Here $$m_3 = \frac{3.5\,Q_3}{h_5 - h_8} = \frac{3.5 \times 10}{353 - 228} = 0.28 \text{ kg/s and}$$

m_3' (mass of refrigerant supplied to flash chamber F_2)

$$= \frac{m_{c_2}(h_4 - h_5)}{h_5 - h_8} = 1.451 \times \frac{(355 - 353)}{353 - 228} = 0.0232 \text{ kg/s}$$

$\therefore \quad m_{c3} = 1.451 + 0.28 + 0.0232 = 1.754$ kg/s

(a) The mass of refrigerant flowing through each evaporator is,

$$m_1 = 0.583 \text{ kg/s}\ ; \quad m_2 = 0.861 \text{ kg/s}$$

$$\text{and } m_3 = 0.28 \text{ kg/s}$$

(b) Power required to drive the compressing units,

$$= \text{work input to compressors}$$

$$= m_{c1}(h_2 - h_1) + m_{c2}(h_4 - h_3) + m_{c3}(h_6 - h_5)$$

$$= 0.583\ (351.5 - 348) + 1.451(355 - 350) + 1.754(368 - 353)$$

$$= 2.04 + 7.25 + 26.31 = 35.6 \text{ kJ/s} = 35.6 \text{ kW}$$

(c) Coefficient of performance of the system,

$$= \frac{\text{total refrigerating effect}}{\text{work input to compressors}} = \frac{3.5(20+30+10)}{35.6} = \mathbf{5.9}$$

EXAMPLE 6.16

A two-stage ammonia system designed for a fish freezing plant, uses flash gas removal and intercooling operation as shown in the adjoining figure. The condensing temperature is 35°C. The saturation temperature of the high-temperature evaporator is 0°C and its capacity is 90 kW. The saturation temperature of the low-temperature evaporator is – 40°C and its capacity is 170 kW of refrigeration

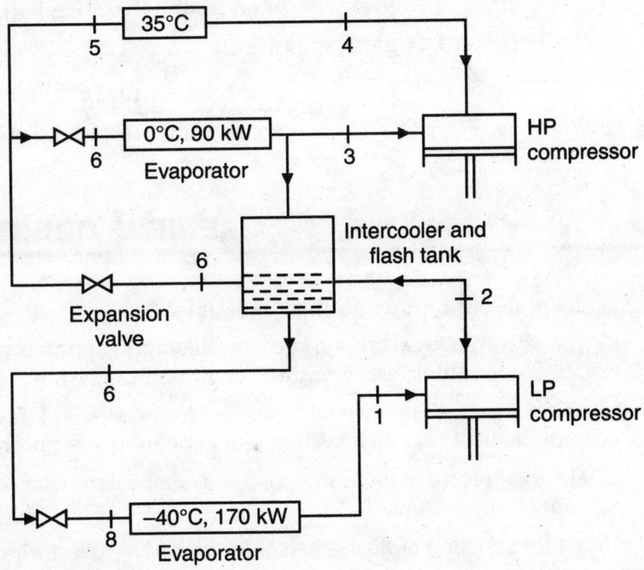

Draw the cycle on a p-h diagram. Determine the following

(a) The rate of refrigerant compressed by the high-stage compressor

(b) The power required by the compressors

(c) The coefficient of performance

Solution : Refer Fig. 6.28 for *p-h* diagram of the processes involved in the given refrigeration system.

The relevant values as read from the *p-h* chart are :

$$h_1 = 1408.5 \text{ kJ/kg}$$

$$h_2 = 1655 \text{ kJ/kg}$$

$$h_3 = 1461.5 \text{ kJ/kg}$$

$$h_4 = 1625 \text{ kJ/kg}$$

$$h_5 = h_2 = 366 \text{ kJ/kg}$$

$$h_7 = h_8 = 200 \text{ kJ/kg}$$

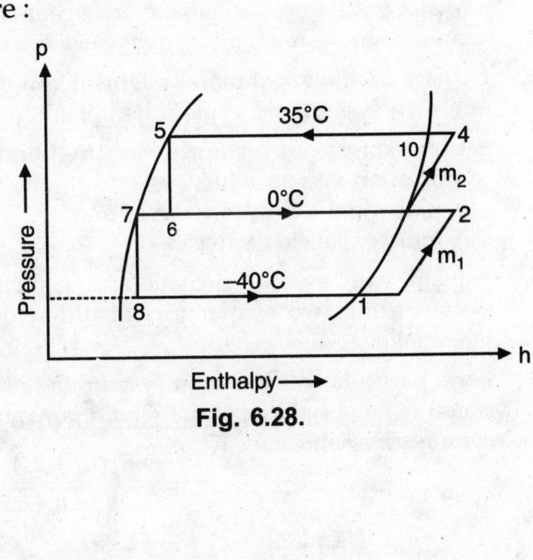

Fig. 6.28.

(a) Let m_1 be the mass flow of refrigerant passing through the evaporator (cooling capacity 170 kW) or low pressure compressor. Then

$$m_1(h_1 - h_8) = \text{cooling capacity}$$

That gives: $m_1 = \dfrac{170}{1408.5 - 200} = 0.1406$ kg/s

Considering energy balance about both the high temperature evaporator and the flash tank,

$$m_5\, h_5 + 90 + m_2\, h_2 = m_3\, h_3 + m_7\, h_7$$

A closer look at the schematic arrangement reveals that

$m_3 = m_5$ and $m_2 = m_7 = m_1 = 0.1406$ kg/s as calculated above.

Then $m_3 \times 366 + 90 + 0.1406 \times 1655 = m_3 \times 1461.5 + 0.1406 \times 200$

$$\therefore \quad m_3 = \frac{90 + 232.69 - 28.12}{1461.5 - 366} = 0.269 \text{ kg/s}$$

This gives the mass of refrigerant being compressed by the high pressure compression.

(*b*) Power required to run the compressors

= work input to low pressure and high pressure compressors

$= m_1\,(h_2 - h_1) + m_3(h_4 - h_3)$

$= 0.1406\,(1655 - 1408.5) + 0.269(1625 - 1461.5)$

$= 34.66 + 43.98 = 78.64$ kJ/s $= 78.64$ kW $=$ **78.64** kW

(*c*) Coefficient of performance

$$= \frac{\text{total cooling produced}}{\text{work input to compressors}} = \frac{90 + 170}{78.64} = \mathbf{3.306}$$

REVIEW QUESTIONS

A. Conceptual and conventional questions.

1. What is meant by compound or multi-stage vapour compression system ? List the advantages of this system over single-stage vapour compression system.
2. Classify multi-stage vapour compression systems. Draw the schematic arrangement of the different components of one such system and explain its operation. Represent the processes involved on *p-h* plot.
3. State the role of liquid intercooler, water intercooler and liquid subcooler in a multi-stage vapour compression system.
4. State the function of flash intercooler provided in a compound vapour compression refrigeration system.
5. Explain, with a neat schematic arrangement and pressure-enthalpy plot, the operation of a two-stage compression system employing water intercooler, liquid intercooler and a liquid flash chamber.
6. Classify multi-evaporation systems. Draw the schematic arrangement of the different components of one such system and explain its operation. Represent the pressures involved on *p-h* plot.
7. Set up expressions for power required and coefficient of performance of a multi-evaporator vapour compression system with
   (*i*) individual expansion valves and
   (*ii*) multi-expansion valves.
8. Explain, with a neat schematic arrangement and *p-h* diagram, the vapour compression refrigeration system using two evaporators at different temperatures with compound compression and perfect intercooling.
9. For a particular refrigeration system, the ratio of condenser temperature to evaporator temperature is stated to be very large. What would be your choice between single stage compression and compound compression, and why ?

10. (*a*) What do you mean by cascade refrigeration ? Explain, with a neat sketch, the operation of this system.
    (*b*) State the difference between overlap temperature and intermediate temperature in the context of cascade refrigeration.
    (*c*) Compare cascade and multi-compressor systems.

11. The following data refers to a compound refrigeration system employing water and flash intercoolers. The system has a load of 20 tons of refrigeration in the evaporator and uses ammonia as refrigerant. The saturation temperatures in the evaporator and condenser are – 40°C and 32°C respectively, and the intercooler pressure is adjusted for minimum power consumption. Assuming no superheating, no subcooling, isentropic compression and 75 percent volumetric efficiency at each stage of compression, calculate :
    (*a*) mass circulation of ammonia through the condenser,
    (*b*) piston displacements for each cylinder
    (*c*) power required to drive the unit, and
    (*d*) coefficient of performance of the system
    (**Ans.** 0.74 kg/s ; 9.52 and 4.25 m^3/min ; 28.2 kW ; 2.48)

12. A vapour compression system with ammonia as refrigerant works between pressure limits of 3 bar (evaporator) and 12 bar condenser. The system is fitted with water and flash intercooling and subcooling. The flash intercooling pressure is 6 bar, saturated vapours enter both the compressors and the limiting temperature for water intercooling and subcooling is 20°C. For a refrigerating load of 15 ton, determine:
    (*a*) power required to drive the two compressors,
    (*b*) coefficient of performance of the plant, and
    (*c*) swept volume of each compressor if each has a volumetric efficiency of 80 percent. Draw schematic arrangement of the system and use *p-h* chart to obtain your solution.
    (**Ans.** 8.631 kW, 6.15, 0.0224 m^3/s, 0.01278 m^3/s)

13. A 10 ton three stage ammonia refrigeration system works between the pressure limits of 2 bar (evaporator) and 12 bar (condenser). The system is fitted with expansion valves and flash chambers such that dry saturated vapours are extracted from the flash chambers at 4 bar and 8 bar and sent to the appropriate compressor suction. Determine :
    (*a*) power required to drive the three compressors,
    (*b*) coefficient of performance of the system.
    How these parameters would be affected if the system is made to operate with a simple saturated cycle between the same pressure limits ?
    Sketch the schematic diagram of the system and use *p-h* chart for obtaining the solution.
    (**Ans.** 7.35 kW, 5.29 ; 8.92 and 4.35).

14. A refrigeration installation using R-717 (ammonia) has one compressor, one condenser and three evaporators of capacity 20 ton each and operating at 5°C, – 10°C and 20°C. The condenser temperature is 40°C. The system uses individual expansion valves and back pressure valves fitted to each evaporator. The vapours leaving the evaporators are superheated by 10°C and the liquid refrigerant is subcooled in the subcooler suction of condenser by 10°C. Assuming isentropic compression, determine :
    (*a*) total mass flow of refrigerant in the unit,
    (*b*) compressor power and
    (*c*) COP of the system. (**Ans.** 5.33 kg/min, 29.95 kW, 3.5)

15. The compound refrigeration system depicted below uses Freon-12 as refrigerant and is used for a multi-load purpose.
    The system employs two evaporators with compound compressors, multi-expansion valves and flash intercooling.

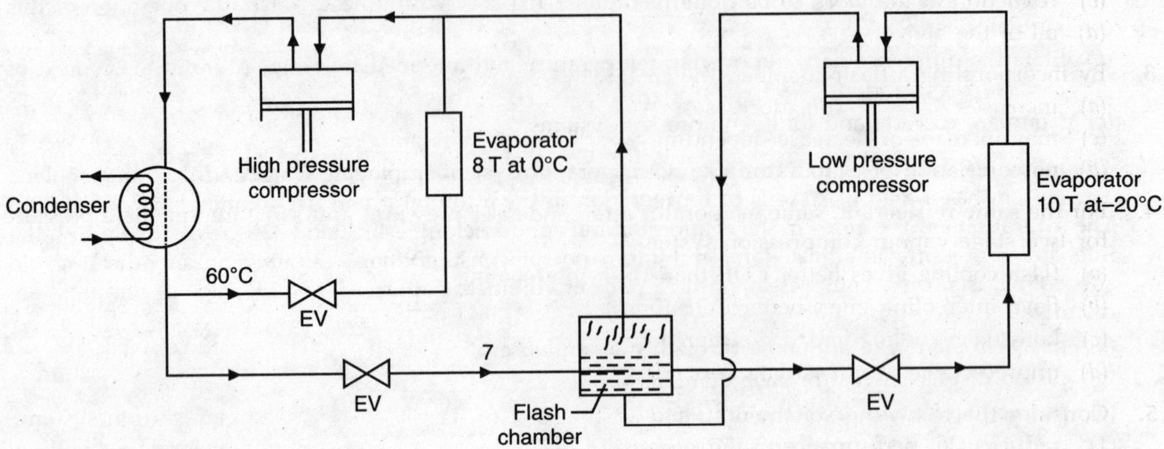

Fig. 6.29.

Draw the cycle on *p-h* chart and estimate coefficient of performance of the plant and power required to run the compressors. **(Ans.** 2.49, 25.62 HP)

B. Fill in the blanks with appropriate word/words

1. A large compression ratio causes a considerable .... in the capacity of a compressor.
2. *COP* of a refrigeration system .... with the employment of multistage compression.
3. In the vapour compression refrigeration system employing two-stage compression with liquid intercooler, the cooling medium is the liquid refrigerant from the ....
4. The temperature reduction in the subcooler of the vapour compression refrigeration system employing two stage compression with water intercooler and liquid subcooler, takes place at ....
5. The pressures in the different evaporators of multi-evaporator system at the same temperature with single compressor and expansion valve is same and equal to that at entry to ....
6. The suction vapour to the high pressure stage in the refrigeration system with multiple evaporator at different temperatures with compound compression and flash intercooler is ....
7. For the same refrigerant, same evaporator pressure, same condenser pressure and same intermediate pressure for two stage vapour compression system, flash intercooling results into .... *COP* when compared that with flash cooling.
8. In a .... refrigeration system, the condenser for the two temperature cycle works as an evaporator for the high temperature cycle.

Answers

1. reduction 2. improves or increases 3. condenser 4. constant pressure 5. compressor 6. dry saturated 7. higher 8. cascade

C. Multiple choice questions.

1. Compound vapour compression refrigeration systems
   1. improve volumetric efficiency
   2. reduce mechanical work input per unit mass
   3. take care of very large pressures.

   Which of the above statements are correct ?

   (*a*) 1 and 2 (*b*) 2 and 3 (*c*) 1 and 3 (*d*) 1, 2 and 3
2. The use of water intercooling in a multistage compression results into
   (*a*) need of a compressor of less stroke volume
   (*b*) decrease into specific volume of the refrigerant

(*c*) reduction in the work to be done in the high pressure compressor
(*d*) all of the above

3. By incorporating a flash chamber, the size of the evaporator
(*a*) increases (*b*) decreases
(*c*) may increase or decrease depending upon refrigeration capacity of the system
(*d*) no correlation of evaporator size with incorporation of flash chamber.

4. For the same refrigerant, same evaporator and condenser pressures and same intermediate pressure for two stage vapour compression system
(*a*) flash cooling gives better *COP* than flash intercooling
(*b*) flash intercoling gives better *COP* then flash cooling
(*c*) both flash cooling and flash intercooling give same *COP*
(*d*) unpredictable, result is data dependent

5. Consider the following statements : The installation of flash chamber in a refrigeration system
1. reduces the pressure losses through the evaporator
2. reduces the size of evaporator by not allowing the vapours to go to the evaporator
3. improves over all heat transfer.

Which of those statements are correct ?
(*a*) 1 and 2 (*b*) 2 and 3 (*c*) 1 and 3 (*d*) 1, 2 and 3

6. The employment of a compound compression system with multiple evaporators at different temperatures results into :
(*a*) decrease in power input to compressor
(*b*) increase in power input to compressor
(*c*) decrease/increase in power input to compressor depending upon operating conditions
(*d*) power requirement of the compressor is not affected in anyway by the given system.

Answers

1. (*d*) 2. (*d*) 3. (*b*) 4. (*b*) 5. (*d*) 6. (*a*)

Chapter – 7

Refrigerants, Driers and Lubricants

Learning objectives : A study of the subject matter included in this chapter will enable the students/readers to

- *understand the meaning of a refrigerant and list the properties of an ideal refrigerant*
- *classify the refrigerants and explain the system of their nomenclature/designation*
- *discuss the environmental effect of refrigerants, green house effect and global warming*
- *know about the necessity of driers and lubricants in a refrigeration system*
- *state the effects of leakage of refrigerant and the methods used for leak detection*

7.1. WHAT IS A REFRIGERANT?

A refrigerant is the working substance or heat carrying medium used in a refrigeration cycle to absorb heat from a low temperature system and discard the heat so absorbed to another system at higher temperature. During heat absorption, the refrigerant evaporates (changes from liquid to vapour state) and during heat rejection it condenses (changes from vapour to liquid state). Obviously, the refrigerant undergoes a phase change from gas to liquid and back. These processes occur in the vapour compression system as well as in the vapour absorption system. A wide variety of refrigerants are available, each having typical properties which limit its efficient use in a particular refrigerating machine.

The selection of refrigerant for an application is essentially based on certain parameters such as pressure-temperature characteristics, density, viscosity, flammability, toxicity, miscibility with oil, and heat absorbing capacity etc.

Early mechanical systems employed sulphur dioxide gas or anhydrous ammonia with small home refrigerators. Being toxic, sulphur dioxide disappeared from the market with the introduction of Freon. Ammonia is still used in large commercial plants, well away from residential areas, where a leak will not cause widespread injuries. Until concerns about depletion of ozone layer arose in 1980s, the most widely used refrigerants were the halomethanes R-12 and R-22, with R-12 being more common in automotive air conditioning and small refrigerators and R-22 being used for residential and light commerical air conditioning, refrigerators and freezers. Some very early systems used R-11 because its low boiling point allows low pressure systems for which low mechanical strength components can be used.

7.2. AN IDEAL REFRIGERANT—DESIRABLE PROPERTIES

The properties of an ideal refrigerant include :

1. Non-flammable, non-explosive, non-viscous, non-toxic and non-corrosive to metals
2. Easy availability, low cost and easy storage and disposal
3. Low specific heat of liquid and high total heat of vaporization

4. Low freezing point, low boiling point and high critical temperature
5. Moderate density in liquid form and a relatively high density in gaseous form
6. High thermal conductivity and electrical insulation
7. Easy to liquefy at moderate pressure
8. Ability to operate on a positive pressure
9. Long life; chemical stability
10. Odorous; easy detection of leakage
11. Mixes well with oil
12. Eco-friendly, and
13. High COP and low power consumption

No single refrigerant satisfies all the qualities desired of a refrigerant and which can be used for all types of applications. Obviously no ideal refrigerant exists.

7.3. CLASSIFICATION OF REFRIGERANTS

The refrigerants are usually classified as follows on the basis of :

• **Working Principle:**

– ***Primary refrigerant*:** These are the heat carrier substances that take part directly in the refrigeration system and cool the product/space by absorption of latent heat. The refrigerant undergoes a cyclic process involving compression, condensation, expansion and evaporation with change in phase.

– ***Secondary refrigerant*:** These are the heat carrier substances that are first cooled by primary refrigerants, and are subsequently used as a medium for producing cooling effect in the product or space being refrigerated. The secondary refrigerants do not undergo any cyclic process; no change in phase and produce the cooling affect by absorption of sensible heat.

Fig. 7.1 shows an air-conditioning system that uses both the primary and secondary refrigerants.

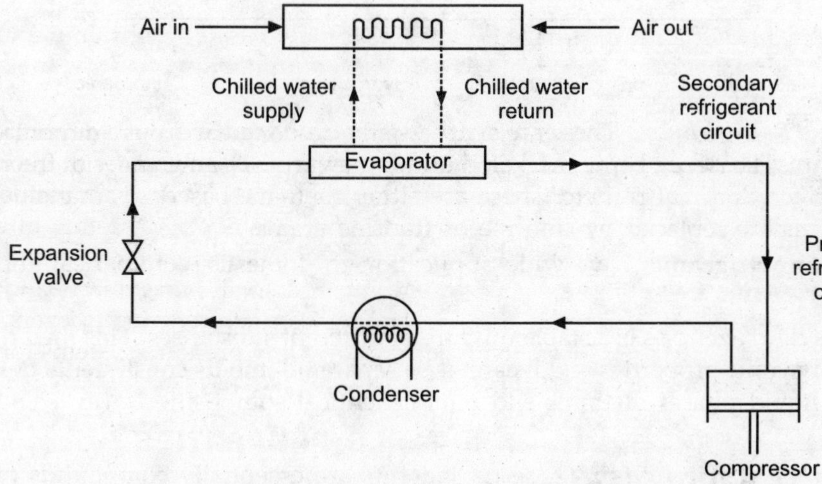

Fig. 7.1. Primary and secondary refrigerant circuits

The distinguishing features of primary and secondary refrigerants are listed below in Table 7.1.

Table 7.1. Primary and Secondary Refrigerants

Parameters	*Primary refrigerant*	*Secondary refrigerant*
• Cyclic/non-cyclic operation	Operates in a cycle; alternatively evaporating and condensing (phase change occurs)	Non-cyclic operation and no phase change
• Heat absorption	Latent heat	Sensible heat
• Refrigerating effect	High	Low
• Cost	Expensive	Comparatively much cheaper
• Quantity of refrigerant required	Less	More
• Corrosive effect	May or may not corrode	Usually corrosive
• Requirement of anti-freeze	No need of anti-freeze	Anti-freeze is required
• Examples	R-11, NH_3 and SO_2	Water and Brines

Water is used as secondary refrigerant for applications above 0°C, *i.e.* its freezing point, and for sub-zero applications brines or glycols are used. Brine is a solution formed by dissolving a soluble substance in water; the soluble substance could be a salt (NaCl, $CaCl_2$) or a glycol. When any soluble substance is mixed in water, the solution so formed has a freezing point lower than that of water.

- **Chemical Composition:**

The generic classification of refrigerants is shown below :

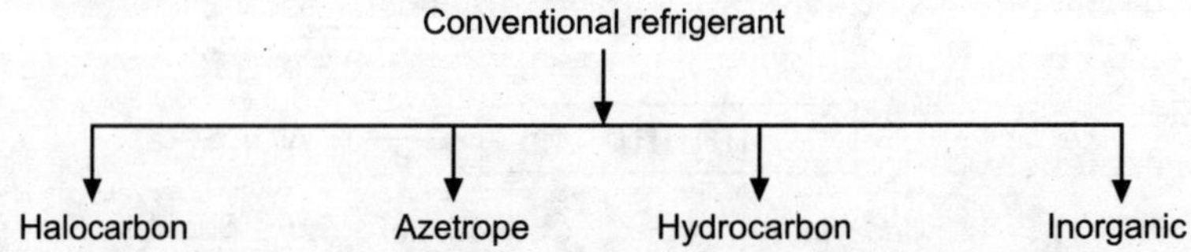

(*i*) ***Halocarbon Refrigerants :*** These are synthetically produced and were developed as Freon family of refrigerants. These are being marketed under the various trade names of freon, genetron, arcton, isotron, frigen and maftron etc. These are either methane-based or ethane-based, where the hydrogen atoms are replaced by chlorine or fluorine atoms.

The halocarbon refrigerants have wide application in domestic, commerical and industrial installations.

(*ii*) ***Azetropes :*** An azetrope is a stable mixture of different substances. This mixed refrigerant has fixed thermodynamic properties and it cannot be separated into its constituents by distillation. The vapour and liquid phases of this blend retain identical composition over a wide range of temperatures.

(*iii*) ***Hydrocarbon Refrigerants :*** These refrigerants are essentially compounds of hydrogen and carbon. These refrigerants have satisfactory thermodynamic properties, but are flammable and thus unsafe.

Examples: ethane, propane, iso-butane and ethylene

(iv) ***Inorganic Refrigerants :*** These are non-hydrocarbon refrigerants and are mostly available in nature. These refrigerants were exclusively used before the advent of halocarbon refrigerants, and are still in use due to their inherent physical and thermodynamic properties.

Examples : air, ammonia, carbon dioxide, sulphur dioxide and water.

- **Nature of Refrigerants :**
  (i) Natural which exist in nature such as water, ammonia etc.
  (ii) Artificial which are man-made and are synthetically produced such as azetropes and halocarbon refrigerants.
- **Safety :**
  (i) Safe refrigerants which are non-toxic and non-flammable.
  Examples : CO_2 , water and Freon family refrigerants.
  (ii) Toxic refrigerants which have irritating, obnoxious smell and are harmful.
  Examples : SO_2 and NH_3.
  (iii) Flammable refrigerants which are strongly prone to catch fire.
  Examples : Butane and isopropane etc.

7.4. DESIGNATION (NUMBERING) SYSTEM FOR REFRIGERANTS

(A) The **halogenated** hydrocarbons constitute the most widely used refrigerants. These are produced and marketed by number of manufacturers using their propriety (trade) names such as Freon, Arcton, Floron, Mafron etc with suffix numerals for identifying different refrigerants. The American Society of Heating, Refrigerating and Airconditioning Engineers (ASHRAE) has standardized the nomenclature system wherein the refrigerants are designated as R followed by a suffix numeral. Here letter R abbreviates for the common name of refrigerant (instead of trade name) and the suffix numeral identifies refrigerant of the same chemical composition. For example, Monochloro-difluro methane is R-22 whether produced by *X*, *Y* or *Z*.

This system was developed by DuPont and systematically identifies the molecular structure of refrigerants made with a single halogenated hydrocarbon. The salient features of this designation system are :

(i) The halogenated hydrocarbons are either methane based or ethane based where the hydrogen atoms are replaced by chlorine or fluorine atoms. Methane based compounds are denoted by a two digit number whereas the three digit number identifies ethane based refrigerant.

(ii) The general formula for these hydrocarbons is $C_m\ H_n\ F_p\ Cl_q$ in which

$$n + p + q = 2m + 2$$

where m = number of carbon atoms,
n = number of hydrogen atoms
p = number of chlorine atoms, and
q = number of fluorine atoms

(iii) The refrigerant is designated by

$$R(m - 1)\ (n + 1)\ (q)$$

The meaning of the code is

- Right most digit → number of fluorine atoms per molecule
- Tens digit → number of hydrogen atoms plus one
- Hundreds digit → number of carbon atoms minus 1. This is however omitted for the compounds which have only one carbon atom.

The following examples illustrate how to

(*i*) workout the number of refrigerant when the chemical formula is known,

(*ii*) find out the chemical formula of a refrigerant if its number is known

- **Dichloro-difluoro methane:** There are two chlorine atoms ($p = 2$), two fluorine atoms ($q = 2$) and no hydrogen atom ($n = 0$).

Then from the identify : $n + p + q = 2m + 2$ we get $m = 1$

$\therefore$ Number of carbon atoms = 1

Thus the chemical formula for dichloro-difluoro methane becomes $C\,Cl_2\,F_2$ and it is designated by the number.

$$R(m - 1)\ (n + 1)\ (q) \rightarrow R(1 - 1)\ (0 + 1)\ (2) \rightarrow R\ 012 \rightarrow R\ 12$$

- **Dichloro-tetrafluoro ethane :** There are two chlorine atoms ($p = 2$), four fluorine atoms ($q = 4$) and no hydrogen atom ($n = 0$).

Then from the identify : $n + p + q = 2m + 2$ we get $m = 2$

$\therefore$ Number of carbon atoms = 2

Thus the chemical formula for dichloro-tetrafluoro ethane is $C_2\,Cl_2\,F_4$ and it is designated by the number.

$$R(m - 1)\ (n + 1)\ (q) \rightarrow R(2 - 1)\ (0 + 1)\ (4) \rightarrow R\ 114$$

It is to be noted that number of chlorine atoms play no part in this designation system. It is really unfortunate since the ozone layer depletion potential of the refrigerant depends on the chlorine content.

- **Chemical formula for R 12 :** Since it is a two digit number, it is derived from methane base. The general formula for a hydrocarbon is $C_m\,H_n\,Cl_p\,F_q$ in which $n + p + q = 2m + 2$. Further, the refrigerant is designated by the number $R(m - 1)\ (n + 1)\ (q)$. Comparing it with the given number R 12 → R 012, we have

$$m = 1, \quad n = 0 \text{ and } q = 2$$

Then from the identity : $n + p + q = 2m + 2$, we get $p = 2$

$\therefore$ Chemical formula for R 12 = $C\,Cl_2\,F_2$

- **Chemical formula for R 114:** Since it is a three digit number, it is derived from ethane base. The general formula for a hydrocarbon is $C_m\,H_n\,Cl_p\,F_q$ in which $n + p + q = 2m + 2$. Further, the refrigerant is designated by the number $R(m - 1)\ (n + 1)\ (q)$. Comparing it with the given number R 114, we have

$$m = 2\,; \quad n = 0 \text{ and } q = 4$$

Then from the identity : $n + p + q = 2m + 2$, we get $p = 2$

$\therefore$ Chemical formula for R 114 = $C_2\,Cl_2\,F_4$

B. The brominated refrigerants are designated by putting the numerical suffix with capital B and a number; the number represents as to how many chlorine atoms are replaced by bromine atoms. Thus R 13 B1 is derived from R 13 (Monochloro-trifluoro methane $C\,Cl\,F_3$) with the replacement of one chlorine atom by a bromine atom. As such, its chemical formula is $C\,F_3\,Br$.

C. The designation of inorganic refrigerants is obtained by adding their molecular mass to 700.

(*i*) CO_2 (molecular mass 44) is designated 744

(*ii*) NH_3 (molecular mass 17), is designated 717

(*iii*) H_2O (molecular mass 18), is designated 718

D. The azeotropes (stable mixture of refrigerants) are given number 5 after R. For example
(*i*) R – 500 for a mixture of 73.8% R 12 and 26.2% R 152
(*ii*) R – 502 for a mixture of 48.8% R 22 and 51.2% R 115

E. Certain organic refrigerants are numbered with digit 6 after R. For example
(*i*) R – 600 for butane C_4H_{10}
(*ii*) R – 600 *a* for isobutane C_4H_{10}. Here subscript *a* denotes an isomer, *i.e.*, structural variant of the compound

F. The unsaturated compounds for which $n + p + q = 2m$ are distinguished by putting the digit 1 before $(a - 1)$. For example.
(*i*) R – 1150 for ethylene C_2H_4
(*ii*) R – 1270 for propylene C_3H_6

7.5. PROPERTIES OF REFRIGERANTS

The refrigerants should have certain chemical, physical and thermodynamic properties to make them safe and energy efficient.

7.5.1. Thermodynamic Requirements

(*i*) Freezing and boiling points : The freezing point of the refrigerant must be below the lower temperature of the cycle. If the refrigerant has high freezing point, it will freeze under operating condition and then hinder the continuous operation of the plant. Except in case of water for which the freezing point is 0°C, all other refrigerants have reasonably low value for the freezing point.

It is desirable that boiling temperature be as low as possible. This reduces the capacity of compressor and overall cost of the system. Moreover, with low boiling point, a lower vacuum is required in the condenser.

(*ii*) Critical temperature and pressure : The critical temperature of a vapour is the temperature above which it cannot be condensed irrespective of the pressure. The critical pressure signifies the pressure that causes liquefaction of a vapour at the critical temperature.

The critical temperature and pressure of the refrigerant must be well above the maximum operating temperature and pressure limits. The refrigerant has maximum heat transfer rate below the critical points.

The approximate values of freezing and boiling points, critical temperature and pressure for some common refrigerants are indicated in Table 7.2.

Table 7.2. FP, BP and Critical temperature and pressure

Refrigerant	*Freezing point (°C)*	*Boiling point (°C)*	*Critical temperature (°C)*	*Critical pressure (bar)*
NH_3	– 77	– 33	132	112
CO_2	– 56	– 73	30	72
SO_2	– 75	– 10	156	77
R-11	– 110	+ 23	197	43
R-12	– 157	– 30	112	40
R-22	– 168	– 40	95	48

***(iii)* Evaporating and condensing pressures :** Pressure of a refrigerant corresponds to saturation pressures during heat abstraction (evaporation) and heat rejection (condensation) processes. The evaporating pressure should be positive and as near atmospheric as possible. A positive pressure is required to restrict the possibility of the entry of air and moisture into the system.

The condensing pressures should be moderately low. This will permit the use of light weight equipment and piping on the high pressure side of the system. Too high evaporator and condenser pressures result in the requirement of high strength compressors, condensers and that increases the capital costs.

Table 7.3 lists the evaporator and condenser pressures and latent heat of vaporization of some common refrigerants.

Table 7.3. Evaporator and condenser pressures

Refrigerant	*Evaporator pressure (bar)*	*Condenser pressure (bar)*	*Latent heat of vaporization (kJ/kg)*
NH_3	2.363	11.67	1316.5
CO_2	22.9	71.93	274.0
SO_2	0.814	4.58	394.7
R-11	0.202	1.261	195.7
R-12	1.826	7.45	159.0
R-22	2.967	12.03	205.1

***(iv)* Specific heat :** The specific heat of the liquid refrigerants should have the lowest possible value. This characteristic reduces losses during throttling of refrigerant, and also improves the COP of the system.

Table 7.4. Specific volume, piston displacement and compression ratio

Refrigerant	*Specific volume (m^3/kg)*	*Piston displacement per ton of refrigeration (m^3/min)*	*Compression ratio*
NH_3	0.51	0.096	4.94
CO_2	0.0167	0.027	3.14
SO_2	0.40	0.254	5.63
R-11	0.77	1.016	6.24
R-12	0.09	0.163	4.08
R-22	0.078	0.101	4.05

***(v)* Latent heat of vaporization :** A high latent heat of vaporization is desirable since it permits greater refrigerating effect (amount of heat absorbed from the system being cooled) per kg of refrigerant circulated. Obviously it reduces mass of refrigerant to be circulated per ton of refrigeration and hence saving in the operating cost. The size of the system also becomes small and hence low initial cost.

***(vi)* Specific volume and compression ratio :** The power input to the compressor depends on the flow rate of the refrigerant. For the volume of the refrigerant dealt to be small, the refrigerant must have a low specific volume. Reciprocating compressors are used with refrigerants having high pressures and small volumes of the suction vapour. Centrifugal compressors are adopted when large volumes of refrigerant at low pressures have to be dealt with.

The compression ratio should be as small as possible. With high compression ratio, there is leakage of refrigerant across the piston and a decrease in the volumetric efficiency of the compressor.

Table 7.4 shows the specific volume of refrigerant vapours, theoretical piston displacement and compression ratio for some common refrigerants.

***(vii)* COP and power consumption :** The refrigerant should give high COP, and the power required per ton of refrigeration should be as low as possible. The power required in the operation of refrigeration plant is influenced by latent heat of vaporization, specific heat, compression ratio and density of the refrigerant vapour. For an ideal refrigeration system operating on Carnot cycle with temperature limits – 15°C (evaporator temperature) and 30°C (condenser temperature), the critical COP is 5.74.

Table 7.5. COP and power consumption

Refrigerant	NH_4	CO_2	SO_2	R-11	R-12	R-22
COP	4.76	2.56	4.87	5.09	4.70	4.66
Power consumption in kW/ton of refrigeration	0.738	1.372	0.724	0.694	0.746	0.753

From the values listed in Table 7.5, it is seen that :

(*i*) COP for R-11 is 5.09 and it is closest to the Carnot value of 5.74.

(*ii*) For most of the commonly used refrigerants, the COP values are fairly high and quite close to each other.

(*iii*) The refrigerant R-744 (CO_2) gives the lowest COP and has the highest power requirement per ton of refrigeration.

The refrigerant with COP closer to Carnot COP are the best for applications needing higher refrigerating effect.

7.5.2. Chemical Requirements

***(i)* Chemical stability :** The refrigerants should be chemically stable and inert. Chemical stability implies that the refrigerant should not decompose or disintegrate at temperatures normally reached in the system. Chemical inertness ensures that the refrigerant does not react with the metals/substances used in the system.

Sulphur dioxide is the most stable refrigerant, while Freons become unstable above 500°C. The recommended material for use for piping with halocarbons is copper. However, NH_3 reacts with copper and in that situation non-corrosive materials such as stainless steel must be used for piping.

***(ii)* Flammability (amenability to catch fire) :** The refrigerant should be non-flammable and and non-explosive when in contact with air or lubricating oil.

Halo-carbon refrigerants, CO_2 and Freon are neither flammable nor explosive. Ammonia becomes explosive when mixed with air in the ratio of 16 to 25 percent by volume of gas (ammonia). Hydrocarbons such as methane, ethane, butane and propane are highly explosive and flammable.

***(iii)* Toxicity :** Toxicity is a measure of harmful nature of a refrigerant and it should be as low as possible. If the degree of toxicity is very high, it will destroy the natural flavour of the eatables stored in the system. Further in case of leakage, it will create problems to human beings. Table 7.6 indicates the duration of exposure to produce serious health effects for some refrigerants.

Table 7.6. Toxicity of refrigerants

Refrigerant	*Concentration in air to produce serious effects (% by volume)*	*Duration of exposure to produce serious effects (minutes)*
NH_3	0.5 to 0.6	30
CO_2	0.7	5
SO_2	2 to 2.5	120
R-11	10	120
R-12	28 to 30	120
R-22	18 to 22	120

Generally, Freons are the safest refrigerants with least toxicity. Ammonia and sulphur dioxide are highly toxic and strong irritants. Obviously these toxic refrigerants are not used in domestic refrigeration and comfort air-conditioning; their use is limited only to cold storages.

(*iv*) Miscibility (ability to mix with oil) : The refrigerant should not react with lubricating oil but be completely miscible with the oil. The miscibility depends upon the temperature of oil and pressure of the refrigerating vapour.

- The Freon group of refrigerants are highly miscible. These refrigerants act as a carrier of oil to the moving parts and obviously then the lubrication becomes better.
- The refrigerants like carbon dioxide, sulphur dioxide and ammonia are relatively non-miscible. These refrigerants require larger heat transfer surfaces due to poor heat conduction properties of oil.

(*v*) Solubility of water : The solubility of water in a refrigerant is dependent both on the nature of refrigerant and its temperature. When the water/moisture in the refrigerant at a particular temperature exceeds the solubility limit at that temperature, the excess separates as free moisture/water. This excess moisture freezes into ice when the temperature of refrigerant falls below 0°C (the freezing point of water) as it normally happens at the throttling device and evaporator. The ice so formed chokes the expansion valve or capillary tube used for throttling in the system. The flow of refrigerant gets restricted and that adversely affects the performance of the refrigerating unit. Free moisture will also have the tendency to react with the refrigerant at high temperature (on the low side of the system) and form acids. The acid attacks the system components and causes corrosion.

Except for ammonia, the solubility of water in the commonly used refrigerants is low..

(*vi*) Effect on commodity being refrigerated : For the refrigeration of perishable foods (dairy products, meats, fruit and vegetables) drinking beverages and products like flowers and furs, the refrigerant chosen should be such that its leakage has no harmful effect on these materials. It is worthwhile to know that :

(*a*) NH_3 has great affinity for water and consequently tendency to become alkaline. Since the fruit and vegetables are generally acidic, NH_3 reacts with these products and spoils their taste.

(*b*) SO_2 destroys flowers, plants and furs but has no effect or food.

(*c*) Freon group of refrigerants is safest to use for the preservation of perishable commodities.

7.5.3 Physical Requirements

(*i*) Viscosity : Low viscosity is desirable both in the liquid and vapour state of the refrigerant. Low viscosity ensures better heat transfer rates, and reduced pressure drops of the refrigerant when passing through liquid and suction lines.

***(ii)* Thermal conductivity :** The refrigerant should have high thermal conductivity for both liquid and vapour states. This aspect improves the heat transfer rates in the condensers and evaporating units of the refrigeration plant. Also higher the thermal conductivity, better the refrigerating effect.

***(iii)* Dielectric strength :** The dielectric strength of a refrigerant is indicative of its electric insulation properly, *i.e.,* a measure of the resistance of the refrigerant to the flow of electric current through it. The dielectric strength of a refrigerant is of concern in hermetically sealed units where the motor is exposed to refrigerant. Table 7.7 lists the relative dielectric strength of a few commonly used refrigerants.

Table 7.7. Dielectric strength

Refrigerant	NH_4	SO_2	CO_2	R-11	R-12	R-13
Relative dielectric strength	0.82	1.9	0.85	3.0	2.4	1.31

Relative dielectric strength is the ratio of dielectric strength of refrigerant vapour to dielectric strength of nitrogen at atmospheric conditions.

***(iv)* Corrosive action :** Corrosion is the ability to corrode metal. Preference is given to the refrigerant which does not cause corrosion of metals and materials used in the system. Freon group can be used with almost all metals/materials (aluminium, steel, synthetic rubber used for gaskets) without fear of corrosion. Ammonia is corrosive to copper and copper based alloys, and can be used only with iron and steel.

***(v)* Odour :** The odour of the refrigerant has both positive and negative aspects :

(*a*) An irritant odour causes panic and headache to the persons working near to the refrigerated place. Further, some eatables are highly sensitive to the odour and that spoils their natural flavour and taste.

(*b*) Any leakage of the refrigerant can be easily detected if it has a distinct odour. Sulphur dioxide has pungent odour which is both irritating and obnoxious. Ammonia is poisonous gas if inhaled in large quantity. The freon group of refrigerants is generally odourless and has no effect on perishable materials.

***(vi)* Leakage tendency :** The leakage tendency should be low, and also the detection of leak should be easy. The leakage may occur due to openings in the joints or flaws in material used for construction: leakage needs extra cost of refrigerant for charging/topping.

Soap solution, halide torch and the electronic leak detector are the methods adopted to trace the leaks of halogenated (Freon group) refrigerants. Ammonia has an irritating smell which becomes indicative of its leakage. Likewise, the obnoxious smell of SO_2 can give a clue to its leakage.

***(vii)* Cost and availability :** These parameters are very important factors in making choice of refrigerant for a particular refrigeration system. The high capacity commercial and industrial systems (e.g. cold storages and ice plants) require large quantities of refrigerant. To keep their operating costs within reasonable limits, the cost of refrigerant should be low. Moreover, the refrigerant should be available in plenty so that its procurement poses no problem.

Ammonia is the cheapest refrigerant. R-12 is slightly cheaper than R-22.

7.6. COMMONLY USED REFRIGERANTS

A brief description is given below of the commonly used refrigerants in the context of their properties and applications.

(A) Halocarbon compounds :

R-11 : Trichloro-monofluoro-methane ($C\ Cl_3\ F$)

Properties:

- Stable, non-flammable and non-toxic
- Low pressure and high boiling point liquid
- Complete miscibility with mineral lubricating oil under all conditions
- High specific volume of suction vapour at any evaporator temperature ; suitable only with centrifugal compressor
- Leakage detected by halide touch or by using an electronic detector
- Boiling point + 23.77°C at atmospheric pressure and latent heat 195.66 kJ/kg at – 15°C

Applications:

- Air-conditioning of office buildings, theatres and departmental stores
- Flushing agent for cleaning the internal parts of the compressor during overhauling
- Flushing out moisture from the system

R-12 : Dichloro-difluro-methane ($C\ Cl_2\ F_2$)

Properties :

- Colourless and almost odourless
- Non-toxic, non-corrosive, non-irritating and non-flammable
- Chemical stability upto 550°C
- No problem of oil return as it has good miscibility with oil
- Boiling point – 29.8°C at atmospheric pressure and latent heat 161.3 kJ/kg at – 15°C. Because of low latent heat, large amount of refrigerant is circulated for a given refrigeration duty. This permits the use of less sensitive and more positive operating and regulatory mechanism.
- Leakage easily detected by a halide torch or by an electronic detector
- Available in abundance and accordingly cheaper than R-22
- Operates at low pressures and provides good volumetric efficiency
- Equally amenable with all types of compressors and condensing units

Applications :

- Most widely used and popular refrigerant of Freon group.
- Equally suitable for low, medium and high temperature applications.
- Used in domestic refrigerators, freezers, water coolers etc.
- Universally used in sealed type compressors because of being excellent insulator.

R-22 : Monochloro-difluro-methane ($CH\ Cl\ F_2$)

Properties :

- Non-toxic, non-flammable, non-corrosive and non-irritating.
- Solubility in water is fairly large (three times greater than that of R-12). Special driers and dessicants are needed to remove water.
- Miscible with oil at condenser temperature but separation occurs at evaporator temperature. Oil separators used to ensure return of oil from the evaporator.
- High discharge temperatures; compressor head and cylinder require water cooling.

- Odour is similar to that of chloroform, leakage can be detected by its odour or by halide torch.

Applications :

- This synthetic refrigerant has been specially developed for refrigeration installations that need a low evaporating temperature. It is successfully used with centrifugal compressor in chillers and window air conditioners.

Other refrigerants of halocarbon groups are mentioned below with a peculiar characteristic/ feature of each.

- R-13 : Monochloro-trifluoro methane ($C\ Cl\ F_3$) → suitable with reciprocating compressor and used for low pressure side of cascade systems.
- R-14 : Carbon tetrafluoride ($C\ F_4$) → an ultra low temperature refrigerant for use in cascade systems.
- R-21 : Dichloro-monofluoro methane ($CH\ Cl_2\ F$) → used for relatively high temperature requirements with centrifugal compressor systems.
- R-113 : Trichloro-trifluoro ethane ($C_2\ Cl_3\ F_3$) → remains liquid at room temperature and pressure, can be carried in sealed tins rather than cylinders, used in industrial and air conditioning units with centrifugal compressor system.
- R-114 : Dichloro-tetra fluoro ethane ($C_2\ Cl_2\ F_4$) → non-toxic and non-corrosive even in the presence of water, used in fractional power household refrigeration systems and drinking water coolers employing rotary vane type compressors.

(B) Azetrope refrigerants

R-500 : It is an azeotropic mixture of 73.8% R-12 ($C\ Cl_2\ F_2$) and 26.2% of R-152 ($CH_3\ CH\ F_2$).

- non-flammable, non-corrosive and low in toxicity
- fairly high solubility with oil
- solubility with water is critical; moisture needs to be kept out of the system by careful dehydration and by using driers
- for the same motor size and when used for the same purpose, the refrigeration capacity with R-500 is about 20% greater than that with R-12
- leaks detected by using soap solution, halide torch or electronic leak detector

This refrigerant is used both for commercial and industrial applications but only with reciprocating compressor systems.

R-502 : It is an azeotropic mixture of 48.8% R-22 ($CH\ Cl\ F_2$) and 51.2% R-15 ($C\ Cl\ F_2\ CF_3$)

- non-flammable, non-corrosive and practically non-toxic
- good refrigerant for obtaining low and medium temperatures
- positive evaporator pressure at low temperature
- lubrication oil retains its viscosity at low temperature and that ensures better lubrication
- less compressor displacement (reduced size and cost) and low power input (reduced running cost)

This refrigerant has replaced R-22 because of difficulties of high discharge temperature and poor oil return experienced with R-22. The blend R-502 has become the standard refrigerant for frozen food lockers, frozen food processing plants, frozen food display cases and in storage units for frozen foods and ice creams.

R-503 : It is an azeotropic mixture of 40.1% R-23 ($CH F_3$) and 59.9% R-13 ($C Cl F_3$). It is non-flammable, non-corrosive and practically non-toxic. It can hold more moisture than some other low temperature refrigerants. Its leak is easily detected with soap solution, halide torch or electronic detector.

This low temperature refrigerant is quite good for use in the low state of cascade liquid which require temperatures in the range – 70°C to – 85°C.

R-504 : It is azeotropic mixture of 48.2% R-32 ($CH_2 F_2$) and 51.8% R-115 ($C Cl F_3$). It is non-flammable, non-toxic and non-corrosive. Its leakage can be easily detected with soap solution, halide torch or electronic detector.

This refrigerant finds favour for use in industrial processes where a low temperature range of – 40°C to – 62°C is desired.

(C) Inorganic refrigerants :

- **Ammonia (NH_3) : R-717**
  - (*i*) excellent thermal properties : high latent heat and high heat transfer coefficient
  - (*ii*) large refrigerating effects with relatively small sized machinery
  - (*iii*) lighter than oil and can be easily separated. However in the presence of moisture, ammonia forms an emulsion with oil and that creates certain operating difficulties
  - (*iv*) low specific volume and so low volumetric displacement
  - (*v*) low mass of liquid to be circulated per ton of refrigeration
  - (*vi*) cheap refrigerant and low running cost
  - (*vii*) leaks easily detectable due to pungent odour. The leak is detected by bringing a sulphur candle near the leak : dense white smoke is given off when ammonia vapour is present.
  - (*viii*) highly volatile and explosive when in contact with air
  - (*ix*) toxic, irritating to eyes, nose and throat
  - (*x*) non-corrosive to ferrous metals but attacks copper and its alloys (brass and bronze)

Ammonia is an ideal refrigerant and is invariably used in ice factories, breweries and cold storage plants

- **Sulphur dioxide (SO_2) : R-764**
  - (*i*) colourles, non-explosive and non-flammable
  - (*ii*) extremely toxic and pungent irritating odour
  - (*iii*) works at low pressure and has small latent heat of vaporisation
  - (*iv*) corrosive when in contact with moisture due to formation of H_2SO_4
  - (*v*) leaks easily detectable on account of its pungent odour. The leakage is detected by bringing aqueous ammonia near the leak; white smoke is given off
  - (*vi*) low refrigerating effect and high specific volume. Therefore, for the same refrigerating effect with this refrigerant, we need larger compressors with higher speeds
  - (*vii*) not readily miscible with oil, liquid SO_2 is heavier than oil and accordingly it floats on the surface of oil. This aspect simplifies the problem of oil removal and return

In earlier days, sulphur dioxide was widely used in household and small commerical units.

- **Carbon dioxide (CO_2) : R-744**
  - (*i*) colourless and odourless
  - (*ii*) non-toxic, non-corrosive and non-flammable
  - (*iii*) higher concentration causes suffocation

(*iv*) leaks difficult to detect ; leak detection done by soap solution
(*v*) low specific volume makes the plant more compact; size of the compressor is small
(*vi*) heavier working parts needed because of high operating pressures
(*vii*) non-miscible with oil ; there is no dilution of oil in the compressor crankcase
(*viii*) low critical temperature and accordingly low temperature cooling medium is required in the condenser
(*ix*) chemically stable ; no reaction with oil and metals in the presence of moisture

The use of this refrigerant has been limited to marine application and to the manufacture of dry ice (solid CO_2).

- **Water (H_2O) : R–118**

(*i*) non-toxic, non-flammable
(*ii*) thermally stable and chemically inert even at high temperatures
(*iii*) very high enthalpy of vaporization
(*iv*) very safe refrigerant ; no leakage problem on health or environment
(*v*) high level transfer coefficient allows small dimensions for condenser and evaporators
(*vi*) available in abundance at no cost
(*vii*) light system components due to very low operating pressures with water

Water is essentially used as ice to produce the cooling effect. Its high freezing point limits its use in vapour compression systems. Water is used as a solvent in vapour absorption system, and also in steam jet refrigeration.

Freon v/s Ammonia

Freon-12 is superceding NH_3 in domestic and commercial plants because of its low toxicity; completely safe refrigerant as it does not affect the food products and is also non-irritant to human body. Home window air conditioners, package units and small central air-conditioners also use F-12 as refrigerant. Being an excellent insulator, it is universally employed in sealed compressor units.

Before the advent of Freon refrigerants, ammonia has been used universally for many applications due to low running cost, easy maintenance (no evacuation and drying), low volumetric displacement, low weight of liquid refrigerant per ton of refrigeration etc. Even today, ammonia is used in cold storages, ice manufacturing plants, skating rinks and breweries without exception.

Since Freon-12 is associated with ozone depletion, its production and use will be completely phased out in the coming years. With its replacement being both costly and time consuming, ammonia will remain ever green and popular due to its ODP (Ozone Depletion Potential) and GWP (Global Warming Potential) being practically zero.

7.7. OZONE DEPLETION AND MONTREAL PROTOCOL

The atmosphere envelops the earth in a series or concentric layers; the two major layers being recognized as troposphere and stratosphere. The troposphere (lowest layer) extends upto height of about 11 km above sea level and beyond 11 km lies troposphere. At 15 to 40 km above the earth's surfaces, a thin veil of chemical compound ozone exists, and this layer is called ***ozonosphere*** or ***mesosphere.*** Ozone layer acts as a protective umbrella for life on earth from the harmful effects of high concentration of sun's ultra violet (UV) radiation. During the last about 30 years, the ozone layer has been threatened by human-made gases such as CFCs released into the atmosphere.

Wayback in 1930, Freon-12 (a CFC) was discovered to replace the flammable and toxic refrigerants like SO_2. Freon-12 was then hailed as a miraculous refrigerant and that paved the way for CFC refrigerants belonging to the Freon family. Being non-toxic, non-corrosive, non-flammable and stable, CFCs enjoy a wide range of applications such as blowing agents for foams and plastics, aerosol propellants, solvents for cleaning precision electronic equipment and as working media in engines for waste heat recovery systems.

The recent findings indicate and confirm that the CFCs have the potential to destroy the stratospheric ozone layer. Once released into the atmosphere, CFCs move upwards by random diffusion from troposphere to stratosphere. Eventually the Cl and Br atoms are liberated from the parent compound by the action of ultraviolet light and destroy the ozone molecules through certain cyclic reactions. In the simplest way, a chlorine atom reacts with an ozone molecule, forms ClO and leaves a normal oxygen molecule. The free oxygen atom then takes away the oxygen from the ClO and the final result is an oxygen molecule and a chlorine atom, which then reinitiates the cycle:

$$Cl + O_3 \rightarrow ClO + O_2$$

$$ClO + O \rightarrow Cl + O_2$$

In sum : $$O_3 + O \rightarrow O_2 + O_2$$

The overall effect is to increase the rate of recombination, leading to an overall decrease in the amount of ozone. The chlorine atoms act as a catalyst and each can break tens of thousands of ozone molecules before being removed form the atmosphere. Each 1% reduction in ozone causes an increase of about 2% in ultraviolet-B radiation reaching the earth's surface. Exposure to increased UV-B reduces the effectiveness of the body's immune system, and so increases disease rates, and produces eye cataracts and skin cancer. Plants and aquatic organisms too get severely damaged by UV-B. Phenomenon of the damaging of the ozone veil is called ***ozone depletion***. The disappearance of ozone on localized region is termed as ***ozone hole***. An ozone hole was first observed in Antarctic stratosphere and that generated a great deal of interest.

Ozone depletion potential : The term refers to the capacity of a refrigerant to cause ozone depletion and is defined as :

(*i*) integrated change in total ozone per unit mass emission of a specific compound relative to integrated change in total ozone per unit mass of R-11.

(*ii*) capacity of a compound to destroy ozone molecules in the atmosphere as compared to R-11.

ODP depends on the chemical composition of the refrigerant. CFCs have more chlorine content than HCFC, and accordingly CFC have higher potential for ozone depletion. CFCs are estimated to contribute nearly 70% of man-made ozone depleting chemicals. Halons (used in fire extinguishers) are the compounds containing bromine, fluorene and carbon. Halons break down in the atmosphere releasing bromine which is even more destructive on ozone layer than chlorine.

Table 7.8. ODP values

Refrigerant	R-11	R-12	R-500	R-502	HCFC-22
ODP	1.0	1.0	0.74	0.34	0.05

The value of ODP for some common refrigerants have been listed in Table 7.8. Lower the ODP, better is the refrigerant.

Montreal Protocol : Consequent to detrimental changes in global environmental resulting from ozone depleting refrigerants, there has been concerted international cooperation to meet the impending threats on human life, plants and aquantic organisms. The Montreal Protocol is an international treaty designed to protect the ozone layer by adopting a two pronged strategy that involves :

(i) to regulate and restrict the release of emissions of CFCs,

(ii) to search and develop suitable ozone-benign refrigerants as substitutes for CFCs.

The treaty was opened for signature on September 16, 1987, entered into force on January 1, 1989 and since then it has under gone five revisions.

The treaty is structured around several groups of halogenated hydrocarbons, and for each group the treaty provides a time table on which the production of those compounds must be phased out and eventually eliminated.

In the original schedule, the participating countries agreed to freeze production of CFCs at 1986 levels and to reduce production by 50% by 1999 with a 10 yeas grace period for developing countries like India. The subsequent ammendments envisage total elimination of all CFCs on the banned list viz R-11, R-12, R-113, R-114 and R-115 by January 2000. There is a slower phase out to zero by 2010 of other substances (halon, 1211, 1301, 2402; and CFCs R-13, R-111 and R-112 etc). The phasing out of less active HCFs will go on till a complete phase out is achieved in 2030.

There are a few exceptions for essential uses where no acceptable substitutes have been found, viz in the metered dose inhalers commonly used to treat asthma and other respiratory problems.

Due to its widespread adoption and implementation, the Montreal Protocol has been hailed as the single most successful international agreement to date.

Substitutes for CFCs : It has been established that chlorine in the CFCs is responsible for chemical reaction with ozone and causing its destruction. Presence of ozone is essential as it protects the earth's surface from ultraviolet radiations being emitted from the sun. Obviously there is the need to replace Freons with alternative refrigerants which should be environmental friendly, and equally effective. The key problem areas in the quest for CFC substitutes are : cost of CFC alternatives and the associated replacement technologies. Replacement of a refrigerant means replacement of compressor and condenser as the saturation pressures are not the same for different refrigerants. Further, volume flow rate of refrigerants need to be adjusted for proper flow of refrigerant in the suction and liquid lines.

The following alternatives have been suggested to replace the CFCs :

(i) the neutral refrigerants such as air, water and ammonia

(ii) the hydrocarbons such as isobutane

(iii) the artificial blends like bromine derivatives of hydrocarbons

Further :

(a) HFCs (hydro-fluoro carbons) which contain no chlorine atoms and have zero ODP.

(b) HCFCs (hydro-chloro-fluoro carbons) have reduced ODP due to their association with one or more hydrogen atoms. This association allows them to dissociate faster in the lower atmosphere of the earth. The rain water absorbs the chlorine atoms which are released due to dissociation. Only a fewer chlorine atoms are able to reach the upper atmosphere and disturb the ozone layer. However, HCFs have a level of GWP and ultimately they too have to be phased out.

There is a dire need to find substitutes for R-11, R-12 and R-113; and the following alternatives have been recommended for these commonly used CFCs :

(a) R-11 ($C\,Cl_3\,F$) : R-11 is the refrigerant of choice in centrifugal chillers. HCFC-123 is the strongest contender to replace R-11 as both are non-flammable and low in toxicity. About 2 percent decrease in efficiency and capacity is expected with this replacement.

HCFC-123 with an ozone depletion factor of 0.02 mostly breaks down before reaching the upper atmosphere and the ozone layer.

• **R-12 ($C\ Cl_3\ F_2$)** : R-12 is the most prevalent and popular refrigerant used in domestic refrigeration and automotive air conditioning. Presently HFC 134a (R-134a) is the most promising alternative to R-12 because both have very similar saturation/temperature properties. However, the mineral oil lubricants used with R-12 are immiscible with R-134a and that causes concern about oil return and heat transfer. Further, with use of R-134a, there will be 5 to 10 percent drop in efficiency and refrigerating capacity of the system.

7.8. GREEN HOUSE EFFECT AND GLOBAL WARMING

These refer to an increase in temperature of earth's surface due to release of certain gases into the atmosphere. The earth's temperature balance is maintained by a subtle interplay of forces :

"Radiation from the sun passes through the atmosphere, warms the earth, and the earth then radiates some of this heat back to space."

The green house gases in the atmosphere (CO_2 and water vapour) trap some of this radiation, maintaining the planet at temperatures that allow life to flourish. The burning of fossil fuels for power and deforestation are increasing the level of CO_2 and so enhancing the natural green house effect.

If no action is taken to regulate green house gas emissions, the earth could heat upto 0.3°C a decade producing a warming of 2 to 5°C over the next century. This phenomenon of increase in temperature of earth's surface due to release of gases like CO_2 into the atmosphere is called ***Global Warming.***

The hydrochlorofluoro carbons (HCFCs) and hydrofluoro carbons (HCFs) are also now thought to contribute towards green house effect. On a molecule-for-molecule basis, these compounds are upto 10,000 times more potent green house gases than CO_2 and their increased use significantly increases the danger that human activity will change the climate. The Montreal Protocol currently calls for a complete phase out of HCFCs by 2030 but does not place any restriction on HCFs.

The global warming would effect climate and agriculture, change rainfall patterns and upset ocean currents. Flora and fauna could either flourish in a more benign climate or perish in a less hospitable one, thus altering the structures of existing eco-systems. As the oceans warm, their waters will expand and ice caps will melt. The resulting rise in sea levels would flood low-lying islands and coastal lands. That would reduce the area of land on which crops can be grown..

Global warming potential (GWP) and Total equivalent warming impact (TEWI) : These are the index of global warming and are defined as :

(*i*) GWP is the capacity of a refrigerant to increase the temperature of atmosphere similar to CO_2, and is defined as the ratio of time integrated temperature increases from the instantaneous releases of 1 kg trace substance relative to that of 1 kg of reference gas. GWP values of a few refrigerants are given in Table 7.9.

Table 7.9. GWP values

Refrigerant	R-11	R-12	R-13	HCFC-22	R-152a	R-600a
GWP value	1.2	3.1	4.5	0.03	0.03	0.04

Less the value of GWP, better the refrigerant.

(*ii*) TEWI is defined as the sum of amount of refrigerants released into the atmosphere and of CO_2 released during energy generation used to drive refrigeration and air conditioning equipment.

This global warming index emphasises the combined environmental effect of the direct emissions of green house gases with indirect effects of CO_2 emissions from energy required to run the equipment using refrigerant.

Kyoto Protocol : This is an international treaty on climate change, assigning mandatory targets for the reduction of green house gas emissions for signatory nations. The objective is the stabilization of green house gas concentrations in the atmosphere at a level that would prevent dangerously anthropogenic interface with the climate system. This is to be achieved by lowering overall emissions of six green house gases: carbon dioxide, methane, nitrous oxide, sulphur hexafluoride, CFCs and HCFC calculated as an average over the five year period of 2008-12.

The treaty was negotiated in Kyoto, Japan in December 1997, and entered into force in February 2005. As on December 2006, a total of 169 countries and other government entities have ratified the agreement. Notable exceptions include United States and Australia. The parties agreed that :

(*i*) the largest share of historical and current global emissions of green house gases has originated in developed countries.

(*ii*) The per capita emissions in developing countries are still relatively low.

China, India and other developing countries are exempted from the requirements of the Kyoto Protocol because they are not the main contributors to the green house gas emission during the industrialization period that is believed to be causing today's climatic change.

7.9. SECONDARY REFRIGERANTS

There are several applications where the direct use of a refrigerant is not desirable and has to be avoided. For example :

- toxic refrigerants (NH_3 and SO_2) cannot be used for the air-conditioning of schools and residential buildings.
- big cold storages need large quantity of refrigerant, and the cost of refrigerant may become higher than that of the refrigeration system itself.

Under such situations, indirect cooling is adopted and use is made of a liquid (water/brine) which is much cheaper than the refrigerant. The liquid is first cooled by the refrigerant and subsequently this chilled liquid is circulated by pumps to the space/products which are to be refrigerated. The liquid becomes a carrier of refrigeration and is known as secondary refrigerant.

The use of secondary refrigerant has the following advantages :

(*i*) easy to handle and easy to control

(*ii*) different temperatures can be maintained in different parts of the building by controlling the flow rate of chilled liquid

(*iii*) reduction in size and overall cost of the system

(*iv*) toxic refrigerants can be kept away from the space and products being refrigerated

(*v*) network of pipeline is reduced in size and length, and low grades materials can be used for piping and heat exchangers

The commonly used secondary refrigerants are :

Water, brines, glycols and halocarbons

The following aspects need consideration while selecting a secondary refrigerant for a particular application :

- low freezing point and high specific heat. Higher the specific heat, lower will be the rate of flow required
- non-inflammable and non-toxic
- compatibility with the materials of the equipment
- neutral (neither acidic nor basic) so as to minimize corrosion
- low viscosity, it should not rise fast with rall in temperatures otherwise pumping power will become high and uneconomical

(A) Water : Water is used as the secondary refrigerant for temperature applications above its freezing point of 0°C such as air-conditioning plants and industrial cooling applications. In air-conditioning plants, air is cooled and dehumidified by using chilled water in water spraying unit.

(B) Brines : For subzero applications, brines are used as secondary refrigerant. Brine is a solution formed by dissolving a soluble substance in water ; the soluble substance could be salt ($NaCl$; $CaCl_2$) or a glycol. When any salt is mixed with water, the solution so formed has a freezing point lower than that of water and it further decreases with increases in salt concentration The strength or concentration of the brine solution is a measure of the mass of salt or glycol in the solution expressed as a percentage of the total mass of the solution.

When the salt concentration is increased beyond a certain level, the freezing point of brine starts increasing rather than decreasing. The lowest freezing point attainable is called ***eutectic temperature*** and the corresponding salt solution is known as ***eutectic solution***. The brine used in a particular application should have a concentration for which the freezing point of brine is at least 5 to 8°C lower than the required temperature.

(*i*) Calcium chloride brine ($CaCl_2$ + water) : This brine has eutectic temperature of – 55°C and this occurs at salt concentration of 30% by mass. It is widely used in industrial cooling and in ice making plants. However, it dehydrates the food products when it comes in contact with them. It has also tendency to corrode the components and as such its contact with air needs to be avoided.

(*ii*) Sodium chloride brine (NaCl + water) : The eutectic temperature of sodium chloride brine is – 20°C and it is obtained with salt concentration of 22% by mass. It is an ideal secondary refrigerant for the freezing of fish and meat.

Both of these brines are very corrosive to brass and copper, and accordingly standard steel pipes are used for brine piping. The brine solution containers are given coatings of corrosion resisting paints (such as lead oxide) at regular intervals ; at least twice a year.

(C) Glycols : Ethylene and propylene glycols are often mixed with water in varying proportions to depress the freezing point of water. The freezing point decreases to – 9.3°C with 15% ethylene glycol (by volume) and further drops to – 41.5°C with 65% strength. Standard steel and copper pipings are used for glycol lines.

These glycols are being extensively used as anti–freeze elements in brewing and dairy industry due to their following advantages over brines :

(*i*) non-corrosive and non-electrolytic even in the presence of water

(*ii*) extremely stable ; no evaporation during service

(*iii*) not much increase in the density of water solutions

(D) Halocarbons : The commonly used halocarbons used as secondary refrigerants are listed below along with their freezing points.

R-11 (–110°C) ; R-12 (– 136°C) and R-30 (– 96°C)

Since these halocarbons are volatile, they need a sealed system.

7.10. MOISTURE : CAUSES, EFFECTS AND REMOVAL

• **Causes :** The presence of moisture in a refrigeration system may be caused by one or more of the followings :

(*i*) incomplete drying of equipment before charging

(*ii*) moisture leaking into the system in dissolved form in the liquid refrigerant

(*iii*) formation of water due to oxidation of certain hydrocarbons in lubricating oil

(iv) leakage of water in the systems using water cooled condensers, and

(v) air leaking in the system on low pressure side has some moisture/water particles dissolved in it.

• **Effects :** The following are the detrimental effects resulting from the presence of water in a refrigeration system :

(i) moisture/water particles may freeze and clog the expansion valve or capillary tube and restrict the flow of refrigerant

(ii) moisture may cause short circuiting of motor windings in hermetically sealed units

(iii) corrosion caused by moisture may damage the tubings, valves and seals etc. and consequently reduce the life of equipment

(iv) deterioration in the quality of lubricant

• **Removal :** The following aspects help in restricting the presence of moisture in a refrigeration system :

(i) heating the entire assembly prior to charging

(ii) circulating dehumidified air through the system

(iii) creation of partial vacuum within the system by mechanical vacuum pump, and removing moisture by the use of CO_2 and alcohol

Quite often, a suitable drier and strainer are installed in liquid line between the condenser and expansion unit. This gives satisfactory results relating to removal of moisture in small quantities.

• **Dehydrators and dessicants :** The dehydrator or drier is a device installed in a refrigeration system to absorb/remove moisture from the refrigerant. This shell-shaped unit is made of steel or brass, is provided with inlet and outlet openings and is fitted in the liquid line.

The drier is filled with a moisture absorbing substance called dessicant. A strainer fitted on the outlet side does not allow any impurity to move towards the evaporator.

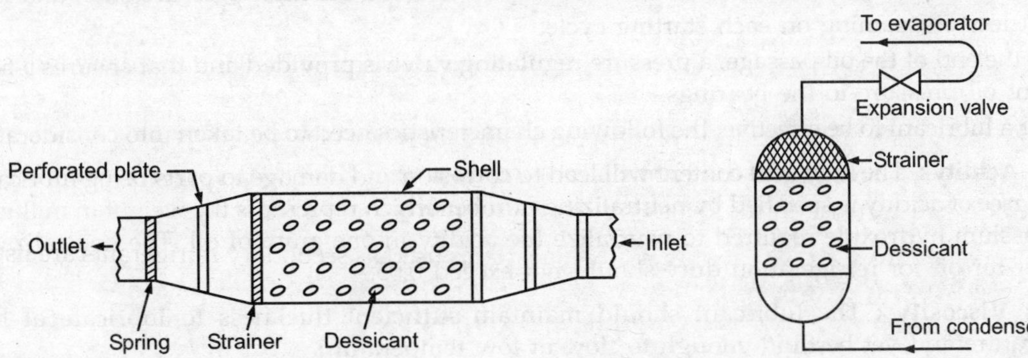

Fig. 7.2. Drier and its installation in liquid line

• **Types of driers :** The drier may be sealed type or refillable type. The shell of the sealed type drier is filled with the dessicant and is permanently sealed. The entire unit has to be replaced if the dessicant loses its effectiveness. The use of such driers is limited to small sealed units. The refillable/rechargeable drier has the provision to recharge/refill it with the dessicant whenever required, and such driers are used with large refrigeration units.

• **Dessicants :** The drying agents which are commonly used to remove moisture from a refrigeration system may be of absorbent type or chemical type.

Silicon dioxide (silica gel) and granular aluminium oxide (activated alumina) constitute the absorbent type of drying agents. The moisture removal process is rather slow and these need to be left in the system for long periods and sometimes permanently.

The chemical type dessicants remove moisture through chemical reaction. The moisture removal process is very quick and as such are never left in the system longer than 24 hours after their installation. The common chemical drying agents are : calcium chloride, calcium oxide (quick lime) and anhydrous calcium sulphate.

7.11. LUBRICATION AND LUBRICANTS

The lubrication system serves to reduce friction by forcing oil film between the mating surfaces of moving parts. The moving surfaces are then separated by a film of lubricating oil and this avoids solid to solid contact. The moving parts ride on a cushion of oil which offers much less resistance and so lower friction. The lubrication also cools the parts which get heated during operation.

Splash and forced feed lubrication are the two methods usually employed for lubricating different parts of the compressor unit of a refrigeration system. In **splash lubrication**, used for slow speed compressors, the crank case is filled with oil upto the bottom of main bearings or to the middle of crankshaft bearing. As the crankshaft rotates, the oil scoops fitted to it at various points pick up oil from the oil sump and pour it into the bearings and splash onto the cylinder walls.

In the **forced feed (pressure)** lubrication a small oil pump, attached to one end of the crankshaft, forces oil under pressure to the bearings through drilled passages in the crankshaft. It is an expensive but effective system for modern compressors designed to run at high speeds. Compared to the splash system, the forced feed system has the following advantages :

(*i*) more accurate control of oil distribution,

(*ii*) not much variation in the amount of oil delivered to the high pressure side, and

(*iii*) less oil foaming on each starting cycle.

At the end of the oil passage, a pressure regulating valve is provided and that ensures a fairly constant oil pressure to the bearings.

For a lubricant to be effective, the following characteristics need to be taken into consideration.

(*i*) Acidity : The high acid content will lead to corrosion and damage to parts being lubricated. The degree of acidity is specified by neutralizing number which represents the weight in milligram of potassium hydroxide required to neutralize the acidity in one gram of oil. The neutralization number for oil for refrigeration duty should not exceed 0.05.

(*ii*) Viscosity : The lubricant should maintain sufficient thickness to lubricate at high temperature and yet be fluid enough to flow at low temperature.

(*iii*) Pour point : The temperature at which oil ceases to flow is its pour point and it depends on wax content and viscosity of oil. With a high pour point, the oil becomes sluggish in movement and tends to cling to the cold evaporator surface affecting oil return and evaporator capacity. The oil must have a pour point low enough to allow flow at any point in the system.

(*iv*) Fire point and Flash point : The fire point of an oil is that temperature at which the vapours support combustion and the oil catches fire. At the flash point temperature, the oil vapours flash when exposed to flame. The flash point should always be higher than the operating temperature on the higher pressure side, *i.e.*, outlet of the compressor.

(*v*) Chemical stability : The oil should be chemically stable at the extremes of temperatures encountered in the system and against chemical reaction by refrigerant in the presence of all the materials such as metals, motor insulation, moisture, contaminants etc in the system.

(*vi*) Floc point : The floc point refers to the temperature at which the wax present in the oil begins to precipitate, *i.e.*, its solubility decreases and it starts separating. The wax thus made free

(*a*) may block orifice in throttling device and restrict flow rate,

(*b*) may deposit on heat transfer surfaces and then reduce the heat transfer rate (wax is insulator) in the evaporating unit.

(*vii*) Dielectric strength : It is a measure of the resistance of oil to current flow, and it is specified in terms of voltage (in kilo volts) to cause an electric spark to penetrate 1 mm thick strip of oil. Good lubricating oil has a rating of 25 kilo volts and above. The dielectric strength is of great importance, particularly in hermetically sealed units where the oil comes in direct contact with the motor.

(*viii*) Sulphur content : The lubricating oil should have low sulphur content. In the presence of moisture, sulphur forms sulphurous acid which has high tendency to corrode the components.

(*ix*) Freedom from dissolved air : Air is non-condensable and when it separates from the oil, it adversely affects the performance of the system.

(*x*) Moisture content : Lubricating oil absorbs moisture when exposed to the atmosphere. This moisture and refrigerant react chemically at high temperature (at high side of the system) and form acids which attack/corrode the components in the refrigeration system. Moisture may also cause sludge formation and freezing which affect the system performance. As such, the lubricant must contain minimum amount of moisture if not free of that.

7.12. LEAK DETECTION

The refrigerant system has to be tight as not to allow the refrigerant to leak out or the surrounding air and moisture to enter the system. The tendency of leakage of refrigerant is more pronounced on the high pressure side and leakage may be due to

(*i*) microscopic holes/flaws in the material used for construction,

(*ii*) openings in the joints, and

(*iii*) sealing material which might have got ruptured.

The leakage in refrigeration plants can lead to

(*a*) extra cost of refrigerant for charging/topping

(*b*) reduction in refrigerating capacity (cooling effect) of the plant

(*c*) excessive heating and drop in viscosity of lubricant

(*d*) breakdown of lubricating medium into acids/sludge and making the system dirty

(*e*) problems like toxic exposure, suffocation and fire hazards

(*f*) improper cooling of the motor windings of hermetically sealed units. This may result into gradual deterioration and ultimate failure of the windings. The repair would be both costly and time consuming.

The methods used for datecting the leakage of commonly used refrigerants are :

(A) Leak detection of SO_2 :

(*i*) Obnoxious smell can give a clue to the leakage

(*ii*) Soap solution is spread with a brush over all the joints. Soap bubbles will appear at the leaking joints.

(*iii*) By keeping aqueous (liquid) ammonia at the spot; a white smoke appears which is indicative of the leakage of SO_2.

(B) Leak detection of NH_3 :

(*i*) Ammonia has an irritating, pungent smell which becomes indicative of its leakage.

(*ii*) Soap solution is spread with brush over all the joints. Soap bubbles form at the leaking joints.

(*iii*) **Sulphur test :** The process consists in melting sulphur flowers with slow heat, then dipping short pieces or heavy strings into melted sulphur and finally drying it. The burning stick is passed around all the joints and suspected leaky points. The dense white fumes of ammonia sulphite (NH_3 S) are seen if NH_3 leak is present.

(*iv*) Solution of hydrochloric acid brought in the vicinity of ammonia leak produces white smoke of ammonium chloride (NH_3 Cl)

(*v*) Moistened strip of phenopthalene paper turns pink when faintist trace of ammonia vapour comes in contact with it. The same paper turns into scarlet colour when it gets exposed to large quantity of ammonia vapour.

(C) Leak detection of halogenated refrigerants :

Soap solution, halide torch and the electronic leak detector are the methods adopted to trace the leaks of halogenated refrigerants.

(*i*) **Soap bubble method :** The soap solution is applied at the joints or suspected places of leak. The leaking gas forms bubbles and their appearance gives an indication of leakage.

This method, however, is to be applied when the refrigeration system operates with positive pressure (pressure above atmospheric). If there is vacuum present, the moisture or soap bubbles will be sucked in and that will be quite detrimental.

(*ii*) **Halide torch :** The unit is like a kerosene lamp ; a slender tank that contains alcohol with a cotton wick. The burning wick has a light blue flame that turns greenish in the presence of slight leakage, and brilliant purple, if the leakage is large.

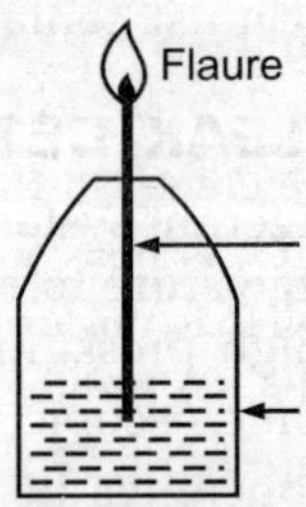

Fig. 7.3. Halide torch

The term halide is used as the torch is a popular apparatus for the leak detection of halides (Freon) group of refrigerants namely R-11, R-12 etc.

The method is, however, slow and relatively insensitive, and in brightly lighted area it is almost impossible to locate small leaks.

(*iii*) **Electronic leak detector :** The unit essentially consists of oppositely charged platinum electrode. When a halogenated refrigerant vapour comes between the electrodes, the current flow varies due to ionization of the leaking refrigerant. This sends a signal which is amplified and displayed on a screen with or without audible warning.

By adjustments and controls provided with the instrument, the magnitude of the leakage can also be detected. The instrument is portable, requires no warm up and its sensitivity can be changed as desired. The instrument can detect leakage upto 10^{-6} cc per second which is 10 times more sensitive than halide torch.

REVIEW QUESTIONS

A. Conceptual and conventional questions.

1. Define a refrigerant and name a few widely used refrigerants.
2. State the reasons for the wide use of Freon group of refrigerants.
3. Distinguish between primary and secondary refrigerants. Give examples of each category.
4. List the characteristics of an ideal refrigerant.

5. Evaluate the following group of refrigerants regrading their properties and applications :
   (*i*) NH_3, CO_2 and SO_2
   (*ii*) R-11, R-12 and R-22
6. Why the use of halogenated hydrocarbons as refrigerants is presently being discarded ?
7. Mention the effect of CFCs on the environment. How do they affect the ozone layer ?
8. Discuss the environmental impact of refrigerants. What is the Montreal and Kyoto Protocol?
9. Discuss the importance of following parameters in the selection of a refrigerant :
   (*i*) freezing point and boiling point
   (*ii*) critical temperature and critical pressure
   (*iii*) specific heat and heat of vaporization
10. Can water be used as a refrigerant ?
11. Discuss the validity of following statements :
    (*i*) A completely odourless refrigerant is not desirable
    (*ii*) A good refrigerant should have a low specific heat and high latent heat
12. What are azetropes ?
13. Name the refrigerants you would recommend for each of the follows applications :
    (*a*) sealed type air-conditioning plant of 5 ton capacity
    (*b*) domestic refrigerator
    (*c*) ice plant
14. Name three refrigerants that are most commonly used for commercial refrigerating plants. Mention for relative merits and demerits.
15. Explain the criteria adopted for selection of refrigerants. Justify use of ammonia as a refrigerant for ice plants.
16. With reference to thermodynamic properties of refrigerants, explain why
    (*i*) CO_2 is not commonly employed as refrigerant in India,
    (*ii*) NH_3 is not used in domestic refrigerator.
17. Describe briefly the method/methods used for detecting the leakage of NH_3, SO_2 and CO_2.
18. How are refrigerants numbered ? Illustrates with an example.
19. (*a*) Discuss nomenclature of refrigerants. Give numbers to :
    (*i*) Dichloro-tetrafluro ethane
    (*ii*) Trichloro-monofluoro methane
    (*iii*) Dichloro-diffuoro methane
    (*iv*) Monochloro-difluoro methane, and
    (*v*) Triichloro-trifluro ethane
    (*b*) Write down the chemical formula and name of the followign refrigerants :
    R-11, R-12, R-22 and R-717.
20. (*a*) Why is it necessary to keep refrigerants free from moisture ?
    (*b*) Compare sulphur dioxide with carbon-dioxide, and ammonia with Freon-12 as refrigerants.
21. (*a*) State the characteristics of a good lubricant for a refrigeration system.
    (*b*) How can moisture be removed from a refrigeration system ?

B. Fill in the blanks with appropriate word/words

1. Monochloro difluoro methane is ....
2. The refrigerant R-744 stands for ....
3. Freon group of refrigerants are .... and ....

4. The weight of refrigerant to be circulated in the refrigeration system for a given duty can be reduced by using a refrigerant having .... as large as possible.
5. .... is used as a secondary refrigerant in milk chilling plants.
6. For vapour compression refrigeration system using ammonia, the pipe material has to be ....
7. The leakage of .... can be easily detected by its pungent odour.
8. Electronic leak detector for refrigerant works on the principle of .... decomposed refrigerant.
9. The leak of Freon vapour in a refrigeration system is checked by using ....
10. Oil separator in a refrigeration cycle is installed between .... and ....
11. The moisture in a refrigeration cycle is removed by ....

Answers

1. carbon dioxide 2. Freon-22 3. non-toxic, non-flamable 4. latent heat of vaporisation 5. glycol 6. iron and steel 7. SO_2 or NH_3 8. ionisation 9. halide torch 10. compressor, condenser 11. driers

C. Multiple choice questions.

1. Which of the followings is not a desirable property of a good refrigerant ?
(*a*) low specific heat
(*b*) high specific volume of vapour
(*c*) large latent heat at evaporator pressure
(*d*) high critical temperature
(*e*) low boiling point

2. Which is not true in respect of an ideal refrigerant ?
(*a*) high latent heat of vaporisation and specific heat
(*b*) critical pressure and temperature well above the maximum operating pressure and temperature limits
(*c*) low value of specific volume
(*d*) high value of thermal conductivity

3. The critical temperature is the temperature
(*a*) above which explosion of gas occurs
(*b*) above which a gas will never be liquefied
(*c*) below which a gas is always liquefied
(*d*) below which a gas does not obey gas laws

4. The critical pressure is the pressure
(*a*) above which explosion of liquid occurs
(*b*) above which a liquid will remain a liquid
(*c*) above which a liquid turns into a vapour
(*d*) below which a liquid remains only in the liquid state

5. Which of the following pair of refrigerants is both non-toxic and non-inflammable ?
(*a*) NH_3 and SO_2
(*b*) CO_2 and SO_2
(*c*) CO_2 and F_{12}
(*d*) F_{12} and NH_3

6. The refrigerant Freon-12 belongs to the family of
(*a*) methanes
(*b*) ethanes
(*c*) ketones
(*d*) aldehydes

7. The freon group of refrigerants are
(*a*) inorganic refrigerants
(*b*) hydrocarbon refrigerants
(*c*) halo-carbon refrigerants
(*d*) azeotrope refrigerants

8. The chemical formula of Freon-12 is
(*a*) $CCI\ F_2$
(*b*) $CCI_2\ F_3$
(*c*) $CCI_2\ F_2$
(*d*) CCIF

9. The refrigerant Freon-12 is a compound consisting of
(*a*) carbon, hydrogen and fluorine
(*b*) carbon, hydrogen and chlorine
(*c*) carbon, chlorine and flourine
(*d*) carbon, hydrogen, chlorine and flourine

10. Which of the following pair of refrigerants is both colourless and odourless ?
(*a*) carbon dioxide and Freon-12 (*b*) sulphur dioxide and carbon dioxide
(*c*) Freon-12 and ammonia (*d*) carbon dioxide and ammonia

11. The refrigerant commonly used for commercial ice plants is
(*a*) Freon-12 (*b*) NH_3 (*c*) CO_2 (*d*) air

12. Ammonia is preferred as a refrigerant in large commercial installations because
(*a*) it is non-toxic (*b*) it has low latent heat
(*c*) it has low working pressure (*d*) it is relatively cheap

13. R-12 is generally preferred over R-22 in deep freezers since
(*a*) it has low operating pressure
(*b*) it gives high coefficient of performance
(*c*) it is miscible with oil over large range of temperatures
(*d*) it is non-toxic and non-inflammable

14. The refrigerant R-22 is
(*a*) monochloro difluoro methane (*b*) dichloro difluoro methane
(*c*) trichloro monofluoro methane (*d*) tetra chloro difluoro methane

15. The refrigerant R-744 is
(*a*) Freon-12 (*b*) NH_3 (*c*) CO_2 (*d*) SO_2

16. Identify the refrigerant used for refrigeration purposes in passenger aircraft.
(*a*) Freon-12 (*b*) carbon dioxide (*c*) air (*d*) sodium chloride

17. Which of the refrigerant from the list given below is least used these days ?
(*a*) ammonia (*b*) carbon dioxide (*c*) sulphur dioxide (*d*) Freon-12

18. Which of the following refrigerants is used in domestic refrigerators and in air-conditioning of cinema halls ?
(*a*) Freon-12 (*b*) Freon-22 (*c*) Methyl chloride (*d*) Carbon dioxide

19. The refrigerant R-22 is
(*a*) monochloro difluoro methane (*b*) dichloro difluoro methane
(*c*) trichloro monofluoro methane (*d*) tetra chloro difluoro methane

20. The refrigerant R-744 is
(*a*) Freon-12 (*b*) NH_3 (*c*) CO_2 (*d*) SO_2

21. Match the refrigerant (column A) with its number (Column B)

Column A	Column B
(*a*) NH_3	(*i*) R-40
(*b*) CO_2	(*ii*) R-717
(*c*) SO_2	(*iii*) R-744
(*d*) Methyl chloride	(*iv*) R-764

22. Suggest the material to be used for the pipes and pipe fittings in an ammonia refrigeration plant
(*a*) aluminium (*b*) brass (*c*) copper (*d*) cast steel or wrought iron

23. Suggest the material to be used for the condenser and evaporator pipes of Freon refrigeration plant
(*a*) aluminium (*b*) brass (*c*) copper (*d*) stainless steel

24. Which of the following refrigerant has the maximum ozone depletion potential in the stratosphere ?
(*a*) Ammonia (*b*) Carbon dioxide (*c*) Sulphur dioxide (*d*) Fluorine

25. Secondary refrigerant is invariably used in
(*a*) water coolers (*b*) deep freezers (*c*) ice plants
(*d*) domestic refrigerators (*e*) window type air conditioners

26. The refrigerant obtained by mixing two refrigerants in proper proportion is called
(*a*) secondary refrigerant (*b*) auxiliary refrigerant
(*c*) azeotrope (*d*) synthetic refrigerant

27. The brine is an aqueous solution of.............. in water
(*a*) sodium chloride (*b*) calcium carbonate
(*c*) sodium bicarbonate (*d*) magnesium sulphate

28. What is an azeotrope?
(*a*) a non-hologenic refrigerant
(*b*) a refrigerant dissolved in alcohol
(*c*) a mixture of refrigerants without phase separation
(*d*) an eco-friendly refrigerant

29. The choice of a particular refrigerant depends upon
1. refrigeration capacity
2. type of compressor used (reciprocating, centrifugal or screw)
3. service required (whether for air conditioning, cold storage or food freezing)
Which of these statements is/are correct?
(*a*) 1 and 3 (*b*) 1 only (*c*) 3 only (4) 1, 2 and 3

30. Environment friendly refrigerant R-134 is used in the new generation domestic refrigerators. Its chemical formula is
(*a*) $CH\ Cl\ F_2$ (*b*) $C_2\ Cl_3\ F_3$ (*c*) $C_2\ Cl_2\ F_4$ (4) $C_2\ H_2\ F_4$

31. Environmental protection agencies advise against the useof chloro-fluro carbon refrigerants because these react with
(*a*) water vapour and cause rain (*b*) plants and cause green house effect
(*c*) ozone layer and cause its depletion (*d*) carbon dioxide and cause global warming

32. When there is leakage of ammonia refrigerant, the sulphur sticks give
(*a*) red smoke (*b*) blue smoke (*c*) white smoke (*d*) black smoke

33. Leakage of halocarbon refrigerants can be best detected by
(*a*) halide torch (*b*) sulphur sticks (*c*) soap and water (*d*) burning candle

34. When there is leakage of Freon refrigerant, the colour of flame in halide torch test changes to
(*a*) yellow (*b*) red (*c*) blue (*d*) violet

35. The function of halide torch is
(*a*) defrosting of cooling coil (*b*) super heating of vapour refrigerant
(*c*) detecting leakage of refrigerant (*d*) facilitating better lubrication in the refrigerator

36. Moisture should be removed from the refrigerant to avoid
(1) failure of compressor seal (2) freezing at the expansion valve
(3) restriction to refrigerant flow (4) corrosion of steel parts

Which of these statements are correct ?
(*a*) 1 and 2 (*b*) 1, 3 and 4 (*c*) 2, 3 and 4 (*d*) 1, 2, 3 and 4

Answers

1. (*b*)	2. (*a*)	3. (*b*)	4. (*b*)	5. (*c*)	6. (*c*)	7. (*c*)	8. (*c*)
9. (*c*)	10. (*a*)	11. (*b*)	12. (*c*)	13. (*c*)	14. (*a*)	15. (*c*)	16. (*c*)
17. (*b*)	18. (*a*)	19. (*a*)	20. (*c*)	21. (*a*)	22. (*d*)	23. (*c*)	24. (*d*)
25. (*c*)	26. (*c*)	27. (*c*)	28. (*c*)	29. (*a*)	30. (*d*)	31. (*c*)	32. (*c*)
33. (*a*)	34. (*c*)	35. (*a*)	36. (*c*)				

Chapter – 8

Vapour Absorption Refrigeration

> **Learning objectives :** A study of the subject matter included in this chapter will enable the students/readers to
> - *understand the basic principle of vapour absorption refrigeration, and comment upon the essential requirements of refrigerant, absorbent and refrigerant-absorbent combination*
> - *explain the working of ammonia – water, water-lithium bromide and electrolux refrigeration systems*
> - *gain knowledge about theory of mixtures, temperature-concentration and enthalpy-concentration diagrams*
> - *make a comparison between the vapour compression and vapour absorption systems*

8.1. VAPOUR ABSORPTION REFRIGERATION SYSTEM

The vapour absorption cycles using water-ammonia systems were quite popular and widely used in the early years of 20th century. However due to its low COP, the absorption system lost much of its importance with the development of vapour compression cycles. Now-a-days, the use of vapour absorption cycle is limited to the places where waste heat is available or where heat is obtained from solar collectors.

Vapour compression systems use a compressor to compress the refrigerant vapour. The compressor requires input of high grade energy (mechanical or electrical) for its operation. In the vapour absorption system, the solar energy or waste heat is the motive power and the compressor is replaced by an absorber, a heat exchanger and a generator. The vapour refrigerant from the evaporator is drawn into an absorber where it is absorbed by the weak (poor) solution of the refrigerant and a strong (rich) solution is formed. The rich solution is pumped to generator raising its pressure to the level of condenser pressure. The liquid solution is practically incompressible and undergoes no change in specific volume. Obviously, very little work is required for raising its pressure. The solution pumped into the generator is heated from an external heat source. Consequently, the refrigerant is vapourized and driven off to the condenser where it is liquefied. The liquid flows by gravity into the liquid receiver and then into the evaporator where refrigerating effect is produced.

8.2. REQUIREMENTS OF REFRIGERANT, ABSORBENT AND REFRIGERANT—ABSORBENT COMBINATION

The pair of substances normally used in the vapour absorption system are :

- Ammonia + Water : Here ammonia is the refrigerant and water is the absorbent. Ammonia dissolves in water to form aqua ammonia.

- Water + Lithium bromide : Here water acts as refrigerant and lithium bromide as absorbent.

A vapour absorption system will work quite effectively if the refrigerant, absorbent and the refrigerant-absorbent combination meet the following requirements:

Refrigerant

(*i*) Low specific heat

(*ii*) High enthalpy of vaporization

(*iii*) High critical temperature and pressure

(*iv*) Chemical stability throughout the cycle

(*v*) High affinity for the absorber at low temperature and less affinity at high temperature.

Absorbent

(*i*) Low specific heat

(*ii*) High boiling point

(*iii*) Chemical stability

(*iv*) High affinity for the refrigerant at low temperature and less affinity at high temperature

(*v*) Minimum heat liberation when refrigerant is absorbed.

Refrigerant-absorbent combination

(*i*) Low specific heat and low viscosity

(*ii*) Non-corrosive

(*iii*) Low freezing point

(*iv*) Large difference in the normal boiling points of refrigerant and absorbent (at least 200°C). The absorbent would then exert negligible vapour pressure at the generator temperature.

(*v*) High affinity of refrigerant for the absorbent at low temperature, and less affinity at high temperature. This will ensures that a strong solution, highly rich in the refrigerant, is formed in the absorbent by the absorption of refrigerant vapour.

(*vi*) Solution to be obtainable with as small a heat of dilution as is compatible with other properties.

8.3. AMMONIA-WATER ABSORPTION SYSTEM

The absorption system is a heat operated system in which the refrigerant (NH_3) is alternately absorbed and liberated from the absorbent (water). The system operates on the principle that NH_3 and water though completely soluble in each other have different boiling points at the same pressure. When the high pressure solution is heated, the NH_3 vapours are driven out because the boiling point of NH_3 is lower than that of water.

Fig. 8.1 shows the schematic arrangement of different components in the water-ammonia (also called aqua-ammonia) absorption system. The various components of the system have been arranged according to their pressures and temperatures.

A strong aqua-ammonia solution is formed by dissolving fairly dry ammonia in cold water. The process of absorption is exothermic *i.e.,* heat is generated. The resulting rise in temperature of the solution adversely affects the capacity of water to absorb ammonia. To have a continuous and much more absorption, the absorber is kept cooled by removing the heat generated to water which is circulated through the cooling coils kept immersed in the absorber.

The high concentrated NH_3 solution known as strong liquor is then circulated through the system by the pump which raises the pressure to about 10.5 bar. For a steady flow process,

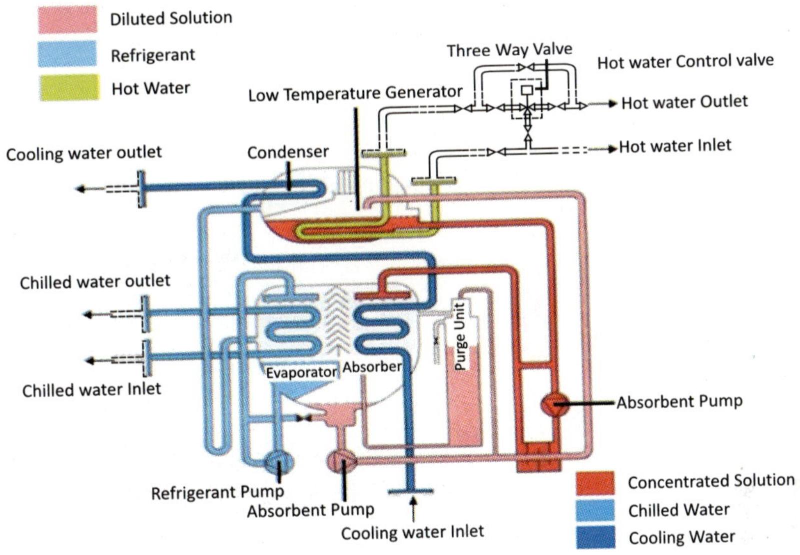

(a)

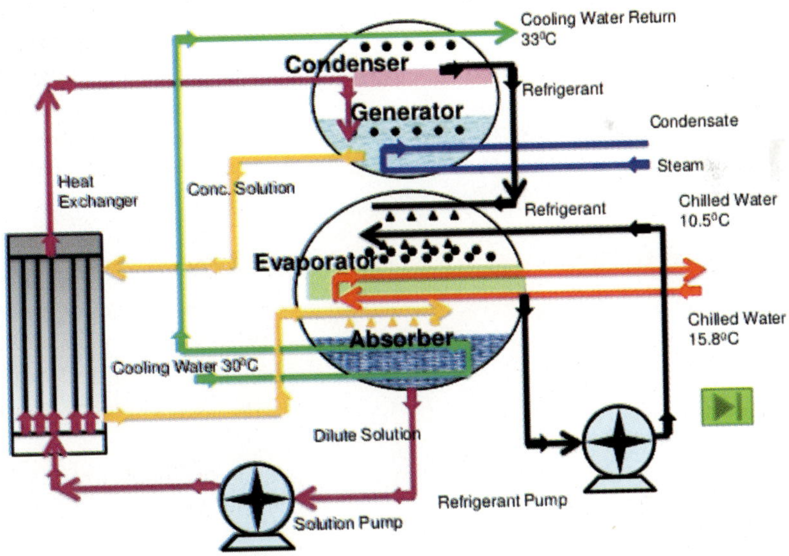

(b)

Vapour Absorption Systems

compression work = $\int_{p_1}^{p_2} V\,dp$ where V is the volume of the fluid to be handled and p_1, p_2 are the lower and higher pressures of the cycle.

The pump requires a very small input due to low volume of the liquid solution.

The solution next passes through heat exchanger where the strong ammonia solution is heated by the hot weak ammonia solution returning from generator to absorber. The warm high pressure solution then enters the generator where it is further heated by steam coils, gas or electricity. At high pressure and temperature existing in the generator, the solubility of ammonia in water is appreciably reduced and consequently NH_3 vapours are driven out of solution. Since the boiling point of NH_3 is less than that of water, the vapour leaving the generator are mainly those of NH_3. The weak NH_3 solution left in the generator called weak aqua is throttled and returned to the absorber via heat exchanger.

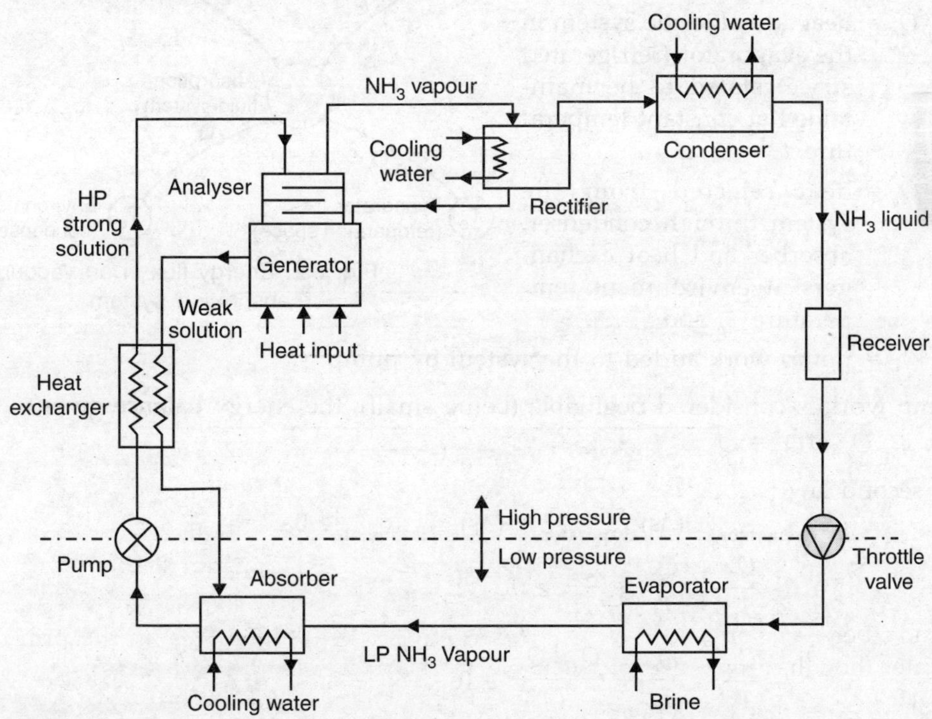

Fig. 8.1. Ammonia-water absorption refrigeration system

Ammonia vapours leaving the generator contain some water vapours. These water vapours if allowed to go to condenser and expansion valve, are likely to freeze resulting in choking of flow. Analyser and rectifier are the devices incorporated to remove water vapours escaping with ammonia vapour. Analyser is essentially an open type of cooler that consists of a series of trays mounted on the top of the generator. Most of the water particles escaping with the hot rising ammonia vapours get cooled, condense into liquid stream and fall back into the generator. A rectifier is water cooled heat exchanger which condenses water vapour and some ammonia and the condensed mixture is drained back to the generator. Thus final reduction or elimination of the water vapour takes place in the rectifier.

The virtually pure ammonia vapours are then led to the condensing unit in which the latent heat of ammonia vapours is rejected to the condenser cooling water and the ammonia vapours condense into liquid ammonia. The high pressure liquid ammonia is then throttled to a pressure of about 1.75 bar. The wet vapours at low pressure and temperature finally enter the evaporator, absorb heat from the brine and get partially dried up. The low pressure ammonia vapours leaving the evaporator are passed on the absorber and the cycle is repeated.

It is to be noted that the absorption system consists of a condenser, throttling or expansion valve and an evaporator similar to vapour compression refrigerator. The combination of absorber, pump, heat exchanger and generator replaces the compressor and this combination can be considered equivalent to a simple heat engine and refrigerator inter-connected for work transfer.

For an absorption refrigeration system (Fig. 8.2), let it be assumed that

Q_g = heat supplied to the system from heating medium at constant temperature T_g

Q_e = heat added to the system in the evaporator (refrigerated space) stated to be maintained at constant temperature T_e

Q_a = heat rejected from the system through condenser, absorber and heat exchangers at environment temperature T_a and

W_p = pump work added to the system by pump.

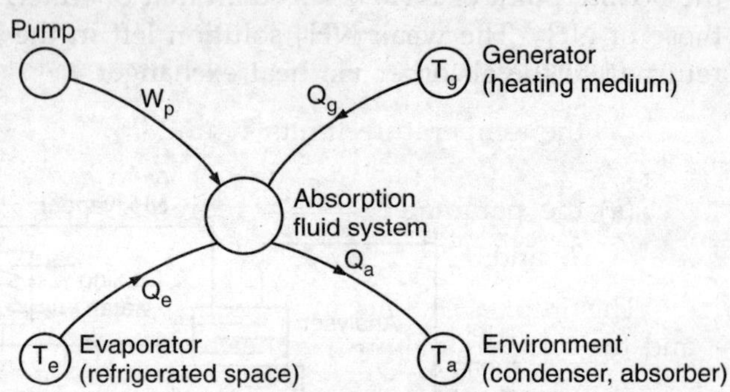

Fig. 8.2. Energy flux in an vapour absorption system

If pump work is considered negligible (being small), the energy balance gives :

$$Q_a = Q_g + Q_e$$

From second law :

$$(\Delta s)_{\text{generator}} + (\Delta s)_{\text{evaporator}} + (\Delta s)_{\text{environment}} \geq 0$$

i.e.,
$$-\frac{Q_g}{T_g} - \frac{Q_e}{T_e} + \frac{Q_a}{T_a} \geq 0$$

or
$$-\frac{Q_g}{T_g} - \frac{Q_e}{T_e} + \frac{Q_g + Q_e}{T_a} \geq 0$$

or
$$Q_g\left[\frac{1}{T_a} - \frac{1}{T_g}\right] + Q_e\left[\frac{1}{T_a} - \frac{1}{T_e}\right] \geq 0$$

or
$$Q_g \frac{T_g - T_a}{T_a T_g} + Q_e \frac{T_e - T_a}{T_a T_e} \geq 0$$

or
$$Q_g \frac{T_g - T_a}{T_a T_g} \leq Q_e \frac{T_a - T_e}{T_a T_e}$$

That gives :
$$\frac{Q_e}{Q_g} \leq \frac{T_g - T_a}{T_g} \times \frac{T_e}{T_a - T_e}$$

The parameter $\frac{Q_e}{Q_g}$ represents the ratio of refrigerant effect to the energy supplied in the generator, *i.e.*, *COP* of the system

$$\therefore \quad (COP)_{max} = \frac{T_g - T_a}{T_g} \times \frac{T_e}{T_a - T_e}$$

For *COP* to be high,

- temperature T_g of the heating medium (generator) be as high as possible.
- temperature T_a of the heat sink (environment and condenser be as low as possible.
- temperature of the refrigerated space (evaporator) be as high as possible.

It is to be noted that :

(*i*) the parameter $\frac{T_g - T_a}{T_a}$ is the thermal efficiency of a reversible engine working between the temperature limits T_g and T_a

(*ii*) the parameter $\frac{T_e}{T_a - T_e}$ represents the ideal *COP* of a refrigerator operating between T_e and T_a

This implies that an absorption system can be considered equivalent to a cyclic heat engine and the refrigerator put together as shown in Fig. 8.3.

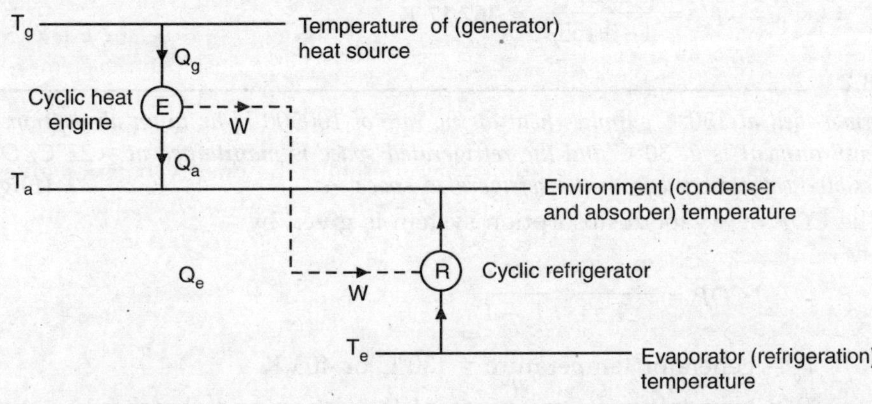

Fig. 8.3. An absoption system as equivalent to a cyclic heat engine and a cyclic refrigerator

Note : In case the condenser and absorber temperatures are not the same and are equal to T_c and T_a respectively, then the maximum possible *COP* is given by

$$(COP)_{max} = \frac{T_g - T_a}{T_g} \times \frac{T_e}{T_c - T_e}$$

EXAMPLE 8.1

A vapour absorption refrigeration system works with generator temperature = 87°C ; absorber temperature = 37°C ; evaporator temperature = – 13°C

Obtain the COP of the system. If the evaporator temperature falls to –18°C, what should be the generator temperature so that the system operates with the same COP? ***(WBUT)***

Solution: The *COP* of an absorption system is given by

$$COP = \frac{T_g - T_a}{T_g} \times \frac{T_e}{T_a - T_e}$$

where T_g = generator temperature = 87°C or 360 K

T_a = temperature of environment (*i.e.*, condenser and absorber) = 37°C or 310 K

T_e = temperature of evaporator (refrigerated space) = – 13°C or 260 K

Then : $$COP = \frac{360-310}{360} \times \frac{260}{310-260} = \mathbf{0.722}$$

(*b*) When the evaporator temperature falls to T_e = – 18 + 273 = 255 K, let the generator temperature be T_g' so that the system operates with the same *COP*. Then

$$0.722 = \frac{T_g' - 310}{T_g'} \times \frac{255}{310-255}$$

or $$1 - \frac{310}{T_g'} = \frac{0.722 \times 55}{255} = 0.1557$$

$$\therefore \quad T_g' = \frac{310}{1-0.1557} = \mathbf{367.17}\ \text{K}$$

EXAMPLE 8.2

A geothermal well at 130°C supplies heat at the rate of 100500 kJ/hr to an absorption refrigeration system. The environment is at 30°C and the refrigerated space is maintained at – 22°C. Determine the maximum possible heat removed from the refrigerated space. ***(Engg. Services)***

Solution : The *COP* of a vapour absorption system is given by

$$COP = \frac{T_g - T_a}{T_g} \times \frac{T_e}{T_a - T_e}$$

Given : T_g = generator temperature = 130°C or 403 K

T_a = temperature of environment (condenser and absorber) = 30°C or 303 K

T_e = temperature of evaporator (refrigerated space) = – 22°C or 251 K

$$\therefore \quad COP = \frac{403-303}{403} \times \frac{251}{303-251} = 1.198$$

The cefficient of performance is also given as

$$COP = \frac{\text{heat removal from the refrigerated space}}{\text{heat added in the generator}}$$

∴ Heat removal from the refrigerated space

$$= 1.198 \times \frac{100500}{3600} = 33.44 \text{ kJ/s} = 33.44 \text{ kW}$$

EXAMPLE 8.3

The operating temperatures of a single stage vapour absorption refrigeration system are : generator 90°C, condenser and absorber 40°C and evaporator 0°C. The system has a refrigeration capacity of 100 kW and the heat input to the system is 160 kW. Neglecting pump work, find the COP of the system and the total heat rejection rate from the system.

An inventor claims that by improving the design of all the components of the system, he could reduce the heat input to the system to 80 kW while keeping the refrigeration capacity and operating temperatures same as before. Examine the validity of the claim. **(MRIU)**

Solution : The *COP* of a vapour absorption system is given by

$$(COP)_{max} = \frac{T_g - T_a}{T_g} \times \frac{T_e}{T_a - T_e}$$

where T_g = generator temperature = 90°C or 363 K

T_a = temperature of environment (*i.e.* condenser and absorber) = 40°C or 313 K

T_e = evaporator temperature = 0°C or 273 K

Then $$COP = \frac{363 - 313}{363} \times \frac{273}{313 - 273} = 0.94$$

Applying 1st law of thermodynamics,

total heat rejection rate from the system

= refrigeration capacity + heat input to the system

= 100 + 160 = 260 kW

(*b*) With operating temperatures remaining the same ; *COP* = 0.94. With improved design:

Heat input to the system = 80 kW

Refrigeration capacity = 100 kW

$$\therefore \quad COP = \frac{100}{80} = 1.25$$

Since *COP* of the proposed system is greater than the maximum possible COP, the claim of the inventoris not valid.

EXAMPLE 8.4

In an aqua-ammonia absorption refrigeration system, heat is supplied to the generator by condensing steam at 0.25 MPa and 90 percent dryness fraction. Determine maximum COP of the system if the refrigerator is to be maintained at – 10°C whilst the ambient temperature is 27°C. What will be the required steam flow rate if the refrigerant load is 15 tons and the relative COP is 40 percent ?

Solution : At 0.25 MPa, i.e., 2.5 bar pressure,

t_{sat} = 127.43°C and h_{fg} = 2181.5 kJ/kg (from steam tables)

The maximum *COP* of an absorption system is given by

$$(COP)_{max} = \frac{T_g - T_a}{T_g} \times \frac{T_e}{T_a - T_e}$$

where T_g = generator temperature = 127.43 + 273 = 400.73 K

T_a = temperature of environment (*i.e.* condenser and absorber) = 27 + 273 = 300 K

T_e = temperature of evaporator (refrigerated space)
= – 10 + 273 = 263 K

$$\therefore \quad (COP)_{max} = \frac{(400.43 - 300)}{400.43} \times \frac{263}{(300 - 263)} = 1.783$$

Now Relative $COP = \dfrac{\text{actual } COP}{\text{maximum possible } COP}$

∴ Actual $COP = 0.4 \times 1.783 = 0.7132$

Since $COP = \dfrac{\text{refrigeration load}}{\text{heat supplied in generator}}$ and 1 ton of regeneration load is equivalent to 3.5 kJ/s, we get

Heat supplied in generator $= \dfrac{15 \times 3.5}{0.7132} = 73.61$ kJ/s $= 265 \times 10^3$ kJ/hr

Heat transferred by 1 kg of steam on condensation

$= x\, h_{fg} = 0.9 \times 2181.5 = 1963.35$ kJ/kg

∴ Required steam flow rate $= \dfrac{265 \times 10^3}{1963.35}$ = **134.97 kg/hr**

EXAMPLE 8.5

In a 18 tonnes absorption refrigeration system,the heating in the generator is carried out by using steam at 3 bar and 85% dry. The refrigeration temperature is –10°C. The condensation of the refrigerant is carried out a 30°C using cooling water. Determine :

(a) maximum possible COP of the system, and

(b) quantity of steam required per hour to run the plant if the steam leaks the generator as saturated water at same pressure.

Assume relative COP = 0.4. **(UPTU)**

Solution : At 3 bar pressure

t_{sat} = 133.5°C and h_{fg} = 2163.5kJ/kg (from steam tables)

The maximum COP of an absorption system is given by

$$(COP)_{max} = \frac{T_g - T_a}{T_g} \times \frac{T_e}{T_a - T_e}$$

where T_g = temperature at generator = 133.5°C or 406.5 K

T_a = temperature of environment (*i.e.* condenser and absorber)
= 30°C or 303 K

T_e = temperature of evaporator (refrigerated space)
= – 10°C or 263 K

$$\therefore \quad (COP)_{\max} = \frac{406.5 - 303}{406.5} \times \frac{263}{303 - 263} = 1.67$$

Now : Relative $COP = \dfrac{\text{actual } COP}{\text{maximum possible } COP}$

Then Actual $COP = 0.4 \times 1.67 = 0.668$

Since $COP = \dfrac{\text{refrigeration load}}{\text{heat supplied to generator}}$ and 1 ton of refrigeration load is equivalent to 3.5 kJ/s, we get

$$\text{Heat supplied to generator} = \frac{18 \times 3.5}{0.668} = 94.31 \text{ kJ/s} = 339520 \text{ kg/hr.}$$

EXAMPLE 8.6

In the generator of an ammonia absorption refrigeration system, 8 kg/min of strong ammonia solution at 80°C is supplied and 1.5 kg/min of dry ammonia vapours evaporate from it. The ammonia vapours leaving the generator have an enthalpy of 1800 kJ/kg and the remaining weak ammonia solution leaves the generator at 100°C. Compute the quantity of heat supplied form the external source per kg of ammonia vapour generated.

Assume mean mass concentration of the aqua ammonia as 0.32 and use the following equation for making calculations for the heat absorption

$$Q_a = 802.5\,(1 - x_m) - 928\,x_m^2$$

where x_m is the mean mass concentration of the aqua ammonia

(b) There is complete condensation of ammonia (generated in the generator) to liquid at 30°C in the condenser and its subsequent evaporation in the evaporator at 2 bar pressure. Using p-h chart, make calculations for the refrigerating effect, COP and refrigerating capacity of the system. How this refrigerating effect compares with that obtained by vapour compression system working within the same pressure limits ? Assume compression to be isentropic, suction is dry and saturated and there is no subcooling.

The other pertinent data is :

pressure in the generator = 12 bar
mean specific heat of solution = 4.85 kJ/kg K
heat loss from the generator to atmosphere = 80 kJ/kg of NH_3
generator thermal efficiency = 18%.

Solution : Heat of absorption in the generator is prescribed by the relation

$$Q_a = 802.5\,(1 - x_m) - 928\,x_m^2$$
$$= 802.5\,(1 - 0.32) - 928 \times (0.32)^2 = \mathbf{450.67} \text{ kJ/kg of } NH_3$$

(*b*) Quantity of strong aqua required per kg of NH_3 generated

$$= \frac{7.5}{1.5} = 5 \text{ kg/kg of } NH_3 \text{ generated}$$

Then, the quantity of weak aqua left in the generator
$= 5 - 1 = 4$ kg/kg of NH_3 generated

The enthalpy of liquid ammonia at 0°C as read from the *p-h* chart is 418 kJ/kg.

Now : Heat coming into the generator
= heat carried by strong aqua
+ heat given by heating source
$= 5[418 + 4.85 \times 80] + Q_s = 4030 + Q_s$

Heat going out of generator
= heat carried by NH_3 vapour
+ heat carried by weak aqua
+ heat of absorption
+ heat lost to surrounding
$= 100 + 4\,(418 + 4.85 \times 100) + 450.67 + 80 \simeq 5943$ kJ

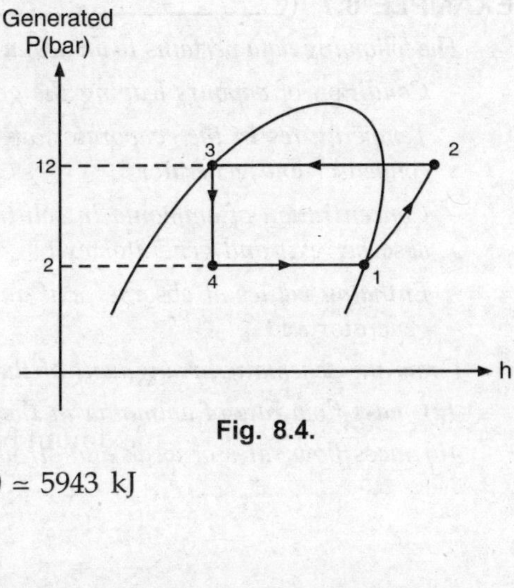

Fig. 8.4.

Then from energy balance of the generator, we have

$$4030 + Q_s = 5943$$

That gives : Heat supplied by the heating source

$$Q_s = 5943 - 4030 = 1913 \text{ kJ/kg of } NH_3$$

From *p-h* chart : $h_1 = 1660$ kJ/kg, $h_2 = 1925$ kJ/kg

$h_3 = h_4 = 560$ kJ/kg

Refrigerating effect : $h_1 - h_4 = 1660 - 560$

$= 1100$ kJ/kg of NH_3

$$COP \text{ of absorption system} = \frac{\text{refrigerating effect}}{\text{heat supplied}} = \frac{1100}{1913} = \mathbf{0.575}$$

Total refrigerating effect = 1100 × 5 = 5500 kJ/min

Since 1 ton of refrigeration = 210 kJ/min,

$$\text{Refrigeration capacity} = \frac{5500}{210} = \mathbf{26.19} \text{ tons}$$

For the vapour compression cycle working with in the same pressure limits,

refrigerating effect = $h_1 - h_4 = 1660 - 560 = 1100$ kJ/kg

work input to compressor = $h_2 - h_1 = 1925 - 1660 = 265$ kJ/kg

With thermal efficiency of 18%, the heat required to produce work equivalent to 265 kJ/kg is

$$= \frac{265}{0.18} = 1472.22 \text{ kJ/kg}$$

Then, *COP* of vapour compression refrigeration system is

$$= \frac{1100}{1472.22} = \mathbf{0.747}$$

For the same refrigeration effect, we have

$$\text{Ratio} = \frac{\text{heat input in vapour compression system}}{\text{heat input in vapour absorption system}}$$

$$= \frac{1472.22}{1913} = \mathbf{0.758}$$

EXAMPLE 8.7

The following data pertains to an acqua-ammonia absorption system of 10 tonnes refrigeration capacity.

- *Condition of vapours leaving the generator* : *100% pure ammonia saturated at 40°C*
- *Temperatures in the evaporator, absorber, condenser and generator* : *– 18°C, 32°C, 40°C and 160°C respectively*
- *Concentration of ammonia in solution at absorber exit and generator exit* : *0.3 and 0.1 respectively*
- *Enthalpy values at absorber exit and generator exit* : *25 kJ/kg and 680 kJ/kg respectively*

Draw the schematic arrangment of the system and make calculations for

(a) mass flow rate of ammonia in the evaporator

(b) mass flow rates of weak and strong solutions

(c) heat rejections in absorber and condenser, heat addition in generator and coefficient of performance of the system.

Solution : Refer Fig. 8.5. for schematic arrangement of the given refrigeration system.

The concentration and enthalpy values at exit from absorber and generator are :

$$x_2 = 0.3 \quad \text{and} \quad h_2 = 25 \text{ kJ/kg}$$
$$x_4 = 0.1 \quad \text{and} \quad h_4 = 680 \text{ kJ/kg}$$

Further from the table of properties for ammonia, we have :

h_3 (saturation vapour at 40°C)
$= 1473.3$ kJ/kg

h_5 (saturation liquid at 40°C)
$= 371.93$ kJ/kg

h_1 (saturated vapour at – 18°C)
$= 1422.72$ kJ/kg

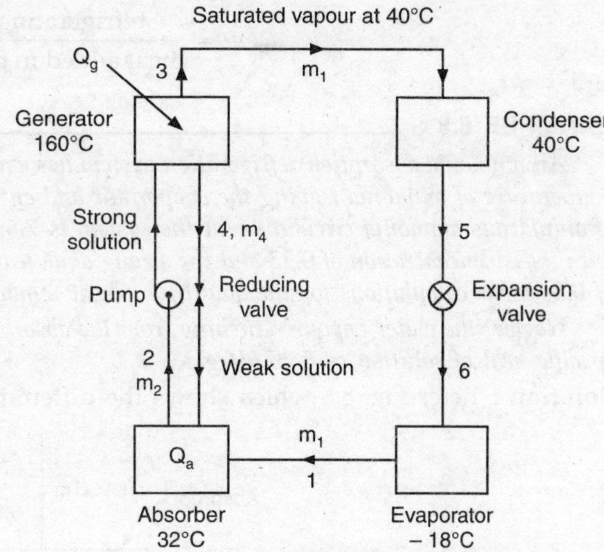

Fig. 8.5.

(*a*) Refrigerating effect

$$= h_1 - h_5$$
$$= 1422.72 - 371.93 = 1050.79 \text{ kJ/kg}$$

1 ton of refrigeration load is equivalent to 3.5 kJ/s. Therefore mass flow rate of ammonia

$$= \frac{3.5 \times 10}{1050.79} = 0.0333 \text{ kg/s}$$

(*b*) Overall mass balance for absorber gives $m_1 + m_4 = m_2$

From ammonia balance from absorber, we have

$$m_1 + 0.1\, m_4 = 0.3\, m_2 = 0.3\,(m_1 + m_4)$$

That gives
$$m_4 = \frac{m_1(1-0.3)}{(0.3-0.1)} = \frac{0.0333 \times 0.7}{0.2} = 0.1165 \text{ kg/s}$$

and
$$m_2 = 0.0333 + 0.1165 = 0.1498 \text{ kg/s}$$

(*c*) From energy balance for the absorber, we obtain

$$m_1 h_1 + m_4 h_4 = m_2 h_2 + Q_a$$

or
$$0.0333 \times 1422.72 + 0.1165 \times 680 = 0.1498 \times 25 + Q_a$$

or
$$47.376 \times 79.22 = 3.745 + Q_a$$

Thus heat rejection in the absorber,

$$Q_a = 47.376 + 79.22 - 3.745$$
$$= 122.85 \text{ kJ/s} = 122.85 \text{ kW}$$

From energy balance for the generator

$$m_2\, h_2 + Q_g = m_4 h_4 + m_1\, h_3$$
$$0.1498 \times 25 + Q_g = 0.1165 \times 680 + 0.0333 \times 1473.3$$

or
$$3.745 + Q_g = 79.22 + 49.06$$

Thus heat addition in the generator,

$$Q_g = 79.22 + 49.09 - 3.745 = 124.56 \text{ kJ/s} = 124.56 \text{ kW}$$

Heat rejected in the condenser $= m_1 (h_3 - h_5)$

$$= 0.0333\ (1473.3 - 371.93) = 36.67 \text{ kJ/s} = 36.67 \text{ kW}$$

$$COP = \frac{\text{refrigerating effect}}{\text{heat added in generator}} = \frac{3.5 \times 10}{124.56} = 0.28$$

EXAMPLE 8.8

An ammonia absorption refrigeration system has evaporator and absorber pressure of 2.5 bar, and the temperature of ammonia leaving the evaporator and entering the absorber is – 10°C. The mass flow rate of anhydrous ammonia circulating in the system is 7.65 kg/min, weak aqua enters the absorber at 48°C with mass concentration of 0.23 and the strong aqua leaves the absorber at 32°C with mass concentration of 0.3. Make calculations for the quantity of heat removed from the absorber.

Neglect the water vapour returning from the absorber and take liquid heat at 0°C as 418 kJ/kg K and specific heat of solution as 6.25 kJ/kg K.

Solution : Refer Fig. 8.6 which shows the different streams entering and leaving the absorber.

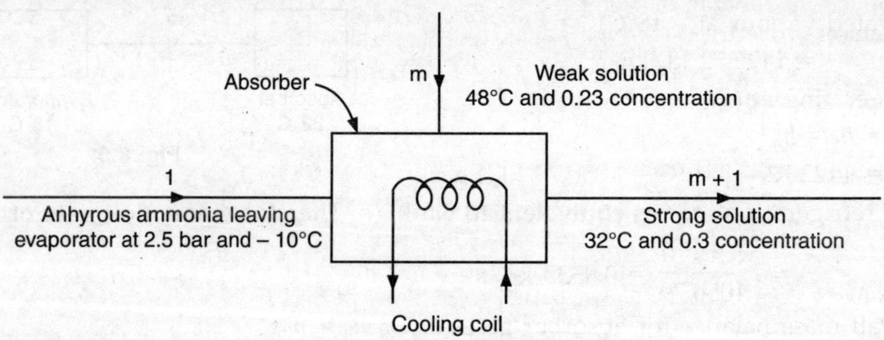

Fig. 8.6.

From mass balance :

anhydrous ammonia coming from evaporator
+ ammonia in weak solutions
= ammonia in strong solution

That gives : $1 + m \times 0.23 = (m + 1) \times 0.3$; $m = \dfrac{-0.3}{0.3 - 0.23} = 10$ kg

where m is the quantity of weak solution per kg of anhydrous ammonia

Heat absorption is prescribed by the relation,

$$Q_a = 802.5\ (1 - x_m) - 928\ x_m$$

where x_m = mean mass concentration $= \dfrac{0.23 + 0.3}{2} = 0.265$

$\therefore \qquad Q_a = 802.5(1 - 0.265) - 928 \times (0.265)^2 = 524.67$ kJ/kg

The total heat of unit mass of anahydrous ammonia at 2.5 bar and – 10°C as read from p-h chart is = 1685 kJ/kg

Enthalpy value at inlet to absorber

= heat of ammonia vapour coming from evaporator
+ heat in weak aqua + heat of absorption

$= 1685 + 10(418 + 625 \times 48) + 524.67$

$= 8389.67$ kJ

Enthalpy value at exit from the absorber

$$= \text{heat in strong ammonia} + \text{heat carried by cooling water}$$

$$= (10 + 1)\ [418 + 6.25 \times 32] + Q_w = 6798 + Q_w$$

Then heat balance gives :

$$Q_w = 8389.67 - 6798$$

$$= 1591.67 \text{ kg/kg of anhydrous ammonia}$$

∴ Heat carried by cooling water

$$= 1591.67 \times 7.65 = \mathbf{12176 \text{ kJ/min}}$$

8.4. VAPOUR COMPRESSION *VERSUS* VAPOUR ABSORPTION SYSTEMS

The salient points of difference between vapour compression and vapour absorption system of refrigeration have been given below :

Vapour compression system	*Vapour absorption system*
1. Energy supplied for running the compressor has to be mechanical or electrical *i.e.,* high grade energy.	The energy consumption of the system is mainly in the generator in the form of heat which is a low grade energy.
2. Energy supply is low being only one-third of the refrigerating effect.	Energy input is high being about one and half times the refrigerating effect.
3. Due to moving parts of the compressor unit, the system is quite noisy in operation and has appreciable wear and tear.	The moving parts are only in the pump which is a small element of the system. As such, the operation of the system is quiet and subjected to little wear.
4. The system has a poor performance at partial loads.	The performance of the system is not affected by load variations. The load variations can be easily met by controlling the quantity of aqua circulated and heat input to generator.
5. Due to reciprocating parts, there are chances of leakage of the refrigerant.	No chances of leakage as there are no reciprocating components.
6. The charging of refrigerant is quite simple.	The charging of refrigerant is rather difficult.
7. The valves and other compressor parts get damaged by the presence of traces of water vapour in the suction lines.	There is no danger from the liquid traces at the exit of the evaporator.
8. The system is used when no cheap source of energy is available.	The system is useful at places where waste heat is available in abundance, *i.e.,* in electric power houses, gas turbines and chemical industries.
9. Equipment is bulky and initial cost is high.	More precise equipment.
10. The space requirements, automatic control needs and high capacity (above 1000 tons) favour the use of vapour absorption systems.	

8.5. WATER – LITHIUM BROMIDE ABSORPTION SYSTEM

In the water – lithium bromide absorption system, the refrigerant is water and the absorbent is the solution of lithium bromide salt in water. Lithium bromide, white powder with a bitter taste, melts at 547°C and is soluble in water, alcohol and glycol. It is used :

– as an operating medium for air conditioning and industrial drying system due to its hygroscopic nature;

— as a sedative and hypnotic in medicine ;
— as brazing and welding fluxes.

The working of water - lithium bromide absorption system involves the following main aspects :

- Boiling point of water is a function of pressure. At atmospheric pressure water boils at 100°C. When maintained at high vacuum, it will boil and subcool itself.
- Lithium bromide (Li Br) has the property to absorb water due to its chemical affinity. At higher concentration and lower temperature, Li Br absorbs water vapour very effectively.
- As Li Br becomes dilute, it loses its capacity to absorb water vapour. It then needs to be reconcentrated using a heat source. Heat source may be steam, flue gases or exhaust gases from automobiles, power plants, steel plants etc.
- The heating causes the solution to release the absorbed refrigerant in the form of vapour. This vapour is cooled in a separate chamber to become liquid refrigerant.

Figure 8.7 shows the schematic arrangement of different components in the water-lithium bromide absorption system. The salient features are :

- The absorber and the evaporator operate almost at the same low pressure and are accordingly in one cylindrical container. Likewise, the generator and condenser are also enclosed in another cylindrical container as they operate at the high pressure of the cycle.

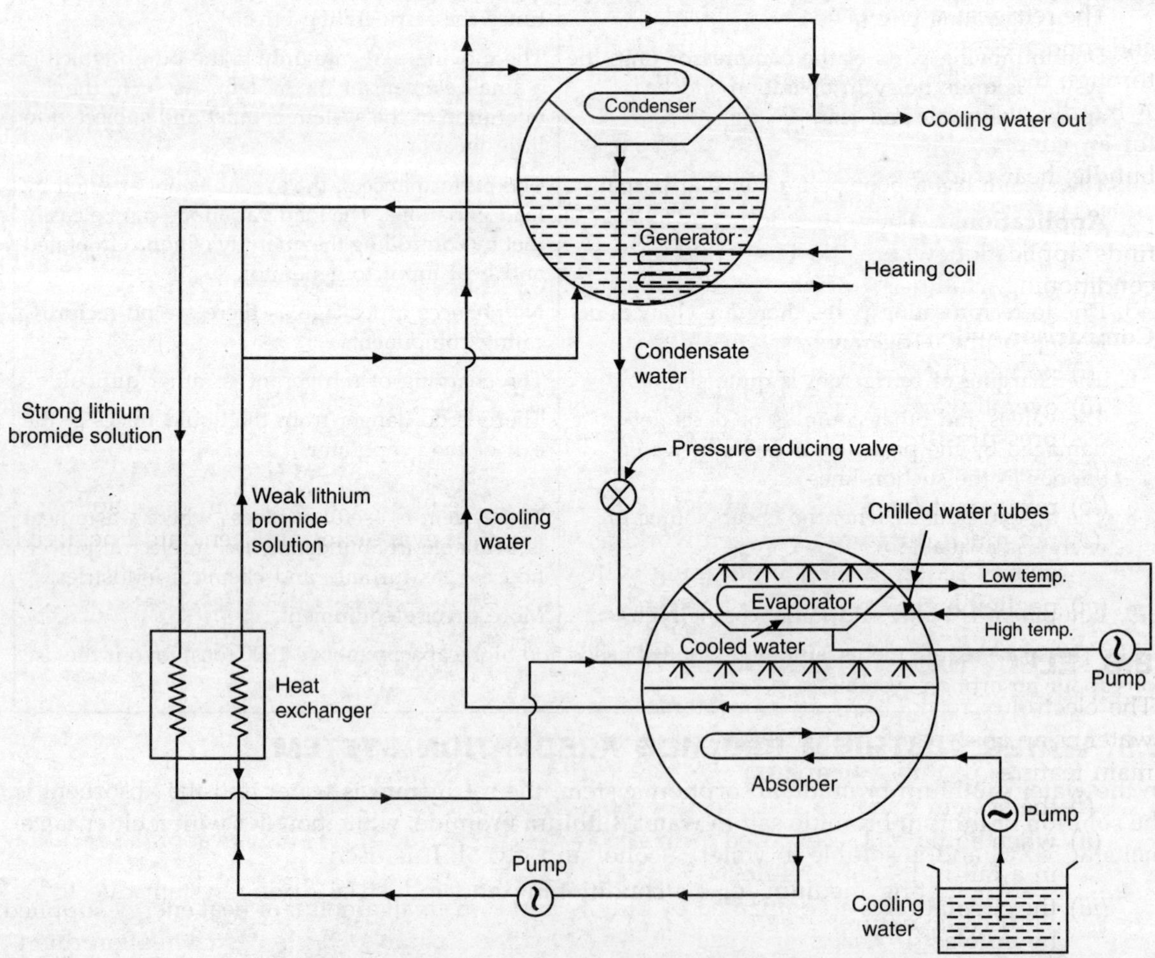

Fig. 8.7. Water-lithium bromide absorption system

- The evaporator contains the refrigerant water and the absorber has charge of lithium bromide. Li Br solution has a very strong affinity for water vapour because of its very low vapour pressure. It absorbs vapour as fast as is released in the evaporator. Li Br is highly corrosive and accordingly an inhibitor (lithium chromate) is often added to protect the metal parts of the system against corrosion.
- The absorber and the condenser have bundle of tubes through which water from the cooling tower is made to circulate. This cooling water lowers the temperature of lithium bromide solution in the absorber where as in the condensing unit it cools and condenses the refrigerant (water) vapour.

Working Circuit : Due to low pressure maintained inside the evaporator, the refrigerant water evaporate and the vapours are absorbed by the strong Li Br solution contained in the absorber. By absorbing water, the solution becomes weak and the weak solution is pumped to the generator where it is heated by using steam or hot water passed through the heating coils. Due to heating, a portion of water evaporates and the solution becomes more strong. The strong solution from the generator flows by gravity through a heat exchanger to the absorber. The weak Li Br solution, on its way from the absorber to the generator, also passes through the heat exchanger and takes up heat from the strong solution and that results in saving of some quantity of heat supplied in the generator.

The refrigerant (water) vapours formed in the generator due to heating of solution are cooled and condensed by the cooling water flowing through the condenser tubes. The condensate passes through the pressure reducing valve where the pressure is reduced to the evaporator pressure. A bundle of tubes passing through the evaporator section carries the water to be cooled either for air conditioning or for process application. The refrigerant water is sprayed over this tube bundle; heat is absorbed and refrigerating effect is produced.

Applications : The water-Li Br absorption system uses water as the refrigerant and as such finds application where the temperatures required are above 4°C. Such applications are air-conditioning, chilling of drinking water and chilling of beverages and fruit juices etc..

Comparison with ammonia-water system

(*i*) no need of analyzer and rectifier units which are essential for ammonia-water system
(*ii*) overall working pressure is barely above 0.15 bar; no need to design the system for high pressures
(*iii*) pressure inside the system is below atmosphere; absolutely no chance of bursting
(*iv*) refrigerant (water) is non-toxic; the system can be used directly for air-conditioning
(*v*) not much difference between working pressure of evaporator and generator; negligible energy required for the pumping system
(*vi*) negligible maintenance cost as there are no moving parts, valves and glands etc.

8.6. ELECTROLUX REFRIGERATOR

The electrolux refrigerator is of ammonia absorption type that uses ammonia as a refrigerant, water as an absorbent and hydrogen gas which reacts neither with ammonia nor with water. The main features of this system are :

(*i*) the system operates at a constant total pressure of about 14 bar throughout the circuit
(*ii*) when liquid NH_3 is exposed to an atmosphere of hydrogen, it readily evaporates resulting in a fall in its temperature
(*iii*) the circulation is accompanied by gravity and by a small amount of heat energy supplied by an electric heater or a gas. Hydrogen, being light, also facilitates circulation due to large difference in densities between the generator and evaporator units

(*iv*) the unit does not incorporate any pump or compressor. Instead, it has a generator, a separator and an absorber

Fig. 8.8 shows the schematic arrangement of the different components of the electrolux refrigerator.

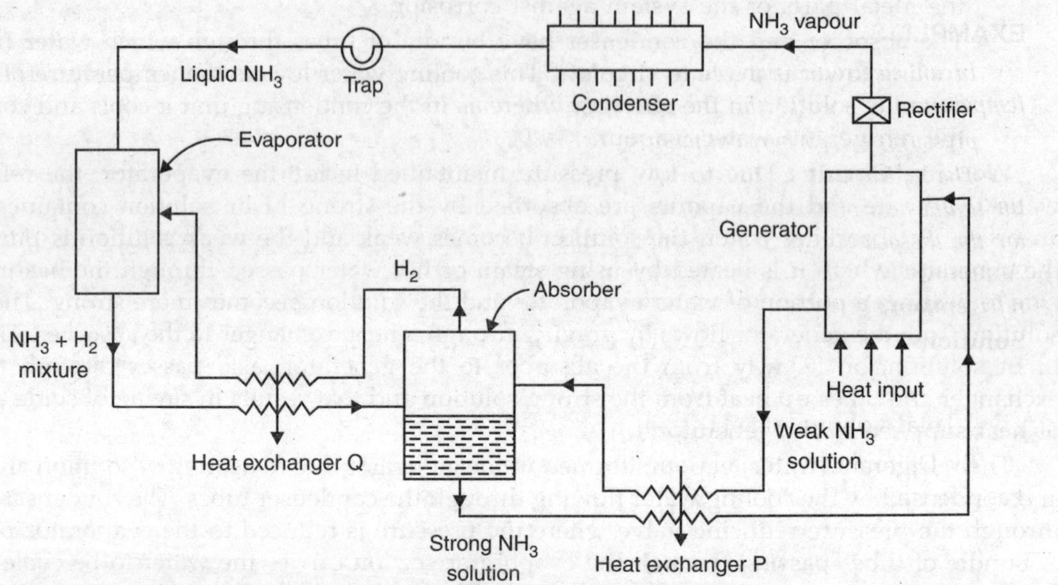

Fig. 8.8. Schematic arrangement of an electrolux refrigerator

The absorber absorbs ammonia vapour contained in (NH_3 + H_2) mixture coming from the evaporator and forms a strong solution of ammonia dissolved in distilled water. Ammonia is much more soluble in water than in hydrogen. As such NH_3 vapour gets dissolved in water while hydrogen is driven upward. Further, the absorption of NH_3 vapour is more and quick in cold water and accordingly the absorber is kept cooled by cooling fins or coils around it.

The strong ammonia solution flows down from the absorber to the generator via heat exchanger P. In the heat exchanger, the strong solution of ammonia is heated by the weak solution of ammonia returning from the generator.

Small heat energy from gas or electric heater is supplied to the generator and this causes the ammonia vapour to be driven out of the solution. The particles of water clinging to ammonia vapour are retained in the rectifier unit and flow by gravity to the absorber.

The pure ammonia vapours rise upwards to the air cooled condensing unit, condense to liquid ammonia which flows by gravity to the evaporator via a liquid trap. The evaporator contains hydrogen at 12 bar pressure. The partial pressure of ammonia vapour in the evaporator gets reduced to 2 bar according to Dalton's law of partial pressure, the corresponding temperature being – 18°C. Ammonia then evaporates at low temperature and the heat required for evaporation is taken from the system being refrigerated.

The mixture of ammonia and hydrogen flows to the absorber where ammonia is absorbed and the separated hydrogen rises to the top. The hydrogen is taken back to the evaporator and on its way is cooled in heat exchanger Q by the ammonia hydrogen mixture flowing from the evaporator to the absorber.

The coefficient of performance of the electrolux refrigerator is given by

$$COP = \frac{\text{heat absorbed in the evaporator}}{\text{heat supplied in the generator}}$$

The electrolux refrigerator is mostly used for household refrigeration in remote places, where there is no electricity. The system does not incorporate any machinery (pump or compressor) and is, therefore, quiet in operation and free from wear and tear. Further, the unit is compact, durable and easy to regulate.

EXAMPLE 8.9

In an electrolux refrigeration system, the total pressure is maintained at 16 bar. The evaporator temperature is – 15°C and the corresponding properties of NH_3 are :

pressure = 2.75 bar, enthalpy = 1675 kJ/kg and specific volume = 0.5 m³/kg

The hydrogen (gas constant 4.16 kJ/kg K and c_p = 12.75 kJ/kg K) enters the evaporator at 25°C and the liquid heat of ammonia entering the evaporator is 330 kJ/kg. The generator is supplied 420 kJ of heat for the dissociation of 1 kg of vapour and 1470 kJ/kg of heat for increasing the total enthlapy of NH_3. Assuming that NH_3 leaves the evaporator in saturated state, compute the coefficient of performance of the refrigeration system.

Solution : Pressure of hydrogen in the evaporator,

$$p_h = p_{total} - p_{NH_3} = 16 - 2.75 = 13.25 \text{ bar}$$

Total volume of $NH_3 = m_{NH_3} \times 0.5 \text{ m}^3$ (m_{NH_3} = mass of NH_3)

By Dalton's law, this is also the volume occupied by hydrogen. Then applying characteristic gas equation

$$m_h = \frac{p_h \times (m_{NH_3} \times 0.5)}{R_h T_h} = \frac{(13.25 \times 10^5) \times (m_{NH_3} \times 0.5)}{(4.16 \times 10^3) \times (273 + 25)} = 0.534\ m_{NH_3}$$

∴ mass of hydrogen per kg of NH_3 = 0.534 kg

Neglecting losses, refrigerating effect per kg of NH_3 evaporated

= heat of (NH_3 + H_2) leaving the evaporator

– heat of (NH_3 + H_2) entering the evaporator

= [1675 + 0.534 × 12.75 × (– 15 – 0)] – [330 + 0.534 × 12.75 (20 – 0)]

= 1572.87 – 466.17 = 1106.7 kJ

Heat supplied in the generator per kg of NH_3 generated

= 420 + 1470 = 1890 kJ

$$\therefore \quad COP = \frac{\text{refrigerating effect}}{\text{heat supplied}} = \frac{1106.7}{1890} = \mathbf{0.585}$$

8.7. THEORY OF MIXTURES

A knowledge of the properties of mixtures is needed for determining the heat transfer for an absorber, a condenser and generator etc. of a vapour absorption refrigeration system.

- A binary mixture is formed by combining two pure substances and its thermodynamic state is defined in terms of pressure p, temperature T and concentration C.
- The binary mixture may be homogenous or heterogenous. The ***homogenous*** mixture is uniform in composition and cannot be separated into its constituents by mechanical means such as settling or centrifuging. A thermal process is required for splitting this mixture into its constituents. Examples : moist air, and a mixture of refrigerant R-12 and compressor lubricating oil.

A ***heterogeneous*** mixture is non-uniform in compositions and its constituents can be easily segregated by mechanical means. Examples : fog and (refrigerant 717 + compressor lubricating oil).

- The binary mixture suitable for vapour absorption system should be mixture perfectly ***miscible*** in all proportions and in both the liquid and vapour phases.
- The ***concentration*** is defined as the ratio of the mass of any arbitrary constituent and the mass of the mixture. The concentration of a binary homogenous mixture remains practically constant (neither increases nor decreases) and the mixture is said to have attained an equilibrium concentration.

Let two liquids A and B with respective masses m_a and m_b be present in different compartments of a container. Both the liquids are at the same temperature, *i.e.*,

$$T_a = T_b = T_1 \qquad ...(i)$$

$$\text{Total mass } m = m_a + m_b$$

and concentrations c_a and c_b of the components A and B in the mixture are :

$$c_a = \frac{m_a}{m} \quad \text{and} \quad c_b = \frac{m_b}{m} \qquad ...(ii)$$

From identities (*i*) and (*ii*)

$$\frac{m_a}{m} + \frac{m_b}{m} = 1 \quad \text{or} \quad c_a + c_b = 1 \quad \text{or} \quad c_b = 1 - c_a$$

The total volume of the two components before mixing is

$$V_1 = V_a + V_b \quad \text{or} \quad mv_1 = m_a v_a + m_b v_b$$

$$\text{or} \quad v_1 = \frac{m_a}{m} v_a + \frac{m_b}{m} v_b = c_a v_a + (1 - c_a) v_b \qquad ...(iii)$$

where v_1 denotes specific volume of the two components before mixing, and v_a and v_b are the specific volumes of components A and B respectively.

Likewise specific enthalpy of the two components before mixing is

$$h_1 = c_a h_a + (1 - c_a) h_b \qquad ...(iv)$$

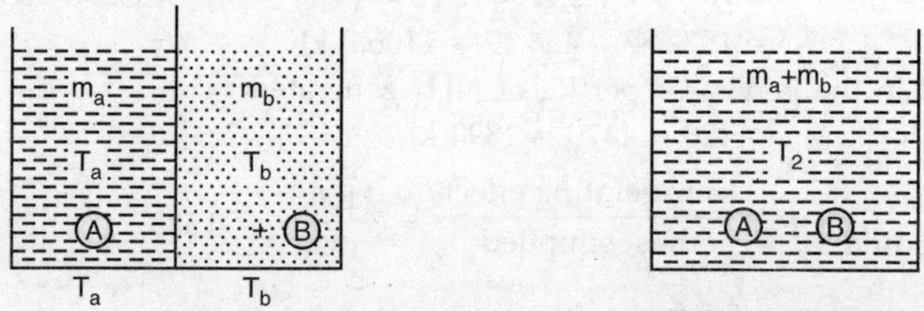

(a) Before mixing $T_a = T_b = T_1$ (b) After mixing $T_2 \neq T_1$

Fig. 8.9.

Let the partition be removed and the liquids A and B be allowed to mix. After thermodynamic equilibrium state is reached, the following conditions will apply depending on the nature of constituent liquids :

(*i*) the mass is conserved

(*ii*) the volume may increase, remain constant or decrease : $v_2 \gtreqless v_1$

(*iii*) the temperature may increase, remain constant or decrease : $T_2 \gtreqless T_1$

For the liquid ammonia-water mixture, the specific volume is worked out from the relation

$$v = 0.84\, c_a\, v_{fa} + (1 - c_a)\, v_{fw}$$

where suffix f stands for liquid states; a and w denote ammonia and water respectively.

8.8. TEMPERATURE CONCENTRATION DIAGRAM

The temperature-concentration diagram represents the variation of temperature with concentration (both the liquid and vapour states) for various pressures.

Withe reference to plot in Fig. 8.10, point 1 represents the initial state of the liquid mixture contained in a vessel filled with a loaded piston which maintains a constant pressure. When heat is supplied, the temperature rises (no change in concentration) and the appearance of a first bubble is observed. The temperature is then noted and the state marked by point 2. With further heating, vaporization starts and the temperature continues rising till the entire mixture has vaporized. The corresponding temperature is noted and marked by state point 3. Obviously the vaporization process of the mixture differs from that of a single component where the temperature remains constant during phase change. The superheating of the mixture vapour would occur if heating continues beyond state 3.

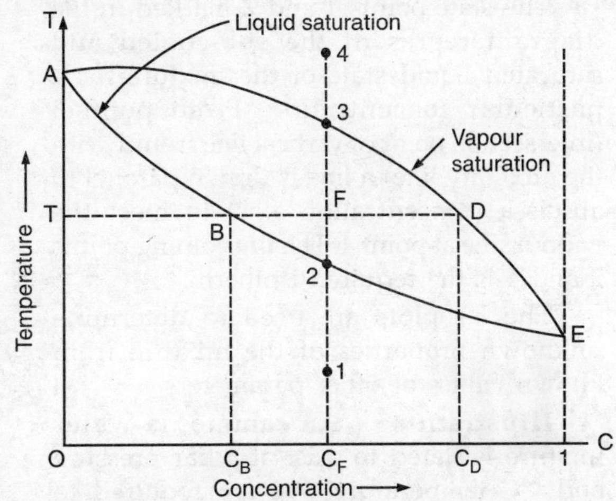

Fig. 8.10. Temperature-concentration diagram

Keeping the pressure same, the above processes can be repeatedly carried out for different concentrations. The locus of the temperature points (like 2) at which the first bubble just appears gives the curve ABE which is known as ***liquid saturation curve***. Likewise, the locus of temperature points (like 3) at which the entire liquid has evaporated gives the ***vapour saturation curve*** ADE. Similar curves can be obtained for other pressures also.

It is to be noted that corresponding to a particular temperature, the mixture concentrations would be different for the liquid and vapour states.

For example, corresponding to temperature $T = T_B = T_D$, the concentration is c_B when the mixture is in the liquid state, and c_D when the mixture is in the vapour state.

For the state F, the mass of the mixture

$$m_F = m_B + m_D$$

where m_B and m_D are the masses of the liquid and vapour corresponding to state F of the mixture. In terms of concentration.

$$m_F \times c_F = m_B\, c_B + m_D\, c_D$$

or $$(m_B + m_D) \times c_F = m_B\, c_B + m_D\, c_D$$

Simplification gives :

$$\frac{m_B}{m_D} = \frac{c_D - c_F}{c_F - c_B} = \frac{\text{length of segment FD}}{\text{length of segment FB}}$$

Obviously ratio of masses of liquid and vapour are inversely proportional to their lengths. The above correlation is referred to as the ***mixing rule*** or the ***lever law***.

8.9. ENTHALPY-CONCENTRATION DIAGRAM

The enthalpy-concentration plot depicts the variation of enthalpy with concentration both for the liquid and vapour state of the mixture for a given pressure.

The *h-c* plot essentially consists of saturated liquid line ABD and a vapour saturation line PQR. To establish the constant temperature lines (isotherms) in the region lying between these curves, use is made of an auxiliary line PSR which is drawn corresponding to the given pressure as per the prescribed procedure outlined in the literature on theory of mixtures.

The state points 1 and 2 marked in the diagram represent the sub-cooled and saturated liquid state of the mixture for a particular concentration. From point 3 (intersection point of vertical line from 1 with the auxiliary line) a line is drawn parallel to abscissa (concentration axis) to meet the vapour line at point 4. The line joining points 2 and 4 is the required isotherm.

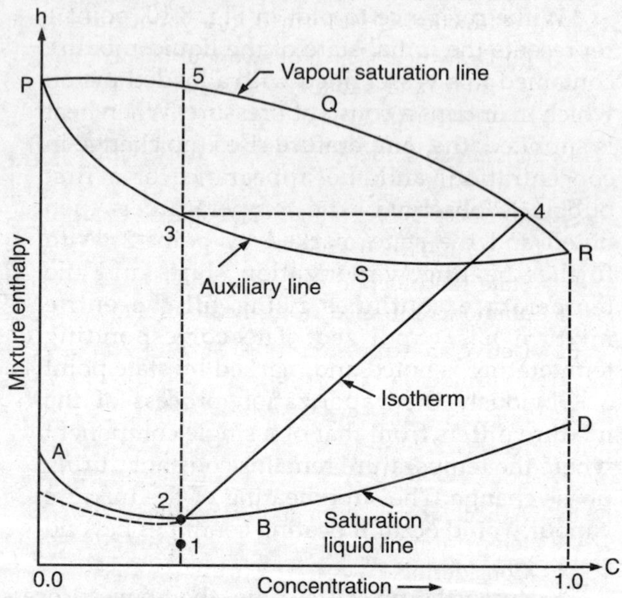

Fig. 8.11. Enthalpy-concentration plot

The *h-c* plots are used to determine unknown properties of the mixture from known values of other parameters.

Illustration : An ammonia-water mixture is stated to exist at *p* bar pressure and *t*°C temperature. We are required to determine the enthalpy and concentration when it is in the (*a*) saturation liquid state (*b*) saturated vapour state..

The procedure for obtaining the solution is outlined below :

(*a*) Locate the state point A on the enthalpy-concentration plot of ammonia-water mixture. The point A is the point of intersection of isotherm *t*°C and the *p* bar saturated liquid line. Projection of point A onto the axes of ordinate and abscissa gives :

Concentration = c_A and

enthalpy = h_A.

These are the concentration and enthalpy values when the mixture is in the saturated liquid state.

(*b*) From point A, draw a vertical line meeting the *p* bar auxiliary line at point B and therefrom draw a horizontal line that intersects the *p* bar saturated vapour line at point D. Projection of point D onto the axes of ordinates gives :

Concentration = c_D and

enthalpy = h_D

These are the concentration and enthalpy values when the mixture is in the saturated vapour state.

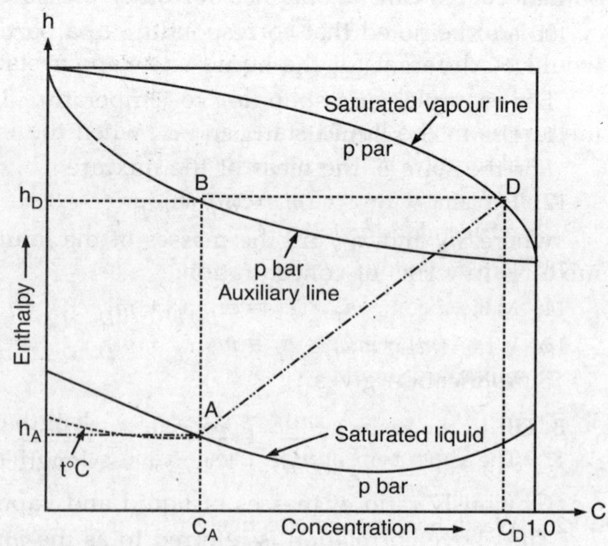

Fig. 8.12.

REVIEW QUESTIONS

A. Conceptual and conventional questions.

1. State the functions of following components in a vapour absorption refrigeration system :
   (*i*) absorber (*ii*) rectifier (*iii*) analyzer, and (*iv*) heat exchanger.
2. What is simple vapour absorption system ? State how its performance can be improved.
3. (*a*) What is the basic function of compressor in vapour compression refrigeration system? How this function is performed in a vapour absorption refrigeration system.

   (*b*) Draw neat sketch of a vapour absorption refrigeration system. Indicate thereon the phase of various fluids and name of component. Also indicate the directions of external energy flow to or from the component.
4. (*a*) Which components of the ammonia – water absorption system perform the function of compressor and how?

   (*b*) Make a comparative list between the different aspects of a vapour absorption system and a vapour compression system.
5. State the function of heat exchanger between the hot weak solution leaving the generator and aqua ammonia pumped from the absorber.
6. Derive an expression for COP of the vapour absorption system in teams of the generator temperature T_g, condenser temperature T_c and the evaporator temperature T_e.
7. In a vapour absorption refrigeration system, the generator is operated by solar heat and the temperature reaches 385 K. The refrigeration temperature is 260 K and the environmental conditions are stated to be at 325 K. Determine the maximum possible COP of the system.
8. A heat driven refrigeration system absorbs heat from low temperature T_E and rejects it to sink at temperature T_C. The supply of heat to the unit is from a high temperature source at temperature T_H. Considering $T_H > T_C > T_E$, use the first and second laws of thermodynamics to derive an expression for maximum *COP* of refrigeration system in terms of these temperatures.
9. In an absorption type of refrigerator, the heat is supplied to NH_3 generator by condensing steam at 2 bar and 90% dry. The temperature in the refrigerator is to be maintained at –5°C. Find the maximum possible *COP*. If the refrigeration load is 20 tonnes and actual *COP* is 70 percent of the maximum *COP*, determine the mass of steam required per hour. Take temperature of atmosphere as 30°C.
   (1.75, 104 kg/hr)
10. (*a*) Name any two pair of substances used in the vapour absorption system.

    (*b*) Mention the requirements to be met by the refrigerant, absorbent and refrigerant absorbent combination for effective working of a vapour absorption system.
11. Draw the schematic arrangement of the various components comprising lithium bromide – water absorption system and explain its working.
12. Explain, with a neat sketch, the principle of working of an electrolux refrigerant. Where are such refrigerators preferred ?
13. Explain, with a neat sketch, how solar energy can be used to obtain refrigeration effect.
14. Make a comparison between a two-fluid and a three fluid vapour absorption system.
15. Write a short note on temperature-concentration and enthalpy-concentration diagrams for a binary mixture of liquids.

B. Fill in the blanks with appropriate word/words

1. The vapour absorption refrigeration system may be considered as a combination of .... and ....
2. Waste heat can be effectively used in a .... refrigeration system.
3. A .... is used in a vapour absorption system to cool the vapours leaving the analyser.

4. In lithium bromide absorption system, lithium bromide is used as a .... and water as an ....
5. The fluids used in electrolux refrigerator are ammonia, water and ....
6. An electrolux refrigerator is called a .... fluids absorption system.
7. Hydrogen is used in electrolux refrigeration system to reduce vapour pressure of ammonia in the ....
8. The .... is defined as the ratio of the mass of any arbitrary constituent and the mass of a mixture of pure substances.

Answers

1. heat engine, heat pump 2. vapour absorption 3. rectifier 4. refrigerant, absorbent 5. hydrogen 6. three 7. evaporator 8. concentration

C. Multiple choice questions.

1. Which of the following components in a vapour absorption plant is a substitute for the compressor of the vapour compression system ?
(*a*) absorber (*b*) heat exchanger and generator
(*c*) aqua pump and generator (*d*) absorber, aqua pump and generator
2. In ammonia absorption system, there is transfer of heat from
(*a*) weak ammonia solution to strong ammonia solution
(*b*) strong solution to ammonia vapour
(*c*) ammonia vapour to strong solution
(*d*) weak solution to ammonia vapour
3. Which of the following components is common between vapour compression and vapour absorption systems ?
(*a*) absorber (*b*) generator (*c*) condenser (*d*) rectifier
4. Aqua ammonia solution used in vapour absorption refrigeration system is a solution of ammonia in
(*a*) water (*b*) hydrogen (*c*) lithium bromide
(*d*) carbon tetrachloride
5. In aqua-ammonia and Li-Br water absorption refrigeration systems, the refrigerants are respectively
(*a*) water and water (*b*) water and Li-Br
(*c*) ammonia and Li-Br (*d*) ammonia and water
6. There is need of cooling water at the following units of vapour absorption refrigeration plant :
(*a*) condenser and rectifier (*b*)condenser, absorber and rectifier
(*c*) condenser, absorber, rectifier and evaporator
(*d*) condenser, absorber and heat exchanger
7. The most distinguishing feature of absorption type refrigerator is
(*a*) uses only low grade of energy in the form of heat
(*b*) small size for the same refrigerating effect
(*c*) higher coefficient of performance
(*d*) quieter operation
8. Which aspect is not true in the context of vapour absorption system when compared to vapour compression system?
(*a*) need for more precise equipment
(*b*) energy input is mainly heat which is a low grade energy
(*c*) energy supply is low; only 1/2 to 1/3 of cooling effect
(*d*) performance of the system is not affected by load variations

9. Lithium bromide is used as absorbent in
   (a) electrolux refrigerator
   (b) vapour absorption refrigeration
   (c) vapour compression refrigeration
   (d) steam jet refrigeration

10. A rectifier is fitted in an ammonia absorption plant to
   (a) superheat ammonia vapour
   (b) remove the unwanted water vapour by heating the vapour mixture
   (c) remove the unwanted water vapour by cooling the vapour mixture
   (d) remove the unwanted water vapour by cooling the vapour mixture and condensing the water vapour

11. There is no need of rectifier in the vapour circuit of lithium bromide refrigeration plant because
   (a) water vapour and lithium bromide cannot be separated
   (b) condensation of vapour mixture can take place in the condenser
   (c) lithium bromide is non-volatile
   (d) vacuum conditions prevail in the lithium bromide plant

12. Which of the following machines can be used to obtain refrigeration at places where there is no electric power ?
   (a) air refrigeration
   (b) steam jet refrigeration
   (c) vapour compression
   (d) vapour absorption

13. In an electrolux refrigerator, there is
   (a) absorption of ammonia in water
   (b) absorption of hydrogen in water
   (c) evaporation of ammonia in hydrogen
   (d) evaporation of water in an atmosphere of some inert gas

14. Electrolux refrigerator essentially works on
   (a) vapour absorption system
   (b) vapour compression system
   (c) thermo-electric system
   (d) vortex tube system

15. Hydrogen is essential in an electrolux refrigeration system, because
   (a) it acts as a catalyst in the evaporator
   (b) it helps in maintaining a low partial pressure for the evaporating ammonia
   (c) the cooled hydrogen leaving the heat exchanger cools the refrigerant entering the evaporator
   (d) the reaction between hydrogen and ammonia is endothermic in evaporators and exothermic in adsorbate

Answers

1. (d)	2. (a)	3. (c)	4. (a)	5. (d)	6. (b)	7. (d)	8. (c)
9. (b)	10. (d)	11. (c)	12. (d)	13. (d)	14. (a)	15. (b)	

❑❑❑

Chapter – 9

Low Temperature Refrigeration (Cryogenics)

Learning objectives : A study of the subject matter presented in this chapter will enable the students/readers to

- *understand the term cryogencis and the applications of low temperature refrigeration.*
- *comment upon the limitations of vapour compression refrigeration system for the production of low temperature.*
- *explain the working of the systems for the production of dry ice.*
- *outline the processes used for the liquefaction of air, hydrogen and helium.*

9.1. WHAT IS CRYOGENICS?

Cryogenics ≡ Kryos (Cold or Frost) + Genic (Formation)

The term cryogenics is derived from the Greek word *Kryos* which means cold or frost and *Genic* meaning formation. The word cryogenics thus literally means "***the production of icy cold***"; however the term is used as a synonym for the low temperature state. It is not well-defined at what point on the temperature scale refrigeration ends and cryogenics begins . However, it has been suggested that the term cryogenics be applied to all temperatures upto –150°C. This is a logical dividing line, since the normal boiling points of the so called permanent gases (such as helium, hydrogen, neon, nitrogen, oxygen and normal air) lie below –180°C while the Froen refrigerants, hydrogen sulphide and other common refrigerants have boiling points above – 180°C.

Cryogenic temperatures are achieved with multistage vapour compression, cascading compression, magnetic cooling system, rapid evaporation of volatile liquid or by expansion of gases confined initially at pressures of 150 to 200 atmospheres. The expansion may be through a valve to a region of low pressure, or it may occur in the cylinder of a reciprocating engine with the gas driving the piston of engine.

The first low temperature refrigerant system was developed around 1875 for solidification of carbon dioxide and liquefaction of oxygen and nitrogen. Now-a-days, a temperature much below 0.001 K has been successfully attained. This fascinating study covers a broad spectrum of applications in the industrial, nuclear, electronic and medical fields. The study also helps in the undrestanding of certain physical phenomenon at temperatures approaching absolute zero.

9.2. LIMITATION OF VAPOUR COMPRESSION CYCLE FOR LOW TEMPERATURE REFRIGERATION : NEED FOR CRYOGENICS

(*i*) Limitation on low temperature imposed by the freezing point of refrigerant. The refrigerant must have a freezing point much below the temperature required to be attained.

From the values of freezing point listed in Table 9.1, it is evident that NH_3, CO_2 and SO_2 cannot be used for obtaining cryogenic temperatures.

(*ii*) Condenser pressure is excessively large when the refrigerant used has a low boiling point.

Table 9.1. Freezing temperature for commonly used refrigerants

Refrigerant	R-11	R-12	R-13	R-21	R-22	R-717 (NH_3)	R-744 (CO_2)	R-764 (SO_2)
Freezing temperature in°C at atmospheric pressure	– 111	– 157.5	– 181	– 135°	– 160	– 77.8	– 56.3	– 75.6

(*iii*) Evaporator pressure is very low when the boiling point of the refrigerant used is high. At low evaporator pressures, the suction volume becomes large and that entails large input to compressor.

(*iv*) System has a low COP when the operating pressure ratio is large.

(*v*) Operation of the mechanical equipment at low temperatures is rather difficult.

9.3. APPLICATIONS OF LOW TEMPERATURE

The science of cryogenics has proven its utility in variety of applications such as :

(*i*) Liquefaction and separation of gases.

(*ii*) Formation of dry ice (solid carbon dioxide).

(*iii*) Cold treatment of metals.

(*iv*) Cooling system for super computers and electronic components.

(*v*) Underground power lines.

(*vi*) Preservation of semen and biological specimens. The semen is used for artificial insemination and for producing test tube babies. The semen is preserved in liquid nitrogen at –195°C.

(*vii*) Crysurgery for treatment of nervous disorders. Cryogenic surgery is being used for the treatment of Parkinson disease, the technique being based on the selective destruction of tissues by freezing it with a small cryogenic probe. A similar technique is employed to destroy brain tumors and to arrest cervical cancer.

(*viii*) Cryogenic engine for missile and space propulsion systems. Cryogenic fuels, mainly oxygen and nitrogen, have been used in rockets and for getting the space shuttles into the orbit.

(*ix*) Cryogenic detectors which need temperatures well below 77K (–196°C) for their operation.

(*x*) Large scale production of oxygen and nitrogen from air.

The oxygen is needed for rocket engines, for cutting and welding torches, for supporting life in space and deep sea vehicles, and for blast furnace operations.

The nitrogen goes into the making of ammonia for fertilizers and it is used to prepare frozen foods by cooling them rapidly enough to prevent destruction of cell tissues. It can also serve as a refrigerant and for transporting frozen foods.

9.4. SOLID CARBON DIOXIDE (DRY ICE)

Fig. 9.1 shows the schematic arrangement of the equipment used and *p-h* plot of the processes involved in the manufacture of dry ice. The equipment includes a compressor, a condenser, an expansion valve and a separating unit (snow chamber).

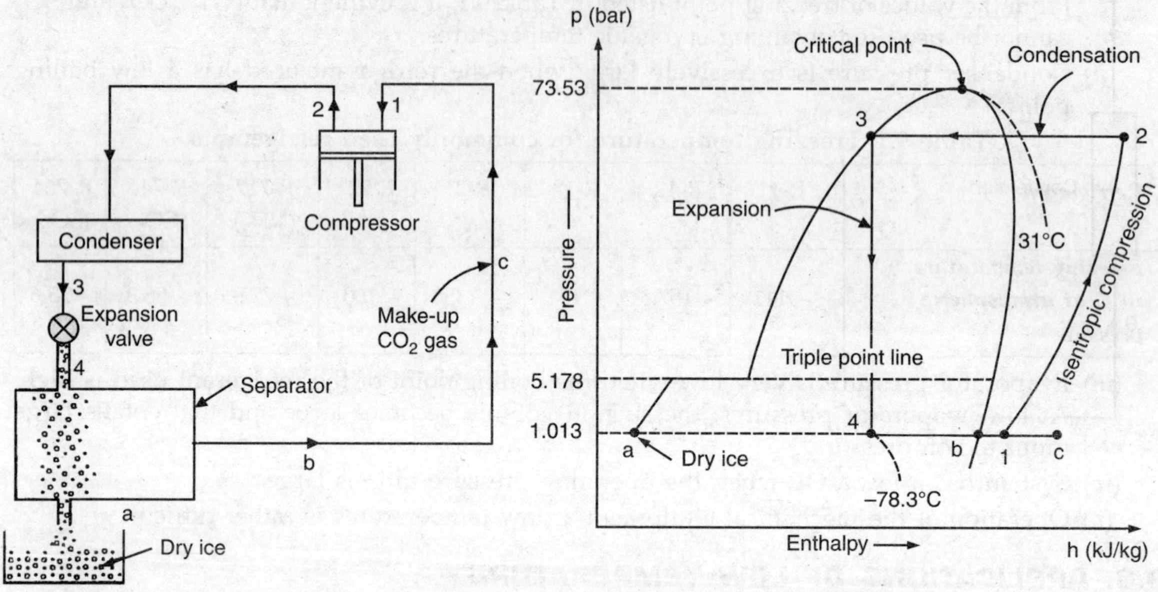

Fig. 9.1. Schematic arrangement of producing dry ice

The sequence of operation is

1-2 : Isentropic compression of CO_2 in the compressing unit to a sufficiently high pressure (60-70 bar)

2-3 : Condensation of compressed CO_2 in the condenser. The condensation is done by a stream of circulating water.

3-4 : Expansion of high pressure liquid CO_2 in an expansion valve. Expansion continues upto a pressure which is below the triple point of CO_2 (5.178 bar)

At state point 4, the CO_2 is in (solid + vapour) form, i.e, a mixture of solid CO_2 and gaseous CO_2. The solid CO_2 is removed at point 'a' and pressed into blocks. The gaseous CO_2 removed at point 'b' is mixed with make up CO_2 gas (state point c) and the mixture attains the state 1; the condition at entry to the compressor. The quantity of make up gas equals the weight of dry ice formed. The solid CO_2 system can produce low temperatures upto – 60°C

The system operates with high pressure ratio (about 70) and as such has high power consumption (nearly 310 kWh per ton of dry ice formed). The following modifications in the system have been suggested to reduce the production cost of dry ice :

– multistaging with intercoolers
– operating of snow chamber at a pressure slightly below the triple point but above the atmospheric pressure
– using flash chambers for sub-cooling the liquid and de-superheating the vapour

The schematic layout of three stage compression system, incorporating water and flash intercoolers, used for the manufacture of dry ice is shown in Fig.9.2 along with the corresponding *p-h* plot. The use of flash intercoolers minimizes the problem of clogging of expansion valve and refrigerant lines due to production of dry ice.

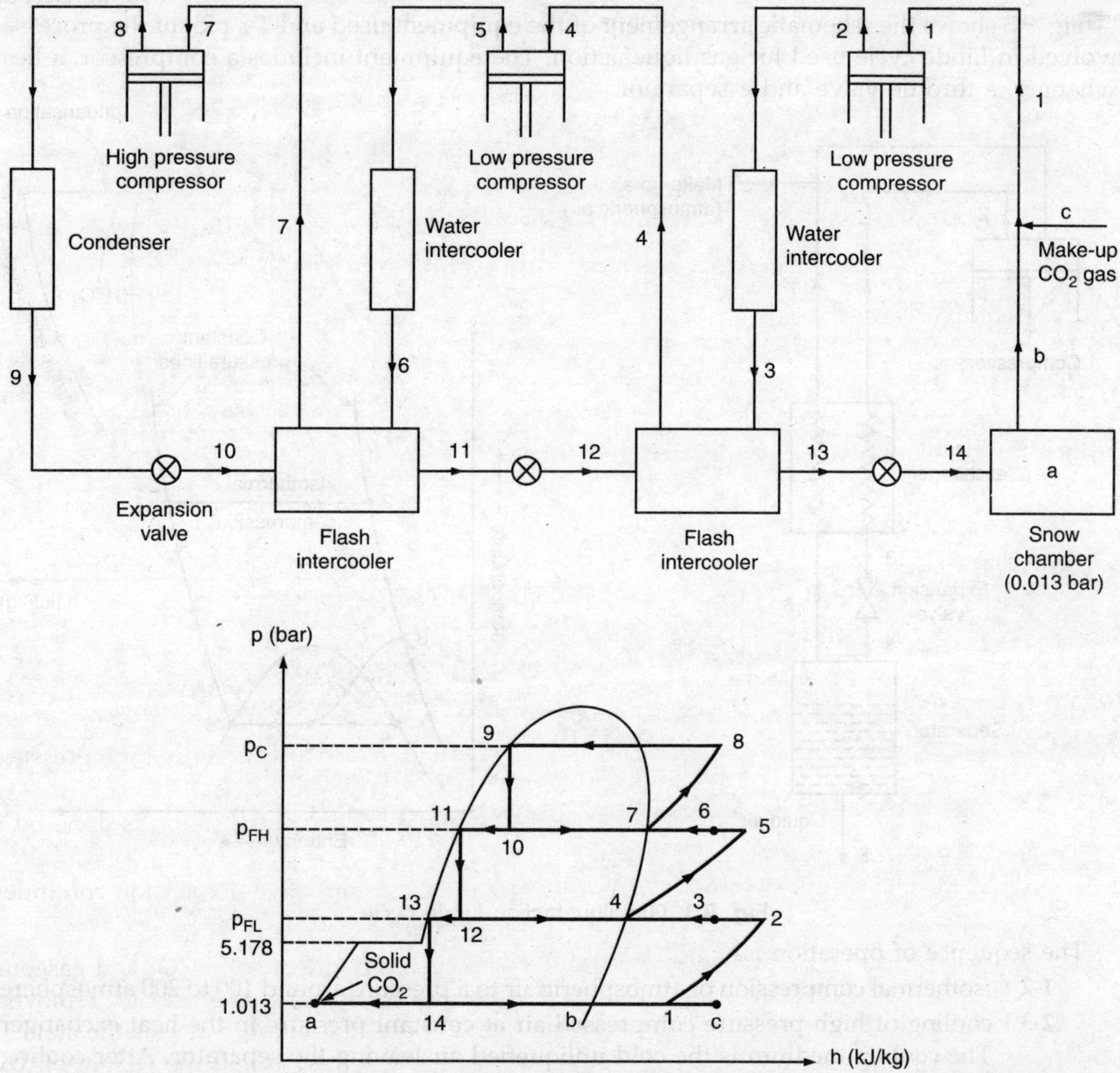

Fig. 9.2. Schematic arrangement and p-h diagram of three stage compression system with water and flash intercoolers

Applications : Dry ice is used as a refrigerant

(*i*) It is non-toxic and non-corrosive.

(*ii*) It is easy to cut and handle.

(*iii*) It requires less space for a fixed cooling capacity.

Dry ice sublimates (changes directly from solid to gas) at – 78.3°C and atmospheric pressure, and so it can be used to maintain wide range of temperatures required for preservation of perishable materials during transportation.

9.5. GAS LIQUEFACTION: LINDE CYCLE

Operation :

Fig. 9.3 shows the schematic arrangement of the equipment used and *T-s* plot of the processes involved in Linde cycle used for gas liquefaction. The equipment includes a compressor, a heat exchanger, a throttle valve and a separator.

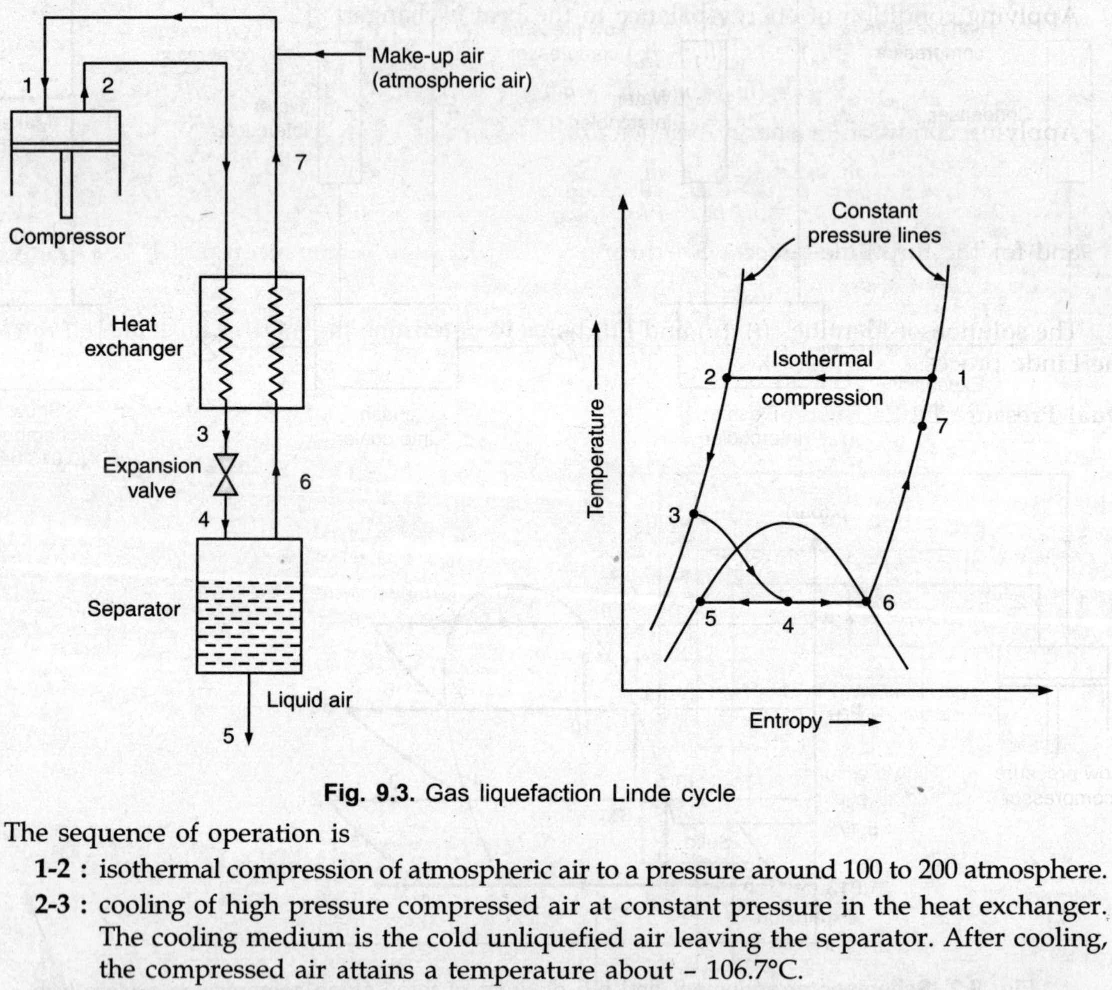

Fig. 9.3. Gas liquefaction Linde cycle

The sequence of operation is

1-2 : isothermal compression of atmospheric air to a pressure around 100 to 200 atmosphere.

2-3 : cooling of high pressure compressed air at constant pressure in the heat exchanger. The cooling medium is the cold unliquefied air leaving the separator. After cooling, the compressed air attains a temperature about – 106.7°C.

3-4 : throttling of cooled compressed air from the heat exchanger to atmospheric pressure. This process is accomplished in the throttle valve and the resulting temperature is – 190°C.

At state point 5, a portion of the air which gets liquefied and is collected in the separator is removed. The remainder of the cold but unliquefied air leaves the separator at state point 6 and returns to the compressor inlet. While returning, it is made to pass through heat exchanger where it provides cooing to the high pressure air which is on its way from the compressor to the throttle valve. At state point 7, the make-up air from the atmosphere joins the air stream leaving the heat exchanger. The quantity of make up air equals the amount of air liquefied.

Analysis :

Let m_2 = mass of air compressed in the compressor (or mass of high pressure air passing through the heat exchanger between the state points 2 and 3)

m_5 = mass of air liquefied and removed from the separator

m_6 = mass of low pressure air that passes the heat exchanger between state points 6 and 1 ($m_6 = m_2 - m_5$)

Applying condition of energy balance to the heat exchanger,

$$m_2 (h_2 - h_3) = m_6 (h_1 - h_6)$$
$$= (m_2 - m_5)(h_1 - h_6) \quad \text{....}(i)$$

Applying condition of energy balance to the separator,

$$m_2 h_4 = m_5 h_5 + m_6 h_6$$
$$= m_5 h_5 + (m_2 - m_5) h_6 \quad \text{...}(ii)$$

and for the throttling process 3-4 during which enthalpy remains constant

$$h_3 = h_4 \quad \text{...}(iii)$$

The solution of identities (*i*), (*ii*) and (*iii*) helps to determine the mass of air liquefied during the Linde process.

Dual Pressure Linde System :

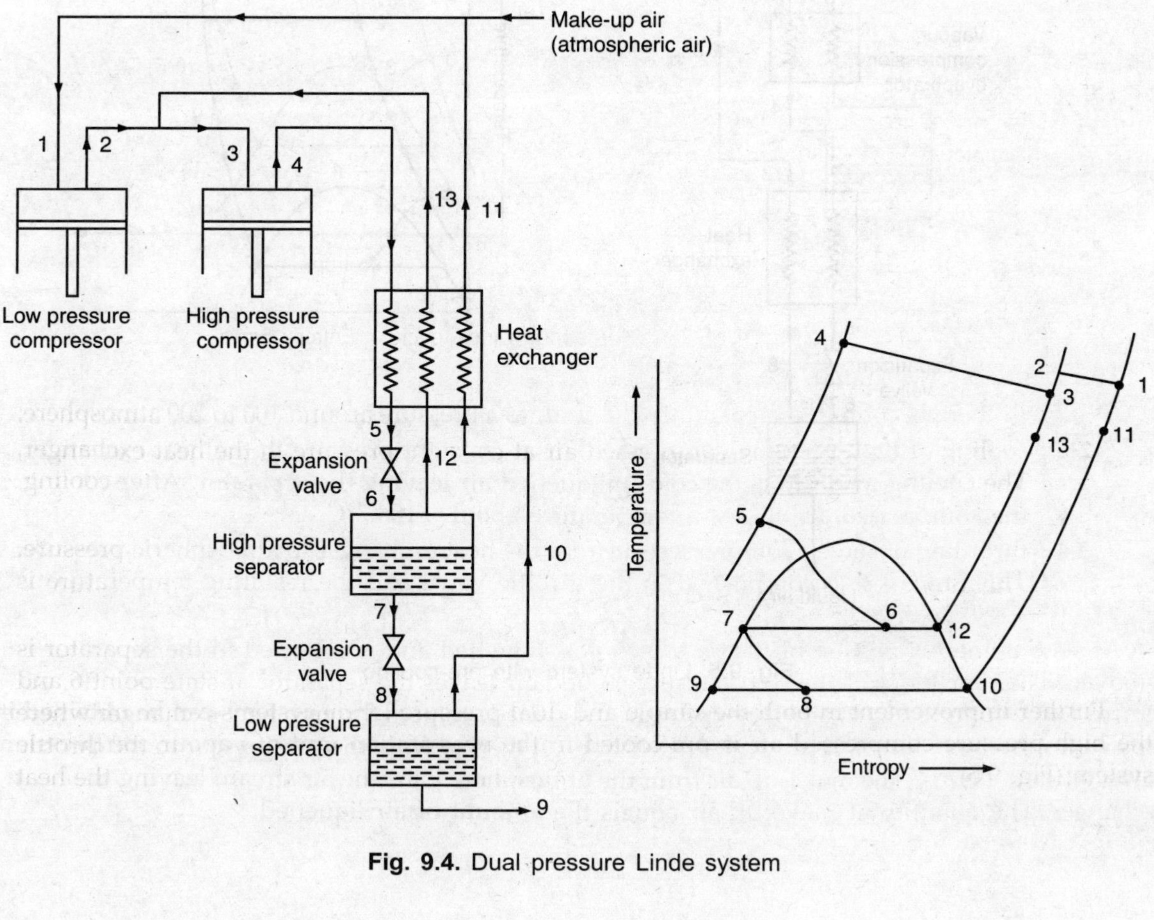

Fig. 9.4. Dual pressure Linde system

The single stage Linde cycle has poor efficiency and is employed only when a small quantity of liquid air is needed. A considerable improvement in efficiency results when compression process is achieved in two stages by using two compressing units. In this dual pressure Linde system (Fig. 9.4), a portion of the air after first throttling (process 5-6) is bled and sent back through heat exchanger into the discharge from first (low pressure) stage of compression. The dual system provides economical operation because there is only one expansion to the intermediate pressure and one compression from this intermediate pressure to the high pressure.

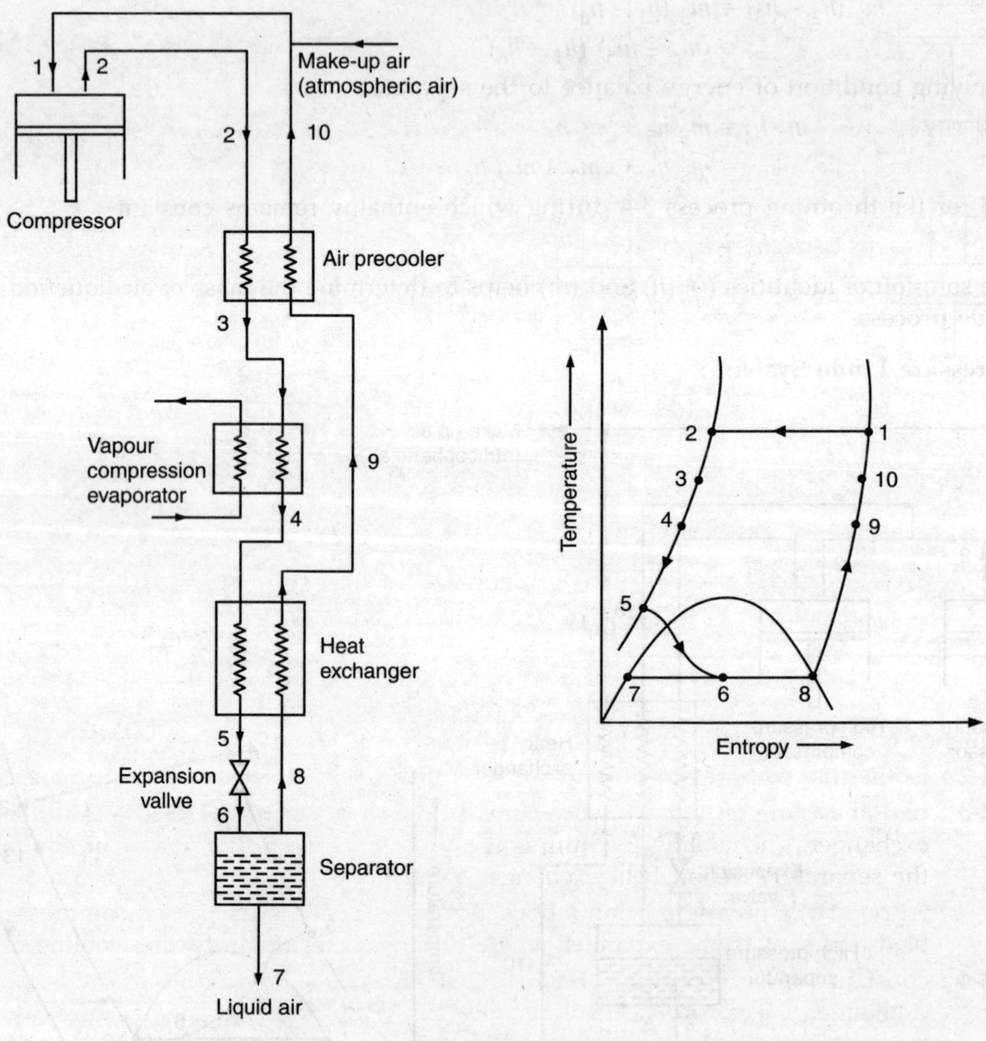

Fig. 9.5. Linde system with pre-cooling

Further improvement in both the simple and dual pressure Linde systems can be obtained if the high pressure compressed air is pre-cooled in the evaporation unit of vapour compression system (Fig. 9.5).

9.6. GAS LIQUEFACTION : (CLAUDE CYCLE OPERATION)

Fig. 9.6 shows the schematic arrangement of the equipment used and *T-s* plot of the processes involved in Claude cycle used for gas liquefaction. The equipment includes a compressor, two heat exchangers, an expander, a throttle valve and a separator.

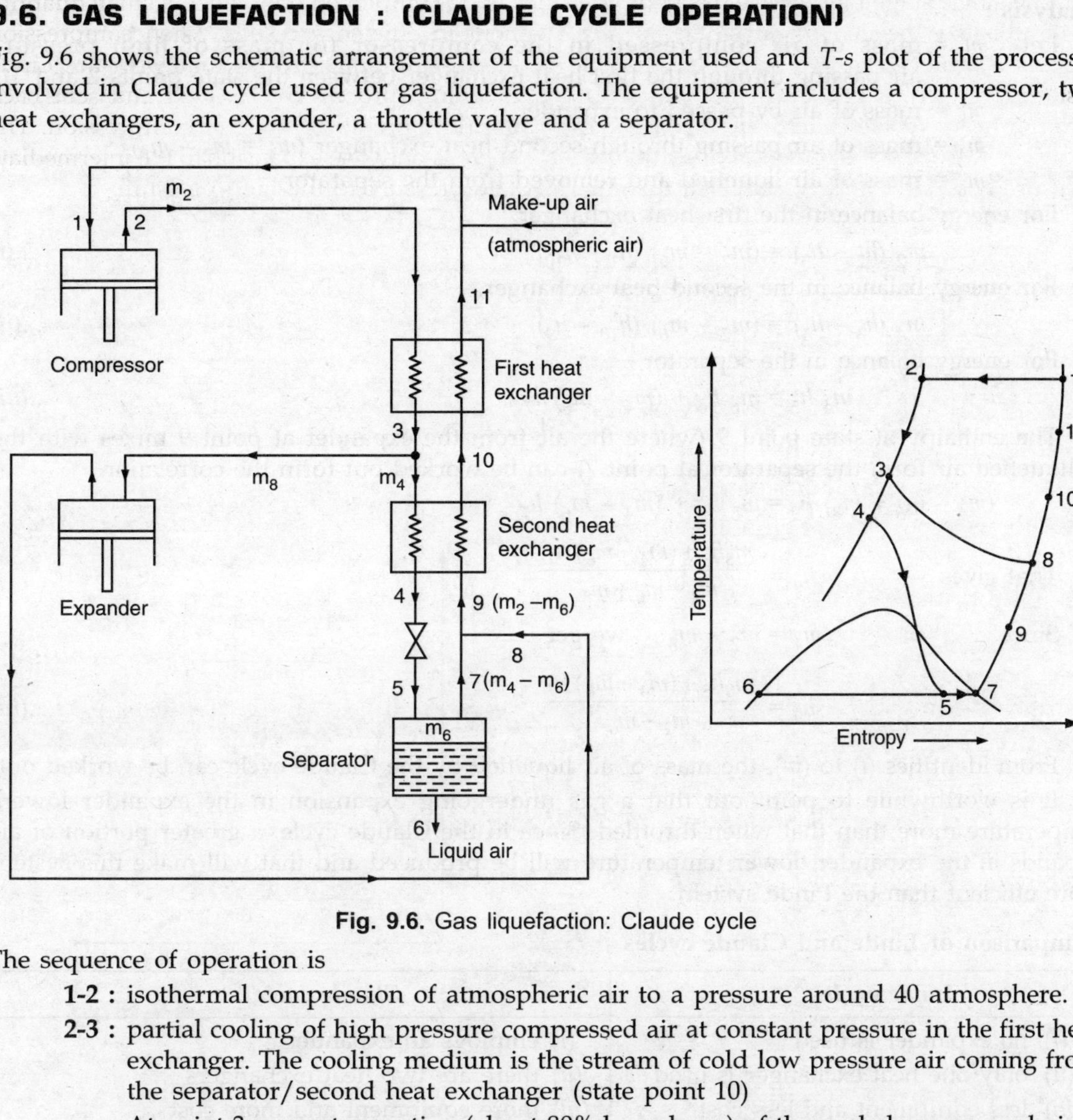

Fig. 9.6. Gas liquefaction: Claude cycle

The sequence of operation is

1-2 : isothermal compression of atmospheric air to a pressure around 40 atmosphere.

2-3 : partial cooling of high pressure compressed air at constant pressure in the first heat exchanger. The cooling medium is the stream of cold low pressure air coming from the separator/second heat exchanger (state point 10)

At point 3, a greater portion (about 80% by volume) of the cooled compressed air is bled and sent to the expander where expansion results in further cooling of air.

3-4 : constant pressure cooling of remaining about 20% by volume in the second heat exchanger. Here too the cooling medium is the cold unliquefied air leaving the separator.

4-5 : throttling of cooled compressed air from the second heat exchanger.

In the separating unit, the liquefied air is separated and removed (state point 6) and is joined by the stream of cold air coming from the expander (state point 8). The mixed air (state point 9) passes through the two heat exchangers and returns back to the compressor. On its way, it meets the stream of make-up air (state point 11) from the atmosphere. The quantity of make up equals the amount of air liquefied.

Analysis :

Let m_2 = mass of air compressed in the compressor (or mass of high pressure air passing through the first heat exchanger between the state points 2 and 3)

m_8 = mass of air by-passed to expander

m_4 = mass of air passing through second heat exchanger ($m_4 = m_2 - m_8$)

m_6 = mass of air liquefied and removed from the separator

For energy balance in the first heat exchanger,

$$m_2 (h_2 - h_3) = (m_2 - m_6)(h_1 - h_{10}) \quad \text{...}(i)$$

For energy balance in the second heat exchanger,

$$m_4 (h_3 - h_4) = (m_2 - m_6)(h_{10} - h_9) \quad \text{...}(ii)$$

For energy balance in the separator,

$$m_4 h_5 = m_6 h_6 + (m_4 - m_6) h_7 \quad \text{...}(iii)$$

The enthalpy at state point 9 (where the air from the expander at point 9 mixes with the unliquefied air form the separator at point 7) can be worked out form the correlation

$$(m_4 - m_6 + m_8) h_9 = m_8 h_8 + (m_4 - m_6) h_7$$

That gives :

$$h_9 = \frac{m_8 h_8 + (m_4 - m_6) h_7}{m_4 - m_6 + m_8}$$

Since $m_4 = m_2 - m_8$, we get

$$h_9 = \frac{m_8 h_8 + (m_4 - m_6) h_7}{m_2 - m_6} \quad \text{...}(iv)$$

From identities (*i*) to (*iv*), the mass of air liquefied by the Claude cycle can be worked out.

It is worthwhile to point out that a gas undergoing expansion in the expander lowers temperature more than that when throttled. Since in the Claude cycle, a greater portion of air expands in the expander, lower temperature will be produced and that will make this system more efficient than the Linde system.

Comparison of Linde and Claude cycles

Linde cycle	*Claude cycle*
(*i*) no expander is used	(*i*) employs an expander
(*ii*) only one heat exchanger is used	(*ii*) there are two heat exchangers
(*iii*) less equipment and less cost	(*iii*) more equipment and more cost
(*iv*) low temperatures are attained	(*iv*) possibility of attaining very low temperatures
(*v*) less efficient	(*v*) more efficient
(*vi*) possibility of operation with dual pressure	(*vi*) dual pressure system not used
(*vii*) used for liquefaction of air and hydrogen	(*vii*) used for liquefaction of air and helium

9.7. LIQUEFACTION OF HYDROGEN

Fig. 9.7 shows the schematic arrangement of the equipment used for liquefying hydrogen. The equipment includes a compressor, four heat exchangers, an expansion valve and a separator.

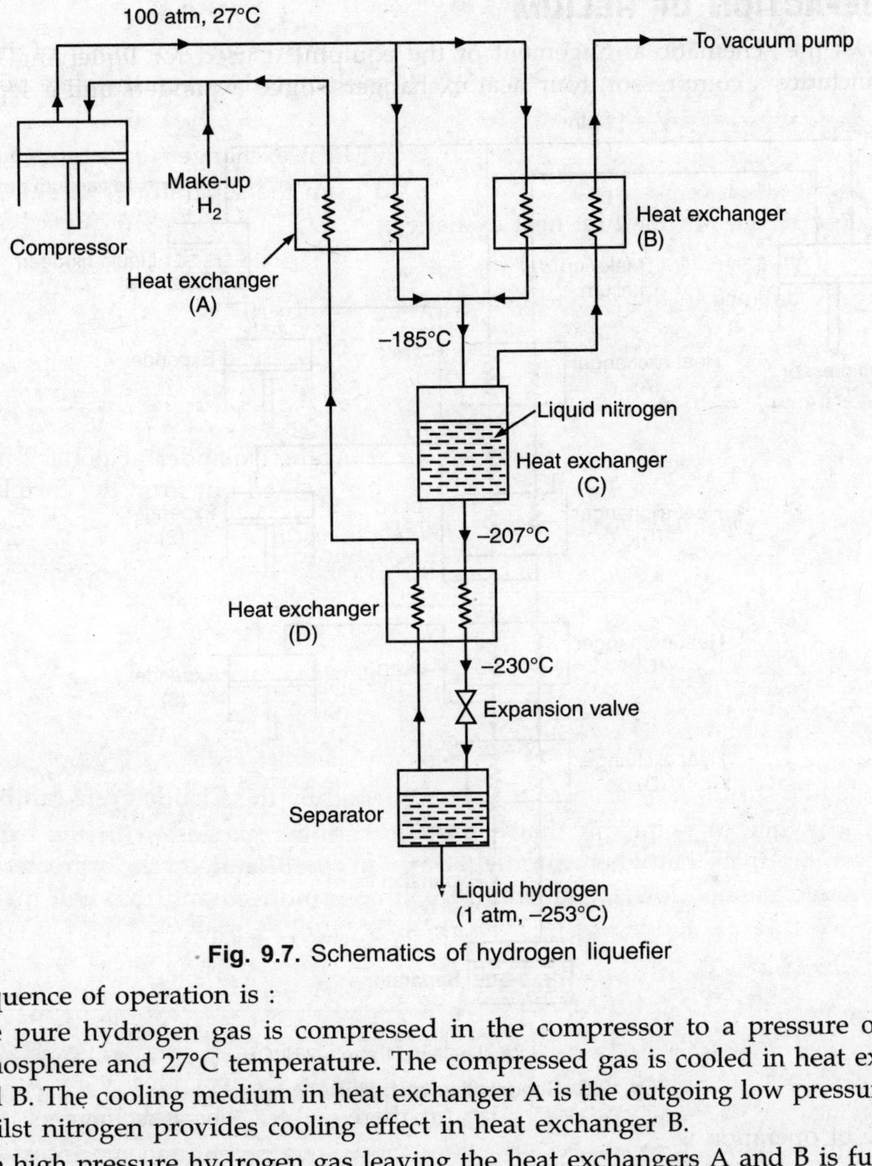

Fig. 9.7. Schematics of hydrogen liquefier

The sequence of operation is :

(*i*) The pure hydrogen gas is compressed in the compressor to a pressure of about 100 atmosphere and 27°C temperature. The compressed gas is cooled in heat exchangers A and B. The cooling medium in heat exchanger A is the outgoing low pressure hydrogen whilst nitrogen provides cooling effect in heat exchanger B.

(*ii*) The high pressure hydrogen gas leaving the heat exchangers A and B is further cooled to about – 207°C in heat exchanger C. Here nitrogen boiling under reduced pressure acts as the cooling medium.

(*iii*) Further cooling of hydrogen gas to – 230°C occurs in heat exchanger D by low pressure hydrogen gas returning from the separating unit.

(*iv*) The hydrogen gas leaving the heat exchanger D at – 230°C passes through the expansion valve where it gets throttled to atmospheric pressure resulting in its liquefaction.

There is also a provision for adding make-up hydrogen to the system. The quantity of make-up hydrogen equals the amount of hydrogen liquefied.

Hydrogen gas has extremely low liquefaction temperature and that makes the hydrogen liquefaction process quite difficult.

9.8. LIQUEFACTION OF HELIUM

Fig. 9.8 shows the schematic arrangement of the equipment used for liquefying helium. The equipment includes a compressor, four heat exchangers, three expanders and a separator.

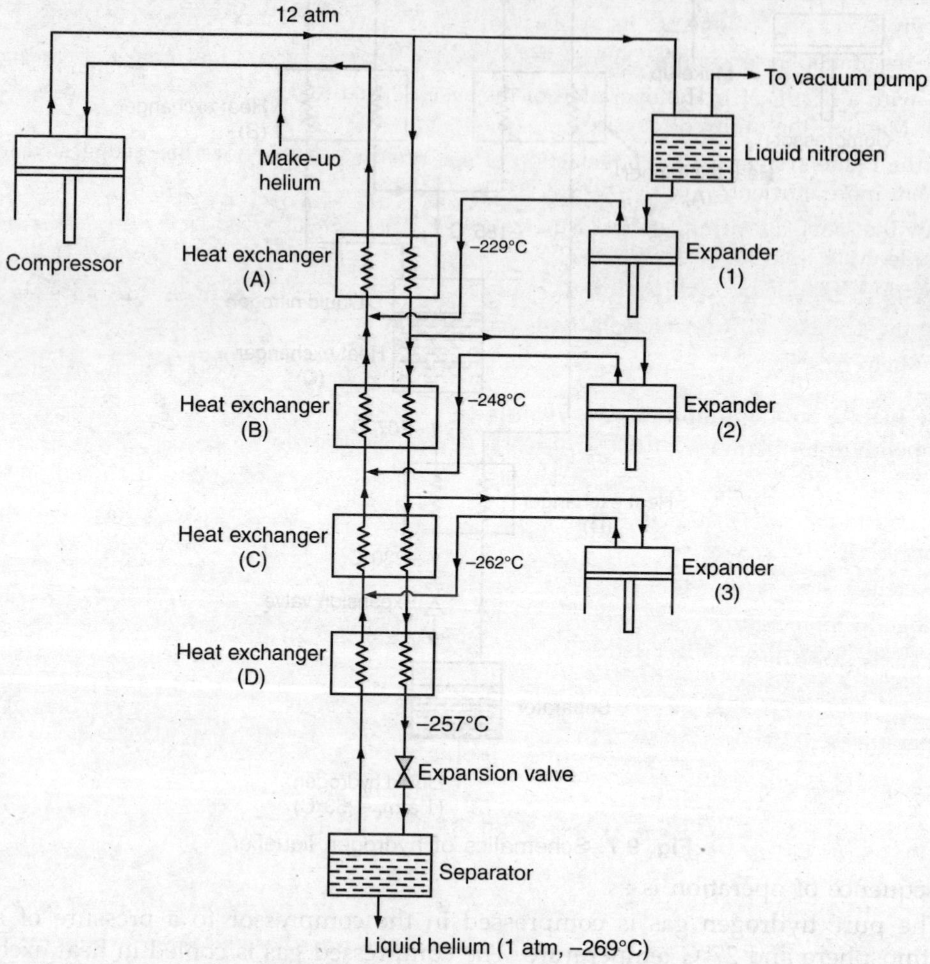

Fig. 9.8. Schematics of helium liquefier

The sequence of operation is

- The helium gas is compressed in a four-stage compressor to a pressure of about 12 atmosphere.
- The compressed helium gas is cooled to about – 257°C by making it to pass through four heat exchangers (A, B, C, D) and three expanders (1, 2, 3) arranged as shown in the set up.
- The helium gas is subsequently throttled to atmospheric pressure; and liquid helium is produced at – 269°C.

There is a also a provision for adding make-up helium to the system. The quantity of make-up helium equals the amount of helium liquefied.

It is worthwhile to point out that

(*i*) the arrangement provides 25 to 30 litres per hour of liquefied helium with a power consumption of about 45 kW.

(*ii*) helium is most difficult of all gases to liquefy.

REVIEW QUESTIONS

A. Conceptual and conventional questions.

1. Define cyrogenics and enumerate its applications.
2. What are the limitations of simple vapour compression cycle in the field of low temperature refrigeration?
3. Explain, with a neat sketch, the operation of the system used for the making of dry ice (solid carbon dioxide). Mention the utility of dry ice.
4. Explain the Linde system for the liquefaction of air. What methods have been suggested for making this system more efficient?
5. (*a*) Draw the schematic arrangement of the equipment used and *T-s* plot of the processes involved in Claude cycle used for liquefaction of air. Explain the operation of the system.
   (*b*) How does Claude cycle differ from Linde cycle in the context of liquefaction of air?
6. Describe the sequence of operation in the liquefaction of hydrogen and helium.
7. With the help of schematic diagram and *p-h* plot, explain how dry ice is produced.

B. Fill in the blanks with appropriate word/words

1. The temperature in the range –100°C to –273°C (absolute zero) are treated as ....
2. At the ...., the solid, liquid and vapour phases are in equilibrium.
3. .... can be easily handled and that makes it an ideal choice for storing and transportation of foods and perishable materials.
4. Linde system for air liquefaction is .... efficient than the Claude system.
5. The change in temperature with drop in pressure at constant enthalpy is called ....
6. .... is most difficult of all gases to liquefy.

Answers

1. low temperatures 2. triple 3. solid CO_2 or dry ice 4. less 5. Joule-Thomson coefficient 6. helium

□□□

Chapter – 10

Non-Conventional Refrigeration Systems

Learning objectives : A study of the subject matter presented in this chapter will enable the students/readers to understand and assimilate the salient aspects (principle, operation and utility) of the following non-conventional refrigration systems :

- *Stream jet refrigeration*
- *Vortext tube refrigeration*
- *Thermo-electric refrigeration, and*
- *Adiabatic demagnetisation*

10.1. STEAM JET REFRIGERATION

Principle : When a liquid contained in a chamber is flashed under vacuum and the flashed vapour is removed, the liquid cools down. Essentially, it is boiling of liquid at low temperature by reducing the pressure on its surface. The process is called ***flash cooling***, and it forms the basis of steam jet refrigeration.

Operation : Fig. 10.1 shows the general arrangement of the different components comprising the steam jet refrigeration system, and the corresponding thermodynamic cycle on *T-s* and *h-s* plots. The main components are :

(*i*) steam jet ejector
(*ii*) flash chamber (evaporator)
(*iii*) condenser, and
(*iv*) chilled water circulating pump, cooling water circulating pump, condensate pump, necessary valves and controls etc.

High pressure steam from the boiler is admitted to a nozzle, it expands and a high speed jet at low pressure issues from the nozzle exit. Suction is created and that entrains the water vapour to be extracted from the flash chamber. The combined stream of flash vapour and steam is diffused in the divergent part of venturi until desired pressure (about 5 cm of mercury) is reached. The kinetic energy of the mixture is converted into an increase in static head and the mass is discharged to the condenser. The condensate from the condenser is cascaded by pump and fedback to the boiler.

In flash chamber, pressure is low and so is the flash point which corresponds to the saturation temperature at the pressure of flash chamber. During flash or being vapour, the water in the flash chamber gets cooled due to absorption of latent heat of vaporization of the portion of water flashed. The chilled water form the chamber is taken to cooling load *i.e.*, the space/products being refrigerated.

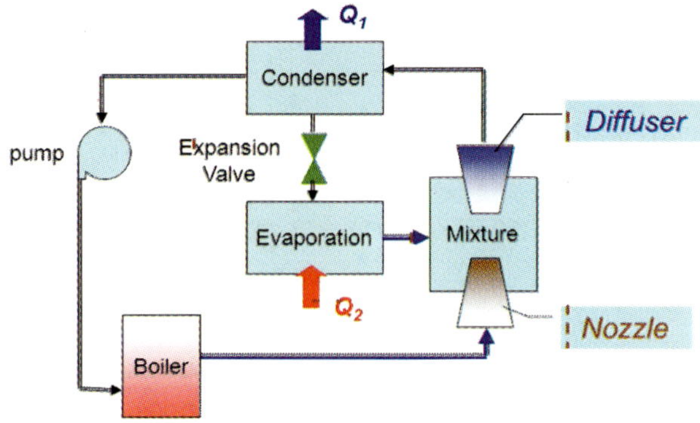

(a) Vapour-Jet Refrigeration System

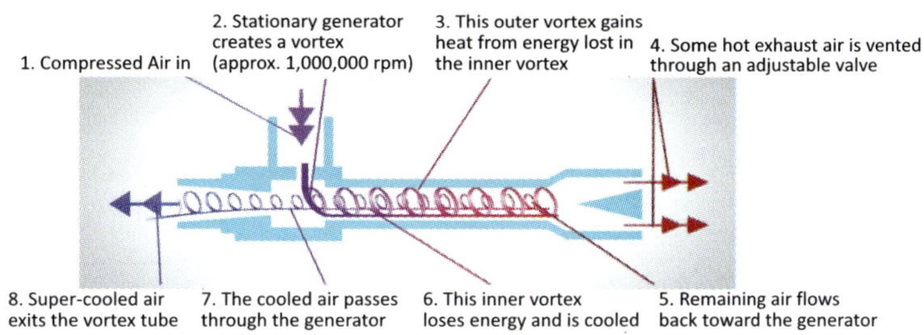

(b) Vortex Tube Refrigeration System

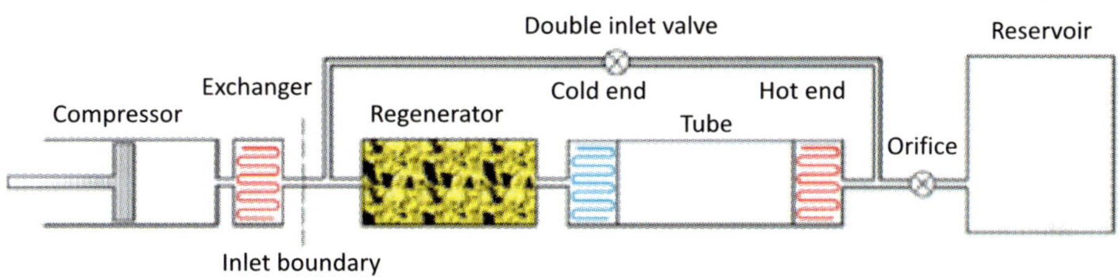

(c) Pulse Tube Refrigeration System

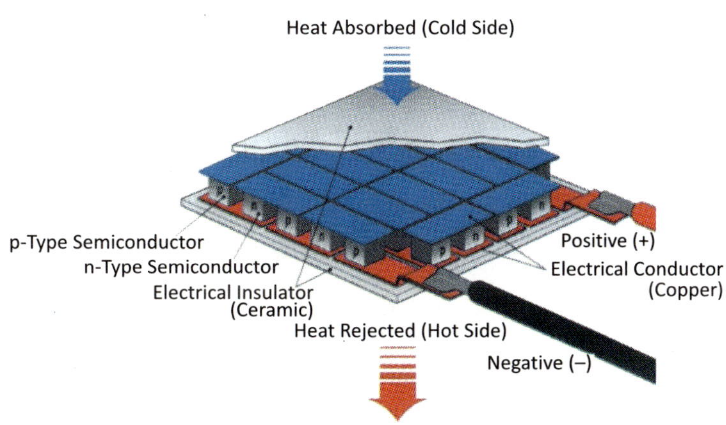

(d) Thermo-electric Refrigeration System

Non-Conventional Refrigeration Systems

The amount of water flashed out from the chamber is replenished by make-up water sprayed into the chamber. The flash chamber is provided with a float valve that maintains water in the chamber at constant load. It is to be noted that the steam jet refrigeration cycle is similar to conventional refrigeration cycle with an evaporator, a compression device and a condenser.

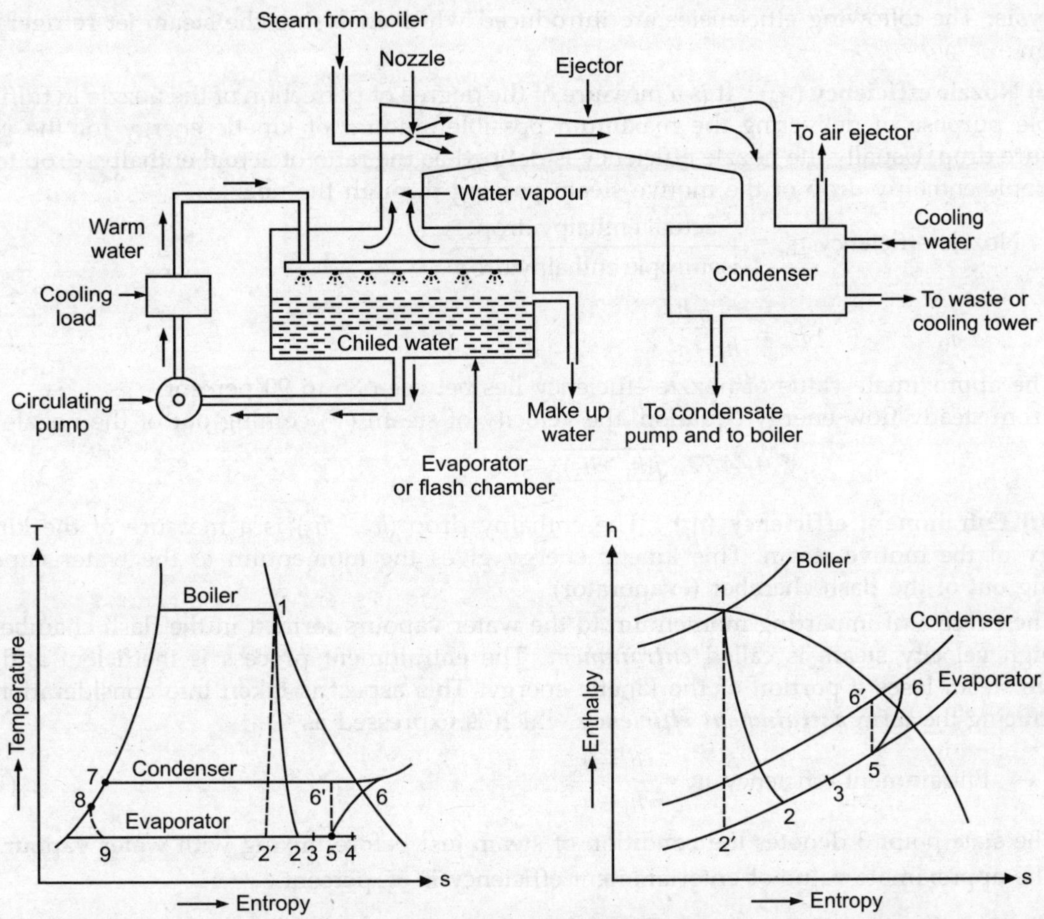

Fig. 10.1. Steam jet refrigeration system

However, instead of mechanical compression, this cycle employes a steam ejector to compress the refrigerant to the level of condenser pressure.

With reference to *T-s* and *h-s* plots

1: initial state of high pressure steam being supplied to the nozzle (assumed dry and saturated)

2′ and 2: state points after isentropic and actual expansion processes of steam from boiler pressure to evaporator pressure

3: condition of steam just before mixing with water vapour

4: initial condition of water vapour in the evaporator (assumed dry and saturated at evaporator pressure)

5: condition of mixture of high velocity steam from the nozzle and entrained water vapour before compression (diffusion)

6′ and 6: state points after isentropic and actual compression of mixture from evaporator pressure to condenser pressure.

It is to be further noted that the make up water is supplied at state point 8 where its temperature is slightly lower than that of condenser (state point 7). The throttling of make up water takes place in the flash chamber (process 8-9) and subsequently it leaves corresponding to the condition at state point 4.

Analysis: The following efficiencies are introduced while analysing the steam jet refrigeration system :

***(i)* Nozzle efficiency (η_n) :** It is a measure of the degree of perfection of the nozzle in fulfilling its sole purpose of delivering the maximum possible amount of kinetic energy for the given pressure drop. Usually the nozzle efficiency is defined as the ratio of actual enthalpy drop to the isentropic enthalpy drop of the motive steam passing through the nozzle.

$$\text{Nozzle efficiency } \eta_n = \frac{\text{actual enthalpy drop}}{\text{isentropic enthalpy drop}}$$

$$= \frac{h_1 - h_2}{h_1 - h_2'} \qquad \text{...(10.1)}$$

The approximate value of nozzle efficiency lies between 85 to 90 percent.

From steady flow energy equation, the velocity of steam (V) coming out of the nozzle is

$$V = 44.72 \sqrt{(h_1 - h_2)}$$

***(ii)* Entrainment efficiency (η_e) :** The enthalpy drop (h_1 - h_2) is a measure of the kinetic energy of the motive steam. This kinetic energy gives the momentum to the water vapours coming out of the flash chamber (evaporator).

The process of imparting momentum to the water vapours formed in the flash chamber by the high velocity steam is called ***entrainment***. The entrainment process is inefficient and the motive steam loses a portion of the kinetic energy. This aspect is taken into consideration by introducing the term ***entrainment efficiency*** which is expressed as

$$\text{Entrainment efficiency } \eta_e = \frac{h_1 - h_3}{h_1 - h_2} \qquad \text{...(10.2)}$$

The state point 3 denotes the condition of steam just before mixing with water vapour.

The approximate value of entertainment efficiency is 65 percent.

***(iii)* Compression efficiency (η_c) :** Assuming no friction, the adiabatic compression or diffusion of the mixture of high velocity steam from the nozzle and entrained water vapour can be represented by the vertical line 5-6′. However due to frictional effects, the actual compression of the mixture is along the curve 5-6 ; the state points 6 and 6′ are on the same constant pressure line. The compression efficiency is then defined as the ratio of the isentropic enthalpy increase to the actual enthalpy increase required for the compression of the mixture (motive steam + water vapours).

$$\text{Compression efficiency } \eta_c = \frac{\text{isentropic enthalpy increase}}{\text{actual enthalpy increase}}$$

$$= \frac{h_6' - h_5}{h_6 - h_5} \qquad \text{...(10.3)}$$

The compression efficiency is also known as diffusion efficiency and it lies in the range 75 to 80 percent.

Steam requirement for the prescribed refrigeration load :

Let m_s = mass of steam supplied

m_v = mass of water vapour entrained from the flash chamber

Then the mass of mixture for compression is

$$m = m_s + m_v$$

Now : energy available for compression = $m_s\,(h_1 - h_3)$

energy required for compression = $(m_s + m_v)\,(h_6 - h_5)$

In accordance with the law of conservation of energy, the energy available for compression equals the energy required for compression. That is

$$m_s\,(h_1 - h_3) = (m_s + m_v)\,(h_6 - h_5) \qquad ...(10.4)$$

From the expression for different efficiencies defined above :

$$\eta_n = \frac{h_1 - h_2}{h_1 - h_2'}\;;\quad h_1 - h_2 = \eta_n\,(h_1 - h_2')$$

$$\eta_e = \frac{h_1 - h_3}{h_1 - h_2}\;;\quad h_1 - h_3 = \eta_e\,(h_1 - h_2)$$

$$\eta_c = \frac{h_6' - h_5}{h_6 - h_5}\;;\quad h_6 - h_5 = \frac{h_6' - h_5}{\eta_c}$$

Substituting the values from above identifies in equation 10.4, we get

$$m_s\,\eta_e\,(h_1 - h_2) = (m_s + m_v) \times \frac{h_6' - h_5}{\eta_c}$$

and $$m_s\,\eta_e\,\eta_n\,(h_1 - h_2') = (m_s + m_v) \times \frac{h_6' - h_5}{\eta_c}$$

or $$m_s\,\eta_n\,\eta_e\,\eta_c\,(h_1 - h_2') = m_s\,(h_6' - h_5) + m_v\,(h_6' - h_5)$$

or $$m_s\,[(h_1 - h_2')\,\eta_n\,\eta_e\,\eta_c - (h_6' - h_5)] = m_v\,(h_6' - h_5)$$

or $$\frac{m_s}{m_v} = \frac{(h_6' - h_5)}{(h_1 - h_2')\,\eta_n\,\eta_e\,\eta_c - (h_6' - h_5)} \qquad ...(10.6)$$

The fraction m_s/m_v represents the mass of motive steam required per kg of water vapour produced in the flash chamber.

It is further noted that :

(*i*) the make up water is supplied at state point 8 where its temperature is slightly lower than that of condenser (state point 7)

(*ii*) the throttling of make up water takes place in the flash chamber (process 8-9) and finally leaves it corresponding to the condition at point 4.

Throttling is a process of constant enthalpy and so enthalpy at state 8 equals the enthalpy at state 9 (*i.e.*, $h_8 = h_9$). Therefore,

$$\text{Net refrigerating effect} = m_v\,(h_4 - h_8)$$

If Q tons is the refrigerating load which the plant is to meet, then

$$m_v = \frac{Q \times 3.5}{(h_4 - h_8)}\ \text{kg/sec} \qquad (\because 1\ \text{ton} = 3.5\ \text{kJ/s})$$

Since each kg of water vapour requires m_s kg of motive steam,
Mass of motive steam for refrigerating load of Q tonnes of refrigeration,

$$= \frac{3.5Q}{h_4 - h_8} \times m_s \quad ...(10.6)$$

The coefficient of performance for the system is given by

$$COP = \frac{\text{refrigerating effect}}{\text{energy input in motive steam}} = \frac{m_v(h_4 - h_8)}{m_s(h_1 - h_8)} \quad ...(10.7)$$

Advantages / Disadvantages:

(*i*) simple and reliable system with quick adjustment to load variation
(*ii*) pump is the only moving part; vibration free and negligible maintenance
(*iii*) water as refrigerant is very safe; non-flammable, non-toxic and non-poisonous
(*iv*) possibility of outdoor installation
(*v*) quite economical when both steam and cooling water are readily available
(*vi*) overall COP of the system is low
(*vii*) high freezing point of water; utility of the system is restricted to temperature above 4°C
(*viii*) proper functioning of the system necessitates high vacuum in the evaporator which is quite difficult to maintain
(*ix*) enormous volume of refrigerant (water) has to handled; 665 m^3/hr per ton of refrigeration compared to 7 m^3/hr for R-12

Applications: The steam jet refrigeration has proven its utility in the applications such as:

(*i*) airconditioning for comfort where steam is available and safety is prime consideration
(*ii*) direct vaporization for the concentration or drying of heat sensitive foods and chemicals
(*iii*) processing of cold water for application in paper and pulp mills, chemical and food processing plants, rubber and vulcanizing industries, breweries and distilleries, refineries and on ships
(*iv*) industrial plants where handling of large quantities of condenser cooling water is relatively inexpensive

EXAMPLE 10.1

The following data has been collected for steam jet refrigeration system installed in an industrial plant :

steam condition supplied to nozzle : 7 bar, dry saturated
flash chamber (evaporator) water temperature : 5°C
temperature at which make up water is supplied : 18°C
pressure in the condenser = 5 cm of Hg absolute
nozzle η = 85%, entertainment η = 63%, compression η = 80%

The quantity of mixture of motive steam and flashed vapour at the beginning of compression is 90%.
Compute the following parameters for the given system :

(a) mass flow of motive steam/kg of flash vapour
(b) quality of vapour flashed from the flash chamber
(c) refrigerating effect per kg of flash vapour
(d) mass of motive steam per tonne of refrigeration
(e) volume flow rate of vapour leaving the flash chamber per tonne of refrigeration.

Solution : Refer Fig. 10.2 for *T-s* and *h-s* plots of the processes involved in the given steam jet refrigeration system.

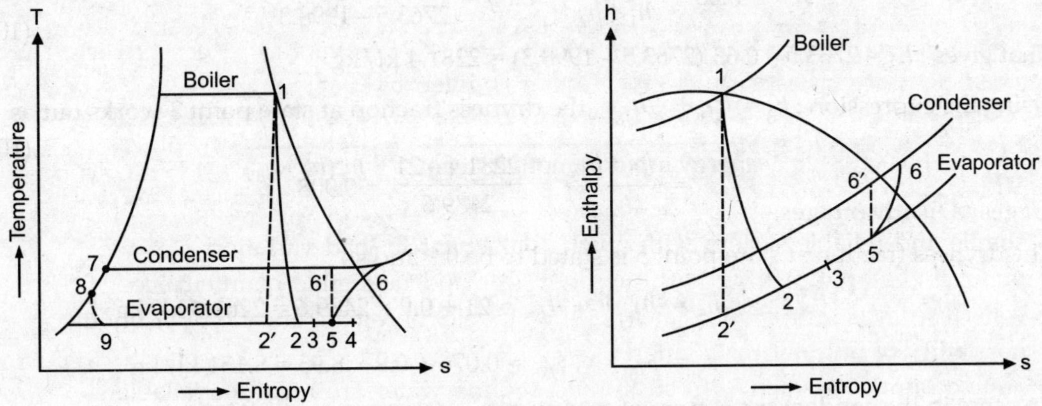

Fig. 10.2. T-s and h-s plots for a steam jet refrigeration system

The solution involves first finding the dryness fraction and enthalpy valves at salient points marked on these plots.

The following values can be read from steam tables :

For 7 bar dry saturated steam

$$h_1 = 2763.5 \text{ kJ/kg} \quad s_1 = 6.708 \text{ kJ/kg K}; \quad t_1 = 164.97°\text{C}$$

At 5°C temperature :

$$h'_{f2} = 21 \text{ kJ/kg K}, \quad h'_{fg2} = 2489.6 \text{ kJ/kg}, \quad s'_{f2} = 0.076 \text{ kJ/kg K}$$

$$\text{and } s'_{f2} = 8.95 \text{ kJ/kg K}$$

For the isentropic expansion process 1–2′ : $s_1 = s'_2$ and that gives

$$6.708 = 0.076 + x'_2 \times 8.95$$

$$\therefore \quad x'_2 = \frac{6.708 - 0.076}{8.95} = 0.74$$

$$\text{and} \quad h'_2 = h'_{f2} + x'_2 h'_{fg2} = 21 + 0.74 \times 2489.6 = 1863.3 \text{ kJ/kg}$$

From the expression for nozzle efficiency,

$$0.85 = \frac{h_1 - h_2}{h_1 - h'_2}; \quad 0.85 = \frac{2763.5 - h'_2}{2763.5 - 1863.3}$$

That gives : $h'_2 = 2763.5 - 0.85\,(2763.5 - 1863.3) = 1998.3$ kJ/kg

Since the points 2′, 2, 3 and 5 lie on the same pressure line corresponding to 5°C temperature, we can write

$$h_{f2} = h'_{f2} = h_{f3} = h_{f5} = 21 \text{ kJ/kg}$$

$$\text{and} \quad h_{fg2} = h'_{fg2} = h'_{fg2} = h'_{fg3} = h_{fg5} = 2489.6 \text{ kJ/kg}$$

$$\text{Further:} \quad h_2 = h_{f2} + x_2\, h_{fg2}$$

$$\text{Then, dryness fraction at state point 2} = \frac{h_2 - h_{f2}}{h_{fg2}} = \frac{1998.3 - 21}{2489.6} = 0.79$$

From the expression for entrainment efficiency

$$0.63 = \frac{h_1 - h_3}{h_1 - h_2}; \quad 0.63 = \frac{2763.5 - h_3}{2763.5 - 1998.3}$$

That gives : $h_3 = 2763.5 - 0.63\,(2763.5 - 1998.3) = 2281.4$ kJ/kg

Using the expression : $h_3 = h_{f3} + x_3 h_{fg3}$, the dryness fraction at state point 3 works out as

$$x_3 = \frac{h_3 - h_{f3}}{h_{fg3}} = \frac{2281.4 - 21}{2489.6} = 0.908$$

The dryness fraction at state point 5 is stated to be 0.9 and so

$$h_5 = h_{f5} + x_5\, h_{fg5} = 21 + 0.9 \times 2489.6 = 2261.6 \text{ kJ/kg}$$

$$s_5 = s_{f5} + x_5\, s_{fg5} = 0.076 + 0.9 \times 8.95 = 8.131 \text{ kJ/kg}$$

Pressure in the condenser = 4.5 cm of mercury

$$= (1000 \times 13.6) \times 9.81 \times \frac{4.5}{100} = 6003.7 \text{ N/m}^2 \simeq 0.06 \text{ bar}$$

At this pressure : $h'_{f6} = 151.5$ kJ/kg and $h'_{fg6} = 2415.9$ kJ/kg

$s'_{f6} = 0.521$ kJ/kg K and $s'_{fg6} = 7.809$ kJ/kg K

For the isentropic compression process 5–6′ : $s_5 = s'_6$ and that gives

$$8.131 = 0.521 + x'_6 \times 7.809$$

$$\therefore \quad x'_6 = \frac{8.131 - 0.521}{7.809} = 0.974$$

and $h'_6 = h'_{f6} + x'_6\, h'_{fg6} = 151.5 + 0.974 \times 2415.9 = 2504.6$ kJ/kg

From the expression for compression efficiency

$$0.8 = \frac{h'_6 - h_5}{h_6 - h_5} = \frac{2504.6 - 2261.6}{h_6 - 2261.6}$$

That gives : $h_6 = \dfrac{2504.6 - 2261.6}{0.8} + 2261.6 = 2565.4$ kJ/kg.

(*a*) Mass of motive steam required per kg of water vapour produced in the flash chamber

$$\frac{m_s}{m_v} = \frac{h'_6 - h_5}{(h_1 - h_2)\eta_n \eta_e \eta_c - (h'_6 - h_5)}$$

$$= \frac{2504.6 - 2261.6}{(2763.5 - 1863.3) \times (0.95 \times 0.63 \times 0.8) - (2763.5 - 1863.3)}$$

$$= \frac{243}{431.02 - 243} = 1.292 \text{ kg/kg of flash vapour}$$

(*b*) The enthalpy at state point 4 can be worked out by writint the following expression for energy balance

$$m_v h_4 + m_s h_3 = (m_s + m_v) \times h_5$$

or $$h_4 + \frac{m_s}{m_v} \times h_3 + \left(\frac{m_s}{m_v} + 1\right) \times h_5$$

$$\therefore \quad h_4 = (1.292 + 1) \times 2261.6 - 1.292 \times 2281.4 = 2236.0 \text{ kJ/kg}$$

From the identity : $h_4 = h_{f_4} + x_4 h_{fg4}$

$$x_4 = \frac{h_4 - h_{f_4}}{h_{fg4}} = \frac{2236.0 - 21}{2489.6} = 0.89 \qquad (h_{f_4} = h_{f_2} \text{ and } h_{fg4} = h_{fg2})$$

(*c*) Refrigerating effect per kg of flash vapour

$$= (h_4 - h_9) = (h_4 - h_8) \qquad (\because h_9 = h_8 \text{ ; throttling process})$$

Further h_8 = enthalpy corresponding to 18°C = 4.186 × 18 = 75.35 kJ/kg

$$RE = 2236.0 - 75.35 = 2160.65 \text{ kJ/kg}$$

(*d*) Mass of motive steam for refrigerating load of 1 ton of refrigeration

$$= \frac{3.5 \times 1}{RE} \times \frac{m_s}{m_v}$$

$$= \frac{3.5 \times 1}{2160.65} \times 1.292 = 2.093 \times 10^{-3} \text{ kg/s per ton of refrigeration}$$

(*e*) Volume of vapour removed from the flash chamber

v_4 = volume of liquid at point 4 + x_4(volume of saturated vapour – volume of liquid)

From steam tables, volume of saturated vapour corresponding to 5°C is 147.1 m^{3}kg.

$$v_4 = 1 + 0.89\,(147.1 - 1) = 131.03 \text{ m}^3/\text{kg}$$

Then volume of vapour removed from the flash chamber for refrigerating load of 1 ton of refrigeration is

$$= v_4 \times \frac{3.5 \times 1}{RE} = 131.03 \times \frac{3.5 \times 1}{2160.65}$$

$$= 0.212 \text{ m}^3/\text{s per ton of refrigeration}$$

(*f*) Coefficient of performance = $\dfrac{\text{refrigerating effect}}{\text{work input}}$

$$= \frac{m_v(h_4 - h_8)}{m_s(h_1 - h_7)} = \frac{1}{1.292} \times \frac{236.0 - 75.35}{2763.5 - 151.5} = 0.64$$

Here h_7 = 151.5 kJ/kg is the enthalpy of mixture at a pressure of 4.5 cm of mercury.

10.2. VORTEX TUBE REFRIGERATION

Principle : When compressed air is released into a tube through vortex generator, streams of hot and cold air appear simultaneously. This phenomenon was noted in 1930 by a French physicist Georges Ranque, is called Ranque effect and it forms the basis of vortex tube refrigeration.

Construction and working : Fig. 10.3 shows the schematic arrangement of the different components comprising the vortex tube refrigeration system. The unit essentially consists of three components namely

(*i*) a nozzle
(*ii*) a diaphragm and valve
(*iii*) a tube with hot and cold sides

A stream of compressed air passes through a nozzle where it expands and attains a high velocity. The nozzle is designed to have higher exit velocity, greater mass flow with minimum inlet losses. Further, the nozzle configuration is such that a vortex (swirl) flow is created in the chamber. The chamber is essentially a portion of the nozzle and it facilitates tangential entry of the high velocity air stream into the tube. The vortex makes the air travel in a spiral like motion along the periphery towards the valve end. The flow is restricted by the valve and the kinetic energy is converted into pressure energy. The resulting stagnation pressure is higher than the pressure in the nozzle plane and consequently reversal in flow takes place. There is then energy transfer between the reversed stream and the forward stream and that leads to cooling of air stream through the core. The cold air passes through diaphragm hole into the cold side. Diaphragm is a cylindrical piece of small thickness and has a small hole of specified diameter at the centre. The hot stream escapes to the atmosphere through opening in the valve. By controlling the valve opening, the quantity of cold air and its temperature can be varied.

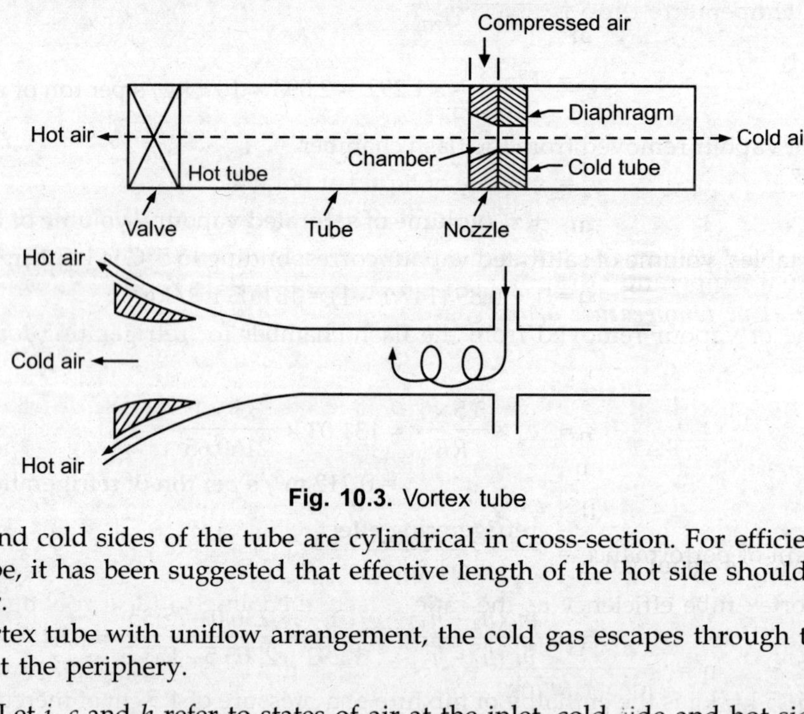

Fig. 10.3. Vortex tube

The hot and cold sides of the tube are cylindrical in cross-section. For efficient operation of the vortex tube, it has been suggested that effective length of the hot side should be 3-times the tube diameter.

In the vortex tube with uniflow arrangement, the cold gas escapes through the central core and hot gas at the periphery.

Theory : Let *i*, *c* and *h* refer to states of air at the inlet, cold side and hot side respectively. Then from:

mass balance : $m_i = m_c + m_h$

energy balance : $m_i\, h_i = m_c\, h_c + m_h\, h_h$ ...(10.8)

Neglecting kinetic energy changes

$$\Delta T_c = T_i - T_c \quad \text{and} \quad T_h = T_h - T_i \qquad ...(10.9)$$

Considering the system to be adiabatic (no heat loss),

heat lost by cold stream = heat gained by hot stream

$$m_c\, c_p\, (T_i - T_c) = m_h\, c_p\, (T_h - T_i)$$
$$= (m_i - m_c)\, c_p\, (T_h - T_i)$$

$$\therefore \quad (T_i - T_c) = \left(\frac{m_i}{m_c} - 1\right)(T_h - T_i) = \left(\frac{1}{\mu} - 1\right)(T_h - T_i) \qquad ...(10.10)$$

where parameter μ defines the ratio of mass of cold air to that of air supplied. Under normal operating condition

$$\mu = 0.3 \text{ to } 0.45$$

Equation (10.10) can be recast as

$$\frac{1}{\mu} = \frac{T_i - T_c}{T_h - T_i} + 1 = \frac{T_i - T_c + T_h - T_i}{T_h - T_i} = \frac{T_h - T_c}{T_h - T_i}$$

or
$$\mu = \frac{T_h - T_i}{T_h - T_c} = \frac{T_h - T_i}{(T_i - T_c) + (T_h - T_i)} = \frac{\Delta T_h}{\Delta T_h + \Delta T_c} \qquad ...(10.11)$$

Presuming isentropic expansion from inlet pressure p_i to atmospheric pressure p_a at the cold end, the isentropic temperature drop is given by

$$\Delta T_c' = T_i - T_c' = T_i \left[1 - \frac{T_c'}{T_i}\right]$$

$$= T_i \left[1 - \left(\frac{p_a}{p_i}\right)^{\frac{\gamma-1}{\gamma}}\right]$$

The ratio of actual temperature drop ΔT_c in the vortex tube to isentropic temperature drop $\Delta T_c'$ is called the ***relative temperature drop***. That is

$$\Delta T_{rel} = \frac{\Delta T_c}{\Delta T_c'} = \frac{\left[\dfrac{\Delta T_c}{T_i}\right]}{\left[1 - \left(\dfrac{p_a}{p_i}\right)^{\frac{\gamma-1}{\gamma}}\right]} \qquad ...(10.12)$$

The value of ΔT_{rel} is about 0.6.

Defining the vortex tube efficiency as the ratio of actual cooling to ideal cooling, we get

$$\eta = \frac{\text{actual cooling}}{\text{ideal cooling}} = \frac{m_c\, c_p\, \Delta T_c}{m_i\, c_p\, \Delta T_c'} = \mu\, \Delta T_{rel}$$

$$= \frac{\mu\, \Delta T_c}{\left[T_i \left\{1 - \left(\dfrac{p_a}{p_i}\right)^{\frac{\gamma-1}{\gamma}}\right\}\right]} \qquad ...(10.13)$$

Taking μ = 0.4 and ΔT_{rel} = 0.6, the vortex tube efficiency works out to be 0.24.

The COP of the vortex tube is defined as the ratio of cooling effect to the work input to the air compressor.

$$\text{Cooling effect} = m_c\, c_p\, \Delta T_c$$

$$\text{Energy supplied to compressor} = \frac{m_i\, c_p T_a}{\eta_c}\left[\left(\frac{p_i}{p_a}\right)^{\frac{\gamma-1}{\gamma}} - 1\right]$$

where η_c is the isentropic efficiency of the compressor.

$$\therefore \quad COP = \frac{m_c\, c_p\, \Delta T_c}{\dfrac{m_i\, c_p T_a}{\eta_c}\left[\left(\dfrac{p_a}{p_i}\right)^{\frac{\gamma-1}{\gamma}} - 1\right]}$$

Since $\dfrac{m_c}{m_i} = \mu$ (the cold air fraction), we can write

$$COP = \frac{\mu\, \Delta T_c\, \eta_c}{T_a\left[\left(\dfrac{p_i}{p_a}\right)^{\frac{\gamma-1}{\gamma}} - 1\right]} \qquad \text{... (10.14)}$$

Using equation 10.13 into equation 10.14 and simplification yields

$$COP = \eta\, \eta_c \left(\frac{p_a}{p_i}\right)^{\frac{\gamma-1}{\gamma}} \qquad \text{... (10.15)}$$

The above identity indicates that COP of a vertex tube is dependent on pressure ratio; the COP decreases as the pressure ratio increases. Further, presuming that η = 0.24, η_c = 0.8 and (p_i/p_a) = 5, the COP works out to be

$$COP = 0.24 \times 0.8\ (1/5)^{\frac{1.4-1}{1.4}} = 0.121$$

Advantages/Disadvantages :

(*i*) Simplicity of design, fabrication and maintenance. These aspects ensure low initial and operating costs and long life.

(*ii*) Light in weight and quite compact; needs only small space for its installation.

(*iii*) No moving parts (except the valves) and noiseless operation.

(*iv*) Low power consumption.

(*v*) Quite economical when dry compressed air is readily available.

(*vi*) Simplicity of operation; no need of the services of any expert attendant.

(*vii*) COP is small and it has considerable dependence on tube length. Moreover, the tubes are quite costly to manufacture.

Applications :

(*i*) Successful use in systems where cooling and heating are required simultaneously.

(*ii*) Cooling of certain commodities at temperatures as low as – 50°C.

(*iii*) Use of cool air from vortex tube for the manufacture of air suits. These suits are used by operators and workers who handle toxic gases during spray painting and work in mines.

(*iv*) Cooling of machine operation such as drilling, milling, turning etc. over small areas. Such cooling increases the tool life, improves surface finish and allows higher cutting speeds.

(*v*) Cooling of rotor blades of gas turbines used for marine and aircraft applications. The cooled air form the tube is made to pass through radial holes drilled in the blades, and effective cooling results.

(*vi*) Spot cooling of supersonic aircraft because of simplicity of construction, controls, light weight and saving in space required.

10.3. THERMO-ELECTRIC REFRIGERATION

Principle : When an electric current is applied across the junction of two dissimilar metals, cooling is produced at one junction and simultaneously there is heating at the other junction. This aspect was noted in 1934 by a French scientist Jean Peltier and is called the Peltier effect. This fact was further enforced in 1939 by a German scientist Lenz who placed a drop of water on a junction of two dissimilar metals and a dc current was made to pass through the circuit. He observed that when the current passed in one direction, the water froze. Subsequently the ice melted when the direction of current was reversed. This concept forms the basis of thermo-electric refrigeration and has commercially materialized with the development of semiconductors.

Thermo electric Effects : The following effects have been observed when a direct current passes through a pair of thermo couples with their junctions maintained at different temperatures.

1. Seeback effect : When the two junctions of a pair of dissimilar metals are maintained at different temperatures, there is generation of electromotive force (emf). Based upon a series of tests conducted by varying the temperatures of the junctions of various combinations of a set of materials, it was found that the open circuit voltage (ΔV) developed is proportional to the temperature difference ΔT of the functions. That is

$$\Delta V \propto \Delta t \ ; \ \Delta V = \alpha_{ab} \Delta T = (\alpha_a - \alpha_b) \Delta T$$

The proportionality constant α_{ab} is called the *seebeck coefficient* or *thermo electric power* and has a value different for a pair of different metals (*A* and *B*).

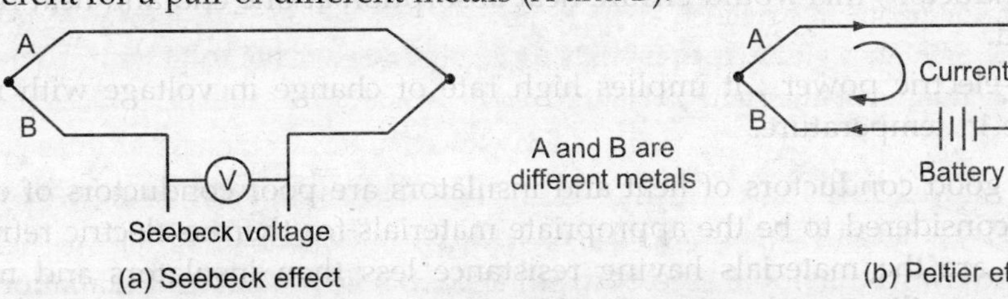

Fig. 10.4. Seebeck and Peltier effects

2. Peltier effect : When a direct current is made to pass through a pair of dissimilar metals, there is heating at one junction and cooling at the other junction set of metals chosen. By varying the current, it was found that heating and cooling rates (heat evolved or absorbed in a unit time) for different set of materials are proportional to the current. That is

$$Q \propto I \ ; \ Q = \pi_{ab} I$$

The proportionality constant π_{ab} is called the *Peltier coefficient* and it has certain value for two different metals. That is

$$\pi_{ab} = \pi_a - \pi_b$$

3. Thomson effect : When a current *I* is made to pass through a single conductor having a temperature gradient (dT/dx), there is heat transfer and it is prescribed by the relation

$$\frac{\delta Q}{dx} = \tau I \left(\frac{dT}{dx} \right)$$

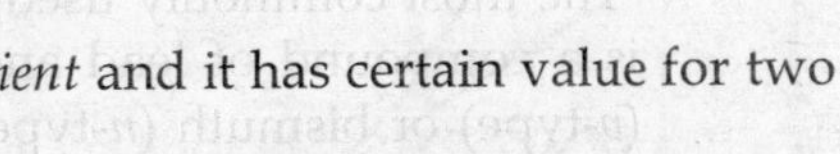

Fig. 10.5. Thomson effect

The factor τ is called the Thomson coefficient. All the three effects (Seebeck, Peltier, Thomson) are reversible pheno-menon.

By applying the 1st and 2nd laws of thermodynamics to a simple thermoelectric circuit, the following two relations have been set up connecting the coefficients α, π and τ

$$\pi_{ab} = \alpha_{ab}\, T \text{ and } \tau_a - \tau_b = T \times \frac{d\alpha_{ab}}{dT}$$

It is to be noted that where as the Seebeck and Peltier coefficients refer to a junction between two materials, the Thomson coefficient is defined for single conductor.

Further combining the Peltier effect $Q = \pi_{ab} I$, and the relation $\pi_{ab} = \alpha_{ab} T$, we get the following expression for the heat evolved or absorbed in a unit time

$$Q = \alpha_{ab}\, I\, T$$

Apparently for a high value of cooling or heating, the combination of materials should be such that the Peltier coefficient is high.

4. Joulean effect : When an electric current flows through a conductor, there is a dissipation of electrical energy.

In terms of current I and electrical resistance R, the dissipation of electrical energy is

$$Q = I^2 R$$

5. Conduction effect : When the ends of any element are maintained at different temperatures, the heat transfer from the hot end (temperature T_h) to the cold end (temperature T_c) is given by

$$Q = U(T_h - T_c) \text{ where } U \text{ is the overall conductance}$$

Thermo-electric materials : The desirable properties of a thermo-electric material are :

(*i*) good conductor of electricity; ohmic losses would then be minimum.

(*ii*) poor heat conductor ; this would ensure heat absorption at one end and heat rejection at the other end.

(*iii*) high thermo-electric power ; it implies high rate of change in voltage with respect to small change in temperature.

Since metals are good conductors of heat and insulators are poor conductors of electricity, semi-conductors are considered to be the appropriate materials for thermo-electric refrigeration. The semi-conductors are the materials having resistance less than insulators and more than conductors. Germanium, silicon and carbon are the common examples of semiconductors.

Thermo-electric materials are denoted by p if their thermo-electric power is positive, and n if thermo-electric power is negative. For the selection of thermo-electric elements, care is taken to have p and n-type combinations. One of the arm consists of purely hole type and the other of a purely electronic type semi-conductor.

- The most commonly used material for thermo-electric converter is ***lead telluride*** which is a compound of lead and tellurium and containing small amounts of either sodium (p-type) or bismuth (n-type).
- ***Bismuth telluride*** is another appropriate material if mechanical strength, suitability under operating conditions and ease of fabrication is considered.
- For high temperature applications, semi-conductors based on silicon germanium and compounds of selenium are preferred.

Further, the thermo-electric material should have good mechanical strength, be amply elastic and be able to resist chemical influences such as oxidation etc.

Operation : Consider a circuit made of materials A and B into which a battery is introduced to provide direct current *I*. The heat evolved or absorbed (*Q*) per unit time at the junctions is proportional to current and is given by

$$Q \propto I, \quad Q = \alpha_{ab}\, I\, T$$

where $\alpha_{ab} = (\alpha_a - \alpha_b)$ is the Peltier coefficient for two different metals.

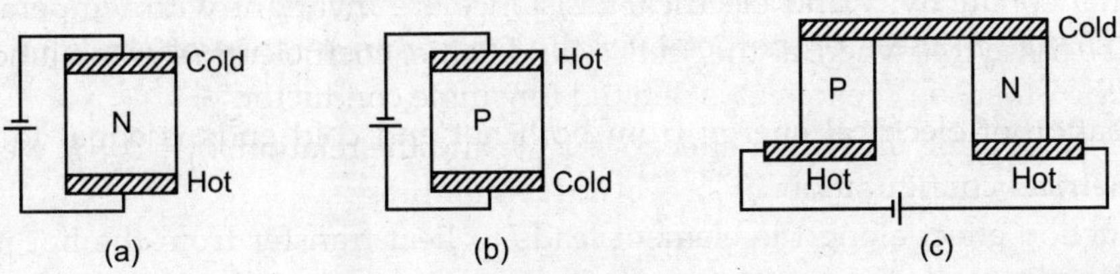

Fig. 10.6.

Refer Fig. 10.6, the *n*-type semiconductor is sandwiched between two metal strips and is connected to a dc voltage source. The junction where the current enters becomes hotter and the other junction gets cooler. When the *p*-type semi-conductor is sandwiched between the metal strips, Fig. 10.6 (*b*), the effect is opposite to that observed in *n*-type semiconductor. Here, the junction where the current enters becomes cold and the other gets hot. The maximum cooling effect can be produced if the *n*-type and *p*-type conductors are coupled, Fig. 10.6 (*c*). A continuous dissipation of heat liberated at hot end will make the temperature of cold end lower than the room temperature. In a typical domestic refrigerator operating on Peltier effect, a large number of couples are put together. These couples are electrically in series and thermally parallel.

Analysis : Refer Fig 10.7 which shows an enlarged view of thermoelectric elements with control volume.

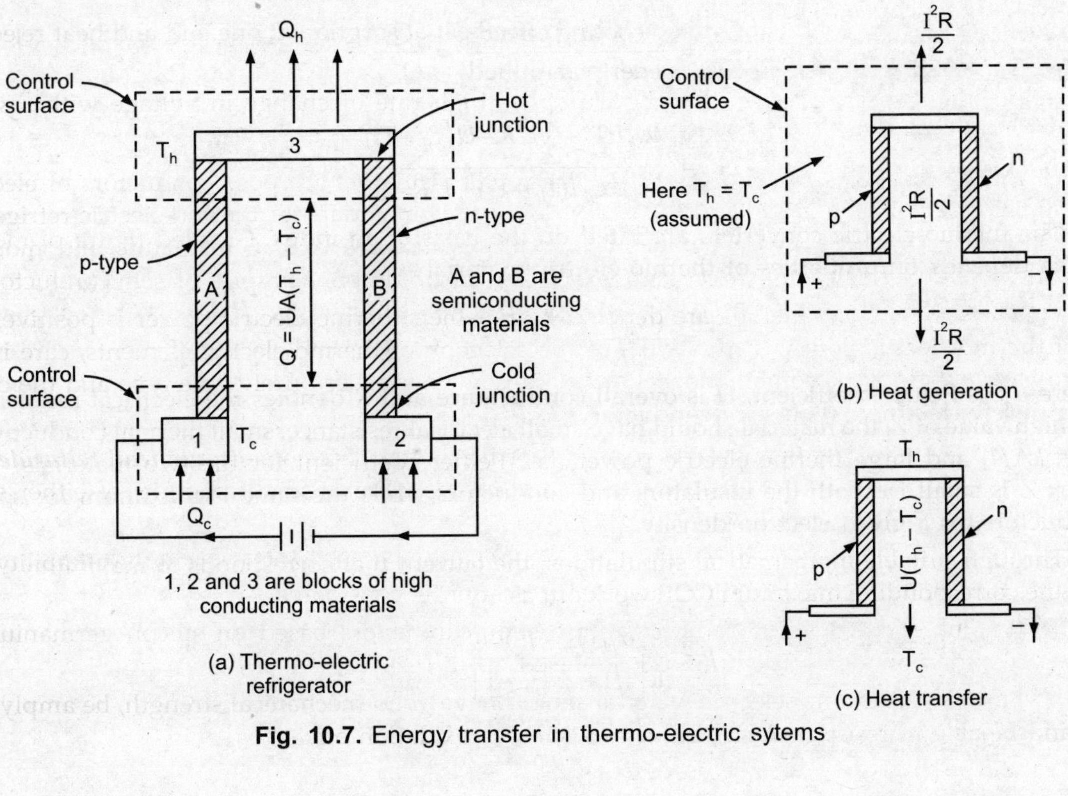

Fig. 10.7. Energy transfer in thermo-electric sytems

An analysis of this system for obtaining the refrigeration effect and coefficient of performances has been worked out by making the following assumptions;

(i) Heat absorption and heat rejection take place through the semi-conductors at the ends only.

(ii) No energy exchange occurs between the elements through the space separating them

(iii) Thermal conductivity and electrical resistance are invariant with temperature, *i.e.* there is no change in these properties over the temperature range of working.

(iv) Dissipation of electrical energy from both hot and cold ends is equal to $\frac{1}{2} I R^2$ due to symmetrical configuration.

(v) Conduction effect along the element leads to heat transfer from the hot junction to the cold junction.

With these stipulations and considering the control volume

Thermo electric cooling $Q_c = \alpha_{ab} I T_c - \frac{1}{2} I^2 R - U(T_h - T_c)$

Thermo electric heating $Q_h = \alpha_{ab} I T_h + \frac{1}{2} I^2 R - U(T_h - T_c)$

Applying 1st law of thermodynamics, the energy input to the system from outside is

$$\oint \delta Q = Q_{net} = - Q_h + Q_c$$

$$= -[\alpha_{ab} I (T_h - T_c) + I^2 R]$$

Then *COP* of th system is

$$= \frac{\text{cooling effect}}{\text{energy supplied}} = \frac{Q_c}{Q_{net}}$$

$$= \frac{\alpha_{ab} I T_c - \frac{1}{2} I^2 R - U(T_h - T_c)}{\alpha_{ab} I (T_h - T_c) + I^2 R}$$

The thermo-electric converters are rated on the basis of an index Z (called figure of merit) which depends on properties of thermo-electric material

$$Z = \frac{(\alpha_{ab})^2}{UR}$$

where α_{ab} is Peltier coefficient, U is overall conductance and R denotes the electrical resistance. For high value of Z, the material should have small electrical resistance, small thermal conductivity ($U = kA/l$) and large thermo-electric power, *i.e.*, Peltier coefficient for the two materials. The index Z is small for both the insulators and conductors while the same is maximum for semi-conductors for a given electron density.

Through suitable mathematical stipulations, the current that corresponds to maximum COP and the corresponding maximum COP work out as

$$I = \frac{\alpha_{ab}(T_h - T_c)}{R\left[\sqrt{1 + Z T_m} - 1\right]}$$

$$(COP)_{max} = \left[\frac{T_c}{T_h - T_c}\right]\left[\frac{\sqrt{1+ZT_m} - \left(\frac{T_h}{T_c}\right)}{\sqrt{1+ZT_m}+1}\right]$$

where T_m is the mean temperature ; $T_m = \frac{T_h + T_c}{2}$

The maximum COP can be rewritten as

$$(COP)_{max} = (COP)_{Carnot} \times (COP)_{ref}$$

where, $(COP)_{Carnot} = \frac{T_c}{T_h - T_c}$ is the Carnot value for a refrigeration system that operates between the same upper and lower temperature limits.

$(COP)_{ref} = \frac{\sqrt{1+ZT_m} - \left(\frac{T_h}{T_c}\right)}{\sqrt{1+ZT_m}+1}$ represents the system performance when the effect of actual thermo-electric behaviour is taken into considerations.

An closer examination of the above expression shows that:

(*a*) $(COP)_{ref}$ tends to unity when figure of merit Z approaches infinity. In that case

$$(COP)_{max} = (COP)_{Carnot} = \frac{T_c}{T_h - T_c}$$

(*b*) For maximum cooling $\frac{dQ_c}{dI} = 0$ and that gives

$$I_{opt} = \frac{\alpha_{ab} T_c}{R} = T_c\sqrt{\frac{ZU}{R}}$$

The corresponding cooling and COP are then prescribed as

$$(Q_c)_{max} = U\left[\frac{T_c^2 Z}{2} - (T_h - T_c)\right]$$

and

$$(COP)_{max} = \frac{ZT_c^2/2 - (T_h - T_c)}{ZT_h T_c}$$

A thermo-electric system gives a reasonable amount of refrigeration effect and COP, when the figure of merit Z has the highest possible value; Z has a maximum value of $0.007/k$ uptil the present investigations

Advantages / Disadvantages :

(*i*) Thermo-electric module is of very compact size; requires very little space for installation.

(*ii*) Noiseless operation as there are no moving parts.

(*iii*) No refrigerant is used and as such no toxication etc. and that allows direct use for air conditioning.

(*iv*) Light in weight, portable and can operate in any position (vertical or horizontal).

(*v*) Easy servicing, most parts can be easily replaced by the end user.

(*vi*) Low maintenance; only dusting and vacuuming needed to ensure good heat dissipation.
(*vii*) Low COP (0.1 to 0.2)
(*ix*) Unavailability of suitable material with high figure of merit Z.
(*x*) High running cost ; system needs more power for its operation.

Applications :

Thermo-electrics have 30 to 40 years of proven record for use in military, aerospace, laboratory and now consumer applications.

(*i*) Peltier refrigerators, water coolers and spot coolers have gained prominence and are being successfully used in Western countries.
(*ii*) Peltier refrigerators, put on a moving trolley, are used by some international airlines to provide cold drinks and hot snacks. This is possible because the system provides cooling and heating effects simultaneously.
(*iii*) Air-conditioning of rooms where large cooling capacities are required but the temperature difference is not to be large.
(*iv*) Thermo-electric cooling is being effectively used in the following applications related to medical and scientific field:
- constant low temperature bath and chambers
- dew point hygrometer for determining absolute humidity
- cooled baffles for oil diffusion in vacuum pumps
- serium coolers for preservation of blood plasma and serums etc.

EXAMPLE 10.2

Why are semiconductor materials preferable to metals in thermo electric refrigeration.

A thermo electric refrigeration removes hat from a refrigerated space at – 6°C at a rate of 125 w and rejects it to an environment at 23°C. Calculate the maximum possible COP, and minimum input required for the working of this refrigeration system.

Solution : Temperature at the hot and cold junctions are:

$$T_h = 23°C = 296 \text{ K} \quad \text{and} \quad T_c = -6°C = 267 \text{ K}$$

$$(COP)_{max} = \frac{T_c}{T_h - T_c} = \frac{267}{296 - 267} = 9.21$$

Now :
$$COP = \frac{\text{refrigerating effect}}{\text{power input}} \; ; \; 9.21 = \frac{125}{\text{power input}}$$

$$\therefore \text{ Minimum required input} = \frac{125}{9.21} = \mathbf{13.57} \textbf{ kW}$$

EXAMPLE 10.3

The hot and cold junctions of a thermoelectric pair are maintained at temperatures of 23°C and – 23°C. the properties relevant to the thermo electric materials forming the pair are:

$$\alpha_a = [10^5 + 100\,t - t^2] \times 10^{-7} \text{ volts/K}$$

where t is the temperature in °C

$$\alpha_b = \alpha_a$$

If the current flow in the circuits limited to 10 amperes, make calculations for the Peltier amount of heat absorbed and heat rejected at the two junctions.

Solution : The temperatures at the hot and cold junctions are:

$$T_h = 23°\text{C} = 296 \text{ K} \quad \text{and} \quad T_c = -23°\text{C} = 250 \text{ K}$$

$$\alpha_{ab} = \alpha_a - \alpha_b = \alpha_a - (-\alpha_a) = 2\alpha_a$$

Peltier heat absorbed at cold junction,

$$Q_c = \alpha_{ab}\, I\, T_c = 2\,\alpha_a\, I\, T_c$$
$$= 2[10^5 + 100 \times (-23) - (-23)^2] \times 10^{-7} \times 10 \times 250$$
$$= 2[10^5 - 2300 - 529] \times 10^{-7} \times 10 \times 250 = \mathbf{48.58} \text{ kW}$$

Peltier heat rejected at hot junction,

$$Q_h = \alpha_{ab}\, I\, T_h = 2\,\alpha_a\, I\, T_h$$
$$= 2[10^5 + 100 \times (23) - (23)^2] \times 10^{-7} \times 10 \times 298$$
$$= 2[10^5 + 2300 - 529] \times 10^{-7} \times 10 \times 296 = \mathbf{60.25} \text{ kW}$$

EXAMPLE 10.4

The following data pertains to a thermo electric cooling system:

Temperature of hot and cooling junctions = 315 K and 275 K

Total electric resistance = 0.0035 Ω

Total thermal conductance = 0.028 W/K

Thermo electric power = 450 × 10^{-6} V/K

For 18 amperes of current flowing through the system, compute figure of merit and cooling capacity in tonnes of refrigerations. The arrangement is stated to use 150 couples.

Solution : The term figure of merit Z is defined as

$$Z = \frac{(\alpha_{ab})^2}{UR} = \frac{(450 \times 10^{-6})^2}{0.028 \times 0.0035} = 2066.3 \times 10^{-6}$$

Heat absorbed at the cooling junction per couple is

$$Q_c = \alpha_{ab}\, I T_c - \frac{1}{2} I^2 R - U(T_h - T_c)$$

$$= (450 \times 10^{-6} \times 18 \times 275) - \left(\frac{1}{2} \times 18^2 \times 0.0035\right) - 0.028(315 - 275)$$

$$= 2.2275 - 0.567 - 1.12 = 0.5405 \text{ J/s}$$

Since 1 tonne ≡ 3.5 kJ/s, the cooing capacity of the arrangement with 150 couples is

$$= \left[\frac{0.5405}{10^3} \times 150\right] \times \frac{1}{3.5} = \mathbf{0.0232} \text{ tonne}$$

10.4. PULSE TUBE REFRIGERATION SYSTEM

Principle : When a gas is released after sudden expansion, the cooling effect is produced. This phenomenon was proposed by Gifford in 1961 and forms the basis of pulse tube refrigeration system.

Operation : Fig. 10.8 shows the schematic arrangement of the different components of a pulse tube refrigeration system. The unit essentially consists of :

(*i*) a high pressure gas source

(*ii*) valve and the associated operating mechanism

(*iii*) pulse tube

(*iv*) cooling system and

(*v*) vacuum system

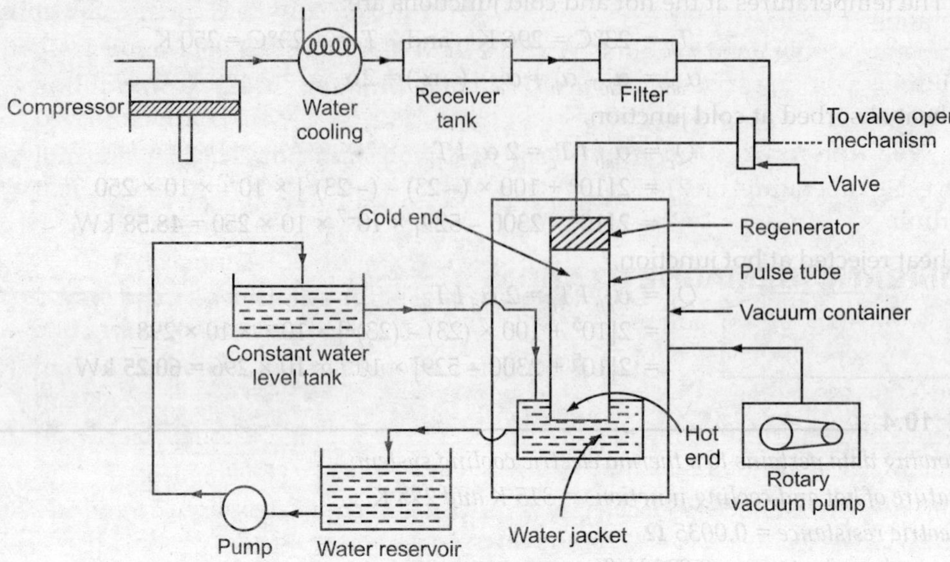

Fig. 10.8. Pulse tube refrigeration system

The operation of the pulse unit involves two processes namely pressurisation and depressurization and these occur alternately.

(*i*) Pressurization and heat rejection: The high pressure gas enters the pulse tube by a system actuated by the valve mechanism. At entry, the gas is cooled in the regenerator to the refrigeration (cold end) temperature. This gas acts as a gas piston and compresses the gas (tube gas) already present in the tube. The tube gas is pushed downwards towards the lower (closed) end of the tube. During compression process, the temperature of the tube gas rises. The heat energy of the gas is rejected to the cooling water and gas attains temperature which is very close to that of the cooling water.

(*ii*) Depressurization (Pressure release) and heat absorption : The gas in the pulse tube is connected to the exhaust by actuating the valve mechanism. The gas is released and during its expansion, the temperature drops at the cold end. The reduced temperature takes care of the refrigeration load.

The radiation and correction heat losses from the system are minimized by placing the pulse tube along with regenerator inside a vacuum container.

Advantages :

(*i*) No moving parts; noiseless operation and no lubrication problems.

(*ii*) Reliable and compact construction.

(*iii*) Low temperature obtainable with moderate pressure.

Applications :

(*i*) Good choice for obtaining very low termperature if the cooling requirements are limited to a few watts.

(*ii*) Most suitable for cryogenic applications where even the employment of multi-stage and cascade vapour compression systems cannot meet the requirements of very low temperature.

It is worthwhile to point out that the following data has been reported in literature on pulse tube refrigeration:

- Temperatures obtained at cold end with the cooling medium temperatures of 300 K and helium as working gas at a pressure of 25 bar absolute.
  124 K with a single stage pulse tube
  79 K with a two stage pulse tube
- Lowest temperature of 214.5 K with a single stage pulse tube using air as the work medium.

10.5. ADIABATIC DEMAGNETIZATION

Fig. 10.9 shows the set up for adiabatic demagnetization of paramagnetic salt.

The set up essentially consists of

(*i*) A chamber filled with low pressure gaseous helium. The paramagnetic substance is suspended by a nylon thread inside this chamber. The gaseous helium provides thermal communication with the surrounding bath of liquid helium.

(*ii*) A bath of liquid hydrogen or nitrogen which surrounds the liquid helium bath.

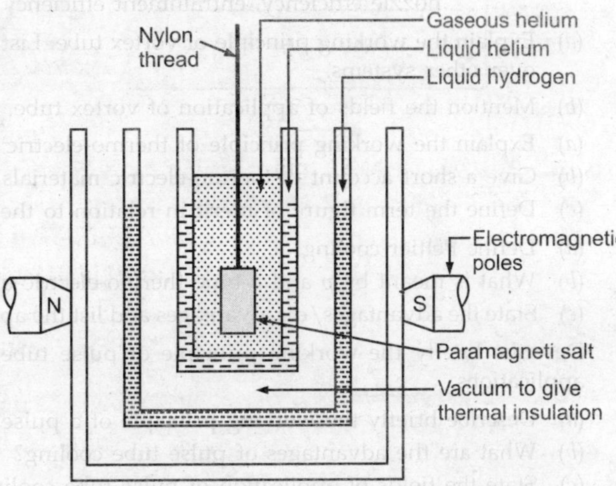

Fig. 10.9. Cooling by adiabatic demagnetization

The following is the sequence of processes that result into the cooling of salt:

(*i*) The liquid helium is cooled by pumping to the lowest practical pressure and usually a temperature of about 1 K (– 227°C) is attained. The salt too attains this temperature by conduction through exchange gas.

(*ii*) A magnetic field of about 50,000 gauss intensity is applied to the set up. The salt gets magnetized and the heat generated during magnetization process is transferred to the liquid helium without any change in the temperature of salt (it remains unchanged at 1 K).

(*iii*) The gaseous helium surrounding the salt is evacuated. That makes the salt thermally isolated and at 1 K temperature under the stress of strong magnetic field.

(*iv*) The magnetic field is released and the temperature of salt decreases almost isentropically as a consequence of adiabatic demagnetization. It is possible to reduce the salt temperature to a value as low as 0.001 K.

Such low temperatures are measured by the magnetic thermometer which essentially measures the susceptibility of salt and correlates that measurement with temperature. According to Curies law, the magnetic susceptibility (χ) is inversely proportional to absolute temperature (T). That is

$$\chi \propto \frac{1}{T} \quad \text{or} \quad \chi = \frac{C}{T}$$

REVIEW QUESTIONS

A. Conceptual and conventional questions.

1. (*a*) State the principle of steam jet refrigeration system.
(*b*) Explain, with a neat sketch, the working of steam jet refrigeration system.

2. (*a*) Set up an expression for determining the mass of motive steam required per kg of water vapour formed in the flash chamber (evaporator) of a steam jet refrigeration system.
(*b*) Enumerable the advantages/disadvantages of steam jet refrigeration system and mention its applications.

3. Define the following terms in relation to steam jet refrigeration system :
nozzle efficiency, entrainment efficiency and compression efficiency

4. (*a*) Explain the working principle of vortex tube. List the merits/demerits of this refrigeration system over other systems.
(*b*) Mention the fields of application of vortex tube.

5. (*a*) Explain the working principle of thermo-electric refrigeration system.
(*b*) Give a short account of thermo-electric materials.
(*c*) Define the term figure of merit in relation to thermo-electric refrigeration system.

6. (*a*) Define Peltier cooling.
(*b*) What is meant by *p* and *n*-type thermo-electric elements?
(*c*) State the advantages/disadvantages and list the applications of thermo-electric refrigeration system.

7. Describe briefly the working principle of pulse tube refrigeration system. State its advantages and applications.

8. (*a*) Describe briefly the working principle of a pulse tube system.
(*b*) What are the advantages of pulse tube cooling?
(*c*) State the fields of application of pulse tube cooling.

9. Explain, with a neat sketch, the arrangement used for producing low temperature by adiabatic demagnetization of a paramagnetic salt.

B. Fill in the blanks with appropriate word/words

1. A cooling effect can be produced by passing a direct current through a junction of two dissimilar metals. This effect is called the ....
2. A .... is the proper material for thermo electric refrigeration.
3. The coefficient of performance of vortex tube lies between .... and ....
4. The lowest temperature produced by adiabatic demagnetisation of a paramagnetic salt is .... kelvin approximately.
5. The magnetic susceptibility is inversely proportional to the absolute temperature. This correlation is referred to as .... law.
6. In steam jet refrigeration system, the motive steam expands in ....
7. .... is the compression device used in a steam jet refrigeration system.
8. In .... refrigeration system, sudden expansion and release of gas are employed to get the refrigerating effe·:.

Answers

1. Peltier effect 2. semi-conductor 3. 0.1 and 0.2 4. 0.001 5. curie's 6. convergent-divergent nozzle 7. Steam ejecter 8. pulse tube

❑❑❑

Chapter – 11

Refrigeration Equipment

Learning objectives : A study of the subject matter presented in this chapter will enable the students/readers to

- *know the main components and their location in vapour compression refrigeration system*
- *explain the various components and devices regarding their function, construction and operation*
- *analyse the reciprocating compressor for work input with and neglecting clearance, and establish the need for multi-stage compression*
- *compare the different types of main components on the basis of their performance*

11.1. MAIN COMPONENT OF REFRIGERATION SYSTEM

The refrigerating effect (cooling) in a refrigeration system is produced when the working substance (refrigerant) undergoes certain changes in its state or condition. Each change is referred to as a process and for every process carried out, the refrigeration system is equipped with a separate component. The main four components are :

(*i*) compressor (*ii*) condenser
(*iii*) expansion device (*iv*) evaporator

Fig 11.1 shows the general arrangement of the location of these components in a vapour compression refrigeration system.

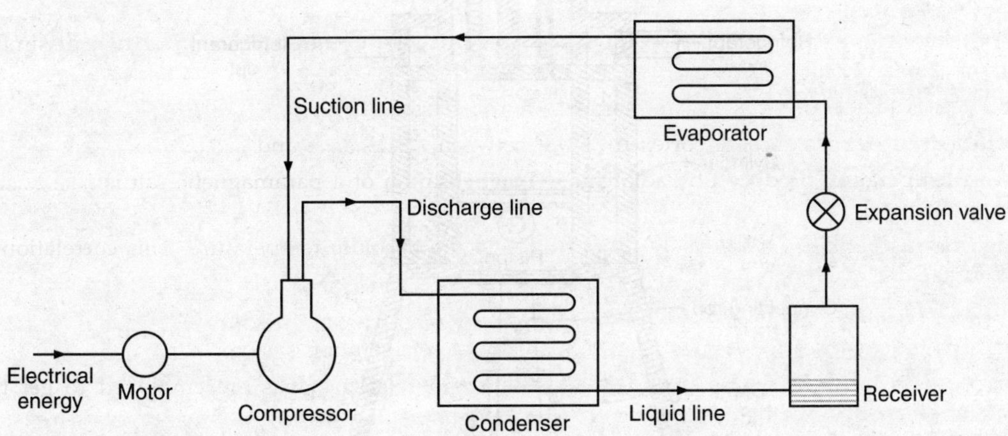

Fig. 11.1. Location of the components in a refrigeration system

Further for effective, economical and safe working of the refrigeration system, certain control devices are also provided. These devices control the flow of refrigerant at various points through out the refrigeration cycle.

11.2. COMPRESSORS

Compressor is the heart of a refrigeration system

The compressor regulates the flow of refrigerant by pumping action and is considered to be the heart of a vapour compression refrigeration system. The compression unit is required to perform the following functions :

- Pump the refrigerant through the system and circulate it again and again in cyclic processes. The human heart pumps and circulates blood through the body. Likewise, the compressor pumps and circulates refrigerant again and again through the refrigeration system
- Produce high pressure (and hence high temperature) to enable the refrigerant to reject its heat in the condenser.
- Produce low pressure in the evaporator to enable the refrigerant to pick up maximum amount of heat from the space being refrigerated.

The refrigerant compressor takes in refrigerant gas (vapour) at low pressure (evaporator pressure), compresses it and then delivers the high pressure gas to condenser. The absolute pressure of refrigerant at the inlet of a compressor is called the *suction pressure,* and the absolute pressure of refrigerant at exit from the compressor is called *discharge pressure.* The compressor acts like a heat pump as it takes heat at low temperature from the evaporator and pumps it at high temperature to the condenser. The compression process requires that work must be done upon the gas and accordingly the compressor has to be driven by some external agency (electric motor or steam turbine). Compressor is the costliest component of a refrigeration system and obviously lies the need for reliable and cheap compressors.

The most common compressors used in the modern vapour compression system are : reciprocating, rotating screw, centrifugal and scroll compressors. Each application prefers one or another due to size, noise, efficiency and pressure aspects.

11.2.1. Reciprocating Compressors

These are the piston style, positive compressors in which the pressure changes are obtained by the co-ordinated action of pistons and valves.

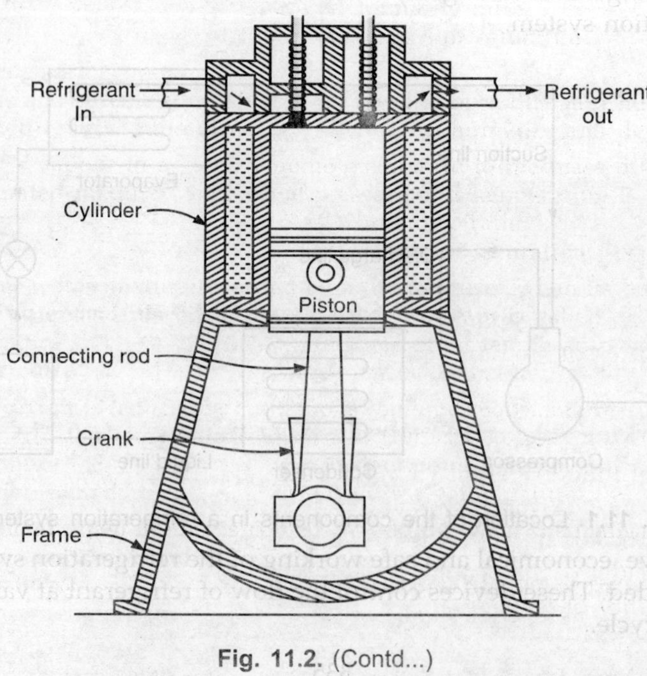

Fig. 11.2. (Contd...)

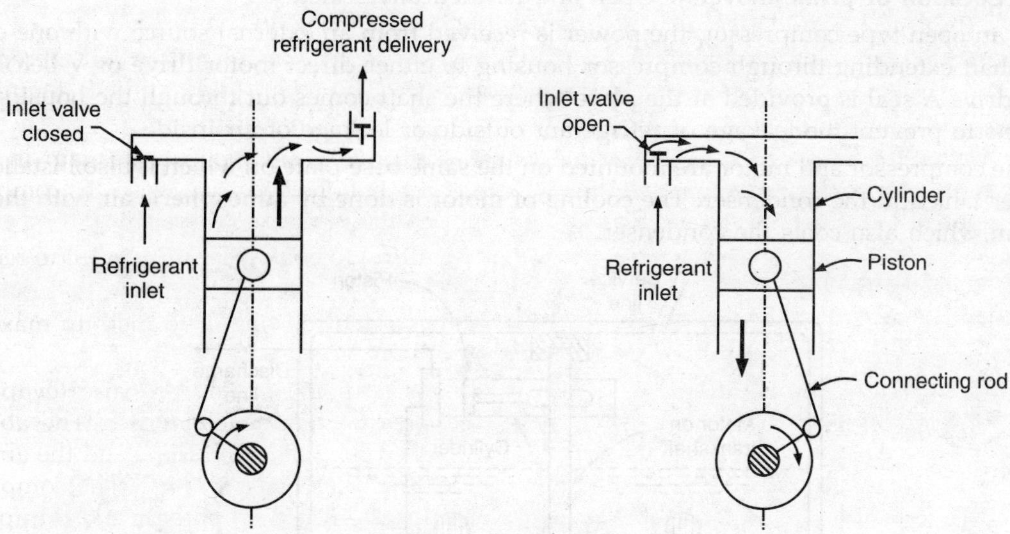

Fig. 11.2. Single stage reciprocating compressor

The piston moves in reciprocating motion to draw the refrigerant gas into the cylinder in one stroke and to compress and discharge it to the condenser in second, *i.e.*, return stroke. The piston derives its motion through a connecting rod and crank mounted in a crankcase. The inlet and outlet valves are provided in the head of cylinder. The valves are mostly of thin steel plates provided with light springs for reducing the inertia. The valves operate because of small pressure difference across them. The head and cylinder of high capacity compressors are cooled by means of water jackets. Low capacity compressors are cooled by providing fins on the cylinder head. Flywheel is used to ensure that uniform turning moment is supplied throughout the cycle of operation.

11.2.2. Classification

The reciprocating compressors are classified on the basis of

1. Arrangement of piston and cylinder : Vertical and horizontal.

The vertical compressors are single acting and the piston is driven directly by connecting rod and crankshaft which are enclosed in a pressure-tight crankcase. They are best used for small size applications as they need less space for their installation.

The horizontal reciprocating compressors essentially operate on double acting principle, and the piston is driven with the mechanism comprising crankshaft, connecting rod, crosshead and piston rod. These compressors require large floor space for their installation and find favour for larger size applications.

2. Number of working strokes : Single acting or double acting.

The compression of the refrigerant vapour in a single acting compressor occurs only on one side of the piston and once during each revolution of the crankshaft. In a double acting unit, the compression of vapour occurs on both sides of the piston and accordingly twice in each revolution of crankshaft. As both sides of the piston are exposed to rapid temperature and pressure changes, there is greater possibility of leakage.

3. Location of prime-movers : Open and hermetically sealed.

In an open type compressor, the power is received from an external source with one end of crankshaft extending through compressor housing to either direct motor drive or V-belt/gear/chain drive A seal is provided at the place where the shaft comes out through the housing. The seal acts to prevent the leakage of refrigerant outside or leakage of air inside.

The compressor and motor are mounted on the same base plate on which is also installed the receiver tank and the condenser. The cooling of motor is done by atmospheric air with the help of a fan which also cools the condenser.

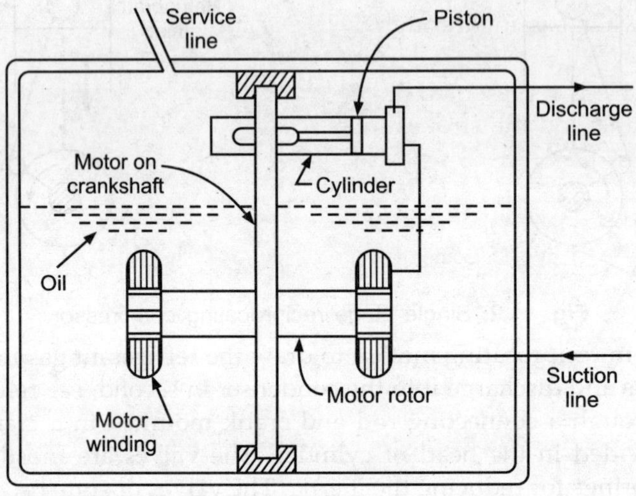

Fig. 11.3. Sealed compressor

The hermetically sealed compressor is a direct drive unit in which both the compressor and the motor driving the compressor use a common shaft and are enclosed in a common casing or a welded steel shell. Since the compressor motor assembly is suspended inside a steel shell, the winding and motor cannot get natural cooling. The cooling of motor and its winding is accomplished by cold refrigerant vapour coming from the evaporator. The unit is dehydrated before charging; otherwise the moisture in the system would damage the motor. The hermetically sealed compressors may be reciprocating or rotary and may be mounted in vertical position. These compressors are generally employed for small domestic refrigerators, home freezers and window air-conditioners, and have the following advantages/disadvantages :

- minimum noise due to absence of external drive
- requires small space for installation as it is compact
- no need of crankshaft seal which is necessary to prevent leakage of refrigerant in ordinary compressors
- cooling is more effective as it needs to be done in low pressure area
- simple lubrication system
- the moving parts are inaccessible and accordingly the maintenance and repair is not easy
- the motor winding is likely to be harmed by the presence of moisture in the refrigerant
- there are chances of lubricating oil mixing with refrigerant and that can deteriorate the quality of refrigerant
- the system requires a separate vacuum pump for evacuation and charging of refrigerant. If the evacuation is done by the compressor, the refrigerant would leave the system but the motor will continue running and creating vacuum. Consequently the motor will get heated up and the winding will burn.

In a modified version, the moving parts are made accessible by bolting the top cover with the shell. The bolted construction permits the assemblies to be opened as and when some repair or servicing needs to be done.

11.2.3. Power Required to Drive the Compressor

To study the operation of a single stage reciprocating compressor and to establish a relation for the work required to drive it, let it be presumed that

(*i*) there is no clearance,

(*ii*) there are no losses due to friction,

(*iii*) the refrigerant when passing through the pipes and valves does not undergo any pressure loss.

The compression cycle is represented on *p-v* and *T-s* plot in Fig. 11.4. The cycle is completed in one revolution of crank and the strokes in order of operation are :

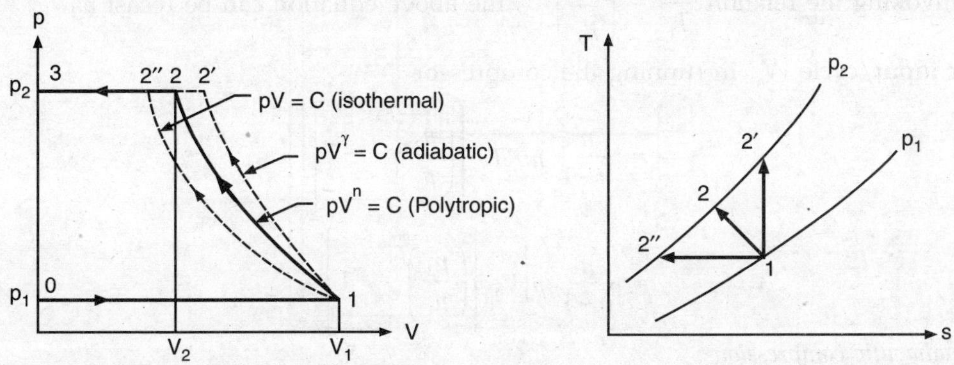

Fig. 11.4. Compression cycle for a single-stage reciprocating compressor

Stroke 0-1 : The piston moves outwards, the valve opens and the refrigerant is sucked into the cylinder. During this operation

(*i*) the pressure and temperature remain constant,

(*ii*) the mass of refrigerant in the cylinder increases,

(*iii*) there is no exchange of heat with the surroundings,

(*iv*) the flow work is done by air

Stroke 1-2-3 : The piston moves inwards, the inlet valve closes and the refrigerant is compressed. The compression law has the following three possibilities :

(*i*) isentropic (adiabatic) compression along 1-2

(*ii*) polytropic compression 1-2′

(*iii*) isothermal compression 1-2″

When the air has been compressed to delivery pressure p_2, the delivery valve opens and the refrigerant is delivered to the receiver at constant pressure p_2. During compression, work is done on air and the temperature of air at the end of compression depends on the law of compression and heat exchanged with the surroundings. During delivery 2-3,

(*i*) pressure and temperature remain constant,

(*ii*) there is no exchange of heat with surroundings,

(*iii*) the mass of refrigerant in the cylinder decreases,

(*iv*) flow work is done on refrigerant in pushing it out of cylinder.

After exhaust, the pressure in the cylinder drops from p_2 to p_1 and the cycle is repeated. Area 0-1-2-3 represents the network when the compression follows the polytropic law.

Network on refrigerant/cycle = area 0-1-2-3

= work done on refrigerant during compression 1-2

+ work done an refrigerant during delivery 2-3

– work done by air during suction 0-1

$$= \frac{p_2V_2 - p_1V_1}{n-1} + p_2V_2 - p_1V_1$$

$$= \frac{n}{n-1}(p_2V_2 - p_1V_1) = \frac{n}{n-1} mR(T_2 - T_1)$$

Invoking the relation $\frac{T_2}{T_1} = \left(\frac{p_2}{p_1}\right)^{\frac{n-1}{n}}$, the above equation can be recast as

Work input/cycle W_a in running the compressor

$$= \frac{n}{n-1} mRT_1 \left[\left(\frac{p_2}{p_1}\right)^{\frac{n-1}{n}} - 1\right]$$

$$= \frac{n}{n-1} p_1V_1 \left[\left(\frac{p_2}{p_1}\right)^{\frac{n-1}{n}} - 1\right] \quad ...(11.1)$$

With adiabatic compression

$$\text{Work input/cycle } W_a = \frac{\gamma}{\gamma-1} mRT_1 \left[\left(\frac{p_2}{p_1}\right)^{\frac{\gamma-1}{\gamma}} - 1\right]$$

$$= \frac{\gamma}{\gamma-1} p_1V_1 \left[\left(\frac{p_2}{p_1}\right)^{\frac{\gamma-1}{\gamma}} - 1\right] \quad ...(11.2)$$

With isothermal compression,

Work input/cycle W_a = area 0-1-2″-3

$$W_a = p_2V_2'' + p_1V_1 \log_e \frac{V_1}{V_2''} - p_1V_1$$

$$= p_1V_1 \log_e \frac{V_1}{V_2''} \qquad (\because p_1V_1 = p_2V_2'')$$

$$= p_1V_1 \log_e \frac{p_2}{p_1} = mRT_1 \log_e \frac{p_2}{p_1} \quad ...(11.3)$$

For single acting air compressor, air power P_a is given by

$$P_a = W_a \times \frac{N}{60}$$

For double acting air compressor

$$P_a = W_a \times \frac{2N}{60}$$

where N is the speed of the compressor in revolutions per minute.

The power input to the compressor can also be determined if there is an arrangement for taking indicator diagram.

$$P_a = k\, p_m\, LA \frac{N}{60} \text{ kJ/s} \quad \text{...(11.4)}$$

where p_m is mean effective pressure in kPa, L is stroke length in meter, A is area of piston in m^2, N is speed in rev/min and k is a factor : 1 for single acting and 2 for double acting.

11.2.4. Best Value for Compression Index

For a given compressor,

(*i*) The pressure ratio p_2/p_1 is fixed which means a fixed height (line 0-3) of *p-v* diagram.

(*ii*) The capacity given by the cylinder volume fixes the stroke length which means line 0-1 of the *p-v* plot has a fixed value.

The only process which can influence the area of the *p-v* diagram (work input) is the path of compression 1-2. The compression index decides the position of line 1-2. The work expended in running the compressor would be least when the area of *p-v* plot is minimum.

Comparison of the expressions derived above for the work required to compress and deliver unit mass of refrigerant for a given pressure ratio would show that isothermal compression requires least amount of energy. However, isothermal compression cannot be achieved in practice but an attempt is made to approach it by cooling the compressor either by providing fins or a water jacket to the compressor cylinder.

The term ***isothermal efficiency*** of a compressor is a measure of the degree to which isothermal compression has been achieved. It is expressed as the ratio of isothermal work to actual work.

$$\text{Isothermal efficiency} = \frac{\text{isothermal work input}}{\text{actual work input}}$$

$$= \frac{p_1 V_1 \log_e \dfrac{p_2}{p_1}}{\dfrac{n}{n-1} p_1 V_1 \left[\left(\dfrac{p_2}{p_1}\right)^{\frac{n-1}{n}} - 1\right]} \quad \text{...(11.5)}$$

11.2.5. Effect of Clearance and Volumetric Efficiency

When the piston is at its extreme left position (Fig. 11.4), it does not flush with the cylinder head. A dead space remains between the cylinder head and the piston. The dead space is called clearance space. The clearance ensures mechanical freedom to the moving parts without much wear and tear and provides the necessary space for valve operation.

The volume occupied by the refrigerant gas when the piston is at its extreme left position is known as clearance volume V_c and the volume swept by the piston as it moves in the cylinder from its extreme position on the left to right is referred to as swept volume or piston displacement V_s.

Swept volume or piston displacement volume

$$V_s = \text{area of piston} \times \text{stroke length}$$

$$= \frac{\pi}{4} d^2 l$$

The pressure-volume diagram for a single stage compressor when clearance is considered is shown in Fig. 11.5. The small quantity of refrigerant at pressure p_2 remaining in the dead space at the end of delivery line 2-3 expands polytropically along 3-4. Theoretically the expansion continues until pressure p_1 is reached at point 4. The inlet valve then opens and the induction of fresh refrigerant into the cylinder begins. The intake continues for the remaining period of stroke. The volume sucked in during suction is $(V_1 - V_4)$. At point 1, the inlet valve closes and the charge is compressed polytropically along the path 1-2 until delivery pressure p_2 is reached at point 2. The delivery valve then drops and the refrigerant is discharged to the condenser. The delivery continues till the piston reaches the extreme position on the left. The cycle is then repeated.

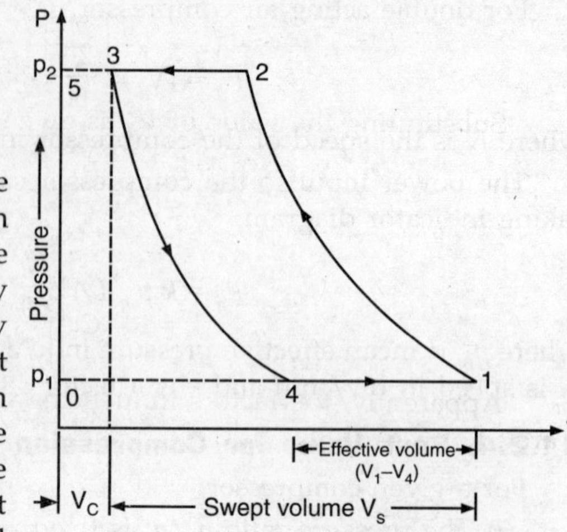

Fig. 11.5. Effect of clearance and volumetric efficiency

Work input/cycle = area 1-2-3-4

= area (0-1-2-5) – area (3-4-0-5)

$$= \frac{n}{n-1} p_1 V_1 \left[\left(\frac{p_2}{p_1}\right)^{\frac{n-1}{n}} - 1\right] - \frac{n}{n-1} p_1 V_4 \left[\left(\frac{p_2}{p_1}\right)^{\frac{n-1}{n}} - 1\right]$$

$$= \frac{n}{n-1} p_1 (V_1 - V_4) \left[\left(\frac{p_2}{p_1}\right)^{\frac{n-1}{n}} - 1\right] \qquad \text{...(11.6)}$$

where $(V_1 - V_4)$ is the effective volume sucked in during the suction stroke. Work input thus depends on effective volume and not on the swept volume.

For a given capacity, increase in clearance volume would necessitate increased swept capacity of the compressor. The compressor would become heavier entailing an increase in mechanical losses. The limits of clearance volume are 5 to 10% of swept volume for good quality machines and 2% of swept volume for sleeve valve machines.

The volumetric deficiency η_v of the reciprocating compressor is defined as the ratio of actual volume sucked in to the piston displacement or swept volume of the compressor. Thus

$$\eta_v = \frac{V_1 - V_4}{V_s} = \frac{V_s + V_c - V_4}{V_s} = 1 + \frac{V_c}{V_s} - \frac{V_4}{V_s} \qquad \text{...}(i)$$

Considering polytropic expansion process 3-4, we have

$$p_3 V_3^{\,n} = p_4 V_4^{\,n} \; ; \quad V_4 = V_3 \left(\frac{p_3}{p_4}\right)^{1/n}$$

But $\quad p_3 = p_2 \; ; \quad p_4 = p_1 \quad$ and $\quad V_3 = V_c$

$$\therefore \quad V_4 = V_c \left(\frac{p_2}{p_1}\right)^{1/n} \qquad ...(ii)$$

Substituting the value of V_4 as given by expression (*ii*) in expression (*i*), we get

$$n_v = 1 + \frac{V_c}{V_s} - \frac{V_c}{V_s}\left(\frac{p_2}{p_1}\right)^{1/n} = 1 + C - C\left(\frac{p_2}{p_1}\right)^{1/n}$$

$$= 1 - C\left[\left(\frac{p_2}{p_1}\right)^{1/n} - 1\right] \qquad ...(11.7)$$

Apparently, the factors influencing the volumetric efficiency of a reciprocating compressor are :

Clearance factor $C = \frac{V_c}{V_s}$; pressure ratio $\frac{p_2}{p_1}$ and the polytropic index of expansion n.

Fig. 11.6 shows the variation in volumetric efficiency when each of these parameters is changed, keeping the other two factors constant. From this plot, it is seen that volumetric efficiency

- decreases as the clearance factor C increases,
- decreases with increases of $\frac{p_2}{p_1}$ and
- increases with increase of index n

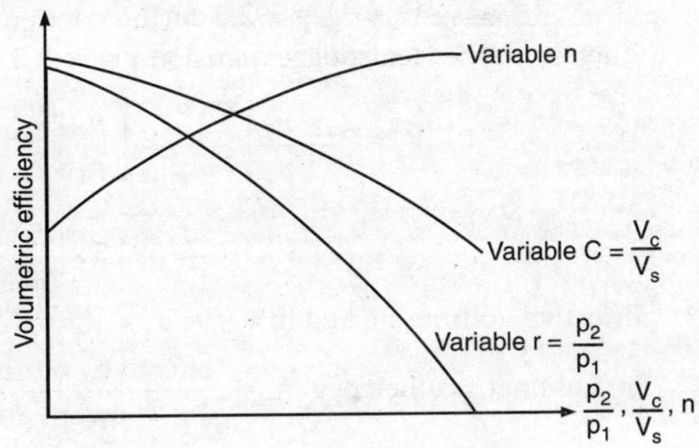

Fig. 11.6.

The volumetric efficiency as expressed by equation 11.7 is further reduced due to :

(*i*) increase in temperature of fresh (sucked) refrigerant when it comes in contact with heated cylinder walls

(*ii*) inertia effect of air in suction pipe

(*iii*) wire drawing or throttling (pressure drops) at suction and delivery valves

(*iv*) leakage in the system, *i.e.*, valve and piston leakage

(*v*) the refrigerant does not get enough time to fill the cylinder fully during suction strokes if the compressor runs at high speeds.

Further in actual operation, the inlet valve opens only when the high pressure refrigerant left in the dead space expands to a pressure below the suction pressure. Some portion of the suction stroke may thus be wasted.

When all these factors are taken into consideration, the actual volumetric efficiency of a reciprocating compressed is worked out from the relation

$$\eta_{av} = (1+C)\left(\frac{p_s}{p_1}\right)^{\frac{1}{n_c}} - C\left(\frac{p_s}{p_1}\right)^{\frac{1}{n_e}} - 0.01\left(\frac{p_d}{p_1}\right) \qquad ...(11.8)$$

where p_1 is the pressure at the flange on the suction side, p_s is suction pressure, p_d is the pressure at exit from the compressor, n_c and n_e are polytropic indices for compression and expansion processes respectively.

EXAMPLE 11.1

A single acting single stage reciprocating compressor of 20 cm bore and 25 cm stroke is so constructed that its clearance volume is 5 percent of active stroke. The compressor receives refrigerant at 1 bar, turns 500 rev/min and delivers the compressed refrigerant at 5 bar. Presuming that the compression and expansion curves follow the polytropic law $pV^{1.3}$ = constant, make calculations for the volumetric efficiency of the compressor and the power required to run it.

Solution : Refer Fig 11.7

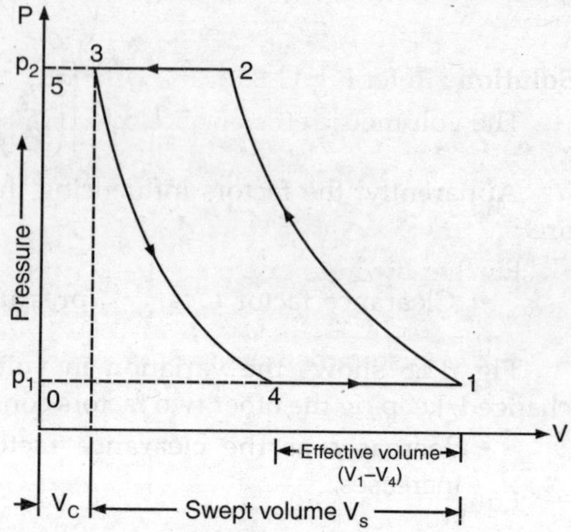

Fig 11.7.

$$\text{Swept volume } V_s = \frac{\pi}{4} d^2 l$$

$$= \frac{\pi}{4} \times (20)^2 \times 25 = 7850 \text{ cm}^3$$

$$\text{Clearance volume } V_c = \frac{5}{100} \times 7850$$

$$= 392.5 \text{ cm}^3$$

Total cylinder volume,

$$V_1 = 7850 + 392.5 = 8242.5 \text{ cm}^3$$

$$V_3 = V_c = 392.5 \text{ cm}^3$$

Then from the isentropic expansion process 3 – 4

$$V_4 = V_3 \left(\frac{p_3}{p_4}\right)^{1/n} = V_3 \left(\frac{p_2}{p_1}\right)^{1/n}$$

$$= 392.5 \times \left(\frac{5}{1}\right)^{1/1.3} = 1353.6 \text{ cm}^3$$

Effective volume sucked in $V_{eff} = V_1 - V_4 = 8242.5 - 1353.6 = 6888.9 \text{ cm}^3$

$$(a) \text{ volumetric efficiency } \eta_v = \frac{\text{effective volume sucked in}}{\text{swept volume}}$$

$$= \frac{6888.9}{7850} \times 100 = \mathbf{87.76\ \%}$$

$$\textit{Alternatively}: \ \eta_v = 1 - \frac{V_c}{V_s}\left[\left(\frac{p_2}{p_1}\right)^{1/n} - 1\right] = 1 - 0.05\left[\left(\frac{5}{1}\right)^{1/1.3} - 1\right]$$

$$= 0.8776 = 87.76\%.$$

$$(b) \text{ Work input to compressor} = \frac{n}{n-1} p_1 V_{eff}\left[\left(\frac{p_2}{p_1}\right)^{\frac{n-1}{n}} - 1\right]$$

$$= \frac{1.3}{1.3-1} \times (1 \times 10^5) \times (6888.9 \times 10^{-6}) \left[\left(\frac{5}{1}\right)^{\frac{1.3-1}{1.3}} - 1\right]$$

$$= 1342.88 \text{ Nm}$$

Since the compressor turns 500 rev/min.

$$\text{Power input} = 1342.88 \times \frac{500}{60} = 11190.6 \text{ Nm/s} \simeq \mathbf{11.19} \text{ kW}$$

EXAMPLE 11.2

Obtain an expression for the capacity of a refrigeration system (tonnage) in terms of the ratio of upper and lower pressure limits, expansion index η, ratio V_c/V_s (V_c = clearance volume and V_s = swept volume), speed N in rpm, specific volume v_1 at inlet to compressor and the refrigeration effect $(h_1 - h_4)$.

Calculation the capacity of the system if pressure ratio is 6.5, index η = 1.1, ratio V_c/V_s = 0.023, speed N = 900 rpm, V_s = 600 cc, v_1 = 0.078 m³/kg and refrigeration effect is 150 kJ/kg

(Engg. Services)

Solution : Refer Fig 11.8 for the *p-v* plot.

The volumetric efficiency of the compressor works out as

$$\eta_v = 1 + \frac{V_c}{V_s} - \frac{V_c}{V_s}\left(\frac{p_2}{p_1}\right)^{1/n}$$

Further by definition :

$$\eta_v = \frac{\text{actual volume sucked in}}{\text{piston displacement or swept volume}}$$

$$= \frac{mv_1}{V_s \times N/60}$$

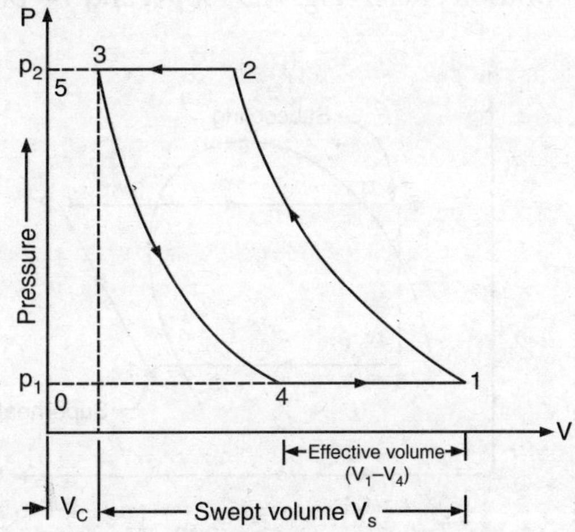

Fig 11.8.

Equating these two identities, we get

Mass of refrigerant

$$m = \left[1 + \frac{V_c}{V_s} - \frac{V_c}{V_s}\left(\frac{p_2}{p_1}\right)^{1/n}\right] \times \frac{V_s}{v_1} \times \frac{N}{60}$$

Then : capacity (tonnage) = $m\,(h_1 - h_4)$

$$= \left[1 + \frac{V_c}{V_s} - \frac{V_c}{V_s}\left(\frac{p_2}{p_1}\right)^{1/n}\right] \times \frac{V_s}{v_1} \times \frac{N}{60} \times (h_1 - h_4)$$

Substituting the given values, we get

$$\text{Capacity} = \left[1 + 0.025 - 0.025 \times (6.5)^{1/1.1}\right] \times \frac{600 \times 10^{-6}}{0.078} \times \frac{900}{60} \times 150$$

$$= 15.37 \text{ kJ/s} = 15.37 \text{ kW}$$

Since 1 ton is equivalent to 3.5 kW,

$$\text{Capacity in tonnage} = \frac{15.37}{3.5} = \mathbf{4.39}$$

EXAMPLE 11.3

In order to keep off food stuff in better condition, a food storage locker requires 20 tons of refrigeration. The evaporator temperature is –8°C and the condenser temperature is 30°C. The refrigerant Fr - 12 used in the plant is subcooled by 5°C before entering the expansion valve and the vapour is superheated to 6°C before leaving the evaporator coil. The compressor employed is of two cylinders single acting having stroke to 1.5 times the bore. The compressor runs at 900 rev/min and the compression process is isentropic. Represent the processes on p-h and T-s diagrams and make calculations for the followings :

(a) Refrigerant effect per kg of refrigerant

(b) mass flow rate of refrigerant being circulated

(*c*) *theoretical piston displacement*
(*d*) *power input to compressor if mechanical efficiency is 0.9*
(*e*) *coefficient of performance*
(*f*) *bore and stroke if volumetric efficiency is 90%*

Given: Enthalpy at inlet to compressor = 356 kJ/Kg
Enthalpy at outlet from compressor = 377 kJ/kg
Enthalpy at exit from subcooler = 225.5 kg
Specify volume at compressor inlet = 0.077 m³/kg **(Engg. Services)**

Solution : Refer Fig. 11.9 for *p-h* and *T-s* plots of the working cycle

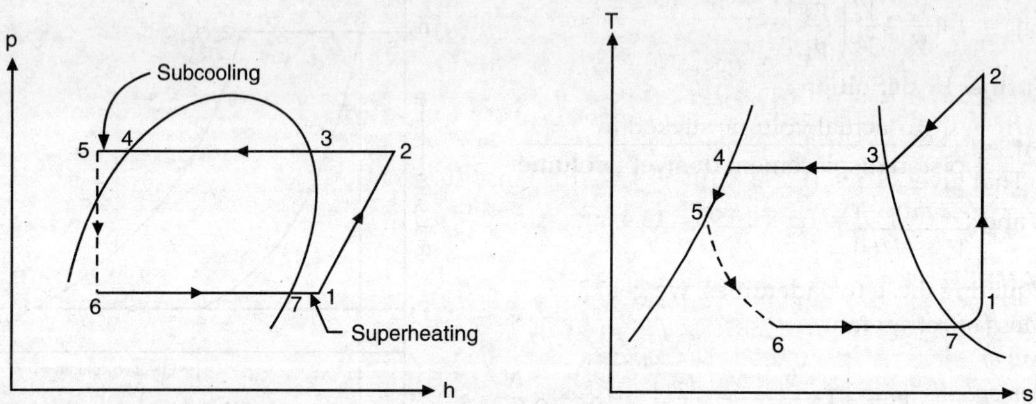

Fig. 11.9

From the given values and points marked on the plots,

$$h_1 = 356 \text{ kJ/kg} \; ; \quad h_2 = 377 \text{ kJ/kg} \; ; \quad v_1 = 0.077 \text{ m}^3/\text{kg}$$

The enthalpy remains constant during throttling process 5-6 and so

$$h_5 = h_6 = 225.5 \text{ kJ/kg}$$

(*a*) Refrigerating effect $= h_1 - h_6 = 356 - 225.5 = 130.5$ kJ/kg

(*b*) 1 ton of refrigration is equivalent of heat absorptions of 3.5 kJ/s of heat in the evaporator. Therefore

$$\text{Mass flow of refrigerant} = \frac{\text{refrigerating capacity}}{\text{refrigerating effect}}$$

$$m_r = \frac{20 \times 3.5}{130.5} = 0.5364 \text{ kg/s}$$

(*c*) Theoretical piston displacement $= m_r \times$ specific volume
$= 0.5364 \times 0.077 = 0.0413 \text{ m}^3/\text{s}$

$$(d) \text{ Power input to compressor} = \frac{m_r(h_2 - h_1)}{\eta_m} = \frac{0.5364 \times (377 - 356)}{0.9}$$

$$= 12.516 \text{ kJ/s} \simeq 12.52 \text{ kW}$$

$$(e) \text{ Coefficient of performance} = \frac{\text{refrigerant effect}}{\text{work input to compressor}}$$

$$= \frac{20 \times 3.5}{12.52} = 5.59$$

(*f*) Actual air intake per cylinder $= \dfrac{\text{total piston displacement}}{\text{number of cylinders}}$

$$= \frac{0.0413}{2} = 0.02065 \text{ m}^3/\text{s}$$

$$\text{Swept volume} = \frac{\pi}{4} d^2 l \times \frac{N}{60} = \frac{\pi}{4} \times d^2 \times 1.5\, d \times \frac{900}{60}$$

$$= 17.6625\, d^3$$

Volumetric efficiency of a compressor is defined as

$$\eta_v = \frac{\text{actual air intake}}{\text{swept volume}} = \frac{0.02065}{17.6625\, d^3}$$

$$\therefore \quad d^3 = \frac{0.02065}{17.6625 \times 0.9} = 1.299 \times 10^{-3}$$

That gives : $d = (1.299 \times 10^{-3})^{1/3} = 0.1093$ m = **10.93** cm

and $l = 1.5 \times 10.93 \simeq$ **16.4** cm

EXAMPLE 11.4

A two cylinder reciprocating compressor with 5% clearance volume is used in a refrigerant plant to take a load of 8 tons of refrigeration. The compression index is 1.15, stroke to diameter ratio l/d is 0.75 and the piston speedis limited to 3 m/s. The pressure drops at suction and delivery are stated to be 0.18 bar and 0.35 bar respectively. The refrigeration plant works between 5°C evaporator temperature, 40°C condenser temperature and uses Freon-12 as refrigerant. Design the system for

(a) power required to run the compressor and coefficient of performance of the cycle

(b) volumetric efficiency, speed and dimensions of the compressor.

Solution : Refer Fig. 11.10 for *p-h* plot of the working cycle; steam dry saturated at entry to compressor and no undercooling of refrigerant in the condenser. From property tables for the refrigerant Freon-12.

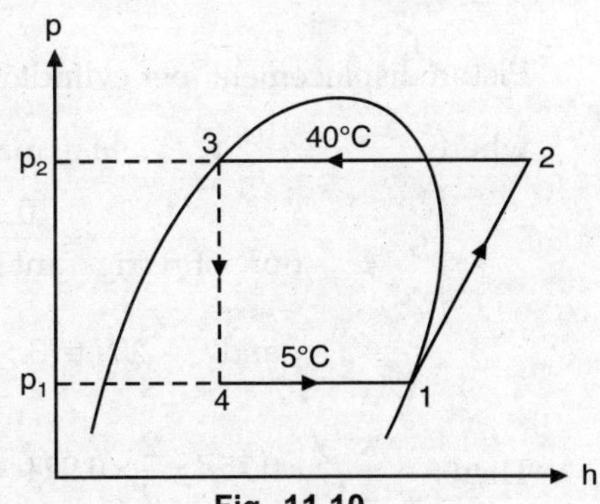

Fig. 11.10

$p_1 = 3.62$ bar and p_2 9.6 bar

Taking pressure drops at suction and delivery into account

$$\text{Suction pressure } p_s = 3.62 - 0.18 = 3.44 \text{ bar}$$

$$\text{Delivery pressure } p_d = 9.6 + 0.35 = 9.95 \text{ bar}$$

The corresponding enthalpy values obtained from the property values (by interpolation) are

$h_1 = 188.42$ kJ/kg and $h_4 = h_3$ (the process 3–4 is at constant enthlapy)

$= 75.61$ kJ/kg

Suction specific volume $v_{s1} = 0.0495 \text{ m}^3/\text{kg}$

(*a*) Total refrigeration effect = 8 tons of refrigeration

$= 8 \times 3.5 = 28$ kJ/s

Refrigeration effect per kg of refrigeration

$$= h_1 - h_4 = 188.42 - 75.61 = 112.81 \text{ kJ/kg.}$$

Then mass flow of refrigerant $m_r = \dfrac{28}{112.81} = 0.248$ kg/s

Work input to compressor per kg refrigerant

$$W = \frac{n}{n-1} p_s v_{s_1} \left[\left(\frac{p_d}{p_s} \right)^{\frac{n-1}{n}} - 1 \right]$$

$$= \frac{1.15}{1.15-1} \times (3.44 \times 10^5 \times 0.0495) \left[\left(\frac{9.95}{3.44} \right)^{\frac{1.15-1}{1.15}} - 1 \right]$$

$$= 19392.6 \text{ J/kg} \simeq 19.39 \text{ kJ/kg}$$

Power input to compressor,

$$= m_r \times W = 0.248 \times 19.39 = 4.809 \text{ kJ/s (or kW)}$$

Coefficient of performance, COP

$$= \frac{\text{refrigerant effect}}{\text{work input}} = \frac{112.81}{19.39} = 5.82$$

(*b*) Volumetric efficiency of a reciprocating compressor is given by

$$\eta_v = 1 - C \left[\left(\frac{p_d}{p_s} \right)^{1/n} - 1 \right]$$

$$= 1 - 0.05 \left[\left(\frac{9.95}{3.44} \right)^{1/1.15} - 1 \right] = 0.924 \quad \text{or} \quad 9.24\%$$

Piston displacement per cylinder $V = \dfrac{\pi}{4} d^2 l \, N \times \eta_v = m v_{s_1}$

where m = mass flow of refrigerant per cylinder

$$= \frac{0.248}{2} = 0.124 \text{ kg/s}$$

and $2lN = 3\,;\quad N = \dfrac{3}{2 \times l} = \dfrac{3}{2 \times 0.75 d} = \dfrac{2}{d}$ rev/s

Then: $\dfrac{\pi}{4} d^2 \times 0.75 d \times \dfrac{2}{d} \times 0.924 = 0.124 \times 0.0495$

$$d = \left[\frac{0.124 \times 0.0495 \times 4}{\pi \times 0.75 \times 2 \times 0.924} \right] = 0.0751 \text{ m} = 7.51 \text{ cm}$$

$$l = 0.75\, d = 0.75 \times 7.51 = 5.63 \text{ m}$$

$$\text{Compressor speed } N = \frac{2}{d} = \frac{2}{0.0751} = 26.63 \text{ rev/s}\ ;\quad 1597.87 \text{ rev/min.}$$

EXAMPLE 11.5

A single stage vapour compression system operates with condensing temperature of 35°C, evaporator temperature of –10°C and use R –134 a as refrigerant for which the relevant properties are

t °C	p bar	v_g m³/kg	h_f kJ/kg	h_g kJ/kg	s_f kJ/kg K	s_g kJ/kg K
35	8.870	—	249.1	417.6	1.168	1.715
–10	2.014	0.0994	186.7	392.4	0.951	1.733

For the single acting compressor used : swept volume = 270 cm³, ratio of clearance volume to swept volume = 0.03, index of expansion = 1.13, speed = 2500 rev/min, compression is isentropic and the vapour at suction is dry saturations.

The specific heats at the condensing pressure are 1.1 kJ/kg K and 1.46 kJ/kg K. Taking the condenser subcooling to be 5°C, compute

(a) capacity of the system in tonnes of refrigeration
(b) power utilized in running the compressor
(c) coefficient of performance of the system
(d) heat rejected to condenser and
(e) ratio of COP of the cycle to that of Carnot cycle operating between the same temperature limits.

Solution : Refer Fig. 11.11 for *p-h* and *T-s* plots of the operating cycle.

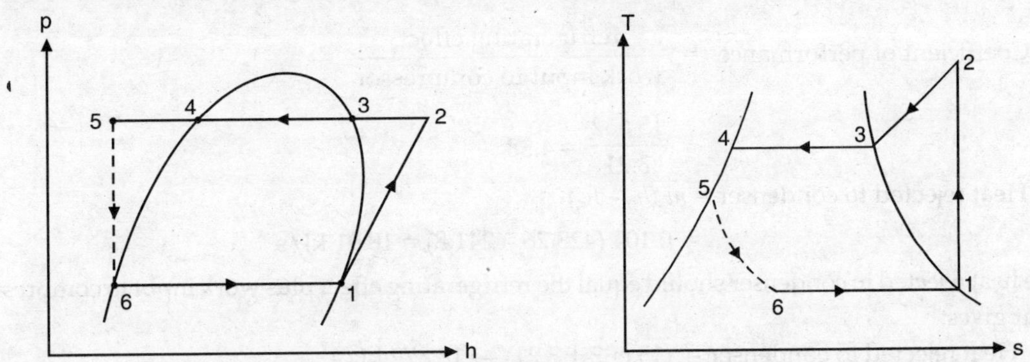

Fig. 11.11.

Given : $h_1 = 392.4$ kJ/kg , $h_3 = 417.6$ kJ/kg ; $h_4 = 249.1$ kJ/kg
$s_1 = 1.733$ kJ/kg K and $s_3 = 1.715$ kJ/kg K

The compression process 1-2 is isentropic and so $s_1 = s_2$. That is

$$s_1 = s_3 + c_p \log_e \frac{T_2}{T_3} \; ; \quad 1.733 = 1.715 + 1.1 \log_e \frac{T_2}{(35+273)}$$

Solution gives : $T_2 = 313.15 \text{ K} = 40.15\ ^\circ\text{C}$

Then : $h_2 = h_3 + c_p(T_2 - T_3) = 417.6 + 1.1(313.15 - 308) \text{ kJ/kg} = 423.26 \text{ kJ/kg}$

$h_5 = h_4 - c_p(T_4 - T_5) = 249.1 - 1.46 \times 5 = 241.8 \text{ kJ/kg}$

The throttling process 5 – 6 is at constant enthalpy and therefore

$$h_6 = h_5 = 241.8 \text{ kJ/kg}$$

$$\text{Volumetric efficiency, } \eta_v = 1 - \frac{V_c}{V_s}\left[\left(\frac{p_2}{p_1}\right)^{1/n} - 1\right]$$

$$= 1 - 0.03\left[\left(\frac{8.870}{2.014}\right)^{1/1.13} - 1\right] = 1 - 0.0814 = 0.9186$$

Then volume of refrigerant admitted to compressor at state point 1.

$$= \eta_v \times V_s = 0.9186 \times (270 \times 10^{-6}) = 248.02 \times 10^{-6}\ \text{m}^3/\text{cycle}$$

$$= (248.02 \times 10^{-6}) \times \frac{2500}{60} = 0.01033\ \text{m}^3/\text{s}$$

∴ Mass of refrigerant being circulated

$$m_r = \frac{\text{volume of refrigerant}}{\text{specific volume}} = \frac{0.01033}{0.0994} = 0.104\ \text{kg/s}$$

(*a*) Refrigerant effect = $m_r(h_1 - h_6) = 0.104\ (392.4 - 241.8)$

$= 15.662$ kJ/s = 15.662 kW

Since 3.5 kW of heat abstraction is equivalent to 1 tonne of refrigeration,

$$\text{Capacity of the system} = \frac{15.662}{3.5} = 4.475\ \text{tonnes}$$

(*b*) Power input to compressor = $m_r(h_2 - h_1)$

$= 0.104(423.26 - 392.4) = 3.21$ kJ/s = **3.21 kW**

$$\text{(c) Coefficient of performance} = \frac{\text{refrigerating effect}}{\text{work input to compressor}}$$

$$= \frac{15.662}{3.21} = \mathbf{4.88}$$

(*d*) Heat rejected to condenser = $m_r(h_2 - h_5)$

$= 0.104\ (423.26 - 241.8) = 18.91$ kJ/s

The heat rejected to condenser should equal the refrigerating effect plus work input to compressor. That gives:

Heat rejected to condenser = $(15.662 + 3.21) \rightarrow 18.872$ kJ/s

(*e*) *COP* of Carnot cycle working within the same temperature limits,

$$(COP)_{\text{Carnot}} = \frac{T_{evo}}{T_{con} - T_{evo}} = \frac{(273 - 10)}{(273 + 35) - (273 - 10)}$$

$$= \frac{263}{45} = 5.84$$

$$\text{Then:} \quad \frac{(COP)_{\text{cycle}}}{(COP)_{\text{Carnot}}} = \frac{4.88}{5.84} = \mathbf{0.8356}$$

EXAMPLE 11.6

A double acting reciprocating compressor with bore 100 mm and storke 80 mm turns 400 rev/min and has a mean effective pressure of 3.2 bar. The pressure limits for the operating cycle are 11 bar and 1.85 bar, and the temperature at inlet and exit from the condenser are 60°C and 30°C respectively. The cooling water for the condenser has a flow rate of 12 kg/min and its temperature rise is limited to 8°C. If the plant is required to produce 45 kg of ice per hour from water available at 25°C, compute,

(a) theoretical and actual COP of the system

(b) mass flow of refrigerant

The refrigerant used is ammonia and its properties are as given below :

Pressure	*Saturation temperature*	*Enthalpy*		*Entropy*
		Liquid	*Vapour*	*Vapur*
bar	°C	*kJ/kg*	*kJ/kg*	*kJ/kg K*
11.0	45	*465*	*595*	*4.75*
1.85	– 15	*405*	*568*	*4.79*

Take latent heat of ice as 335 kJ/kg and assume specific heat for liquid refrigerant as 0.96 kJ/kg K and for superheated refrigerant as 0.65 kJ/kg K.

Solution : With specific heats given for liquid and superheated refrigerant, it is good to assume that the refrigerant is superheated at entry to compressor and is subcooled at exit from the condenser. The operating cycle would then be as depicted in Fig. 11.12

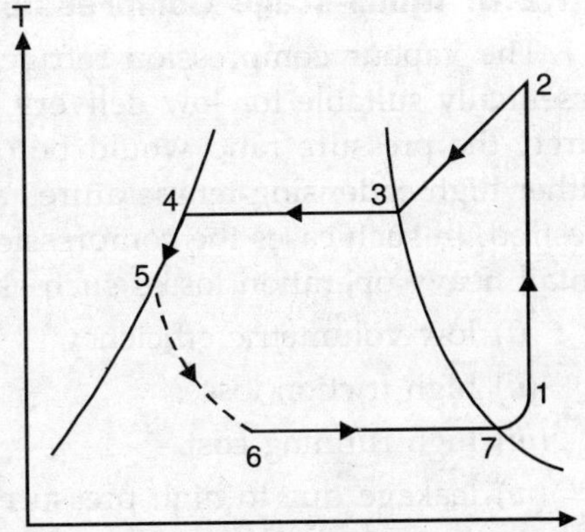

Fig. 11.12.

$$h_2 = h_3 + c_{pv}(t_2 - t_1)$$
$$= 595 + 0.96\,(60 - 45) = 609.4 \text{ kJ/kg}$$
$$h_5 = h_4 - c_{pl} \times (t_4 - t_5)$$
$$= 465 - 0.65\,(45 - 30) = 455.3 \text{ kJ/kg}$$

From isentropic compression process 1–2 , $s_1 = s_2$ and that gives

$$4.79 + 0.65 \log_e \left[\frac{T_1}{-15+273}\right] = 4.75 + 0.65 \log_e \left[\frac{60+273}{45+273}\right]$$

Solution gives : $T_1 = 265$ K or -8°C

Then $h_1 = 568 + 0.65\,[-8 - (-15)] = 572.5$ kJ/kg

(*a*) Theoretical $COP = \dfrac{\text{refrigerating effect}}{\text{work input}} = \dfrac{h_1 - h_6}{h_2 - h_1}$

It is being assumed that superheating occurs in the evaporator (space being cooled). Further, $h_6 = h_5$ as the throttling process 5–6 is at constant enthalpy. Then

$$COP = \frac{h_1 - h_5}{h_2 - h_1} = \frac{572.5 - 455.3}{609.4 - 572.5} = \mathbf{3.18}$$

Actual refrigeration effect = heat absorbed for converting water into ice

$$= \frac{45}{3600} \times [4.186 \times (25 - 0) + 335] = 5.49 \text{ kJ/s}$$

Swept volume per revolution $= 2 \times \frac{\pi}{4} d^2 l = 2 \times \frac{\pi}{4} (0.1)^2 \times 0.08 = 1.256 \times 10^{-3}$ m^3

The factor 2 accounts for the compressor to be double acting.

$$\text{Work done/s} = p_m \times \text{swep volume per second}$$

$$= 3.2 \times 10^5 \times \left(1.256 \times 10^{-3} \times \frac{400}{60}\right)$$

$$= 2679.5 \text{ J/kg} = 2.679 \text{ kJ/kg}$$

Then actual $COP = \frac{5.49}{2.679} = \mathbf{2.05}$

(*b*) Let m_r be the mass flow of refrigerant. Then

heat lost by Freon-12 = heat gained by cooling water

$$m_r (h_2 - h_5) = m_w c_{pw} \Delta t_w$$

$$\therefore \quad m_r = \frac{12 \times 4.186 \times 8}{609.4 - 455.3} = \mathbf{2.61} \text{ kg/min}$$

11.2.6. Multi-stage Compression

The vapour compression refrigeration systems employing a single stage compression are essentially suitable for low delivery pressures. In situations where delivery pressure has to be large, the pressure ratio would become high. High pressure ratios (more than 4) occur when either high codensing temperatures are required and/or very low evaporator temperatures are desired. In such cases the compression of the vapour refrigerant in a single compression would entail heavy operation losses such as :

(*i*) low volumetric efficiency

(*ii*) high friction losses

(*iii*) high running cost

(*iv*) leakage due to high pressure potential across the piston.

Moreover, the temperature of refrigerant at the end of compression becomes high and that is likely to heat up the cylinder head, and also damage the lubrication system.

To overcome these aspects, it is considered advantageous to have ***compound*** or ***multistage compression*** system where compression is carried out in more than one cylinder. The cylinders are arranged in series one after the other. The gas after compression in the first cylinder and before entering the next cylinder is cooled at constant pressure by some suitable method ; the cooling process is called ***inter-cooling***. Multistage compression with inter-stage cooling is an effective method of reducing compression work by working on isentropics close to saturation curve.

Compared to single state compression, the multistage compression provides the following main advantages :

1. Improvement in volumetric efficiency of the compressor
2. Reduction in work input to compressor. This is achieved by cooling of compressed refrigerant vapour before their compression in the succeeding stage.
3. Low operating temperatures and consequently there is little chance of lubrication problems.

4. Leakage loss is minimum due to better cylinder lubrication and reduced pressure difference across the two sides of the piston and valve.
5. When refrigerant vapour is being compressed in one cylinder, there is suction in the other cylinder. This provides more uniform torque with consequent reduction in the size of flywheel. For the same reasons, the mechanical balance is more perfect.
6. The compressing machine has a lighter construction. Further since the operating temperatures are lower, cheaper material can be used for construction. These aspects result in saving of the material cost of construction.
7. Increase in refrigerating effect and COP of the system.

The multi-stage compression has increased number of parts and the unit becomes complicated. That requires more initial investment and maintenance costs. However, this gets neutralized by considerable saving in operation cost in terms of energy consumed in running the compressor.

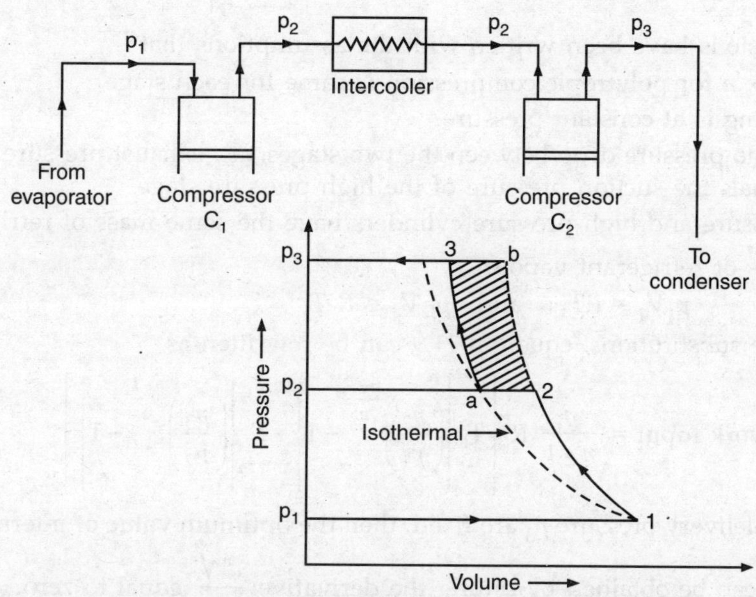

Fig. 11.13. Multi-stage compression with intercooling

Fig. 11.13 shows the schematic arrangement and *p-v* diagram of a two-stage compression with intercooler.

The refrigerant enters the low pressures cylinder at point 1 with volume V_1 and pressure p_1. The refrigerant is compressed polytropically to point 2. This compressed refrigerant at intermediate pressure p_2 and temperature T_2 is then cooled at constant pressure by passing it through the intercooler. Cooling is represented by line 2-*a*. The air then enters the high pressure cylinder where it is further compressed polytropically to pressure p_3 along the path *a*-3.

The cooling is said to be perfect if the refrigerant after compression in the low pressure cylinder is cooled to the original temperature T_1. For perfect cooling, the state points 1 and *a* would lie on the same isothermal curve. Without intercooling, the compression would have followed the path 1-2-*b*. The shaded area *a*-2-*b*-3 represents the saving in work input due to intercooling.

Work input to low pressure cylinder

$$= \frac{n}{n-1} p_1 V_1 \left[\left(\frac{p_2}{p_1} \right)^{\frac{n-1}{n}} - 1 \right]$$

Work input to high pressure cylinder

$$= \frac{n}{n-1} p_2 V_a \left[\left(\frac{p_3}{p_2} \right)^{\frac{n-1}{n}} - 1 \right]$$

∴ Total work input to compressor

$$= \frac{n}{n-1} \left[p_1 V_1 \left\{ \left(\frac{p_2}{p_1} \right)^{\frac{n-1}{n}} - 1 \right\} + p_2 V_a \left\{ \left(\frac{p_3}{p_2} \right)^{\frac{n-1}{n}} - 1 \right\} \right] \quad ...(11.9)$$

These expressions have been written with the assumptions that :

(*i*) The index n for polytropic compression is same for each stage.

(*ii*) Intercooling is at constant pressure.

(*iii*) There is no pressure drop between the two stages, *i.e.*, exhaust pressure of low pressure stage equals the suction pressure of the high pressure stage.

(*iv*) Low pressure and high pressure cylinders have the same mass of refrigerant.

For unit mass of refrigerant vapour

$$p_1 V_1 = R T_1 \quad \text{and} \quad p_i V_a = R T_a$$

Making these substitutions, equation 11.9 can be rewritten as

$$\text{Total work input} = \frac{n}{n-1} R \left[T_1 \left\{ \left(\frac{p_2}{p_1} \right)^{\frac{n-1}{n}} - 1 \right\} + T_a \left\{ \left(\frac{p_3}{p_2} \right)^{\frac{n-1}{n}} - 1 \right\} \right]$$

If p_1, T_1 and delivery pressure p_3 are fixed, then the optimum value of intermediate pressure p_2 for minimum can be obtained by setting the derivative $\frac{dW}{dp_2}$ equal to zero.

That would give

$$\frac{p_2}{p_1} = \sqrt{\left(\frac{T_a}{T_1} \right)^{\frac{n}{n-1}} \frac{p_3}{p_1}}$$

When intercooling is perfect : $T_a = T_1$ and so the above identity reduces to

$$\frac{p_2}{p_1} = \sqrt{\frac{p_3}{p_1}} \quad \text{or} \quad p_2 = \sqrt{p_1 p_3} \quad ...(11.11)$$

The above correlation suggests that for minimum work input with perfect intercooling, the intermediate pressure p_2 for a two stage compressor is the geometric mean of the supply pressure p_1 and the delivery pressure p_3.

Equation 11.11 can be rewritten as

$$\frac{p_2}{p_1} = \frac{p_3}{p_2} = \sqrt{\frac{p_3}{p_1}} \quad ...(11.12)$$

Apparently for a two stage compressor with perfect intercooling, the pressure ratios for each stage are equal and are given by square root of the overall pressure ratio.

Further, with perfect intercooling $T_a = T_1$ and therefore the expression 11.10 for work input can be rewritten as

$$\text{Work input} = \frac{n}{n-1} RT_1 \left[\left(\frac{p_2}{p_1}\right)^{\frac{n-1}{n}} + \left(\frac{p_3}{p_2}\right)^{\frac{n-1}{n}} - 2 \right]$$

The expression for minimum work input can be determined by substituting $\frac{p_2}{p_1} = \frac{p_3}{p_2} = \sqrt{\frac{p_3}{p_1}}$ in the above identity.

$$\text{Minimum work input} = \frac{n}{n-1} RT_1 \left[\left(\frac{p_3}{p_1}\right)^{\frac{n-1}{2n}} + \left(\frac{p_3}{p_1}\right)^{\frac{n-1}{2n}} - 2 \right]$$

$$= \frac{2n}{n-1} RT_1 \left[\left(\frac{p_3}{p_1}\right)^{\frac{n-1}{2n}} - 1 \right] \quad ...(11.13)$$

In N-stage compressor with perfect intercooling,

$$\frac{p_2}{p_1} = \frac{p_3}{p_2} = \frac{p_4}{p_3} .... \left(\frac{p_{N+1}}{p_1}\right)^{\frac{1}{N}}$$

and

$$W_{min} = \frac{Nn}{n-1} RT_1 \left[\left(\frac{p_{N+1}}{p_1}\right)^{\frac{n-1}{Nn}} - 1 \right] \quad ...(11.14)$$

where p_{N+1} si the final delivery pressure and p_1 is the inlet suction pressure.

11.2.7 Performance Characteristics

The performance parameters of a reciprocating compressor are :

- Refrigeration capacity which indicates whether the system is capable or not to meet the cooing requirements of a given space. The refrigeration capacity is the product of mass flow of refrigerant and the heat absorbed in the evaporator (*i.e.* refrigerating effect).
- Power input in kW; this parameter decides whether the prime-mover is capable to take the load or not. The power required by an ideal compressor is given by

  $$P = m_r \times W_{isen}$$

  where P is power (kW), m_r is the refrigerant mass flow rate (kg/s) and W_{isen} is the isentropic compressor work (kJ/kg).
- kW per tonne of refrigeration which indicates whether the system is operating economically or not.

The refrigerant reciprocating compressors are designed for prescribed values of evaporator (or suction) and condenser temperature. The effect of these design variables on the compressor performance parameters has been depicted in Fig. 11.14.

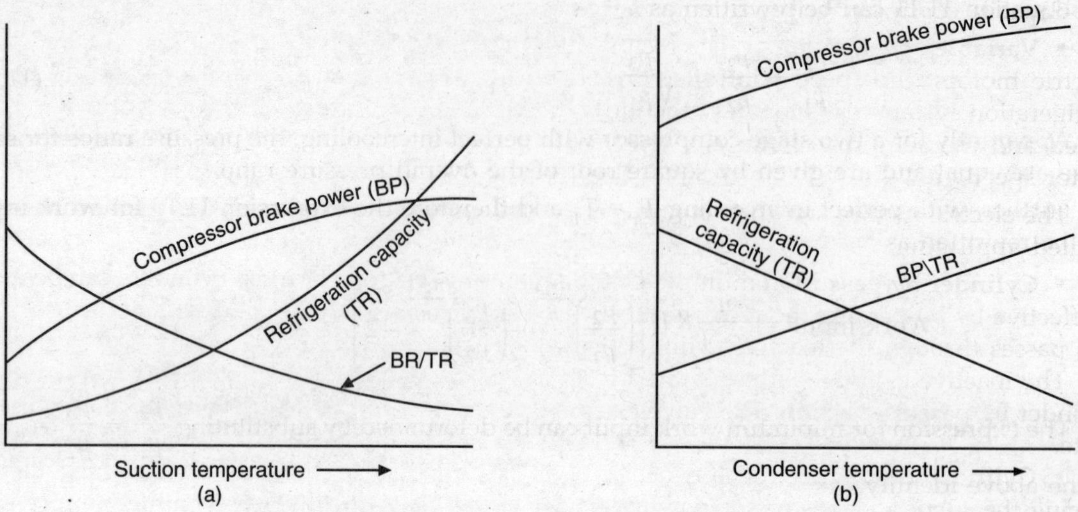

Fig. 11.14. Effect of suction and condenser temperature

These plots are typical of refrigerant compressors and show the general trend; numerical values have not been assigned because of the fact that curves would vary for different refrigerants and compressor configurations.

The notable information as exhibited by these plots is :

(*i*) The refrigerating capacity increases with an increase in the evaporator or suction temperature. This stems from the fact that evaporator temperature affects the refrigerant density. Higher the evaporating temperature, higher will be vaporizing pressure as well as the density of the suction vapour. Hence, the corresponding mass flow rate would be higher and that would contribute towards increased refrigerating capacity.

(*ii*) The power required to run the compressor increases with rise in evaporator temperature

$$P = m_r \times W_{isen}$$

The compressor work is less and the mass flow of refrigerant is more at high evaporating temperature. The increase in the mass of refrigerant circulated predominates over work of compression. Accordingly the net effect is to increase the power input to compressor with the increase in suction temperature.

(*iii*) Brake power per tonne of refrigeration

$$= \frac{\text{power input to compressor}}{\text{refrigerating capacity}}$$

With increase in suction temperature, there is increase in both the power input to compressor and the refrigerating capacity. It has been observed that the effect of increased refrigerating capacity is more pronounced and accordingly kW/ton of refrigeration decreases with increase in evaporator temperature.

(*iv*) With increasing condenser temperature (and keeping evaporator temperature constant), the refrigeration capacity decreases and there occurs an increases both in the compressor power and the kW/ton of refrigeration.

11.2.8. Capacity Control

There are many refrigeration applications in which the refrigeration load is not constant. It is then necessary for both economy and operation that the system must have some means for the capacity control.

Capacity control in reciprocating compressors is generally achieved by the following methods:

• **Variable speed motors :** The reciprocating compressors may be driven by variable speed electric motors, the speed controlled by electrical relays. When the suction pressure of the refrigeration system reads low (*i.e.*, the refrigeration load has decreased), the control is set at low speed and the compressor capacity decreases. This is because the capacity is proportional to motor speed.

The electric motors and their controls are quite expensive and as such this method has only limited application.

• **Cylinder by-pass :** On multiple-cylinder compressors, one or more cylinders can be made ineffective by by-passing the refrigerant gas from the discharge to the intake of compressor. The gas passes through the inactive cylinders but is not compressed.

The inactive cylinders are separated from the active ones by installing a check valve in the cylinder by-pass connection. The valve has automatic operation by either thermostat or pressure state.

• **Cylinder unloading :** The capacity of multi-cylinder compressors can be controlled by forcing the suction valves of some cylinders to remain open and thus preventing compression. With the suction valve in open position, the piston draws refrigerant gas from the suction manifold on the outward stroke. When the piston moves inward, it returns the gas to the suction line without compressing it. The suction valves remain open according to load requirement.

• **On and off :** The simple on-off or start-stop control in conjunction with a thermostat is used effectively in small unitary equipments such as refrigerators, air-conditioners and water coolers etc.

The method essentially finds favour where there is sudden large demand followed by periods of small or no demand on refrigeration capacity.

• **Multiple units :** The system employs multiple units of same capacity with common suction and discharge valves. The operation of all units provides the maximum desired capacity and operation of various combinations permits efficient capacity reduction.

Some units can also serve the purpose of stand-by equipment. That allows partial load operation in case some unit breaks down.

• **Hot gas by-pass :** The method involves putting an artificial load on the compressor and this is done by passing the hot gas from the discharge side to the suction side through a constant pressure expansion valve. When a capacity reduction is needed, the valve opens and permits some of the hot gas discharge by the compressor to be returned to the suction line.

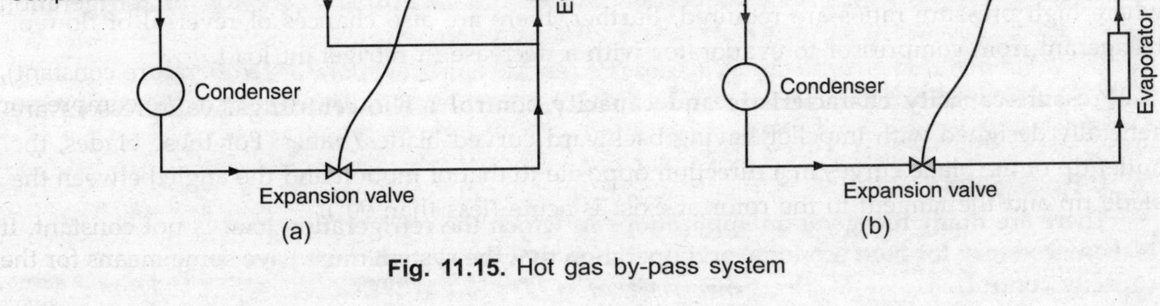

Fig. 11.15. Hot gas by-pass system

(*i*) The hot gas by-pass may be at entrance to or at exit from the evaporator.

(*ii*) The power input to compressor remains constant irrespective of load on the evaporator.

(*iii*) The method is least preferred for capacity control as it only loads the system instead of unloading. However, the method prevents the evaporator surface from frosting up and avoids the compressor working at excessive back pressure.

11.2.9. Centrifugal Compressors

Centrifugal compresors are dynamic units; they use centrifugal force to increase the pressure of low pressure refrigerant.

The centrifugal compressor unit essentially consists of an inlet section, a rotate impeller, a vaneless diffuser and a volute casing. The impeller is a disc to which are fitted a number of radial or approximately radial vanes. When the impeller rotates, the refrigerant vapour is drawn in form the evaporator and pushed from the centre of impeller to its periphery by centrifugal force. The rotating impeller imparts a high velocity to the refrigerant vapour and thus increases its kinetic energy. Subsequently the kinetic energy gained by the vapour is converted to static pressure in the diffuser. The volute casing (an expanding section) collects the refrigerant from the diffuser, further increases the pressure and makes the compressed vapour leave through the outlet.

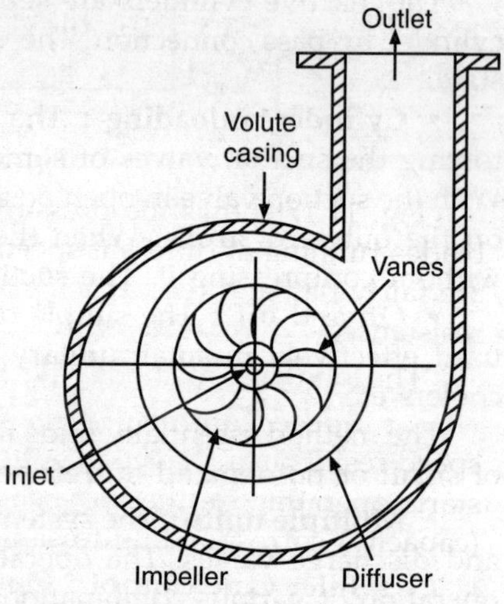

Fig. 11.16. Centrifugal compressor

Main advantages of a centrifugal compressor are :

(*i*) high speed running and as such can be directly coupled with the prime-movers (electric motor or steam turbine)

(*ii*) a continuous supply of high pressure stream of refrigerant is maintained to the condenser

(*iii*) simple lubrication system as there are no reciprocating parts

(*iv*) low maintenance cost because there are no piston rings or valves to wear out

(*v*) operation is free from noise and vibrations

(*vi*) compactness ; need small space even fro large capacity refrigeration units

(*vii*) can handle large volume flow of refrigerants

(*viii*) ease of capacity control and part load efficiency

However, the centrifugal compressors need a large number of stages to deliver the refrigerant where high pressure ratios are required. Further, there are also chances of reversal of flow of refrigerant from compressor to evaporator with a decrease in refrigerant load.

Pressure-capacity characteristic and capacity control : The centrifugal compressors are generally designed with impeller having backward curved blades/vanes. For these blades, the outlet tip of the blade curves in a direction opposite to that of motion and the angle between the blade tip and the tangent to the rotor at exist is acute (less than 90°).

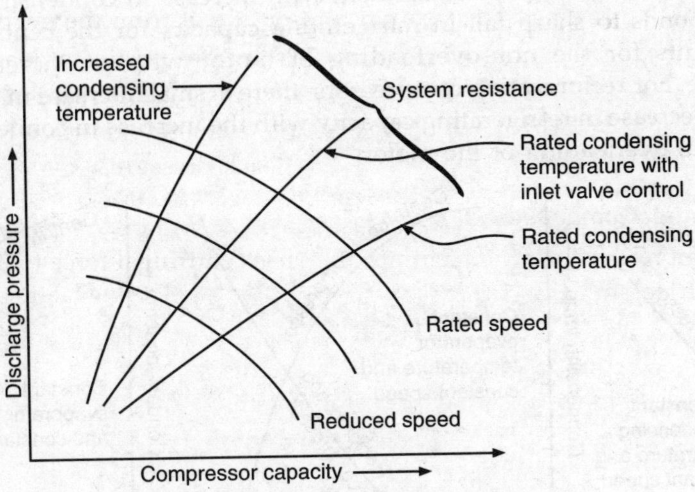

Fig. 11.17. Pressure-capacity characteristic

Fig. 11.17 shows the pressure-capacity characteristic for an impeller with backward curved blades running at different speeds and the different system resistances. The characteristic curves are fairly flat, and the capacity of the compressor is seen to decrease with an increase in system resistance.

The following methods have been suggested for the capacity control of centrifugal compressors:

- **Speed control :** A change in the velocity of flow caused by variation in compressor speed results in a change in system resistance. When the compressor speed decreases, the system starts operating on the low head-capacity characteristic. That provides a low volume flow rate (capacity) corresponding to same pressure ratio.
- **Inlet vane control :** Quart often, the capacity is varied by changing the angle at which the refrigerant vapour enters the impeller. The vapour than enters with pre-rotation and that results in a decrease in flow.
- **Condenser water control :** When the quantity of cooling water passing through the condenser is reduced, the condensation rate decreases and that gives rise to increase in condenser pressure and temperature. This aspect forces the compressor to self-adjust to the part load capacity.

Note:

(*i*) The inlet vane control is cheaper than speed control and condenser water control is cheapest amongst all.

(*ii*) The condenser water control is least efficient and the speed control is most efficient amongst all.

11.2.10. Comparison of Performance of Reciprocating and Centrifugal Compressors

Fig. 11.16 depicts the general trend in the difference between the performance of reciprocating and centrifugal compressors :

The following variations are evident from these plots :

(*i*) The centrifugal compressor has a flat head-capacity characteristic compared to reciprocating compressor. For the same change in cooling load, the variation in evaporative temperature for centrifugal compressor is very small as compared to that for reciprocating compressor.

(*ii*) There is reduction in power requirement with increase in condensing temperature and this corresponds to sharp fall in refrigerating capacity for the centrifugal compressor. This accounts for the non-overloading of motor with an increase in condensing temperature. For reciprocating compressors, there is small increase in power requirement and small decrease in refrigerating capacity with the increase in condensing temperature. This leads to overloading of the motor.

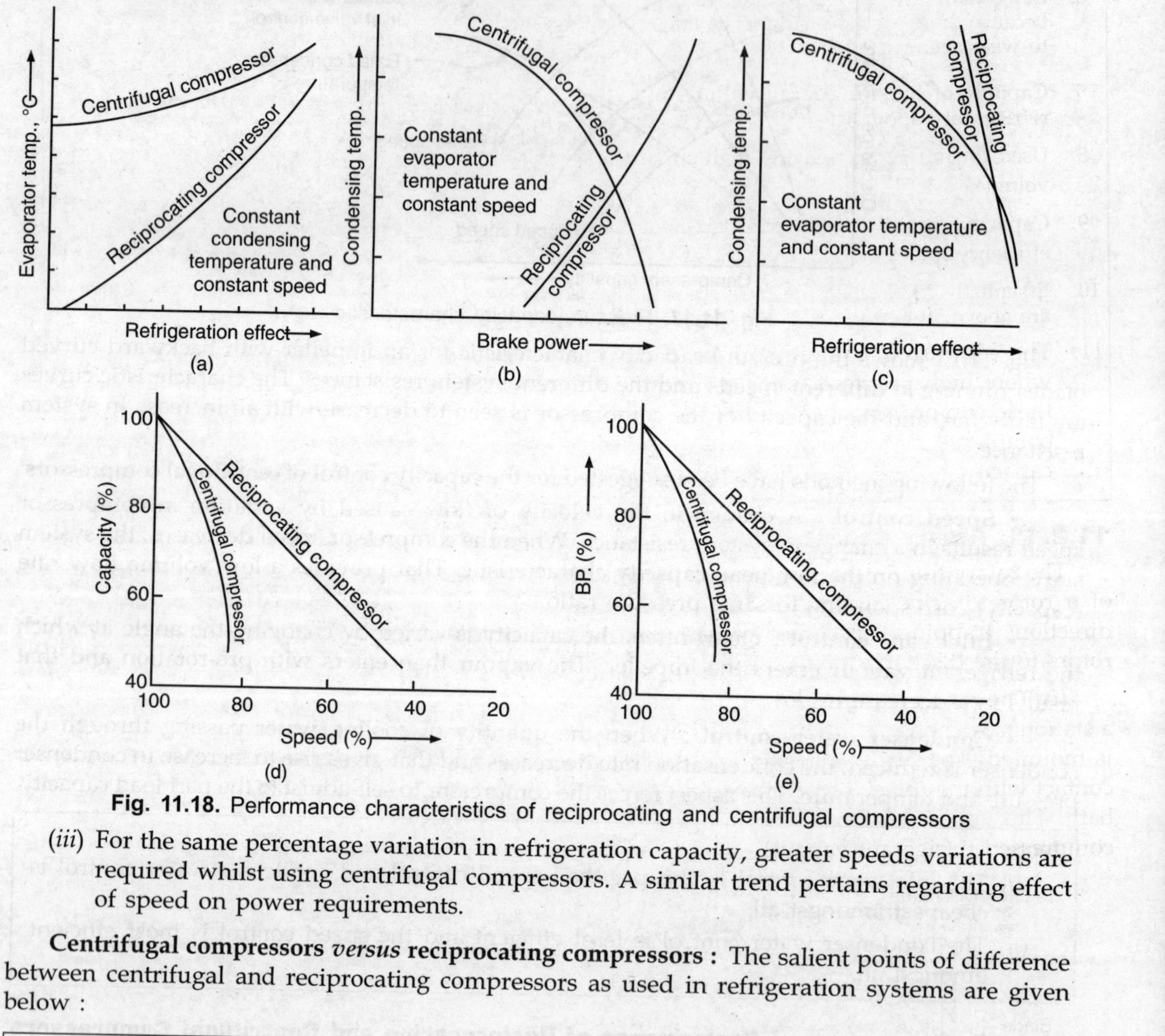

Fig. 11.18. Performance characteristics of reciprocating and centrifugal compressors

(*iii*) For the same percentage variation in refrigeration capacity, greater speeds variations are required whilst using centrifugal compressors. A similar trend pertains regarding effect of speed on power requirements.

Centrifugal compressors *versus* reciprocating compressors : The salient points of difference between centrifugal and reciprocating compressors as used in refrigeration systems are given below :

Centrifugal compressors	*Reciprocating compressors*
1. Pressure increase is due to dynamic action; centrifugal force is imparted to increase the pressure of low pressure refrigerant.	1. Pressure changes are obtained by co-ordinated action of pistons and valves; reciprocating motion of a piston compresses the refrigerant.
2. System maintains a continuous supply of high pressure stream of refrigerant.	2. System provides an intermittent output of compressed vapour.

3. Operation is free from noise and vibrations.	3. High vibrations.
4. High speed running unit.	4. Low speeds due to heavy reciprocating parts.
5. Compactness; need less floor area.	5. Bulky ; need large floor area and heavy foundations.
6. Longer life and low maintenance cost because there are no piston rings and valves to wear act.	6. High wear and tear and so shorter life.
7. Capable of handling large volume of refrigerant vapour.	7. Volume of refrigerant handled is low.
8. Used for refrigerants having high specific volume.	8. Used for refrigerant with low specific volume.
9. Capacity control is easy and part load efficiency is high.	9. System has a poor efficiency at partial loads.
10. Essentially used when the refrigerant loads are above 50 tons	10. Use is restricted to refrigeration loads below 50 tons.
11. The centrifugal compressors need a large number of stages to deliver the referigerant where high pressure ratios are required (increase in pressure per stage is less). Further, these compressors have a surging problem; there is reversal of flow of refrigerant from compresssor to evaporator where refrigeration load decreases to about 35% of the rated capacity.	

11.2.11. Rotary Compressors

The rotary compressors operate on the positive displacement principle, and use rotary motion of a rotor to achieve compression of the refrigerant. Two meshing rotors rotate in opposite directions trapping refrigerant vapour, and reducing the volume of the refrigerant along the rotors to the discharge point.

***(i)* The rotary compressor with single fixed blade and eccentric rotor** essentially consists of a stationary cylinder, cylindrical roller and a shaft. The shaft has an eccentric on which the roller is mounted. The blade is spring loaded, is mounted in the slot of cylinder and always makes contact with the roller. The entire unit is placed in an air tight housing and submerged in the oil bath. The upper portion of the housing is, however, kept free from oil to accommodate the compressed refrigerant vapour.

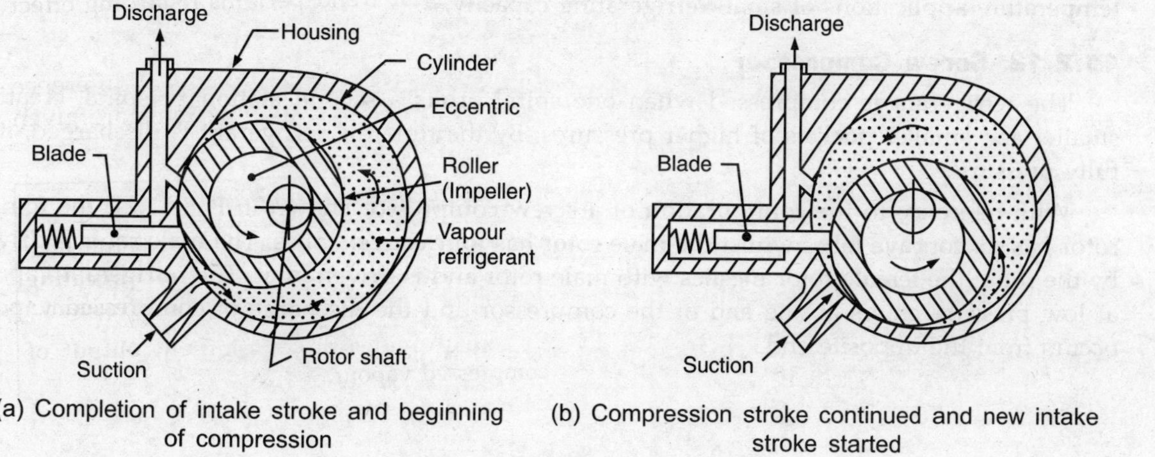

(a) Completion of intake stroke and beginning of compression

(b) Compression stroke continued and new intake stroke started

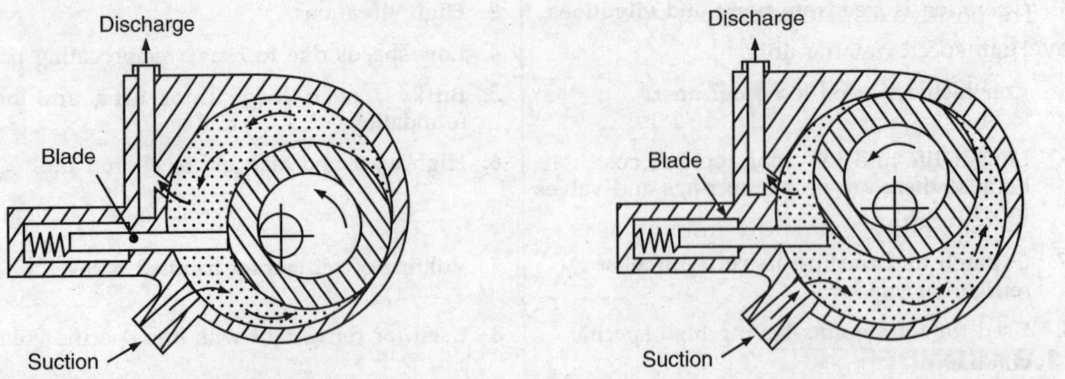

(c) Compression continued and new intake stroke continued

(d) Compressed vapour discharged to condenser and new intake stroke continued

Fig. 11.19. Stationary single blade rotary compressor

Fig. 11.19 shows the various positions of roller as the vapour refrigerant is compressed. When the motor shaft revolves the rollers, one of its point makes continuous contact with the cylinder wall. The refrigerant enters the open space from the lower side of the blade and is finally pushed through the discharge valve located at the upper side of the blade.

***(ii)* The rotating vane/blade type** rotary compressor consists of a cylinder and a slotted rotor containing a number of vanes. The rotor is arranged co-axially with the motor shaft which is eccentric with the cylinder. The vanes move in and out of slots in the rotor to make sliding contact with the cylinder. Essentially, the vanes are forced against the cylinder wall by the centrifugal action during the motor rotation. The vapours are compressed as the vanes rotate from the point of maximum rotor clearance to minimum rotor clearance.

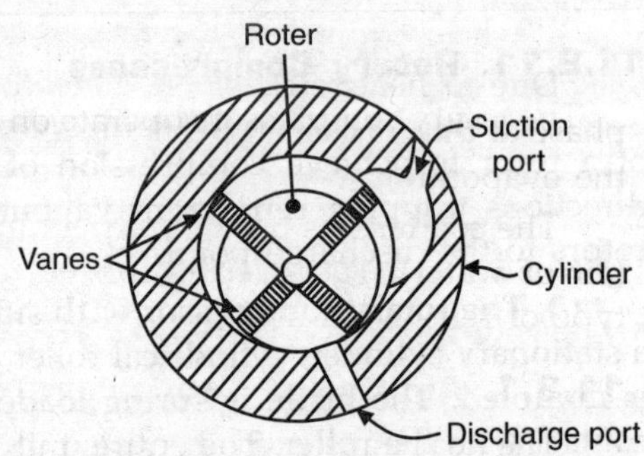

Fig. 11.20. Rotating blade type rotary compressor

The rotary compressors are silent in operation and are free from vibrations. These compressors are quite suitable to handle refrigerants with low specific volume at low suction pressure. Further, they are used for low temperature applications of small refrigerating capacity.

11.2.12. Screw Compressor

The refrigerant is compressed when one spiral orbits a second stationary spiral, creating smaller and smaller pockets of higher pressures. By the time the refrigerant is discharged, it is fully pressurized.

With reference to the configuration of a screw compressor shown in Fig. 11.21, the female rotor has six concave lobes whilst the male rotor has four convex lobes. The male rotor is driven by the rotor; the female rotor meshes with male rotor and is driven by it. The refrigerant vapour at low pressure enters at one end of the compressor and the discharge of compressed vapour occurs from the opposite end.

The positive displacement com-pressors have gained popularity due to

– high volumetric efficiency
– less wear and tear due to few rubbing parts
– less vibrations
– capability to handle ammonia as well as halocarbon refrigerants.

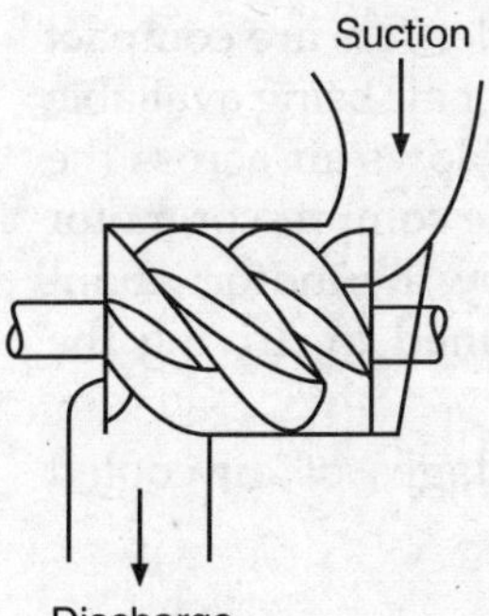

Fig. 11.21. Screw compressor

11.3. CONDENSERS

The condenser has been defined as

– a heat exchanger in which the refrigerant compressed to a hot gas is condensed to liquid by rejecting heat.
– component of a refrigeration system where the refrigerant gives up heat and changes from vapour to liquid state.

The condenser is essentially a heat exchange device in which the refrigerant vapour changes to liquid phase by transferring its heat to a cooling medium flowing through it. The heat rejected consists of

- heat picked up by the refrigerant from the space/product being cooled, and
- heat equivalent to work of compression in driving the compressor unit.

Due to heat rejection in the condenser, the refrigerant vapour is converted into the liquid phase at the condenser pressure, and the liquid refrigerant is made available for re-expansion in the evaporator.

The condensers used in refrigerant system are classified on the basis of cooling medium used (air or water). The selection of the condenser depends on the capacity of refrigeration system, type of refrigerant and the cooling medium available.

11.3.1. Air Cooled Condensers

The heat transfer in air-cooled condensers is caused by air circulation around their surface; the air circulation may be caused by free convection (caused by buoyancy) or by forced convection (caused by the action of fan element). These condensers are made of steel, copper and aluminium tubings with fins attached and that increases the effective area for heat dissipation. The fins are made from aluminium because of its light weight, and the fin spacing is quite wide to reduce dust clogging.

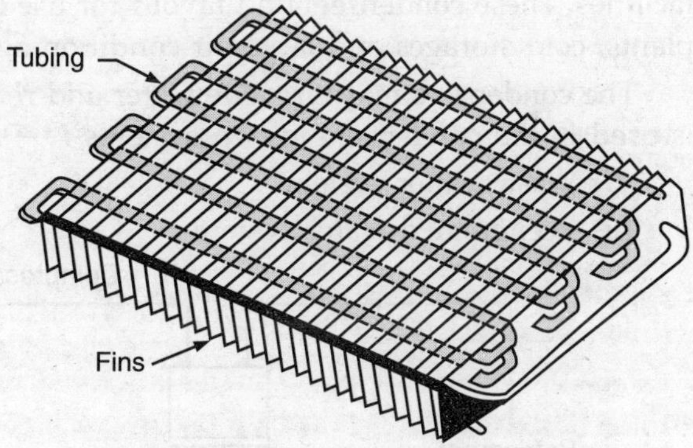

(a) Air condenser having tubing provided with fins

In natural convection condensers, the surrounding air picks up heat and becomes warmer. The warm air (being lighter) rises up and the cool air flows to take its place. These condensers have a limited cooing capacity and as such their use is restricted only to domestic refrigerators, freezer cabinets and room air-conditioners.

The forced convection condensers are compact in design due to more quantity of air being available for circulation. The fan which blows air across the condenser coil is mounted on the compressor motor pulley and is directly driven by the motor; about 25% of motor power is consumed in driving the fan.

The advantages/disadvantages of air-cooled condensers are :

- simplicity of design
- low installation and maintenance cost
- no requirement for any piping arrangement for flow of air
- hot air can be easily disposed off
- less fouling effect
- poor heat carrying capacity
- requirement of high volume of air flow
- variable cooling capacity due to variation in atmospheric temperature.

There is also the possibility of dirt, dust and other foreign material lying on floor being picked up and carried to the condenser by moving air.

The air-cooled condensers are generally designed for condensing temperatures of 15 to 20°C above the atmospheric temperature.

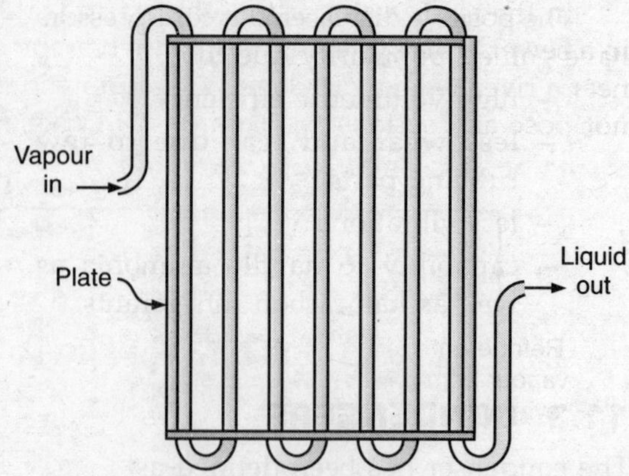

(b) Plate and tube air condender

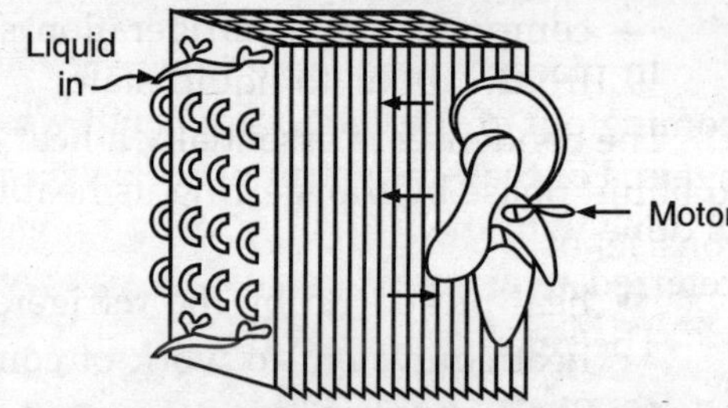

(c) Forced circulation type

Fig. 11.22. Air cooled condensers

11.3.2. Water Cooled Condensers

Water is used as the cooling medium in water cooled condensers, and these condensers operate with condensing temperature 5 to 15°C above the temperature of entering water. They are always preferred where adequate supply of cooling water is available along with disposal facilities. These condensers find favour for use in large capacity refrigeration systems such as ice plants, cold storages and large air-conditioning installations.

The condensers may circulate water and reuse it or discharge it to a sewer. Obviously water is used in the condensers in two systems :

(*i*) Waste water system

(*ii*) Recirculated water system

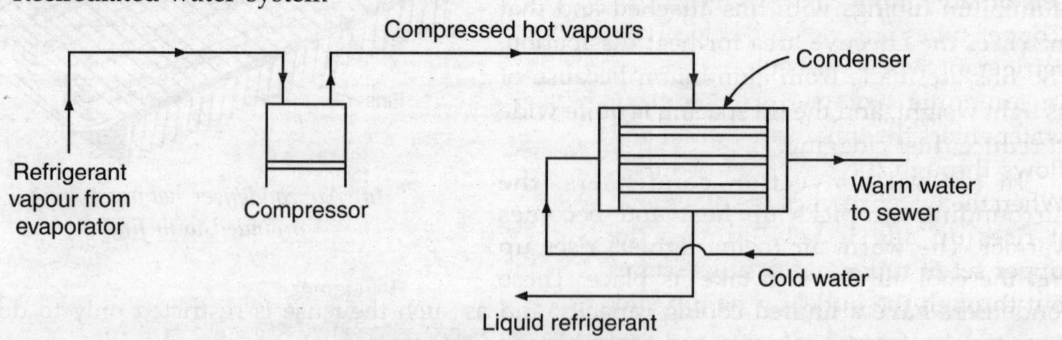

Fig. 11.23. Waste water system

In the *waste water system*, the warm water which comes out of the condenser is disposed of in a sewer. This system is adopted when the refrigeration and air-conditioning plants are located near a river or lake or when water supply is cheaply available in abundance and its disposal does not pose any serious problem.

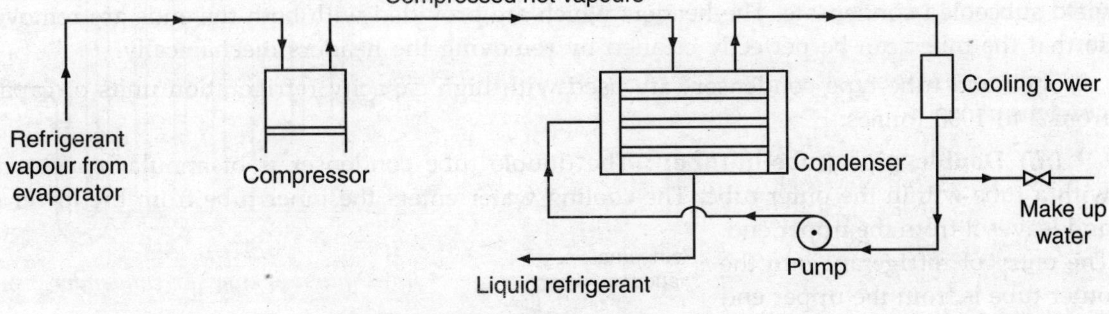

Fig. 11.24. Recirculating water system

In places where there is scarcity of clean water, the warm water coming out of the condenser is not wasted but is circulated again and again. For that, the warm water needs to be cooled every time and this is done with the help of a cooling tower or pond. The arrangement is referred to as *recirculating water system.*

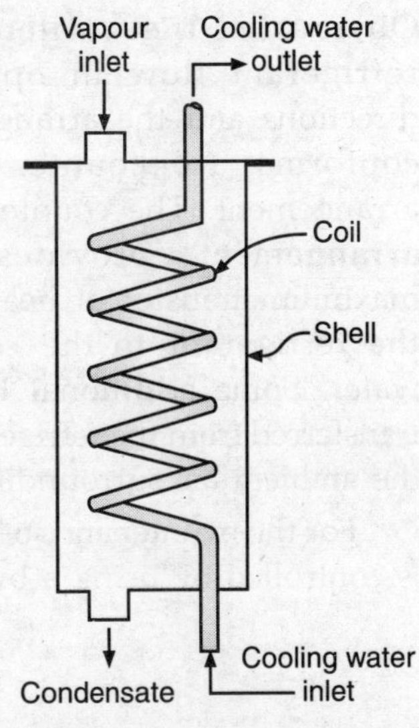

Fig. 11.25. Shell and coil type

The main types of water cooled condensers are :

***(i)* Shell and coil type :** The unit consists of a welded steel shell with a continuous copper coil inside. The cooling water passes through the coil; entering at the lower end and leaving from the upper end. The refrigerant vapour enters near the top of shell and while travelling downward, it comes in contact with the comparatively cooled surface of the coil. The refrigerant loses heat, gets condensed and collects at the bottom of shell.

For satisfactory operation, the water used has to be clean because cleaning of the inside surface of the coil is quite difficult. Such condensers have been used with refrigeration systems upto 50 tonne capacity

***(ii)* Shell and tube type :** The unit comprises a cylindrical shell and two sets of large number of parallel tubes. Copper tubes are used for halocarbon refrigerant while steel tubes are used in ammonia systems. The cooling water enters the unit from bottom and flows through the lower set of tubes. When the water reaches the other end, it rises up and travels through the upper set of tubes and finally escapes out through the outlet at the top end.

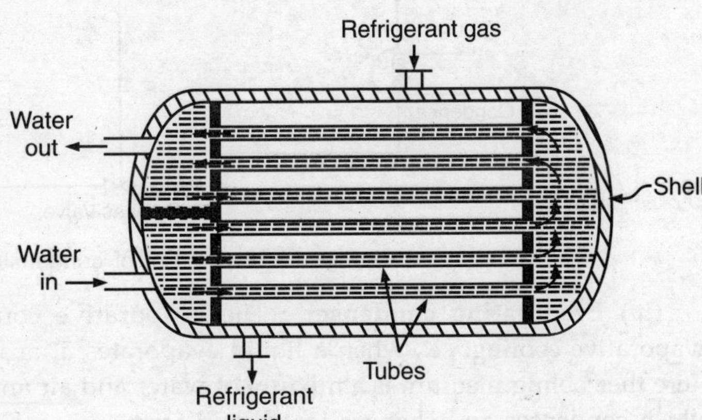

Fig. 11.26. Shell and tube type

The travel of water is thus almost two times the length of condenser shell. The entry of refrigerant is from the top and its exit is from the bottom. While flowing downwards, it loses its heat to the comparatively cool outside surface of the water tubes and gets condensed.

Copper tubes are used for halocarbon refrigerants while steel tubes are used in ammonia systems. In such condensers, the upper tubes act as desuperheaters while the tubes at the bottom yield subcooled condensate. The headers which are provided with both the ends are removable so that the tubes can be perfectly cleaned by removing the headers mechanically.

Shell and tube type condensers are used with high capacity refrigeration units of capacity from 2 to 1000 tonnes.

***(iii)* Double pipe (tube-in-tube)** : The double tube condenser is of annular construction with a tube within the outer tube. The cooling water enters the inner tube from the lower end and leaves it from the upper end. The entry of refrigerant into the outer tube is from the upper end and exit is from the lower end. Obviously the water and refrigerant flow in opposite directions and the arrangement conforms to counter flow arrangement. The counter flow arrangement provides the maximum transfer of heat from the refrigerant to the cooling water. Some additional heat is transferred from the refrigerant to the ambient air surrounding the condenser, resulting in a relatively compact size.

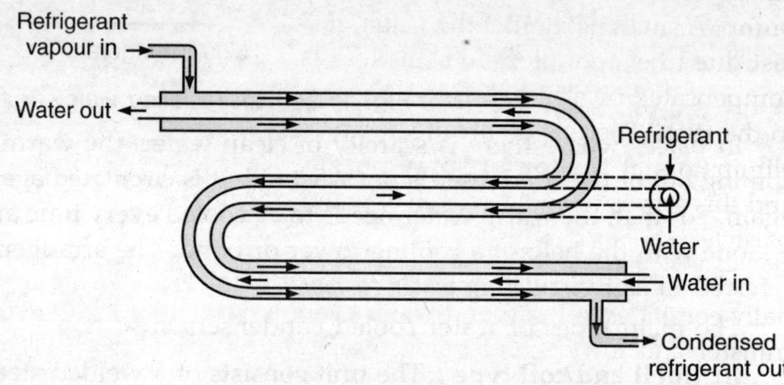

Fig. 11.27. Double pipe condenser

For the maintenance of condenser pressure in water cooled condensers, the water temperature is controlled by using a by-pass valve between the condenser discharge and inlet. (Fig. 11.28)

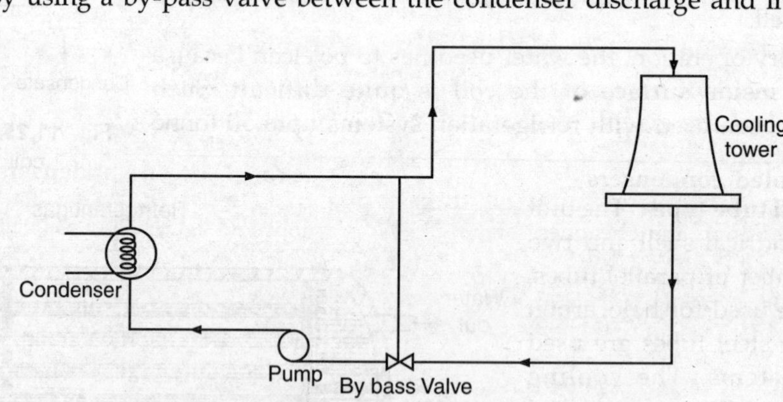

Fig. 11.28. Control of condenser pressure

***(iv)* Evaporating condenser** : The evaporative condenser operates on the principle of evaporative cooling, *i.e.*, when a liquid evaporates, it absorbs heat equivalent to its latent heat. Here the cooling medium is a mixture of water and air and heat is absorbed from the refrigerant which condenses and changes into liquid form.

The air is drawn from the opening provided near the bottom with the air flow caused by a fan installed at the top of the unit. While moving upwards, the air crosses the refrigerant coils over which water is being sprayed. The refrigerant enters the coil from its upper end, loses heat to the air and water, condenses and the condensate is taken out from the lower end of the coil. The water collects at the bottom in a tank and is recirculated with the help of a pump. A small portion of the water gets lost due to evaporation and this loss is compensated by adding make-up water to the tank. There is an arrangement of eliminators at the top of spray nozzle, and this serves to avoid escape of water particles and mist with air.

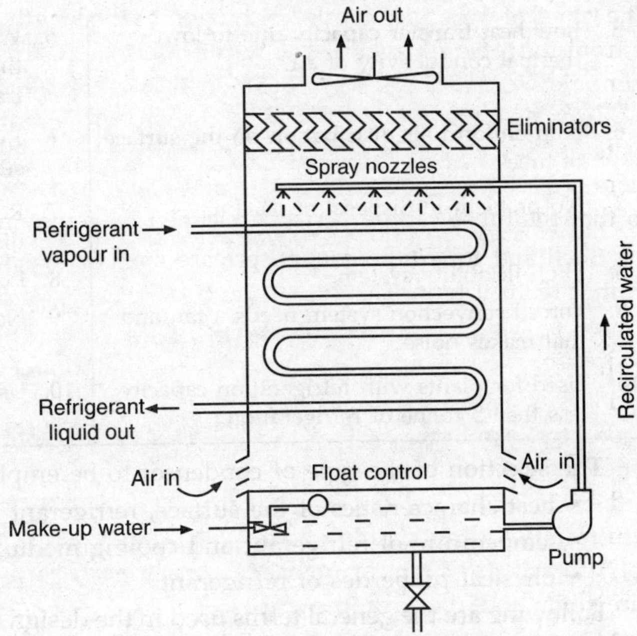

Fig. 11.29. Evaporative condenser

The evaporative condenser essentially combines both air and water cooling with advantages over air cooling both in terms of heat transfer and lower condensing temperature. In comparison with water cooling condenser, the evaporative condenser has much less water consumption. There is no need to dispose water and there is saving in space required for installation.

The evaporative condensers are used where supply of cooling water is limited and the construction of cooling towers and spray ponds is uneconomical. Evaporative condenser is essentially a water conservative device ; here a condenser and a cooling tower are combined together in a single unit. Large ammonia plants are generally fitted with evaporative condensers.

11.3.3. Water Cooled *versus* Air Cooled Condensers

The salient aspects of difference between air cooled and water cooled condensers as used in refrigeration systems are given below :

Air cooled condensers	Water cooled condensers
1. Simplicity of design and cheapness of construction.	1. Complicated construction and high initial cost.
2. Cost of repair, servicing and maintenance is quite low.	2. High cost of repair, servicing and maintenance.
3. No piping arrangement is required for carrying the air.	3. Pipes are required to take water to and from the condenser.
4. System is free from corrosion and therefore fouling effect is low.	4. Corrosion does occur on the inside surface of tubes carrying water and therefore the system suffers from fouling effects.

5. Low heat transfer capacity due to low thermal conductivity of air.	5. Water has high thermal conductivity and therefore the system operates with high heat transfer rates.
6. Non-uniform air distribution on the surface of tubes.	6. Even distribution of water on the condensing surface.
7. No problem in disposal of used air.	7. Recirculation system has to be adopted for the disposal of used water.
8. High flexibility system.	8. Low flexibility system.
9. Forced convection system needs a fan and that makes noise.	9. No fan and no noise.
10. Used for plants with refrigeration capacity less than 5 tonne of refrigeration.	10. Used for large capacity plants.

The selection of the type of condenser to be employed depends on :

- heat characteristics of the surface, refrigerant and cooling medium
- temperature of refrigerant and cooling medium
- physical properties of refrigerant.

Following are the general terms used in the design and application of condensers :

1. Heat rejection factor (HRF) : The function of condenser in a refrigeration system is to reject the heat absorbed at the evaporator (refrigerating effect) and the heat of compression that is added to the system by the compressor. Obviously then load on the condenser is given by

$$Q_c = \text{refrigerating effect } (N) + \text{compressor work } (W)$$

The term *heat rejection factor* (*HRF*) for the condenser is defined as the ratio of heat rejected at the condenser to the refrigerating effect. That is

$$HRF = \frac{Q_c}{N} = \frac{N+W}{N} = 1 + \frac{W}{N} = 1 + COP$$

It is worth while to comment that

(*i*) *HRF* depends upon *COP* which in turn is dependent upon the condenser and evaporator temperatures.

(*ii*) There is increase in heat load on the condenser with increase in heat of compression. Heat of compression is the heat that develops from work input to the compressor needed for compressing the suction gas. Work input is more with increased compression ratio.

(*iii*) *HRF* is about 1.25 in air-conditioner equipment which operates at condenser temperature of 40°C, evaporator of 5°C with *R*-12 and *R*-22 as refrigerants.

2. Fouling factor and rate of heat exchange : The water used for cooling purposes in a water cooled condenser contains a certain amount of minerals and other foreign matter. These materials form deposits (scale) inside the condenser tubes. The phenomenon of deposition of fluid impurities is called *fouling*. The surface deposits increase thermal resistance with a corresponding drop in the performance of the heat exchange equipment. Since the thickness and thermal conductivity of the scale deposits are difficult to ascertain, the effect of scale on heat flow is considered by specifying an equivalent scale heat transfer coefficient h_s. If h_{si} and h_{so} denote the heat transfer coefficients for the scale formed on the inside and outside surfaces respectively, then the thermal resistances due to scale formed on the inside and outside surfaces are

$$R_{si} = \frac{1}{A_i h_{si}} \text{ and } R_{so} = \frac{1}{A_o h_{so}}$$

Then for a tube wall of outside and inside radii r_o and r_i respectively, the overall resistance is

$$R = \frac{1}{A_i h_i} + \frac{1}{A_i h_{si}} + \frac{1}{2\pi kl}\log_e\frac{r_o}{r_i} + \frac{1}{A_o h_{so}} + \frac{1}{A_o h_o}$$

where h_i and h_o are convection resistances due to fluid film at the inside and outside surface respectively

Then overall heat transfer coefficients based on the inside and outside surface area are :

$$U_i = \frac{1}{\frac{1}{h_i} + \frac{1}{h_{si}} + \frac{r_i}{k}\log_e\frac{r_o}{r_i} + \frac{r_i}{r_o h_{so}} + \frac{r_i}{r_o h_o}}$$

and $$U_o = \frac{1}{\frac{1}{h_o} + \frac{1}{h_{so}} + \frac{r_o}{k}\log_e\frac{r_o}{r_i} + \frac{r_o}{r_i h_{si}} + \frac{r_o}{r_i h_i}}$$

The structure of heat exchange transfers the heat from the hot fluid to the coolant.

$$Q_{ex} = U_i A_i \theta_m = U_o A_o \theta_m$$

where θ_m denotes the logarithmic temperature difference defined as,

$$\theta_m = \frac{\theta_1 - \theta_2}{\log_e\left(\frac{\theta_1}{\theta_2}\right)}$$

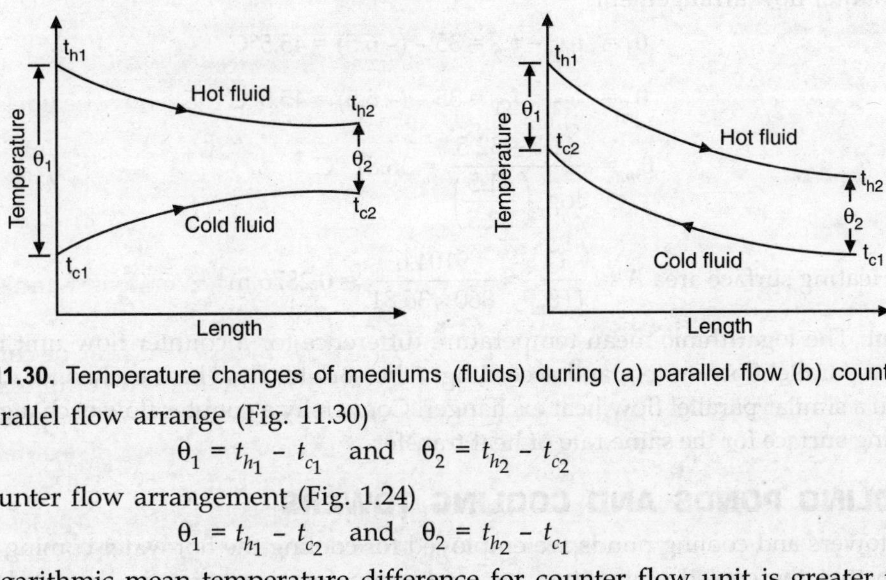

Fig. 11.30. Temperature changes of mediums (fluids) during (a) parallel flow (b) counter flow

For parallel flow arrange (Fig. 11.30)

$$\theta_1 = t_{h_1} - t_{c_1} \quad \text{and} \quad \theta_2 = t_{h_2} - t_{c_2}$$

For counter flow arrangement (Fig. 1.24)

$$\theta_1 = t_{h_1} - t_{c_2} \quad \text{and} \quad \theta_2 = t_{h_2} - t_{c_1}$$

The logarithmic mean temperature difference for counter flow unit is greater than that of parallel flow system and accordingly the counterflow heat exchanger can transfer more heat than a similar parallel flow heat exchanger. Conversely, a counter flow heat exchanger needs a smaller heating surface for the same rate of heat transfer.

EXAMPLE 11.7

In a food processing plant, a brine solution is heated from –12°C to 6.5°C in a double pipe heat exchanger by water entering at 35°C and leaving at 20.5°C at the rate of 9 kg/hr. If the overall heat transfer coefficient is 860 W/m² K, what heat exchanger area is required for (a) parallel flow and (b) counter flow arrangement. For water c_p *= 4.186 kJ/kg K.*

Solution : Refer Fig. 11.30.

The heat transfer from the water is

$$Q = m\,c_p\,\Delta_t = \frac{9}{60}(4.186\times10^3)\times(35-20.5)$$

$$= 9104.6\ \text{J/s}$$

Log mean temperature difference $\theta_m = \dfrac{\theta_1-\theta_2}{\log_e\left(\dfrac{\theta_1}{\theta_2}\right)}$

(*a*) For parallel fow arrangement,

$$\theta_1 = t_{h_1} - t_{c_1} = 35 - (-12) = 47\ °\text{C}$$

$$\theta_2 = t_{h_2} - t_{c_2} = 20.5 - (-6.5) = 27\ °\text{C}$$

The subscripts *h* and *c* represent the hot and cold fluids respectively.

$$\theta_m = \frac{47-27}{\log_e(47/27)} = 36.10\ °\text{C}$$

Heat exchange $Q = UA\,\theta_m$

$\therefore$ Heating surface area $A = \dfrac{Q}{U\,\theta_m} = \dfrac{9104.6}{860\times36.10} = \mathbf{0.293\ m^2}$

(*b*) For counter flow arrangement

$$\theta_1 = t_{h_1} - t_{c_2} = 35 - (-6.5) = 45.5°\text{C}$$

$$\theta_2 = t_{h_2} - t_{c_1} = 35 - (-6.5) = 45.5°\text{C}$$

$$\theta_m = \frac{41.5-32.5}{\log_e\left(\dfrac{41.5}{32.5}\right)} = 36.81\ °\text{C}$$

$\therefore$ Heating surface area $A = \dfrac{Q}{U\,\theta_m} = \dfrac{9104.6}{860\times36.81} = 0.2876\ \text{m}^2$

Comment: The logarithmic mean temperature difference for a counter flow unit is greather than that for a parallel flow system and accordingly the counter flow heat exchanger can transfer more heat tan a similar parallel flow heat exchanger. Conversily a counter flow exchanger needs a smaller heating surface for the same rate of heat transfer.

11.4. COOLING PONDS AND COOLING TOWERS

The cooling towers and cooling ponds are employed for cooling the hot water coming out of the condenser so that the cooled water may be used again for circulation in the condenser. The cooling is affected by exposing the hot water to an atmosphere of air. These units become a necessity when the supply of cooling water is scarce and the same water has to used over and over again for cooling purposes.

11.4.1. Cooling Ponds

• A cooling pond is a man-made body of water primarily formed for the purpose of providing cooling to the condenser circulating water. Pond cooling is always more efficient when the pond is located in a naturally windy area. However, the water losses by windage are high

; ranging between 1 to 5 percent of the recirculation rate. Impurities from surrounding area can easily enter the pond leading to extremely difficult deposit control problems in the system.

- A spray pond consists of an arrangement of piping and spray nozzles suspended over an outdoor open pond. The hot water coming out of condenser is sprayed through the nozzles to expose maximum surface area of water to air for effective cooling. Consequently the rate of evaporation increases and faster cooling results.

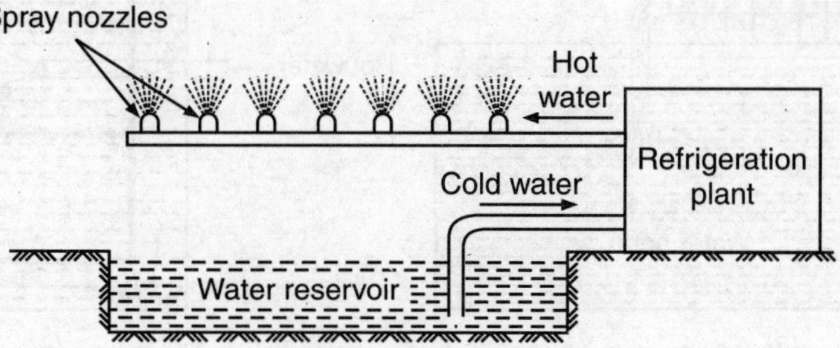

Fig. 11.31. Spray pond

The following aspects have been suggested for the design of cooling/spray ponds :

(*i*) The long dimension should be placed broad side to the prevailing wind.

(*ii*) For maximum cooling, spray nozzles are to the placed 1 to 2 m above the water surface.

(*iii*) There is not to be any interference between two sprays.

(*iv*) The distributing pipes are to be spaced 6 to 7 m apart.

(*v*) There has to be sufficient space (15 to 20 m) between the last nozzle and the edge of pond; this prevents excessive spray loss.

(*vi*) For the atomization of water, the pressure in the nozzles is limited to 1.5 bar. The nozzle pressure should break up the water into droplets but not into a mist which would easily drift off.

(*vii*) A surface area of 2.5 to 3 m^2 has to be provided per 1 litre/sec flow of water.

(*viii*) Louvered fence needs to be provided to reduce the carry over losses.

For a given cooling capacity, the size of spray pond is much less than that of a natural or open pond.

The cooling ponds may be of :

(*i*) Single and double deck system type : In the single deck system, the spray nozzles are arranged at the same level where as the double deck system has nozzles at different elevations.

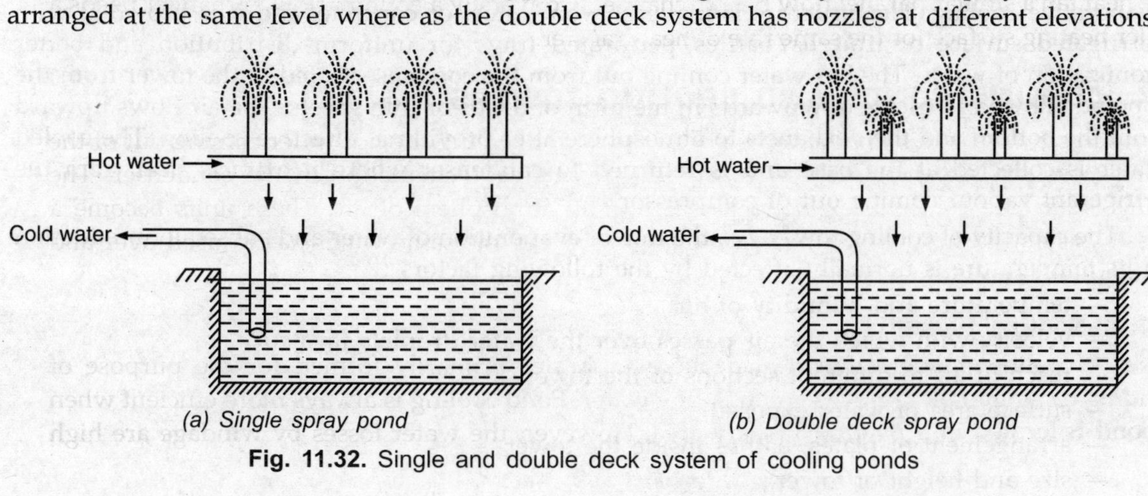

Fig. 11.32. Single and double deck system of cooling ponds

***(ii)* Natural and directed flow system type :** In a natural flow system the warm water coming out from the condenser is just allowed to flow into the pond where its cooling occurs.

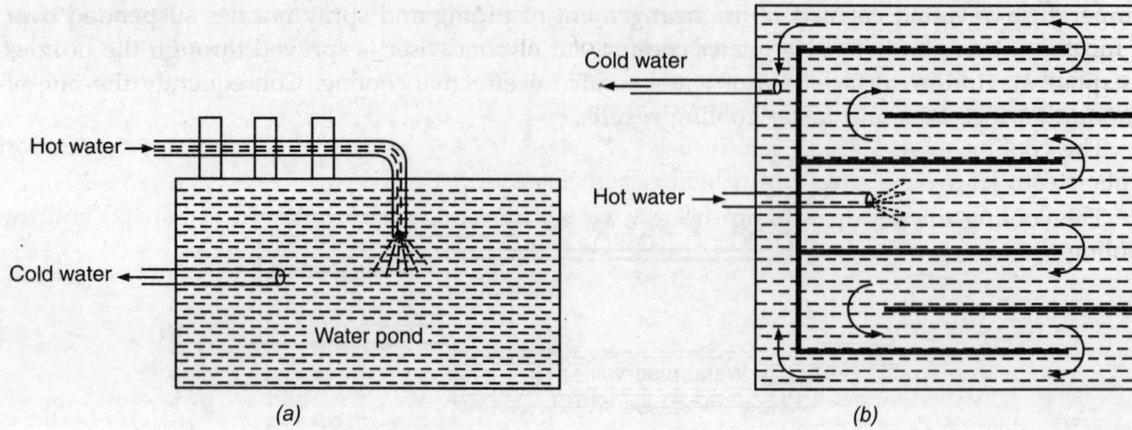

Fig. 11.33. Natural and directed flow system

In the directed flow system, the pond is divided into a number of flow channels by providing baffles. The hot water coming out from the condenser enters the middle channel and on reaching the far end divides into two streams. These streams are then directed by the baffles and made to traverse several times and finally exit at the outlet point. The water passes over more surface area and that ensures thorough mixing and optimum cooling.

When the wind velocity is high, the drift losses (water particles carried out by wind) become more in case of an open spray pond. This can be avoided by providing louvre fencing. Further, the use of cooling ponds is disadvantageous due to :

– large area requirement
– low cooling efficiency
– spray losses because of water evaporation
– no control over temperature of cooled water.

The system of removing heat from the condenser cooling water in a cooling/spray pond is restricted to small condensers only.

11.4.2. Cooling Towers

A cooling tower is a semi-closed tower like structure that operates on the principle of evaporative cooling. The structure is made of wood, steel or concrete, and it is provided with corrugated surface or troughs/baffles/perforated trays for uniform distribution and better atomization of water. The hot water coming out from the condenser is fed to the tower from the top and allowed to trickle downwards in the form of drops or thin sheets. The air flows upward from the bottom and then exhausts to atmosphere after providing effective cooling. The cooled water is collected at the base and is returned to condenser where it extracts heat from the refrigerant vapour coming out of compressor.

The capacity of cooling tower, *i.e.,* the rate of evaporation of water and subsequent reduction in its temperature is normally affected by the following factors :

– temperature and humidity of air
– velocity with which the air passes over the water droplets formed
– reach of air to different sections of the tower
– surface area of water exposed
– arrangement of plates/baffles inside the tower
– size and height of tower.

Some of water during its fall is evaporated and is carried along by heated air. The loss of water due to evaporation is made up by fresh water called make-up-water. The reduction in temperature of water is called ***range***. The temperature difference between wet bulb temperature of incoming air and the outgoing temperature of water is called ***cooling tower approach***. Higher the quantity of water circulated lower would be the cooling tower approach. Lower the cooling tower approach, better and effective would be the cooling in the cooling tower.

The cooling towers are generally used with the large installations, absorption refrigeration units, steam jet systems and centrifugal refrigeration units.

The cooling towers operate on draught which may be natural, induced or forced, and the cooling towers are classified as shown below :

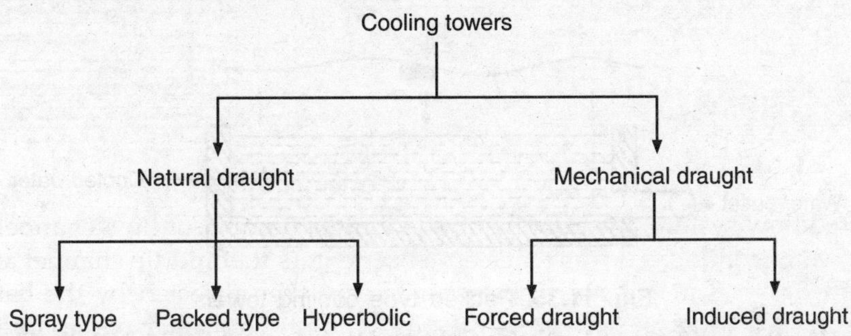

• **Natural draught cooling towers :** The air flow through the shell is created by density differences between the atmospheric air and the air inside the tower which has been warmed by the hot water coming from the condenser. These towers are categorized as spray type, packed type and hyperbolic cooling tower.

In the ***spray type*** natural draught cooling tower, the air enters through the louvered sides and flows across the unit in a transverse direction. The capacity of the tower varies from 50 to 100 litres/min per m^2 of base area depending on the wind velocity. These towers are suitable only for small plants as their cooling range is limited and there is no control over outlet temperature of water.

Quite often, the natural draught cooling towers are provided with water distributing packing or troughs which help to break the water spray into smaller droplets and thereby increase the amount of wetted surface. Such towers are referred to as ***packed-type*** cooling towers. Like spray type towers, they are not used with large installations because of high installation cost, high pumping head, greater height and anchoring required to withstand high winds.

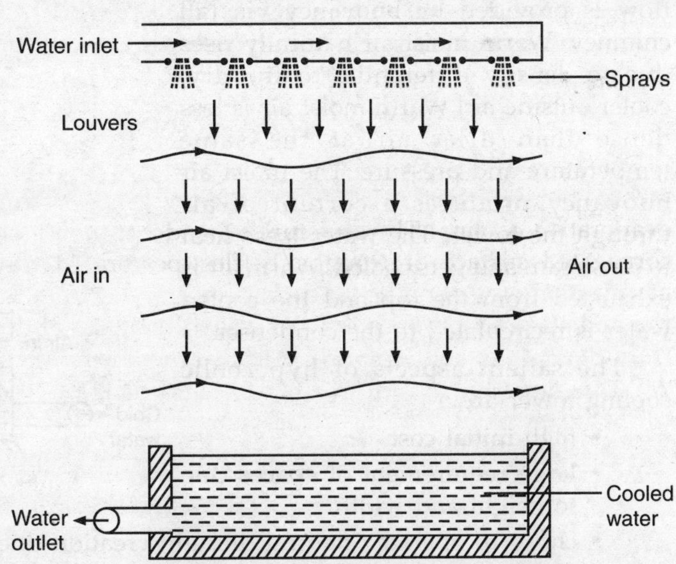

Fig. 11.34. Spray filled cooling tower

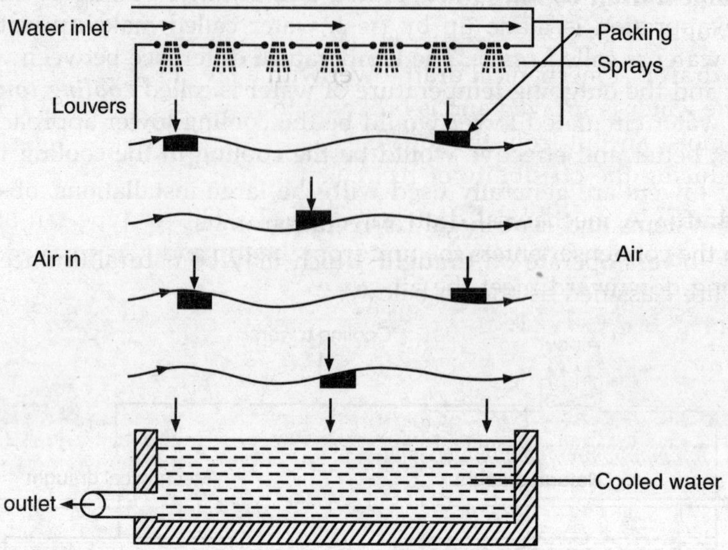

Fig. 11.35. Packed type cooling tower

The ***hyperbolic tower*** is a steel reinforced stack of concrete which is hollow from inside, circular in plan and hyperbolic in shape. The hot water coming form the condenser enters from the bottom through a vertical inlet pipe and the water sprays falling downwards are broken into droplets by the packing. The natural draft causing air flow is provided by buoyancy via tall chimney. Warm moist air naturally rises due to density differential to the dry, cooler outside air. Warm moist air is less dense than drier air at the same temperature and pressure. The moist air buoyancy produces a current of air through the tower. The water loses heat to air stream and gets cooled. Warm air is exhausted from the top and the cooled water is recirculated to the condenser.

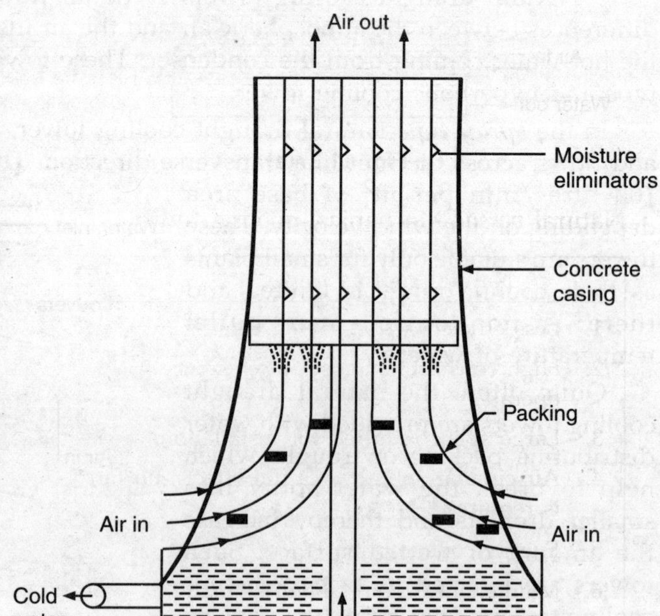

Fig. 11.36. Hyperbolic tower

The salient aspects of hyperbolic cooling tower are :

- high initial cost
- large requirement of spaces due to height and width
- cheap to operate ; natural draught creation due to chimney shape of tower.
- self-supported structure; can withstand wind velocities upto 150 km/hr.

In general, the performance of natural draught cooling towers is affected by variations in atmospheric conditions.

• **Mechanical draft cooling towers :** These towers use power driven fan motors to force or draw air through the tower

Induced draft : A mechanical draft tower with a fan at the discharge which pulls air through the tower. The water is sprayed and is broken into droplets by pads and is cooled by air. The fan induces hot moist air out from the top. This aspect produces low entering and high exiting air velocities, reducing the possibility of recirculation (back flow of hot air back towards intake).

Forced draft : A mechanical draft tower with a blower type fan at the intake. Hot water coming from the condenser enters the unit from the top and gets sprayed through nozzles. Water droplets falling downward meet the air.

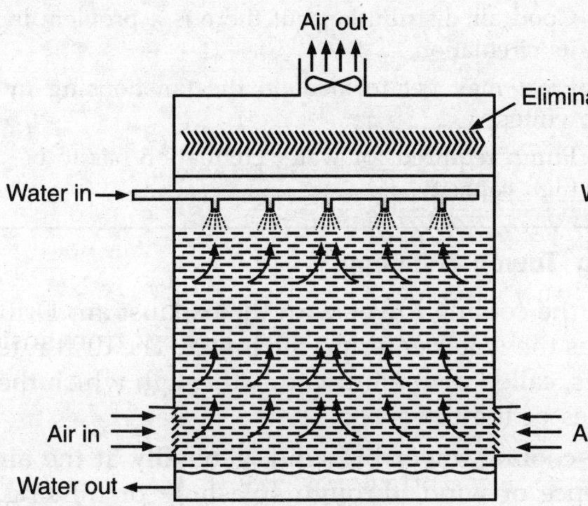

Fig. 11.37(a)

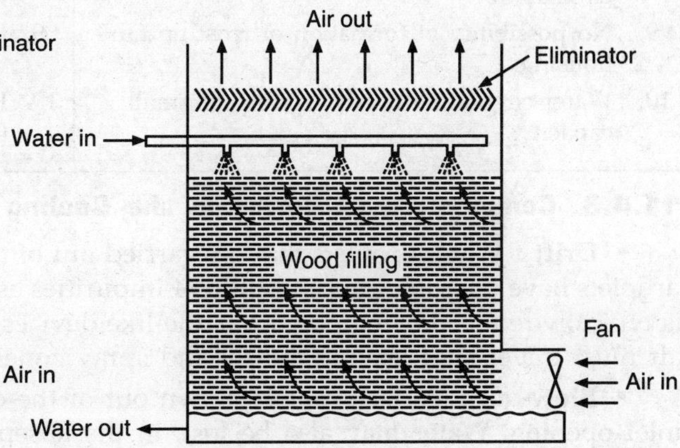

Fig. 11.37(b)

Natural *versus* Mechanical cooling tower :

Natural cooling tower	Mechanical cooling tower
1. Less efficient in providing cooling.	1. More efficient
2. High construction cost but cheaper to operate.	2. Low construction cost but high operation cost.
3. Large space required for installation.	3. Space requirement is less.
4. Air circulation is by natural effect ; no fan is required.	4. System requires fan for air circulation.
5. Energy consumption is low.	5. High energy consumption.
6. Maintenance requirement is small.	6. System needs proper maintenance.
7. Safe operation.	7. Possibility of failure of mechanical equipment.
8. Preferred in valley locations.	8. Preferred in inland locations.
9. Used for high capacity plants.	9. Used for low capacity plants.

Induced *versus* Forced Draft Cooling Tower :

Induced draught cooling tower	Forced draught cooling tower
1. Fan is installed at the top.	1. Fan is installed near the base.
2. Fan size is limited by the weight of tower.	2. Fan size is limited by size of tower.
3. Fan blades have high chances of erosion.	3. Blade erosion is less.
4. Vibration and noise level is high.	4. Minimum noise and vibrations.
5. Less efficient.	5. More efficient.
6. Unsafe operation.	6. Operation is safe.
7. Low in initial cost and power consumption.	7. High initial cost and power consumption.
8. Poor air distribution but no problem in its circulation.	8. Good air distribution but there is a problem in its circulation.
9. No possibility of formation of frost on fan housing.	9. Frost may get formed on the fan housing in winter.
10. Water circulation needs a pump of small capacity.	10. Pump required for water circulation has to be high capacity.

11.4.3. Commonly used Terms in the Cooling Tower Systems

- **Drift :** Water droplets that are carried out of the cooling tower with the exhaust air. Drift droplets have the same concentration of impurities as the water entering the tower. The drift rate is typically reduced by employing baffle-like devices, called drift eliminators, through which the air must travel after leaving the fill and spray zones of the tower.

- **Blow-out :** Water droplets blown out of the cooling tower by wind, generally at the air inlet opening. Water may also be lost, in the absence of wind, through splashing or missing. Devices such as wind screens, louvers, splash deflectors and water diverters are used to limit these losses.

- **Plume :** The stream of saturated exhaust air leaving the cooling tower. The plume is visible when water vapour it contains condenses in contact with cooler ambient, like the saturated air in one's breath fogs on a cold day. Under certain conditions, a cooling tower plume may present fogging or icing hazards to its surroundings.

- **Blow-down :** The portion of the circulating water flow that is removed in order to maintain the amount of dissolved solids and other impurities at an acceptable level. It may be noted that higher TDS (total dissolved solids) concentration in solution results in greater potential cooling tower efficiency. However, higher the TDS concentration, the greater the risk of scale, biological growth and corrosion.

- **Leaching :** The loss of wood preservative chemicals by the washing action of the water flowing through a wood structure cooling tower.

- **Noise :** Sound energy emitted by a cooling tower and heard (recorded) at a given distance and direction. The sound is generated by the impact of falling water, by the movement of air by fans, the fan blades moving in the structure, and the motors, gearboxes or drive belts.

- **Approach :** The approach is the difference in temperature between the cooled-water temperature and the entering-air wet bulb temperature. Since the cooling towers are based on the principles of evaporative cooling, the maximum cooling tower efficiency depends on the wet bulb temperature of the air.

- **Range :** The range is the temperature difference between the water inlet and water exit.

• **Fill :** Inside the tower, fills are added to increase contact surface as well as contact time between air and water. Thus they provide better heat transfer. The efficiency of the tower also depends on them. There are two types of fills that may be used :

– Film type fill (causes water to spread into a thin film)

– Splash type fill (breaks up water and interrupts its vertical progress)

Both in the induced and forced draft cooling towers, spray eliminators are provided at the top and these entrap the water which is likely to go along with hot air leaving from the top. Compared to natural draft cooling tower, the mechanical draft cooling systems are considerably small in size of tower and pipes, have high efficiency but have high running cost.

11.5. EXPANSION DEVICES

Expansive devices are the control systems used for regulating the expansion of liquid refrigerant from condenser (high) pressure to evaporator (low) pressure in proportion suiting to its rate of evaporation in the evaporator. Essentially the main purpose of an expansion device is two fold:

(*i*) Reduce the pressure of liquid refrigerant at constant enthalpy before being fed to the evaporator.

(*ii*) Control and regulate the flow of refrigerant to the evaporator.

Further, they expand the refrigerant by throttling process and maintain pressure difference between high and low pressure sides of the refrigeration system.

A brief description of the different types of expansion devices used in refrigeration systems has been given in the following paragraphs.

11.5.1. Capillary Tube

The capillary tube is a length of small diameter tubing (1 to 6 m long with the inside diameter generally from 0.5 to 2 mm) located between the condenser (HP side) and evaporator (LP side) of a refrigeration system. When the liquid refrigerant flows through the tube, the pressure drops because of friction (high resistance) and fluid acceleration through the narrow passage. The resistance increases with decrease in diameter and increase in length of the tube.

The length and the small diameter reduce the pressure from condensing pressure to evaporator pressure. With drop in pressure and temperature, the liquid refrigerant flashes into vapour. Vapour has a large specific volume and consequently the fluid speeds up. A choked flow results if the pressure drops low wnough. Further drop in pressure will then have no effect on the flow rate as then the sonic velocity occurs at the end of tube. The flow of refrigerant through the capillary tube depends on the load on condenser; it increases with increase in load (summer) and decreases with decrease in load (winter).

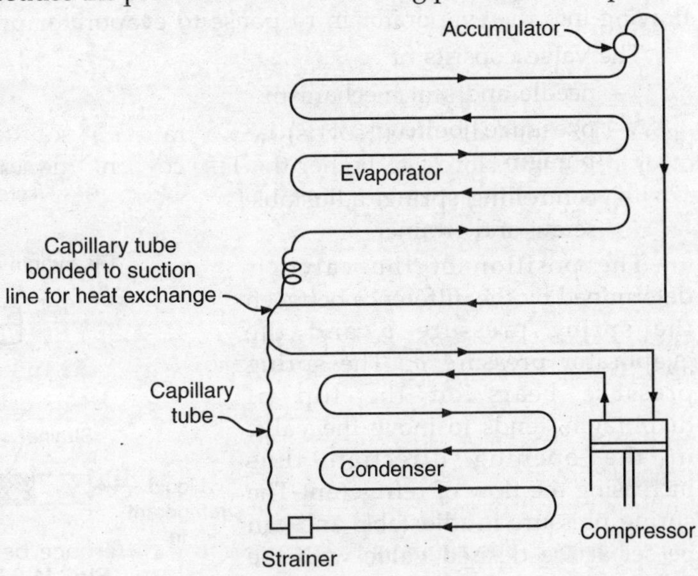

Fig. 11.38. Capillary tube

The salient aspects of a capillary tube are :

(i) simplicity of design and construction. Numerous combinations of bore and length are available to obtain the desired pressure drop.

(ii) cheap in cost and reliable in operation.

(iii) long life; approximately 20-30 years.

(iv) flow of refrigerant continues (even when compressor stops) until pressure becomes equal between the high and low pressure sides. The motor is then under no load condition. This allows the compressor to restart in unloaded condition with a less expensive motor as starting torque is low.

(v) capillary tubes give optimum performance at one set of operating conditions. Once a capillary tube has been selected, it will be suitable only for design pressure drop.

(vi) mass flow rate is determined by a balance between the tube and compressor performance. If the tube resistance is too high, the refrigerant flow becomes less than the pumping capacity of the compressor and the evaporator starves. Starved evaporator means that there is not enough refrigerant to provide enough cooling capacity.

(vii) Capillary tube may get choked due to moisture (which may freeze) in the system, kinked tubing or foreign particles lodged in the lines. These aspects lead to stoppage in the flow of refrigerant; the compressor keeps running continuously but no cooling effect is produced. Under such a condition, the capillary tube should be replaced.

If the tube resistance is too low, the refrigerant flow becomes more than the pumping capacity of the compressor and the evaporator gets overfed. This results in slugging of the compressor; liquid drops enter the compressor.

Usually, the refrigerant charge in the capillary tube is at critical level (just sufficient) and accordingly no receiver need to be provided for storage.

The capillary tubes are used with small capacity units such as domestic refrigerators, freezers and room coolers.

11.5.2. Automatic Expansion Valve

The automatic expansion valve works on the principle of maintaining a set pressure in the evaporator. The valve opens and closes to either increase or decrease the amount of refrigerant flowing into the evaporator in response to evaporator pressure.

The valve consists of

- needle and seat mechanism
- pressure bellows or diaphragm
- controlling spring, adjustable screw and strainer

The position of the valve is determined by the difference between the spring pressure p_s and the evaporator pressure p_e. The spring pressure bears on the top of diaphragm, tends to move the valve in the opening direction thus increasing the flow of refrigerant The spring pressure is adjustable and can be set at the desired value with the help of adjustable screw. The

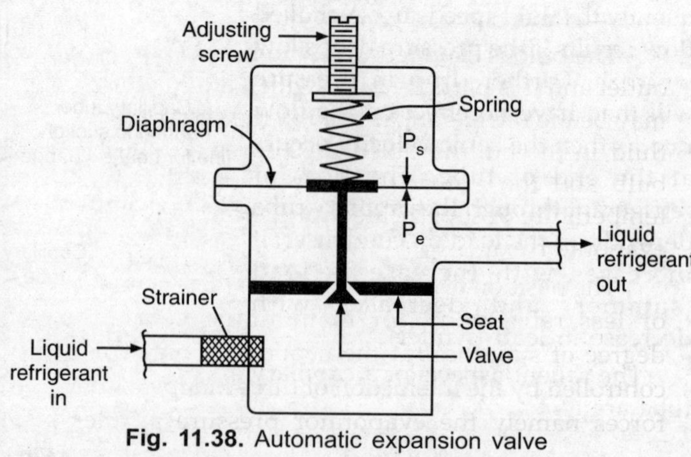

Fig. 11.38. Automatic expansion valve

evaporator pressure acts beneath the diaphragm, attempts to move the valve in closing direction thus reducing the flow of refrigerant. At a given setting, the diaphragm is held in equilibrium position by the interaction between the evaporator pressure and the spring force that counterbalance each other. A constant pressure results from the interaction of the two opposing forces.

When the compressor starts operating, the refrigerant gas is sucked from the evaporator and the evaporator pressure falls below the set value. The diaphragm is pushed downwards and the downward motion is transmitted to the valve through push rods. The valve opens more and that increases the liquid flow to the evaporator. With entry of more refrigerant, the evaporator pressure builds till it balances the spring pressure. In the situation when the evaporator pressure becomes more than the set value, the diaphragm is pushed upwards. The valve then tends to close and less refrigerant liquid flows to the evaporator. This results in reduction of evaporator pressure till it again equals the spring pressure and equilibrium state is restored.

When the plant is shut down, the liquid refrigerant continues to flow into the evaporator and the resulting further buildup pressure closes the valve. The valve remains closed until the compressor starts again and reduces the evaporator pressure.

The major disadvantages of the automatic expansion valve are :

– poor efficiency
– changes of over flooding the evaporator at lower loads
– it limits the capacity of the refrigerant system due to its constant pressure, characteristics
– large pressure fluctuations disallow its use with low pressure motor controls

Due to low efficiency at heavy load, this valve is used only in small equipment (having relatively constant load) such as domestic refrigerators, freezers and ice-cream cabinets.

11.5.3. Thermostatic Expansion Valve

The thermostatic expansion valve ensures complete filling of evaporator with refrigerant irrespective of load and also no spilling over to the suction line to the compressor. Further, the refrigerant flow has to be so regulated that the vapour at the evaporator exit has a constant degree of superheat.

This expansion device essentially consists of

– needle and seat assembly
– pressure bellows or diaphragm
– controlling spring with tension adjusting screw
– fluid charged remote bulb

The bulb is clamped firmly to the evaporator outlet and it is partially filled with a liquid which may be the refrigerant as used in the system. The fluid in the bulb is called the power fluid. The bulb and the power fluid closely assume the temperature of superheated refrigerant gas at the evaporator exit.

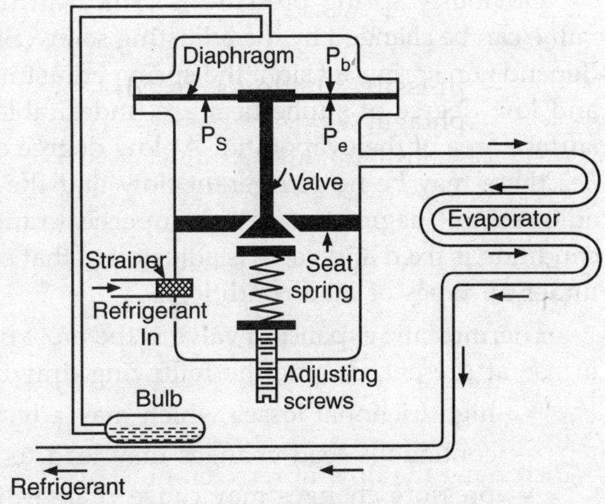

Fig. 11.39. Thermostatic expansion valve

The position of valve that allows flow of more or less refrigerant (for maintaining constant degree of superheat at the evaporator outlet) is controlled by the interaction of three independent forces namely the evaporator pressure p_e, the spring pressure p_s and the bulb pressure p_b. The pressure exerted by the fluid in the bulb equals the pressure of vapour in the suction line at the point where the bulb is attached to the evaporator outlet. The bulb pressure attempts to open the valve, whereas the spring pressure added to the evaporator pressure provides the total closing pressure for the valve. Thus under normal operating condition

$$p_b = p_e + p_s$$

That is: ***the pressure exerted by the vapour in the bulb on the diaphragm is balanced by spring tension and pressure in the evaporator.***

To understand the working of thermostatic expansion valve, let it be presumed that Freon-12 is the refrigerant and the evaporator temperature is – 5°C. Corresponding to this temperature, the evaporator pressure p_e = 2.612 bar. If the spring tension is set at p_s = 0.477 bar, then the combined pressure tending to close the valve becomes

$$p_e + p_s = 2.612 + 0.477$$
$$= 3.089 \text{ bar}$$

Then for the valve to be in equilibrium position, the bulb pressure should be 3.089 bar. This is also the pressure of refrigerant at exit from the evaporator; the corresponding saturation temperature is 0°C. Thus the refrigerant vapours at evaporator exit are superheated with degree of supherheat equal to 0 – (– 5) = 5 degree. The diaphragm is on balance point and remains in equilibrium condition till there is change in the superheat of suction vapour at bulb location, *i.e.*, at evaporator exit. Any change in superheat of suction vapour causes the valve to take compensatory action and re-establish equilibrium.

(*i*) When the degree of superheat decreases, *i.e.*, becomes less than 5-degree, the valve tends to move towards the closing position. That throttles the flow to evaporator till again the same degree of supherheat is attained.

Conversely

(*ii*) When the degree of superheat increases, *i.e.*, becomes more than 5-degree, the valve tends to move towards the opening position. This causes increase in the flow of refrigerant, thereby reducing the degree of superheat.

Obviously spring pressure is critical in the operation of thermostatic expansion valve; its value can be changed by the adjusting screw. Since the equilibrium condition degree of superheat depends on spring tension, the spring adjustment is also called superheat adjustment. Both high and low degree of supherheat are undesirable. High degree of superheat reduces the effective surface area of the evaporator. At low degree of superheat, the valve loses control of refrigerant, *i.e.*, there may be no refrigerant flow and the evaporator may starve. These valves are usually adjusted to 4-degree to 6-degree superheat range by the manufacturer. Once the superheat range condition is fixed and valve is adjusted to that condition, the set value of supherheat is maintained under all types of load conditions.

Thermostatic expansion valve is the most popular and very efficient type of expansion device in use at present, despite the following drawbacks :

– high frictional losses which may adversely affect its control characteristics
– continuous load changes may lead to failure of bellows/diaphragm
– pressure changes may cause changes in flow pattern across the valve
– stability of superheat condition may be affected by wide fluctuations in load

The salient points of difference between capillary tube and thermostatic expansion valve are enumerated below :

Capillary tube	Thermostatic expansion valve
1. Simple construction, absence of moving parts and low cost.	1. Complicated construction and high cost.
2. A small diameter tubing which expands the refrigerant by throttling.	2. An expansion valve which controls the flow of refrigerant by the valve opening.
3. Pressure drop is low.	3. High pressure drop.
4. Motor driving the compressor has low starting torque.	4. Starting torque of motor driving the compressor is high.
5. Low efficiency.	5. High efficiency.
6. No requirement of a receiver.	6. Receiver is required.
7. Used in a small refrigeration capacity systems.	7. Used in large refrigeration capacity installations.

The distinguishing aspects between thermostatic expansion valve and automatic expansion valve are enumerated below :

Thermostatic expansion valve (TEV)	Automatic expansion valve (AEV)
1. Operation of the valve is controlled by spring pressure, evaporator pressure and bulb pressure.	1. Spring pressure and evaporator pressure control the operation of the valve.
2. Pressure drop across the valve is large.	2. Pressure drop across the valve is small.
3. Large friction loss.	3. Small friction loss.
4. TEV maintains constant superheat of vapour refrigerant.	4. AEV maintains constant pressure of refrigerant.
5. External equalization is required.	5. Equalization is not required.
6. Flow pattern affected by a change in pressure.	6. Change of load affects the valve efficiency.
7. More popular for industrial applications.	7. Less popular for industrial applications.
8. Used with flooded evaporators	8. Used with dry expansion evaporators.

EXAMPLE 11.8

A straight charged thermostatic expansion valve is designed to operate at an evaporator temperature of 7 °C with a degree of superheat of 5 °C. R-134 a is the refrigerant used in the refrigeration system. The saturation temperature and corresponding pressure of R-134 a are as given below:

T_{sat} °C	–23	–13	–12	–11	2	7	12	13
Psat kPa	116.39	177.92	185.24	192.8	314.62	374.63	443.01	457.76

(a) Write down the chemical formula for R-134 a. What does 'a' signify.

(b) Calculate the required spring pressure at the design condition.

(c) Find the degree of superheat if the same TEV operates at an evaporator temperature of –23°C. It may be assumed that the spring pressure remains constant.

Solution : Chemical formula of R-134 *a* is $CF_3\,CH_2\,F_1$.

The symbol '*a*' stands for isomer

(*b*) Pressure at 7°C = 374.63 kPa. The pressure 5°C will have to be interpolated from the values given in the table. That is,

$$P \text{ at } 5\,°\text{C} = \frac{374.63 - 314.62}{7 - 2} \times (5 - 2)$$

$$= 314.62 + 36.002 \simeq 350.63 \text{ kPa}$$

Then : Spring pressure = P at 7 °C – P at 5 °C

$$= 374.63 - 350.63 = \mathbf{24} \text{ kPa}$$

(*c*) Corresponding to evaporative temperature of –23°C,

Pressure at the evaporator inlet = 116.39 kPa

For the same spring pressure of 24 kPa, the equilibrium pressure of the charge is,

$$P_{eq} = 116.39 + 24 = 140.39 \text{ kPa}$$

The corrosponding saturation temperature of charge is,

$$T = -23 + \frac{-13 + (-23)}{177.92 - 116.39} \times (140.39 - 116.39)$$

$$= -23 + 3.9 = -19.1\ °\text{C}$$

Then : Degree of superheat = – 19.1 – (– 23) = **3.9** °C

11.6. EVAPORATORS

Component of a refrigerating mechanism in which the refrigerant vaporizes and absorbs heat

Evaporator is a component of the refrigeration system which produces the refrigeration (cooling) effect by vaporization of liquid refrigerant at low temperature and low pressure. The liquid refrigerant at exit from the throttling unit/expansion valve is fed to the evaporator where low pressure is maintained by the suctioning effect of the compressor. At low pressure, the liquid refrigerant boils and changes to vapour. During this phase change, it absorbs large quantum of heat from itself and cools. The cooled refrigerant subsequently cools the evaporator coils and the cooled evaporator surface causes cooling of the space/product or application and refrigerating effect is produced.

Evaporator works as a heat exchanger but its function is opposite to that of a condenser ; in a condenser heat is released where as in a evaporator heat is absorbed by the refrigerant. Evaporators are also called the chillers, chilling coils or cooling coils.

The variation in temperature of the refrigerant within the evaporator is as reflected in Fig. 11.40.

(*i*) During evaporation, the refrigerant temperature remains constant as only latent heat is absorbed (0 – 1′)

(*ii*) Beyond saturation point 1′, there is sensible heat gain by the refrigerant, and accordingly there is increase both in temperature and enthalpy (1′ – 1).

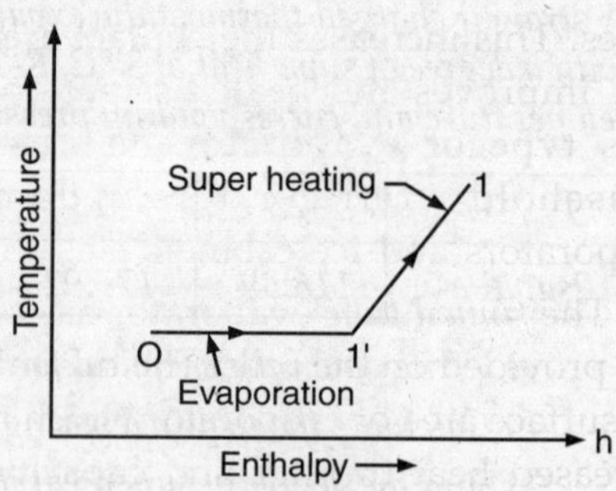

Fig. 11.40. T-h plot for an evaporator

The capacity of an evaporator refers to the heat absorbed by the evaporator coils, and is given by

$$Q = UA\ \Delta t \qquad \text{Watt (J/s)}$$

where U is overall heat transfer coefficient in W/m^2-deg ; A is area of evaporator in m^2 ; Δt is temperature difference between the evaporator coils and the product from which heat is being absorbed.

The heat transfer in an evaporator is given by :

– material used for fabrication of evaporator and its thermal conductivity,
– temperature difference between the evaporator coils and the product/application,
– surface area of evaporator coil,
– thickness of evaporator coil and
– flow velocity of refrigerant in the evaporator coil. Higher the flow velocity of refrigerant, greater will be the heat absorbing capacity of the evaporator.

11.6.1. Classification of Evaporators

Evaporators are of various types and the primary difference between the different types are the ways in which the heat exchanger and the vapour body are connected together, and their physical relationship to each other.

Based on the type of construction, the evaporators are classified as bare tube, plate and finned tube evaporators.

The ***bare tube*** evaporator coil (Fig. 11.41) is laid bare without any covering and is generally made of steel pipe (for ammonia refrigerant) or copper tube (for Freon refrigerant). Bare tube coils are available in a number of sizes, shapes and designs, and are usual custom made for a particular application. The use is further limited to application where the evaporator operates at a temperature below 0°C and in liquid cooling. Because of low operating temperatures, the tubes are likely to be covered with a layer of frost. However, the bare coils are easy to clean and defrost when compared to those provided with fins.

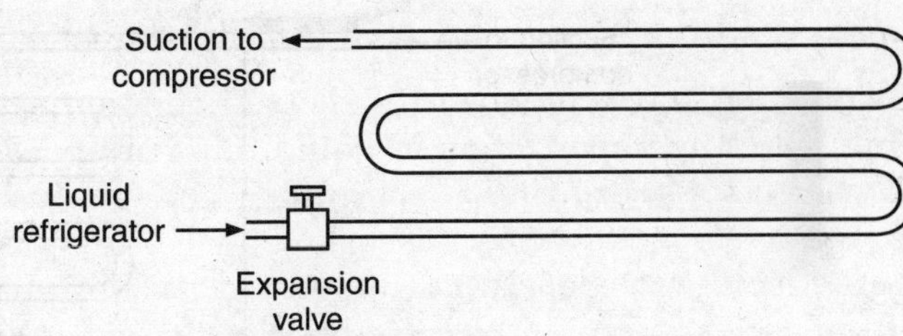

Fig. 11.41. Bare tube evaporator

The ***plate surface*** evaporator has coils which are either welded on one side of plate or between two plates welded together at the edges. This increases the contact area and that improves the heat dissipation rate. This type of evaporators are used in household refrigerators, domestic evaporators and ice cabinets etc.

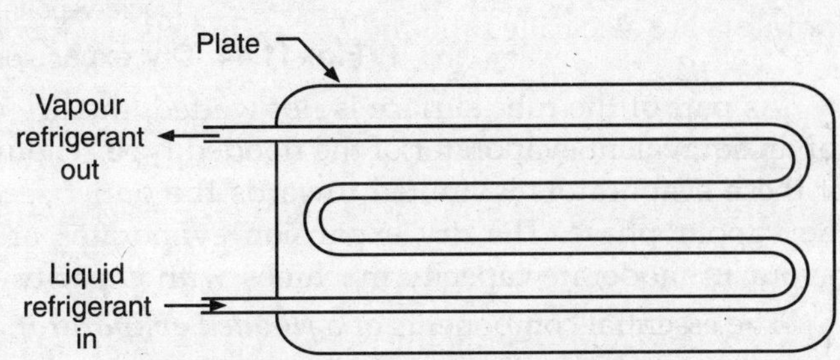

Fig. 11.42. Plate surface evaporator

The ***finned tube*** evaporator has several fins provided on the coils. The fins increase the surface area of evaporator which means increased heat transfer and capacity. The fins are made of high thermal conductivity materials, and their shape-size and spacing is adapted to provide optimum rate of heat transfer for a given application. It also needs to be ensured that no frost develops on the finned coil because accumulation of frost between the fins reduces the heat transfer capacity. The finned evaporators are primarily designed for window type air-conditioners and other air-cooling applications where operating temperatures are above 0°C.

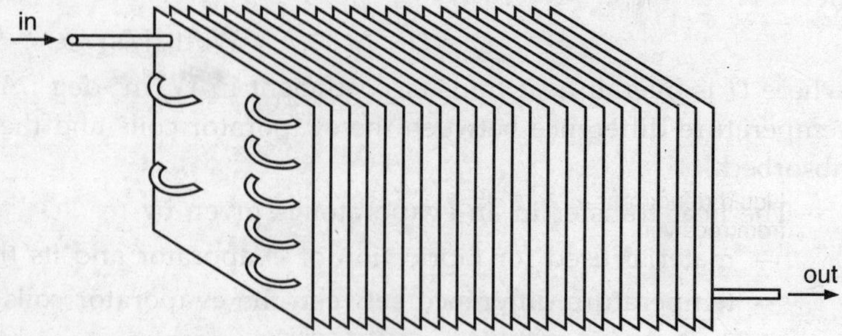

Fig. 11.43. Finned tube evaporator

Based on the operating conditions, the evaporators and classified as dry expansion type and flooded type evaporator.

In the ***dry expansion evaporator***, the liquid refrigerant from the receiver enters the bottom end of evaporator coil through a thermostatic expansion valve. The expansion device regulates the refrigerant flow such that the whole of it gets vaporized before it leaves the evaporator coil. The vapour leaving the evaporator is usually slightly superheated to ensure that there is no risk of any refrigerant liquid reaching the compressor. Further the amount of refrigerant varies with load on it; greater the load, more has to be the flow of refrigerant.

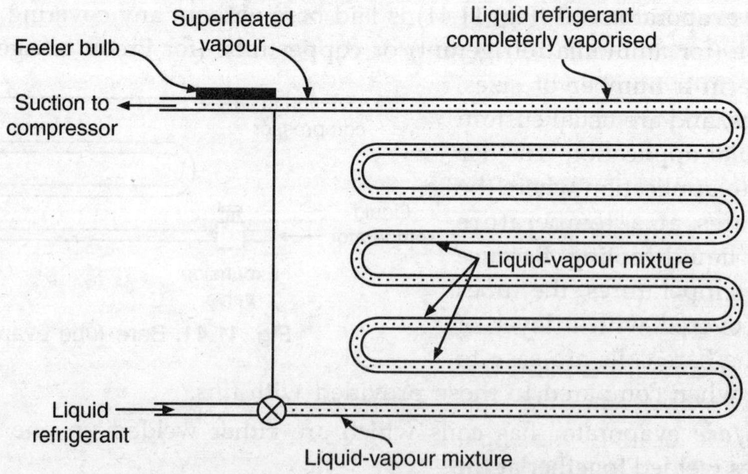

Fig. 11.44. Dry expansion evaporator

As part of the tube surface is not wetted, the dry expansion evaporators need more surfaces for an equivalent evaporator of the flooded type. Though simple to operate, the thermal efficiency of these evaporators is limited towards the outlet section where the refrigerant is essentially in the vapour phase. The dry expansion evaporatos are most widely used and particularly find favour in moderate capacity machines with capacity upto 150 tonne.

The essential components of a ***flooded evaporator*** are : evaporator coil, accumulator and float control mechanism.

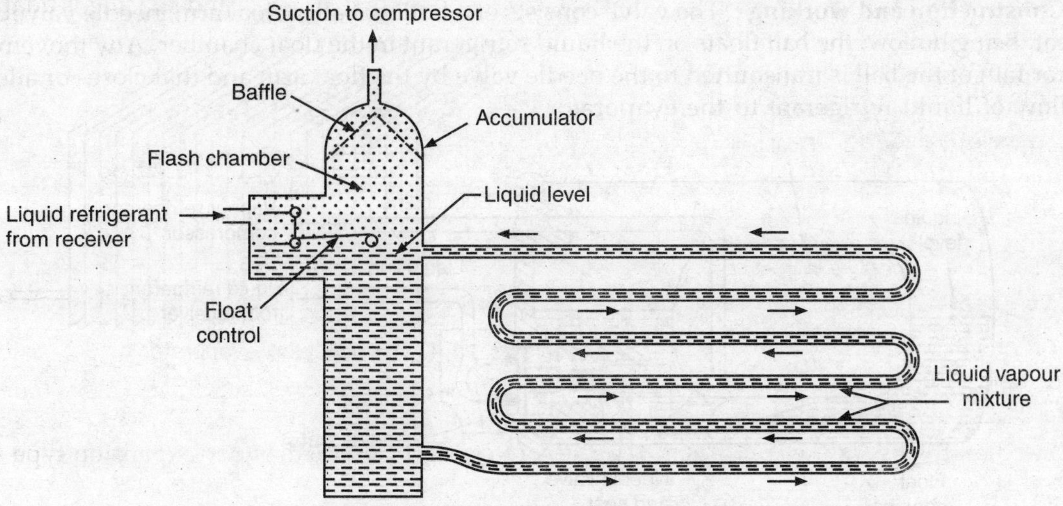

Fig. 11.45. Flooded type evaporator

These evaporators evaporate only a part of the refrigerant volume delivered. The evaporator coil is completely filled with liquid refrigerant and the level of liquid refrigerant is maintained constant by a float control mechanism. The control valve is fitted in a drum called accumulator/ liquid separator. The liquid refrigerant from the receiver enters the accumulator and therefrom to the evaporator where it absorbs heat and vaporizes. The liquid-vapour mixture is taken to the liquid separator where occurs the separation of vapour from the liquid. The vapour collected in the upper part of the separator is drawn off by the compressor suction. The unevaporated liquid is fed back to the evaporator inlet.

In the flooded evaporator, the entire internal exchange remains in contact with the liquid refrigerant under all load conditions. Accordingly these evaporators have high heat exchange efficiency compared to dry expansion evaporators of the same size. However, they are bulky and require more charge, *i.e.*, large flow rate of refrigerant. They are essentially used in large installations, especially in chemical and food processing industries.

11.6.2. Float Valves

Float valves are a type of expansion devices used with flooded evaporators for maintaining liquid refrigerant at constant level in the evaporator. The valve mechanism comprises a system that actuates as the refrigerant liquid level changes (rises or falls) in the float chamber.

Depending on the location of the float chamber, these valves are categorized as low side float valve and high side float valve. The low side float valve has the float and float chamber in the low pressure side of the system, and the high side float valve has its float and float chamber in the high pressure side of the system.

- **Low Side Float Valve**

Location : The valve is located on the low pressure side of the refrigeration system. It is fixed in a chamber (float chamber) which is connected to the evaporator.

Function : The valve acts to maintain a constant level of liquid in the evaporator by regulating flow of liquid refrigerant into it according to the rate of vaporization. It maintains a constant liquid level in the evaporator under all loading conditions regardless of the evaporator pressure and temperature.

Construction and working : The valve consists of a hollow ball, a float arm, needle valve and its seat. Being hollow, the ball floats on the liquid refrigerant in the float chamber. Any movement (rise or fall) of the ball is transmitted to the needle valve by the float arm and that closes or allows the flow of liquid refrigerant to the evaporator.

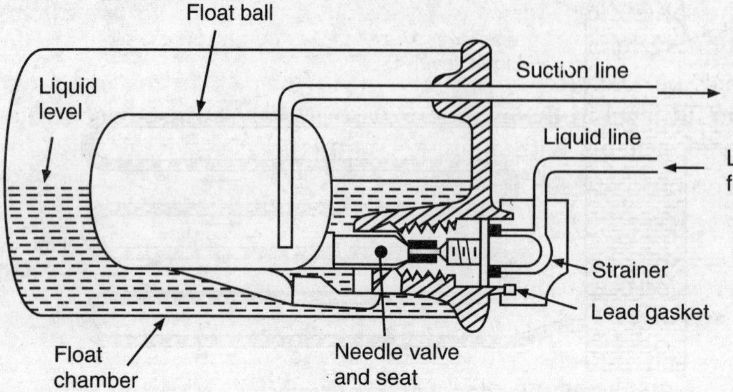

Fig. 11.46. Low side float valve

With an increase in the refrigeration load on the system, the evaporation rate of refrigerant becomes more and that causes the liquid level to drop. The float also then comes down and that pulls the needle away from its seat. The opening of the valve allows entry of more refrigerant into the chamber. When enough liquid has entered, the float ball rises and ultimately the valve closes when desired liquid level is reached.

Conversely, when the evaporation decreases with decrease in load on the system, the ball goes up due to increase in the liquid level. The valve then closes and that stops the inflow of refrigerant into the chamber.

- **High Side Float Valve**

Location : The valve is located on the high pressure side (*i.e.*, between the condenser and evaporator) of the refrigeration system. It is fixed in a chamber (float chamber) which is connected to the evaporator.

Function : The valve acts to regulate flow of liquid refrigerant to evaporator at a rate commensurate with rate of vaporization regardless of load on the system. The control of liquid level in the evaporator is obtained indirectly by maintaining liquid level in the float chamber.

Construction and working : The valve assembly consists of a hollow ball, a float arm, needle valve and its seat. Being hollow, the ball floats on the liquid refrigerant in the float chamber. Any movement (rise or fall) of the ball is transmitted to the needle valve by the float arm and that closes or allows the flow of liquid refrigerant to the evaporator.

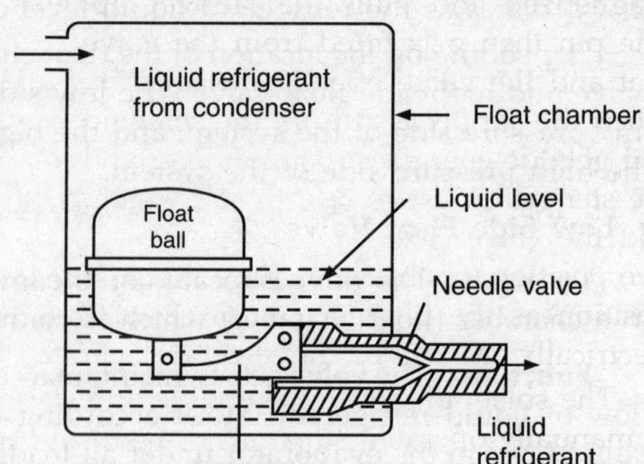

Fig. 11.47. High side float valve

The refrigerant from the condenser flows into the chamber and that raises the liquid level. The ball also then moves up and pulls the needle away from its seat. That allows the entry of liquid in the evaporator which replaces the liquid evaporated. Such a situation occurs when there is increase in load; more of liquid evaporates and sub-sequently condenses. A decrease in load would decrease the rate of evaporation (and condensation) and consequently liquid level would fall in the chamber and the float valve would tend to close the seat orifice. The rate of condensation of refrigerant vapour in the condenser equals the rate at which liquid refrigerant evaporates in the evaporator; and the float chamber receives and feeds liquid to the evaporator at the same rate. Thus the valve continuously and automatically works as per load on the refrigeration system.

High side float valve may be installed at any level with respect to the evaporator as its working is independent of liquid level in the evaporator. These valves find their greatest use in domestic and small commercial systems with a single compressor, evaporator and condenser.

The salient points of difference between low side and high side float valves are :

Low side float valve	*High side float valve*
1. The valve and its chamber is located on the low pressure side of the system.	1. The valve and its chamber is located on the high pressure side of the system.
2. The valve opens and allows entry of refrigerant into the evaporator on a fall in the liquid level in the chamber. The valve would close with a rise in the liquid level.	2. The valve opens on a rise in liquid level in the chamber and then the liquid refrigerant is allowed to enter into the evaporator. The valve would close with a drop in the liquid level.
3. The system can be completely vacuumized.	3. No possibility of complete vacuum in the system.
4. The machine fitted with low side float valve may get oil choked.	4. There is least possibility of oil choking in a machine fitted with high side float valve.

11.6.3. Solenoid Valve

The solenoid valve is an electrically operated valve installed ahead of thermostatic expansion valve to prevent the flow of refrigerant liquid from the evaporator when the compressor is not working.

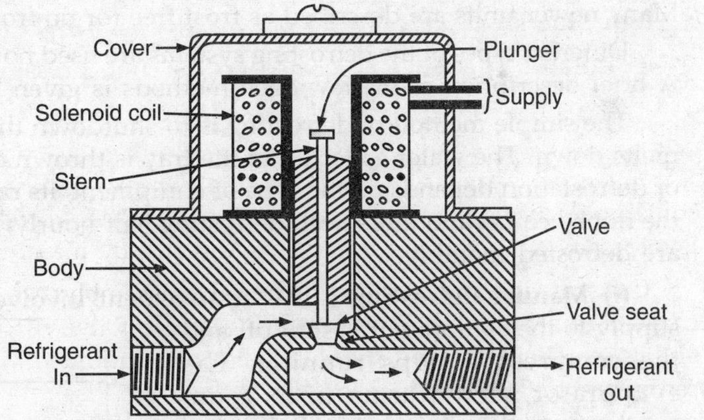

Fig. 11.48. Solenoid valve

The valve consists of a coil of insulated copper wire (called solenoid) and an iron core (called plunger) to which a pin and a stem are attached. When a current is passed through the coil, it gets magnetized and pulls the plunger up. The pin then gets lifted from the valve seat and the valve opens. Subsequently when the supply is cut off, the solenoid coil gets de-energized. The plunger and the stem drop down due to self weight and the valve closes. Obviously, it is a two position (open and close) uni-directional valve. These valves are always fitted in the vertical position as the plunger comes down due to gravity. The operation of the valve is adjusted electrically and can be conveniently operated through a suitable electrical controlled mechanism.

The solenoid valves have wide applications in refrigeration, water and brine lines in place of manually operated stop valves as they provide automatic operation. They are also frequently used to control the flow of condensing water. They automatically stop the flow of refrigerant as soon as the compressor stops. The supply restarts only with the start of compressor. This aspect provides considerable saving in water cost.

11.7. FROST AND DEFROSTING

The accumulation of ice on the cooling coils of an evaporator is called ***frost***. Frost is essentially a condition that occurs when low temperature and humidity combine and cause freezing. The presence of moisture in air forms dew on the surface of cooing coil/evaporator when its temperature reaches the dew point temperature. With further drop in temperature to freezing point, the dew is frozen to ice. Frosting is common to domestic as well as industrial refrigerators. The frost acts as a thermal insulation and inhibits (retards) the heat transfer between air and the cooing coil.

The condition becomes severe when the frost bridges the gap between adjacent fins and greatly reduces the surface area available for heat transfer. With a reduced airflow rate, the value of overall heat transfer coefficient for the coil decreases, and the mean temperature difference between the air and refrigerant must increase in order to transfer the same rate of heat flow.

Both these aspects cause the compressor to operate at lower suction pressure which reduces the refrigerating capacity and COP of the system.

Obviously it becomes necessary to periodically remove the frost from the surface of coils. The process of frost removal is called ***defrosting***. Defrost means to make or become free of ice or frost. Defrosting is a procedure performed periodically on refrigerators and freezers to maintain their operating efficiency. Defrosting of domestic refrigerators is normally achieved by :

- temporarily removing all food from the cabinet
- turning off power supply to the unit
- leaving the doors of the unit open
- waiting for the ice to melt and draining it appropriately.

The process may be speeded up by mechanical removal of ice, or by the introduction of a gentle heat into the cabinet. Placing a pan of hot water in the cabinet and closing is an effective method. Any mechanical removal of ice should be done gently so that the coils are not damaged. Many newer units are described as frost free (or no frost) and do not require manual defrosting.

Different types of the defrosting systems are used now a days for different types of installations. A brief description of the few such methods is given below :

The simple method of defrosting is to shutdown the system for few minutes, till all the frost melts down. The water collected in the tray is thrown out and the system is cleaned. The period of defrostation depends on the type of equipment, its rating and also on the season. For example, the finned coil evaporators need defrosting half-hourly, where as evaporators used in cold storage are defrosted monthly.

(*i*) Manual defrosting : The arrangement involves in opening the switch so that power supply to the compressor is shut off and the compressor stops running. The evaporator then warms upto room temperature, thereby causing the frost to melt. The manual defrosting is employed in large installations, even though it has the drawbacks of

– waiting time till the frost melts, and

– refrigeration equipment has to be out of operation and that may cause losses.

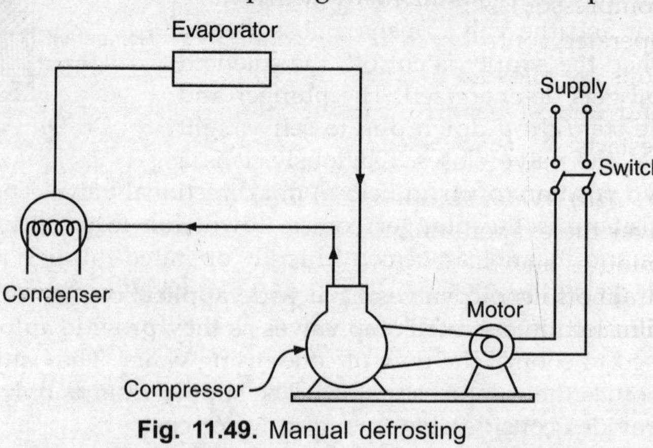

Fig. 11.49. Manual defrosting

***(ii)* Automatic period defrosting :** Like the manual defrost method, the automatic period defrosting also involves the shutting down of the refrigeration equipment for melting the frost. However, the shutting down and restarting the system is carried out automatically as per change in evaporator temperature. Due to frost formation and its accumulation on the coils, the temperature of the evaporator drops. At a pre-determined low temperature, the thermostat comes into action; it passes the signal to a controller. The controller then disconnects the motor switch through an activator. The power supply to the compressor is shut and the equipment stops functioning. The frost melts due to rise in tempeature of evaporator. Once the temperature reaches the set point, the controller activates the switch. The power supply to the compressor motor is restored and the equipment starts functioning.

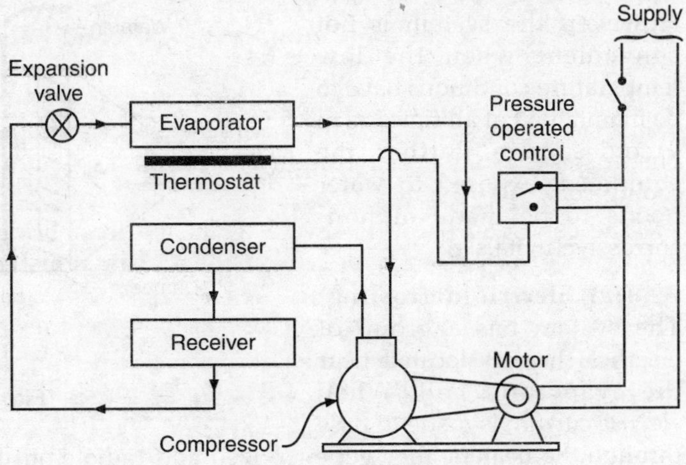

Fig. 11.50. Temperature control defrosting

As the temperature controls the system, the arrangement is known as temperature defrosting. The method, though quite popular with domestic refrigerators, is not reliable in hot weather or at high loads.

Sometimes, pressure sensor can also be used to sense evaporator pressure instead of temperature. The frosting on the coil adversely affects the heat transfer between the air and cooling coil, and this reduces the suction pressure of the compressor. When the pressure falls below the pre-determined value, the running of compressor is stopped by a pressure operated control system. The cooing coils are allowed to warm upto 0°C and defrost before the compressor restarts.

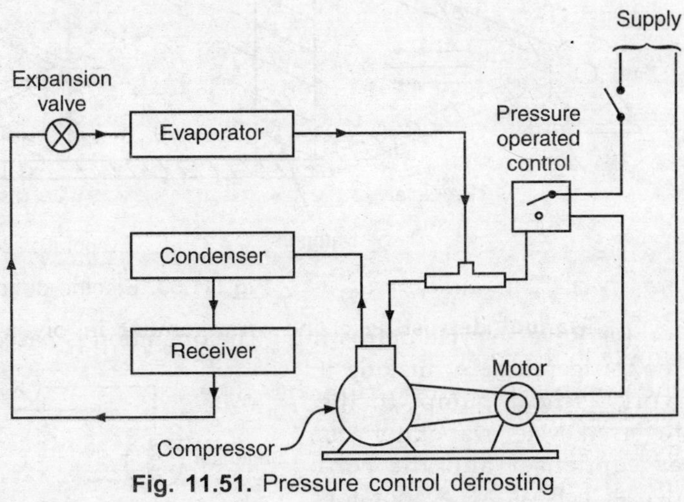

Fig. 11.51. Pressure control defrosting

***(iii)* Water defrosting:** Water at about 20°C is sprayed in sufficient quantity and velocity over the evaporator coils where the frost has accumulated. The process continues for about 4-5 minutes and the frost melts away. After the spray has been turned off, the water is allowed to drain off into a pan from where it can be pumped back to the water source for reuse. The water film left on the surface of coil is dried by using air blown from a fan.

This is a simple and economical defrosting method. However, the system is not convenient when the low temperature conditions have to be maintained at all times as in a cold storage. Further, the equipment exposed to water needs to be made of non-corrosive metals.

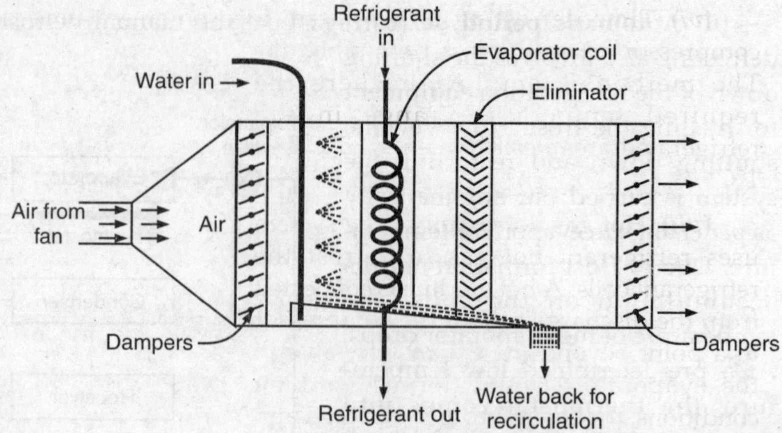

Fig. 11.52. Water defrosting

(iv) Electric defrosting: The system has a group of electrical heaters located near the evaporator coil. When electric current is made to pass through the heaters, they get energized and radiate out heat which melts the frost accumulated on the evaporator coils.

The electric defrosting finds favour for use in finned evaporators. The arrangement requires sufficient long time (more than half hour) for defrost as most of the heat from heaters is lost. This drawback is overcome by using the evaporator coil itself as resistance by connecting it with the supply.

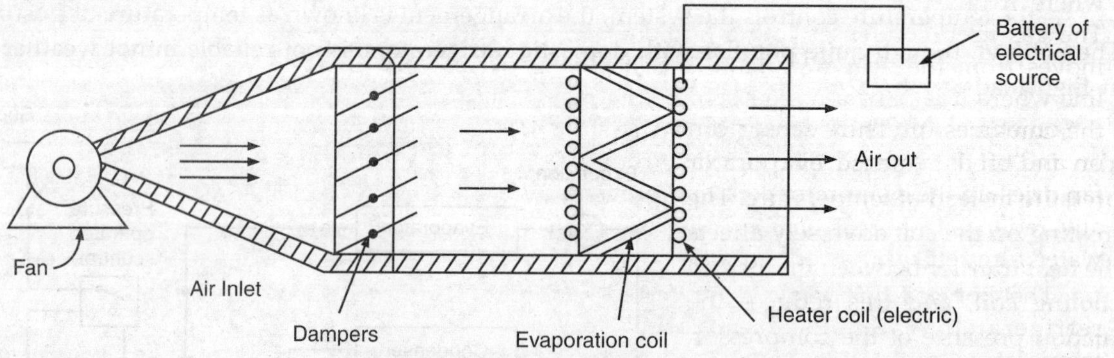

Fig. 11.53. Electric defrosting

(v) Reverse cycle defrosting : The operation is based on reversing the cycle of operation of the refrigerator, *i.e.,* making it work as heat pump. In the reversed cycle, evaporator acts as condenser and the condenser acts as an evaporator. The hot gas from the compressor then flows to the evaporator where it melts the frost.

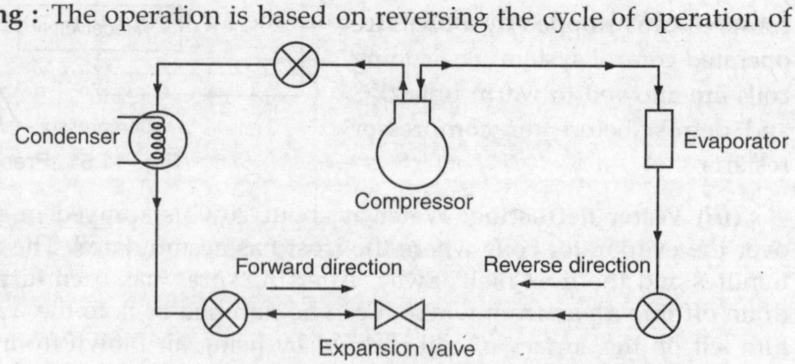

Fig. 11.54. Reverse cycle defrosting

(vi) Time defrosting : The system incorporates a time switch which stops and starts the compressor according to a time schedule. The method is employed where the required temperature range in the refrigerated area lies between – 4.5°C and 1°C.

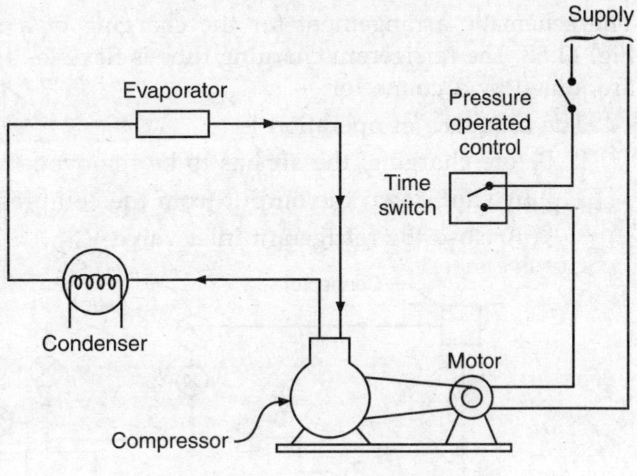

Fig. 11.55. Time defrosting

(vii) Hot gas defrosting : The system uses refrigerant hot gas to defrost the refrigerant coils. A hot gas line is connected from the discharge side of the compressor to a point beyond the expansion valve in the evaporator. Under normal working conditions, the solenoid valve is closed and the refrigerant gas flows from compressor to the evaporator via condenser, receiver, heat exchanger and expansion valve. Finally the gas returns to the compressor from the evaporator.

At regular intervals, the defrosting timer starts defrosting cycle by opening the solenoid valve. The hot refrigerant gas from the compressor is led directly to the evaporator where it gives off heat and melts the frost. The gas itself condenses into a liquid which moves from the evaporator into the suction line where it is re-evaporated before entering the com-pressor. The evaporator fan is shut on and off as required. The air blast from the fan dries up the evaporator coils and prevents any redeposit of frost on the coils from the water film left on the coil surface.

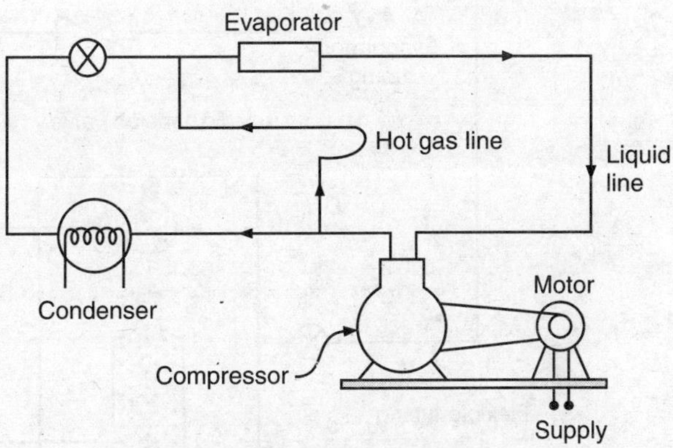

Fig. 11.56. Hot gas defrosting

The hot gas defrosting is preferred for refrigerated warehouses and other low temperature applications.

(viii) Solid absorbent defrosting : The system uses a solid dehydrant (silica gel or activated alumina) which absorbs moisture from the air before it is passed over the evaporator coils.

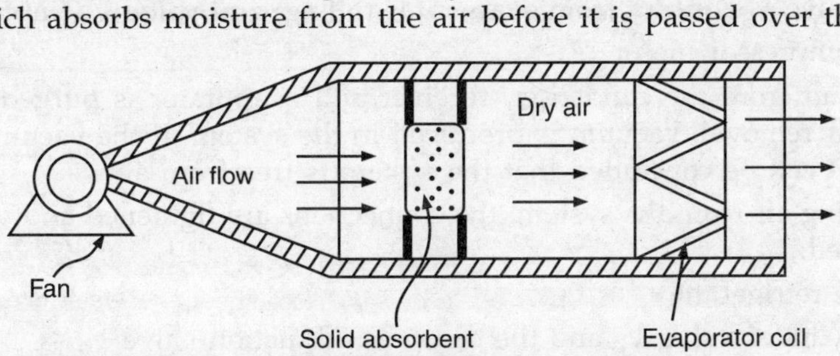

Fig. 11.57. Solid absorbent defrosting

The solid drying agent can be reactivated by blowing hot air over it. This defrosting system has high initial cost and is used only when humidity of the air is not important.

11.8. CHARGING THE REFRIGERANT

The schematic arrangement for the charging of a refrigerant unit has been shown below in Fig. 11.58. The refrigerant charging tube is flexible, the main refrigerant piping is rigid and these are joined by a connecter.

The sequence of operation is :

1. Before charging, the air has to be removed from the refrigerant equipment and for that
   - (*i*) shut down the output from the compressor by closing valve V_1
   - (*ii*) close the refrigerant inlet valve V_2

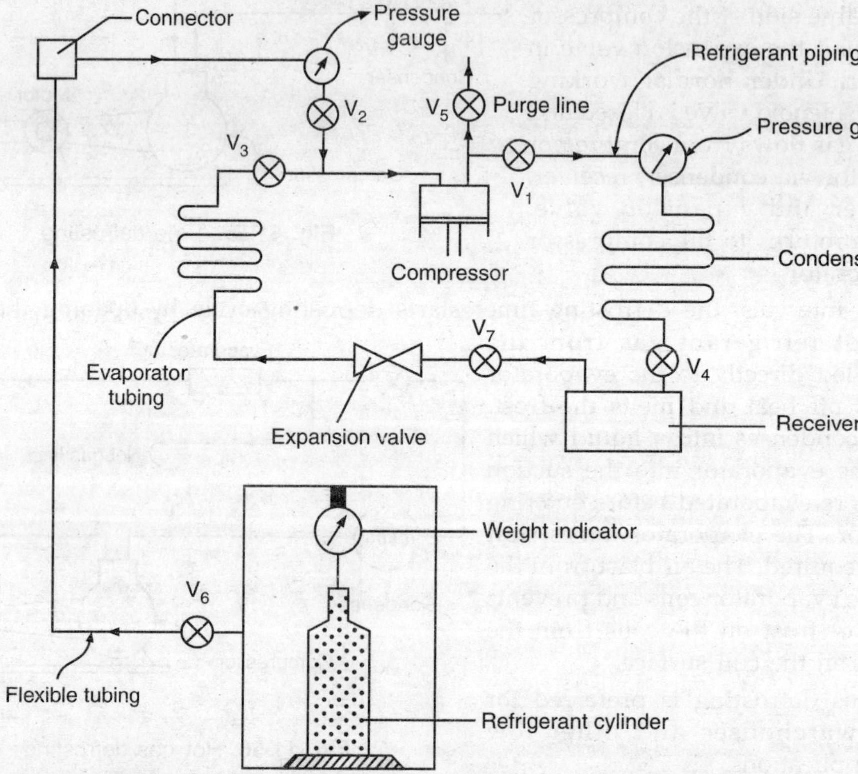

Fig. 11.58. Schematic arrangement for refrigerant charging

   - (*iii*) open the valve V_3 (output from evaporator tubing) and valve V_4 (input to receiver)
   - (*iv*) start the compressor motor.

Consequently, the air from the condenser, receiver and evaporator is purged out through open valve V_5. After air removal, vacuum is produced in the system. If the vacuum is retained for about 30 minutes, it can be concluded that the system is free from air.

   - (*v*) After purging air from the system, the connections are tightened and valves V_3 and V_4 are closed.

2. For charging the refrigerant,
   - (*i*) open the cylinder valve V_6 and the compressor suction valve V_2
   - (*ii*) stop the compressor motor when sufficient quantity of refrigerant has entered the system.
   - (*iii*) close the valves V_2 and V_6. That will disconnect the charge cylinder from the system.

The quantity of refrigerant charged in the system is observed by noting the difference in weight of the refrigerant cylinder. The amount of refrigerant to be charged is dependent on the size and type of installation. For small installations, like water cooler, room cooler and ice cream cabinets etc., the exact quantity is prescribed by the manufacturer. If the recommended weight of refrigerant is not known, the refrigerant is charged in the system till no flashing or bubbles are visible in the sight glass in the liquid line.

Proper charging of the refrigerant is essential for smooth functioning of the refrigeration system. Over-charging results in high temperature and pressure in the system and that may lead to a damage in the compressor. Under-charging results in insufficient cooling and frequent tripping of the compressor motor.

REVIEW QUESTIONS

A. Conceptual and conventional questions.

Compressors

1. State the functions of compressor in a refrigeration system.
2. Describe, with neat sketches, the construction and working principle of a (*i*) reciprocating compressor and (*ii*) centrifugal compressor.
3. What are hermetically sealed compressors ? Point out their salient features.
4. The centrifugal compressors are normally used in large capacity refrigeration systems. Why ?
5. List the different types of compressors. Specify the application of each in refrigeration systems.
6. What is meant by the volumetric efficiency of a reciprocating compressor ? Set up an expression for the volumetric efficiency in terms of pressure ratio, clearance factor and expansion index.
7. (*a*) What is multistage compression ?

   (*b*) A two stage reciprocating compressor works between the evaporator pressure p_1 and condenser pressure p_2 with perfect intercooling. Show that the optimum intermediate pressure for minimum compressor work is given by

$$p_i = \sqrt{p_1 p_2}$$

8. Name and describe the methods of capacity control in :

   (*a*) reciprocating compressor (*b*) centrifugal compressor

   Mention the relative advantages and disadvantages of each method.
9. Explain how does the compressor power, capacity and coefficient of performance vary with the evaporating temperature in an ideal compressor operating at constant condensing temperature.
10. Discuss the influence of condensing tempeature on compressor power, capacity and COP of reciprocating refrigerant compressor when suction (evaporator) temperature is fixed.

Condensers

1. List the advantages/disadvantages of air cooled, condenser over water cooled condenser.
2. Explain, with neat sketches, the construction and working of

   (*a*) shell and coil condenser

   (*b*) shell and tube condenser

   (*c*) double pipe condenser

   Mention the fields for the use of each.

3. (*a*) Sate the function of condenser in a refrigeration system.
   (*b*) Name the different types of condensers used with refrigeration systems.
   (*c*) Describe, with a new sketch, the working of an evaporator condenser. Name the application for which an evaporator condenser is recommended.
4. (*a*) What are the different types of water cooled condensers ?
   (*b*) With the aid of line diagram, describe the working principle of a shell-and-tube condenser.
5. (*a*) What purpose is served by a cooling tower in a condensing unit ?
   (*b*) Name the different types of cooling towers and give a brief description of each one of them.
6. Give a brief description of the waste water system and the recirculated water system used in water cooled condensers.
7. What is fouling factor ? How does it affect the proper functioning of a condenser.
8. State the difference between :
   (*a*) Natural draft cooling tower and mechanical draft cooling tower.
   (*b*) Induced cooling tower and forced draft cooling tower.
9. Explain the following terms as referred to cooling towers :
   (*a*) cooling range
   (*b*) approach.

Capillary Tube and Expansion Valves

1. Explain the salient aspects of a capillary tube. How this tube controls the flow of refrigerant in response to load changes.
2. State the purpose of refrigerant flow control device and name the different types of flow control devices.
3. (*a*) Sketch a thermostatic expansion valve. Name its different parts and explains how it operates.
   (*b*) How a thermostatic expansion valve controls the flow of refrigerant in response to load changes?
4. Distinguish between :
   (*a*) capillary tube and thermostatic expansion valve
   (*b*) thermostatic expansion valve and automatic expansion valve
5. What is the function of a high pressure side float valve ? How is that achieved ?
6. Differentiate between a low-side float valve and a high-side float valve.
7. What is solenoid valve ? Where these valves are used in refrigeration system ?
8. Explain the working of automatic expansion valve. Will it work satisfactorily under changing load conditions ?

Evaporators

1. (*a*) State the purpose of an evaporator in a refrigeration system.
   (*b*) Mention the main requirements which an evaporator must fulfil ?
2. Describe, with a neat diagram, the working principle of :
   (*a*) dry expansion evaporator
   (*b*) flooded type evaporator
3. How is the capacity of an evaporator defined ? List the main factors which effect the heat capacity of an evaporator.
4. (*a*) Define frost and state the reason of frost formation. How does it affect the performance of an evaporator coil?
   (*b*) Explain, with a neat sketch, the working of
   (*i*) hot gas defrost system (*ii*) automatic defrost system.

B. Fill in the blanks with appropriate word/words

1. With increase in clearance volume, the ideal work of compressing unit mass of refrigerant ....
2. The clearance factor is the ratio of .... volume to .... volume of the cylinder.
3. Multi-stage compression with interstage cooling is an effective method of reducing ....
4. Rotary compressors are used mostly for .... refrigerators.
5. For ammonia refrigeration systems, the tubes of shell and tube condenser are made of .... steel.
6. The heat rejection factor for a condenser is defined as the ratio of heat rejected at the condenser to the ....
7. The .... used as expansion device in vapour compression system works on the principle of flow through a pipe with friction causing pressure drop.
8. The thermostatic expansion valve maintains a fixed .... for the vapours at entry to the compressor.
9. In .... condenser, both air and water are employed as a condensing medium.
10. The thermostatic expansion valve is also called the constant .... valve.
11. The .... changes the low pressure liquid refrigerant from the expansion valve into low pressure vapour refrigerant.
12. A solenoid valve may be .... acting or .... operated.
13. The .... on the evaporator acts as an insulation and retards the heat transfer.

Answers

1. remains same 2. clearance, swept 3. power consumption 4. domestic 5. steel 6. refrigerant effect 7. capillary tube 8. degree of superheat 9. evaporative 10. superheat valve 11. evaporator 12. direct, pilot 13. frost.

C. Multiple choice questions.

1. In reciprocating compressors, clearance is provided to
   (*a*) improve the volumetric efficiency of the compressor
   (*b*) accommodate valves
   (*c*) reduce power consumption of the compressor
   (*d*) increase the isentropic efficiency
2. The work input to reciprocating air compressor will be least if the exponent n in the expression pv^n = constant equals
   (*a*) $n = 1$ (*b*) $n = 1.2$
   (c) $n = Y = 1.4$ (adiabatic compression) (*d*) n takes infinite valve
3. Consider the following factors
   1. cylinder size 2. clearance ratio
   3. delivery pressure 4. compressor shaft power

   The factors which affect the volumetric efficiency of a single stage compressor will include.
   (*a*) 1 and 2 (*b*) 3 and 4 (*c*) 2 and 3 (*d*) 1 and 4
4. An intercooler is used with multi-stage reciprocating compressors to cool the
   (*a*) refrigerant gas before it is sucked inside the compressor cylinder
   (*b*) refrigerant gas before it is discharged to the receiver
   (*c*) refrigerant gas after one stage of compression before it is sent to the next stage
   (*d*) compressor cylinder
5. Which aspect is not true for hermetically sealed type compressors?
   (*a*) compact unit and less requirement of space
   (*b*) difficult maintenance as the welded joint has to be broken for repairs
   (*c*) no need of separate pump for evacuation
   (*d*) more effective cooling of motor

6. Natural convection air cooled condensers are used for
   1. domestic refrigerators 2. domestic air conditioners
   which of these statements are correct ?
   (*a*) 1 only (*b*) 2 only (*c*) both 1 and 2 (*d*) none
7. The heat rejection factor (ratio) for a condenser depends upon *COP* of the refrigeration system. If *COP* of the system is 5, the *HRF* would be
   (*a*) 0.8 (*b*) 1.2 (*c*) 2.236 (*d*) 2.45
8. Fouling or scaling factor is the term used with
   (*a*) evaporators (*b*) compressors
   (*c*) condensers (*d*) expansion valves
9. Capillary tube used as expansion device in vapour compression refrigeration systems works on the principle of drop in pressure because of
   (*a*) flow through pipe with friction (*b*) isothermal expansion
   (*c*) adiabatic expansion (*d*) throttle expansion
10. Thermostatic expansion valve operates on the changes in
    (*a*) temperature in the evaporator
    (*b*) pressure in the evaporator
    (*c*) degree of superheat exit from the evaporator
    (*d*) degree of subcooling in the evaporator
11. Hunting of thermostatic expansion valve refers to
    (*a*) alternate over-feeding and starving of refrigeration flow in the evaporator
    (*b*) variation of evaporator load with degree of superheat
    (*c*) variation of pressure in the evaporator with variation of load
    (*d*) there is no such term in relation to expansion valve
12. Periodic defrosting is required in domestic refrigerators because frost
    (*a*) overcools food stuff (*b*) reduces heat extraction
    (*c*) causes corrosion of metals (*d*) blocks flow of refrigerant
13. The performance of which of the following units in a vapour compression system is adversely affected by the presence of moisture ?
    (*a*) condenser (*b*) compressor (*c*) evaporator (*d*) expansion valve
14. The presence of frost on the cooling coils of the cvaporator of a refrigeration system
    (*a*) increases the compressor pressure
    (*b*) acts as the insulator and decreases the heat transfer rate
    (*c*) makes the compressor run for comparatively short durations
    (*d*) reduces the life of cooling coils
15. Which of the following valves maintains a constant degree of superheat at the end of evaporator coil?
    (*a*) automatic expansion valve (*b*) thermostatic expansion valve
    (*c*) low side float valve (*d*) high side float valve
16. Moisture should be removed from the refrigerant to avoid
    (1) failure of compressor seal
    (2) freezing at the expansion valve
    (3) restriction to refrigerant flow
    (4) corrosion of steel parts

Which of these statements are correct?

(*a*) 1 and 2 (*b*) 1, 3 and 4 (*c*) 2, 3 and 4 (*d*) 1, 2, 3 and 4

17. The charging of ammonia in a vapour compression system is done on
    (*a*) low pressure side near the evaporator
    (*b*) low pressure side near the compressor
    (*c*) high pressure side close to receiver
    (*d*) any where in the cycle
18. The compressor of a domestic refrigerator has been noted to run for comparatively large durations. That indicates
    (*a*) highly efficient operation
    (*b*) over loading of the machine
    (*c*) capillary tube choked
    (*d*) thermostat gone defective
19. A clicking sound was heard in the expansion valve of a newly charged ammonia compressor system. This could be due to
    (*a*) leakage of ammonia
    (*b*) less charge of ammonia
    (*c*) excessive charge of ammonia
    (*d*) evaporator coils choked

Answers

1. (*b*)	2. (*a*)	3. (*c*)	4. (*c*)	5. (*c*)	6. (*c*)	7. (*b*)	8. (*c*)
9. (*a*)	10. (*c*)	11. (*a*)	12. (*b*)	13. (*b*)	14. (*b*)	15. (*b*)	16. (*c*)
17. (*c*)	18. (*b*)	19. (*b*)					

□□□

Chapter – 12

Psychrometry: Moist Air Properties

Learning objectives : A study of the subject matter presented in this chapter will enable the students/readers to

- *define psychrometry and discuss the important terms related to it*
- *learn how different properties of moist air are laid on a typical psychrometric chart*
- *use psychrometric chart for understanding the different air-conditioning processes*

12.1. WHAT IS PSYCHROMETRY?

Psychrometry is the science of studying the thermodynamic properties of moist air and the use of these properties to analyse the conditions and processes involving moist air.

For many purposes, the composition of real air can be assumed to be a mixture of two components:

• **Standard dry air :** Dry air is a mixture of number of gases such as oxygen, nitrogen, carbon dioxide, hydrogen, argon etc. Oxygen and nitrogen are the main constituents with the following composition

21% oxygen and 79% of nitrogen ..... by volume

23% oxygen and 77% nitrogen ..... by mass

The molecular mass of dry air is taken is 20.966 and gas constant equal to 287 J/kgK. For psychrometric purpose, dry air is assumed to be a pure substance and not a mixture.

• **Water vapour:** Air has affinity for water and consequently the atmospheric air always contains some water; water vapour content varies from 0 to 3% by mass.

The moist air is essentially a mixture of dry air and water vapour. The amount of water vapour present depends on the absolute pressure and temperature of the mixture. This two-component model is more than adequate for many practical thermo-physical calculations such as:

- analysis of combustion products in the cylinder of an internal combustion engine which contains CO_2, CO, N_2 and water vapours.
- processing plants involving heating, cooling, drying and humidity control systems.
- defining performance criteria and designs for cooling tower systems.

Treating air as a binary mixture makes it relatively simple to define the relationships between various properties such as temperature, pressure, density, water content (humidity) and internal energy or heat content (enthalpy).

12.2. DALTON'S LAW OF PARTIAL PRESSURE

"the total pressure of a mixture of gases is equal to the sum of the partial pressures exerted by each gas when it occupies the mixture volume at the mixture temperature".

For the moist air, which is a mixture of dry air and water vapour

$$p_t = p_a + p_v$$

where p_t = total pressure of moist air

p_a = partial pressure of dry air

p_v = partial pressure of water vapour

Refer Fig.12.1., for the three states on *T-s* diagrams for water vapour and dry air respectively,

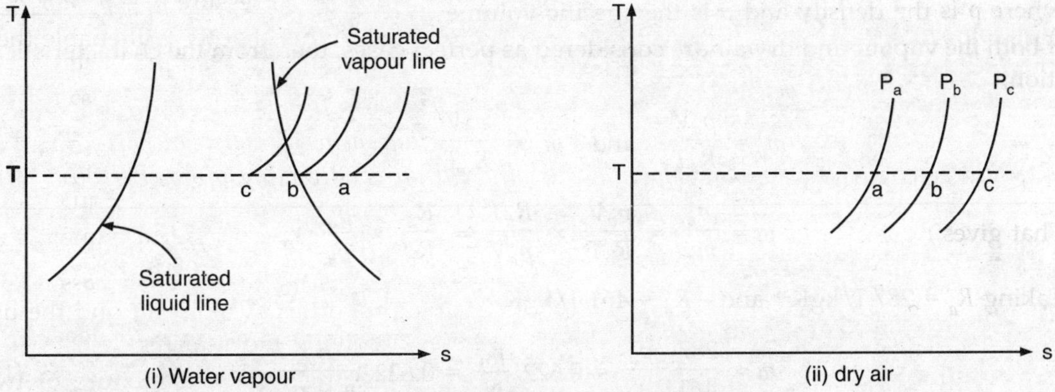

Fig. 12.1. States for water vapour and dry air

It is to be noted that at constant temperature, the partial pressure of air decreases with an increase in the partial pressure of water vapour.

The total pressure p_t is usually the barometric (atmospheric) pressure. Then

$$\text{mole fraction of dry air} = \frac{p_a}{p_t}$$

$$\text{mole fraction of water vapour} = \frac{p_v}{p_t}$$

12.3. SATURATED/UNSATURATED AIR MIXTURE AND SUPER-SATURATED VAPOUR

Saturated air-mixture: Mixture of dry air and water vapour in which the partial pressure of the water vapour is equal to its saturation pressure corresponding to the temperature of the mixture.

Unsaturated air mixture: Mixture of dry air and water vapour in which the partial pressure of the water vapour is less than its saturation pressure corresponding to the temperature of the mixture.

Super saturated air mixture: Mixture of dry air and water vapour in which the partial pressure of the water vapour is greater than its saturation pressure corresponding to the temperature of the mixture.

12.4. SPECIFIC HUMIDITY, RELATIVE HUMIDITY AND DEGREE OF SATURATION

Humidity refers to the dampness, *i.e.*, the water content of air. ***Absolute humidity*** represents the amount of water vapour actually present in the air, expressed as gram per cubic meter of air.

The ***specific humidity*** or ***humidity ratio*** or ***moisture content*** is the ratio of mass of water vapour to the mass of dry air in a given volume of the mixture. Consider a mixture consisting of m_a kg of dry air and m_v kg of water vapour contained in a vessel of volume V at total pressure p_t and temperature T. Then

$$\text{Specific humidity } \omega = \frac{m_v}{m_a}$$

m_a kg of dry air
m_v kg of water vapour
Volume V
Temperature T
Pressure $p_t = p_a + p_v$

Since both masses occupy volume V,

$$\omega = \frac{m_v/V}{m_a/V} = \frac{\rho_v}{\rho_a} = \frac{v_a}{v_v}$$

where ρ is the density and v is the specific volume

If both the vapour and dry air are considered as perfect gases, then from the characteristic gas equation,

$$m_a = \frac{p_a V}{R_a T} \quad \text{and} \quad m_v = \frac{p_v V}{R_v T}$$

That gives :

$$\omega = \frac{m_v}{m_a} = \frac{p_v V}{R_v T} \times \frac{R_a T}{p_a V} = \frac{R_a}{R_v} \times \frac{p_v}{p_a}$$

Taking R_a = 287 J/kgK and R_v = 461 J/kgK

$$\omega = \frac{287}{461}\frac{p_v}{p_a} = 0.622\frac{p_v}{p_a} = 0.622\frac{p_v}{p_t - p_v} \quad ...(12.1)$$

The above relation shows that if the total pressure remains constant, the specific humidity is a function of partial pressure of water vapour only. This relationship has been obtained by assuming that behaviour of water vapour is identical with that of an ideal gas. Such an assumption is quite valid at low pressure and normal humidity conditions.

The ***relative humidity*** is the ratio (expressed as a percentage) of the amount of water vapour actually present in a given nolume of air to the maximum amount that the air could hold under the same pressure and temperature conditions.

$$\text{Relative humidity } \phi = \frac{\text{mass of water vapour in a given volume}}{\text{mass of water vapour in the same volume if saturated at the same temperature}}$$

With the assumption that vapours behave as perfect gases, we have

$$m_v = \frac{p_v V_v}{R_v T_v} \quad \text{and} \quad m_{vs} = \frac{p_{vs} V_{vs}}{R_{vs} T_{vs}}$$

where *vs* subscript is used for saturated vapour.

$$\therefore \quad \phi = \frac{m_v}{m_{vs}} = \frac{p_v V_v}{R_v T_v} \times \frac{R_{vs} T_{vs}}{p_{vs} V_{vs}}$$

Also $V_v = V_{vs}$; $R_v = R_{vs}$ and $T_v = T_{vs}$. That gives

$$\phi = \frac{p_v}{p_{vs}} \quad ...(12.2)$$

where p_{vs} is the saturation pressure at the temperature of the mixture. This saturation pressure is obtained from the steam tables corresponding to the given temperature.

Apparently, the relative humidity can also be defined as the ratio of the partial pressure of water vapour in a given volume of mixture to the partial pressure of water vapour when the same volume of mixture is saturated at the same temperature. This implies that ϕ equals unity for saturated air; 100 percent relative humidity means that air contains the maximum moisture it can hold.

Essentially, the term relative humidity compares the humidity of the given air with the humidity of saturated air at the same pressure and temperature. For saturated air, the relative humidity is 100 percent.

The ***degree of saturation*** μ represents the ratio of mass of water vapour associated with unit mass of dry air to the mass of water vapour associated with unit mass of dry air saturated at the same temperature.

$$\mu = \frac{\text{mass of water vapour with unit mass of dry air}}{\text{mass of water vapour with unit mass of dry saturated air}}$$

This relation implies that μ represents the ratio of specific humidity of moist air to specific humidity of saturated air at the same temperature. That is

$$\mu = \frac{\omega}{\omega_s} = \frac{0.622\dfrac{p_v}{p_t - p_v}}{0.622\dfrac{p_{vs}}{p_t - p_{vs}}} = \frac{p_v}{p_{vs}}\,\frac{p_t - p_{vs}}{p_t - p_v} = \frac{p_v}{p_{vs}}\left[\frac{1-\dfrac{p_{vs}}{p_t}}{1-\dfrac{p_v}{p_t}}\right] \qquad \text{...(12.3)}$$

The following observations can be made from the above relation:

(*i*) If the air is dry, then $p_v = 0$ and therefore $\mu = 0$.

If the relative humidity $\phi = \dfrac{p_v}{p_{vs}} = 1$, then $p_v = p_{vs}$ and accordingly $\mu = 1$.

Thus the degree of saturation varies between 0 and 1 .

(*ii*) The degree of saturation is a measure of the capacity of the air to absorb moisture.

When $\mu = 1$, then $\omega = \omega_s$

which implies that air is holding the maximum amount of water vapour.

(*iii*) The expression for the degree of saturation can be recast as

$$\mu = \frac{p_v}{p_{vs}}\left[\frac{1-\dfrac{p_{vs}}{p_t}}{1-\dfrac{p_v}{p_{vs}}\times\dfrac{p_{vs}}{p_t}}\right] = \phi\left[\frac{1-\dfrac{p_{vs}}{p_t}}{1-\phi\dfrac{p_{vs}}{p_t}}\right]$$

or $$\phi - \phi\,\frac{p_{vs}}{p_t} = \mu - \mu\,\phi\,\frac{p_{vs}}{p_t}$$

or $$\phi\left[1-\frac{p_{vs}}{p_t}+\mu\frac{p_{vs}}{p_t}\right] = \mu$$

or $$\phi \left[1-(1-\mu)\frac{p_{vs}}{p_t} \right] = \mu$$

$$\therefore \quad \phi = \frac{\mu}{1-(1-\mu)\dfrac{p_{vs}}{p_t}} \qquad \text{...(12.4)}$$

The difference between relative humidity and degree of saturation is usually less than 2%.

12.5. DRY BULB TEMPERATURE AND WET BULB TEMPERATURE

The ***dry-bulb temperature*** (*dbt*) is the normal temperature of an air-vapour mixture as indicated or recorded by any temperature measuring device placed in the mixture. This temperature is not affected by the moisture content in the mixture.

The ***wet-bulb temperature*** (*wbt*) of an air-vapour mixture is the temperature measured by a thermometer whose bulb is covered by a wick soaked in water. When the air passes over the wet wick, the moisture contained in the wick tends to evaporate. That produces cooling effect at the bulb and an equilibrium temperature lower than that of the air stream is recorded.

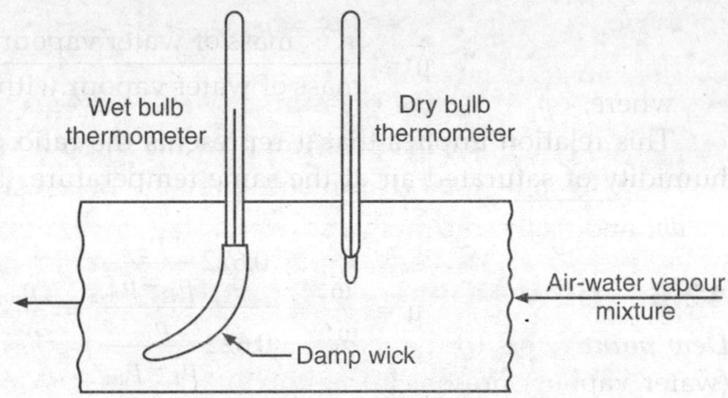

Fig. 12.2. Dry and wet bulb temperature

It is worthwhile to note that:

(*i*) The wet-bulb temperature is lower than the dry-bulb temperature and the difference is known as the ***wet-bulb depression.***

(*ii*) The wet-bulb depression is greatest when the air is initially completely dry, *i.e.*, capable of absorbing a maximum amount of moisture.

(*iii*) When the air is initially saturated, there will be no evaporation of water and hence the two thermometers will record equal temperatures. This implies that the depression is zero with 100 percent relative humidity.

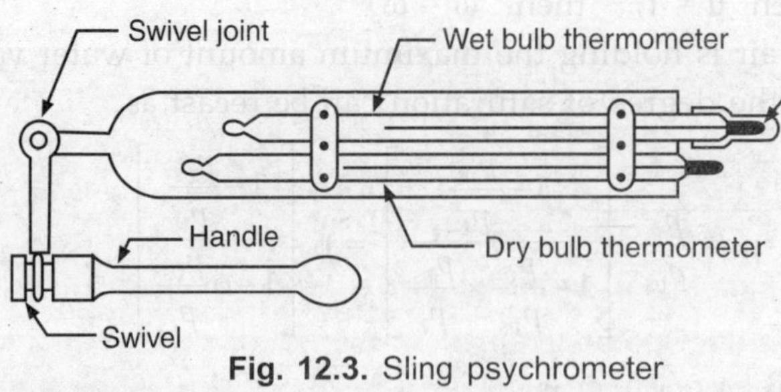

Fig. 12.3. Sling psychrometer

(*iv*) The dry and wet-bulb temperatures are simultaneously measured by instruments called ***psychrometers.*** A sling psychrometer consists of two identical mercury-in-glass themometers mounted on a suitable frame and arranged with a swivel-mounted handle as shown in Fig. 12.3. The temperature sensing bulb of one of the thermometers is covered with a knitted or woven cotton wick which is wetted with pure clean water. For better and accurate measurements of the wet bulb temperature, a fast movement of air past the moistened wick is necessary. This is to ensure that the surrounding air does not cling to the moistened wick and that the air at the wet-bulb thermometer is always in immediate contact with the wet wick. The necessary air motion, 5 m/s to 10 m/s, is provided by rotating the psychrometer frame with the swivel mounted handle. The readings are taken after swinging the psychrometer in a smooth circular path for 15 to 20 seconds. With a too short duration, the temperature will not be depressed to its proper value. If the swinging period is too large, the wick will dry and the bulb temperature will not remain at its minimum value.

(*v*) When dry and wet bulb temperatures are known, the other psychrometric properties can be determined by calculations.

(*vi*) Many investigation have suggested different expressions to determine the partial pressure of water vapour in air from the wet and dry bulb temperature readings. Carrier's equation, as given below, is most widely used.

$$p_v = (p_{vs})_{wb} - \frac{(p_t - p_{vs})(dbt - wbt)}{1544 - 1.44\, wbt} \qquad ...(12.5)$$

where, p_v = partial pressure of water vapour

p_{vs} = partial pressure of water vapour when air is fully saturated

p_t = total (barometric pressure of moist air)

dbt and *wbt* = dry bulb and wet bulb temperature respectively in °C.

12.6. DEW POINT AND ADIABATIC SATURATION TEMPERATURE

Dew point refers to the temperature at which the mixture becomes saturated, *i.e.*, the moisture (water vapour) present in the mixture begins to condense consequent to continuous cooling at constant pressure.

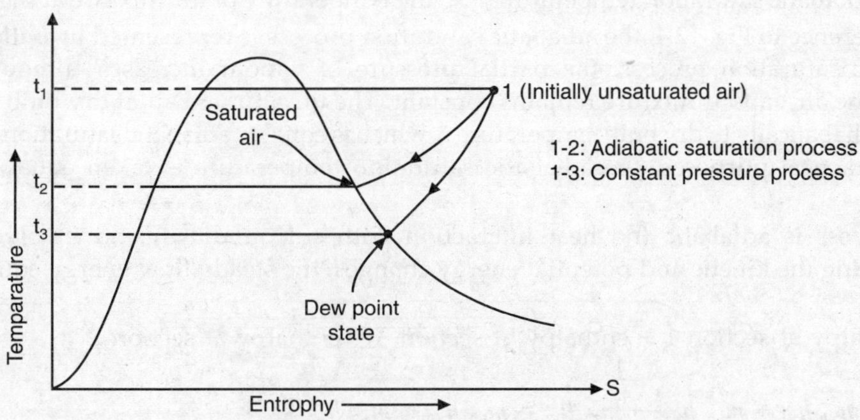

Fig. 12.4. Concept of dew point temperature

With reference to Fig. 12.4, let point 1 represent the initial state of air-water vapour mixture. The vapour is superheated, the pressure equals the partial pressure of the superheated vapour and temperature corresponds to the temperature of the mixture. When the mixture is cooled at constant pressure along the path 1-3, the cooling continues until the vapour attains the saturated state at point 3. With further cooling, the vapour condenses, *i.e.*, the moisture is released. The state point 3 represents the dew point temperature (*dpt*) of the air-water vapour mixture. The dew point temperature thus corresponds to the saturation temperature of steam at the partial pressure of water vapour in the mixture. For saturated air, the dry bulb temperature, the wet bulb temperature and the dew point temperature is same.

The adiabatic saturation temperature (or the thermodynamic wet bulb temperature) refers to the temperatures at which the air can be brought adiabatically to saturation state by the evaporation of water into a flowing stream.

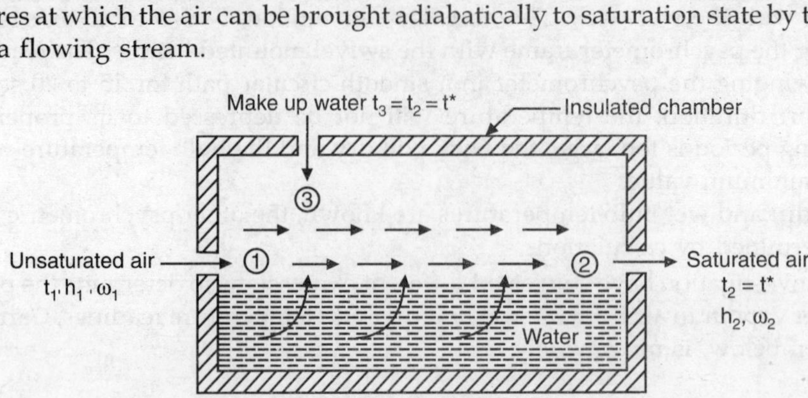

Fig. 12.5. Concept of adiabatic saturation process

Consider a stream of unsaturated air-vapour mixture flowing over a surface of water contained in a chamber which is sufficiently long and is perfectly insulated. Due to intimate contact between the unsaturated air and liquid water, some of the water evaporates and is carried by air resulting into an increase in its humidity. The heat required for evaporation comes both from the air-vapour mixture and the liquid water in the chamber. The process continues until the energy transferred from the air to the water is equal to the energy required to vaporise the water. By the time, air reaches the exit section, it becomes saturated and an equilibrium is established. The temperature of the saturated air at the exit section is known as ***adiabatic saturation temperature*** or the ***thermodynamic wet bulb temperature***. The steady state thermal equilibrium conditions are maintained by adding make up water steadily at the rate of evaporation. This make up water has to be at the adiabatic saturation temperature, *i.e.*, the temperature of the mixture at the exit section.

With reference to Fig. 12.4, the adiabatic saturation process is represented by path 1-2. During the adiabatic saturation process, the partial pressure of vapour increases, although the total pressure of the air-vapour mixture remains constant. The unsaturated air at dry bulb temperature t_1 is cooled adiabatically to dry bulb temperature t_2 which is equal to adiabatic saturation temperature t^*. For all practical purposes, the adiabatic saturation temperature is taken equal to wet bulb temperature.

The process is adiabatic (no heat interaction with surroundings) and involves no work. Then neglecting the kinetic and potential energy changes, the steady flow energy equation can be written as:

enthalpy at section 1 + enthalpy at section 3 = enthalpy at section 2

That is

$$(m_{a1}\, h_{a1} + m_{v1}\, h_{v1}) + m_{f3}\, h_{f3} = m_{a2}\, h_{a2} + m_{v2}\, h_{v2} \qquad \text{...}(i)$$

where m_a, m_v = mass of dry air and water vapour respectively

h_a, h_v = specific enthalpy of dry air and water vapour respectively

Suffix 1 and 2 refer to the inlet (unsaturated) and outlet (saturated) conditions respectively.

m_{f3}, h_{f3} = mass and specific enthalpy of water (fluid) added to make up.

From the principle of mass conservation:

$$m_{a1} = m_{a2} = m_a \text{ for dry air}$$

and $$m_{v1} + m_{f3} = m_{v2}$$

$$\text{or} \qquad \frac{m_{v1}}{m_{a1}} + \frac{m_{f3}}{m_a} = \frac{m_{v2}}{m_{a2}}$$

$$\text{or} \qquad \omega_1 + \frac{m_{f3}}{m_a} = \omega_2$$

$$\text{or} \qquad m_{f3} = m_a\,(\omega_2 - \omega_1) \qquad \text{...(ii)}$$

Dividing expression (*i*) by $m_a = m_{a1} = m_{a2}$, we get

$$\frac{m_{a1}\,h_{a1}}{m_{a1}} + \frac{m_{v1}\,h_{v1}}{m_{a1}} + \frac{m_{f3}\,h_{f3}}{m_a} = \frac{m_{a2}\,h_{a2}}{m_{a2}} + \frac{m_{v2}\,h_{v2}}{m_{a2}}$$

$$\text{or} \qquad h_{a1} + \omega_1\,h_{v1} + \frac{m_{f3}\,h_{f3}}{m_a} = h_{a2} + \omega_2\,h_{v2}$$

where ω is the specific humidity.

Substituting $m_{f3} = m_a\,(\omega_2 - \omega_1)$ from identity (*ii*), we get

$$h_{a1} + \omega_1\,h_{v1} + (\omega_2 - \omega_1)\,h_{f3} = h_{a2} + \omega_2\,h_{v2}$$

$$\text{or} \qquad \omega_1(h_{v1} - h_{f3}) = (h_{a2} - h_{a1}) + \omega_2(h_{v2} - h_{f3})$$

$$\text{or} \qquad \omega_1 = \frac{(h_{a2} - h_{a1}) + \omega_2(h_{v2} - h_{f3})}{(h_{v1} - h_{f3})}$$

The liquid water at section 3 enters the chamber at the same temperature as that of saturated air leaving the chamber at the exit section 2. In that case $h_{f3} = h_{f2}$. Further at low pressures, water vapours behave like an ideal gas even near the saturated vapour state. Hence $h_{v1} = h_{g1}$ and $h_{v2} = h_{g2}$ where suffix *g* stands for gas. Accordingly,

$$\omega_1 = \frac{(h_{a2} - h_{a1}) + \omega_2(h_{g2} - h_{f2})}{(h_{g1} - h_{f2})}$$

If air is considered to behave like an ideal gas, then

$$(h_{a2} - h_{a1}) = c_p\,(t_2 - t_1)$$

$$\text{Hence} \qquad \omega_1 = \frac{c_p(t_2 - t_1) + \omega_2\,h_{fg2}}{(h_{g1} - h_{f2})}$$

where h_{fg2} is the latent heat of vaporization corresponding to adiabatic saturation temperature $t^* = t_2$. Normally, the properties at adiabatic saturation temperature are marked with a star and so we have

$$\omega_1 = \frac{c_p(t^* - t_1) + \omega^*\,h^*_{fg2}}{(h_{g1} - h^*_f)} \qquad \text{...(12.6)}$$

The above equation is a convenient approach for evaluating the specific humidity ω_1 by measuring t_1 (dry bulb temperature) and t^* (adiabatic saturation temperature).

The process of adiabatic saturation occurs in desert coolers employed for the evaporative cooling of air in dry climate.

12.7. ENTHALPY OF MOIST AIR

The enthalpy of moist air is obtained by adding the enthalpies of dry air and associated water vapour. Thus

$$mh = m_a h_a + m_v h_v \qquad ...(12.7)$$

where m , m_a and m_v are masses of moist air, dry air and associated water vapour.

h , h_a , h_v are specific enthalpies of moist air, dry air and associated water vapour.

Then enthalpy of moist air per unit mass of dry air is

$$h = \frac{m_a h_a + m_v h_v}{m_a} = h_a + \frac{m_v}{m_a} h_v = h_a + \omega h_v$$

$$= c_{pa} t_{db} + \omega h_v \qquad ...(12.8)$$

Further, enthalpy of vapour

$$h_v = (h_f)_{dp} + (h_{fg})_{dp} + c_{pv} (dbt - dpt) \qquad ...(12.9)$$

$$= c_{pf} \times dpt + (h_{fg})_{dp} + c_{pv} (dbt - dpt)$$

where *dpt* is the dew point temperature corresponding to *dbt*

c_{pf} is specific heat of water (4.186 kJ/kgK)

$(h_{fg})_{dp}$ is latent heat of vaporisation of water at *dpt*

c_{pv} is specific heat of super heated vapour (1.88 kJ/kgK)

$(dbt - dpt)$ is degree of superheat of the vapour

$$\therefore \quad h = c_{pa}\, dbt + \omega\, [c_{pf}\, dpt + (h_{fg})_{dpt} + c_{pv} (dbt - dpt)]$$

$$= 1.005\, dbt + \omega\, [4.186\, dpt + (h_{fg})_{dpt} + 1.88\, (dbt - dpt)]$$

$$= (1.005 + 1.88\, \omega)\, dbt + \omega\, (h_{fg})_{dpt} + 2.306\, \omega\, dpt$$

$$= c_{pm}\, dbt + \omega\, [(h_{fg})_{dpt} + 2.306]\, dpt \qquad ...(12.10)$$

where c_{pm} is called the *humid specific heat* or heat capacity of (1 + ω) kg per kg of dry air.

At low pressures normally encountered in psychrometry, the superheated vapour behaves as a gas and its enthalpy can be considered only as a function of temperature. In that case, the enthalpy of superheated vapour at dew point temperature *dpt* is approximately equal to the enthalpy of saturated vapour at saturation temperature equal to dry bulb temperature *dbt*. Hence

$$h_f + h_{fg} + c_{pv} (dbt - dpt) = \text{enthalpy of saturated vapour at } dbt$$

$$\therefore \quad h = c_{pa}\, dbt + \omega\, (h_g)_{at\ dbt} \qquad ...(12.11)$$

A better approximation is given by the relation: $h_v = 2500 + 1.88\, dbt$. Then enthalpy of moist air

$$h = c_{pa}\, dbt + \omega\, (2500 + 1.88\, dbt) \qquad ...(12.12)$$

It is noted that here 2500 kJ/kg is the enthalpy of saturated vapour at 0°C and 1.88 kJ/kgK is the specific heat of superheated vapour.

Equation 12.12 is a convenient empirical relation for calculating the enthalpy of moist air when steam tables are not available.

EXAMPLE 12.1

A metal beaker contains water initially at room temperature and the water is cooled by gradually adding ice water to it. When the water temperature reaches 13°C, the moisture from room air begins to condense on the beaker. Make calculations for the specific humidity and parts by mass of water vapour in the room air. Take:

room air temperature = 22°C and barometric pressure = 1.01325 bar

Solution: From steam tables, the partial pressure of water vapour at *dpt* of 13°C

$$p_v = 1500 \text{ N/m}^2$$

(*a*) Specific humidity or humidity ratio,

$$\omega = \frac{m_v}{m_a} = 0.622\frac{p_v}{p_a} = 0.622\frac{p_v}{p_t - p_v}$$

$$= 0.622 \times \frac{1500}{101325 - 1500} = \mathbf{0.009346} \text{ kg/kg of dry air}$$

(*b*) Parts of mass of water vapour,

$$\frac{m_v}{m} = \frac{\omega}{1+\omega} = \frac{0.009346}{1 + 0.009346} = \mathbf{0.00926} \text{ kg/kg of mixture}$$

EXAMPLE 12.2

The following data pertains to a sample of moist air:

dry bulb temperature = 30 °C ; dew point temperature = 15° C

total pressure = 1 bar

Compute the values for relative humidity, specific humidity and degree of saturation.

Solution : Corresponding to *dbt* = 303 K and *dpt* = 288 K, the saturation pressures read from the steam tables are:

$$p_{vs} = 0.04246 \text{ bar} \quad \text{and} \quad p_v = 0.01705 \text{ bar}$$

Then: Relative humidity $\phi = \frac{p_v}{p_{vs}} = \frac{0.01705}{0.04246} = 0.4015$

$$\text{Specific humidity } \omega = 0.062\frac{p_v}{p_t - p_v}$$

$$= 0.622 \times \frac{0.01705}{1 - 0.01705} = \mathbf{0.01079} \text{ kg of vapour per kg of dry air}$$

$$\text{Degree saturation } \mu = \frac{p_v}{p_t - p_v} \times \frac{p - p_{vs}}{p_s}$$

$$= \frac{0.01705}{1 - 0.01705} \times \frac{1 - 0.04246}{0.04246} = \mathbf{0.3911}$$

EXAMPLE 12.3

(a) A mixture of dry air and water vapour is at a temperature of 21°C under a total pressure of 736 mm of mercury. The partial pressure of water vapour is 1707.5 N/m² and its saturation pressure is 2489.8 N/m². Compute the relative humidity specific humidity and degree of saturation.

(b) The atmospheric conditions are 20°C and specific humidity of 0.0095 kg/kg of dry air. Calculate partial pressure of vapour and relative humidity. ***(WBTU)***

Solution : (*a*) Total pressure $p_t = \rho g h = (1000 \times 13.6) \times 9.81 \times (736 \times 10^{-3})$

$$= 98194 \text{ N/m}^2$$

Given : $p_v = 1707.5 \text{ N/m}^2$ and $p_{vs} = 2489.8 \text{ N/m}^2$

(*i*) Relative humidity $\phi = \frac{p_v}{p_{vs}} = \frac{1707.5}{2489.8} = \mathbf{0.686}$

(*ii*) Specific humidity $\omega = 0.622\left(\frac{p_v}{p_t - p_v}\right)$

$$= 0.622\left[\frac{1707.5}{98194 - 1707.5}\right] = \mathbf{0.011}\ \text{kg/kg of dry air}$$

(*iii*) Degree of saturation $\mu = \frac{p_v(p_t - p_{vs})}{p_{vs}(p_t - p_v)}$

$$= \frac{1707.5\,(98194 - 2489.8)}{2489.8\,(98194 - 1707.5)} = \mathbf{0.68}$$

(*b*) The specific humidity is prescribed by the relation,

$$\omega = 0.622\frac{p_v}{p_t - p_v}$$

Substituting the given values for ω and *p*, we have

$$0.0095 = 0.622\left(\frac{p_v}{1.0132 - p_v}\right)$$

or $\quad 0.0095\,(1.0132 - p_v) = 0.622\,p_v$

That gives : $p_v = \frac{1.0132 \times 0.0095}{0.622 + 0.0095} = 0.01524$ bar

(*ii*) Corresponding to 20°C, we read $p_{vs} = 0.0234$ bar steam tables.

Then : Relative humidty $\phi = \frac{p_v}{p_{vs}} = \frac{0.01524}{0.0234} = 0.65$ or 65%.

EXAMPLE 12.4

***(a)** A room contains 35 kg of dry air and 0.5 kg of water vapour. The total pressure and temperature of air in the room are 100 KPa and 25°C respectively. Given that the saturation pressure of water at 25°C is 3.17 kPa, compute the relative humidity of air in the room.*

***(b)** The pressure, dry bulb temperature and relative humidity of dry air in a room are 1 bar, 30°C and 70% respectively. If the saturated steam pressure at 30°C is 4.25 kPa, Calculate the specific humidity of the room air in kg of water per kg of dry air.*

Solution : (*a*) The specific humidity or humidity ratio is defined as the ratio of water vapour to the mass of dry air in a given volume of mixture. That is

$$\omega = \frac{m_v}{m_a} = \frac{0.5}{35} = 0.01428$$

The specific humidity is also given as

$$\omega = 0.622 \times \frac{p_v}{p_t - p_v} \quad ; \quad 0.01428 = 0.622 \times \frac{p_v}{100 - p_v}$$

The solution gives: $p_v = 2.244$ kPa

Relative humidity $\phi = \dfrac{p_v}{p_{vs}} = \dfrac{2.244}{3.14} = 0.708$ or **70.8 %**

(*b*) The relative humidity ϕ is defined as : $\phi = \dfrac{p_v}{p_{vs}}$ where p_v is the parlial pressure of vapour and p_{vs} is the satruation pressure. That gives

$$p_v = \phi \times p_{vs} = 0.7 \times (4.25 \times 10^3) = 2975 \text{ N/m}^2$$

Futher, specific humidity or water vapour content is given by

$$\omega = 0.622 \frac{p_v}{p_t - p_v}$$

$$= 0.622 \times \frac{2975}{10^5 - 2975} = 0.0191 \text{ kg/kg of dry air}$$

$$= 19.1 \text{ g/kg of dry air}$$

EXAMPLE 12.5

(a) A certain volume of moist air with 50% relative humidity is isothermally reduced to half its original volume. What will be the change in relative humidity?

(b) A certain air has 35°C dry bulb temperature and 20°C dew point temperature, and the corresponding saturation pressure of water being 5.628 kPa and 2.33 kPa. Assuming the atmospheric pressure to be 1.0132 bar, calculate the specific humidity of air.

(c) In a sample of moist air at standard atmospheric pressure of 101.325 kPa and 26°C, the partial pressure of water vapour is 1.344 kPa. If the saturation pressure of water vapour is 3.36 kPa at 26°C, compute the humidity ratio and relative humidity of the moisture air sample.

(Engg. Services)

Solution : (*a*) The relative humidity is defined as : $\phi = \dfrac{p_v}{p_{vs}}$, where p_v is the partial pressure of water vapour and p_{vs} is the saturation pressure. During isothermal process, temperature remains constant and so will be the saturation pressure. As such $\phi \propto p_v$. Assuming the water vapour to behave as a perfect gas, we have

$$p_{v1} V_1 = RT \quad \text{and} \quad p_{v2} V_2 = RT \qquad (T_1 = T_2 = T)$$

That gives :

$$\frac{\phi_1}{\phi_2} = \frac{p_{v1}}{p_{v2}} = \frac{RT/V_1}{RT/(V_1/2)} = \frac{1}{2}$$

$$\therefore \quad \phi_2 = 2\,\phi_1 = 2 \times 50 = \mathbf{100}\ \%$$

(*b*) Specific humidity of humidity ratio of air

$$\omega = 0.622 \frac{p_v}{p_t - p_v}$$

where p_v is saturation pressure of water and p_t is the atmospheric (total) pressure

$$\therefore \quad \omega = 0.622 \times \frac{2.33}{1.0132 - 2.33} = 0.01464 \text{ kg of vapour/kg of dry air}$$

(*c*) Humdity ratio (specific humidity) of air

$$\omega = 0.622 \frac{p_v}{p_t - p_v}$$

where, p_v is the saturation pressure of water and p_t the atmopshereic (total) pressure.

$$\therefore \qquad \omega = 0.622 \times \frac{1.344}{101.325 - 1.344} = 0.00836 \text{ kg of vapour/kg of dry air}$$

Relative humidity of moist air sample,

$$\phi = \frac{p_v}{p_{vs}} = \frac{1.344}{3.36} = 0.4 \quad \text{or} \quad 40\%$$

EXAMPLE 12.6

The air supplied to an air-conditioned room is noted to be at temperature 20°C and specific humidity 0.0085. Corresponding to these conditions, detetrmine the partial pressure of vapour, relative humdiity and dew point temperature.

Take barometric or total pressure = 1.0132 bar

Solution: Specific humidity $\omega = 0.622 \frac{p_v}{p_a} = 0.622 \frac{p_v}{p_t - p_v}$

Thus: $0.0085 = 0.622 \frac{p_v}{1.0132 - p_v}$

∴ Partial pressure of water vapour $p_v = \frac{1.0132 \times 0.0085}{0.622 + 0.0085} = \mathbf{0.01366}$ bar

(*b*) The relative humidity is defined as $\phi = \frac{p_v}{p_{vs}}$. From steam tables, the saturation vapour pressure at 20°C = 0.0234 bar.

Then: $\phi = \frac{0.01366}{0.0234} = 0.5837$ or **58.38%**

(*c*) The dew point temperature is the saturation temperature of water at a pressure of 0.01366 bar

DPT (from steam tables by interpolation)

$$= 11 + (12 - 11) \times \frac{(0.01366 - 0.01312)}{0.01401 - 0.01312}$$

$$= 11 + 0.607 = \mathbf{11.607°C}$$

EXAMPLE 12.7

10 gm of water vapour per kg was removed from a given sample of atmospheric air at 40°C and 60% relative humidity. The temperature of air after the removal of moisture reduced to 30°C. Determine the humidity ratio, partial pressure of vapour, relative humidity and dew point temperature of this air (air after removal of moisture).

Take total atmospheric pressure as 101.325 kPa.

Solution: The relative humidity is defined as $\phi = \frac{p_v}{p_{vs}}$. From steam tables, the saturation vapour pressure p_{vs} at 40°C *dbt* = 7.384 kPa.

That gives : $p_v = \phi\, p_{vs} = 0.6 \times 7.384 = 4.43$ kPa

Specific humidity or humidity ratio $\omega = 0.622 \dfrac{p_v}{p_t - p_v}$

$$= 0.622 \times \frac{4.43}{101.325 - 4.43} = 0.0284 \text{ kg/kg of dry air}$$

$$= 28.4 \text{ gm/kg of dry air}$$

(*a*) Since 10 gm of water vapour per kg has been removed,

Specific humidity = 28.4 – 10 = 18.4 gm/kg = **0.0184** kg/kg of dry air

(b) At this state, the air is at 30°C with specific humidity 0.0184 kg/kg of dry air. Then

$$0.0184 = 0.622 \frac{p_v}{101.325 - p_v}$$

That gives: partial pressure of water vapour,

$$p_v = \frac{101.325 \times 0.0184}{0.0184 + 0.622} = \mathbf{2.92} \text{ kPa}$$

(*c*) At 30°C *dbt*, p_{vs} = 4.246 kPa (from steam tables)

$$\therefore \quad \text{Relative humidity } \phi = \frac{p_v}{p_{vs}} = \frac{2.92}{4.246} = 0.688 \text{ or } 68.6\%$$

(*d*) The dew point temperature is the saturation temperature at the pressure of 2.92 kPa. Then from steam tables

DPT = 24°C

EXAMPLE 12.8

Calculate all the psychrometric properties of air at 1 bar, 25°C dbt and 15° wbt. Assume that the following properties of water hold good.

Temperature 0°C	*Saturation pressure bar*	*Specific volume of vapour m³/kg*	*Enthalpy kJ/kg*	
			Saturated liquid	*Saturated vapour*
25	0.03166	43.40	–	2547.3
15	0.017307	77.98	62.94	–
10	0.01078	–	–	–
0	0.01002	–	–	–

The following expression may be used if necessary

$$p_v = (p_{v\,\text{sat}})_{wbt} - \frac{(p_t - p_{v\,\text{sat}})\,(dbt - wbt) \times 1.8}{2854 - 1.325\ (1.8\ dbt + 32)}$$ **(RTU, Engg. Services)**

Solution : Substituting the given values *dbt* = 25°C, *wbt* = 1° C and p_t = 1 bar in the given expression, we get

$$p_v = 0.017307 - \frac{(1 - 0.03166)\,(25 - 15) \times 1.8}{2854 - 1.325\,(1.8 \times 25 + 32)}$$

$$= 0.017307 - \frac{17.43}{2751.97} = \mathbf{0.01097} \text{ bar}$$

The different psynchorometic properties of air and then worked out as:

(*a*) Specific humidity $\omega = 0.622 \dfrac{p_v}{p_t - p_v} = 0.622 \times \dfrac{0.01097}{1-0.01097}$

$= 0.006899$ kg vapour/kg of dry air

(*b*) Relative humidity $\phi = \dfrac{p_v}{p_{vs}} = \dfrac{0.01097}{0.03166}$

$= 0.3465$ or **34.65 %**

(*c*) Degree of saturation $\mu = \phi \dfrac{p - p_{vs}}{p_t - p_v} = 0.3465 \left[\dfrac{1-0.03166}{1-0.01097}\right]$

$= 0.34565 \left(\dfrac{0.96834}{0.98903}\right) = 0.339$ or **33.9 %**

(*d*) Enthalpy of moist air $h = c_{pa}\, t + \omega\,(2500 + 1.88\, t)$

$= (1.005 \times 25) + 0.006899\,(2500 + 1.88 \times 25)$

$= 25.125 + 17.571 =$ **42.698** kJ/kg of dry air

EXAMPLE 12.9

A mixture of air and water-vapour occupies a volume of 650 m³ at 1 bar pressure and 35°C temperature. If its relative humidity is 78 percent, calculate the specific humidity, the dew point and the masses of air and vapour in the mixture.

Solution: Relative humidity $\phi = \dfrac{m_v}{m_{vs}} = \dfrac{p_v}{p_{vs}}$

From steam tables at 35°C: $p_{vs} = 0.0563$ bar

∴ partial pressure of vapour $p_v = \phi \times p_{vs} = 0.78 \times 0.0563 = 0.0439$ bar

(*a*) Specific humidity or humidity ratio,

$$\omega = 0.622 \frac{p_v}{p_a} = 0.622 \frac{p_v}{p_t - p_v}$$

Taking total (barometric) pressure equal to 1.01325 bar,

$$\omega = 0.622 \times \frac{0.0439}{1.01325 - 0.0439} = \mathbf{0.0282} \text{ kg/kg of dry air}$$

(*b*) The dew point refers to the saturation temperature corresponding to the vapour pressure

Dew point $= t_{sat}$ at 0.0439 bar

From steam tables: $t_{sat} = 29$°C at 0.04 bar and $t_{sat} = 31$°C at 0.045 bar

Then by interpolation, the saturation temperature at 0.0439 bar is

$$(t_{sat})_{0.0439\text{ bar}} = 29 + (31 - 29) \times \frac{0.0439 - 0.04}{0.045 - 0.04} = \mathbf{30.56°C}$$

(*c*) Partial pressure of air $p_a = p_t - p_v$ = 1.01325 – 0.0439 = 0.9693 bar
Apply characteristic gas equation, $pV = mRT$, we have

$$\text{Mass of air } m_a = \frac{p_a V}{R_a T} = \frac{0.9693 \times 10^5 \times 650}{287 \times (273 + 35)} = 712.75 \text{ kg}$$

Since $\omega = \dfrac{m_v}{m_a}$, we get

Mass of water vapour $m_v = \omega \times m_a$ = 0.0282 × 712.75 = **20.1 kg**

EXAMPLE 12.10

An air-water vapour mixture at 20°C and 760 mm mercury has a relative humidity of 70 percent. Determine (a) partial pressures of vapour and air (b) humidity ratio (c) saturation ratio (d) dew point and (e) density of mixture.

Solution: Relative humidity $\phi = \dfrac{m_v}{m_s} = \dfrac{p_v}{p_{vs}}$

From steam tables at 20°C: p_{vs} = 0.0234 bar

∴ partial pressure of vapour $p_v = \phi \times p_{vs}$ = 0.70 × 0.0234 = 0.01638 bar

Corresponding to 760 mm mercury, the barometric pressure p_b of the mixture is 1.01325 bar

∴ p_t = 1.01325 bar

and partial pressure of air $p_a = p_t - p_v$
= 1.01325 – 0.01638 = 0.996879 bar

(*b*) Humidity ratio or specific humidity,

$$\omega = 0.622 \frac{p_v}{p_a} = 0.622 \times \frac{0.01638}{0.99687} = 0.01022 \text{ kg/kg of dry air}$$

$$\text{(c) Saturation ratio, } \mu = \phi \frac{p_t - p_{vs}}{p_t - p_v} = 0.7 \times \frac{1.01325 - 0.0234}{1.01325 - 0.01638} = \mathbf{0.695}$$

(*d*) The dew point refers to the saturation temperature corresponding to the vapour pressure.

Dew temperature = t_{sat} at 0.01638 bar

From steam tables: t_{sat} = 13°C at 0.015 bar and t_{sat} = 17°C at 0.02 bar

Then by interpolation, the saturation temperature at 0.01638 bar is

$$(t_{sat})_{0.01638 \text{ bar}} = 13 + (17 - 13) \times \frac{0.01638 - 0.015}{0.02 - 0.015} = 14.104°\text{C}$$

∴ Dew point = **14.104°C**

(*e*) Density of mixture = $\rho_a + \rho_v = \rho_a + \omega\, \rho_a = \rho_a\, (1 + \omega)$

$$= \frac{p_a}{R_a T_a} (1 + \omega)$$

$$= \frac{0.99687 \times 10^5}{287 \times 293} (1 + 0.01022) = \mathbf{1.197 \text{ kg/m}^3}$$

EXAMPLE 12.11

An air-water vapour mixture at 25°C and 1 bar has a relative humidity of75 percent. Determine:
(a) the partial pressures of water vapour and dry air
(b) the specific volume of each constituent
(c) the dew point, the specific humidity and the saturation ratio
(d) the density of the mixture.

Also find the amount of water vapour condensed, if the mixture is cooled at constant pressure to a temperature of 10°C.

Solution: Corresponding to 25°C, the saturated vapour pressure p_{vs} as read from the steam tables is

$$p_{vs} = 0.0317 \text{ bar}$$

∴ partial pressure of vapour $p_v = \phi\, p_{vs} = 0.75 \times 0.0317 = \mathbf{0.0237\ bar}$

and partial pressure of air $p_a = p_t - p_v = 1 - 0.0277 = \mathbf{0.9762\ bar}$

(*b*) Applying the perfect gas equation $pv = RT$ to both air and water vapour,

For air: $v_a = \dfrac{R_a T_a}{p_a} = \dfrac{287 \times 298}{0.9762 \times 10^5} = \mathbf{0.876\ m^3/kg}$

For vapour: $v_v = \dfrac{R_v T_v}{p_v} = \dfrac{(8314/18) \times 298}{0.02377 \times 10^5} = \mathbf{57.91\ m^3/kg}$

(*c*) The dew point refers to the saturation temperature corresponding to vapour pressure

Dew point = t_{sat} at 0.02377 bar

From steam tables: $t_{sat} = 17$°C at 0.020 bar and $t_{sat} = 21.1$°C at 0.025 bar

Then by interpolation, the saturation temperature at 0.02377 bar is

$$t_{sat} \text{ at } 0.02377 \text{ bar} = 17 + (21.1 - 17) \times \frac{0.02377 - 0.020}{0.025 - 0.20} = \mathbf{20.09°C}$$

$$\text{Specific humidity } \omega = 0.622\ \frac{p_v}{p_a} = 0.622 \times \frac{0.02377}{0.9762} = \mathbf{0.0151\ kg/kg\ of\ dry\ air.}$$

$$\text{Saturation ratio } \mu = \phi\, \frac{p_t - p_{vs}}{p_t - p_v} = 0.75 \times \frac{1 - 0.0317}{1 - 0.02377} = \mathbf{0.744}$$

(*d*) Density of the mixture,

$$\rho = \rho_a + \rho_v = \frac{1}{v_a} + \frac{1}{v_v}$$

$$= \frac{1}{0.876} + \frac{1}{57.91} = 1.1415 + 0.01727 = 1.1588 \text{ kg/m}^3$$

Alternatively: $\rho = \rho_a\,(1 + \omega) = \dfrac{1}{0.876}\,(1 + 0.0151) = 1.1588 \text{ kg/m}^3$

(*e*) Since 10°C is below the dew point temperature of 20.09°C, there will be condensation of some of the vapour and the mixture will have a relative humidity of 100 percent.

At 10°C: p_{vs} = 0.01227 bar

$$p_v = \phi\, p_{vs} = 1 \times 0.01227 \text{ bar} = 0.01227 \text{ bar}$$

$$\omega = 0.622\,\frac{p_v}{p_a} = 0.622 \times \frac{0.01227}{1-0.01227} = 0.007727 \text{ kg/kg of dry air}$$

∴ Amount of vapour condensed per kg of dry air

$$= 0.0151 - 0.007727 = \mathbf{0.007373\ kg}$$

EXAMPLE 12.12

Moist air at a total pressure of 760 mm of mercury occupies 3.2 m³ at 35°C; the relative humidity being 0.75. It is subsequently cooled at constant pressure to 5°C. Using the property values given below, determine the mass of moisture that gets deposited during the process.

Saturation temperature °C	*Pressure* bar	*Specific volume* m³/kg
35	0.0563	25.22
5	0.0087	147.1

Solution: Barometric pressure p_b = 760 mm of Hg = 1.01325 bar

For initial state at 35°C,

$$\text{mass of water vapour present} = \frac{m}{v}\phi = \frac{3.2}{25.22} \times 0.75 = 0.09516 \text{ kg}$$

partial pressure of water vapour $p_v = \phi \times p_{vs} = 0.75 \times 0.0563 = 0.04222$ bar

The barometric pressure is the total pressure and therefore

partial pressure of air $p_a = p_t - p_v = 1.01325 - 0.04222 = 0.97103$ bar

For final state at 5°C

Here the air would be saturated and therefore partial pressure of air

$$p_a = 1.01325 - 0.0087 = 1.00455 \text{ bar}$$

Applying characteristic gas equation to initial and final states, we have

$$\frac{p_1V_1}{T_1} = \frac{p_2V_2}{T_2} \qquad \text{(mass and gas constant same)}$$

$$\therefore \quad \text{Volume of air } V_2 = \frac{p_1V_1}{T_1} \times \frac{T_2}{p_2} = \frac{0.97103 \times 3.2}{273+35} \times \frac{273+5}{1.00455} = 2.792 \text{ m}^3$$

$$\text{Mass of water vapour} = \frac{2.792}{147.1} = 0.01898 \text{ kg}$$

∴ Mass of moisture deposited = 0.09516 – 0.01898 = **0.07618 kg**

EXAMPLE 12.13

A sample of moist air (a mixture of dry air and water vapour) has a dry bulb temperature of 20°C and a dew point temperature of 15°C, when the total pressure is 740 mm of mercury. Calculate the following by using only the steam tables (use of psychrometric chart is not allowed).

(a) partial pressure of water vapour

(b) relative and specific humidity

(c) specific enthalpy of water vapour, and

(d) specific volume and enthalpy of air per kg of dry air.

Solution : From steam tables, the partial pressure of water vapour at dpt of 15°C is

$$p_v = \mathbf{1707}\ \text{N/m}^2$$

(*b*) Saturation pressure of water vapour at 20°C dbt

$$p_{vs} = 2340\ \text{N/m}^2$$

$$\text{Relative humidity } \phi = \frac{p_v}{p_{vs}} = \frac{1707}{2340} = 0.729 \text{ or } \mathbf{72.9\%}$$

Barometer pressure p_t = 740 mm of Hg

$$= 740 \times \frac{101325}{760} = 98658\ \text{N/m}^2$$

Specific humidity or humidity ratio,

$$\omega = 0.622\frac{p_v}{p_a} = 0.622 \times \frac{p_v}{p_t - p_v}$$

$$= 0.622 \times \frac{1707}{98658 - 1707} = \mathbf{0.01095}\ \text{kg/kg of dry air}$$

(*c*) Specific enthalpy of water vapour can be determined by using the following empirical relation

$$h_v = 1.88\ dbt + h_{fg}$$

where h_{fg} is the latent heat of vaporisation of water at 0°C = 2500 kJ/kg

$$\therefore \quad h_v = 1.88 \times 20 + 2500 = 2537.6\ \text{kJ/kg of mixture}$$

Another relation suggested for determining the specific enthalpy of water is

$$h_v = c_{pw} \times dbt + (h_{fg})_{\text{at } dbt}$$

From steam tables, the latent heat of vaporisation of water at dbt of 20°C is 2454 kJ/kg

$$\therefore \quad h_v = 4.186 \times 20 + 2454 = 2537.7\ \text{kJ/kg of mixture}$$

(*d*) Partial pressure of air $p_a = p_t - p_v = 98658 - 1707 = 96951\ \text{N/m}^2$

$$\text{Specific volume of air } v = \frac{RT}{p} = \frac{287 \times (273 + 20)}{96951} = \mathbf{0.867}\ \text{m}^3\text{/kg of dry air}$$

Specific enthalpy of air

$$h = h_a + \omega h_v = c_{pa}\ dbt + \omega h_v$$

$$= 1.005 \times 20 + 0.01095 \times 2597.6$$

$$= 20.1 + 27.786 = \mathbf{47.886}\ \text{kJ/kg of dry air}$$

EXAMPLE 12.14

A sample of moist air initially at 20°C dbt and 15°C wbt and at barometric pressure of 1.04325 kPa is heated by a battery through 20°C. Find the load on battery if the hot water at 80°C is used and its temperature drops to 70°C. Also find the flow rate of water required to accomplish the task.

Take mass flow rate of moist air as 1.5 m³/s.

(b) In another set up, this moist air is required to be cooled by 5°C. What will be the load on cooling coil and the flow rate of cooling water. The rise in temperature of chilled water used for cooling is limited to 5°C.

Solution: The pressure of water vapour in the mixture is obtained by using the Carrier's equation

$$p_v = (p_{vs})_{wt} - \frac{(p_t - p_{vs})[dbt - wbt]}{1544 - 1.44\, wbt}$$

where p_t is total (barometric) pressure of moist air and p_{vs} is the saturation pressure at wet bulb temperature (p_{vs} at 15°C wbt = 0.017 bar as read from property tables).

$$p_v = 0.017 - \frac{1.04325 - 0.017}{1544 - 1.44 \times 15}(20 - 15) = 0.01363 \text{ bar}$$

$$\text{Specific humidity } \omega = 0.622\frac{p_v}{p_a} = 0.622\frac{p_v}{p_t - p_v}$$

$$= 0.622 \times \frac{0.01363}{1.04325 - 0.01363}$$

$$= 8.234 \times 10^{-3} \text{ kg/kg} = 8.234 \text{ gm/kg}$$

$$p_a \text{ (partial pressure of air)} = p_t - p_v = 1.04325 - 0.01363$$

$$= 1.02962 \text{ bar}$$

Applying characteristic gas equation,

$$\text{Mass of dry air } m_a = \frac{p_a V}{RT} = \frac{(1.02962 \times 10^5) \times 1.5}{287 \times (20 + 273)} = 1.837 \text{ kg/s}$$

(*a*) Heating of air

$$h_1 \text{ (before heating)} = (1.005 + 1.88\,\omega)\, dbt_1 + \omega\,(2500 + 1.88\, dbt_1)$$

$$h_2 \text{ (after heating)} = (1.005 + 1.88\,\omega)\, dbt_2 + \omega\,(2500 + 1.88\, dbt_2)$$

Then : $\Delta h \text{ (heat added)} = h_2 - h_1$

$$= (1.005 + 1.88\,\omega)\,(dbt_2 - dbt_1) + \omega[1.88\,(dbt_2 - dbt_1)]$$

$$= (1.005 + 1.88 \times 0.008234) \times 20 + 0.008234\,(1.88 \times 20)$$

$$= 20.41 + 0.31 + \mathbf{20.72} \text{ kJ/kg of air}$$

Now : heat lost by hot water = heat gained by air

or $\quad m_{w_1} c_{p_w} \Delta t_w = m_a \Delta h$

$$\therefore \quad m_{w_1} = \frac{1.837 \times 20.72}{4.186 \times (80 - 70)} = \mathbf{0.9093} \text{ kg/s}$$

(*b*) Cooling of air

From the identities given in part (*a*), heat removed

$$\Delta h' = (1.005 + 1.88\,\omega)\,(dbt_1 - dbt') + \omega[1.88(dbt_1 - dbt')]$$

$$= (1.005 + 1.88 \times 0.008234) \times 5 + 0.008234 \times 1.88 \times 5$$

$$= 5.10 + 0.077 = 5.177 \text{ kJ/kg of air}$$

Now : heat gained by chilled water = heat lost by air

or $\quad m_{w_2} c_{p_w} \Delta t_w = m_a \Delta h'$

$$\therefore \quad m_{w_2} = \frac{1.837 \times 5.177}{4.186 \times 5} = \mathbf{0.454} \text{ kg/s}$$

EXAMPLE 12.15

For a given sample of moist air, the wet bulb and dry bulb temperatures have been measured as 30°C and 45°C respectively. Assuming total pressure of the mixture to be 1 bar, determine the following parameters for this air sample:

(a) *vapour pressure* *(b)* *dew point temperature*
(c) *specific humidity, relative humidity and the degree of saturation*
(d) *vapour density of air, and* *(e)* *enthalpy of the mixture.*

Solution: The following data can be read from the steam tables:

At 45°C *dbt*: p_{vs} = 0.0958 bar; v_v = 15.28 m^3/kg and h_v = 2582 kJ/kg

At 30°C *wbt*: p_{vs} = 0.0425 bar; v_v = 32.93 m^3/kg and h_v = 126 kJ/kg.

(a) We may use the Carrier equation to determine the partial pressure of water vapour in air

$$p_v = (p_{vs})_{wbt} - \frac{[p-(p_{vs})_{wbt}][dbt-wbt]}{1547-1.44\,wbt}$$

$$= 0.0425 - \frac{(1-0.0425)(45-30)}{1547-1.44\times 30} = 0.0425 - 0.00955 = \mathbf{0.03285\ bar}$$

(b) The dew point refers to the saturation temperature corresponding to the vapour pressure.

Dew point temperature = t_{sat} at 0.03285 bar

From steam tables: t_{sat} = 24.1°C at 0.030 bar and t_{sat} = 26.7°C at 0.035 bar

Then by interpolation, the saturation temperature at 0.03285 bar is

$$= 24.1 + (26.7 - 24.1) \times \frac{0.03285-0.030}{0.035-0.030} = \mathbf{25.582°C}$$

(c) Specific humidity or humidity ratio

$$\omega = 0.622\,\frac{p_v}{p_t - p_v}$$

$$= 0.622 \times \frac{0.03285}{1-0.03285} = \mathbf{0.02113}\ \text{kg/kg of dry air}$$

$$\text{Relative humidity } \phi = \frac{p_v}{p_{vs}} = \frac{0.03285}{0.0958} = \mathbf{0.3429}$$

$$\text{Saturated ratio } \mu = \phi\,\frac{p_t - p_{vs}}{p_t - p_v} = 0.3429 \times \frac{1-0.0958}{1-0.03285} = \mathbf{0.32}$$

(d) Vapour density of saturated air is the reciprocal of specific volume of saturated air

$$= \frac{1}{15.28} = 0.0654\ \text{kg/m}^3$$

For the moist air with relative humidity 0.3429, the vapour density would be

$$= 0.0654 \times 0.3429 = \mathbf{0.02243\ kg/m^3}$$

(e) Enthalpy of mixture,

= enthalpy of dry air + enthalpy of vapour associated with dry air

Enthalpy of water vapour can be determined by using the following reation

$$h_v = 1.88\ dbt + h_{fg}$$

where h_{fg} is the latent heat of vaporisation of water at 0°C = 2500 kJ/kg. That gives

$$h_v = 1.88 \times 45 + 2500 = 2584.6\ \text{kJ/kg}$$

This is approximately equal to enthalpy of saturated vapour at saturation temperature equal to dry bulb temperature.

Then enthalpy of vapour associated with dry air

$$= \omega\, h_v = 0.02113 \times 2584.6$$
$$= 54.61 \text{ kJ/kg of dry air}$$

Enthalpy of dry air $h_a = c_{pa}\, dbt = 1.005 \times 45 = 45.22$ kJ/kg of dry air

∴ Enthalpy of mixture = 45.22 + 54.61 = **99.83** kJ/kg of dry air

EXAMPLE 12.16

Atmospheric air at 0.965 bar enters the adiabatic saturation where the wet bulb temperature is 20°C and the dry bulb temperature is 31°C. Determine:

(i) humidity ratio of entering air

(ii) vapour pressure and relative humidity at 31°C, and

(iii) dew point temperature

Solution : At DBT = 31° C

Saturation pressure p_s = 0.04496 bar,

Enthalpy of saturated vapour h_{g1} = 2558.1 kJ/kg

Enthalpy of make up water h_{f3} = 84.0 kJ/kg

Enthalpy of unsaturaed air,

$$h_{a_1} = m\, c_p \times DBT = 1 \times 1.005 \times 31 = 31.55 \text{ kJ/kg}$$

At WBT = 20° C, p_{v2} = 0.0233 a bar ; $h_{s_2} = h_{g2}$ = 2538.1 kJ/kg

$$h_{a_2} = m\, c_p \times WBT = 1 \times 1.003 \times 20 = 20.1 \text{ kJ/kg}$$

Humidity ratio of saturated air,

$$\omega_2 = 0.622 \frac{p_{v2}}{p_t - p_{v2}} = 0.622 \times \frac{0.02339}{0.965 - 0.02339}$$
$$= 0.01545 \text{ kg/kJ of dry air}$$

(*a*) Humidity of entering air,

$$\omega = \frac{\omega_2 (h_{g1} - h_{f3}) + (h_{a_2} - h_{a_1})}{h_{g1} - h_{f3}}$$

$$= \frac{0.01545(2538.1 - 84) + (20.1 - 31.55)}{2558.1 - 84} = 0.01085 \text{ kg/kg of dry air}$$

(*b*) Let p_{v1} be the vapour pressure at 31°C. Then

$$\omega_1 = 0.622 \times \frac{p_{v1}}{p_t - p_{v1}} \quad ; \quad 0.01085 \left[\frac{p_{v1}}{0.965 - p_{v1}}\right]$$

Solution gives : $\phi = \dfrac{p_{v1}}{p_s} = \dfrac{0.0165}{0.0449} = 0.3669$ or **36.69** %

(*c*) Dew point is the saturation temperature corresponding to the partial pressure of water vapour (p_{v1}). Then from steam tables the dew point temperature corresponding to p_{v1} = 0.0165 bar is read as **14.5°** C.

EXAMPLE 12.17

A sling psychrometer in a laboratory test recorded the following readings of an atmospheric air with 0.98 bar pressure:

Dry bulb temperature = 30°C ; Wet bulb temperature = 20°C

This air is made to enter into an adiabatic saturator where it undergoes adiabatic saturation process. Make calculations for:

(a) humidity ratio for the flowing stream at entry and exit,
(b) vapour pressure and relative humidity at 30°C,
(c) vapour density and enthalpy of mixture at entry and
(d) dew point temperature.

Solution: From steam tables at 20°C, the saturation pressure of vapour is

$$p_{v2} = 0.0234 \text{ bar}$$

(*a*) Humidity ratio,
$$\omega_2 = 0.622 \frac{p_{v2}}{p_t - p_{v2}} = 0.622 \times \frac{0.0234}{0.98 - 0.0234}$$
$$= 0.0152 \text{ kg/kg of dry air}$$

The humidity ratio at entry (unsaturated state) is given by:

$$\omega_1 = \frac{c_p(t^* - t_1) + \omega^* h_{fg}^*}{h_{g1} - h_f^*}$$

where, t^* = adiabatic saturation temperature = *wbt* = 20°C
t_1 = temperature of unsaturated stream at inlet = *dbt* = 30°C
ω^* = ω_2 = humidity ratio at saturated state
h_{fg}^* = latent heat corresponding to t^* (20°C) = 2454 kJ/kg
h_{g1} = enthalpy of vapour corresponding to *dbt* (30°C) = 2556 kJ/kg
h_f^* = sensible heat of water corresponding to t^* (*wbt* = 20°C) = 84 kJ/kg

$$\therefore \quad \omega_1 = \frac{1.005(20-30) + 0.0152 \times 2454}{2556 - 84} = \mathbf{0.01084} \text{ kg/kg of dry air}$$

(*b*) Now,
$$\omega_1 = 0.622 \frac{p_{v1}}{p_t - p_{v1}}\ ; \quad 0.01084 = \frac{0.622 \times p_{v1}}{0.98 - p_{v1}}$$

or
$$0.98 \times 0.01084 - 0.01084\, p_{v1} = 0.622\, p_{v1}$$

∴ Vapour pressure at inlet section,

$$p_{v1} = \frac{0.98 \times 0.01084}{0.622 + 0.01084} = 0.01679 \text{ bar}$$

From steam tables, the saturation pressure corresponding to dry bulb temperature of 30°C is 0.0425 bar

$$\therefore \text{ Relative humidity } \phi_1 = \left(\frac{p_v}{p_{vs}}\right)_1 = \frac{0.01679}{0.0425} = \mathbf{0.395 \text{ or } 39.5\%}$$

(c) The specific volume of dry and saturated vapour at 30°C is 32.89 m^3/kg

$$\text{Vapour density} = \frac{1}{32.89} = 0.0304 \text{ kg/m}^3 \text{ of saturated vapour}$$

∴ Vapour density at 39.5% relative humidity is

$$= 0.0304 \times 0.395 = \mathbf{0.012} \text{ kg/m}^3$$

Enthalpy of the mixture is

= enthalpy of dry air + enthalpy of vapour associated with dry air

Now, enthalpy of dry air $h_a = c_{pa}\ dbt = 1.005 \times 30$

$= 30.15$ kJ/kg of dry air

The enthalpy of vapour can be taken approximately equal to the enthalpy of saturated vapour at saturation temperature equal to dry bulb temperature. That is

$$h_v = 2556 \text{ kJ/kg}$$

Then enthalpy of vapour associated with dry air

$$= \omega\, h_v = 0.01084 \times 2556 = 27.71 \text{ kJ/kf of dry air}$$

∴ Enthalpy of mixture = 30.15 + 27.71 = **57.86** kJ/kg of dry air

(d) The dew point refers to the saturated temperature corresponding to the vapour pressure.

Dew point = t_{sat} at 0.01679 bar

From steam tables: t_{sat} = 14°C at 0.016 bar

t_{sat} = 15°C at 0.017 bar

Then by interpolation, the saturation temperature at 0.01679 bar is

$$(t_{sat})_{0.01679 \text{ bar}} = 14 + (15 - 14) \times \frac{0.017 - 0.01679}{0.017 - 0.016} = \mathbf{14.21°C}$$

EXAMPLE 12.18

The properties of air in a room are: temperature 35°C and relative humidity 70%. Determine (i) partial pressure of air (ii) humidity ratio (iii) saturation ratio (iv) dew point temperature (v) density of air-vapour mixture (vi) enthalpy of mixture per kg of dry air. Assume room pressure to be 1.0132 bar and the specific heat for superheated vapour as 1.863 kJ/kgK.

Solution: From steam tables,

Saturation pressure of vapour at 35°C, p_{vs} = 0.05628 bar

Relative humidity ϕ = 70% = 0.7

The partial pressure of vapour $p_v = \phi\ p_{vs}$ = 0.7 × 0.5628 = 0.394 bar

Partial pressure of air = 1.0132 – 0.394 = **0.6192 bar**

Humidity of ratio, ω = weight of water vapour/kg of dry air

$$= 0.622\ \frac{p_v}{p_t - p_v} = 0.622 \times \frac{0.394}{1.0132 - 0.394}$$

$$= \mathbf{0.0252 \text{ kg/kg of air}}$$

Saturation ratio (degree of saturation)

$$\mu = \frac{\omega}{\omega_s} = \frac{0.622\ \dfrac{p_v}{p_t - p_v}}{0.622\ \dfrac{p_{vs}}{p_t - p_{vs}}} = \frac{p_v}{p_{vs}}\left(\frac{p_t - p_s}{p_t - p_v}\right)$$

$$= \phi\left(\frac{p_t - p_{vs}}{p_t - p_v}\right) = 0.7\left(\frac{1.0132 - 0.05628}{1.0132 - 0.394}\right) = \mathbf{0.6879}$$

Dew point temperature corresponds to saturation temperature at partial pressure of water-vapour. From steam tables:

DPT corresponding to partial pressure of water vapour (p_v = 0.394 bar) is **28°C** (approx)

$$\text{Density of air (dry) } \rho_{air} = \frac{p_a}{R_a(t+273)} = \frac{(1.0132 - 0.0394)\times 10^5}{287\times(35+273)} = 1.102 \text{ kg/m}^3$$

$$\text{Density of water vapour } \rho_v = \frac{p_v}{R_v(t+273)} = \frac{0.0394\times 10^5}{461\times(35+273)} = 0.028 \text{ kg/m}^3$$

Density of mixture $\rho_{mix} = \rho_{air} + \rho_v$

$= 1.102 + 0.028 = \mathbf{1.13}$ kg/m^3

Enthalpy of mixture

= enthalpy of dry air + enthalpy of vapour associated with dry air

Now, enthalpy of dry air = c_{pa} dbt = 1.0005 × 35 = 35.17 kJ/kg of dry air

Enthalpy of superheated water vapour can be determined by using the following relation

$$h_v = (h_v)\, dbt + c_{ps}\, (dbt - dpt)$$

where $(h_v)_{dbt}$ is the enthalpy of saturated vapour at saturation temperature corresponding to dry bulb temperature. From steam tables $(h_v)_{dbt}$ = 2565.4 kJ/kg. Then

$$h_v = 2565.4 + 1.863\,(35 - 28) = 2578.44 \text{ kJ/kg}$$

Then enthalpy of water vapour associated with dry air

$= \omega\, h_v = 0.0252 \times 2578.44$

= 64.98 kJ/kg of dry air

∴ Enthalpy of mixture = 35.17 + 64.98 = **100.15** kJ/kg of dry air

EXAMPLE 12.19

The DBT and WBT of atmospheric air are 35°C and 23°C respectively when the barometer reads 75 cm of Hg. Determine (i) relative humidity (ii) humidity ratio (iii) dew point temperature (iv) density (v) enthalpy of atmospheric air. Use the Carrier equation as given below:

$$p_v = (p_{vs})_{wbt} - \frac{[p_t - (p_{vs})_{wbt}]\,[dbt - wbt]\times 1.8}{2800 - 1.3\,(1.8\, dbt + 32)}$$

where pressure air is in mm of Hg

Solution: $(p_{vs})_{wbt}$ = 21.05 mm of Hg (from steam tables)

$$p_v = 21.05 - \frac{(750 - 21.05)\,(35 - 23)\times 1.8}{2800 - 1.3(1.8\times 35 + 32)}$$

$= 21.05 - 5.85 = 15.2$ mm of Hg

$(p_{vs})_{dbt}$ = 42.5 mm of Hg (from steam tables)

(*i*) Relative humidity $\phi = \dfrac{p_v}{p_{vs}} = \dfrac{15.2}{42.5} = 0.36 = \mathbf{36\%}$

(*ii*) Humidity ratio $\omega = 0.622 \dfrac{p_v}{p_t - p_v}$

$$= 0.622 \times \frac{15.2}{(750-15.2)} = \mathbf{0.01285\ kg/kg\ of\ dry\ air}$$

(*iii*) DPT, saturation temperature corresponding to partial pressure of water vapour = **17.7°C** (from steam tables)

(*iv*) Density of air $\rho_a = \dfrac{p_a}{R_a T} = \dfrac{(750-15.2)\times 133.5}{287\times(35+273)} = 1.1085\ \text{kg/m}^3$

$$\rho_v = \frac{p_a}{R_a T} = \frac{15.2\times 133.5}{461\times(35+273)} = 0.0145\ \text{kg/m}^3$$

$$\therefore \quad \rho_{mix} = \rho_a + \rho_v = 1.1085 + 0.0145 = \mathbf{1.123\ kg/m^3}$$

(*v*) Enthalpy of mixture = enthalpy of dry air + enthalpy of vapour associated with dry air

Now, enthalpy of dry air = $c_{pa}\ dbt$

= 1.005 × 35 = 35.17 kJ/kg

The enthalpy of vapour can be taken approximately equal to the enthalpy of saturated vapour at saturation temperature equal to dry bulb temperature. That is

h_v = 2565.4 kJ/kg (from Steam tables)

Then enthalpy of vapour associated with dry air

= ωh_v = 0.01285 × 2565.4 = 32.96 kJ/kg of dry air

∴ Enthalpy of mixture = 35.17 + 32.96 = **68.13** kJ/kg of dry air

EXAMPLE 12.20

A stream of moist air (mass flow rate = 10.1 kg/s) with humidity ratio of 0.01 kg/kg of dry air mixes with a second stream of superheated water vapour flowing at 0.1 kg/s. Assuming proper and uniform mixing with no condensation, what will be the humidity ratio of the final stream.

Solution : Humidity ratio of final stream is,

$$\omega = \frac{m_1\omega_1 + m_2\omega_2}{m_1 + m_2} = \frac{10.1\times 0.01 + 0.01\times 1}{10.1 + 0.1}$$

The specific humidity of superheated water vapour is taken as unity.

$$\omega = \frac{0.101 + 0.1}{10.2} = 0.0197\ \text{kg/kg of dry air}$$

EXAMPLE 12.21

Air at state 1 (DPT 10°C, ω = 0.040 kg/kg) mixes with air at state 2 (DPT 18°C, ω = 0.0051 kg/kg) in the ratio 1 to 3 by weight. Workout the degree of saturation in percentage of the mixture. Take specific humidity of saturated air at 13.6°C as 0.01 kg/kg.

Solution : Mass of moisture actually confained in the mixture

$$= \frac{m_1\omega_1 + m_2\omega_2}{m_1 + m_2} = \frac{1\times 0.004 + 3\times 0.0051}{1+3}$$

$$= \frac{0.004 + 0.0153}{4} = 0.0048$$

Mass of moisture in saturated state of mixture = 0.01 kg/kg of dry air

$$\therefore \quad \text{Degree of saturation} = \frac{0.0048}{0.01} = 0.48 \text{ or } 48\%.$$

EXAMPLE 12.22

An air stream at 3°C dbt and 80% RH mixes with another air stream at 30°C dbt and 10°C dew point in the ratio 1 to 2 by mass. The mixed air subsequently passes through cooling coils and leaves at 15°C dry bulb. For 1 kg/s of mixture, calculate:

(a) temperature and specific humidity of air after mixing, and

(b) the cooling load on the coils in kW.

The following data may be assumed:

Barometeric pressure = 1.01325 bar

Specific heats of dry air and water vapour at constant pressure

= 1.005 and 1.885 kJ/kg K respectively

Solution: *For stream 1:*

Saturation pressure p_{vs} at 3°C = 0.0076 bar

Partial pressure of vapour $p_v = \phi\, p_{vs}$ = 0.8 × 0.0076 = 0.00608 bar

$$\text{Humidity ratio } \omega = \frac{0.622\, p_v}{p_t - p_v} = \frac{0.622 \times 0.00608}{1.01325 - 0.00608} = 0.003755 \text{ kg/kg dry air}$$

$$\text{Enthalpy } h = c_p \times dbt + \omega\,(h_v)_{dbt}$$

$$= 1.005 \times 3 + 0.003755 \times 2506.9 = 12.428 \text{ kJ/kg dry air}$$

$$\text{Ratio } \frac{\text{mass of dry air}}{\text{mass of moist air}} = \frac{1}{1.003755} = 0.996$$

For stream 2:

Partial pressure of vapour = saturation pressure corresponding to dew point 10°C

= 0.0123 bar

$$\text{Humidity ratio } \omega = \frac{0.622\, p_v}{p_t - p_v} = \frac{0.622 \times 0.0123}{1.01325 - 0.0123} = 0.007643 \text{ kg/kg dry air}$$

$$\text{Enthalpy } h = c_p \times dbt + \omega\,(h_v)_{dbt}$$

$$= 1.005 \times 30 + 0.007643 \times 2556.3 = 49.688 \text{ kJ/kg dry air}$$

$$\text{Ratio } \frac{\text{mass of dry air}}{\text{mass of moist air}} = \frac{1}{1.007643} = 0.9924$$

$$(a)\ \text{Enthalpy of mixture} = \frac{1}{3}\left[\frac{12.428}{1.003755} + \frac{49.688 \times 2}{1.007643}\right]$$

$$= \frac{1}{3}\,[12.381 + 98.622] = 37 \text{ kJ/kg of moist air}$$

Mass of vapour per unit mass of moist air

$$= \frac{1}{3}\left[\frac{0.003755}{1.003755} + \frac{0.007643 \times 2}{1.007643}\right]$$

$$= \frac{1}{3}[0.003741 + 0.01517] = 0.006304 \text{ kg/kg moist air}$$

$$\text{Special humidity of mixture} = \frac{0.006304}{1-006304} = 0.006344 \text{ kg/kg of dry air}$$

Using the empirical relation for the enthalpy of an air sample, we have:

$$h = c_p\, t + \omega\, (2500 + 1.884t)$$

$$37 = 1.005\, t + 0.006344\, (2500 + 1.884t)$$

$$= 1.005\, t + 15.86 + 0.01195\, t$$

$$\therefore \text{ Temperature of mixture } t = \frac{37 - 15.86}{1.005 + 0.01195} = \mathbf{20.79°C}$$

(*b*) The enthalpy of the mixture after it has passed through the battery of cooling coils (t = 15°C) can also be obtained from the relation

$$h = c_p\, t + \omega(2500 + 1.884\, t)$$

$$= 1.005 \times 15 + 0.006344\, (2500 + 1.884 \times 15)$$

$$= \mathbf{31.11} \text{ kJ/kg dry air}$$

$$\therefore \text{ Load on the cooling coils} = \frac{1 \times (37 - 31.11)}{1 + 0.006344} = 5.85 \text{ kJ/s} = \mathbf{5.85\ kW}$$

EXAMPLE 12.23

One kg of air at 35°C DBT and 60% RH is mixed with 2 kg of air at 20°C dbt and 13°C dew point temperature. Calculate the specific humidity, temperature and enthalpy the mixture.

Assume specific heat of steam as 1.88 kJ/kg K and use the property values as given below:

t → °C	*13*	*20*	*26*	*27*	*35*
Saturation pressure bar	*0.0150*	*0.0234*	*0.0336*	*0.0356*	*0.0563*
Enthalpy of saturated steam kJ/kg	*2524.4*	*2538.2*	*2549.1*	*2550.9*	*2564.4*

(Engg. Services)

Solution : *For air at 35°C DBT and 60% RH (Stream 1)*

Corresponding to 35°C : p_{vs} = 0.0563 bar (from steam tables)

$$\text{Relative humidity } \phi_1 = \frac{p_v}{p_{vs}} \quad \therefore \quad p_v = 0.6 \times 0.563 = 0.0338 \text{ bar}$$

Specific humidity $\omega_1 = 0.622\dfrac{p_v}{p_i - p_v}$

$$= 0.622\left(\frac{0.0338}{1.0132 - 0.0338}\right) = 0.0214 \text{ kg/kg of dry air}$$

Corresponding to 0.0338 bar saturation pressure, the dew point temperature *dpt* is worked out by interpolation from the given values. That is

$$dpt = 26 + (27 - 26) \times \frac{0.0338 - 0.0336}{0.0356 - 0.0336} \simeq 26.1° \text{ C}$$

The enthalpy value is given as

$$h_1 = c_p \times DBT + \omega_1\, h_{vapour}$$
$$= (1.005 \times 35) + 0.0214\, [2564.4 + 1.88(35 - 26.1)]$$
$$= 35.175 + 55.236 = 90.41 \text{ kJ/kg of dry air}$$

For the air at 20°C DBT and 13°C DPT (Stream 2)

$$p_v = 0.015 \text{ bar (from the given table)}$$

$$\omega_2 = 0.622\frac{p_v}{p - p_v} = 0.622 \times \left(\frac{0.015}{1.0132 - 0.015}\right)$$
$$= 0.00935 \text{ kg/kg of dry air}$$

$$h_2 = c_p \times DBT + w_2\, h_{vapour}$$
$$= (1.005 \times 20) + 0.00935\, [2538.2 + 1.88\,(20 - 13)]$$
$$= 20.1 + 23.85 = 43.95 \text{ kJ/kg}$$

Then,

Mass of vapour/kg of moist air

$$= \frac{1}{3}\left[\frac{\omega_1}{1+\omega_1} + \frac{2\omega_2}{1+\omega_2}\right] = \frac{1}{3}\left[\frac{0.0214}{1.0214} + \frac{2\times 0.00935}{1.00935}\right] = 0.01316 \text{ kg of dry air}$$

$$\omega_{mixture} = \frac{0.01316}{1 - 0.01316} = \mathbf{0.01333} \text{ kg/kg of dry air}$$

$$h_{mixture} = \frac{1}{3}\left[\frac{h_1}{1+\omega_1} + \frac{2h_2}{1+\omega_2}\right] = \frac{1}{3}\left[\frac{90.43}{1.0214} + \frac{2\times 43.95}{1.00935}\right] = 54.54 \text{ kJ/kg of moist air}$$

$$t_{mixture} = \frac{1}{3}\left[\frac{(DBT)_1}{1+\omega_1} + \frac{2\times(DBT)_2}{1+\omega_2}\right] = \frac{1}{3}\left[\frac{35}{1.0214} + \frac{2\times 20}{1.00935}\right] = \mathbf{24.63° \text{ C}}$$

12.8. PSYCHROMETRIC PROPERTIES

The subject which deals with the behaviour of moist air is known as psychrometry, and the properties of moist air are called ***psychrometric properties***.

Water vapour in atmosphere exists at very low partial pressure. At the low partial pressure and at atmospheric temperature, water vapour behaves like a perfect gas. The partial pressure of dry air is also less than atmospheric pressure. Thus, the mixture of dry air and water vapours follow Dalton's law of partial pressures.

12.9. PSYCHROMETRIC CHART AND PROCESSES

Humidity calculations can be made by using the equations relating the dry and wet bulb temperatures to the humidity. The method, however, tends to be tedious, cumbersome and time consuming. The key to humidity calculations is then provided by the ***Psychrometric*** or ***Hygrometric chart*** which graphically describes the relationship between the properties of moist air, *i.e.,* the dry bulb, the wet bulb and dew point temperatures of the mixture and its humidity. Fig. 12.6 shows how these parameters are laid out on a typical psychrometric chart.

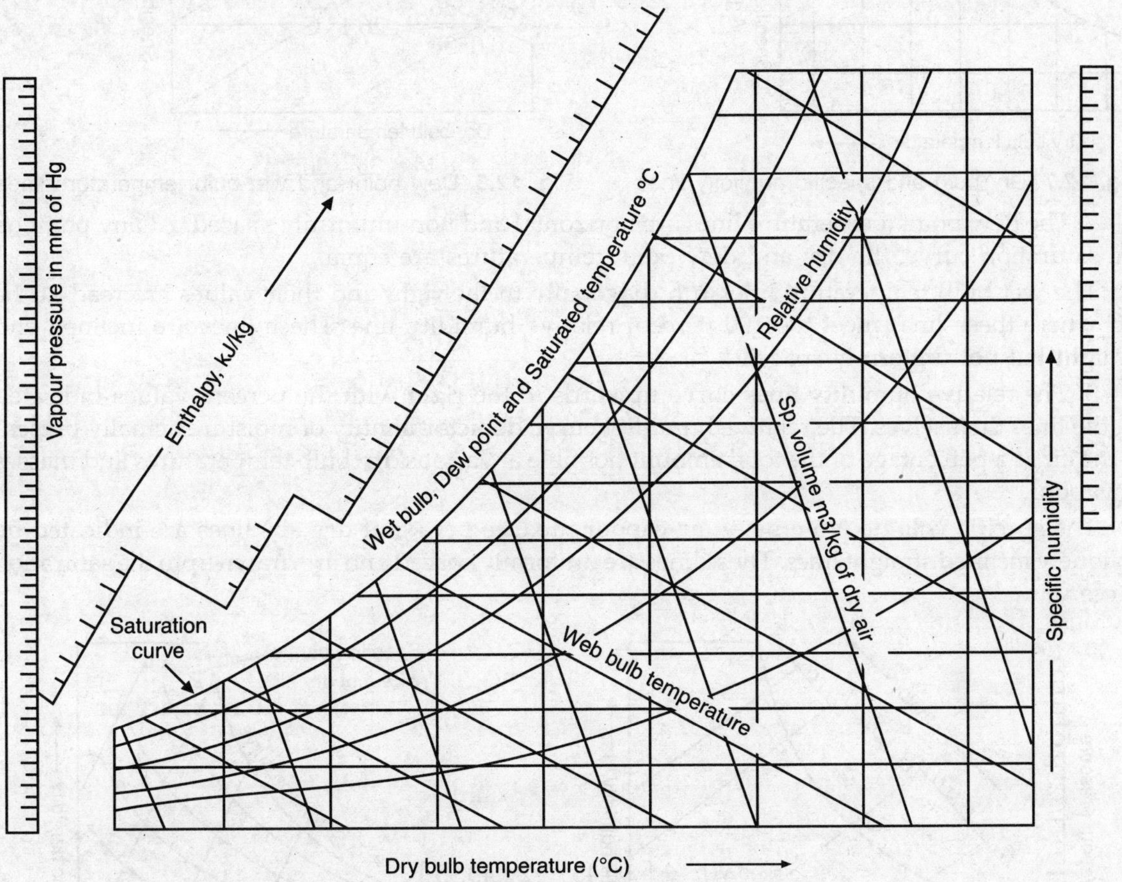

Fig. 12.6. Psychrometric chart

The psychrometic chart has the number of details and its salient aspects are:

1. The dry bulb temperature is taken as abscissa and specific humidity (*i.e.* moisture content) as ordinate.

The dry bulb temperature lines are vertical and uniformly spaced. The specific humidity lines are horizontal and also uniformly spaced. The saturation curve is drawn by plotting the various saturation points at corresponding dry bulb temperatures. The saturation curve represents 100 percent relative humidity at various dry bulb temperatures. It also indicates the wet bulb and dew point temperatures.

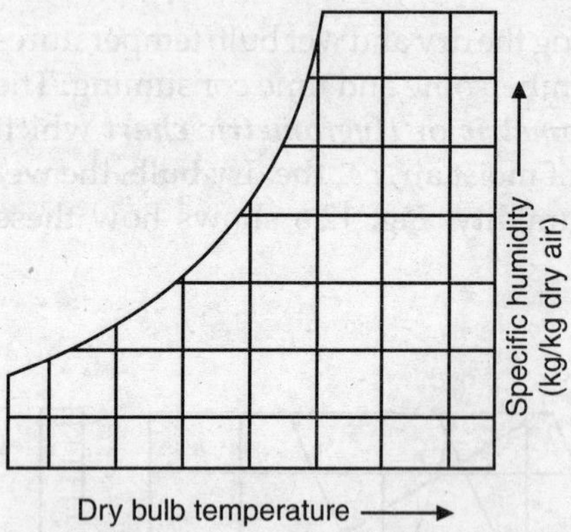

Fig. 12.7. Dry bulb and specific humidity lines

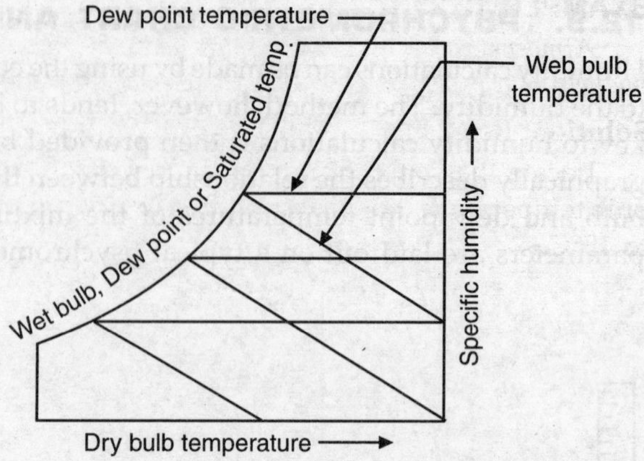

Fig. 12.8. Dew point and wet bulb temperature lines

2. The dew point temperature lines are horizontal and non-uniformly spaced. At any point on the saturation curve, the dry and dew point temperatures are equal.

The wet bulb temperature lines run diagonally to the right and their values are read at the left where these lines meet the 100 percent relative humidity line. These lines are inclined and straight but not uniformly spaced.

3. The relative humidity lines curve upwards to the right with the percent values indicated on the lines themselves. The relative humidity curve depicts quantity of moisture actually present in the air as a percentage of the total amount possible at various dry bulb temperatures and masses of vapour.

The specific volume (volume of air-vapour mixture per kg of dry air) lines are indicated by obliquely inclined straight lines. These lines are uniformly spaced and are drawn upto the saturation curve.

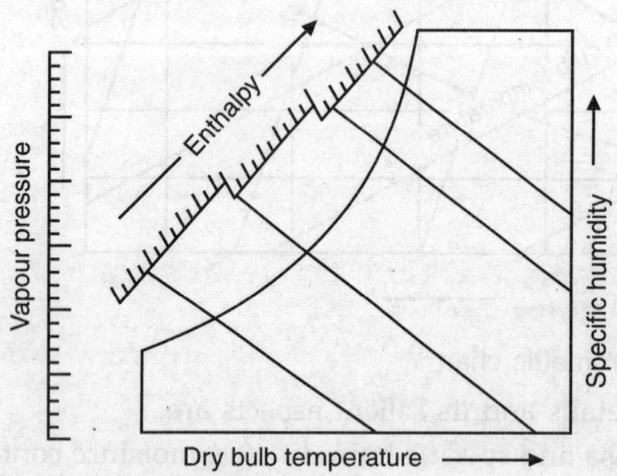

Fig. 12.9. Enthalpy and vapour pressure lines

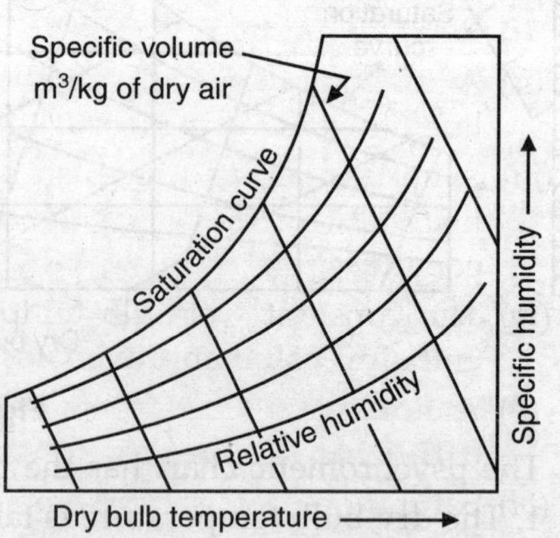

Fig. 12.10. Relative humidity and specific volume lines

4. The vapour pressure and enthalpy (total heat) lines are also scaled on the chart. The total heat at saturation temperature is represented by a diagonal system of co-ordinates. These inclined straight lines are uniformly spaced and are parallel to the wet bulb temperature lines. The scale on the diagonal lines is separate from the body of the chart and is indicated above the saturation line.

Pressure of water vapour is shown in the scale on left and is the absolute pressure of steam in mm of mercury.

EXAMPLE 12.23

Atmospheric air at 1 bar pressure has 15°C wet-bulb temperature and 25°C dry-bulb temperature. With the help of a psychrometric chart, determine the salient psychrometric properties of the air.

Solution: Refer Fig. 12.11

Locate the state point *S* on the psychrometric chart. This is the point where a constant dry-bulb temperature line corresponding to 25°C intersects the wet-bulb temperature line corresponding to 15°C.

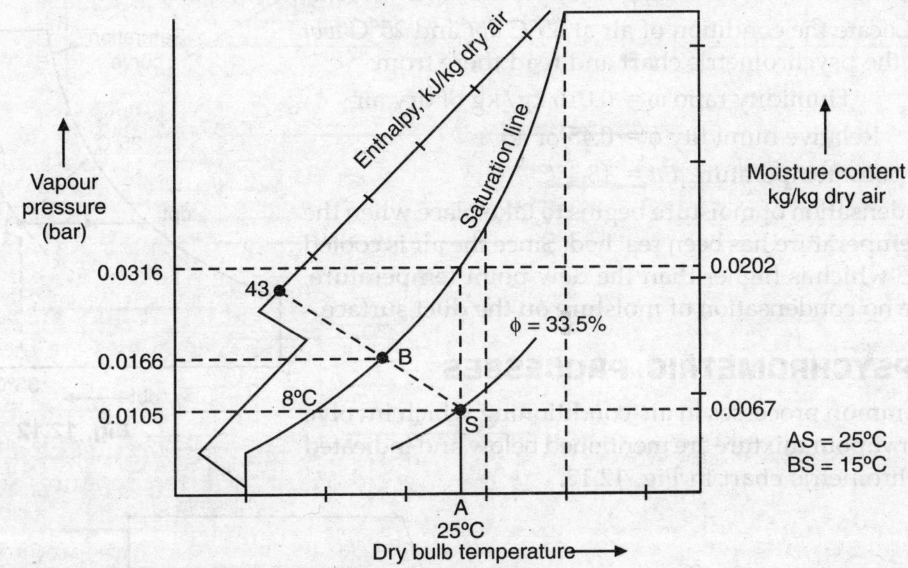

Fig. 12.11

At this point the following properties can be read:

(*i*) Relative humidity = 33.5%

(*ii*) A horizontal line to the right from state point *S* would represent a constant humidity ratio line; its value being ω = 0.0067 kg of water/kg of dry air.

(*iii*) A horizontal line to the left drawn from the state point *S* would intersect the saturation line; the temperature at the intersecting point gives the value of dew point temperature equal to *dpt* = 8°C.

(*iv*) The constant wet-bulb temperature line is produced to meet the enthalpy line ; the enthalpy value approximates to 43 kJ/kg of dry air.

(*v*) A horizontal line drawn from the state point *S* meets the vapour pressure scale at 0.0105 bar. This represents the partial pressure of the vapour.

(*vi*) Saturation pressure corresponding to *dbt* = 25°C can be obtained be drawing a horizontal line to the left from the point where *dbt* = 25°C line cuts the 100% relative humidity line. The saturation pressure value is 0.0316 bar.

(*vii*) Saturation pressure corresponding to the wet-bulb temperature is obtained by drawing a horizontal line to the left from the point where *wbt* = 15°C line cuts the 100% relative humidity line. Here the pressure is scaled as 0.0166 bar.

(*viii*) Humidity ratio corresponding 25°C dry-bulb temperature can be obtained by drawing a horizontal line to the right from the point where *dbt* = 25°C line cuts the 100% relative humidity line. The corresponding value is scaled as 0.0202 kg of water/kg of dry air.

The degree of saturation is then calculated as:

$$\mu = \frac{\omega}{\omega_s} = \frac{0.0067}{0.0202} = 33\%$$

EXAMPLE 12.24

The atmospheric air at 35°C dbt and 25°C wbt is cooled to 20°C. Using psychrometric chart, find (i) humidity ratio, (ii) relative humidity and (iii) dew point temperature. Comment whether or not there will be any condensation on the duct that supplies this cooled air to the conditioned space.

Solution : Locate the condition of air at 35°C *dbt* and 25°C *wbt* (Point 1) on the psychrometric chart and read there from:

Humidity ratio ω = 0.016 kg/kg of dry air

Relative humidity ϕ = 0.45 or 45%

Dew point temperature *dpt* = 18.4°C

The condensation of moisture begins to take place when the dew point temperature has been reached. Since the air is cooled only to 20°C which is higher than the dew point temperature, there will be no condensation of moisture on the duct surface.

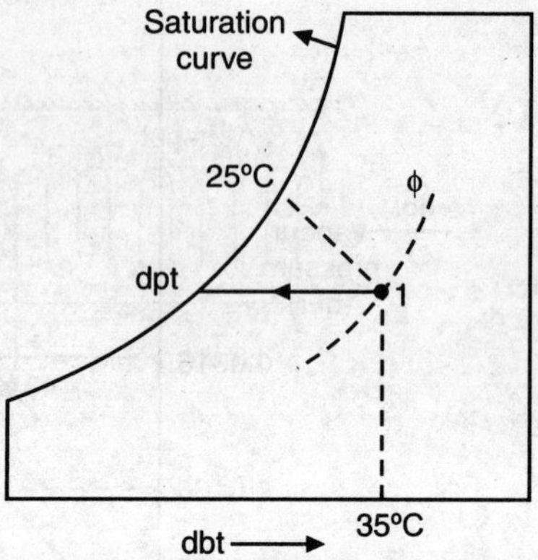

Fig. 12.12

12.10. PSYCHROMETRIC PROCESSES

The most common processes in air-conditioning which involve the air-water vapour mixture are mentioned below and indicated on the psychrometric chart in Fig. 12.12.

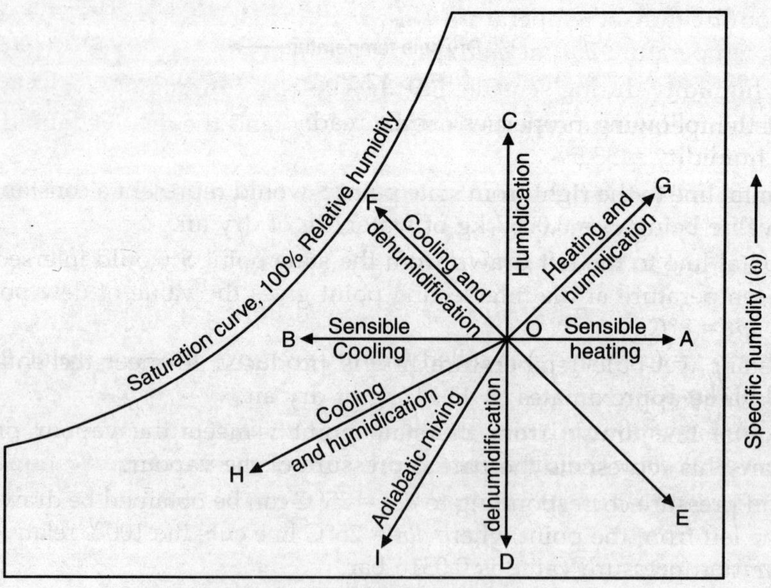

Fig. 12.13. Psychrometric processes

The initial state is represented by point 0. Heating and cooling processes are represented by horizontal lines while humidification and dehumidification plot as vertical lines. Humidification represents the process wherein the moisture is added to air but its dry bulb temperature is maintained constant. In dehumification process, the moisture is removed from air without changing its dry bulb temperature. The combination of these processes are set between them.

12.10.1. Sensible Heating

The mixture is heated without any change in its moisture content. The process results when the mixture is made to pass over a surface whose temperature is above the dry bulb-temperature of the mixture. The heating surface may be the electric resistance heating coils or steam passed through the coils or hot water passed through the coils.

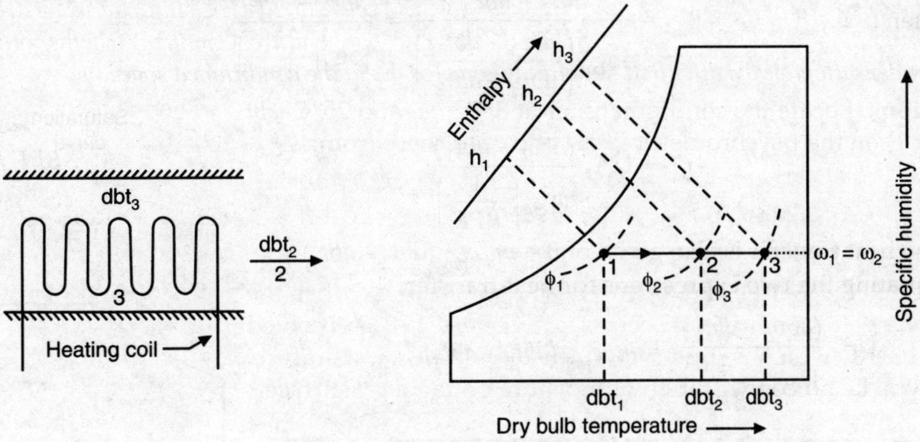

Fig. 12.14. Sensible heating

With reference to Fig. 12.14, the process of sensible heating is represented by horizontal line 1-2 that extends from left to right. The surface temperature of the heating coil is indicated by point 3. Further, it may be noted that:

(*i*) Air-vapour mixture at temperature dbt_1 passes over a heating coil of temperature dbt_3 and acquires temperature dbt_2 at exit; dbt_2 is less than dbt_3.

(*ii*) Specific humidity during sensible heating remains constant, *i.e.*, $\omega_1 = \omega_2$

(*iii*) Dry bulb temperature increases from dbt_1 to dbt_2 and the relative humidity reduces from ϕ_1 to ϕ_2.

(*iv*) The amount of heat absorbed by the mixture can be obtained from the psychrometric chart by the enthalpy difference $(h_2 - h_1)$. Alternatively, it can be calculated from the relation:

$$h = h_2 - h_1 = c_{pa}\ (dbt_2 - dbt_1) + \omega\, c_{pv}\ (dbt_2 - dbt_1)$$
$$= (c_{pa} + \omega\, c_{pv})\ (dbt_2 - dbt_1) = c_{pm}\ (dbt_2 - dbt_1)$$

The factor $c_{pm} = (c_{pa} + \omega\, c_{pv})$ is called ***humid specific heat*** and its value is generally taken as 1.022 kJ/kg K.

(*v*) At exit, the air-vapour mixture ideally should attain temperature dbt_3 of the heating coil but it leaves at dbt_2. The difference is due to inefficiency of the heating coil and this is a prescribed by the terms ***by-pass factor*** and ***coil-efficiency*** defined as

$$\text{By-pass factor } (BPF) = \frac{dbt_3 - dbt_2}{dbt_3 - dbt_1}$$

and Coil efficiency $\eta_c = 1 - BPF$

The coil efficiency is also called ***contact factor.***

The sensible heat given out by the coil can be prescribed by the relation

$$Q = U A_s\, \theta_m$$

where U is the overall heat transfer coefficient, A_s is the surface area of the coil and θ_m is the logarithmic mean temperature difference.

$$\theta_m = \frac{\theta_1 - \theta_2}{\log_e \dfrac{\theta_1}{\theta_2}}$$

where $\theta_1 = dbt_3 - dbt_1$ and $\theta_2 = dbt_3 - dbt_2$

Then :
$$\theta_m = \frac{dbt_2 - dbt_1}{\log_e\left(\dfrac{dbt_3 - dbt_1}{dbt_3 - dbt_2}\right)} = \frac{dbt_2 - dbt_1}{\log_e\left(\dfrac{1}{BPF}\right)}$$

$$\therefore \quad Q = UA_s \frac{(dbt_2 - dbt_1)}{\log_e\left(\dfrac{1}{BPF}\right)} \qquad \left(\because BPF = \frac{dbt_3 - dbt_2}{dbt_3 - dbt_1}\right)$$

The heat transfer is also given of $\theta = m_a\, c_{pm}\, (dbt_2 - dbt_1)$

Equating the two expressions for heat transfer,

$$UA_s \frac{(dbt_2 - dbt_1)}{\log_e\left(\dfrac{1}{BPF}\right)} = m_a\, c_{pm}\, (dbt_2 - dbt_1)$$

or
$$\log_e\left(\frac{1}{BPF}\right) = \frac{UA_s}{m_a\, c_{pm}}\ ;\ \log_e (BPF) = -\frac{UA_s}{m_a\, c_{pm}}$$

That gives :
$$BPF = e^{-\left[\frac{UA_s}{m_a c_{pm}}\right]} = e^{-\left[\frac{U n A_r}{m_a c_{pm}}\right]} = e^{-k}$$

where n is the number of coils and A_r is surface area per row of the coil.

The value of factor k is constant for the given coil surfaces and given air flow quantity. The above expression shows that the by-pass factor decreases with an increase in surface area, increase in depth of the coil and with the decrease in air velocity ($m_a = \rho AV$). Further, the following correlation exists between B_1(by-pass factor of single depth coil) and B_n (by-pass factor of n-depth coil) : $B_n = (B_1)^n$

The factor (1 – BPF) is the mass actually coming in perfect contact with the coil. The by-pass factor is an important tool for measuring the performance of the heating coils. The coil gives better performances with lower by-pass factor.

EXAMPLE 12.24

For a single depth coil, the by-pass factor is stated to be 0.75.

(a) What will be the by-pass factor of a 4-depth coil ?

(b) What number of depths of coil will be required if the by-pass factor in an air conditioner is 0.2 ?

Solution : Invoking the relation : $B_n = B_1^n$

(*a*) By-pass factor of 4-depth coil works out as

$$B_4 = (0.75)^4 = \mathbf{0.3164}$$

(*b*) The number of depths for a by-pass factor of 0.2

$$0.2 = (0.75)^n\ ;\ \log_e 0.2 = n \log_e 0.75$$

or
$$n = \frac{\log_e 0.2}{\log_e 0.75} = \frac{-1.6094}{-0.2877} = \mathbf{5.594}$$

EXAMPLE 12.25

Calculate the by-pass factor and the contact factor for the air-conditioner coil used to bring the condition of outdoor air (40°C dbt, 30% RH) to the required comfort conditions (22°C dbt, 60% RH). The data pertinent to the coil is :

Overall heat transfer coefficient = 850 kJ/m²-hr-deg
Surface area of coil provided per row = 1.25 m²
Depth of coil = 4
Volume of free air passing through the coil = 60 m³/min.

Solution : At outdoor air conditions of 40°C *dbt* and 30% *RH*, the specific volume of air,

$$v_s = 0.908 \text{ m}^3/\text{kg} \qquad \text{(Read from psychrometric chart)}$$

$$\text{Mass flow rate of air } m_a = \frac{V}{v_s} = \frac{60 \times 60}{0.908} = 3964.76 \text{ kg/hr}$$

The by-pass factor of the coil is prescribed by the relation,

$$BPF = e^{-\left(\frac{U n A_r}{m_a c_{pm}}\right)}$$

Substituting the appropriate values, we get

$$BPF = e^{-\left(\frac{850 \times 4 \times 1.25}{3964.76 \times 1.022}\right)} = e^{-1.049} = 0.35$$

Contact factor for the coil (coil efficiency),

$$= 1 - BPF = 1 - 0.35 = \mathbf{0.65}$$

EXAMPLE 12.26

The moist air at 10°C and 50% relative humidity enters a steam heating coil at the rate of 50 kg/s and the temperature at exit is noted to be 30°C. Determine :

(a) sensible heat transfer,

(b) mass flow rate of steam if it enters saturated at 100°C and the condensate leaves at 65°C.

Solution : Refer Fig. 12.15

(*i*) Locate the initial state point 1 on psychrometric chart as per $dbt_1 = 10°\text{C}$ and $\phi_1 = 50\%$.

(*ii*) The state point 2 is located by drawing a constant specific humidity line from point 1 to intersect the vertical line drawn through 30°C *dbt* at point 2.

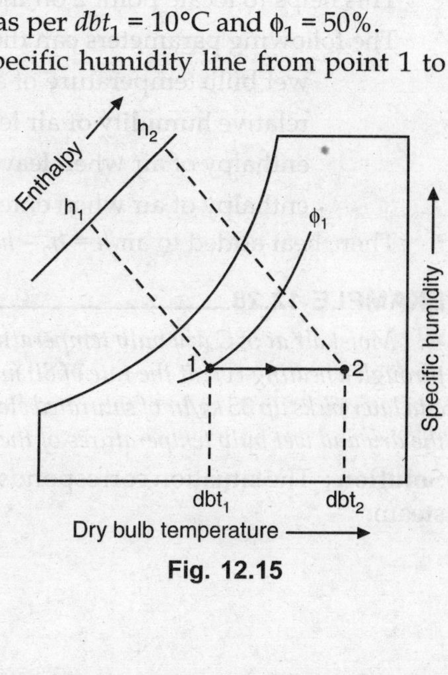

Fig. 12.15

The line 1-2 represents sensible heating of air.

From the chart :

$$h_1 = 19.5 \text{ kJ/kg of dry air and}$$
$$h_2 = 40 \text{ kJ/kg of dry air}$$

Then :

Sensible heat transfer $= m_a (h_2 - h_1)$

$$Q = 50(40 - 19.5) = \mathbf{1025} \text{ kJ/s}$$

(*b*) From steam tables :

Enthalpy of saturated steam at 100°C : $h_g = 2676$ kJ/kg

Enthalpy of condensate at 65°C : $h_f = 272$ kJ/kg

Then : steam mass flow rate $m_s = \dfrac{Q}{h_g - h_f} = \dfrac{1025}{2676 - 272} = \mathbf{0.426}$ **kg/s**

EXAMPLE 12.27

A sample of atmospheric air having 15°C dry bulb temperature and 10°C wet bulb temperature is made to enter a heating coil whose surface temperature is 40°C. Using psychrometric chart, determine :

(a) dry bulb temperature, wet bulb temperature and relative humidity of air at exit from the heating coil, and

(b) heat added to the air.

It may be assumed that the heating coil has a by-pass factor of 0.5.

Solution : Refer Fig. 12.16. The situation corresponds to sensible heating, *i.e.*, heating of air without any change in specific humidity.

(*i*) Mark point 1 corresponding to dry bulb temperature of 15°C and wet bulb temperature of 10°C.

(*ii*) Draw a horizontal line (constant specific humidity) 1-3 that extends from left to right and intersects the vertical line drawn through 40°C (the surface temperature of the heating coil).

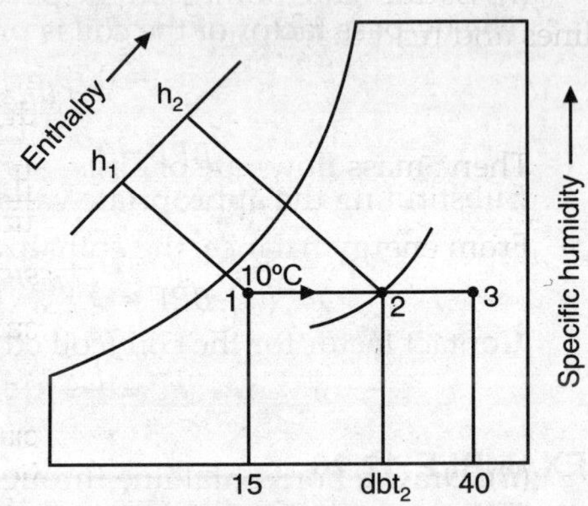

Fig. 12.16

The point 2, representing state of the air on leaving the coil, lies on this line.

We know that by-pass factor in sensible heating is

$$BPF = \frac{dbt_3 - dbt_2}{dbt_3 - dbt_1} \quad \text{or} \quad 0.5 = \frac{40 - dbt_2}{40 - 15}$$

Therefore, the dry bulb temperature of air at exit from the heating coil is

$$dbt_2 = 40 - 0.5\,(40 - 15) = 27.5°\text{C}$$

This helps to locate point 2 on the sensible heating line 1-3.

The following parameters can then be read from the psychrometric chart :

wet bulb temperature of air leaving the coil at point 2, $wbt_2 = 16.5$°C

relative humidity of air leaving the coil at point 2, $\phi_2 = 29.5\%$

enthalpy of air when leaving the coil at point 2, $h_2 = 45$ kJ/kg dry air

enthalpy of air when entering the coil at point 1, $h_1 = 31.5$ kJ/kg dry air

Then, heat added to air $h = h_2 - h_1 = 45 - 31.5 = 13.5$ kJ/kg of dry air.

EXAMPLE 12.28

Moist air at 5°C dry bulb temperature and 2.5°C thermodynamic wet bulb temperature is made to pass through a heating coil at the rate of 80 m³/min. While passing through the coil, the air absorbs 40 kW of heat and later picks up 35 kg/hr of saturated steam at 110°C. Assuming barometric pressure of 1.0132 bar, determine the dry and wet bulb temperatures of the leaving air.

Solution : The situation corresponds to simple heating and humidification of air by the addition of steam.

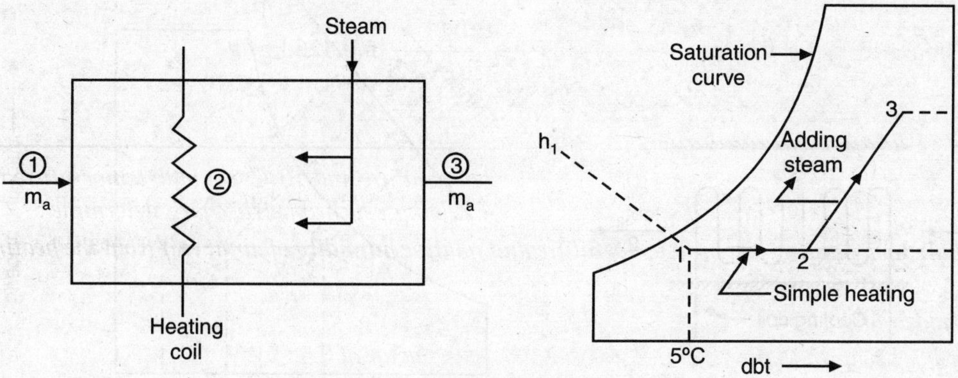

Fig. 12.17

(*i*) Locate state point 1 on the psychrometric chart at the intersection of 5°C dbt and 2.5°C wbt lines and read therefrom

$$v_{s_1} = 0.79 \text{ m}^3/\text{kg}; \quad \omega_1 = 0.0035 \text{ kg/kg, and } h_1 = 13.8 \text{ kJ/kg}$$

Then : mass flow rate of air $m_a = \dfrac{V_1}{v_{s_1}} = \dfrac{80/60}{0.79} = 1.688$ kg/s

From energy balance, the enthalpy of air h_2 at exit from the heating coil is given by

$$m_a (h_2 - h_1) = Q_{abs}; \quad Q_{abs} \text{ is heat absorbed}$$

or $$h_2 = h_1 + \frac{Q_{abs}}{m_a} = 13.8 + \frac{40}{1.688} = 37.5 \text{ kJ/kg}$$

(*ii*) Draw a horizontal line through point 1 till it meets enthalpy line $h_2 = 37.5$ kJ/kg and then mark point 2.

By moisture balance.

$$m_a (\omega_3 - \omega_2) = m_s$$

or $$\omega_3 = \frac{m_s}{m_a} + \omega_2 = \frac{35/3600}{1.688} + 0.0035 = 0.00925 \text{ kg/kg}$$

By energy balance : $m_a (h_3 - h_2) = m_s h_{fg}$

where $h_{fg} = 2691$ kJ/kg is the enthalpy of saturated steam at 110°C.

$$\therefore \quad h_3 = \frac{m_s}{m_a} h_{fg} + h_2 = \frac{35/3600}{1.688} \times 2691 + 37.5 = 53 \text{ kJ/kg}$$

From the above obtained values of ω_3 and h_3, locate the state point 3 on the chart and read the following values pertaining to air at exit

Dry bulb temperature = 27°C

Wet bulb temperature = 18.5°C

12.10.2. Sensible Cooling

The mixture is cooled without any change in its moisture content. The process results when the mixture is made to pass over a surface whose temperature is below the dry bulb temperature of the mixture. The cooling surface may be the cooled water or gas flowing through coils or the refrigerant at low temperature in the coils of the evaporator of a vapour refrigeration system.

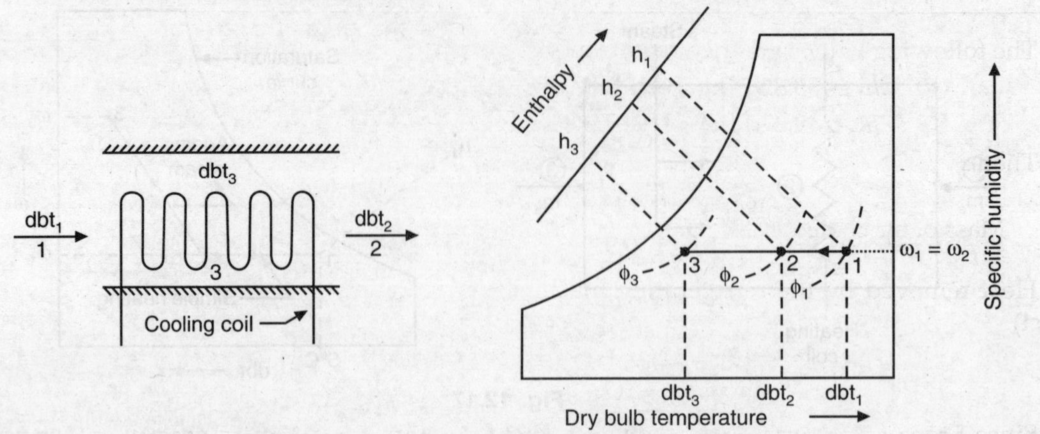

Fig. 12.18. Sensible cooling

With reference to Fig. 12.18, the process of sensible cooling is represented by horizontal line 1-2 that extends from right to left. The surface temperature of the cooling coil is indicated by 3. Further, it may be noted that

(*i*) Air-vapour mixture at temperature dbt_1 passes other a cooling coil of temperature dbt_3 and acquires temperature dbt_2 at exit; dbt_2 is more than dbt_3.

(*ii*) Specific humidity during sensible cooling remains constant, *i.e.*, $\omega_1 = \omega_2$.

(*iii*) Dry bulb temperature decreases from dbt_1 to dbt_2 and the relative humidity increases from ϕ_1 to ϕ_2.

(*iv*) The amount of heat rejected by the mixture can by obtained from the psychrometric chart by the enthalpy difference $(h_1 - h_2)$. Alternatively, it can be calculated from the relation:

$$h = h_1 - h_2 = c_{pa}\,(dbt_1 - dbt_2) + \omega\, c_{pv}\,(dbt_1 - dbt_2)$$
$$= (c_{pa} + \omega\, c_{pv})\,(dbt_1 - dbt_2) = c_{pm}\,(dbt_1 - dbt_2)$$

The factor $c_{pm} = (c_{pa} + \omega\, c_{pv})$ is called ***humid specific heat*** and its value is generally taken as 1.022 kJ/kg K.

(*v*) At exit, the air-vapour mixture should ideally attain temperature dbt_3 of the cooling coil but it leaves at dbt_2. The difference is due to inefficiency of the cooling coil and this is prescribed by the terms ***by-pass factor*** and ***coil efficiency*** defined as

$$\text{By-pass factor } (BPF) = \frac{dbt_2 - dbt_3}{dbt_1 - dbt_3}$$

and Coil efficiency $\eta_c = 1 - BPF$

EXAMPLE 12.29

Atmospheric air at 35°C dbt and 55% relative humidity has been cooled to a state corresponding to 25°C dbt and 50% relative humidity. Determine the capacity of the refrigerating machine if it is to handle 125 m^3 *of air per minute at initial state.*

Solution : Refer Fig. 12.19

(*i*) Set the initial state 1 according to dbt_1 = 35°C and ϕ_1 = 55°C

(*ii*) Locate point 2 by drawing a horizontal line through point 1 such that dbt_2 = 25°C and ϕ_2 = 50°C

The following values are then read from the chart:

$$h_1 = 85.8 \text{ kJ/kg}; \quad v_{s_1} = 0.9 \text{ m}^3\text{/kg}$$
$$h_2 = 49.8 \text{ kJ/kg}; \quad v_{s_2} = 0.85 \text{ m}^3\text{/kg}$$

Then :

$$\text{Mass of air cooled } m_a = \frac{V}{v_{s_1}} = \frac{125}{0.9} = 138.38 \text{ kg/min}$$

Heat removed by the refrigerating machine (refrigerating effect)

$$= m_a (h_1 - h_2)$$
$$= 138.38 (85.8 - 49.8) = 4981.68 \text{ kJ/min}$$

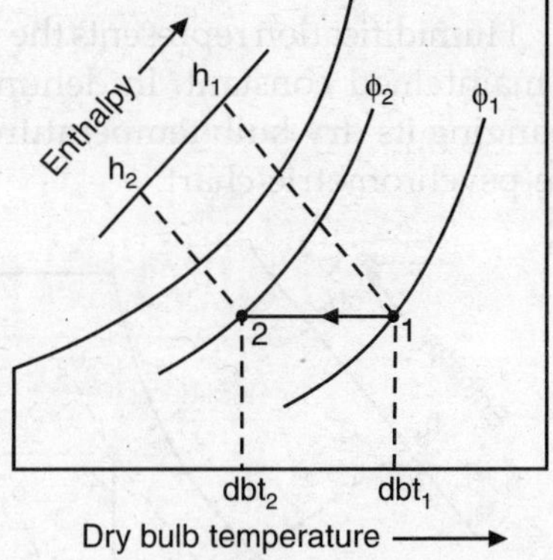

Fig. 12.19

Since 1 ton of refrigeration is equivalent to 210 kJ/min,

$$\text{Capacity of refrigerating machine} = \frac{4981.68}{210} = 23.72 \text{ ton}$$

EXAMPLE 12.30

To study the performance of a cooling coil, a sample of moist air at 30° C dry bulb temperature and 50% relative humidity of 50% was made to enter it at a rate of 5 m³/s. If the air leaves the coil at 10° C in just saturated state, determine the amount of moisture added or removed and tons of refrigeration required for the accomplishment of process.

Solution : Refer Fig. 12.20 which shows the given process laid on psychrometric chart.

(*i*) Set the initial state point 1 according to *dbt* = 30°C and ϕ = 50%, and the find state point 2 at *dbt* = 30°C and ϕ = 50%, and the final state point 2 at *dbt* = 10°C and lying on the saturation curve.

The following values are then read from the chart:

h_1 = 80 kJ/kg of dry air,

ω_1 = 0.0135 kg/kg of dry air

v_1 = 0.875 m³/kg of dry air

h_2 = 47 kJ/kg of dry air, and

ω_2 0.0075 kg/kg of dry air

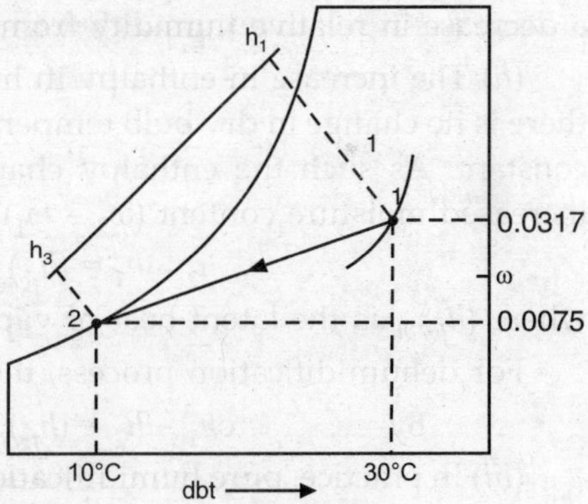

Fig. 12.20

$$(a) \text{ Mass flow of air} = \frac{5}{0.875} = 5.714 \text{ kg/s}$$

The specific humidity decreases from 0.0135 to 0.0075.

Therefore, the amount of moisture removed is

$$= m(\omega_1 - \omega_2) = 5.714 (0.0135 - 0.0075)$$
$$= \mathbf{0.0354} \text{ kg/s}$$

(*b*) The enthalpy decreases during the process and that indicates that heat transfer is from the system to the surroundings.

∴ Heat absorbed or cooling effect is

$$m(h_1 - h_2) = 5.714 (80 - 47)$$
$$= 188.56 \text{ kJ/s} = 188.56 \text{ kW}$$

Since 1 ton of refrigeration is equivalent of 3.5 kW, the capacity of the cooling coil in tonnage is

$$= \frac{188.56}{3.5} = \mathbf{53.87}$$

12.10.3. Humidification and Dehumidification

Humidification represents the process wherein the moisture is added but its dry bulb temperature is maintained constant. In dehumidification process, the moisture is removed from air without changing its dry bulb temperature. These processes are obviously represented as vertical lines on the psychrometric chart.

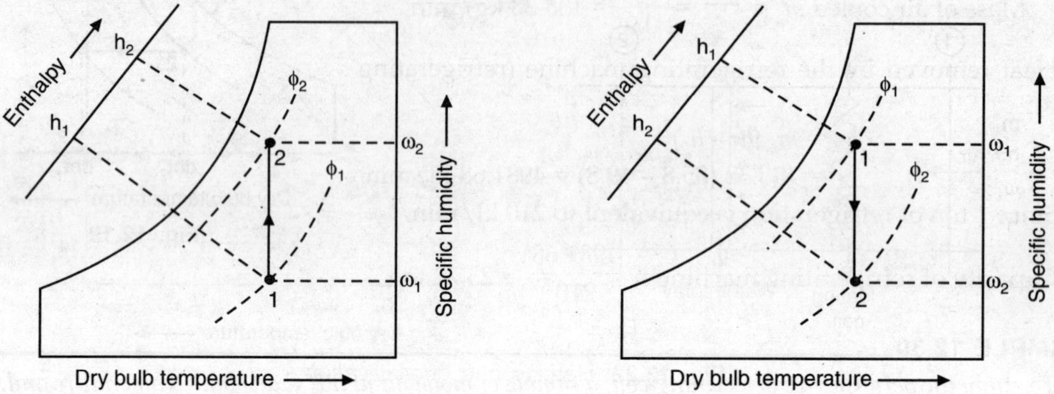

Fig. 12.21. Humidification and dehumidificaion processes

With reference to Fig. 12.21, it may be noted that

(*i*) In humidification the specific humidity increases from ω_1 to ω_2 and the relative humidity also increases from ϕ_1 to ϕ_2.

However in dehumidification, there is decrease in specific humidity from ω_1 to ω_2 and also a decrease in relative humidity from ϕ_1 to ϕ_2.

(*ii*) The increase in enthalpy in humidification is represented by the intercept $(h_2 - h_1)$. Since there is no change in dry bulb temperature during humidification, the sensible heat would remain constant. As such the enthalpy change $(h_2 - h_1)$ equals the latent heat of vaporisation of the increased moisture content $(\omega_2 - \omega_1)$.

$$\therefore \quad h_2 - h_1 = (h_{fg})_{db}\,(\omega_2 - \omega_1)$$

where $(h_{fg})_{db}$ is the latent heat of vaporisation at the dry bulb temperature *dbt*

For dehumidification process, the decrease in enthalpy can be written as

$$h_1 - h_2 = (h_{fg})_{db}\,(\omega_1 - \omega_2)$$

(*iii*) In practice, pure humidification and dehumidificaton processes are not possible. These are always accompanied by heating or cooling.

12.10.4. Heating and Humidification

The process is achieved when the moist air is made to pass through spray water whose temperature is maintained at a temperature higher than the dry bulb temperature of the air. The unsaturated air tends to become saturated and the heat of vaporisation is absorbed from the spray water.

With reference to Fig. 12.22, the process of heating and humidification is represented by line 1-2.

The air-vapour mixture enters at condition 1 and leaves at condition 2. It is to be noted that during the heating and humidification process, (*i*) There is an increase in specific humidity, dry and wet bulb temperatures, dew point temperature and enthalpy.

(*ii*) The relative humidity may either increase or decrease.

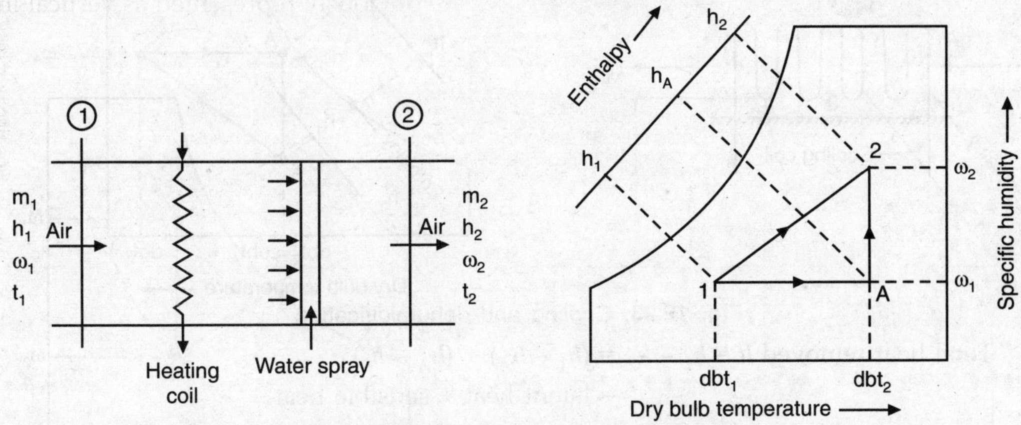

Fig. 12.22. Heating and humidification

(*iii*) The process may be assumed to follow the paths 1-A and A-2. Path 1-A represents sensible heating during which the dry bulb temperature increases. Path A-2 represents humidification during which, there is no change in dry bulb temperature and latent heat of vaporisation is supplied with increased moisture content $(\omega_2 - \omega_1)$. Then

$$\text{Total heat added } h = h_2 - h_1 = (h_2 - h_A) + (h_A - h_1)$$
$$= \text{latent heat + sensible heat}$$

The ratio of sensible heat to total heat is known as ***sensible heat factor*** (*SHF*) or ***sensible heat ratio*** (*SHR*)

SHF can be read from a scale on the psychrometric chart located in tne right corner of the chart

$$SHF = \frac{\text{sensible heat}}{\text{total heat}} = \frac{\text{sensible heat}}{\text{sensible heat + latent heat}}$$

The process of heating and humidification has practical application in winter air conditioning.

12.10.5. Cooling and Dehumidification

The process takes place when the moist air is passed through a cooling coil whose effective surface temperature is lower than the dew point temperature of mixture. The surface temperature of the cooling coil is called the ***apparatus dew point*** (**ADP**)

With reference to Fig. 12.23, the process of cooling and dehumidification is represented by line 1-2. The moist air enters the cooling coil at condition 1 and leaves at condition 2. It is to be noted that during the cooling and dehumidification process,

(*i*) the dry bulb temperature decreases from dbt_1 to dbt_2

(*ii*) the air is cooled and condensation of moisture takes place, *i.e.*, it is dehumidified

(*iii*) the specific humidity decreases from ω_1 to ω_2

(*iv*) the relative humidity at outlet is generally higher than that at inlet

(*v*) the process may be assumed to follow the path 1-A and A-2. Path 1-A represents sensible cooling during which the dry bulb temperature decreases. Path A-2 represents dehumidification during which there is no change in the dry bulb temperature and latent heat condensation is removed with decreased moisture content $(\omega_1 - \omega_2)$

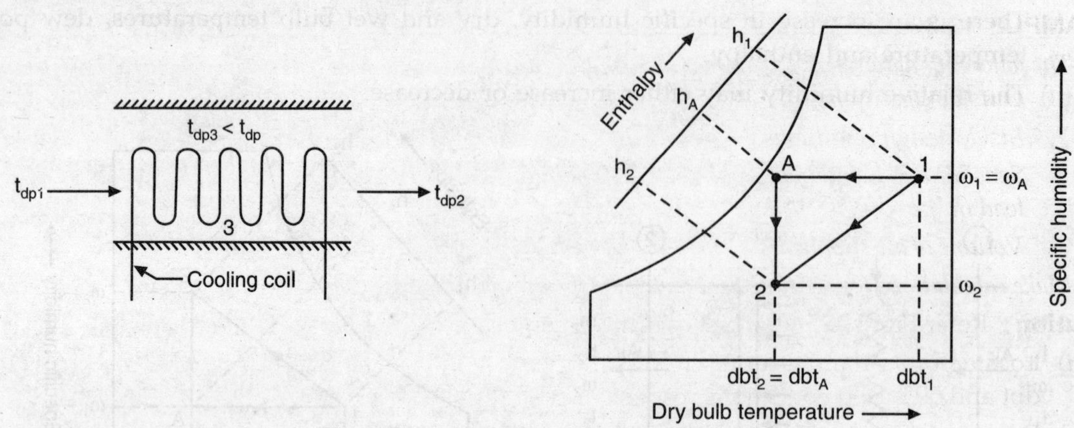

Fig. 12.23. Cooling and dehumidification

Total heat removed $h = h_1 - h_2 = (h_1 - h_A) + (h_A - h_2)$

$= \text{latent heat} + \text{sensible heat}$

The ratio of sensible heat to total heat is known as ***sensible heat factor*** (*SHF*)

$$SHF = \frac{\text{sensible heat}}{\text{total heat}} = \frac{\text{sensible heat}}{\text{sensible heat + latent heat}}$$

The cooling and dehumidification process has practical application in summer air-conditioning.

EXAMPLE 12.31

Air at 15°C dry bulb temperature and 25% relative humidity is heated and humidified at 30°C dry bulb temperature and 50% relative humidity. Using psychrometric chart, calculate the heat and moisture added to air and the sensible heat factor for the process.

Solution : Refer Fig. 12.24 which shows the given process laid on psychrometric chart.

(*i*) Set the initial state point 1 according to dbt_1 = 15°C and ϕ_1 = 25% and state final state point 2 according to dbt_2 = 30°C and ϕ_2 = 50%.

(*ii*) Locate the point 3 by drawing horizontal line through point 1 and vertical line through point 2.

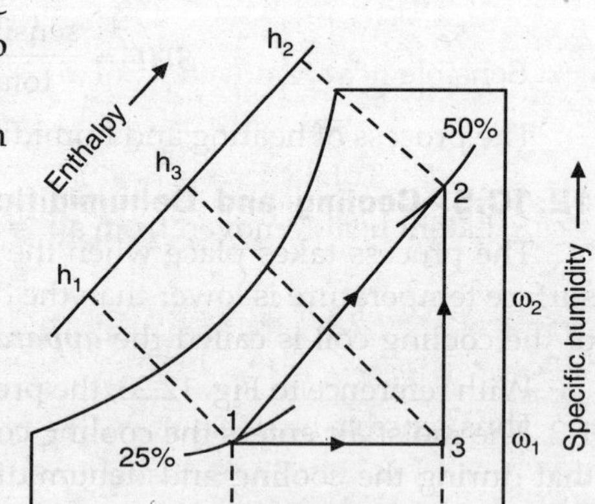

Fig. 12.24

The following values are then read from the chart :

h_1 = 22 kJ/kg of dry air

h_2 = 65 kJ/kg of dry air

h_3 = 38 kJ/kg of dry air

ω_1 = 0.0027 kg/kg of dry air

ω_2 = 0.0134 kg/kg of dry air

Then :

(*a*) Heat added to air $= h_2 - h_1$

$= 65 - 22 =$ **43** kJ/kg of dry air

(*b*) Moisture added to air $= \omega_2 - \omega_1 = 0.0134 - 0.0027 =$ **0.0107** kg/kg of dry air

(*c*) Sensible heat factor $SHF = \frac{\text{sensitble heat}}{\text{total heat}}$

$$= \frac{h_3 - h_1}{h_2 - h_1} = \frac{38 - 22}{65 - 22} = \mathbf{0.372}$$

EXAMPLE 12.32

The following data refers to summer air-conditioning of a small office hall with capacity for 30 persons :

Outside conditions	*: 34°C dbt and 28°C wbt*
Inside design conditions	*: 24°C dbt and 50% RH*
Sensible and latent heat load in the room	*: 2100 kJ/min and 700 kJ/min*
Volume of air supplied	*: 0.4 m³/min per person.*

Make calculations for sensible heat factor for the air-conditioning plant.

Solution : Refer Fig. 12.25 which shows the given process laid on psychrometric chart.

(*i*) Locate points 1 and 2 representing outdoor air (34°C *dbt* and 28°C *wbt*) and room air (24°C dbt and 50% RH) conditions respectively.

(*ii*) Draw a horizontal line towards right from point 2 and a vertical line downwards from point 1 and mark point 3 where these lines intersect.

The horizontal vertical lines represent cooling and dehumidification respectively.

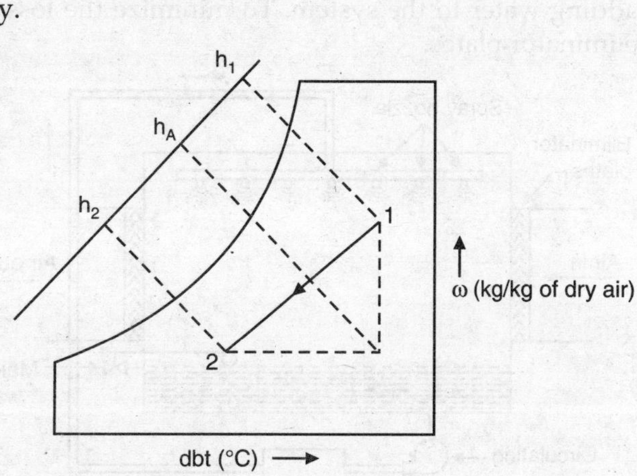

Fig. 12.25

From the chart :

$$h_1 = 90 \text{ kJ/kg of dry air}$$
$$v_1 = 0.9 \text{ m}^3\text{/kg of dry air}$$
$$h_2 = 48 \text{ kJ/kg of dry air}$$
$$h_3 = 58.8 \text{ kJ/kg of dry air}$$

Then, mass of air supplied $m_a = \frac{0.4 \times 30}{0.9}$

$= 13.33$ kg/min

Sensible heat removed from air $= m_a (h_3 - h_2)$

$= 13.33 (58.8 - 48) = 143.96$ kJ/min

$\therefore$ Total sensible heat $SH = 2100 + 143.96 = 2243.96$ kJ/min

Latent heat removed from air $= m_a (h_1 - h_3)$

$= 13.33 (90 - 58.8) = 415.9$ kJ/min

$\therefore$ Total latent heat $LT = 700 + 415.9 = 1115.9$ kJ/min

Thus sensible heat factor $SHF = \frac{SH}{SH + LT} = \frac{2243.96}{2243.9 + 1115.9} = \mathbf{0.668}$

12.10.6. Methods for Humidification and Dehumidification

Humidification (an increase in moisture content of air) is achieved by making the air stream to flow through a spray of cold water/hot water or steam. The humidification may be :

(*i*) **Direct :** Where the hot water/cold water or steam is atomized and sprayed directly into the space being air-conditioned.

(*ii*) **Indirect :** Where an air washer is used to introduce water into the air in the air-conditioning plant. The conditioned air is then supplied, by a system of ducts, to the space which is to be air-conditioned.

The humidification of air with an air washer is carried out by any one of the following methods:

(*a*) by using a spray of heated water,

(*b*) by using re-circulated spray water with no preheating of air, and

(*c*) by preheating the air and then washing it with re-circulated water

An air washer has been referred to as a water spray system or device for cooling and cleaning air in which the entering warm, moist air is cooled below its dew point by refrigerating water so that although the air leaves close to saturation with water, it has less moisture per unit volume than when it entered. In general, the air washer involves the flow of air through a spray of water, and depending on the temperature of water the air may get cooled or heated, humidified or dehumidified or simply adiabatically saturated. The water to be sprayed is externally heated or cooled or simply circulated by a pump. Any loss of water during humidification of air is replenished (made-up) by adding water to the system. To minimize the loss of water droplets, the setup is provided with eliminator plates.

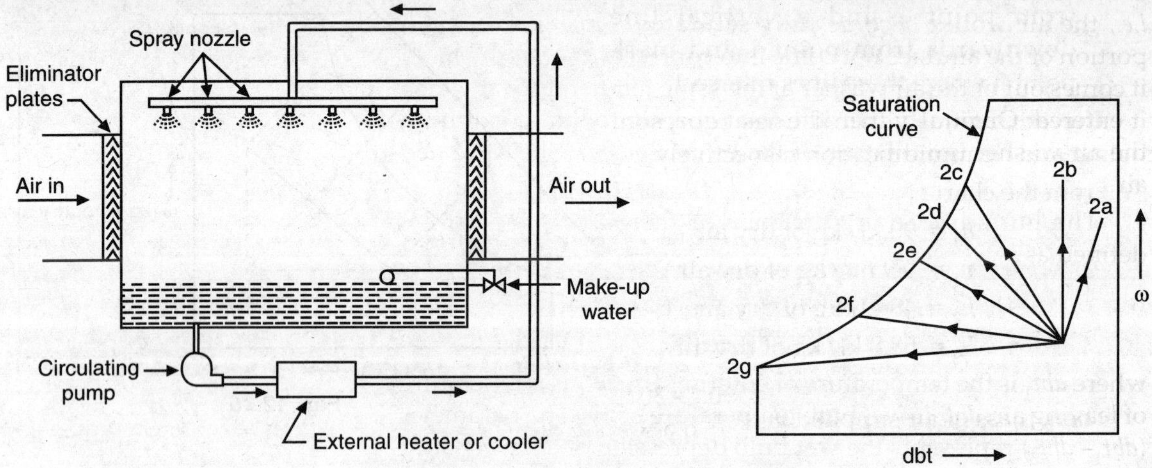

Fig. 12.26. Schematic of an air washer

Fig. 12.26 depicts the thermodynamic changes in state of air along the path 1-2 in an air washer; the path followed depends on the mean surface temperature (t_s) of water. There occur both sensible and latent heat transfers and the direction of heat transfer depends on the temperature and vapour pressure potential. Depending on the spray water temperature, the following processes are possible:

(1) 1-2a : Heating and humidification (t_s > *dbt*) : The water is externally heated and the mean surface temperature of water is greater than the dry bulb temperature of water.

(1) 1-2b : Humidification (t_s = *dbt*) : The mean surface temperature of water is equal to the dry bulb temperature of water and the water is required to be heated externally.

(3) 1-2c : Cooling and dehumidification (*wbt* < t_s > *dbt*) : The spray water is maintained at a constant surface temperature which is higher than the entering wet bulb temperature but lower than the inlet dry bulb temperature of air. This is achieved by heating of spray water before pumping back to spray nozzles. The air gets cooled and humidified but its enthalpy increases.

(4) 1-2d : Adiabatic saturation : There is no heat gain or loss from the spray water, and the same water is circulated again and again. If there is intimate contact between water and air, the process corresponds to that of adiabatic saturation. The temperature of spray water becomes equal to the wet bulb temperature of entering air and as such the process lies along the constant wet bulb temperature line. The cooling and humidification of air occurs at constant enthalpy.

(5) 1-2e : Cooling and humidification ($dpt < t_s < wbt$) : The surface temperature of water is less than the wet bulb temperature of air but more than the dew point temperature. The spray water needs to be cooled externally by coolers before pumping it back to the spray nozzles. During this cooling and humidification process, the enthalpy of air is reduced.

(6) 1-2f : Cooling ($t_s = dpt$) : The temperature of spray water is further reduced by external cooling, and maintained equal to the dew point temperature of entering air. During this sensible cooling process, the water vapour contents of air remain constant.

(7) 1-2g : Cooling and dehumidification ($t_s < dpt$) : The temperature of spray water is further lowered and maintained lower than entering dew point temperature. There is simultaneous cooling and dehumidification of the air stream.

It is worthwhile to note that air washer affords means for year-round air-conditioning system.

Air-washer efficiency : If the heat exchange between the air and water spray in the air-washer is 100%, the dry bulb temperature of the entering air would drop down to its wet bulb temperature. *i.e.*, the air would become fully saturated. However, a small portion of the air does not come into contact with the spray and it comes out of the air washer at the same temperature at which it entered. Obviously then the total quantity of air at exit from the air-washer is a mixture of the by-passed and the saturated air.

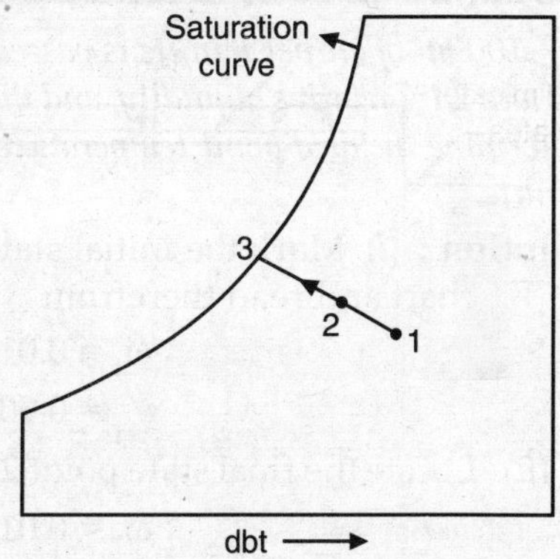

Fig. 12.27

The humidification efficiency of the air-washer is then defined as :

$$\eta = \frac{dbt_1 - dbt_2}{dbt_1 - dbt_3}$$

where dbt_1 is the temperature of entering air, dbt_2 is temperature of leaving air, dbt_3 is wet bulb temperature of air. The parameter ($dbt_1 - dbt_3$) represents the wet bulb depression of entering air.

The extent to which humidification is affected depends on the by-pass factor, fineness and spread of spray, and the velocity of air.

12.10.7. Chemical Dehumidification (Heating and Dehumidification)

The process involves in passing air over a bed of chemicals which have affinity for moisture. When the air comes in contact with the chemicals, the moisture condenses and latent heat is released. That leads to decrease in specific humidity of air. Further, the heat of condensation provides sensible heat for heating the air and accordingly its dry bulb temperature decreases. The path followed during the chemical dehumidification process is along the constant wet bulb temperature line or constant enthalpy line (Fig. 12.28).

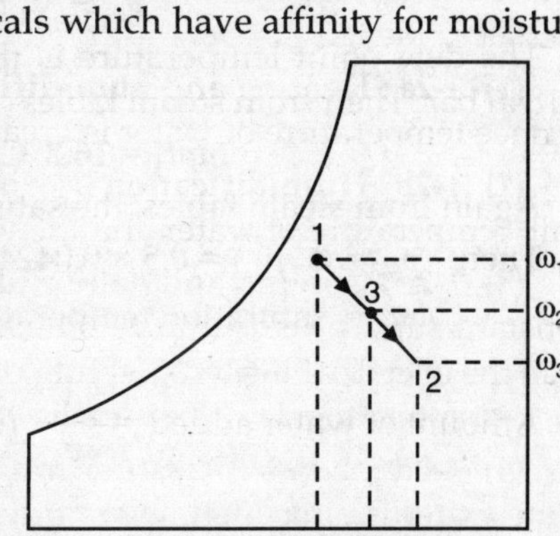

Fig. 12.28

The chemical reagents used for dehumidification process are of two types :

(*a*) Absorbent such as water solutions or brines of calcium chloride, lithium chloride, lithium bromide and ethylene glycol. These substances take up moisture causing a physical or chemical change.

(*b*) **Adsorbent** such as silica gel and activated alumina. The silica gel is a form of silicon dioxide prepared by mixing fused sodium silicate and sulphuric acid, and the activated alumina is a porous form of aluminium oxide. These adsorbents are in the solid state, they take up moisture from air but do not undergo any physical or chemical change :

The effectiveness of the chemical dehumidifier is prescribed by the relation :

$$\eta = \frac{\text{actual increase in dry bulb temperature}}{\text{ideal increase in dry bulb temperature}}$$

$$= \frac{dbt_3 - dbt_1}{dbt_2 - dbt_1}$$

This process finds application in industry for processes where low humidify is needed or precise control over humidify is desired.

EXAMPLE 12.33

100 m³ of air per minute is passed through an adiabatic humidifier. The condition of air at inlet is 40°C dbt and 25% relative humidity, and the corresponding values at exit are 30°C dbt and 50% relative humidity. Determine the dew point temperature, wet bulb temperature and amount of water vapour added to air per minute.

Solution : (*i*) Mark the initial state point 1 as per $dbt_1 = 40°C$ and $\phi_1 = 25\%$ on the psychrometric chart and read therefrom

$$\omega_1 = 0.011 \text{ kg/kg of dry air ;}$$

$$v_{s_1} = 0.90 \text{ m}^3/\text{kg and } h_1 = 69.5 \text{ kJ/kg}$$

(*ii*) Locate the final state point 2 as per $dbt_2 = 30°C$ and $\phi_2 = 50\%$ on the chart and read there from

$$\omega_1 = 0.011 \text{ kg/kg of dry air ;}$$

$$v_{s_2} = 0.88 \text{ m}^3/\text{kg and } h_2 = 64 \text{ kJ/kg and } wbt = 23°C$$

From steam tables, the saturation pressure of water vapour at 40°C is $p_{vs_1} = 0.7385$ bar

$$\text{Now : Relative humidity } \phi_1 = \frac{p_{v_1}}{p_{vs_1}}$$

$$\therefore \quad p_{v_1} = \phi_1 \, p_{vs_1} = 0.25 \times 0.07385 = 0.01846 \text{ bar}$$

The dew point temperature is the saturation temperature corresponding to the pressure of 0.01846 bar. Then from steam tables

$$dpt_1 = \mathbf{16.2°C}$$

Again from steam tables, the saturation pressure of water vapour at 30°C is $p_{vs_2} = 0.04246$ bar

Then $p_{v_2} = \phi_2 \, p_{vs_2} = 0.5 \times 0.04246 = 0.02123$ bar

$dpt_2 =$ saturation temperature corresponding to 0.02123 bar = **19°C**

$$\text{Amount of water added} = \frac{V}{v_{s_1}} (\omega_2 - \omega_1) = \frac{100}{0.9} (0.0135 - 0.011)$$

$$= \mathbf{0.278 \text{ kg/min}}$$

EXAMPLE 12.34

The air-conditioning plant for a hall of 800 seating capacity has been designed with reference to the following conditions :

Outdoor conditions : 40°C dbt and 20°C wbt

Required indoor conditions : 20°C dbt and 50% RH

Free air circulation : 0.25 m³/min per person

The required conditions are achieved by adiabatic humidification followed by cooling. Using psychrometric chart, make calculations for the following parameters :

(a) Capacity of the humidifier in kg/hr,

(b) Capacity of the cooling coil and its surface temperature if it has a by-pass factor of 0.3.

Solution : Refer Fig. 12.29 which depicts the processes involved drawn on psychrometric chart as per give data.

(*i*) Locate points 1 and 2 corresponding to outdoor (40°C *dbt*, 20°C *wbt*) and required indoor (20°C *dbt*, 50% RH) conditions respectively.

(*ii*) Mark point 3 where the constant enthalpy line drawn through state point 1 intersects the horizontal line drawn through state point 2.

Process 1-3 represents adiabatic humidification and process 3-2 represents cooling.

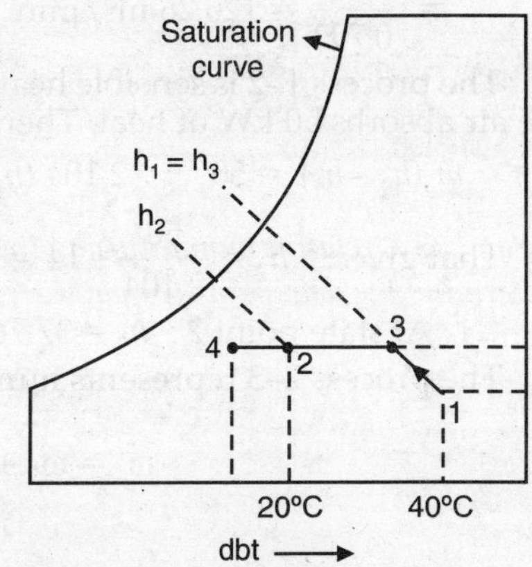

Fig. 12.29

From the psychrometric chart

$$h_1 = h_3 = 57.6 \text{ kJ/kg}; \quad h_2 = 39 \text{ kJ/kg};$$

$$v_{s1} = 0.896 \text{ m}^3/\text{kg}; \quad \omega_1 = 0.0064 \text{ kg/kg}$$

$$\text{and } \omega_3 = 0.0074 \text{ kg/kg}$$

Then :

$$\text{Mass of air } m_a = \frac{V_1}{v_{s_1}} = \frac{800 \times 0.25}{0.896} = 223.2 \text{ kg/min}$$

(*a*) Capacity of the humidifier $= m_a(\omega_3 - \omega_1) = 223.2\,(0.0074 - 0.0064)$

$$= 0.232 \text{ kg/min} = \mathbf{13.93} \text{ kg/hr}$$

(*b*) Capacity of the cooling coil $= m_a(h_3 - h_2)$

$$= 223.2\,(57.6 - 39) = 4151.52 \text{ kJ/min}$$

$$= \frac{4151.52}{210} = 19.77 \text{ tons} \qquad (\because 210 \text{ kJ/min} = 1 \text{ ton})$$

(*c*) $dbt_3 = 38°\text{C}$ (from chart)

$$\text{By-pass factor of cooling coil} = \frac{dbt_2 - dbt_4}{dbt_3 - dbt_4} \quad \text{or} \quad 0.3 = \frac{20 - dbt_4}{38 - dbt_4}$$

That gives : surface temperature of the coil,

$$dbt_4 = \frac{20 - 38 \times 0.3}{1 - 0.3} = 12.28°\text{C}$$

EXAMPLE 12.35

Air enters a chamber at 5°C dry bulb temperature and 2.5°C wet bulb temperature at the role of 100 m³/min and at a pressure of 1 bar ; while passing through the chamber, the air absorbs 50 kW heat and picks up 35 kg/hr of saturated steam at 110°C. Show the process on psychrometric chart and find the dry and wet bulb temperatures of the leaving air. At 110° C, enthalpy of saturated steam is 2691.3 kJ/kg.

Solution : Refer Fig. 12.30 which shows the process on a psychrometric chart.

(*i*) Mark point 1 corresponding to dry bulb temperature of 5°C and wet bulb temperature of 2.5° C. From the chart

$$h_1 = 14 \text{ kJ/kg and } v_{s1} = 0.792 \text{ m}^3/\text{kg}$$

and ω_1 = 0.035 kj/kg of dry air

The mass flow rate of air,

$$m_a = \frac{\text{volume flow rate}}{\text{specific volume}}$$

$$= \frac{100}{0.792} = 126.26 \text{ m}^3/\text{min} \quad \text{or} \quad 2.104 \text{ m}^3/\text{s}$$

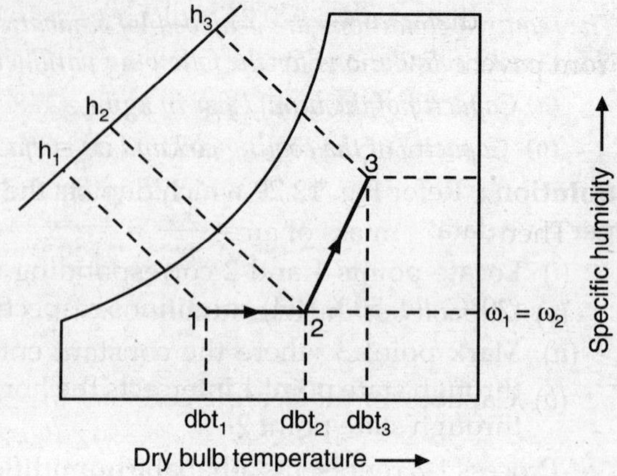

Fig. 12.30

The process 1–2 is sensible heating during which the air absorbs 50 kW of heat. Then

$$m_a(h_2 - h_1) = 50 \quad ; \quad 2.104\,(h_2 - 14) = 50$$

That gives: $h_2 = \dfrac{50}{2.104} + 14 = 37.76$ kJ/kg

∴ At state point 2 : h_2 = 37.76 kJ/kg and $\omega_2 = \omega_1$ = 0.0035 kj/kg of dry air

The process 2–3 represents humdification by steam injection : Then

$$\omega_3 = \omega_2 + \frac{m_s}{m_a} = 0.0035 + \frac{35/3600}{2.104} = 0.00812 \text{ kg/kg of dry air}$$

$$h_3 = h_2 + \frac{m_s}{m_a} \times h_v = 37.76 + \frac{35/3600}{2.104} \times 2691.3$$

$$= 37.76 + 12.436 \simeq 50.2 \text{ kJ/kg}$$

The point 3 is then located corresponding to the above calculated values for w_3 and h_3. Then at state point 3,

$$dbt_3 = 31\ °\text{C} \quad \text{and} \quad wbt_3 = 18.3°\ \text{C}$$

EXAMPLE 12.36

Air at 28°C dbt and 18°C wbt is made to pass through an air washer in which water circulates and has an efficiency of 15 percent. Subsequently the air passes over a heating coil which has a surface temperature of 30°C and a by-pass factor of 0.25. If the mass flow rate of air is 80 m³/min determine :

(*a*) *capacity of the humidifier,*
(*b*) *capacity of the heating coil, and*
(*c*) *condition of air leaving the heating coil.*

Solution : Refer Fig. 12.31 which depicts the processes involved drawn on psychrometric chart as per data given :

(*i*) Locate state point 1 at the intersection of 28°C *dbt* and 18°C *wbt*.

(*ii*) Mark point 2 where the constant enthalpy line drawn from point 1 meets the saturation curve.

(*iii*) Since the air washer has an efficiency of 75%, locate point 3 on line 1-2 such that

distance 1-3 = 0.75 × distance 1-2

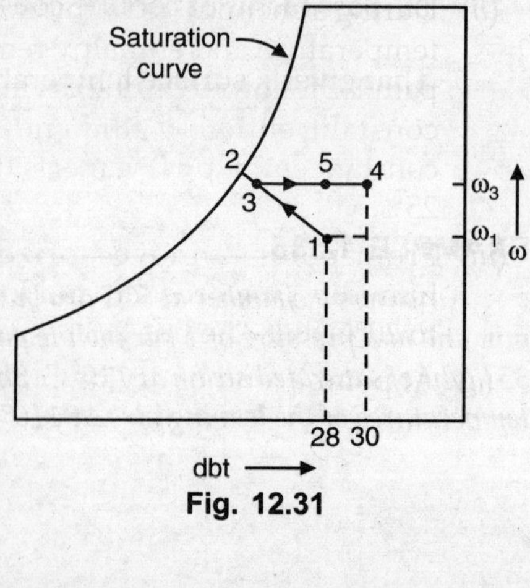

Fig. 12.31

(*iv*) Move horizontally from point 3 and locate point 4 where it intersects the vertical line corresponding to 30°C (surface temperature of the coil).

(*v*) Since the heating coil has a by-pass factor of 0.25, locate point 5 on line 3-4 such that distance 4-5 = 0.25 × distance 4-3

From psychrometric chart :

$$h_1 = h_3 = 50.8 \text{ kJ/kg} ; \quad v_{s1} = 0.865 \text{ m}^3/\text{kg} ; \quad \omega_1 = 0.009 \text{ kg/kg}$$

$$h_5 = 56.5 \text{ kJ/kg} \quad \text{and} \quad \omega_5 = 0.018\ 18 \text{ kg/kg}$$

Then :

$$\text{mass of air} = \frac{V_1}{v_{s_1}} = \frac{80}{0.865} = 92.48 \text{ kg/min}$$

(*a*) Capacity of the humidifier $= m_a\,(\omega_3 - \omega_1)$

$$= 92.48(0.018 - 0.009) = \mathbf{0.832} \text{ kg/min}$$

(*b*) Capacity of the heating coil $= m_a\,(h_5 - h_3)$

$$= 92.48\,(56.5 - 50.8) = 527.14 \text{ kJ/min}$$

$$= \frac{527.14}{60} = \mathbf{8.78 \text{ kJ/s}} \text{ or } \mathbf{8.78 \text{ kW}}$$

(*c*) The state point 5 represents the condition of air leaving the heating coil. Corresponding to point 5,

$$dbt = 27.8°\text{C} \quad \text{and} \quad wbt = 19.8°\text{C}$$

EXAMPLE 12.37

Saturated air at 20°C is made to undergo a chemical dehumidification process in a drier till its relative humidity becomes 25 percent. Subsequently the dried air enters a cooling coil where it gets restored to the initial temperature of 20°C by following a constant specific humidity process. Show the processes on psychrometric chart and obtain therefrom

(*a*) *temperature of air and the dew point temperature at the end of drying process,*

(*b*) *relative humidity at the end of cooling process,*

(*c*) *heat rejected during cooling process, and*

(*d*) *moisture removed during the drying process.*

Solution : Refer Fig. 12.32 which depicts the processes involved drawn on psychrometric chart as per the data given.

(*i*) Set point 1 (20°C) on the saturation line.

(*ii*) During a chemical dehumidification process, the wet bulb temperature and enthalpy remain constant. That sets the path 1-2 followed during drying. From point 1, move along constant enthalpy line and locate point 2 where the constant enthalpy line meets the relative humidity line of 25%.

(*iii*) From point 2, move horizontally (constant specific humidity) towards left and locate point 3 corresponding to 20°C dbt. The line 2-3 represents line cooling process.

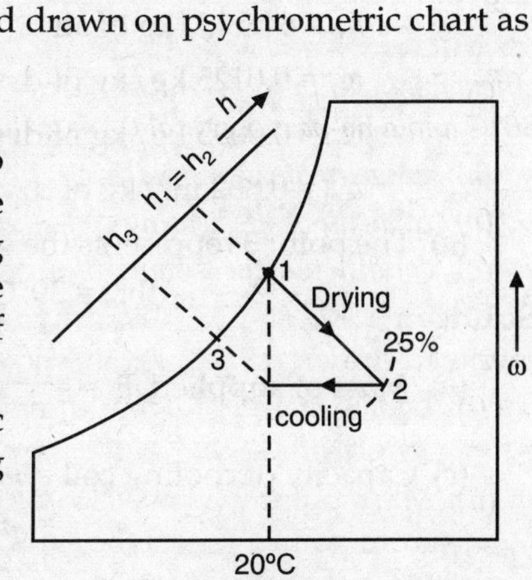

Fig. 12.32

(*a*) At point 2 (end of drying process) :

$$dbt_2 = 35°\text{C and } dpt_2 = 20°\text{C}$$

(*b*) At point 3 (end of cooling process) :

$$\phi_3 = 58\%$$

(*c*) Heat rejected during cooling process

$$= (h_2 - h_3) = 57.5 - 41.8 = 15.7 \text{ kJ/kg}$$

(*d*) Moisture removed during drying process,

$$= \omega_1 - \omega_2 = 0.015 - 0.0085 = 0.0065 \text{ kg/kg of dry air}$$

EXAMPLE 12.38

The design conditions for a summer air-conditioning system for a class room of 65 seating capacity are as follows :

Outdoor conditions	: *32°C dbt and 22°C wbt*
Inside comfort conditions	: *22°C dbt and 55% RH*
Outside air to meet ventilation requirement	: *0.5 m³/min per student*

The comfort conditions are achieved first by chemically dehumidifying air and then cooling it. Analyze the problem on a psychrometric chart and determine the following :

(*a*) *dry bulb temperature of air leaving the dehumidifier,*
(*b*) *capacity of the dehumidifier,*
(*c*) *capacity of the cooling coil, and*
(*d*) *surface temperature of the cooling coil if its by-pass factor is 0.28.*

Solution : Refer Fig. 12.33 which shows the various processes of the given system on psychrometric chart :

(*i*) Locate points 1 and 2 representing outdoor air (32°C *dbt*, 22°C *wbt*) and inside room air (22°C *dbt*, 55% RH).

(*ii*) Draw a constant enthalpy line (chemical dehumidification) from point 1 and a constant humidity (horizontal line) from point 2 and let these meet at point 3.

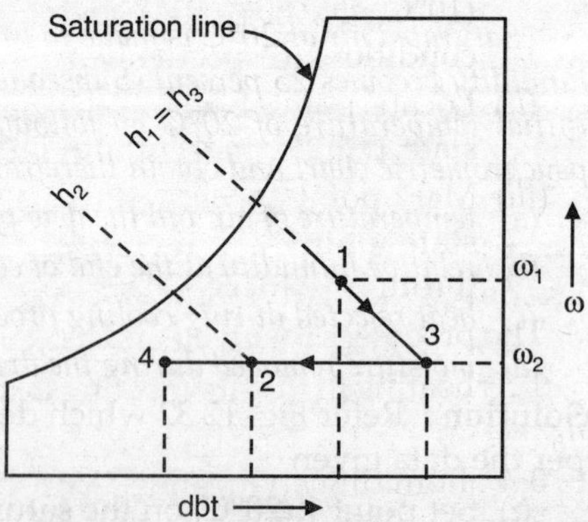

Fig. 12.33

From the chart :

$$h_1 = 64.5 \text{ kJ/kg of air} ; \quad h_2 = 44.8 \text{ kJ/kg of air}$$
$$\omega_1 = 0.0125 \text{ kg/kg of dry air}$$
$$\omega_2 = 0.0095 \text{ kg/kg of dry air}$$
$$v_{s1} = 0.882 \text{ m}^3\text{/kg of dry air.}$$

(*a*) The point 3 represents the state of air leaving the dehumidifier. Corresponding to this point

$$\text{dbt}_3 = 39°\text{C} \quad \text{(read from the chart)}$$

(*b*) Mass of supplied $m_a = \dfrac{V_1}{v_{s_1}} = \dfrac{0.5 \times 65}{0.882} = 36.95 \text{ kg/min}$

(*c*) Capacity of cooling coil $= m_a(h_3 - h_2)$

$$= m_a(h_1 - h_2) = 36.85\ (64.5 - 44.85) = 724 \text{ kJ/min}$$

$$= \frac{724}{210} = 3.45 \text{ tonnes} \qquad (\because\ 210 \text{ kJ/min} = 1 \text{ tonne})$$

(*d*) Let dbt_4 be the surface temperature of the coil. Then

$$\text{By-pass factor} = \frac{dbt_2 - dbt_4}{dbt_3 - dbt_4}; \quad 0.28 = \frac{22 - dbt_4}{39 - dbt_4}$$

$$\text{Solution gives}: \ dbt_4 = \frac{22 - 39 \times 0.28}{1 - 0.28} = \mathbf{14.2°C}$$

EXAMPLE 12.39

The winter air-conditioning system of a theatre of 800 seating capacity is to be designed. The design is based on the following information.

Outside design condition	*: 10°C dbt and 60% RH*
Inside design condition	*: 22°C dbt and 60% RH*
Outside air to meet ventilation	*: 0.95 m³/min per person*

The desired conditions are achieved by heating, humidifying and then again by heating. The air coming out of the humidifier has a relative humidity of 75 percent. Represent the processes on psychrometric chart and make calculation for the following aspects :

(*a*) *capacity and surface temperature of first heating coil if its by-pass factor is 0.25,*

(*b*) *capacity of the humidifier, and*

(*c*) *capacity and by-pass factor of the second heating coil if its surface temperature is 25°C.*

Solution : Refer Fig. 12.34 which depicts the processes involved drawn on psychrometric chart as per given data.

(*i*) Locate points 1 and 2 corresponding to outdoor air (10°C *dbt*, 60% RH) and inside (22°C *dbt*, 60% RH) conditions respectively.

(*ii*) Locate point 4 where the horizontal line drawn from state point 2 meets the relative humidity line 75%.

(*iii*) Mark point 3 where constant humidity line drawn from state point 4 intersects the horizontal line drawn through point 3.

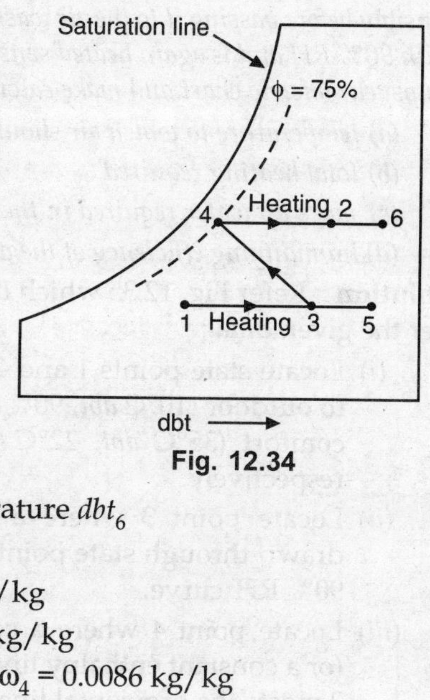

Fig. 12.34

The processes involved and as represented on the chart are :

1-3 : heating in the first heating coil with surface temperature dbt_5

3-4 : humidification, and

4-2 : heating in the second heating coil with surface temperature dbt_6

From the psychrometric chart :

$$h_1 = 21.4 \text{ kJ/kg of air}; \quad v_{s1} = 0.81 \text{ m}^3/\text{kg}$$
$$h_2 = 44.8 \text{ kJ/kg of air}; \quad \omega_1 = 0.0044 \text{ kg/kg}$$
$$h_3 = h_4 = 43 \text{ kJ/kg}; \quad h_2 = 47 \text{ kJ/kg}, \quad \omega_4 = 0.0086 \text{ kg/kg}$$
$$dbt_3 = 31.5°\text{C}; \quad dbt_4 = 18.3°\text{C}$$

$$\text{Then}: \quad \text{Mass of air } m_a = \frac{V_1}{v_{s_1}} = \frac{800 \times 0.95}{0.81} = 938.27 \text{ kg/min}$$

(*a*) Capacity of the first heating coil $= m_a (h_3 - h_1)$

$$= 938.27 (43 - 21.4) = 20266.6 \text{ kJ/min}$$

$$= \frac{20266.6}{60} = \mathbf{337.78 \ kW}$$

By-pass factor of first heating coil $= \dfrac{dbt_5 - dbt_3}{dbt_5 - dbt_1}$ or $0.25 = \dfrac{dbt_5 - 31.5}{dbt_5 - 10}$

∴ Surface temperature of first heating coil,

$$dbt_5 = \frac{31.5 \times 0.25 \times 10}{1 - 0.25} = \mathbf{38.67°C}$$

(*b*) Capacity of the humidifier $= m_a\,(\omega_4 - \omega_3)$

$= 938.27\,(0.0086 - 0.0044) = \mathbf{3.94}$ kg/min

(*c*) Capacity of the second heating coil $= m_a(h_2 - h_4)$

$= 938.27\,(47 - 43) = 3753.08$ kJ/min

$= \dfrac{3753.08}{60} = \mathbf{62.55}$ kW

By pass factor of the second heating coil

$$= \frac{dbt_6 - dbt_2}{dbt_6 - dbt_4} = \frac{25 - 22}{25 - 18.3} = \mathbf{0.448}$$

EXAMPLE 12.40

The air-conditioning system of a certain premises necessities that outside air at 10°C dbt and 90% RH be heated and humidified to indoor comfort conditions of 35°C dbt and 22°C wbt. The air is first preheated sensibly before passing it to the air washer which has a system of water circulation. The air exits the air washer with 90% RH and is again heated sensibly till the final desired condition is attained. Represent the processes on psychrometric chart and make calculations for the following parameters :

(*a*) *temperature to which air should be preheated,*

(*b*) *total heating required,*

(*c*) *make up water required in the air washer, and*

(*d*) *humidifying efficiency of the air washer.*

Solution : Refer Fig. 12.35 which depicts the processes involved drawn on psychrometric chart as per the given data :

(*i*) Locate state points 1 and 2 corresponding to outdoor (10°C *dbt*, 90% *RH*) and indoor comfort (35°C *dbt*, 22°C *wbt*) conditions respectively.

(*ii*) Locate point 3 where a horizontal line drawn through state point 2 intersects the 90% RH curve.

(*iii*) Locate point 4 where a constant *wbt* line (or a constant enthalpy line) through point 3 meets the hroizontal line drawn through state point 1.

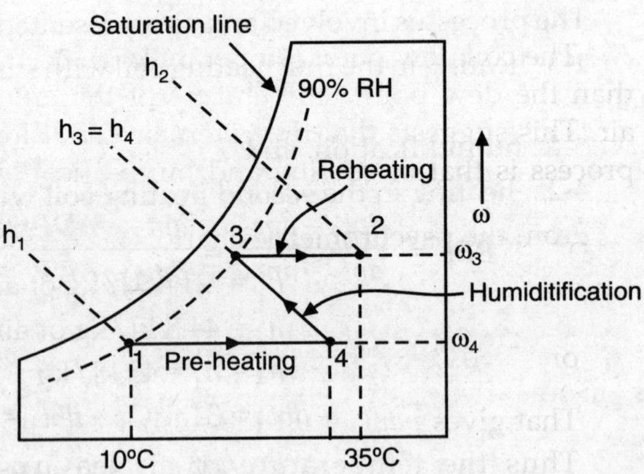

Fig. 12.35

Then 1-4 is the preheating of air, 4-3 represents humidification and 3-2 represents reheating to final condition.

From the chart :

$$h_1 = 27 \text{ kJ/kg}; \quad h_2 = 67 \text{ kJ/kg}; \quad h_3 = h_4 = 49 \text{ kJ/kg}$$

$$\omega_1 = \omega_4 = 0.0068 \text{ kg/kg and } \omega_3 = \omega_2 = 0.012 \text{ kg/kg}$$

(*a*) Air is preheated upto state point 4

$$dbt_4 = \mathbf{31°C} \quad \text{(from chart)}$$

(*b*) Heat required for preheating the air = $(h_4 - h_1) = (49 - 27) = 22$ kJ/kg

Heat required for reheating the air = $(h_2 - h_3) = (67 - 49) = 18$ kJ/kg

∴ Total heat required = 22 + 18 = **40** kJ/kg

(*c*) Make up water required in the air washer

$$= \omega_3 - \omega_4 = 0.012 - 0.0068 = 0.0052 \text{ kg/kg}$$

(*d*) Humidification efficiency of air washer,

$$\eta = \frac{\text{actual drop in } dbt}{\text{ideal drop in } dbt} = \frac{dbt_4 - dbt_3}{dbt_4 - dbt_{3'}}$$

From the chart : $dbt_3 = 18.5°C$ and $dbt_{3'} = 17.5°C$

The point 3′ on the saturation curve is located by extending line 4-3 to intersect the saturation curve.

$$\eta = \frac{31 - 18.5}{31 - 17.5} = 0.926$$

EXAMPLE 12.41

The atmospheric air at 30°C dry bulb temperature and 75% relative humidity enters a cooling coil at the rate of 200 m³/min. The coil dew point temperature is 14°C and the by-pass factor of the coil is 0.1. Determine.

(a) temperature of air leaving the cooling coil.

(b) capacity of the cooling coil in tonnes of refrigeration and in kilowatt,

(c) amount of water vapur removed per minute, and

(d) sensible heat factor for the process. ***(MDU, MRIU)***

Solution : Refer Fig. 12.36 for plot of the air-conditioning process on psychrometirc chart.

Locate state point 1 on psychrometric chart corresponding to initial condition of air, *i.e.*, 30° *dbt* and 75% *RH*. The dew point temperature at this state as read from the chart is

$$dbt_1 = 25°C$$

The coil dew point temperature (*ADP*) is less than the dew point temperature of the entering air. This suggests that the given air conditioning process is that of cooling and humidification

$$BPF = \frac{dpt_2 - dpt_3}{dpt_1 - dpt_3} = \frac{dpt_2 - ADP}{dpt_1 - ADP}$$

or $$0.1 = \frac{dpt_2 - 14}{30 - 14}$$

That gives : $dpt_2 = 0.1\,(30 - 14) + 14 = 15.6°C$

Thus the temperature of air leaving the cooling coil is 15.6°C.

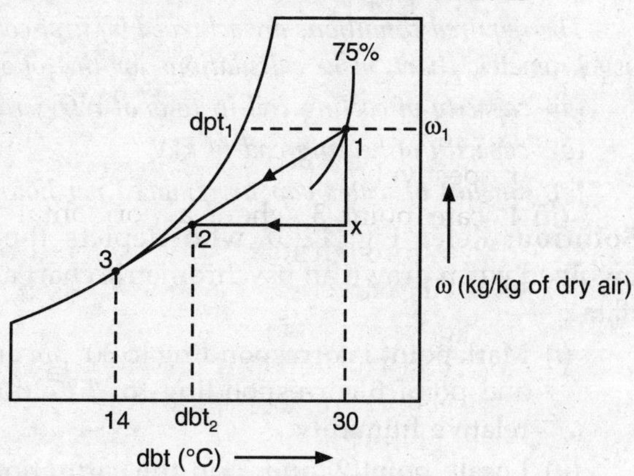

Fig. 12.36

(*b*) Join the points 1 and 3. The point 2 on this line represents the condition of air coming out of the coil. The line 1–2 represents the cooling and dehumidification process which can be assumed to follow the dehumidification process 1–*x* and cooling process *x*–2.

From the chart :

$$\omega_1 = 0.0202 \text{ kg/kg of dry air} \quad \text{and} \quad \omega_2 = 0.011 \text{ kg/kg of dry air}$$

$$v_{s_1} = 0.885 \ m^3/\text{kg of dry air}$$

$$h_1 = 82 \text{ kJ/kg} \ ; \ h_x = 58 \text{ kJ/kg} \quad \text{and} \quad h_2 = 44 \text{ kJ/kg}$$

Mass of air flowing through the cooling coil

$$m_a = \frac{V_1}{v_{s_1}} = \frac{200}{0.885} = 226 \text{ kg/min}$$

Capacity of the cooling coil (refrigerating effect)

$$= m_a (h_1 - h_2) = \frac{226}{60}(82 - 44) = \mathbf{143.13} \text{ kJ/s or kW}$$

1 tonne of refrigeration is equivalent to 3.5 kW of heat absorption

∴ capacity of cooling coil in tonnes

$$= \frac{143.13}{3.5} = \mathbf{40.9} \text{ tonnes}$$

(*c*) Amount of water vapour removed,

$$= m_a (\omega_1 - \omega_2)$$

$$= 226\ (0.0202 - 0.011) = 2.08 \text{ kg/min}$$

(*d*) Sensible heat factor for the process,

$$SHF = \frac{h_x - h_2}{h_1 - h_2} = \frac{58 - 44}{82 - 44} = \mathbf{0.368}$$

EXAMPLE 12.42

The following conditions apply to an air-conditioning system of an industrial process:

outdoor conditions: 30° dbt and 65% relative humidity

inlet conditions: 25°C dbt and 60% relative humidity

free air circulation: 250 m³/min

The required conditions are achieved by first cooling and dehumidification and then by heating. Using psychrometric chart, make calculations for the following parameters:

(a) capacity of cooling coil in tons of refrigeration,

(b) capacity of heating coil in kW

(c) amount of water vapour removed per hour

Solution: Refer Fig. 12.37 who depicts the processes involved when drawn an psychrometric chart as per given data :

(*i*) Mark point 1 corresponding to 30° *dbt* and 65% RH and point 3 corresponding to 25°C dbt and 60% relative humidity.

(*ii*) Locate point 2′ and 2 on the saturation curve by drawing horizontal lines through points 1 and 3 respectively.

The path 1-2′-2, then represents the cooling and dehumidifying process and line 2-3 represents the heating process.

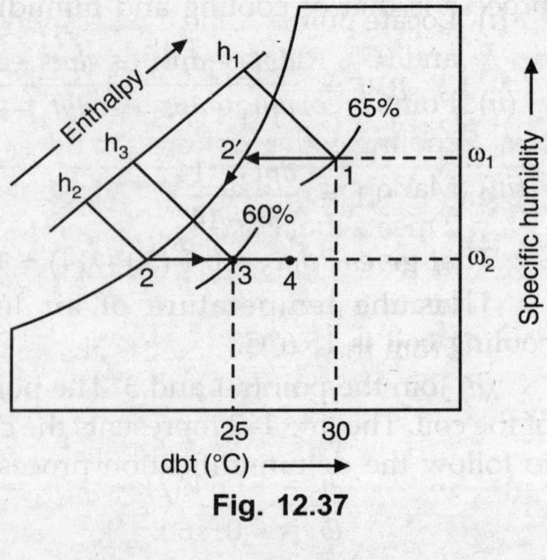

Fig. 12.37

The following values may then be read from the psychrometric chart

$h_1 = 82$ kJ/kg $\quad h_2 = 46$ kJ/kg

$h_3 = 56$ kJ/kg $\quad \omega_1 = 0.0195$ kg/kg

$\omega_2 = 0.0118$ kg/kg $\quad v_{s1} = 0.89$ m³/kg

Mass of air supplied $m_a = \dfrac{V}{v_{s1}} = \dfrac{250}{0.89} = 280.9$ kg/min

(*a*) Capacity of cooling coil $= m_a (h_1 - h_2)$

$= 280.9 (82 - 46) = 10112.4$ kJ/min

$= \dfrac{10112.4}{210} =$ **48.15** ton of refrigeration

(*b*) Capacity of heating coil $= m_a (h_3 - h_2)$

$= 280.9 (56 - 46) = 2809$ kJ/min

$= \dfrac{2809}{60} = 46.81$ kJ/s = **46.81 kW**

(*c*) Amount of water removed $= m_a (\omega_1 - \omega_2)$

$= 280.9 (0.0195 - 0.0118)$

$= 2.163$ kg/min = **129.78** kg/hr

EXAMPLE 12.43

The following information is available for the air-conditioning system of a particular premises :

Outdoor conditions : 30°C dbt and 65% RH

Inside desired condition : 23°C dbt and 65% RH

The specification require that 150 m³/min of air is delivered to the premises after conditioning it by cooling and dehumidification and subsequently heating it in a heating coil. If the dew point temperature of the cooling coil is 25°C, analyze the problem on psychrometric chart and determine :

(a) capacity and by-pass factor of cooling coil,

(b) mass of water vapour removed, and

(c) capacity and surface temperature of heating coil if its by-pass factor is 0.24.

Solution : Refer Fig. 12.38 which depicts the processes involved drawn on psychrometric chart as per given given data :

(*i*) Locate points 1 and 2 corresponding to outdoor (30°C *dbt* and 65% RH) and inside (23°C *dbt* and 65% RH) conditions respectively.

(*ii*) Point 3 corresponds to dew point temperature of the cooling coil, and it lies on the saturation curve.

(*iii*) Mark point 4 where the horizontal line drawn from point 2 meets the line 1-3. This point 4 indicates the condition at which the air leaves the cooling coil and enters the heating coil.

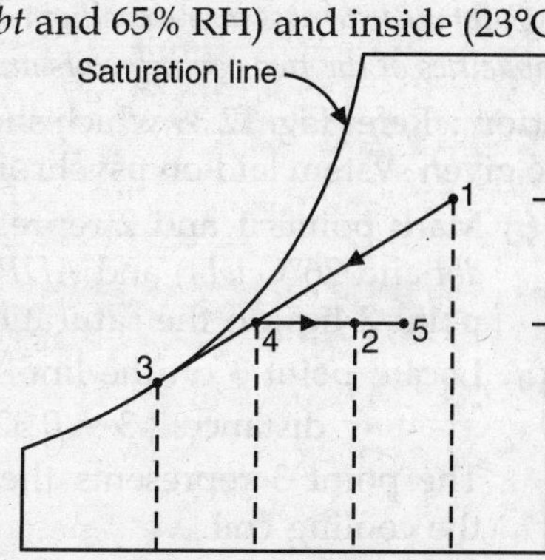

Fig. 12.38

From the psychrometric chart :

$h_1 = 75$ kJ/kg ; $\quad v_{s_1} = 0.882$ m³/kg

$\omega_1 = 0.0175$ kg/kg

$h_2 = 51.5$ kJ/kg ; $\quad h_4 = 46$ kJ/kg

$\omega_4 = 0.0116$ kg/kg and $dbt_4 = 16.5$°C

Then : Mass of air $= \frac{V_1}{v_{s_1}} = \frac{150}{0.882} = 170.06$ kg/min

(*a*) Capacity of the cooling coil $= m_a (h_1 - h_4)$

$= 170.06\ (75 - 46) = 493.174$ kJ/min

$= \frac{4931.74}{210} = \mathbf{23.48}$ **tons** (∵ 210 kJ/min = 1 ton)

By-pass factor of the cooling coil $= \frac{dbt_4 - dbt_3}{dbt_1 - dbt_3} = \frac{16.5 - 15}{30 - 15} = \mathbf{0.1}$

(*b*) Mass of water removed in the humidifier

$= m_a (\omega_1 - \omega_4)$

$= 170.06(0.0175 - 0.011) = \mathbf{1.003}$ **kg/min**

(*c*) The air leaving the cooling coil at condition 4 passes over the heating coil and gets heated to condition represented by state point 2.

Capacity of the heating coil $= m_a (h_2 - h_4)$

$= 17.06\ (51.5 - 46) = 935.33$ kJ/min

$= \frac{935.33}{60} = 15.59$ kW

If surface temperature of the heating coil is represented by state point 5, then

By-pass factor of the heating coil $= \frac{dbt_5 - 23}{dbt_5 - 16.5}$

∴ Surface temperature of the heating coil,

$$dbt_5 = \frac{23 - 0.24 \times 16.5}{1 - 0.24} = \mathbf{25.05°C}$$

EXAMPLE 12.44

Air from outdoor conditions of 40°C dbt and 28°C wbt is supplied to an air-conditioned room at the rate of 250 m³/min. The dehumidification of air is done in two stages : first by a cooling coil whose by-pass factor is 0.32 and dew point temperature is 25°C. and then by a chemical dehumidifier. The air leaving the dehumidifier at 30°C dbt made to pass over a cooling coil having surface temperature 15°C and by-pass factor 0.26. Determine the capacities of the two cooling coils and the dehumidifier.

Solution : Refer Fig. 12.39 which shows the various processes of the given system laid on psychrometric chart.

(*i*) Mark points 1 and 2 representing outdoor air (40°C *dbt* and 26°C *wbt*) and *ADP* = 15°C respectively. The point 2 lies on the saturation curve.

(*ii*) Locate point 3 on the line 1-2 such that

distance 2-3 = 0.32 × distance 2-1

The point 3 represents the condition of air leaving the cooling coil.

(*iii*) The dehumidification process is along the constant wet bulb temperature line or the constant enthalpy line.

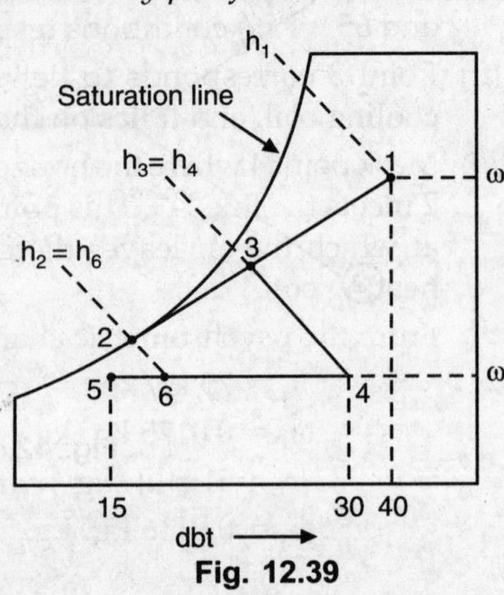

Fig. 12.39

From point 3, draw the constant enthalpy line to meet the 30°C *dbt* line at point 4.

(*iv*) From point 4 move horizontally towards left and locate point 5 where this line meets the vertical line drawn upwards from 15°C *dbt* point.

Since by-pass factor of cooling coil is 0.26, locate point 6 on line 4-5 such that

distance 5-6 = 0.26 × distance 5-4.

The point 6 represents the condition of air leaving the second cooling coil.

From the chart :

h_1 = 80.7 kJ/kg of dry air ; ω_1 = 0.155 kg/kg of dry air

v_{s1} = 0.915 m^3/kg of dry air ; $h_3 = h_4$ = 54.5 kJ/kg of dry air

$h_2 = h_6$ = 42.2 kJ/kg of dry air ; w_4 = 0.0092 kg/kg of dry air

Then : mass of supplied $m_a = \dfrac{V_1}{v_{s_1}} = \dfrac{250}{0.915} = 274.12$ kg/min

Capacity of first cooling coil $= m_a (h_1 - h_3)$

$$= \frac{274.12}{60} (80.7 - 54.5) = 119.7 \text{ kW}$$

Capacity of second cooling coil $= m_a (h_4 - h_6)$

$$= \frac{274.12}{60} (54.5 - 42.2) = 56.2 \text{ kW}$$

Capacity of the dehumidifier $= m_a (\omega_1 - \omega_4)$

$= 274.12(0.0155 - 0.0092) = \mathbf{1.727}$ **kg/min**

12.10.8. Adiabatic Mixing

Let two streams of moist air having different specific humidities and enthalpies be allowed to mix without the addition or rejection of either heat or moisture, *i.e.*, adiabatically and at constant total moisture content.

With reference to Fig. 12.39, the state of the resultant mixture lies on the straight line that joins the state of two streams on psychrometric chart. The location of final state on the straight line depends on the masses involved, and on the enthalpy and specific humidity of each stream.

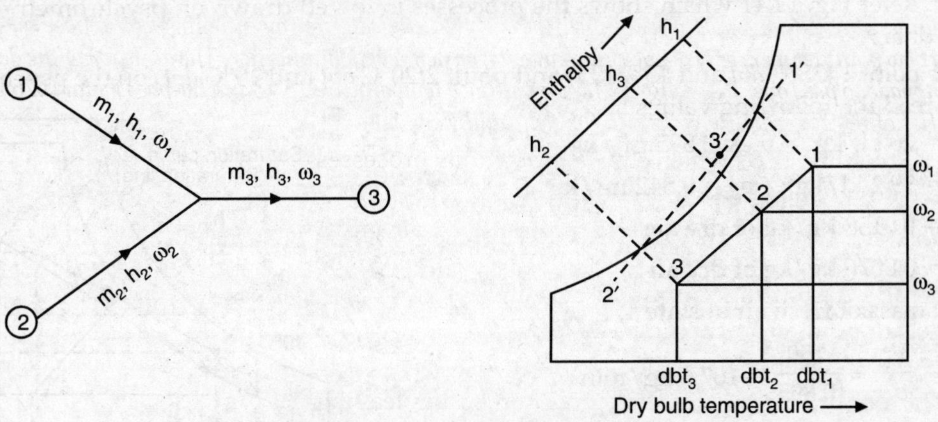

Fig. 12.40. Adiabatic mixing of two air streams

The fundamental equations of the process are:

Mass balance: $m_1 + m_2 = m_3$ ...(*i*)

Energy balance: $m_1 h_1 + m_2 h_2 = m_3 h_3$ ...(*ii*)

Mass balance of water vapour: $m_1 \omega_1 + m_2 \omega_2 = m_3 \omega_3$ ...(*iii*)

where *m*, *h*, ω represent the mass, enthalpy and specific humidity of moist air at a particular state point.

From expressions (*i*) and (*ii*),

$$m_1 h_1 + m_2 h_2 = (m_1 + m_2) h_3 = m_1 h_3 + m_2 h_3$$

or $$m_1 (h_1 - h_3) = m_2 (h_3 - h_2)$$

That gives: $$\frac{m_1}{m_2} = \frac{h_3 - h_2}{\omega_1 - \omega_3} \quad ...(a)$$

Likewise from expressions (*i*) and (*iii*)

$$\frac{m_1}{m_2} = \frac{\omega_3 - \omega_2}{h_1 - h_3} \quad ...(b)$$

Thus from identities (*a*) and (*b*), we have

$$\frac{m_1}{m_2} = \frac{h_3 - h_2}{h_1 - h_3} = \frac{\omega_3 - \omega_2}{\omega_1 - \omega_3}$$

Obviously, the final state point 3 divides the line 1-2 in the inverse ratio of the mixing masses

Note : A condition may exist when the state of mixture lies outside the saturation curve, *i.e.*, point 3′ on line 1′-2′. This state is referred to as *fog condition*.

EXAMPLE 12.45

400 m³/min of air at 20°C DBT and 10°C DPT coming out from air conditioned hall is mixed with 150 m³/min of fresh air at 35°C DBT and 45% relative humidity. Determine :

(a) enthalpy (b) specific humidity (c) specific volume and (d) DPT of the mixture

(JNTU)

Solution : Refer Fig. 12.41 which shows the processes involved drawn on psychrometric chart as per given data :

Locate point 1 (35°C *dbt* and 45% RH) and point 2(20°C *dbt* and 10°C *dpt*) on the psychrometric chart and read the following values :

h_1 = 76 kJ/kg ; v_{s_1} = 0.895 m³/kg

h_2 = 39.2 kJ/kg ; v_{s_2} = 0.842 m³/kg

ω_1 = 0.0158 kg/kg of dry air

ω_2 = 0.0076 kg/kg of dry air

Then : mass of fresh air at state 1,

$$m_1 = \frac{V_1}{v_{s_1}} = \frac{150}{0.895} = 167.6 \text{ kg/min}$$

Mass of recirculated air at state 2,

$$m_2 = \frac{V_2}{v_{s_2}} = \frac{400}{0.842} = 475.0 \text{ kg/min}$$

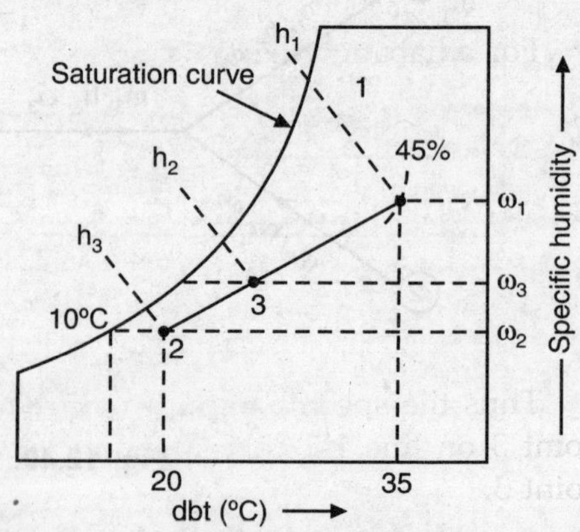

Fig. 12.41

For adiabatic mixing :

$$\frac{m_1}{m_2} = \frac{h_3 - h_2}{h_1 - h_3} \quad ; \quad \frac{167.6}{475.0} = \frac{h_3 - 39.2}{76 - h_3} \quad ; \quad 0.35 = \frac{h_3 - 39.2}{76 - h_3}$$

or $\quad 0.35 \times 76 - 0.35\, h_3 = h_3 - 39.2$

$$\therefore \quad h_3 = \frac{0.35 \times 76 + 39.2}{1 + 0.35} = 48.74 \text{ kJ/kg of dry air}$$

Thus the specific enthalpy of the resulting mixture is 48.74 kJ/kg of dry air. Now locate point 3 on the line 1-2 corresponding to h_3 = 48.74 kJ/kg of dry air. Then the following values may be read from psychrometric chart :

Specific volume v_{s3} = 0.854 m^3/kg of dry air
Humidity ratio ω_3 = 0.0097 kg/kg of dry air
Dew point temperature dpt_3 = 13.8°C

Alternatively : For adiabatic mixing

$$\frac{m_1}{m_2} = \frac{\omega_3 - \omega_2}{\omega_1 - \omega_3} \quad ; \quad \frac{134.23}{385.98} = \frac{\omega_3 - 0.0076}{0.0158 - \omega_3}$$

or $\quad 0.348 \times 0.0158 - 0.348\, \omega_3 = \omega_3 - 0.0076$

$$\therefore \quad \omega_3 = \frac{0.348 \times 0.0158 + 0.0076}{1 + 0.348} = 0.0097 \text{ kg/kg of dry air}$$

The point 3 can then be located on line 1-2 and the relevant data read from the chart.

EXAMPLE 12.46

A stream of moist air at 10°C dry bulb and 90 percent relative humidity mixes with another stream of moist air at 40°C dry bulb and 30 percent relative humidity. The flow rates of these streams are in the ratio of one part of first to two parts of the second stream. Presuming barometric pressure equal to 1 bar and considering adiabatic mixing, determine the temperature, specific enthalpy, humidity ratio and relative humidity of the air after mixing.

Solution: Locate state 1 corresponding to dbt_1 = 10°C and ϕ_1 = 90% and state 2 corresponding to dbt_2 = 40°C and ϕ_2 = 30%. Draw the line joining states 1 and 2. From psychrometric chart:

ω_1 = 0.007 kg/kg of dry air and ω_2 = 0.014 kg/kg of dry air

For adiabatic mixing

$$\frac{m_1}{m_2} = \frac{\omega_3 - \omega_2}{\omega_1 - \omega_3} \quad ; \quad \frac{1}{2} = \frac{\omega_3 - 0.014}{0.007 - \omega_3}$$

or $\quad 0.007 - \omega_3 = 2\omega_3 - 2 \times 0.014$

$$\therefore \quad \omega_3 = \frac{0.007 + 2 \times 0.014}{1 + 2} = \mathbf{0.0117 \text{ kg/kg of dry air}}$$

Thus the specific humidity of the resulting mixture is 0.0117 kg/kg of dry air. Now locate point 3 on line 1-2 corresponding to ω_3 = 0.0117 kg/kg of dry air. Then we may find that at point 3,

dry bulb temperature dbt_3 = **29.5°C**
relative humidity ϕ_3 = **45%**
specific enthalpy h_3 = **55** kJ/kg of dry air

EXAMPLE 12.47

A stream of air at 20°C, 90% relative humidity and flowing at the rate of 5 m³/min mixes adiabatically with another air with corresponding parameters of 30°C, 50% and 20 m³/min. Presuming that both the streams are at 760 mm of mercury as pressure, determine the following parameters for the mixed stream:

(a) dry and wet bulb temperature,

(b) relative humidity and specific humidity.

Solution: Let suffix 1, 2 and 3 refer to first, second and mixed streams respectively.

Locate point 1 for first air stream according to dbt_1 = 20°C and ϕ_1 = 90% and point 2 for second air stream according to dbt_2 = 30°C and ϕ_2 = 50%.

The following values are then read from the chart:

$$\omega_1 = 0.0132 \text{ kg/kg of dry air} \quad \text{and} \quad \omega_2 = 0.0134 \text{ kg/kg of dry air}$$

From characteristic gas equation: $pV = mRT$. Since pressure p and gas constant R are same for both the streams, we have

$$\frac{m_1}{m_2} = \frac{V_1}{T_1} \times \frac{T_2}{V_2} = \frac{V_1}{V_2} \times \frac{T_2}{T_1} = \frac{5}{20} \times \frac{273+30}{273+20} = 0.2585$$

For adiabatic mixing :
$$\frac{m_1}{m_2} = \frac{h_3 - h_2}{h_1 - h_3} = \frac{t_3 - t_2}{t_1 - t_3}$$

or
$$0.258 = \frac{t_3 - 30}{20 - t_3}; \quad \text{or} \quad 0.2585 \times 20 - 0.258\, t_3 = t_3 - 30$$

$$\therefore \quad t_3 = \frac{30 + 0.2585 \times 20}{1 + 0.2585} = \mathbf{27.95°C}$$

Again, for adiabatic mixing

$$\frac{m_1}{m_2} = \frac{\omega_3 - \omega_2}{\omega_1 - \omega_3}; \quad 0.2585 = \frac{\omega_3 - 0.0134}{0.0132 - \omega_3}$$

or
$$0.0132 \times 0.2585 - 0.2585\, \omega_3 = \omega_3 - 0.0134$$

$$\therefore \quad \omega_3 = \frac{0.0134 + 0.0132 \times 0.2585}{1 + 0.2585} = \mathbf{0.01336} \text{ kg/kg of dry air}$$

Now, locate point 3 on the line joining points 1 and 2 corresponding to

$$\omega_3 = 0.01336 \text{ kg/kg of dry air} \quad \text{or} \quad t_3 = 27.95°C$$

Then we find that at point 3,

$$\text{Relative humidity } \phi_3 = 58\% \quad \text{and} \quad \text{Wet bulb temperature} = 21°C$$

EXAMPLE 12.48

A mixture of recirculated air and outside air enters a cooling coil at 30°C dry bulb temperature and 18°C wet bulb temperature. The surface temperature of the coil is maintained at 4.5°C, and the cooling capacity is equivalent to 12 kW of refrigeration with the given entering state and flow rate of 40 m³/min. Determine :

(a) dry bulb and wet bulb temperatures of the air leaving the coil, and

(b) coil by-pass factor.

Solution : Refer Fig. 12.42.

(*i*) Mark the state point 1 as per dbt_1 = 30°C and wbt_1 = 18°C on the psychrometric chart and read therefrom :

$$v_{s1} = 0.87 \text{ m}^3/\text{kg}$$
$$\omega_1 = 0.008 \text{ kg/kg of dry air}$$
$$h_1 = 51.2 \text{ kJ/kg}$$

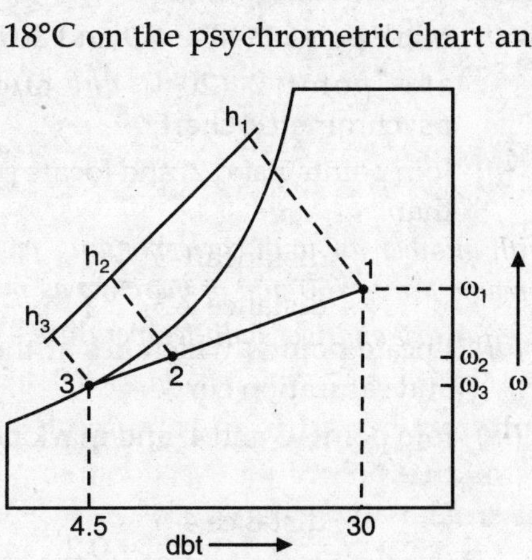

Fig. 12.42

(*ii*) Draw a vertical line through 4.5°C *dbt* which meets the saturation line at point 3. The line 1-3 then represents the path followed by air while passing through the cooling coil.

Corresponding to state point 3,

$$\omega_3 = 0.005 \text{ kg/kg of dry air, and}$$
$$h_3 = 17.8 \text{ kJ/kg}$$

(*iii*) Mass flow of air $m_a = \dfrac{V_1}{v_{s_1}} = \dfrac{40/60}{0.87} = 0.766$ kg/s

From energy balance : $m_a(h_1 - h_2) = Q$ where Q is cooling capacity.

$$\therefore \quad h_2 = h_1 - \frac{Q}{m_a} = 51.2 - \frac{12}{0.766} = 35.53 \text{ kJ/kg}$$

Equation for the condition line 1-3 is

$$\frac{h_1 - h_2}{h_1 - h_3} = \frac{\omega_1 - \omega_2}{\omega_1 - \omega_3}$$

or

$$\frac{51.2 - 35.53}{51.2 - 17.8} = \frac{0.008 - \omega_2}{0.008 - 0.005}; \quad \frac{15.67}{33.4} = \frac{0.008 - \omega_2}{0.003}$$

$$\therefore \quad \omega_2 = 0.008 - 0.003 \times \frac{15.67}{33.4} = 0.0066 \text{ kg/kg of dry air}$$

From the above calculated values of h_2 = 35.53 kJ/kg and ω_2 = 0.0066 kg/kg, locate the state point 2 on the chart and read the following

$$dbt_2 = 18.5°\text{C} \quad \text{and} \quad wbt_2 = 12.5°\text{C}$$

By-pass factor of the coil $BPF = \dfrac{h_2 - h_3}{h_1 - h_3} = \dfrac{35.53 - 17.8}{51.2 - 17.8} = \mathbf{0.53}$

EXAMPLE 12.49

For a certain room air-conditioning system, room air at 20°C dbt and 50% RH is mixed with outside air at 40°C dbt and 30% RH in the ratio of 4 : 1. The mixture is then made to pass through a cooling coil whose surface is kept at a constant temperature of 10°C. If the coil has a by-pass factor of 0.2, determine the condition of air at entry to and at exit from the coil. Also work out the refrigeration load on the coil if the system is designed for air flow rate of 200 m³/min.

Solution : Refer Fig. 12.43

(*i*) Mark point 1 (40°C *dbt* and 30% RH) and point 2 (20°C *dbt* and 50% RH) on the psychrometric chart.

(*ii*) Join points 1 and 2, and locate point 3 on line 1-2 such that

$$\frac{\text{distance 1-3}}{\text{distance 2-3}} = \frac{4}{1}$$

(*iii*) Locate point 4 which lies at the intersection of 10°C and saturation curve.

(*iv*) Join points 3 and 4, and mark point 5 on line 3-4 such that

$$\frac{\text{distance 4-5}}{\text{distance 4-3}} = 0.2$$

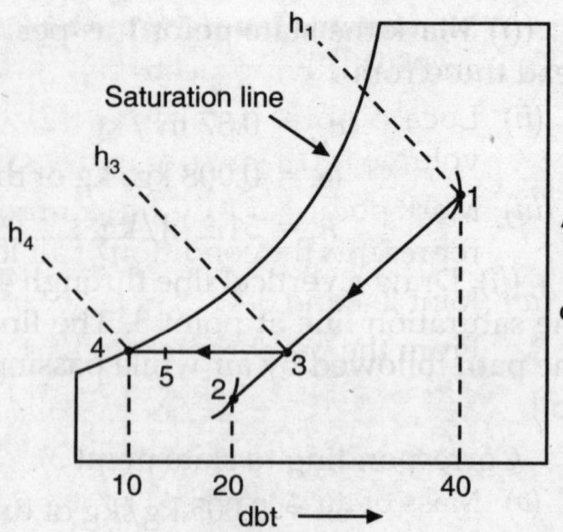

Fig. 12.43

(*a*) The point 3 represents the condition of air at entry to the coil, and corresponding to this point

$$dbt = 24°\text{C} \quad \text{and} \quad \text{RH} = 45.6\%.$$

The point 5 represents the condition of air at exit from the coil and corresponding to this point

$$dbt = 12.8°\text{C} \quad \text{and} \quad \text{RH} = 82.8\%.$$

(*b*) Load on the coil = $m_a (h_3 - h_5)$

Read the following values from the chart :

$$v_{s_1} = 0.84 \text{ m}^3/\text{kg}, \quad h_3 = 45.8 \text{ kJ/kg} \quad \text{and} \quad h_5 = 32.8 \text{ kJ/kg}$$

Then, load on the coil = $\frac{200/60}{0.84}(45.8 - 32.8) = 51.59$ kJ/s

Since 1 ton is equivalent to 3.5 kJ/s, we have

load on the coil = $\frac{51.59}{3.5}$ = **14.74** tons

EXAMPLE 12.50

In a particular premises

Outdoor conditions : 40°C dbt and 27°C wbt

Inside design conditions : 25°C dbt and 50% RH

The specifications require that 4000 m³/min is to be delivered to the room by the air-handling unit of the air-conditioning plant. Of this amount, 25% is to be the outside fresh air and 75% return air from the room. If the air is to leave the cooling coil at 13°C saturated, what is the total cooling load and room heat gain ? Use the data from the table given below :

Condition	*dbt* °C	*wbt* °C	*RH* %	*Specific humidity* kg/kg of dry air	*Enthalpy* kJ/kg of dry air
Outside	40	27	–	0.0172	85
Inside	25	–	50	0.01	50.8
ADP	13	–	100	0.0094	37

Further, the specific volume of air entering the cooling coil may be taken as 0.87 m³/kg.

Solution : Refer Fig. 12.44 which shows the various processes of the given system on psychrometric chart.

(*i*) Mark points 1 and 2 representing outdoor air (40°C *dbt* and 27°C *wbt*) and inside air (25°C *dbt* and 50% RH) respectively.

(*ii*) Locate point 3 on line 1-2 where specific volume is 0.87 m^3/kg of dry air.

(*iii*) Mark point 4 at 13°C on saturation curve. This represents the condition of air leaving the coil.

(*iv*) Joint 2-4 and 3-4.

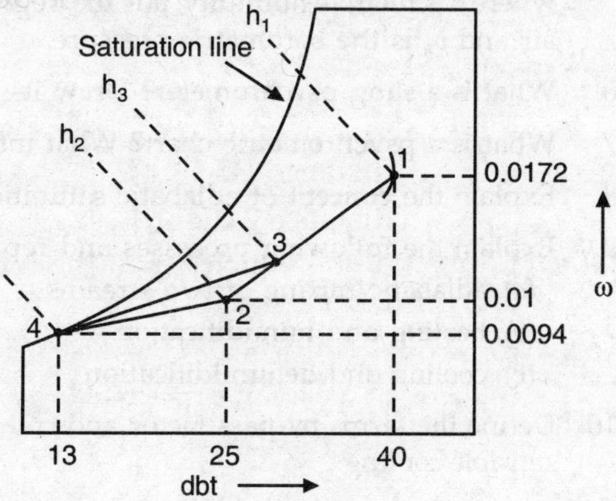

Fig. 12.44

From the psychrometric chart,

$$h_3 = 58.2 \text{ kJ/kg} \quad \text{and}$$

$$\omega_3 = 0.0116 \text{ kg/kg of dry air}$$

(*a*) Mass of air entering the cooling coil,

$$m_{a3} = \frac{V_3}{v_{s_3}} = \frac{4000}{0.87} = 4597.7 \text{ kg/min}$$

$$\text{Total cooling load} = m_{a3}(h_3 - h_4)$$

$$= 4597.7\,(58.2 - 37) = 97471 \text{ kJ/min}$$

Since 1 ton of refrigeration is equivalent to 210 kJ/min,

$$\text{Total cooling load} = \frac{97471}{210} = 464 \text{ tonnes}$$

(*b*) Mass of fresh air = 25% of total mass of air

$$= 0.25 \times 4597.7 = 1149.4 \text{ kg/min}$$

$$\text{Fresh air load} = m_a(h_1 - h_2)$$

$$= \frac{1149.4(85 - 50.8)}{210} = 187.2 \text{ tonnes}$$

Then : room heat gain = total cooling load – fresh air load

$$= 464 - 187.2 = 276.8 \text{ tonnes}$$

REVIEW QUESTIONS

A. Conceptual and conventional questions.

1. What is moist air and saturated air?
2. Define and explain the following terms in relation to psychrometry
   (*a*) dry bulb, wet bulb and dew point temperatures
   (*b*) relative humidity and specific humidity.
3. Write the Carrier's equation for partial pressure of vapour and state the significance of the terms included in it.
4. Establish the following expression for air-vapour mixture

$$\text{specific humidity } \omega = 0.622 \times \frac{p_v}{p_b - p_v}$$

where p_v is partial pressure of water vapour and p_b is the barometric pressure?

5. Set up the following expression for the relative humidity

$$\phi = \frac{\mu}{1-(1-\mu)p_s / p_b}$$

where ϕ is relative humidity, μ is the degree of saturation, p_s is the saturation pressure of vapour in moist air and p_b is the barometric pressure.

6. What is a sling psychrometer? Draw its neat sketch and explain its use.
7. What is a psychrometric chart? What information does it provide?
8. Explain the concept of adiabatic saturation temperature and adiabatic dew point temperature.
9. Explain the following processes and represent these on a psychrometric chart:
   (*a*) adiabatic mixing of two streams
   (*b*) heating and humidification
   (*c*) cooling and dehumidification
10. Define the terms by-pass factor and coil efficiency in relation to the processes of sensible heating and sensible cooling.
11. Explain the process of chemical dehumidification.
12. Name any five psychrometric processes and represent them on the psychrometric chart.
13. A certain sample of moist air exists at 35°C dry bulb temperature and 20°C dew point temperature. The atmospheric pressure is 760 mm of mercury. Calculate the relative humidity and saturation ratio.
(Ans. 41.6%; 40.2%)
14. A sample of air has dry and wet bulb temperatures of 35°C and 25°C respectively as recorded by a sling psychrometer. The barometric pressure is 1.0132 bar. Make calculations for (*a*) the vapour pressure (*b*) the relative humidity (*c*) the specific humidity (*d*) the degree of saturation, (*e*) the vapour density of air (*f*) the dew point temperature and (*g*) the enthalpy of the mixture.
(Ans. p_v = 0.0252 bar; ϕ = 44.7%
ω = 0.01586 kg/kg of dry air; μ = 43.35
ρ_v = 0.0177 kg/m³; t_{dp} = 21.2°C
h = 76.25 kJ/kg of dry air)
15. An air-water vapour mixture at 35°C dry bulb temperature and 20 percent relative humidity is passed through an adiabatic humidifier at the rate of 150 m³/min. If the outlet conditions are noted to the 20°C dry bulb temperature and 15°C wet bulb temperature, make calculations for the following parameters
   (*a*) dew point temperature,
   (*b*) relative humidity of exit air, and
   (*c*) amount of water vapour added to the air per minute. **(Ans. 8.7°C, 58.5%, 0.26 kg/min)**
16. A sample of moist air has a relative humidity of 75 percent at 1 bar and 40°C. Determine the humidity ratio, the dew point, the masses of water vapour and dry air in 100 m³ of the mixture
(Ans. 0.0364 kg/kg dry air, 34.7°C, m_v = 3.83 kg and m_a = 105 kg)
17. Air at 17°C and 60 percent relative humidity has been supplied to an air-conditioned room. Calculate the specific humidity and dew point of the air under these conditions. Assume barometric pressure to be 1.01325 bar.
What amount of vapour would condense if this air is made to pass over a cooling coil with surface temperature 6°C at the rate of 0.5 m³/s? Assume that the air leaving the coil is saturated.
(Ans. 0.007213 kg/kg dry air, 9.38°C, 3.082 kg/hr.)
18. The following data is available for a room on a particular day.
Temperature = 25°C, barometer reading = 760 mm of Hg and relative humidity = 75 percent.
Calculate: (*a*) partial pressure of air and water vapour (*b*) specific humidity (*c*) dew point and (*d*) density of mixture. **(Ans. 0.9895 bar, 0.02375bar, 0.01493 kg/kg dry air, 20.25°C, 1.174 kg/m³)**

19. Air at 27°C *dbt* and 50% RH is cooled and dehumidified to 10°C *dbt* and 100% RH. Determine the amount of sensible and latent heat removed from 0.5 m^3/s of this air. **(Ans. 5.23 kW, 10.05 kW)**

20. The air at 20°C and 30% RH is passing at the rate of $0.6 m^3/s$ through the duct of an air-conditioning system. The air is heated and humidified to raise it to 38°C and 40% RH. Determine the heat and moisture added to the air undergoing the above said process. **(Ans. 35.8 kW, 32.6 kg/hr)**

21. Air at 26°C *dbt* and 18°C *wbt* is made to ener a space with flow rate of 100 m^3/min. The air stream picks up 50 kg/hr of dry saturated steam at 120°C. Taking barometric pressure as 760 mm of mercury, determine the specific humidity and enthalpy of air at exit. Take enthalpy of dry saturated stream at 120°C as 2714.7 kJ/kg.

22. 250 litres/sec of outside air at 35°C *dbt* and 24°C *wbt* is mixed with 750 litres/sec of return air at 27°C *dbt* and 50% RH. Determine the *dbt*, *wbt*, *dpt* and specific humidity of the mixture. **(Ans. 28.9°, 19.5°C, 15°C and 0.011 kg/kg)**

23. Atmospheric air at 40°C *dbt* and 55° RH is passed over a cooling coil at the rate of $600 m^3$/min. At exit from the coil, the air is saturated and its temperature is 8°C. Determine the amount of condensate leaving the coil per minute and the refrigeration required in kW. **(Ans. 12.62 kg/min, 890 kW)**

24. Air at 35°C dry bulb temperature and 60% relative humidity is conditioned to 22°C dry bulb temperature and 55% relative humidity. The task is accomplished by cooling and dehumidifying and then heating. If the air flow rate is $50 m^3$/min, determine :
    (*a*) amount of water vapour drained per hour,
    (*b*) capacity of cooling coil in tons of refrigeration, and
    (*c*) capacity of heating coil in kW **(Ans. 43.33 kg/hr, 14.42 tons, 8.148 kW)**

25. Certain air-conditioning system, saturated air at 21°C *dbt* undergoes chemical dehumidification till its relative humidity becomes 20%. Thereafter, the air is sensibly cooled to the initial temperature of 21°C *dbt*. Depict the processes on pschrometric chart and determine :
    (*a*) dry bulb and dew point temperature at the end of drying (chemical humidification) process,
    (*b*) moisture removed during the drying process
    (*c*) heat removed during cooling and relative humidity at the end of cooling process.
    (Ans. 38.5°C; 0.0073 kg/kg; 18 kJ/kg, 53%)

26. The following information is available for the air-conditioning system for an industrial process :

Outside ambient condition	:	32°C *dbt* and 65% RH
Inside required contition	:	25°C *dbt* and 60% RH
Outside air to meet ventilation	:	300 m^3/min
Dew temperature of the coil	:	13°C

The task is accomplished by first cooling and dehumidifying and then by heating. Analyse the system on psychrometric chart and determine the following parameters :
    (*a*) refrigeration capacity of the cooling coil and its by-pass factor
    (*b*) heating capacity of the heating coil and its surface temperature if its contact factor is 0.72, and
    (*c*) amount of water vapour drained **(Ans. 56 tons, 0.239; 45.4 kW, 27.88° C; 157.38 kg/hr)**

B. Fill in the blanks with appropriate word/words

1. The .... air is a mixture of dry air and water vapour when the air has diffused the maximum amount of water vapour into it.
2. Saturated air is the moist air in which the partial pressure of water vapour equals the .... of stream corresponding to air temperature.
3. Dew point is the temperature at which .... of moisture starts in the unsaturated air.
4. The difference between the dry bulb and wet bulb temperature is called ....
5. Specific humidity is also called ....

6. During adiabatic humidification, ... remains constant.
7. The addition of moisture of air without change in its dry bulb temperature is known as ....
8. During chemical dehumidification process of air dry bulb temperature .... and specific humidity ....
9. The vertical and uniformly spaced lines on a psychromatric chart indicate ....
10. The ratio of sensible heat to total heat is known as ....

Answers

1. saturated 2. saturation pressure 3. condensation 4. wet bulb 5. depression 6. enthalpy 7. humidification 8. increases, decreases 9. dry bulb temperature 10. sensible heat factor

C. Multiple choice questions.

1. Absolute humidity of air-vapour mixture at a particular temperature is defined as
   (*a*) mass of water vapour per kg of dry air in the mixture
   (*b*) mass of water vapour per m^3 of mixture
   (*c*) mass of water vapour per kg of saturated air
   (*d*) ratio of actual particle pressure of water vapour to the saturation pressure of water vapour at the same temperature
2. If m_a = mass of dry air and m_w = mass of water vapour in the air-water vapour mixture, then humidity ratio is given by
   (*a*) $\frac{m_w}{m_a}$ (*b*) $\frac{m_a}{m_w}$ (*c*) $\frac{m_w}{m_a + m_w}$ (*d*) $\frac{m_a}{m_a + m_w}$
3. The degree of saturation is defined as
   (*a*) mass of water vapour per unit volume of air-water vapour mixture
   (*b*) mass of water vapour per kg of dry air in the mixture
   (*c*) the ratio of mass of water vapour in a given volume of moist air to the mass of water vapour in the same volume of saturated air at the same temperature and pressure
4. The relation $0.622 \frac{p_w}{p - p_w}$ where p and p_w are the atmospheric pressure and partial pressure of water vapour respectively, is used for calculating
   (*a*) absolute humidity (*b*) humidity ratio
   (*c*) relative humidity (*d*) degree of saturation
5. Which of the followings are synonyms ?
   (*a*) absolute humidity and specific humidity (*b*) specific humidity and humidity ratio
   (*c*) humidity ratio and relative humidity (*d*) relative humidity and degree of saturation
6. Ratio of mass of water vapour in a given volume of mixture to the mass of water vapour in the same volume of saturated mixture at the same pressure and temperature is known as
   (*a*) specific humidity (*b*) humidity ratio
   (*c*) relative humidity (*d*) degree of saturation
7. The temperature of air-water mixture recorded by a thermometer when its bulb is covered with a wick saturated with water and placed in the air - water stream is called
   (*a*) wet bulb temperature (*b*) dew point temperature
   (*c*) saturation temperature (*d*) critical temperature
8. When the temperature of air is reduced by continuous cooling at constant pressure, the condensation of water starts at a particular temperature known as
   (*a*) wet bulb temperature (*b*) dew point temperature
   (*c*) adiabatic saturation temperature (*d*) critical temperature
9. The temperature attained when the air is cooled by adding water adiabatically to it in a steady flow until it becomes saturated is called
   (*a*) wet bulb temperature (*b*) dew point temperature
   (*c*) adiabatic saturation temperature (*d*) wet bulb depression

10. The wet bulb depression is zero when relative humidity equals
(a) zero (b) 0.5 (c) 0.75 (d) 1.0

11. The moisture content of air is indicated by
(a) wet bulb temperature and dew point temperature
(b) wet bulb temperature and dry bulb temperature
(c) adiabatic saturation temperature and dew point temperature
(d) enthalpy and wet bulb temperature

12. The wet bulb temperature is a measure of
(a) absolute humidity (b) specific humidity
(c) relative humidity (d) degree of saturation

13. A sling psychrometer can measure
(a) absolute humidity (b) specific humidity
(c) wet bulb temperature (d) dry as well as wet bulb temperature

14. Consider the following statements :
A psychrometer measures
(1) wet bulb temperature (2) dry bulb temperature (3) dew point temperature
Which of these statements are correct ?
(a) 1 alone (b) 1 and 2 (c) 2 and 3 (d) 1, 2 and 3

15. Consider the following properties of air-water vapour mixture
I. dry bulb temperature, II. wet bulb temperature,
III. specific humidity IV. relative humidity
The psychometric chart shows the relation between
(a) I, II only (b) II, III only (c) I, II, IV only (d) all

16. In a psychrometric chart, the vertical lines parallel to the ordinate indicate
(a) dry bulb temperature (b) wet bulb temperature
(c) specific humidity (d) enthalpy of saturation

17. The uniformly spaced horizontal lines running parallel to the abscissa in a psychrometric chart indicate
(a) absolute humidity (b) specific humidity
(c) dew point temperature (d) volume

18. On a psychrometric chart, the relative humidity lines are represented by
(a) uniformly spaced horizontal lines (b) uniformly speed inclined lines
(c) non-uniformly spaced inclined lines (d) curved lines

19. By pass factor of a cooling coil will decrease with
(a) increase in fin spacing and increase in number of rows
(b) increase in fin spacing and decrease in number of rows
(c) decrease in fin spacing and decrease in number of rows
(d) decrease in fin spacing and increase in number of rows

20. During sensible cooling of air, the coil efficiency is given by
(a) 1 – BF (b) 1 + BF (c) $\sqrt{1-BF}$ (d) 1/BF
where BF is the by pass factor

21. Air at 20°C dry bulb temperature and 40% relative humidity is heated upto 40°C using an electric heater whose surface temperature is maintained uniformly at 45°C. The bypass factor of the heater is
(a) 0.2 (b) 0.25 (c) 0.88 (d) 1.0

22. The atmospheric air at dry bulb temperature 15° C enters a heating coil maintained at 40° C. If the air leaves the heating coil at 30° C, the coil efficiency would be equal to
(a) 0.2 (b) 0.4 (c) 0.6 (d) 0.75

23. Which is not true in the context of sensible heating and sensible cooling ?
    (*a*) no moisture is added to or removed from the air
    (*b*) dew point temperature decreases
    (*c*) vapour pressure remains constant
    (*d*) these processes plot as straight horizontal line on the psychrometric chart

24. All of the following statements pertaining to humidification process are correct, except
    (*a*) moisture is added
    (*b*) dry bulb temperature remains unchanged
    (*c*) specific humidity and relative humidity increase
    (*d*) humidification process plots as an inclined line on the psychrometric chart

25. Which is false statement in respect of dehumidification process?
    (*a*) moisture is removed
    (*b*) dry bulb temperature remains unchanged
    (*c*) specific humidity and relative humidity decrease
    (*d*) enthalpy of saturation increases

26. During combined cooling with humidification process
    (*a*) air passes through air washer
    (*b*) specific humidity and relative humidity increase
    (*c*) dry bulb temperature decreases
    (*d*) dew point and wet bulb temperatures increase

27. The process generally used for summer air conditioning is
    (*a*) cooling with humidification (*b*) cooling with dehumidification
    (*c*) adiabatic cooling or evaporative process (*d*) chemical dehumidification

28. The mixing of two or more streams of moist air in air conditioning is
    (*a*) an isothermal process (*b*) an adiabatic process
    (*c*) a constant pressure process (*d*) a general polytropic process

29. Quite often atmospheric air is made to pass through a wet khas-khas or bamboo mat to render a surface cool in summer. Thermodynamically this process may be called
    (*a*) sensible cooling (*b*) cooling and humidification
    (*c*) cooling and dehumidification (*d*) adiabatic saturation

30. It is desired to condition the outside air from 70% relative humidity and 45° C dry bulb to 50% relative humidity and 25° C dry bulb condition. The practical arrangement would be
    (*a*) cooling and dehumidification (*b*) cooling and humidification
    (*c*) cooling alone (*d*) dehumidification alone

Answers

1. (*b*)	2. (*a*)	3. (*d*)	4. (*d*)	5. (*b*)	6. (*c*)	7. (*a*)	8. (*b*)
9. (*c*)	10. (*d*)	11. (*a*)	12. (*a*)	13. (*c*)	14. (*a*)	15. (*d*)	16. (*a*)
17. (*b*)	18. (*d*)	19. (*d*)	20. (*a*)	21. (*a*)	22. (*c*)	23. (*b*)	24. (*d*)
25. (*d*)	26. (*a*)	27. (*b*)	28. (*b*)	29. (*d*)	30. (*a*)		

❑❑❑

Chapter – 13

Air-conditioning and Human Comfort

Learning objectives : A study of the subject matter presented in this chapter will enable the students/readers to have an overview of

- *air conditioning, its applications and relationship with refrigeration*
- *comfort air conditioning and factors affecting human comfort*
- *thermal interchange and thermal balance of human body*
- *concept of effective temperature, comfort chart and comfort zone*

Air-conditioning is an artificial process that involves cooling as well as heating coupled with ventilation, filtration and air circulation. It is essentially the process of treating air to control simultaneously its temperature, humidity, cleanliness and distribution to meet the comfort requirements of the occupants of the conditioned space. The functioning of an air-conditioning system can be conceived as depicted in Fig 13.1.

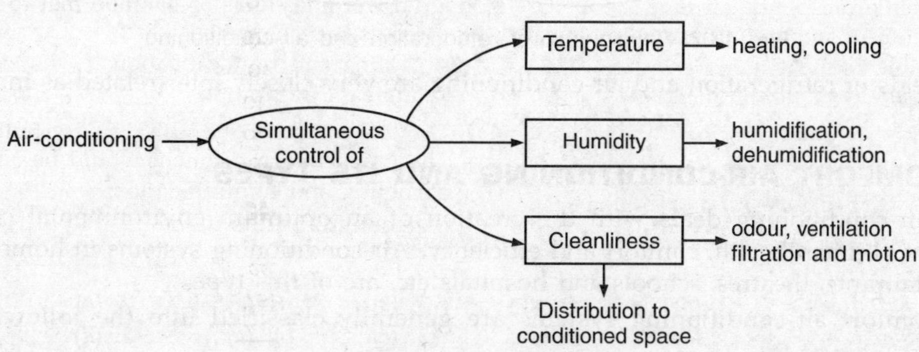

Fig. 13.1. Functioning of air-conditioning

Apart from the creation of an acceptable thermal environment (controlled temperature), control of humidity is of great importance both in humid and arid climates. Further, the air inside the conditioned space gets fouled due to absorption of pollutants from different sources and for human comfort, the indoor air has be purified.

13.1. APPLICATIONS OF AIR-CONDITIONING

Air-conditioning which was once considered as luxury, has now become a necessity in our day-to-day life. The air-conditioning has applications in diverse fields such as

(*i*) Residential and office buildings
(*ii*) Hospitals, cinema halls and departmental stores
(*iii*) Libraries, museums, computer centres and research laboratories
(*iv*) Transport vehicles :
  (*a*) cars, buses and rail coaches
  (*b*) aircrafts, space shuttles and rockets
  (*c*) submarines
(*v*) Printing, textile and photographic products
(*vi*) Food and process industries
(*vii*) Production shop laboratories, manufacture of materials and precision devices.

Air-conditioning essentially performs three services in the manufacture of precision metal parts. These services are :

(*a*) maintenance of uniform temperatures so that the metals neither expand nor contract
(*b*) control of humidity so that the rusting of metals is prevented
(*c*) filtration of air so as to minimize dust. Cleanliness of air conditioned space is absolutely essential where electronic components are being manufactured.

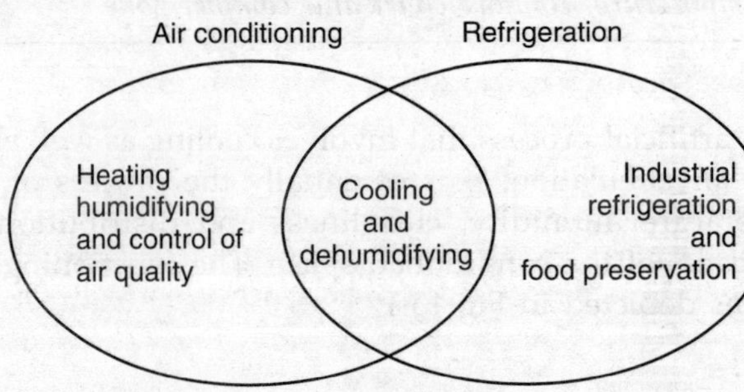

Fig. 13.2. Relationship of refrigeration and air-conditioning

The fields of refrigeration and air-conditioning are very closely inter-related as indicated in Fig. 13.2.

13.2. COMFORT AIR-CONDITIONING AND ITS TYPES

Comfort air-conditioning deals with the creation of an optimum environmental conditions conducive to human health, comfort and efficiency. Air-conditioning systems in homes, offices, stores, restaurants, theatres, schools and hospitals etc. are of this type.

The comfort air-conditioning systems are generally classified into the following three categories :

- **Summer air-conditioning :** These systems when properly designed and installed maintain the temperature and humidity of indoor air to a level at which persons feel comfortable. Essentially it involves reducing the air temperature and humidity (in humid tropics) by the process of cooling and dehumidification.
- **Winter air-conditioning :** These system are meant for the control of environmental conditions of indoor air so as to provide comfort in winter. Essentially it involves an increase in sensible heat and water content of air by the process of heating and humidification. The heating is done by furnaces or boilers fired with solid, liquid or gaseous fuels.
- **Year-round air-conditioning :** This system manifests in the control of temperature and humidity in an enclosed space for all times of the year ; this is despite a change in the atmospheric

conditions. Essentially the system comprises the heating and cooling equipment with associated components and automatic controls.

13.3. HUMAN COMFORT

Thermal comfort is a condition of mind which expresses satisfaction with thermal environment. It is the state where the person is entirely unaware of his surroundings; no consideration whether the space is too hot or too cold. Dissatisfaction with the thermal environment may be caused by the body as a whole (being too hot or cold) or by the unwanted heating or cooling of a particular part of the body (local discomfort).

Human comfort refers to the control of temperature and humidity of air and its circulation so that the resulting environment becomes human friendly; the state of environment where persons feel comfortable. Comfort is however a subjective quality; it is dependent on the preferences of an individual and varies with the age, sex, state of health and clothing etc. of a person.

13.4. THERMAL INTERCHANGE WITH ENVIRONMENT

The human body is continually gaining and producing heat as well as losing heat to its surroundings to maintain temperature equilibrium. Human body can be likened to a machine which receives fuel in the form of food, converts a fraction of energy in the fuel into work and rejects the remainder as heat to the surroundings. It is the continuous process of heat rejection which requires a thermal balance.

The various mechanisms by which temperature control is accomplished and a person feels comfortable are shown in Fig. 13.3.

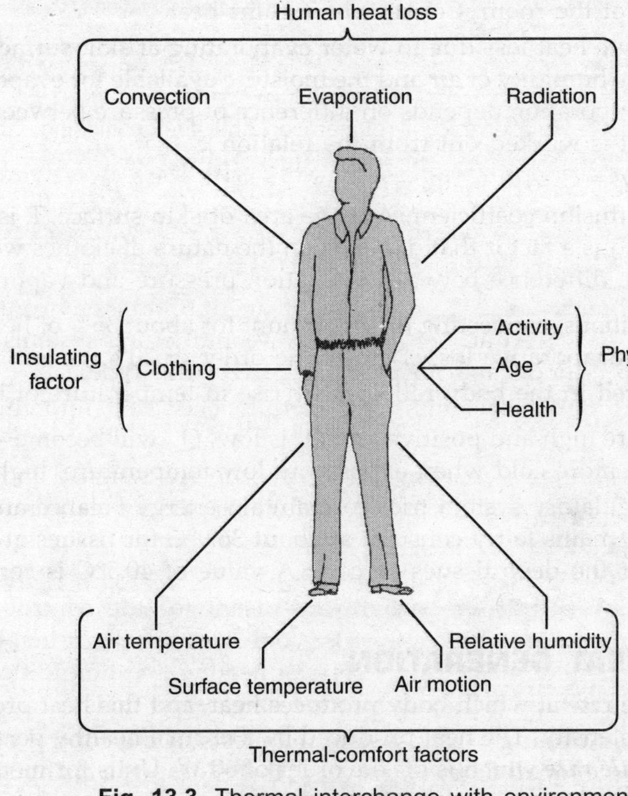

Fig. 13.3. Thermal interchange with environment

The generation of heat occurs in the body cells, circulatory system carries this heat to skin where it is released to environment. The physical basis of comfort lies in the thermal balance of the body, *i.e.*, the heat produced by the body's metabolism must be dissipated to the environment, otherwise the body would overheat.

The thermal balance of the body can be expressed as

$$Q_b = W \pm Q_c \pm Q_r \pm Q_e \pm Q_s \quad \text{... (13.1)}$$

where :

(*i*) Q_b is the heat produced within the body itself as a result of metabolic processes. The rate at which the body produces heat is referred to as metabolic rate. The metabolic heat produced depends on the rate of food consumption which itself is dependent on individual's health, his physical activity and environment. The metabolic heat production will be low for a person who is sick or weak or is fasting.

The heat generated is transported from the warm body core to the body surface partly by conduction through the tissues and partly by blood flow to the skin.

(*ii*) Q_c is the heat gain or loss by convection; it is prescribed by the relation

$$Q_c = UA\ (t_b - t_s)$$

where U is overall heat transfer coefficient the body, A is area of body surface, t_b is body temperature and t_s is temperature of surroudings. Q_c is +ve when heat is lost and –ve when heat is gained from the surroundings.

(*iii*) Q_r is the heat gain or loss by radiation; it depends on the mean radiant temperature which refers to the average surface temperature of the surrounding objects in the conditioned space. The body loses heat if the radiant temperature is more than the dry bulb temperature of the room. Conversely it gains heat.

(*iv*) Q_e is the evaporative heat loss due to water evaporating at skin surface; the evaporation rate depends on the humidity of air and the moisture available for evaporation. Essentially the heat loss by evaporation depends on difference of pressure between skin surface and surrounding air. It is worked out from the relation

$$Q_e = C_d\ A\ L\ C_c\ (p_s - p_v)$$

where C_d is the diffusion coefficient, A is the area of skin surface, L is the latent heat of vaporization and C_c is a factor that depends on the nature of clothes worn by the person, and $(p_s - p_v)$ is the difference between saturation pressure and vapour pressure.

In typical indoor conditions, convecting losses account for about 35% of heat losses, radiative losses are about 45% and evaporative losses are of the order of 20%.

(*v*) Q_s is the heat stored in the body reflected by rise in temperature of body tissues.

When Q_c, Q_r and Q_e are high and positive and Q_b is low, Q_s will become –ve. Hence a weak or fasting person will feel more cold when exposed to low temperature, high draft air.

The body's thermo-regulatory system tries to maintain energy balance and accordingly the temperature of the body remains fairly constant at about 36.9°C for tissues at the surface or the skin, and about 37.2°C for the deep tissues or core. A value of 40.5°C is considered high and 43.5°C is certainly fatal.

13.5. METABOLIC HEAT GENERATION

Metabolic rate refers to the rate at which body produces heat, and this heat production increases in proportion to exercise intensity. The heat produced by a normal healthy person while sleeping is called the ***basal metabolic rate*** which is of the order of 60 W. Units for metabolic heat output are often quoted in ***met***, and one met represents the average heat produced by a sedentary average person at normal mean radiant temperature and 1 met = 58 W/m^2. The appropriate area is the surface area of the body (approximately 2 m^2 for an adult male).

The typical values of metabolic hot generation for various activities have been listed below in Table 13.1.

Table 13.1. Metabolic heat generation for various activities

Activity	*W/m²*	*met*
• **Resting**		
Sleeping	40	0.7
Reclining	45	0.8
Seated and quiet	60	1.0
Standing and relaxed	70	1.2
• **Walking**		
3.5 km/hr	115	2.0
4.5 km/hr	150	2.6
6.5 km/hr	220	3.8
• **Office work**		
Seated and reading	55	1.0
Writing	60	1.0
Typing	65	1.1
Filing (seated, standing)	70, 80	1.2, 1.4
Walking	100	1.7
Lifting/packing	120	2.1

Note : The values listed above pertain only to sedentry and slightly active, normal clothed (0.6 clo) persons at low air velocities when the mean radiant temperature is equal to air temperature.

13.6. TEMPERATURE REGULATION MECHANISMS

Whenever heat exerts a stress on the human body, the thermo-regulatory systems come into operation and try to maintain the energy balance and keep the body core at about 37°C. The body essentially involves two types of regulatory mechanisms :

(*i*) automatic systems, e.g., dilation of blood vessels and sweat production

(*ii*) behavioural regulation, e.g., adjustment/changing of clothing and activity levels

• **Vasomotor control :** It relates to the nerves and muscles that cause blood vessels to constrict or dilate. When the body temperature soars, the blood vessels dilate/expand (vasodilation) increasing the blood circulation. The increased blood supply to the peripheral blood vessels augments the convective transport of heat from the body interior to its surface. The corresponding increase in thermal conductance of skin minimizes further heating of the body. Conversely a decrease in the body temperature causes contraction of the peripheral blood vessels (vasoconstriction) and the decreased blood circulation reduces body cooling.

This mechanism is self-sufficient in controlling the body temperature and is called the *first line of defence* against rise/fall in body temperature.

• **Sudomotor control :** It pertains to the nerves that control the activity of sweat glands. When the temperature of atmosphere becomes high compared to the body temperature, the body gains heat both by convection and radiation. Then only evaporation is the heat transfer mode left to dissipate heat. The sudomotor control then comes into action and the sweat glands start the sweat production.

The capacity of sweat glands to produce sweat differs from person to person, and also diminishes with time in conditions of severe heat. Also, with evaporation of sweat the body gets further cooled. The perspiration is also likely to result in body odour and spoiling of clothes.

13.7. PHYSIOLOGICAL HAZARDS (Impact of Heat and Cold Extremes)

Extremes of heat and cold have a broad and far-reaching set of impact on human beings. The effects vary both in severity and duration of impacts.

The hazards due to extreme heat are :

(*i*) **Heat exhaustion :** A milder form of heat related illness that can develop after several days of exposure to high temperatures and inadequate or unbalanced replacement of fluids. The condition is caused by an inability of the cardiovascular system to satisfy all the requirements of the body's tissues when the blood is shifted to the skin for cooling. Sufferers of heat exhaustion have a normal body temperature but their pulse rate is accelerated; their skin is cold and sweaty, and they often experience headache, weakness and tiredness, drowsiness and vomiting. Those most prone to heat exhaustion are elderly people, people with high blood pressure, and people working in a hot environment.

Heat exhaustion can be relieved/prevented by rest and cooling, and drinking plenty of water preferabley with salt added. However, if the condition persists there is risk of heat neurasthenia. Heat neurasthenia is a progressive condition characterized by feelings of apathy, hysteria, aggression and it can lead to chronic behaviour disorders.

(*ii*) **Heat syncope :** The exposure to heat causes swelling of blood vessels and that results into pooling of blood in lower limbs. Consequently, there is reduction in blood pressure and the person experiences dizziness. The skin appears pale and sweaty but is generally moist and cool.

(*iii*) **Heat stroke :** A medical condition caused by elevated body temperature where the body's internal organs begin to shut down because of excessive exposure to intense heat. A person with heat stroke has a body temperature above 40°C. The body does not perspire (no sweating) to reduce temperature. Other symptoms include confusion, combativeness, faintness, strong and rapid pulse and possible delirium or coma. Heat stroke is a serious medical condition; may result in tissue damage and is potentially fatal and therefore requires immediate medical care.

(*iv*) **Heat cramps :** These are muscle pains or spasm – usually in the abdomen, arms or legs. Heat cramps usually affect people who sweat a lot during strenuous activity. The profuse sweating depletes the body's salt and moisture. The low salt level in the muscles causes painful cramps. Drinking plenty of watery fluids and taking adequate salt in the diet should prevent heat cramps.

The hazards due to extreme cold are :

(*a*) **Cold cramps :** An unpleasant, often painful sensations due to contraction of muscles resulting from exposure to cold. Muscle cramps can be treated by applying a soft massage on the cramped muscle, stretching the muscle and applying heat which would improve superficial blood circulation and make the muscle more flexible.

(*b*) **Frost bite :** A medical condition whereby damage is caused to skin and other tissues due to extreme cold for long periods. Extreme cold causes the small blood vessels in the extremities to constrict dangerously low levels. The combination of cold temperature and poor blood flow can cause severe tissue injury by freezing the tissue. Frost bite is most likely to happen in body parts farthest from the heart, and those with a lot of surface area exposed to cold. The toes, fingers, ears and nose are usually affected first becoming cold, hard, white and bloodless. Generally frost bite is accompanied with discoloration of the skin along with burning sensations, partial or complete numbness and possibly intense pain. To treat frost bite, move the victim to a warm place, place the affected areas in warm water until the areas become soft and the sensation returns.

(*c*) **Freezing :** A severe hazard caused by a prolonged exposure to extreme cold weather. There occurs a drop in the blood pressure of the victim leading to loss of consciousness and finally death.

13.8. EFFECTIVE TEMPERATURE

The degree of warmth or cold felt by a human body depends on three factors namely dry bulb temperature, relative, humidity and air velocity. The combined effect of these factors is evaluated and expressed in terms of ***effective temperature*** which is defined as :

- A sensory index or scale that takes into account the effects of temperature humidity and air movement on human comfort in a noise free pure air environment.
- An index which combines into a single figure the effects of temperature, humidity and air movement on the sensation of warmth or cold felt by a human body; numerically equal to the temperature of still saturated air which induces an identical sensation.
- The temperature at which motionless, saturated air would induce in a sedentary worker (a person who sits down a lot of time and does not much exercise) wearing ordinary indoor clothing the same sensation of comfort as that induced by the actual conditions of temperature, humidity and air movement.

If air at different dry bulb temperatures and relative humidity conditions carries the same amount of heat as the heat carried by air at a particular temperature *T* and 100 percent relative humidity, then *T* represents the effective temperature. A few combinations of dry bulb temperature and relative humidity are given below for the same ***effective temperature.***

Dry bulb temperature (DBT)	*Relative humidity (RH)*	*Effective temperature (ET)*
°C	%	°C
21	100	21
23	88	21
25	76	21

Again with reference to Fig. 13.4, at low relative humidity values the dry bulb temperature of air can be higher for the same effective temperature and for the same feeling of comfort. This implies that as the moisture content in air decreases, the higher dry bulb-temperature has to be used in order to maintain the same level of comfort. This stems from the fact that at high dry bulb temperature (air is dry), the body loses more heat due to evaporation of perspiration. Then to keep the same heat balance, the convective and radiative modes of heat transfer have to be decreased. This would be taken care if the room temperature increases.

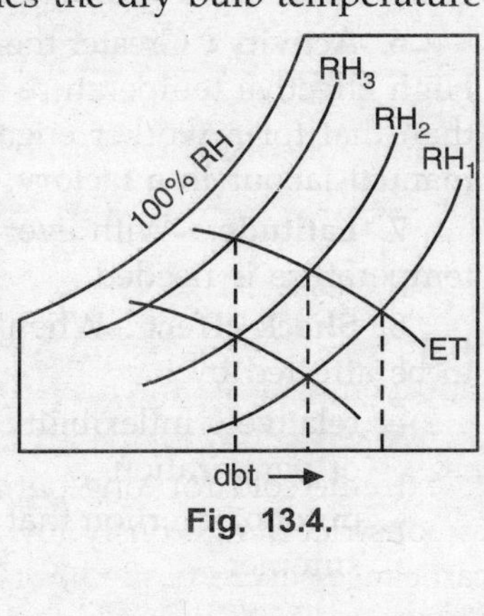

Fig. 13.4.

Quite often the combined effect of temperature and humidity is evaluated in terms of ***heat stress index*** which represents the ratio of total evaporative heat loss required for thermal comfort to the maximum evaporative heat loss possible from the environment for steady state conditions. For steady state conditions

– skin temperature is held constant at 35°C in order to limit the rise in body temperature.
– sweat rate is not to exceed one litre per hour to limit the loss of body fluid

$$\text{Thus: heat stress index} = \frac{Q_e}{(Q_e)_{max}} \times 100$$

where Q_e is the actual evaporation heat loss and $(Q_e)_{max}$ is the maximum evaporative heat loss with skin temperature at 35°C.

The optimum effective temperature is known to be affected by the following factors :

1. Climate and seasonal differences : People living in warmer regions feel comfortable at higher effective temperature than those living in colder climates. It is generally agreed that optimum effective temperatures for summer and winter are 22°C and 19°C respectively.

2. Age and sex : Children need higher effective temperature than adults, and women of all ages prefer an effective temperature about one-degree higher than that preferred by men. Similar is the case with men and women over 40 years age; they require effective temperature about one-degree higher than desired by young people.

3. Clothing : The surface temperature of the body is affected by the amount of heat transferred from the body core, the heat losses from the body, and also by the insulation value of the clothing. The thermal resistance of clothing is given by the ***clo-value*** which is a measure of the ratio of thermal resistance of clothing to a standard of 0.155 m^2K/W (representative of a business suit). The clo-value is 0.5 for light trousers + shift, and 2.0 for heavy suit + overcoat + gloves and hat.

The type of clothing depends on weather conditions and is recommended as follows :

– Loose fitting and white clothes for the hot climate, as they provide high evaporative capacity.
– Light and porous material clothes for the humid climates, where evaporation rate is small. The porous clothes will permit sweating and free movement of air.
– Thick and black clothes for cold climates; these will restrict heat loss from the body to cold surroundings.

In general, less effective temperatures are needed for persons wearing light clothings.

4. Air velocity : At higher velocities, there is requirement for less difference between the outdoor effective temperature and indoor effective temperature.

5. Density of occupants and duration of stay : Higher the density of occupants, lower is the desired effective temperature. Also high value of effective temperature is needed when the stay of a person in a conditioned space is only for a short duration.

6. Activity : Greater the degree of activity, lower the effective temperature needed for comfort. High effective temperature is needed by a sedentary worker (seated at rest, watching a movie) than that for a worker engaged with cumbersome or hectic activity such as performing heavy manual labour in a factory.

7. Latitude : With every 5-degree reduction in latitude, half-a-degree increase in effective temperature is needed.

8. Shock effect : When a person suddenly enters or leaves a conditioned space, he is likely to be affected by :

– relatively inflexibility in the amount of clothing that produces chilling due to evaporation of perspiration.
– outdoor exertion that may provide beneficial warmth in winter and unpleasant heating in summer.

13.9. COMFORT CHART AND COMFORT ZONE

Comfort chart is a diagram showing curves of relative humidity (RH) and effective temperature (ET) superimposed upon rectangular co-ordinates of dry bulb temperature (DBT) and wet bulb temperature (WBT). Essentially it serves to correlate ET, DBT, WBT and air movement.

Comfort zone is the temperature zone in which the body does not shiver or sweat, but has an idiomatic sense of a place where people feel comfortable and can avoid the physical or mental worries of the world. The comfort zone has effective temperatures between 19 to 22°C and relative humidities from 30 to 90%.

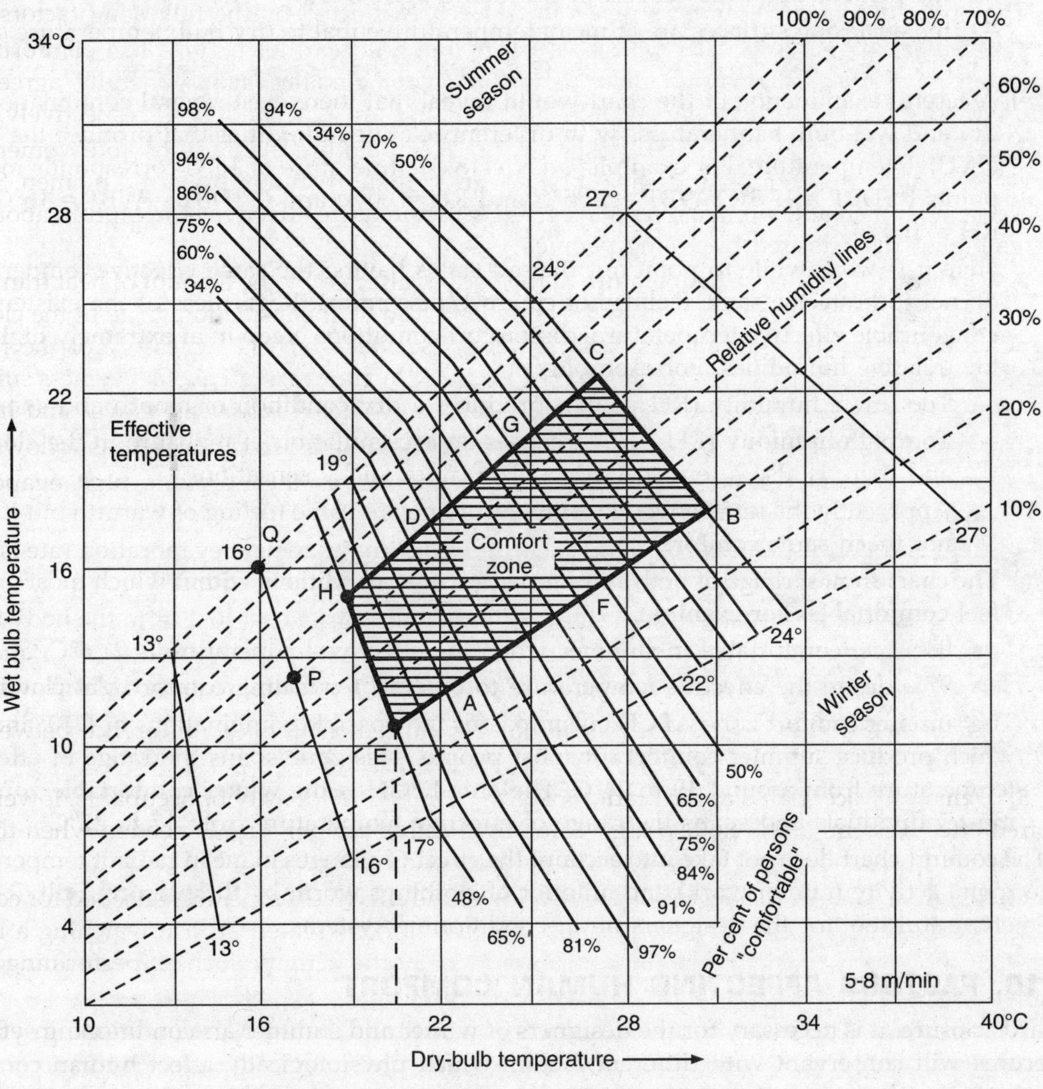

Fig. 13.5. Comfort chart and comfort zone

The salient features of the comfort chart shown in Fig. 13.5 are :

(*i*) The chart has been drawn by taking dry bulb temperature as abscissa and wet bulb temperature as ordinate. The relative humidity lines too are replotted from the psychromatic chart. The plots having effective temperature as abscissa and percentage of people feeling comfortable both in summer and winter seasons too have been superimposed.

(*ii*) The graphs having effective temperature and percentage of people feeling comfortable as coordinate have been constructed by American Society of Heating, Refrigeration and Air-conditioning Engineers (ASHRAE) by subjecting different kinds of people to a wide range of environmental temperature, relative humidity and air movement. These charts apply to the situations where

– the air is reasonably still; the air movement is within 5 to 8 m/min velocity,
– the persons are either at rest or are doing light work,
– the enclosing surfaces are at mean temperature equal to dry bulb temperature of the air.

(*iii*) A careful examination of the chart would reveal that there exist several combinations of dry and wet bulb temperatures with different relative humidities that produce the same effective temperature. For example, an effective temperature of 16°C corresponds to both points P (DBT = 17.5°C, WBT = 12.5°C and RH = 60%) and Q (DBT = WBT = 16°C and RH = 100%).

Here it is worthwhile to point out that all states having the same effective temperature do not indicate the same feelings of comfort/discomfort. Regardless of the existence of a favourable effective temperature, discomfort conditions creep in at extremely high and low relative humidities. For example :

– Too low a humidity (RH < 30%) produces a dry condition of mouth and skin.
– Too high humidity (RH > 70%) causes an accumulation of moisture in the clothing and there is also a tendency for a clammy or sticky sensation.

Apparently the same effective temperature means same feeling of warmth but it does not mean same comfort.

(*iv*) The chart shows range of both summer and winter conditions within which most people feel comfortable. For example

- 98% feel comfortable in summer when the effective temperature is 21.6°C, and
- 97% desire the effective temperature to be 20°C for feeling comfortable in winter.

(*v*) The summer comfort zone ABCD encompasses the possible combinations of DBT and RH which produce summer comfort for most people. This zone spans the range of effective temperature from around 19 to 24°C. The area EFGH is the winter comfortable zone for most individuals and spans the range of effective temperature from 17 to 22°C.

The comfort chart does not take into account the effect of changes in mean radiant temperature in the room, activity (rate of work) and amount of clothings worm by the persons. Still, it is an extremely useful tool for the designers of air-conditioning systems.

13.10. FACTORS AFFECTING HUMAN COMFORT

Towards closure, it is necessary for the designers of winter and summer air-conditioning systems to become will conservant with different factors which physiologically affect human comfort. These are :

- Environmental factors : air temperature, mean radiant temperature, air speed and air humidity.
- Non-environmental factors : activity and clothing level, age and sex, seasonal rhythms and day-to-day fluctuation of temperature
- Psychological factors : stress and mental effort.

Local thermal discomfort can be caused by :

– vertical air temperature differences

– asymmetrical thermal radiation (front to back or head to foot)

– draught, *i.e.*, air speed greater than 0.15 m/s at an air temperature of 20°C..

Some of these factors affecting human comfort are discussed below :

(1) Air temperature : Energy production in the body takes place continuously through metabolic processes which oxidise food into energy. This energy is partly converted into external mechanical work (negligible for most activities and increasing to about 20% for heavy manual labour) and the remainder is dissipated to surroundings. Consequently the temperature of surroundings increases and the objective of an air-conditioning system is to remove heat and maintain the room air temperature within the prescribed limits for human comfort.

(2) Air humidity : Humidity affects the rate at which the body loses heat by evaporation. In summer, the moisture content of air in places near sea is quite high and that makes the environment sultry and hot. High humidity increases discomfort by making it more difficult to evaporate perspiration into the air. Obviously moisture needs to be removed from the inside air (conditioned space) by a dehumidifying process so that a relative humidity of around 60% is attained.

In the context of dry places in winter, the comfortable conditions ask for an increase in moisture content of air in the conditioned space. Relative humidity around 30 to 40% is considered suitable for comfortable conditions in winter. Within these limitations of humidity level in summer and winter, the necessary dry bulb temperature for air can be determined from the comfort chart.

(3) Air motion : When a low velocity breeze contacts the human skin, there is dissipation of heat from the skin surface and one gets a feeling of pleasureness. However, when the breeze is at high velocity and there is turbulence in the air motion, heat dissipation becomes more and that too at a faster rate. The skin temperature then falls below the ambient temperature and the person feels cold. For comfortable state, the air should be uniformly distributed and its speed should not exceed 8-10 m/min. Further, the flow of air should be preferably towards the faces of the individuals rather than from the rear. It has also been suggested that for proper and perfect distribution of air in the conditioned space, downflow should be preferred instead of upflow.

(4) Quality and quantity of air : The atmospheric air generally contains certain toxic gases, bacteria, small dust particles, smoke fumes and viruses, and all these are harmful to human health. Obviously, the air needs to be sprayed with deodorants and freshness chemicals. That imparts pleasant smell to air and makes the conditioned environment comfortable. In short : proper filtration, cleaning and purification of air is necessary to keep it free from dirt dust, and toxic and disagreeable fumes.

For general applications, 15 percent of fresh air mixed with 85% of recirculated air is considered good.

(5) Air stratification : The heated air is less dense and being lighter rises towards the ceiling of the room. The cold air from the top moves downward to take its place. Obviously, there is variation in temperature from floor to ceiling and that makes the occupants uncomfortable. Air stratification is the term used to designate the movement of air that produces temperature gradient (variation).

The air-conditioning system is to be so designed that air stratification is avoided and that is made possible by introducing sufficient air changes per hour.

(6) Hot and cold surfaces : The ceilings get warm in summer and the glass windows get cooled in winter. The level of discomfort of occupants caused by heat exchange between these surfaces and the occupants can be minimized by having regulated air supply over the surfaces.

(7) Body temperature mechanisms and effective temperature : The concept of these parameters has already been discussed in sections 13.6 and 13.8 respectively.

REVIEW QUESTIONS

A. Conceptual and conventional questions.

1. Define air-conditioning and show diagrammatically the relationship between the refrigeration and air-conditioning fields.
2. State a few applications of air-conditioning.
3. How does comfort air-conditioning differ from industrial air-conditioning ? Illustrate your answer with suitable examples.
4. Name the three major types of comfort air-conditioning.
5. Define human comfort and explain the factors which affect it.
6. Which are the main factors affecting thermal comfort, and how can these be categorized ?
7. What is the thermal interaction of a human body with the environment ?
8. Write down the body heat balance equation and explain the different terms included in it.
9. Describe briefly the physical hazards (effect of extreme heat and cold) on human body.
10. What is the main function of the human thermo-regulatory system, and which important pre-requisite needs to be fulfilled to achieve that ?
11. Name and describe briefly the temperature regulation mechanisms in a human body.
12. What is air stratification ?
13. (*a*) Draw a comfort chart and indicate the various quantities incorporated in the chart.
    (*b*) What information is provided by the comfort chart ?
    (*c*) What is comfort zone and where is it located on the comfort chart ?
14. (*a*) Define effective temperature and state the factors which affect it
    (*b*) Mention the range of effective temperature and relative humidity for which most of the people feet comfortable.
15. How does the body respond to following factors in order to stay within the comfort zone ?
    (*a*) muscle activity (*b*) sweating and (*c*) blood flow
    Hint : Increased muscle activity and a higher metabolic rate increase internal heat production. The sweat moisture moves to the skin surface to be evaporated to lower body temperature. Reduced blood flow to the hands, feet and skin surface in the winter to reduce heat loss to surroundings and an increase in blood flow to these areas in the summer to augment heat loss.

B. Fill in the blanks with appropriate word/words

1. .... is the index which correlates the combined effects of temperature, relative humidity and air velocity.
2. The optimum effective temperature for human comfort is .... in winter than that in summer.
3. At the same dry bulb temperature, effective temperature .... with decrease in relative humidity.
4. The rate at which human body produces heat is known as ....
5. In comfort chart, the dry bulb temperature plots along the *x*-axis where as .... plots along the *y*-axis.

Answers

1. effective temperature 2. lower 3. decreases 4. metabolism rate 5. wet bulb temperature

□□□

Chapter – 14

Air-Conditioning Systems and Processes

Learning objectives : A study of the subject matter included in this chapter will enable the students/readers to

- *understand the different air conditioning systems with regard to their purpose, schematic arrangement of the associated components and working.*
- *assimilate and explain the different terms and parameters relevant to air conditioning systems.*
- *analyse, with the help of psychrometric chart, the processes involved in the air conditioning systems*

14.1. AIR-CONDITIONING SYSTEM

Air-conditioning is a process of maintaining range of temperature, humidity and cleanliness of air for human comfort and for proper working of some industrial processes. It is very old technology and has achieved fast growth due to urbanization of culture and industrialization of economy after second world war. The proper maintenance of temperature, cleanliness, and humidity is required due to global warming, industrial pollutions, and changes in season round the year.

The air-conditioning systems are normally categorized according to

(*a*) ***Purpose of systems:*** Comfort air-conditioning and industrial air-conditioning.

(*b*) ***Season of the use of system:*** Summer, winter and year-around air-conditioning systems.

(*c*) ***Working substance used in the system:*** All-air systems, chilled water systems and air water systems.

(*d*) ***Volumes of air handled by the system:*** Constant volume systems and variable air volume systems.

(*e*) ***Arrangement of equipment in the system:*** Central air-conditioning system and unitary air-conditioning system.

Factors for selection of air conditioning system are

- purpose, location, orientation and shape of the space or building.
- temperature, humidity, wind, exposure to sun and shade of environments of the space or building.
- occupancy, lighting and other heat exchanges, and heat storage capability.

An air-conditioning system refers to an assembly of different parts/components that produces a specified condition of air within a required space of building.

Irrespective of the form or type of an air-conditioning system, its basic elements are :

- *Fans :* for producing air movement

- *Filters* : for cleaning air which may be either fresh, recirculated or both
- *Refrigerating plant* : connected with heat exchange surface such as finned coils or chilled water sprays.
- *Warming set up* : hot water or steam heated coils or electrical elements.
- *Humdification/dehumidification system*
- *Control elements* : For automatic regulation of the amount of cooling or warming.

14.2. COMFORT AND INDUSTRIAL AIR-CONDITIONING SYSTEMS

Comfort air-conditioning systems deal with the creation of an optimum environmental conditions conducive to human health, comfort and efficiency. Air-conditioning systems in homes, offices, stores, restaurants, theatres, schools and hospitals etc. are of this type.

There are some manufacturing processes which need to be done under controlled environmental conditions. Even some of the sophisticated and precision instruments need controlled conditons for their effective working and upkeep. The conditioning done for this purpose is called industrial air-conditioning and is concerned with production of environment suitable for

(*i*) Metrology laboratories, precision machine shops, computer centers: Here constantancy of temperature is the main consideration and a variation of ± 10 to 20 per cent in relative humidity does not have much effect on performance.

(*ii*) Paper and textile mills: The temperature requirements for such spaces are not severe but the relative humidity has to be maintained at a high valve (70 to 75%).

(*iii*) Chemical and biological process industry: These systems need a strict control on both temperature and relative humidity.

(*iv*) Photographic processing: The quality of film relies on uniformity of the layers of sensitized photographic films and this depends on careful control of temperature. For processing film, a temperature of 21°C to 30°C is recommended with humidity 50 to 55 percent.

(*v*) Libraries and museums: The recommended conditions are 5 to 25°C temperature and 40 to 50 percent relative humidity. The air-conditioning of these places is essential to prevent the spoilage of antiques of museum, and to have long life of books and journals which is likely to be affected by dust and corrosive vapours with excessive moisture.

14.3. SUMMER AIR-CONDITIONING

These systems, when properly designed and installed, maintain the temperature and humidity of indoor air to a level at which persons feel comfortable during summer. Essentially it involves reducing the air temperature and humidity (in humid tropics) by the processing of cooling and dehumidification.

Figure 14.1 shows the schematic arrangement of the different components incorporated in a summer air-conditioning system.

The outside hot and humid air passes through a damper and mixes with circulated air coming from the conditioned space. The mixed air is pulled through a filter where airborne particles such as dust and lint (small unwanted threads/fibres) are removed. Sophisticated filters may remove microscopic pollutants as well. The filtered air is routed to a cooling coil where it gets cooled due to refrigerant action. Cold air next passes through a perforated membranes where it is made free from moisture. The moisture so entrapped is collected in the sump at the base.

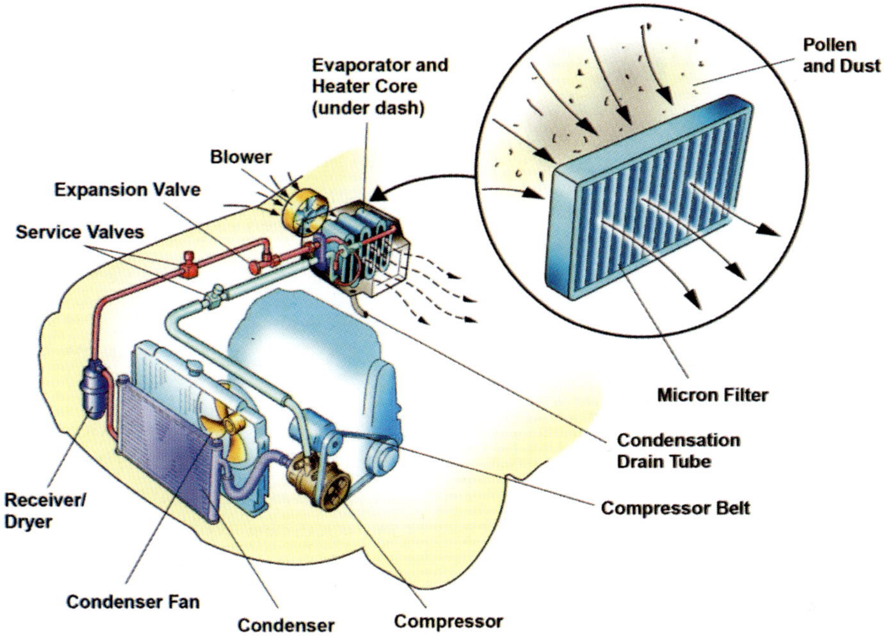

(a)

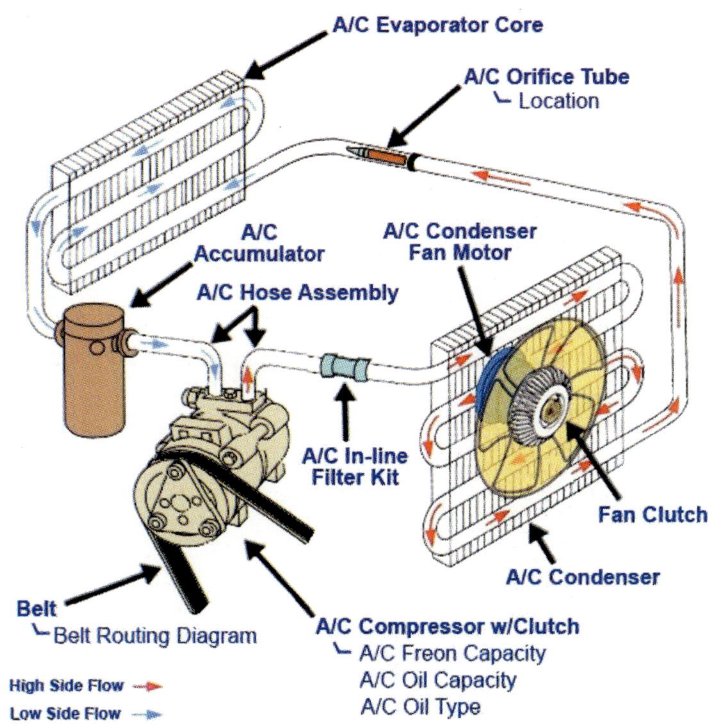

(b)

Air-Conditioning Systems

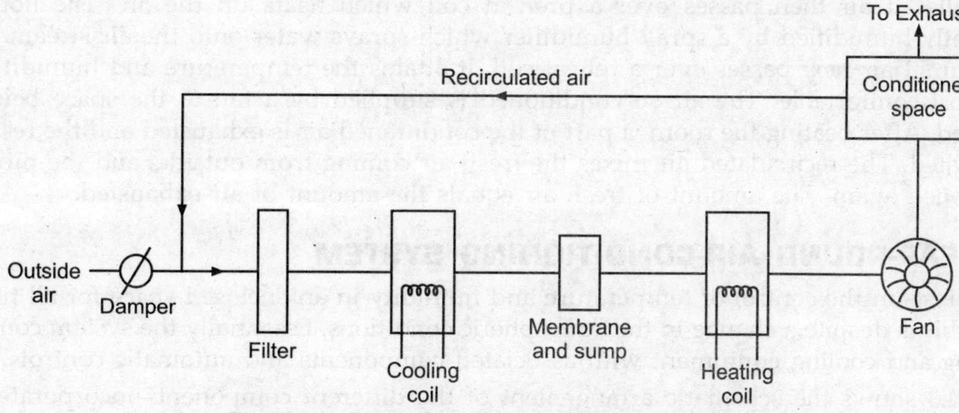

Fig. 14.1. Schematics of summer air-conditioning

When the dry cold air next passes over a heating coil, it attains the temperature and humidity level that is most comfortable and healthful. The air so conditioned is supplied by a fan to the space being air-conditioned. The air picks up heat from the space ; a portion of this hot air is exhausted to atmosphere and the remainder is recirculated. The recirculated air is mixed with the incoming fresh air; the amount of fresh air equals the amount of air exhausted.

The cooling and dehumidifying process is essentially carried out by the use of cooling coil (apparatus dew point temperature lower than the dew point of air entering the coil) or by air washer in which chilled water is sprayed with temperature again lower than dew point temperature of air entering.

14.4. WINTER AIR-CONDITIONING

These systems are meant for the control of indoor air conditions so as to provide human comfort. Essentially it involves an increases in sensible heat and water content of air by the processes of heating and humidification. Essentially, the winter air-conditioning system heats and humidifies air when the surrounding is cold and dry.

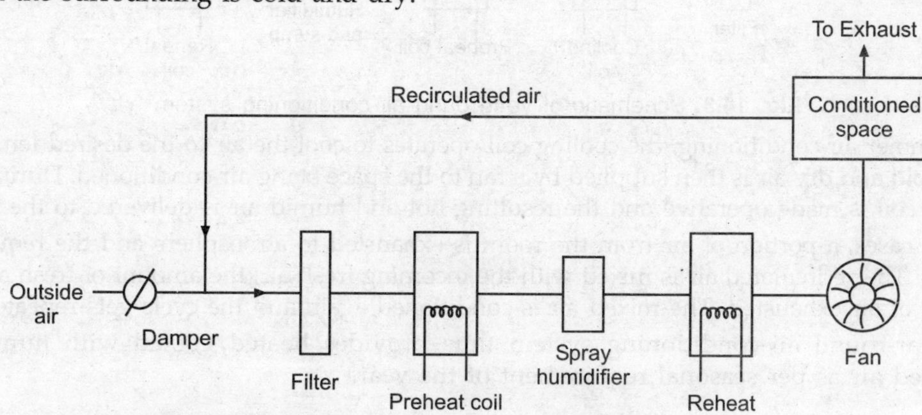

Fig. 14.2. Schematics of winter air-conditioning system

Fig. 14.2 shows the schematic arrangement of the different components incorporated in a winter air-conditioning system. The outside cool air passes through a damper and mixes with circulated air coming from the conditioned space. The mixed air is passed through a filter which serves to purify the air. In the process, dirt/dust/lint; soot and pollens are filtered out and the air may be sterilized as is sometimes done in hospital and public places.

The filtered air then passes over a preheat coil which heats up the air. The hot air is subsequently humidified by a spray humidifier which sprays water onto the air stream. When the hot humid air next passes over a reheat coil, it attains the temperature and humidity level that is most comfortable. The air so conditioned is supplied by a fan to the space being air-conditioned. After heating the room, a part of the conditioned air is exhausted and the rest of air is recirculated. The recirculated air mixes the fresh air coming from outside, and the mixed air is conditioned again. The amount of fresh air equals the amount of air exhausted.

14.5. YEAR-ROUND AIR-CONDITIONING SYSTEM

This manifests in the control of temperature and humidity in an enclosed space for all times of the year; this is despite a change in the atmospheric conditions. Essentially the system comprises the heating and cooling equipment with associated components and automatic controls.

Fig. 14.3 shows the schematic arrangement of the different components incorporated in a year-round air-conditioning systems. The outside air passes through a damper and mixes with the circulated air coming from the conditioned space. The mixed air is pulled through a filter where airborne particles such as dirt, dust and lint (small unwanted threads/fibres) are removed. The pollens also get filtered out and the air may be sterilized as is sometimes done in hospitals and public places. Depending upon weather, the filtered air is made to pass over a cooling coil or a heating coil. At one time, only one coil is in operation.

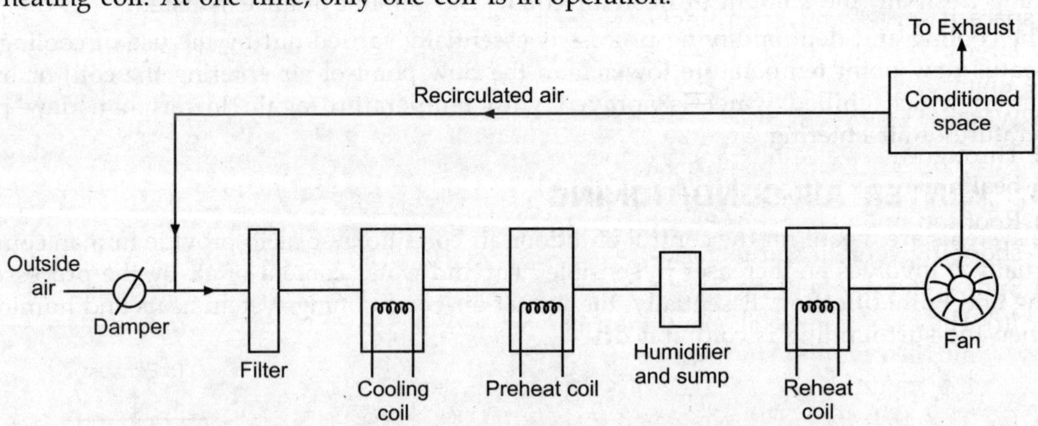

Fig. 14.3. Schematic of year-round air-conditioning system

For summer air-conditioning, the cooling coil operates to cool the air to the desired temperature level. The cold and dry air is then supplied by a fan to the space being air-conditioned. During winter the heating coil is made operative and the resulting hot and humid air is delivered to the room.

In both cases, a portion of air from the room is exhausted to atmosphere and the remainder is recirculated. The recirculated air is mixed with the incoming fresh air; the amount of fresh air equals the amount of air exhausted. The mixed air is conditioned again and the cycle gets repeated.

The year-round air-conditioning system thus provides heated/cooled with humidified/dehumidified air as per seasonal requirement of the year.

14.6. UNITARY AND CENTRAL AIR-CONDITIONING SYSTEMS

The unitary air-conditioning systems consist of a single unit used to condition air in a single conditioned space. These are factory assembled units located in or near the conditioned space. These systems are available as a single package containing all the necessary components inside a casing and suitable for installaiton on windows or wall openings. The room (or cooling) side and the outdoor (or heat rejection) side of the unit are separated by an insulated partition within the casing. They may also be split units with the compressor and condenser located remotely. A single unit may be sufficient for a room of small area. A large area may be divided into several zones; each being served by a conditioner of small to medium capacity.

The salient aspects of unitary air-conditioning systems are:

(*i*) They are mass produced and factory assembled units including internal wiring, controls and piping.

(*ii*) The installation is quite easy and there is considerable flexibility in shifting the unit from one place to another.

(*iii*) The initial and operating cost is less when selection and control of the system is done properly. However, the capital cost per ton of refrigeration is quite high.

(*iv*) These units are preferred for low capacity requirements upto 15 tons and about 200 m^3/min air flow.

(*v*) The system does not need any duct work for air distribution.

(*vi*) Failure of the unit puts off conditioning in one room only.

(*vii*) The units are to be kept running only at the places where cooling is required (units in other rooms can be stopped) and accordingly these systems have high load factor.

However these systems have relatively few options with regard to sizing the evaporator, condenser, fans, compressor and controls. Further, each unit is required to meet the peak loads of the space it serves and accordingly the installed capacity and connected electrical load are usually larger.

The unitary systems are available as:

(*a*) Window units which are primarily found in residential applications.

(*b*) Through-the-wall units are designed into the buildings and find application in motels, health care facilities, schools and offices.

(*c*) Roof top units are primarily applied in low rise buildings with flat roofs such as stores, shopping centres and factories.

(*d*) Split systems where the unit is bifurcated into two parts:

  (*i*) The evaporator coil, evaporator blower with its motor, capillary tube, control panel, air filter, supply and return air grilles etc. are mounted in a cabinet and kept inside the room.

  (*ii*) The compressor, air cooled condenser, condenser fan and motor are installed outside the room in a suitable casing.

The distance between the indoor and outside units has to be kept as small as possible. With increase in distance, the pressure drop in the suction and liquid lines becomes more and that results in a reduction of the capacity of the unit. The split systems enable noise, vibration and maintenance to be kept away from the conditioned space and as such are quite acceptable for office suite, residences, stores and restaurants.

Central air-conditioning, commonly referred to as central air or air-con is an air-conditioning system in which the compressor, condenser, chilled evaporator along with pumps and control panels are located in one place called the plant room. The cooling tower used for cooling the condenser water is installed either on the roof or on the ground at a place where there is no obstruction to the movement of air. The conditioned air from system is distributed to the different places through a system of ducts and grilles. Ducts are suitably designed to give an economical and efficient air-conditioning system.

The salient aspects of central air-conditioning systems are:

(*i*) The components comprising the system (*e.g.*, cooling and heating coils, humidifiers and dehumidifiers, air filters, blowers with motors, and control devices) are assembled at site rather than in the factory.

(*ii*) These units are preferred for large capacity units above 25 tons and about 2500 m^3/min of air flow.

(*iii*) The systems are easy to control, have the less power consumption and are quite economical for large tonnage.

(*iv*) Installation charges are more due to the requirement of duct work.

(*v*) The control plant is located in unoccupied areas, hence facilitating operation, maintenance and repair.

(*vi*) Noise and vibration levels are fairly low.

(*vii*) All rooms are affected by the system failure, and it is also difficult to meet the needs of individual rooms.

(*viii*) Low load factor as the plant has to be kept running even when only few rooms require conditioning.

The central air-conditioning systems can be either of the direct or indirect type

• **Direct type:** The air from the space to be air-conditioned is circulated over the coil in which low pressure refrigerant liquid is boiling. The liquid refrigerant takes its latent heat of vaporization from the air and that gets cooled.

• **Indirect type:** Chilled water or brine for the refrigeration plant is circulated through the cooling coil located in the air handling unit to cool and dehumidify the room air.

The central air-condiitioning systems are generally preferred for auditoriums, departmental stores and large multi-storeys buildings.

14.7. ALL-AIR AIR-CONDITIONING SYSTEM

An all-air system serves to provide complete sensible and latent cooling to the air supplied by air handling unit (AHU) to various parts of the building through duct work. AHU is a package assembly of air conditioning components (fan, filter, cooling and/or heating coil) which provides necessary treatment to the air before it is distributed. The air does not need any additional cooling after leaving the AHU.

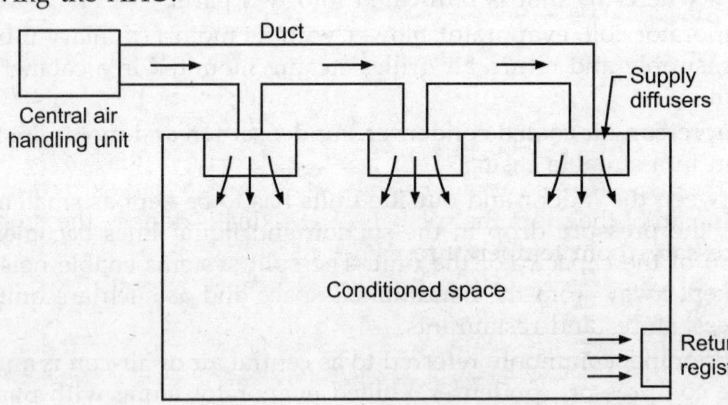

Fig. 14.4. All-air air-conditioning system (elementary form)

The conditioned air is distributed to the space through supply diffusers. Inside the room, the air picks up heat and returns back to AHU through return registers.

Different combinations of the duct networks, coil locations and controls are in use. However, the most commonly employed all-air systems are: constant air volume system and variable air volume system.

14.7.1. Constant Air Volume (CAV) System

The volume of air handled is constant in the CAV system of air-condiitioning. However, the supply air temperature varies in response to the space load. The main features of the system are:

– simple to design, install and operate
– relatively low installation cost
– noise and vibration free as all the mechanical equipment is remotely located.

A brief description of the commonly used CAV systems is given below:

• **Single zone system:** The unit has one single supply air duct that branches to different areas in the building. The air supplied to different areas is at the same temperature and humidity.

This system has applications to premises where load varies uniformly or where temperature control is not critical. The notable applications are residential installations, offices, class rooms and small departmental stores.

• **Dual duct and multi-zone systems:** In the dual duct systems, the control station supplies air to the conditioned spaces through two parallel ducts; one duct is for cold air and the other duct is for warm air.

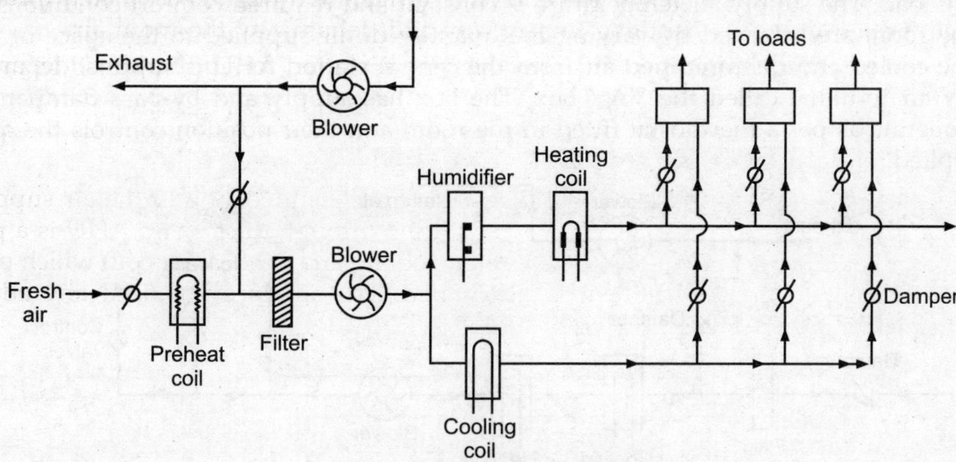

Fig. 14.5. Dual duct system

The mixing of the two steams for a particular room takes place in the mixing box which operates under command of the room thermostat. The mixing is done in the proportions required to maintain the necessary room temperature.

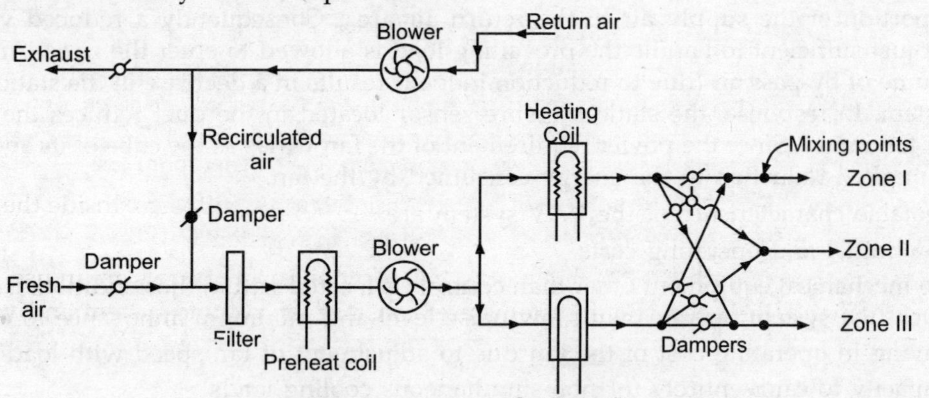

Fig. 14.6. Multi-zone system

Since the warm and cold air streams are simultaneously available at all times at each terminal unit, the dual system provides prompt temperature response and has greater flexibility in satisfying multiple loads. However, the system has high initial cost as it has two supply air-ducts and both have to be large enough to handle the air flow. To reduce the duct size and space requirements, the system is usually designed for high velocity air movement. This increases the operation cost due to increased fan power.

The multi-zone system has a single supply duct and that distributes the fully conditioned and filtered air through out the building. The mixing of the cold and hot air streams is done at the central AHU instead of being mixed at each space served. The mixing process is in response to room or zone thermostats.

This system provides excellent control of temperature and humidity in individual spaces and is well adopted when there are enough different load conditions which cannot be served by a single zone system. It may, however, be pointed that the use of both the dual duct and multi-zone systems is presently being discouraged from the energy conservation viewpoint.

14.7.2. Variable Air-volume (VAV) System

This system controls the room conditions by controlling the air quantity to match or balance the cooling load. The supply air temperature is constant and required comfort conditions as per load in the room are achieved by varying the quantity of air supplied to the space or zone it serves. The cooled and dehumidified air from the central station AHU is supplied to the space via supply air terminal called the VAV box. The box has supply and by-pass dampers; these dampers operate as per a thermostat fitted in the room and their position controls the quantity of air supplied.

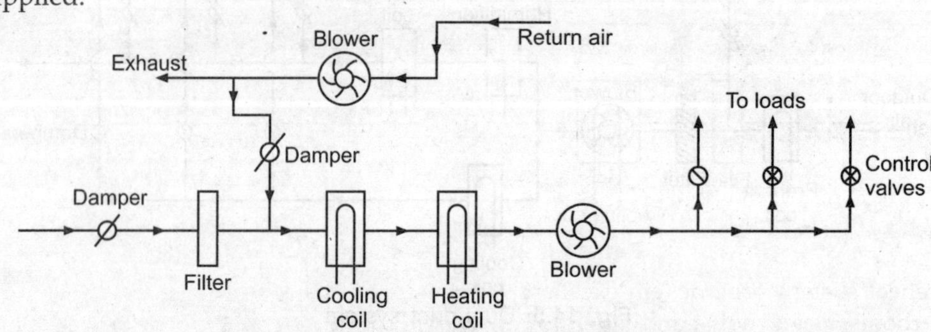

Fig. 14.7. Variable air-volume (VAV) system

When peak load prevails, the supply air damper opens to allow full volume of supply air to match the load in the zone. As the load comes down, the by-pass damper opens and that by-passes a portion of the supply air to the return air area. Consequently a reduced volume of supply air just sufficient to handle the prevailing load is allowed to enter the zone. An increase in the volume of by-pass air (due to reduction in load) results in a decrease in the static pressure in the system. In response, the static pressure sensor located in the duct reduces the speed of supply air fan motor. Since the power requirement of the fan varies as the cube of its speed, there will be a marked reduction in the energy consumed by the fan.

The notable characteristics of the VAV system are:

– low initial and operating costs
– no mechanical equipment other than controls is located within the conditioned space. As such, the system operates with low noise level and minimum annoyance to occupants
– saving in operating cost of the fan due to adjustment of fan speed with load
– capacity to cater entirely for non-simultaneous cooling loads

– complexity of design
– lack of fresh air at part load
– dampers tend to produce noise and this increases proportionally with the amount of throttling. Further, throttling down to 50 to 60% of the maximum design flow may lead to loss of air motion in the room with considerable drafts and stratification.

Being energy effecient, the VAV systems are considered ideally suited to interior zones of single, multi-storey and multi-room buildings that require cooling whenever they are occupied.

• **Terminal reheat system:** The fully conditioned and filtered air is supplied from a control AHU at fixed cold air temperature designed to offset the maximum cooling load in the space. Secondary heating coils, called the reheat coils, are added to individual control zone. The medium for heating may be electricity, hot water or steam. The reheat system is adopted to serve the following purposes:

– to permit space control for areas of unequal loading
– to provide heating or cooling of process or comfort applications requiring closed control of space conditions.

The reheat coils get activated by a thermostat when the temperature falls below the set temperature.

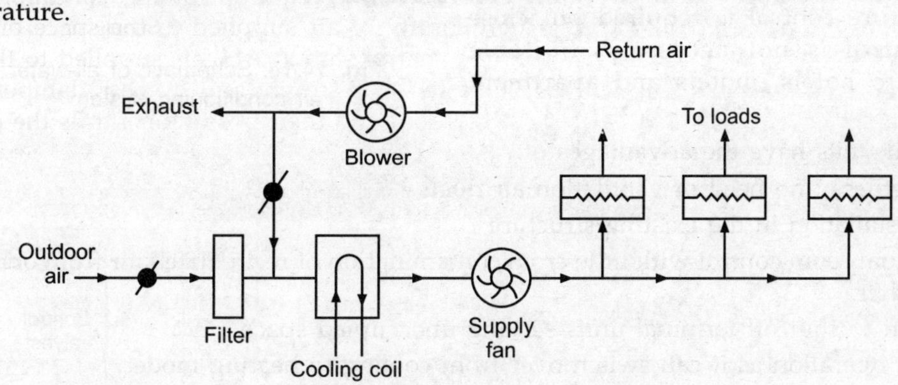

Fig. 14.8. Reheat control system

The reheat system provides an excellent temperature control and is generally used in hospital operation rooms, clean rooms and laboratories where there is possibility of wide variations in load.

• **Face and by-pass control system:** The system adopts an arrangement whereby some fraction of air is by-passed around the cooling coil. The amount of by-passed air is controlled by the by-pass dampers, and the face dampers control the intake of outside air. The resulting mixed air stream at controlled temperature and humidity level is then supplied by the fan to all zones.

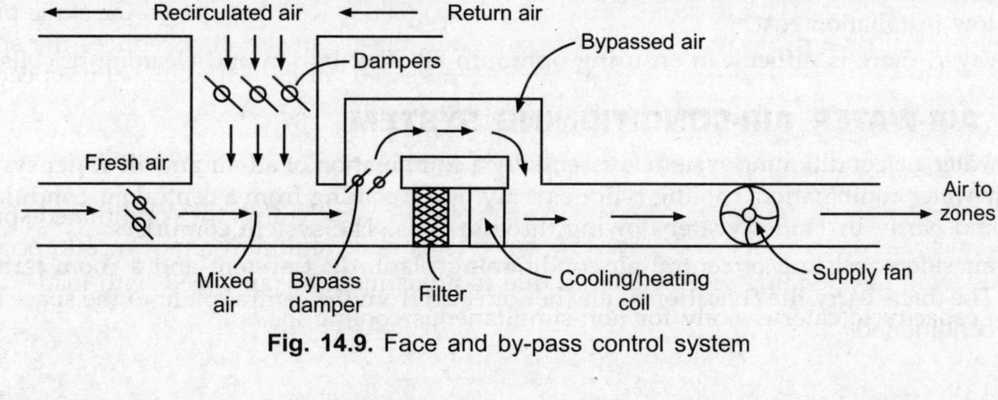

Fig. 14.9. Face and by-pass control system

The sizing of return, by-pass and exhaust duct is such that the outdoor air does not get circulated through the by-pass connection.

14.8. ALL-WATER AIR-CONDITIONING SYSTEM

The unit has coils through which circulates the hot (during winter) or chilled water (during summer) from a remote control source. The coils are housed in terminal units located in the various spaces in the building. The final heat transfer is, however, to or from air.

The all-water air-conditioning systems are generally classified as fan coil units (FCU) and valence units. A fan coil unit consists of a fan, a cooling coil, and a filter. The system also employs a multi-speed motor to control the fan capacity. There may also be a water side control to regulate the flow of water into the coil. The fan blows air which passes over the coils and gets heated or cooled before entering the space being conditioned. The fan coil units are best applied where individual space temperature control is required but close humidity control is not necessary. Suitable applications are hotels, motels and apartment buildings.

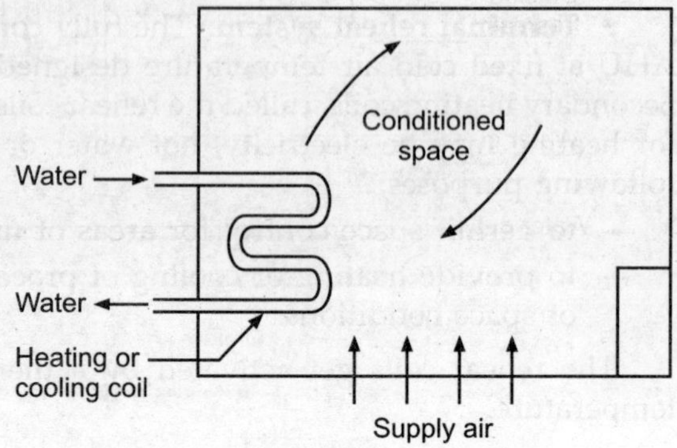

Fig. 14.10. Schematic of all-water air-conditioning system

The fan coil units have the advantages of:

- compactness ; no need of ventilation air ducts
- easy installation in the existing structure
- individual room control with little cross contamination of recirculated air from one space to another.
- freedom to shut off terminal units serving unoccupied space.
- flexible operation as it can switch over from cooling to heating mode.

The disadvantages are: coils are difficult to clean, maintenance required is more, drain pans are needed and there is lack of humidity control.

The valance units do not employ any fan system; the cooling or heating of room air is by convection. The coil having fins is mounted in an insulated sheet metal enclosure and is so arranged that air can pass over it by convection.

The advantages of valance units are:

- freedom from noise,
- provide reasonably god cooling and dehumidification
- low installation cost.

However, there is difficult in ensuring optimum air distribution and cleaning of coils.

14.9. AIR-WATER AIR-CONDITIONING SYSTEM

The air-water air-conditioning system is essentially a combination of all-air and all-water systems. In the air-water combination, cooling is done partly by air coming from a central air-conditioning system and partly by chilled water flowing through coils. The system comprises:

(*i*) air side consisting of central air-conditioning plant, duct system and a room terminal. The ducts carry the conditioned air (hot or cold) from the central plant to the space being conditioned.

(*ii*) Water side consisting of a pump and piping arrangement for the circulation of water. The hot or cold water circulates through coils located at appropriate places in the conditioned space.

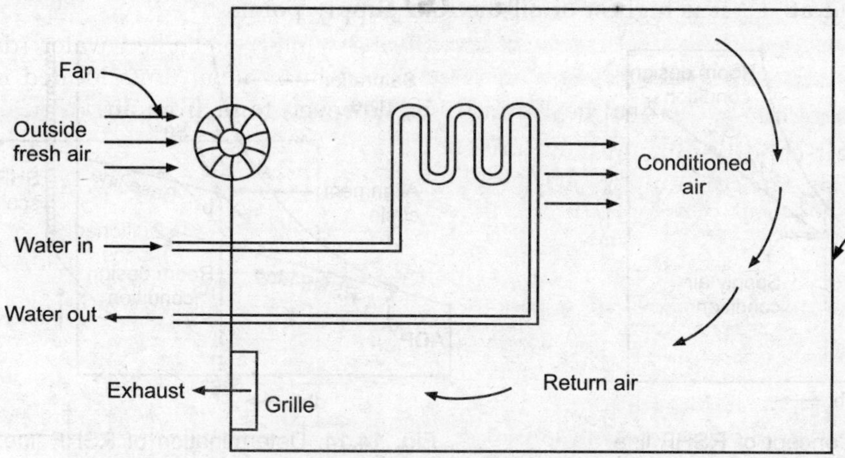

Fig. 14.11. Schematic of air-water air-conditioning system

The room temperature can be controlled by varying the air flow or flow rate of water through the coils. The system has the advantage that less quantity of air needs to be supplied compared to an all-air system and this is a critical factor for high rise buildings. Further, the reduced quantity of air is usually combined with a high velocity method of air distribution to minimise the space required.

14.10. ROOM SENSIBLE HEAT FACTOR

With reference to Fig. 14.12, the thermal properties of air for the heating and humidification process 1-2 can be separated into two parts namely sensible heating (*SH*, process 1-*a*) and latent heating (*LH*, process *a*-2). Then, the term sensible heat factor (*SHF*) is defined as ratio of sensible heat (*SH*) to total heat (*TH*). The total heat (*TH*) is the sum of sensible heat and latent heat. Thus:

$$SHF = \frac{SH}{TH} = \frac{SH}{SH + LH}$$

If these quantities pertain to room load, then one may write

$$RSHF = \frac{RSH}{RTH} = \frac{RSH}{RSH + RLH}$$

where *RSHF* is the room sensible heat factor, *RSH* is the room sensible heat, *RLH* is the room latent heat and *RTH* is the room total heat.

The air capacity supplied to the conditioned space has to be such that offsets simultaneously both the room sensible heat and room latent heat.

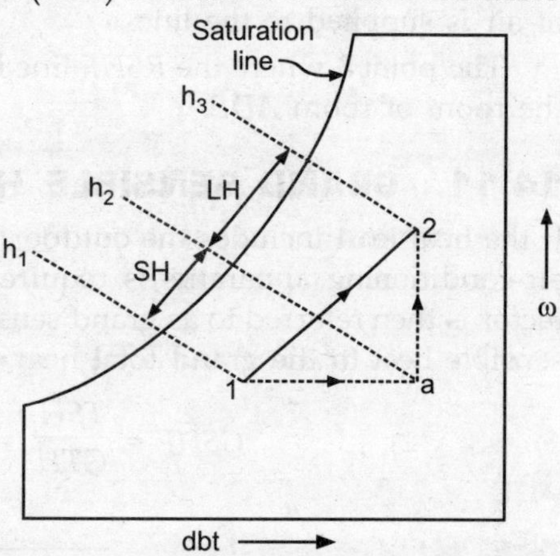

Fig. 14.12

If the points 1 and 2 denote the conditions of supply air and that of room design respectively, then the straight line 1-2 respresents the psychrometric process of the supply air to the conditioned space, and is called the ***room sensible heat factor line***. The slope of the line is a measure of the ratio of sensible heat to latent heat within the room. The air properties of dry bulb and wet bulb temperatures corresponding to any point on the line will satisfy the room requirements with adequate supply of air. However, the quantity of air will be different for different supply air points. For the supply air with properties as at 1, 1′, 1″ and 1‴, the supply condition at 1 requires less air than at 1′ and at 1‴ it is highest of all the four supply points.

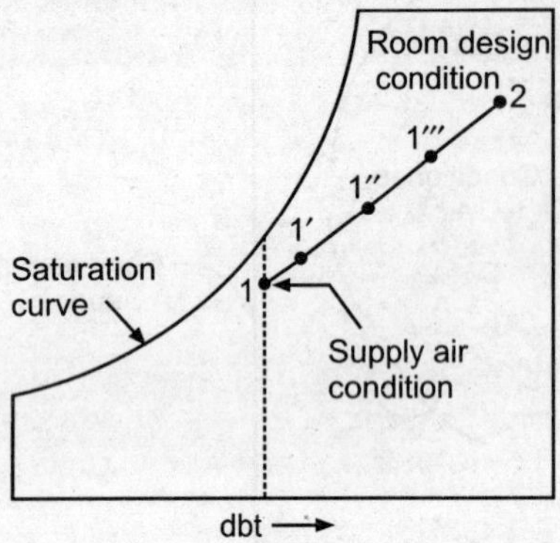

Fig. 14.13. Concept of RSHF line

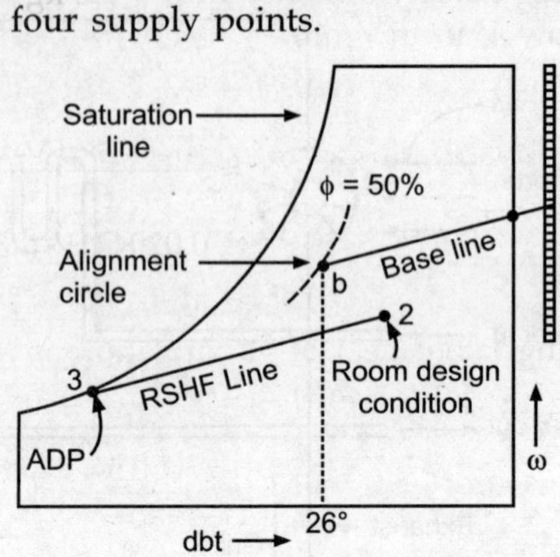

Fig. 14.14. Determination of RSHF line

The supply air conditions are normally not known and these are essentially required to be found and for that:

(*i*) mark point *a* on the *SHF* scale corresponding to the calculated value of *RSHF*. The *SHF* scale is drawn in the upper right corner of the psychrometric chart

(*ii*) draw a line that joins point *a* to the standard reference point *b* on the chart corresponding to 26°C *dbt* and 50% relative humidity. This standard reference point is called the alignment circle and the line *a-b* is called the base line.

(*iii*) mark point 2 on the chart corresponding to room design condition. Through point 2 draw line 2-3 parallel to the base line such that 3 is the saturation point.

The line 2-3 is the *RSHF* line for the room. The supply conditions of air lying at any point on line 2-3 will meet the sensible and latent heat loads of the room provided adequate quantity of air is supplied to the line.

The point 3 where the *RSHF* line intersects the saturation line is the apparatus dew point for the room or room *ADP*.

14.11. GRAND SENSIBLE HEAT FACTORS

If the heat load includes the outdoor load in addition to room load, then the cooling coil or the air-conditioning apparatus is required to handle the grand total heat load. The sensible heat factor is then referred to as grand sensible heat factor (*GSHF*) and it is defined as the ratio of total sensible heat to the grand total heat. That is

$$GSHF = \frac{TSH}{GTH} = \frac{TSH}{TSH + TLH}$$

$$= \frac{RSH + OASH}{(RSH + OASH) + (RLH + OALH)}$$

where *TSH* (total sensible heat) = room sensible heat (*RSH*) + outside air sensible heat (*OASH*)

TLH (total latent heat) = room latent heat (*RLH*) + outside air latent heat (*OALH*)

and, *GTH* (grand total heat) = total sensible heat (*TSH*) + total latent heat (*TLH*)

The outside air sensible heat and outside air latent heat are worked out from the relations:

$$OASH = m\, c_{pm}\, (dbt_1 - dbt_2) = \rho\, V\, c_{pm}\, (dbt_1 - dbt_2)$$

Taking density of moist air $\rho = 1.2$ kg/m^3, humid specific heat $c_{pm} = 1.022$ kJ/kg K and rate of air flow V in m^3/min.

$$OASH = 1.2\, V \times 1.022\, (dbt_1 - dbt_2) \text{ kJ/min}$$

$$= \frac{1.2\, V \times 1.022}{60} (dbt_1 - dbt_2)$$

$$= 0.02044\, V\, (dbt_1 - dbt_2) \text{ kJ/s or kW}$$

$$OALH = m\, h_{fg}\, (\omega_1 - \omega_2) = \rho\, V\, h_{fg}\, (\omega_1 - \omega_2)$$

Taking latent heat of vaporization as $h_{fg} = 2500$ kJ/kg

$$OALH = 1.2\, V \times 2500\, (\omega_1 - \omega_2) = 3000\, V\, (\omega_1 - \omega_2) \text{ kJ/min}$$

$$= \frac{3000\, V}{60} (\omega_1 - \omega_2) = 50\, V\, (\omega_1 - \omega_2) \text{ kJ/s kW}$$

In the above relations V, dbt and ω stand for ventilation volume/min, dry bulb temperature and specific humidity respectively.

Key:

1 = outside air condition

2 = room air condition

3 = mixture air condition

4 = supply air from cooling coil

6 = ADP of coil

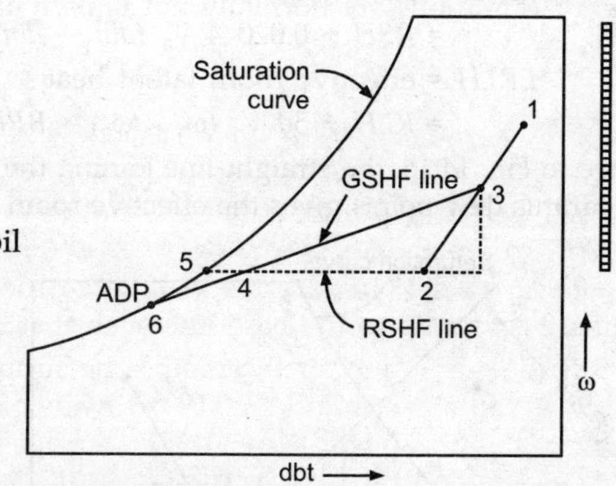

Fig. 14.15. Grand sensible heat factor

Let the point 1 represent the outside air condition and the point 2 represent the room air condition on the psychrometric chart (Fig. 14.15).

The notable respects concerning *GSHF* are:

(*i*) The air supplied to the air-conditioning plant is generally a mixture of fresh air (outside air or ventilation) and the recirculated air having the properties of the room air.

(*ii*) The mixture, represented by state point 3, enters the cooling coil or conditioning apparatus, gets cooled and dehumidified, and leaves the coil with reduced temperature and moisture content. The fall in temperature and moisture content depends on the total sensible heat and total latent heat to be handled by the unit.

(*iii*) When the point 3 (representing the condition of air entering the cooing coil) is joined with the point 4 (representing the condition of air leaving the coil), we get the grand sensible heat factor line (*GSHG* line). The line 3-4 represents the psychrometric process undergone by the mixed stream as it passes over the cooling coil. Any point on this line will give sensible heat and the latent heat removed in a fixed proportion.

(*iv*) When the *GSHF* line is extended, it meets the saturation curve at point 6 which gives the *ADP* (apparatus dew point) or the effective surface temperature of the cooling coil. The effective surface temperature represents that temperature of the coil which would produce the same leaving air conditions as the non-uniform temperature of coil would have produced.

(*v*) If *GSHF* for the system is known, the *GSHF* line can be drawn by following the steps as outlined for the *RSHF* line (Fig. 14.15).

(*vi*) The ideal supply conditions to the room are given by the point 4 which lies at the intersection of *GSHF* line and *RSHF* line.

14.12. EFFECTIVE ROOM SENSIBLE HEAT FACTOR

The effective room sensible heat factor (*ERSHF*) is the term that relates by-pass factor, *ADP* and effective sensible heat, and provides a simplified method for the determination of air quantity and selection of cooling coil.

ERSHF is defined as the ratio of the effective room sensible heat (*ERSH*) to the effective room total heat (*ERTH*). That is

$$ERSHF = \frac{ERSH}{ERTH} = \frac{ERSH}{ERSH + ERLH}$$

where:

$$ERSH = RSH + OASH \times BPF$$
$$= RSH + 0.02044\ V_1\ (dbt_1 - dbt_2) \times BPF$$
$$ERLH = \text{effective room latent heat}$$
$$= RLH + 50\ V_1\ (\omega_1 - \omega_2) \times BPF$$

With reference to Fig. 14.16, the straight line joining the state point 2 (room design condition) with point 6 (apparatus dew point) gives the effective room sensible heat factor line (ERSHF line).

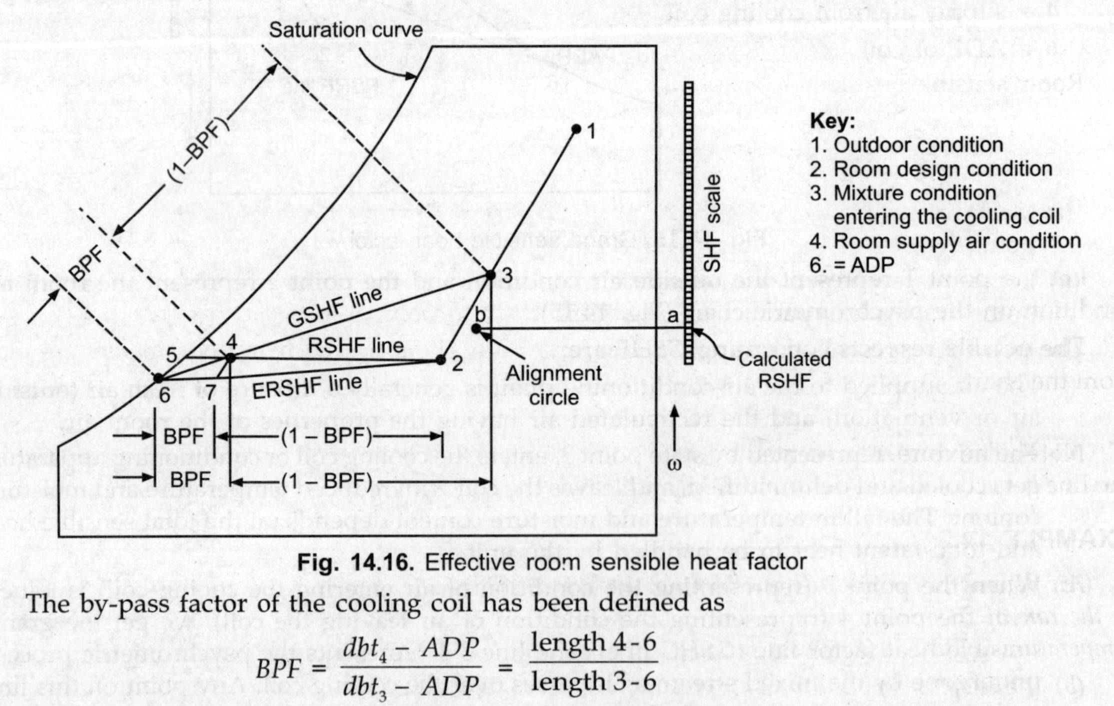

Fig. 14.16. Effective room sensible heat factor

The by-pass factor of the cooling coil has been defined as

$$BPF = \frac{dbt_4 - ADP}{dbt_3 - ADP} = \frac{\text{length } 4\text{-}6}{\text{length } 3\text{-}6}$$

From point 4, let a line 4-7 be drawn parallel to line 3-2. Then from the similarly of triangles 6-4-7 and 6-3-2,

$$\frac{\text{length } 4\text{-}6}{\text{length } 3\text{-}2} = \frac{\text{length } 7\text{-}6}{\text{length } 2\text{-}6} = \frac{dbt_7 - ADP}{dbt_2 - ADP}$$

Obviously, then *BPF* can be re-written as:

$$BPF = \frac{dbt_4 - ADP}{dbt_3 - ADP} = \frac{dbt_7 - ADP}{dbt_2 - ADP}$$

The following expressions can be used to determine the effective room sensible heat (*ERSH*), the effective room latent heat (*ERLH*) and the effective room total heat (*ERTH*)

$$ERSH = 0.02044\ V\ (dbt_2 - ADP)\ (1 - BPF)$$

$$ERLH = 50\ V\ (\omega_2 - \omega_{adp})\ (1 - BPF)$$

EXAMPLE 14.1

The sensible heat gain of a room is 4.8 kW and the latent heat gain is 1.4 kW. A conditioned air supply of 0.5 m³/s is to be delivered to the room. If the room is to be maintained at 25 dbt and 50% RH, find the required wet and dry bulb temperatures of air supply. Take specific heat of air = 1.005 kJ/kg K.

Solution : Refer Fig. 14.17 for plot of the air conditioning process on psychrometric chart.

Applying the relation,

$$R_{SH} = 0.02044\ V\ (dbt_1 - dbt_2)$$

where V is the air supply in m³/min. Then

$$4.8 = 0.02044 \times (0.5 \times 60)\ (25 - dbt_2)$$

∴ Dry bulb temperature of supply air,

$$dbt_2 = 25 - \frac{4.8}{0.02044 \times (0.5 \times 60)} = 17.2°\ C$$

Room sensible heat factor,

$$RSHF = \frac{RSH}{RSH + RLH} = \frac{4.8}{4.8 + 1.4} = 0.77$$

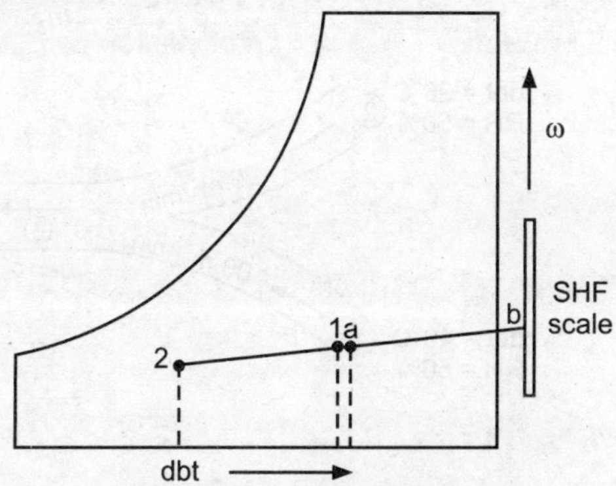

Fig. 14.17

(*i*) Mark the calculated value of *RSHF* = 0.77 on the *SHF* scale as point *b* and join it with point *a* which is the alignment circle (26°C *dbt* and 50% *RH*).

(*ii*) Locate state point 1 corresponding to room conditions (25°C and 50% RH). From this point draw a line parallel to *ab* and extend it to point 2 such dbt_2 = 17.2°C.

The point 2 represents the supply condition of air. Then, the following parameters are read from the chart,

$$wbt = 14°\ C \quad \text{and} \quad \phi = 72\%$$

Note : The condition for points *a* and 1 are identical and as such they merge. Obviously then the line *b* – *a*(1) – 2 becomes a single continous line.

EXAMPLE 14.2

Air flowing at the rate of 100 m³/min at 40°C dbt and 50% RH is mixed with another stream flowing at the rate of 20 m³/min at 26°C dbt and 50% RH. The mixture flows over a cooling coil whose ADP temperature is 10°C and by-pass factor is 0.2. Find dbt and RH of air leaving the coil.

(b) If this air is supplied to an air-conditioned room where dbt of 26°C and RH of 50% are maintained, estimate the room sensible heat factor and cooling capacity of the coil in tons of refrigeration.

(JNTU)

Solution: Refer Fig. 14.18 for the flow diagram of the system and presentation of the processes involved on psychrometric chart.

(*i*) Locate state point 1 (*dbt* = 40°C, *RH* = 50%) and state point 2 (*dbt* = 26°C and *RH* = 50%) or psychrometric chart. Join points 1 and 2.

From chart: $h_1 = 99.5$ kJ/kg ; $v_{s1} = 0.92$ m³/kg

$h_2 = 54$ kJ/kg ; $v_{s2} = 0.86$ m³/kg

Masses of the two air streams are:

$$m_1 = \frac{V_1}{v_{s_1}} = \frac{100}{0.92} = 108.69 \text{ kg/min}$$

$$m_2 = \frac{V_2}{v_{s_2}} = \frac{20}{0.86} = 23.26 \text{ kg/min}$$

Then: mass of air flowing through the coil,

$$m_3 = m_1 + m_2 = 108.69 + 23.26 = 131.95 \text{ kg/min}$$

From energy conservation: $m_1 h_1 + m_2 h_2 = m_3 h_3$

$$\therefore \quad h_3 = \frac{m_1 h_1 + m_2 h_2}{m_3} = \frac{108.69 \times 99.5 + 23.26 \times 54}{131.95} = 91.48 \text{ kJ/kg}$$

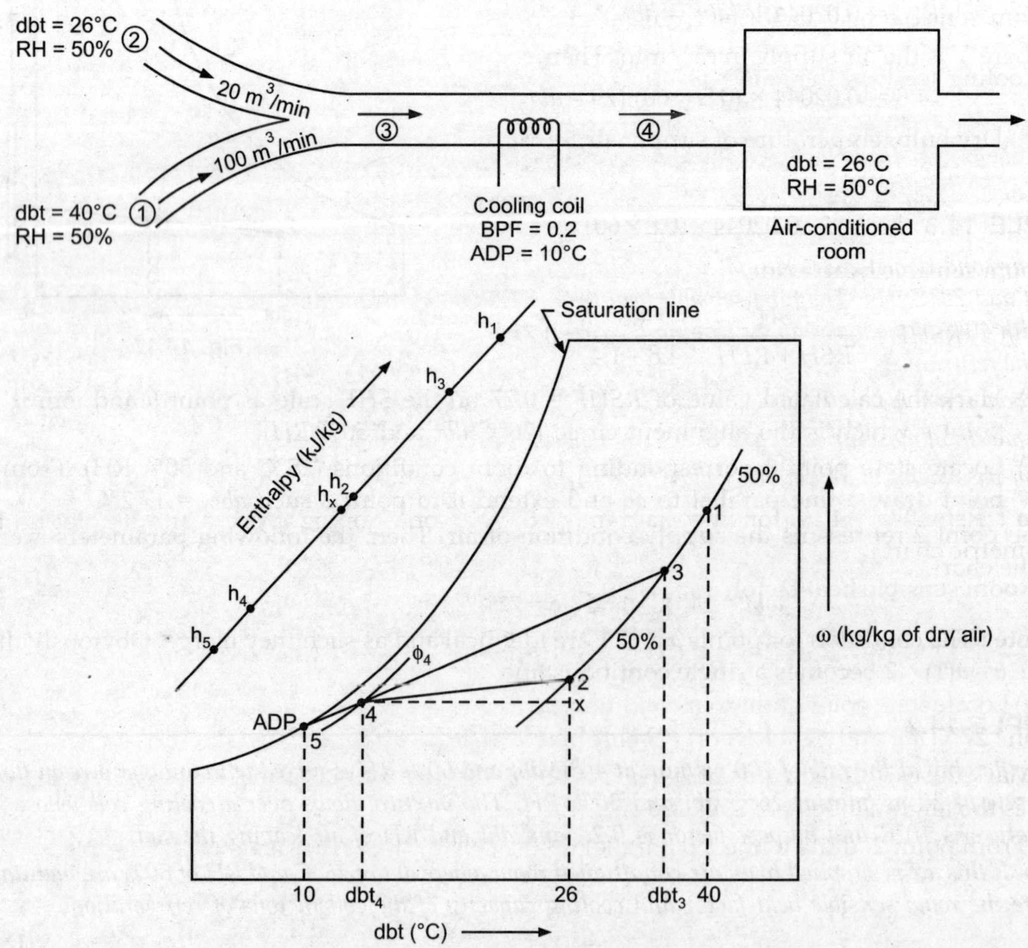

Fig. 14.18

(*ii*) Locate state point 3 on the line 1-2 such that h_3 = 91.48 kJ/kg. The point 3 represents condition of mixed air stream before the coil.

Here dbt_3 = 37.8°C

(*iii*) Locate ADP = 10°C on the saturation curve (state point 5)

Now: $$BPF = \frac{dbt_4 - ADP}{dbt_3 - ADP} \;;\quad 0.2 = \frac{dbt_4 - 10}{37.8 - 10}$$

That gives: dbt_4 = 10 + 0.2 (37.8 – 10) = 15.56°C

Corresponding to this temperature value, locate point 4 on the straight line joining points 3 and 5. The point 4 represents the condition of mixed air stream leaving the coil. From the chart, *RH* at point 4 is noted to be 92%.

∴ Condition of air leaving the coil,

dbt = 15.56°C and RH = 92%

(*b*) From the psychrometric chart:

h_2 = 54 kJ/kg ; h_4 = 42 kJ/kg and h_x = 52.5 kJ/kg

The point x is the intersection of horizontal line from point 4 and vertical line from point 2.

$$\text{Room sensible heat factor} = \frac{RSH}{RTH} = \frac{h_x - h_4}{h_2 - h_4} = \frac{52.5 - 42}{54 - 42} = 0.87$$

$$\text{Cooling load on the coil} = m_3\,(h_3 - h_4)$$

$$= \frac{131.95}{60}(91.48 - 42) = 108.81 \text{ kJ/s} = \frac{108.81}{3.5} = \mathbf{31.09} \text{ TR}$$

EXAMPLE 14.3

An air conditioned auditorium is to be maintained at 27° C dbt and 55% RH. The ambient condition is 39°C dbt and 28°C wbt. The total sensible heat load is 120000 kJ/hr, the total latent air is recirculated and mixed with 40% of the make-up air leaving the cooling coil at 17°C. Determine heat load is 45000 kJ/hr and 60% of the return

(a) room sensible heat factor
(b) condition of air entering the auditorium
(c) amount of make-up air
(d) apparatus dew point, and
(e) by-pass factor of the cooling coil ***(WBUT)***

Solution : Refer Fig. 14.19. for line diagram of the air conditioning process and its plot on the psychrometric chart.

(*a*) Room sensible heat factor,

$$RSHF = \frac{RSH}{RSH + RLH} = \frac{120000}{120000 + 45000} = \mathbf{0.727}$$

(*b*) (*i*) Locate state point 1 (outdoor condition : 39°C *dbt* and 28°C *wbt*) and state point 2 (auditorium condition : 27°C *dbt* and 55% *RH*) on psychrometric chart.

(*ii*) Mark the calculated value of *RSHF* = 0.727 on the *SHF* scale as point *b* and join it with point *a* which is the alignment circle; 26°C *dbt* and 50% *RH*.

(*iii*) From point 2, draw a line 2-5 parallel to line *ab*. This line represents the *RSHF* line.

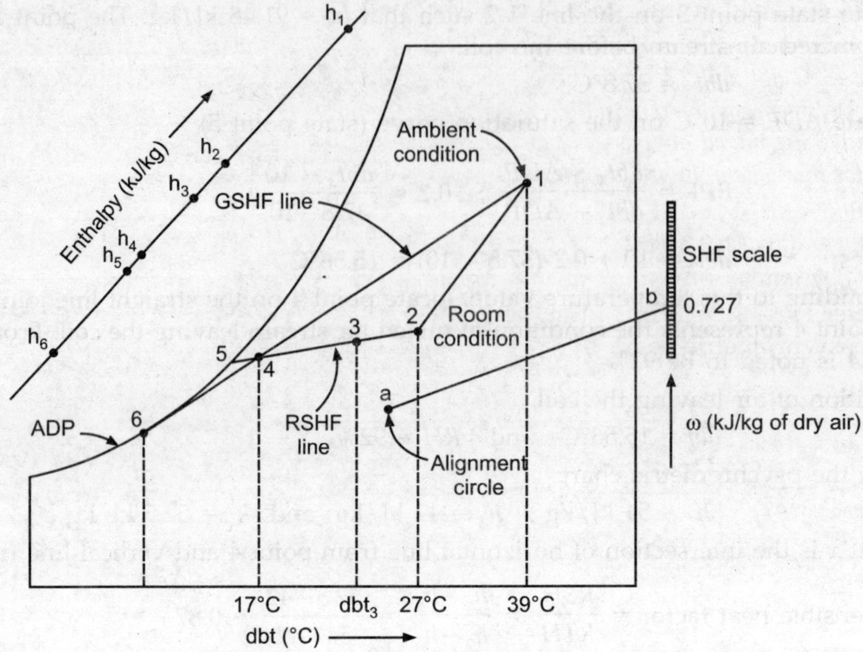

Fig. 14.19.

Since the condition of air leaving the cooling coil is 17°C, mark point 4 such that $dbt_4 = 17°C$. Join point 1 and 4 and extend it to point 6 on the saturation curve. The line 1 – 4 – 6 is the *GSHF* line.

(*iv*) Given that 60% of air from the auditorium is recirculated and mixed with 40% of make-up air after the cooling coil, locate point 3 on *RSHF* line such that,

$$\frac{\text{length } 2-3}{\text{length } 2-4} = 0.4$$

Tthe point 3 represents the condition of air entering the auditorium. From psychrometric chart,

$$dbt_3 = 23°C \quad ; \quad wbt_3 = 18.4°C \text{ and } \phi_3 = 66\%$$

(*c*) Again from the chart

$$h_2 = 58 \text{ kJ/kg of dry air and } h_3 = 52.5 \text{ kJ/kg of dry air}$$

Mass supply of air to auditorium

$$m_a = \frac{\text{room total heat}}{h_2 - h_3} = \frac{R_{SH} + R_{LH}}{h_2 - h_3}$$

$$= \frac{120000 + 45000}{58 - 52.5} = 30000 \text{ kg/kr}$$

Since the make-up air is 40% of supply air,

$$\text{Mass of make-up air} = 0.4 \times 30000 = 12000 \text{ kg/hr}$$

(*d*) From the psychrometric chart, it can be noted that the apparatus dew point of the cooling coil at point 6 is,

$$ADP = dbt_6 = 11°C$$

(e) By-pass factor of the cooling coil,

$$= \frac{dbt_4 - ADP}{dbt_1 - ADP} = \frac{17-11}{39-11} = \mathbf{0.214}$$

EXAMPLE 14.4

An air-conditioned room is maintained at 25°C dbt and 50% relative humidity whilst the ambient conditions are 43°C dbt and 27°C wbt. The sensible and latent heat gains for the room are specified to be 24 kW and 6 kW respectively. The return air from the room mixes with the outside air in the ratio of 4 : 1 and thereafter the mixed air passes over the cooling coil which has a by-pass factor of 0.2. Draw the flow diagram for the system. Represent the processes involved on psychrometric chart and make calculations for the following:

(a) coil apparatus dew point *(b) condition of air at entry to and at exit from the cooling coil,*
(c) mass flow rate of fresh air, and *(d) tonnage of the cooling coil.*

Solution: Refer Fig. 14.20 for the flow diagram for the system and presentation of the processes involved on psychrometric chart:

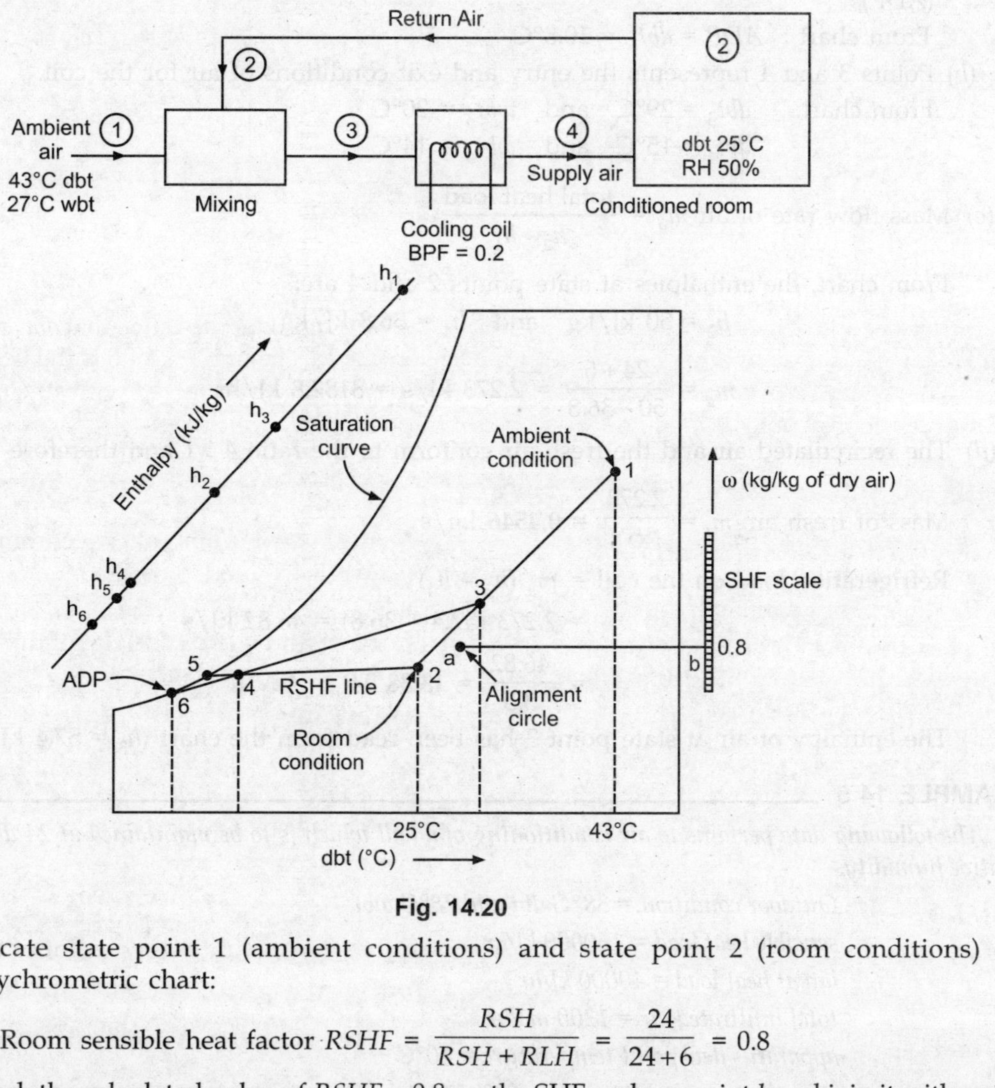

Fig. 14.20

(*i*) Locate state point 1 (ambient conditions) and state point 2 (room conditions) on psychrometric chart:

$$\text{Room sensible heat factor } RSHF = \frac{RSH}{RSH + RLH} = \frac{24}{24+6} = 0.8$$

(*ii*) Mark the calculated value of $RSHF = 0.8$ on the SHF scale as point *b* and join it with point *a* which is the alignment circle (*i.e.*, 26°C *dbt* and 50% *RH*)

(*iii*) From point 2, draw a line 2-5 parallel to line *a*–*b*. The line 2-5 is the *RSHF* line.

(*iv*) The return air from the room and fresh air are mixed in the ratio 4: 1 before entering the cooling coil. Mark point 3 on the line 1-2 such that

$$\text{length 2-3} = \frac{\text{length 1-2}}{5}$$

From point 3, draw a line 3-6 in such a way that its intersection point 4 with RSHF line 2-5 is so located that

$$\text{length 4-6} = 0.2 \times \text{length 3-6}$$

where 0.2 is the by-pass factor of the coil.

The line 3-6 is the *GSHF* line

(*a*) Point 6 where the *GSHF* line meets the saturation curve is the apparatus dew point (*ADP*)

From chart : $ADP = dbt_6 = 10.8°C$

(*b*) Points 3 and 4 represents the entry and exit conditions of air for the coil.

From chart: $dbt_3 = 29°C$ and $wbt_3 = 20°C$

$dbt_4 = 15°C$ and $wbt_4 = 14°C$

(*c*) Mass flow rate of air $m_a = \dfrac{\text{total heat load}}{h_2 - h_4}$

From chart, the enthalpies at state points 2 and 4 are:

$$h_2 = 50 \text{ kJ/kg} \quad \text{and} \quad h_4 = 36.8 \text{ kJ/kg}$$

$$\therefore \quad m_a = \frac{24+6}{50-36.8} = 2.273 \text{ kJ/s} = 8182.8 \text{ kJ/hr}$$

(*d*) The recirculated air and the fresh air conform to the ratio 4 : 1 and therefore

$$\text{Mass of fresh air } m_f = \frac{2.273}{5} = \mathbf{0.4546} \text{ kg/s}$$

$$\text{Refrigeration load on the coil} = m_a (h_3 - h_4)$$

$$= 2.273 (57.4 - 36.8) = 46.82 \text{ kJ/s}$$

$$= \frac{46.82}{3.5} = \mathbf{13.38} \text{ TR}$$

The enthalpy of air at state point 3 has been read from the chart ($h_3 = 57.4$ kJ/kg).

EXAMPLE 14.5

The following data pertains to air conditioning of a hall which is to be maintained at 24 dbt and 60% relative humidity.

Outdoor condition = 38°C dbt and 28°C wbt

sensible heat load = 160000 kJ/hr

latent heat load = 40000 kJ/hr

total infiltrated air = 1200 m³/hr

apparatus dew point temperature = 10°C

quantity of recirculated air = 60% of total

If the recirculated air mixes with condition air coming from the cooling coil, compute the followings:

(a) condition of air leaving the conditioner coil and before mixing with recirculated air

(b) *condition of air before entering the hall*
(c) *mass of air entering the cooler*
(d) *mass of total air passing through the hall*
(e) *by-pass factor of the cooling coil* **(RTU)**

Solution : Refer Fig. 14.21 for line diagram of the air conditioning process and its plot on the psychrometric chart.

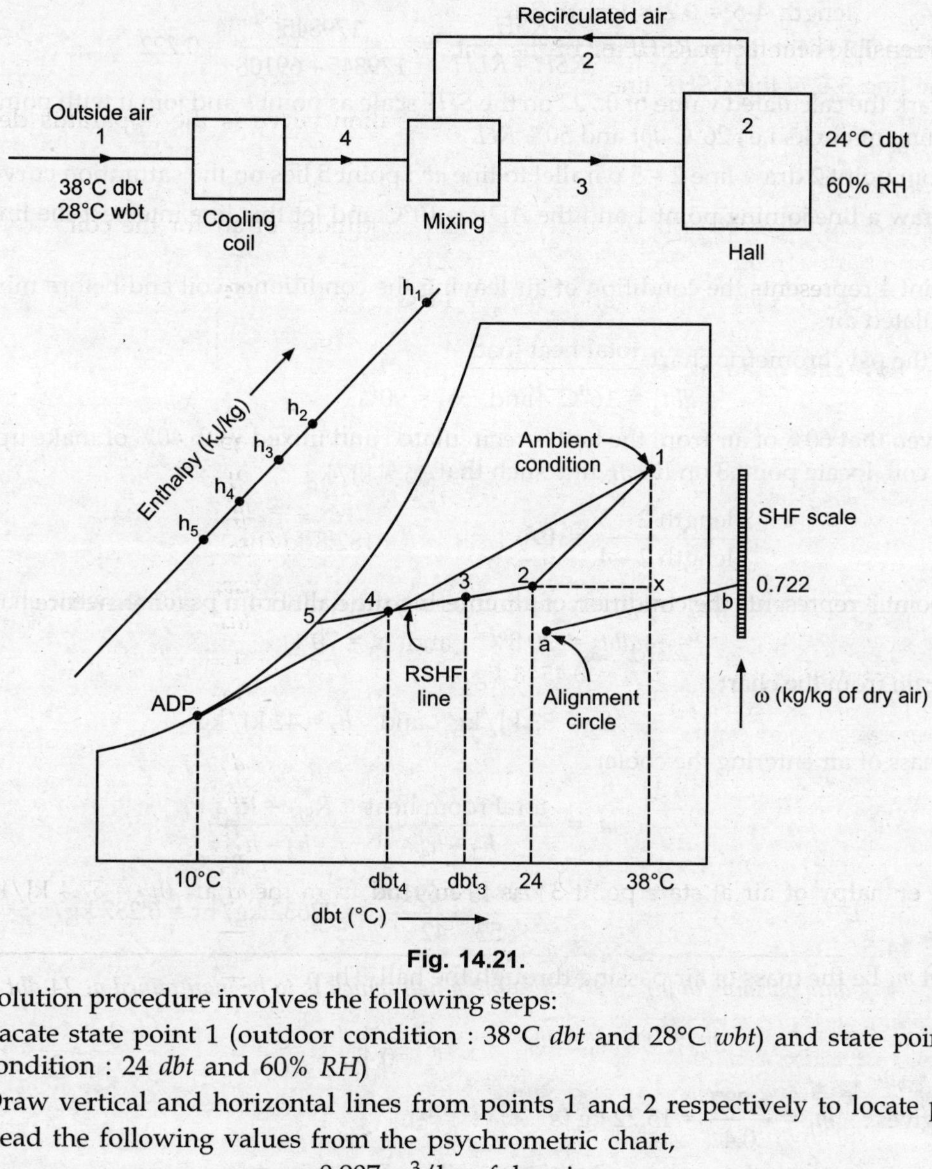

Fig. 14.21.

The solution procedure involves the following steps:

(*i*) Lacate state point 1 (outdoor condition : 38°C *dbt* and 28°C *wbt*) and state point 2 (hall condition : 24 *dbt* and 60% *RH*)

(*ii*) Draw vertical and horizontal lines from points 1 and 2 respectively to locate point *x*.

(*iii*) Read the following values from the psychrometric chart,

$$v_{s1} = 0.907 \text{ m}^3/\text{kg of dry air}$$

$$h_1 = 90 \text{ kJ/kg}, \quad h_2 = 53 \text{ kJ/kg} \quad \text{and} \quad h_x = 68 \text{ kJ/kg}$$

The mass of infiltrated air,

$$m_1 = \frac{V_1}{v_{s1}} = \frac{1200}{0.907} = 1327 \text{ kg/hr}$$

Sensible heat gain due to infiltrated air

$$= m(h_x - h_2) = 1323(68 - 53) = 19845 \text{ kJ/hr}$$

Latent heat gain due to infiltrated air

$$= m(h_1 - h_x) = 1323\ (90 - 68) = 29108 \text{ kJ/hr}$$

Then : $RSH = 19845 + 160000 = 179845$ kJ/hr

$RLH = 29108 + 40000 = 69108$ kJ/hr

$$\text{Room sensible heat factor } RSHF = \frac{RSH}{RSH + RLH} = \frac{179845}{179845 + 69108} = 0.722$$

(*iv*) Mark the calculated value of 0.722 on the *SHF* scale as point *b* and join it with point *a* which is the alignment circle, i.e., 26°C *dbt* and 50% *RH*.

(*v*) From point 2 draw line 2 – 5 parallel to line *ab* ; point 5 lies on the saturation curve.

(*vi*) Draw a line joining point 1 and the *ADP* = 10°C and let this line intersect the line 2 – 5 at point 4.

(*a*) Point 4 represents the condition of air leaving the conditioner coil and before mixing with the recirculated air.

From the psychrometric chart.

$$dbt_4 = 16°\text{C} \quad \text{and} \quad \phi_4 = 90\%$$

(*b*) Given that 60% of air from the hall is recirculated and mixed with 40% of make up air from condition coil, locate point 3 on *RSHF* line such that

$$\frac{\text{length } 2-3}{\text{length } 2-4} = 0.4$$

The point 3 represents the condition of air entering the hall. From psychrometric chart,

$$dbt_3 = 20.8°\text{C} \quad \text{and} \quad \phi_3 = 70\%$$

(*c*) Again from the chart

$$h_2 = 53 \text{ kJ/kg} \quad \text{and} \quad h_4 = 42 \text{ kJ/kg}$$

The mass of air entering the cooler

$$m_c = \frac{\text{total room heat}}{h_2 - h_4} = \frac{R_{SH} + R_{LH}}{h_2 - h_4}$$

$$= \frac{179845 + 69108}{53 - 42} = 22632 \text{ kg/hr} = 6.287 \text{ kg/s}$$

(*d*) Let m_h be the mass of air passing through the hall. Then

$$m_c + 0.6\, m_h = m_h \quad ; \quad m_h = \frac{m_c}{0.4}$$

That gives : $m_h = \dfrac{6.287}{0.4} = 15.72$ kg/s

(*e*) By-pass factor of the cooling conditioner coil is given by,

$$BPF = \frac{dbt_4 - ADP}{dbt_1 - ADP} = \frac{16 - 10}{38 - 10} = \mathbf{0.214}$$

EXAMPLE 14.6

The required inside conditions for an air-conditioned restaurant are 27°C dry bulb temperature and 21°C wet bulb temperature whilst its sensible and latent heat loads are 56 kW and 20 kW respectively. The ambient air at 38°C dbt and 27°C wbt is supplied in two ways: (i) directly into the room through ventilators and by infiltration at the rate of 30 m³/min; and (ii) conditioned by passing it through a cooling coil with 15°C apparatus dew point and thereafter mixed with 60% of recirculated air before entering the restaurant.

Draw the flow diagram for the system. Analyze the problem on psychrometric chart and determine:

(a) condition of air after the cooling coil before mixing with recirculated air,

(b) condition of air entering the restaurant,

(c) mass of fresh air entering the cooling coil,

(d) by-pass factor of the cooling coil, and

(e) refrigeration load on the cooling coil. **(WBUT)**

Solution: Refer Fig. 14.22 for the flow diagram for the system and presentation of processes involved on the psychrometric chart.

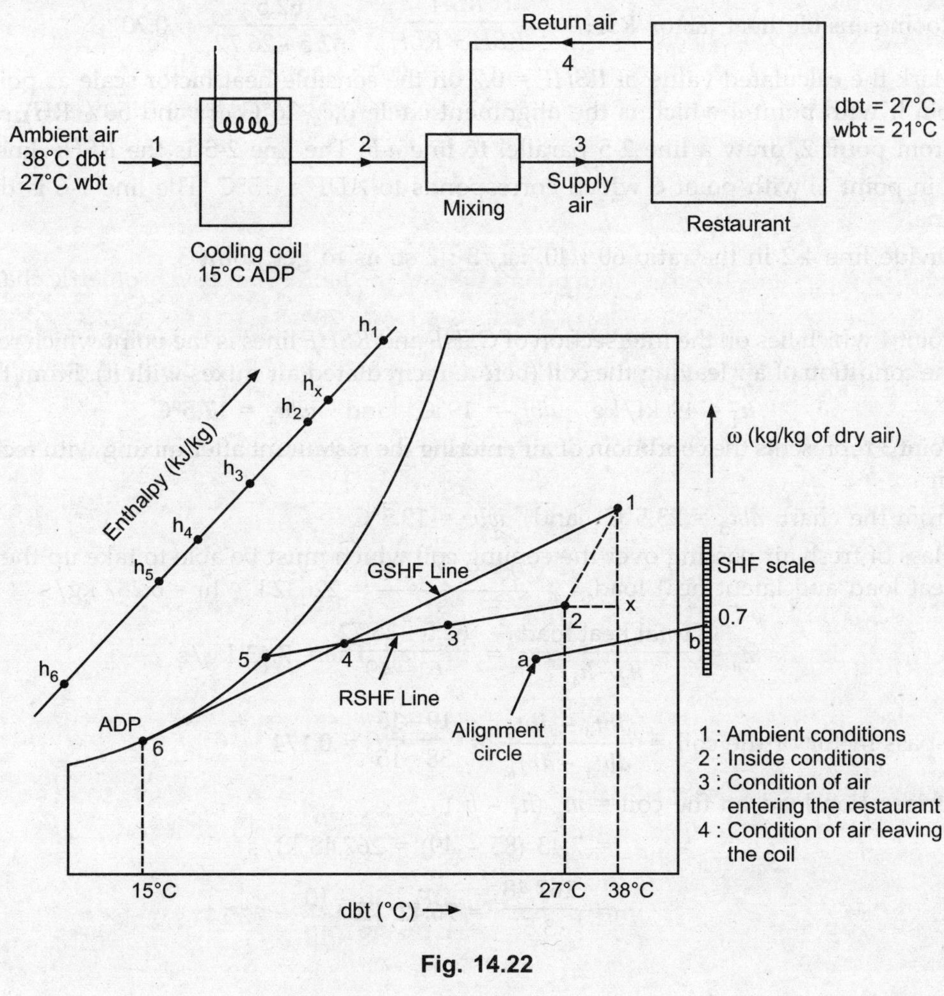

Fig. 14.22

(*i*) Locate state point 1 (ambient conditions) and state point 2 (inside conditions) on psychrometric chart. Also obtain point x at the intersection of a vertical line through 1 and horizontal line through 2. From the chart

$$h_1 = 85 \text{ kJ/kg}; \quad h_2 = 61 \text{ kJ/kg}; \quad h_x = 72.8 \text{ kJ/kg}$$
$$\text{and} \quad v_1 = 0.908 \text{ m}^3\text{/kg}.$$

Then, mass of infiltrated which directly enters the restaurant

$$m_a = \frac{V}{v} = \frac{30}{0.908} = 33.04 \text{ kg/min} = 0.551 \text{ kg/s}$$

Sensible heat load due to infiltrated air $= m_a (h_x - h_2)$
$= 0.551 (72.8 - 61) = 6.5$ kW

Latent heat load due to infiltrated air $= m_a (h_1 - h_x)$
$= 0.551 (85 - 72.8) = 6.72$ kW

∴ Room sensible heat load $RSH = 56 + 6.5 = 62.5$ kW

Room latent heat load $RLH = 20 + 6.72 = 26.72$ kW

$$\text{Room sensible heat factor } RSHF = \frac{RSH}{RSH + RLH} = \frac{62.5}{62.5 + 26.72} = 0.70$$

(*ii*) Mark the calculated value of $RSHF = 0.7$ on the sensible heat factor scale as point b and join it with point a which is the alignment circle (*i.e.*, 26°C *dbt* and 50% *RH*).

(*iii*) From point 2, draw a line 2-5 parallel to line *a-b*. The line 2-5 is the *RSHF* line.

(*iv*) Join point 1, with point 6 which corresponds to $ADP = 15$°C. The line 1-6 is the *GSHF* line.

(*v*) Divide line 4-2 in the ratio 60 : 40, *i.e.*, 3 : 2 so as to get point 3.

Now:

(*a*) Point 4 which lies on the intersection of *GSHF* and *RSHF* lines is the point which represents the condition of air leaving the coil (before recirculated air mixes with it). From the chart:

$$h_4 = 49 \text{ kJ/kg}, \quad dbt_4 = 19°\text{C} \quad \text{and} \quad wbt_4 = 17.5°\text{C}$$

(*b*) Point 3 represents the conditioin of air entering the restaurant after mixing with recirculated air.

From the chart: $dbt_3 = 23.5$°C and $wbt = 19.5$°C

(*c*) Mass of fresh air passing over the cooling coil which must be able to take up the sensible heat load and latent heat load,

$$m_a = \frac{\text{total heat load}}{h_2 - h_4} = \frac{62.5 + 26.72}{61 - 49} = 7.43 \text{ kg/s}$$

(*d*) By-pass factor of the coil $= \dfrac{dbt_4 - dbt_6}{dbt_1 - dbt_6} = \dfrac{19 - 15}{38 - 15} = \mathbf{0.174}$

(*e*) Refrigeration load on the coil $= m_a (h_1 - h_4)$
$= 7.43 (85 - 49) = 267.48$ kJ/s

$$= \frac{267.48}{3.5} = \mathbf{76.42} \text{ TR}$$

EXAMPLE 14.7

The design conditions for the summer air-conditioning system of a room are as follows:

Outside conditions	*: 35°C dbt and 28°C wbt*
Inside conditions	*: 25°C dbt and 50% RH*
Sensible and latent heat gains	*: 12 kW and 8 kW respectively.*
Cooling coil by-pass factor	*: 0.1*
Coil apparatus dew point	*: 8°C*

The return air from the room mixes with outside air before entering the cooling coil in the ratio 4 : 1, and the return air from the room also mixes with air leaving the coil in the same ratio 4 : 1. The air may be reheated, if necessary, before supplying to the room.

Draw the flow diagram for the conditioned space, represent the processes involved on psychrometric chart and determine :

(a) *supply air condition to room*

(b) *fresh air mass flow rate and*

(c) *refrigeration load on the cooling coil.*

Solution : Refer Fig. 14.23 for the flow diagram for the conditioned space and representation of the processes involved on psychrometric chart.

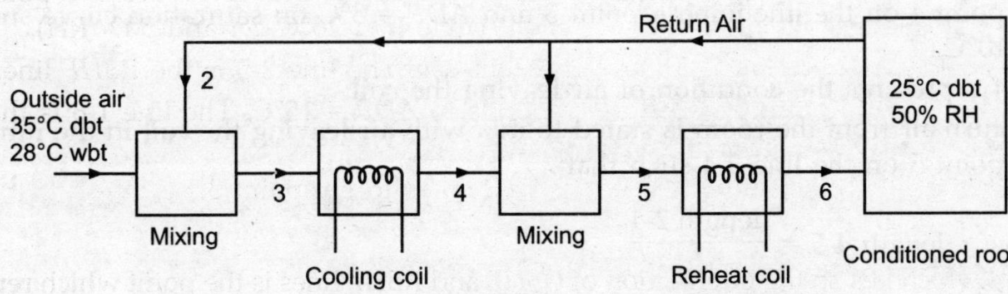

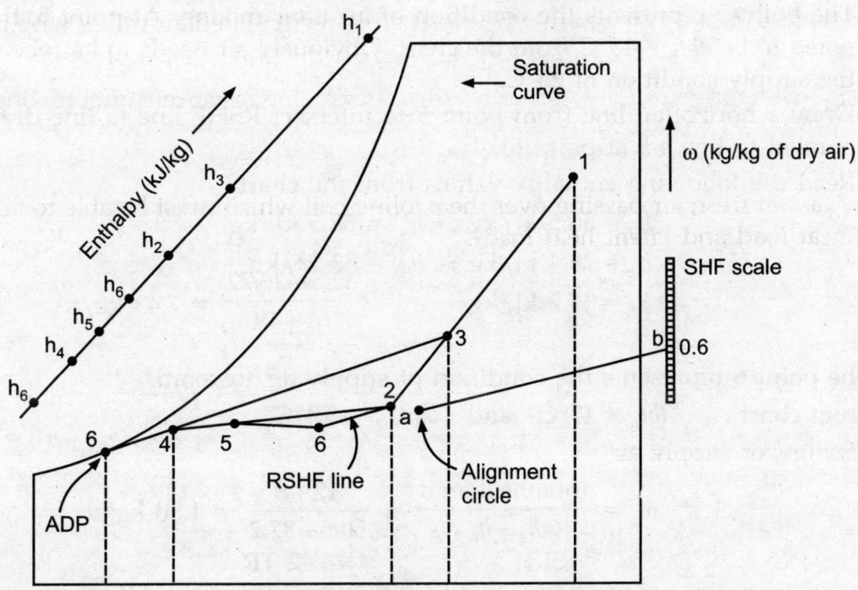

Fig. 14.23

(*i*) Locate state point 1 (outdoor conditions) and state point 2 (room conditions) on psychrometric chart.

$$\text{Room sensible heat factor } RSHF = \frac{RSH}{RSH + RLH} = \frac{12}{12+8} = 0.6$$

(*ii*) Mark the calculated value of $RSHF = 0.6$ on the *SHF* scale as point *b* and join it with point *a* which is the alignment circle (*i.e.*, 26°C *dbt* and 50% *RH*).

(*iii*) The return air from the room and the outside air are mixed in the ratio 4 : 1 before entering the cooling coil. Mark point 3 on the line 1-2 such that

$$\text{length 2-3} = \frac{\text{length 1-2}}{5}$$

The point 3 represents the condition of air before entering the cooling coil. At point 3, the temperature is noted to be $dbt_3 = 28°C$.

If the air leaves the cooling coil at temperature dbt_4, then

$$BPF = \frac{dbt_4 - ADP}{dbt_3 - ADP} \; ; \quad 0.1 = \frac{dbt_4 - 8}{28 - 8}$$

$$\therefore \quad dbt_4 = 20 \times 0.1 + 8 = 10°C$$

Mark point 4 on the line joining point 3 and $ADP = 8°C$ on saturation curve such that $dbt = 10°C$.

Point 4 represents the condition of air leaving the coil.

(*iv*) The return air from the room is stated to mix with air leaving the coil in the ratio 4 : 1. Mark point 5 on the line 2-4 such that

$$\text{length 4-5} = \frac{\text{length 2-4}}{5}$$

The point 5 represents the condition of air after mixing. At point 5, the temperature is noted to be $dbt_5 = 13°C$ from the chart. Obviously air needs to be reheated to bring it to the supply condition of 25°C.

(*v*) Draw a horizontal line from point 5 to intersect RSHF line (a line drawn from point 2 parallel to line *a-b* at point 6).

(*vi*) Read the following enthalpy values from the chart:

$$h_2 = 50.5 \text{ kJ/kg} \; ; \quad h_3 = 58.5 \text{ kJ/kg}$$
$$h_4 = 28.4 \text{ kJ/kg} \; ; \quad h_5 = 33 \text{ kJ/kg}$$
$$h_6 = 37.2 \text{ kJ/kg}$$

Now:

(*a*) The point 6 represents the condition of supply air to room.

From chart: $dbt_6 = 17°C$ and $wbt_6 = 13.4°C$

(*b*) Amount of supply air,

$$m_a = \frac{\text{total heat load}}{h_2 - h_6} = \frac{12+8}{50.5-37.2} = 1.50 \text{ kg/s}$$

$$\text{Amount of dehumidified air} = 1.5 \times \frac{4}{5} = 1.2 \text{ kg/s}$$

Then: mass of fresh air supplied $= \frac{1.2}{5} = 0.24$ kg/s = 864 kg/hr

(c) Refrigeration load = mass of dehumidified air × $(h_3 - h_4)$

$$= 1.2 \times (58.5 - 28.4) = 36.12 \text{ kJ/s} = \frac{36.12}{3.5} = \mathbf{10.32\ TR}$$

EXAMPLE 14.8

The following data refer to a bank for 100 persons in the premises:

Ambient conditions	*: 37°C dbt and 27°C wbt*
Required inside conditions	*: 22°C dbt and 60% RH*
Sensible heat	*: 110 kW*
Latent heat	*: 55 kW*
Ventilation requirement	*: 0.0047 m³/sec per person.*

If the by-pass factor for the coil is 0.15, make calculations for: (a) grand total heat, (b) effective sensible heat factor, (c) apparatus dew point, and (d) volume flow rate of humidified air.

Solution: Refer Fig. 14.24 which shows the processes involved drawn on psychrometric chart as per given data.

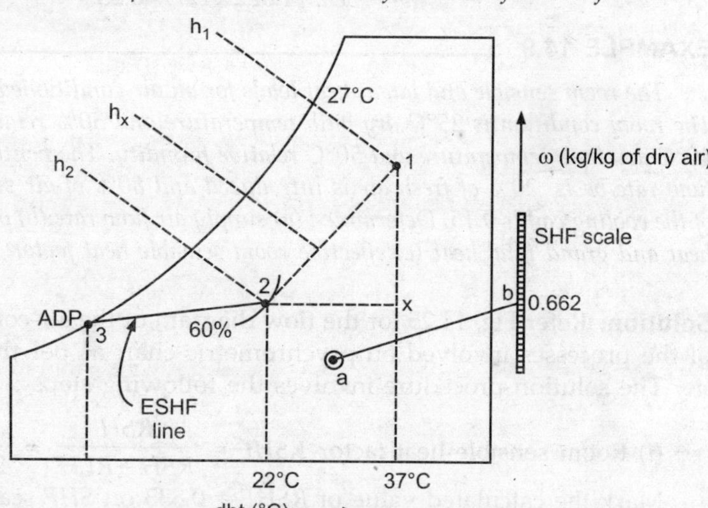

Fig. 14.24

From the chart:

$$h_1 = 85.8 \text{ kJ/kg};$$
$$v_{s_1} = 0.905 \text{ m}^3/\text{kg}$$
$$h_2 = 48 \text{ kJ/kg};$$
$$h_x = 63.5 \text{ kJ/kg}.$$
$$V_{inf} = 0.0047 \times 100 = 0.47 \text{ m}^3/\text{s}$$

Mass of air $m_a = \frac{V_{inf}}{v_{s_1}}$

$$= \frac{0.47}{0.905} = 0.519 \text{ kg/s}$$

Then: *OASH* (outside air sensible heat)

$$= m_a(h_x - h_2) = 0.519\ (63.5 - 48) = 8.045 \text{ kW}$$

OALH (outside air latent heat)

$$= m_a(h_1 - h_x) = 0.519\ (85.8 - 63.5) = 11.574 \text{ kW}$$

(a) *TSH* (total sensible heat) = 110 + 8.045 = 118.045 kW

TLH (total latent heat) = 55 + 11.574 = 66.574 kW

and *GTH* (grand total heat) = 118.045 + 66.574 = 184.619 kW

(b) Effective room sensible heat

$$ERSH = RSH + BPF \times OASH$$
$$= 110 + 0.15 \times 8.045 = 111.207 \text{ kJ/kg}$$

Effective room latent heat

$$ERLT = RLH + BPF \times OALH$$
$$= 55 + 0.15 \times 11.574 = 56.736 \text{ kJ/kg}$$

Then: Effective sensible heat factor (*ESHF*)

$$= \frac{ERSH}{ERSH + ERLH} = \frac{111.207}{111.207 + 56.736} = 0.662$$

(*c*) Mark the calculated value of *ESHF* = 0.662 on the *SHF* scale as point *b* and join it with point *a* which is the alignment circle (*i.e.*, 26°C *dbt* and 50% *RH*)

From point 2, draw a line parallel to line *a-b* and let it meet the saturation curve at point 3. The line 2-3 is the *ESHF* line.

Point 3 represents the apparatus dew point and its value as red from the chart is

$$ADP = 9.5°C$$

(*d*) Let *V* = Quantity of air supply in m^3/min

Then: $$ERSH = m\,c_p\,(dbt_2 - ADP)\,(1 - BPF) \text{ kJ/s}$$

or $$111.207 = \frac{1.2 \times V}{60} \times 1.022\,(22 - 9.5)\,(1 - 0.15) \qquad (\because m = \rho V)$$

$$\therefore \quad V = \frac{111.207 \times 60}{1.2 \times 1.022 \times 12.5 \times 0.85} = 512.06 \text{ m}^3/\text{min}$$

EXAMPLE 14.9

The room sensible and latent heat loads for an air-conditioned space are 25 kW and 5 kW respectively. The room condition is 25°C dry bulb temperature and 50% relatively humidity. The outdoor cndition is 40°C dry bulb temperature and 50°C relative humdity. The ventilation requirement is such that on mass flow rate basis, 20% of fresh air is introduced and 80% of air supply is recirculated. The by-pass factor of the cooling coil is 0.15. Determine : (a) supply air flow rate (b) outside air sensible heat, outside air latent heat and grand total heat (c) effective room sensible heat factor.

(MRIU)

Solution: Refer Fig. 14.25 for the flow diagram of the air-conditioning system and representation of the processes involved on psychrometric chart as per the given data

The solution procedure involves the following steps :

(*i*) Room sensible heat factor $RSHF = \frac{RSH}{RSH + RLH} = \frac{25}{25+5} = 0.833$

Mark the calculated value of *RSHF* = 0.833 on *SHF* scale as point *b* and join it with point *a* which is the alignment circle, *i.e.*, 26°C *dbt* and 50% *RH*.

(*ii*) Locate state point 1 (outdoor conditions; 40°C *dbt* and 50% *RH*) and state point 2 (room conditions : 25° *dbt* and 50% *RH*)

(*iii*) From point 2, draw line 2–5 parallel to *a–b* ; point 5 lies on the saturation curve

The line 2–5 is called RSHF line

(*iv*) Since 20% fresh or outside air is mixed with 80% of supply air, mark point 3 on the line 1–2 such that

$$\frac{\text{length } 2\text{–}3}{\text{length } 1\text{–}2} = \frac{20}{100} = 0.2$$

The point 3 represents the condition of air entering the coil after mixing process.

(*v*) Through point 3, draw the line 3–6 (point 6 lies on saturation curve) and let it intersect the RSHF line in a way that

$$\frac{\text{length } 4-6}{\text{length } 3-6} = BPF = 0.15$$

The line 3–6 is the GSHF line.

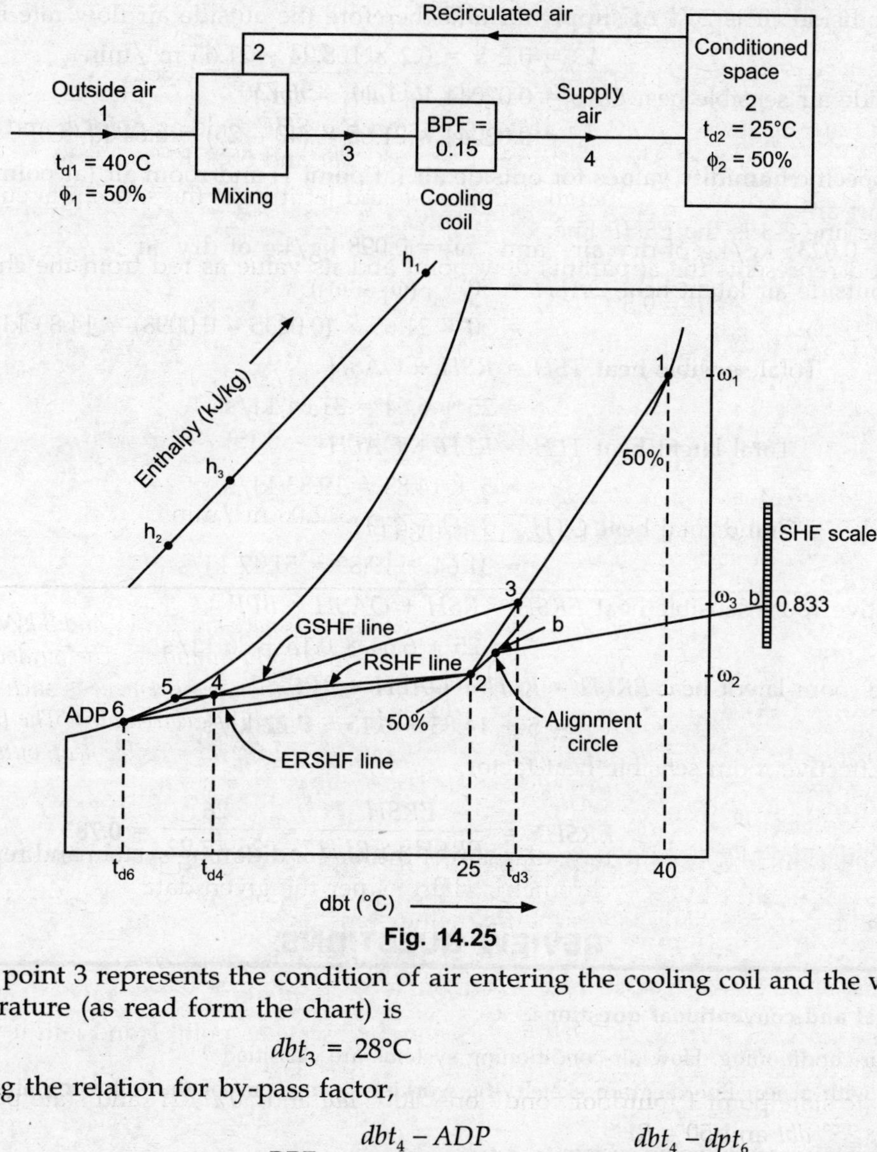

Fig. 14.25

(*v*) The point 3 represents the condition of air entering the cooling coil and the value of dry bulb temperature (as read form the chart) is

$$dbt_3 = 28°C$$

Invoking the relation for by-pass factor,

$$BPF = \frac{dbt_4 - ADP}{dbt_3 - ADP} \; ; \quad 0.15 = \frac{dbt_4 - dpt_6}{28 - dpt_6}$$

Here dbt_4 is dry bulb temperature of air leaving the cooling coil and dpt_6 is the temperature at apparatus dew point (ADP).

The following values are obtained by trial and carrier

$$dbt_4 \simeq 13.7°C \text{ and } dpt_6 = 11.2°C$$

(*a*) The room sensible heat load is give as

$$RSH = 0.02044\ V(dbt_2 - dbt_4)\ , \quad 25 = 0.02044\ V(25 - 13.7)$$

∴ Volume flow rate of air,

$$V = \frac{25}{0.02044 \times (25 - 13.7)} = \mathbf{108.24}\ m^3/kg$$

(*b*) The outside air is 20% of supply air and therefore the outside air flow rate is

$$V_0 = 0.2\ V = 0.2 \times 108.24 = 21.65\ m^3/min$$

Outside air sensible heat $OAS = 0.02044\ V_0\ (dbt_1 - dbt_2)$

$$= 0.02044 \times 21.65 \times (40 - 25) = 6.64\ kJ/s$$

(*c*) The specific humidity values for outside air (at point 1) and room air (at point 2) as read from the chart are :

$$\omega_1 = 0.0235\ kg/kg\ \text{of dry air} \quad \text{and} \quad \omega_2 = 0.098\ kg/kg\ \text{of dry air}$$

Then : outside air latent heat $OALH = 50\ V_0\ (\omega_1 - \omega_2)$

$$= 50 \times 21.65 \times (0.0235 - 0.0098) = 14.83\ kJ/s$$

(*d*) Total sensible heat $TSH = RSH + OASH$

$$= 25 + 6.64 = 31.64\ kJ/s$$

Total latent heat $TLH = RLH + OALH$

$$= 5 + 14.83 = 19.83\ kJ/s$$

Grand total heat $GTH = TSH + TLH$

$$= 31.64 + 19.83 = 51.47\ kJ/s$$

(*e*) Effective room sensible heat $ERSH = RSH + OASH \times BPF$

$$= 25 + 6.64 \times 0.15 \simeq 26\ kJ/s$$

Effective room latent heat $ERLH = RLH + OALH \times BPF$

$$= 5 + 14.83 \times 0.15 \simeq 7.22\ kJ/s$$

Then : Effective room sensible heat factor,

$$ERSHF = \frac{ERSH}{ERSH + ERLH} = \frac{26}{26 + 7.22} = 0.783$$

REVIEW QUESTIONS

A. Conceptual and conventional questions.

1. Define air-conditioning. How air-conditioning systems are classified ?
2. Explain, with a neat line diagram sketch, the working of any one of the following air-conditioning systems:
   (*i*) Summer air-conditioning system
   (*ii*) Winter air-conditioning system
   (*iii*) Year-round air-conditioning system.
3. Compare and contrast the central air-conditioning system and unitary air-conditioning system.
4. Write short notes on:
   (*i*) All-air air-conditioning system
   (*ii*) All-water air-conditioning system
   (*iii*) Air-water air-conditioning system.
5. (*a*) Sketch the schematic of VAV system. Label the different parts and describe its working.
   (*b*) Why is a static pressure controller used in such a system ?
6. Draw the neat line diagram sketch of a constant volume system and point out the salient aspects of such a system.
7. What is meant by dual duct and multi-zone system ?

8. Explain the following terms

(*a*) RSHF and RSHF line

(*b*) GSHF and GSHF line

With reference to these lines, which state of air gives the ideal supply conditions to the room.

9. What do you mean by alignment circle? Mention its utility in psychrometry.

10. The following data refer to summer air-conditioning of a certain premises:

Room design conditions	:	*dbt* = 27°C and *RH* = 60%
Ambient conditions	:	*dbt* = 40°C and *wbt* = 30°C
Room cooling load	:	1,40,000 kJ/hr
By-pass factor for the cooling coil	:	0.14

If the apparatus dew point is 4.5°C lower than that of the dew point temperature of corresponding to room design conditions, make calculations for the following parameters:

(*a*) ADP,

(*b*) Volume of the air supplied to the room and

(*c*) Refrigeration capacity of the cooling coil. **(Ans. 14°C, 103.8 m^3/min, 15.6 TR)**

11. The conditioning plant of a hall consists of fresh air intakes, a cooling coil followed by a mixing chambers for the cooled fresh air and recirculated room air and a supply fan. The cooling coil handles all fresh air and has an ADP of 5°C. The other data is as follows:

Inside conditions	:	20°C *dbt* and 60% *RH*
Outside conditions	:	40°C *dbt* and 26° *wbt*
Heat gains	:	*RSH* = 19.5 kW ; *RLH* = 6 kW
Quantity of infiltrated air	:	30 m^3/min.

If 60% of the total air required is recirculated and mixed with the conditioned air after the cooling coil, determine the following:

(*a*) the condition of air leaving the coil and before entering the hall

(*b*) volume of fresh air passing through air-conditioner

(*c*) by-pass factor of conditioner coil, and

(*d*) refrigeration load on conditioner coil in tons of refrigeration.

(Ans. 8°C, 90% and 13°C, 72% ; 127 m^3/min ; 0.09, 38.7 TR)

12. The design of the summer air-conditioning system for a room is based on the following information:

Outdoor conditions	:	35°C *dbt* and 24°C *wbt*
Indoor conditions	:	24°C *dbt* and 50% *RH*

Room sensible and latent heat gains: 58 kW and 14.5 kW respectively : 55 m^3/min

If the system employs a cooling coil with a by-pass factor of 0.15, determine:

(*a*) out air load and grand total heat

(*b*) effective room sensible heat factor

(*c*) coil apparatus dew point, and

(*d*) dehumidified air quantity. **(Ans. 26.95 kW, 99.45 kW, 0.786, 10°C, 247.9 m^3/min)**

B. Fill in the blanks with appropriate word/words

1. An assembly of different components that produces specified condition of air within a space is called .... system.
2. For human comfort in winter, the air needs to be so conditioned that its relative humidity is not less than .... percent.
3. In winter air conditioning, the air is .... and ....
4. In .... air conditioning system, cooling is done partly by air coming from central air conditioning system and partly by chilled water flowing through coils.
5. The .... air conditioning systems consists of a single unit for the conditioning of air in a single conditioned space.

6. Summer air conditioning essentially involves reducing the air temperature and humidity by the process of .... and ....
7. The .... air conditioning system is adopted when different zones in a building are to be air conditioned.
8. A point on the psychrometric chart representing 26 *dbt* and 50% *RH* is referred to as the .... point.
9. The slope of .... line gives the ratio of the room sensible heat to the room total heat.
10. A point where the R_{SHF} line intersects the saturation line is the .... for the room.
11. .... is defiend as the ratio of the effective room sensible heat (*ERSH*) ot the effective room total heat (*ERTH*).
12. The effective room sensible heat factor line is obtained by joining the room design condition point with the .... point.

Answers

1. air conditioning 2. 40% 3. heated, humidified 4. air-water 5. unitary 6. cooling, dehumidified 7. central 8. alignment circle 9. room sensible heat factor 10. apparatus dew point 11. *ERSHF* 12. apparatus dew point

C. Multiple choice questions.

1. For human comfort in summer, the air needs to be so conditioned that its relative humidity is not less than .... percent
   (*a*) 40 (*b*) 60 (*c*) 75 (*d*) 90
2. In winter air conditioning, the air is
   (*a*) heated and dehumidified (*b*) heated and humidified
   (*c*) cooled and dehumified (*d*) cooled and humidified
3. For room sensible heat = 50 kW and room latent heat = 25 kW, the room sensible heat factor is
   (*a*) 0.33 (*b*) 0.50 (*c*) 0.67 (*d*) 0.75
4. If the by-pass factor of each depth in a coil is 0.8, then the combined by-pass factor of a 4 depth coil would be
   (*a*) 0.41 (*b*) 0.62 (*d*) 0.8 (*d*) 095
5. The alignment circle is marked on the psychrometric chart at
   (*a*) 20°C *dbt* and 60% *RH*
   (*b*) 26°C *dbt* and 50% *RH*
   (*c*) 30°C *dbt* and 75% *RH*
   (*d*) 15°C *wbt* and 50% *RH*
6. The state of supply air from the cooling air is represented by a point of intersection of .... with ....
   (*a*) *RSHF* line, saturation curve
   (*b*) *GSHF* line, saturation curve
   (*c*) *RSHF* line, *GSHF* line
   (*d*) point of intersection divides the *RSHF* line in proportion of *BPF* and (1 – *BPF*) where *BPF* is the by-pass factor of the coil.

Answers

1. (*b*) 2. (*b*) 3. (*c*) 4. (*a*) 5. (*b*) 6. (*c*)

Chapter – 15

Cooling Load Estimation

Learning objectives : A study of the subject matter presented in this chapter will enable the students/readers to

- *understand the meaning of the term load in the context of air-conditioning and make distinction between sensible and latent cooling loads*
- *explain the heat sources which contribute to summer air conditioning systems*
- *analyse, with the help of psychrometric chart, the processes involved in the air conditioning systems*

In the context of air-conditioning, the term load refers to the heat removed in summer or the heat added in winter to any space/room in a given time. The heat removed (in summer) is called the ***cooling load***, and the heat added (in winter) is termed as the ***heating load***. Considering air-conditioning primarily as the science of cooling, we are essentially concerned with the cooling load which may be defined as :

- The amount of heat that must be removed from a building to maintain a comfortable temperature for its occupants.
- The total amount of heat energy that is required to be pumped out from the system by a cooling mechanism in a unit time, equal to the rate at which heat is generated by people, machinery and processes, plus the net flow of heat into the system not associated with the cooling machinery.

To bring down the temperature of space/room to the required comfort conditions, attention needs to be directed to the nature of heat sources which contribute to the air-conditioning system used for summer. That helps to determine the capacity of refrigeration plant or the specifications of air-conditioning equipment required to maintain the room/space at design conditions. Such details are used for comfort air-conditioning, industrial air-conditioning, product storage and product freezing systems.

15.1. SPACE HEAT GAIN

The space heat gain is the rate at which heat enters into and/or is generated within a space/room. The space gains heat from external as well as internal sources, and is generally classified by the mode in which it enters the space, and whether it is sensible or latent heat gain.

15.1.1. Mode of Entry

The heat gain from external heat sources occurs by way of

(*i*) heat conduction through exterior walls and roofs

(*ii*) solar radiation through transparent sources

(*iii*) heat transfer as a result of ventilation and infiltration of outside air

The internal heat sources are

(*a*) heat conduction through interior partitions, ceiling and floors

(*b*) heat generated within the space by occupants, lights, appliances and products

The various components of heat gains from external and internal sources have been shown schematically in Fig. 15.1.

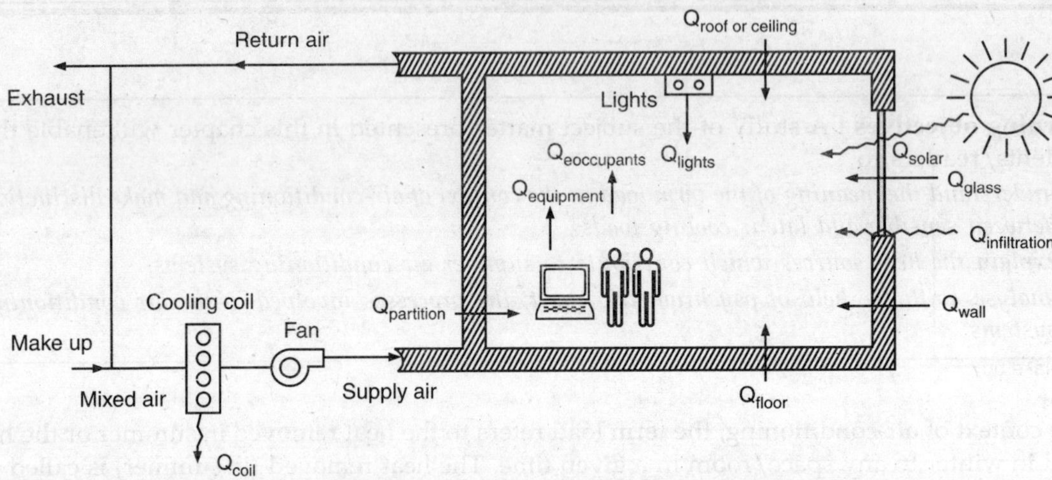

Fig. 15.1. Heat gains in space from internal and external sources

15.1.2. Sensible and Latent Heat Gain

The sensible cooling load refers to the dry bulb temperature, and the latent cooling load refers to wet bulb temperature of the building.

The ***sensible heat gain*** manifests when the heat enters the enclosed space by any one or all the modes of heat transfer (conduction, convection and radiation) and is indicated by the sensible temperature rise. Important components of sensible heat gain are :

(*i*) heat transmitted by conduction through the exterior walls, floors, ceilings, doors and windows due to temperature difference across the two sides.

(*ii*) heat transferred into the enclosed space by solar radiation in the form of

(*a*) direct heat flow through doors, windows and ventilators, and

(*b*) heat conducted and radiated through the roofs, walls and glass windows exposed to solar radiation.

(*iii*) heat liberated by occupants in the building.

(*iv*) heat given off by equipment and appliances operated in the summer.

(*v*) heat given off by electrical fittings such as lamps and tubes.

(*vi*) heat brought in by infiltration (leaking) air passing through doors, windows/ventilators, cracks/openings etc.

(*vii*) heat gain by ducts carrying the conditioned air when the duct passes through unconditioned space in the building.

The ***latent heat gain*** manifests without temperature rise and occurs essentially due to addition of water vapours to the air in the enclosed space. The following aspects contribute towards latent heat gain :

(*i*) heat gain due to moisture in outside air entering by infiltration

(*ii*) heat gain due to condensation of moisture from the occupants.

(*iii*) heat gain due to condensation of moisture from products or processes such as cooking, baking and boiling etc.

(*iv*) heat gain due to moisture that passes directly into the conditioned space through permeable walls where the water vapour pressure is higher.

A brief summary of the sensible and latent heat (moisture) gains has been indicated in Fig. 15.2.

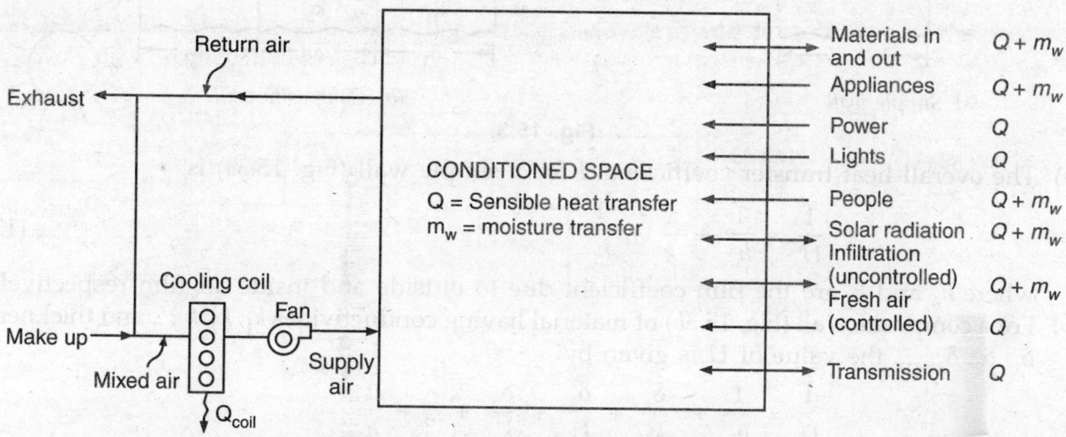

Fig. 15.2. Sensible and latent heat (moisture) gains

The total heat gain is then the sum of sensible and latent heat loads, and the extra load added by the following sources :

(*a*) fresh air taken from outside adds to the sensible and latent heat loads.

(*b*) sensible heat added to fresh air from the fan work, if the fan is installed before the air-conditioner.

The total heat so gained then contributes the cooling load for the air-conditioning and refrigeration equipment. This load determines the size of equipment which will operate to maintain the summer temperature conditions that are :

(*i*) comfortable to persons, and

(*ii*) required by a product or processes

The equipment cost will be very high if it is designed for maximum cooing loads, and there will be long periods of discomfort if design is on the basis of average loads. Hence, a value which gives physical comfort and also cares for economics of the system.

15.2. SENSIBLE HEAT GAIN THROUGH BUILDING STRUCTURE

The heat gain through building structures such as walls, ceiling, floors, doors and windows is a sensible load. Heat is conducted through the building due to difference of temperature between the inside and outside conditions and essentially depends on wall thickness (δ), cross-sectional area of wall (*A*) and thermal conductivity (*k*) of the wall material. Presuming steady state heat condition, the heat gain through a wall is worked out by invoking the heat transfer equation :

$$Q = UA\,(T_0 - T_i) \qquad ...(15.1)$$

where *A* is cross-sectional area of the wall, *U* is overall heat transfer coefficient, T_0 is the outside ambient temperature and T_i is the inside temperature at the design condition.

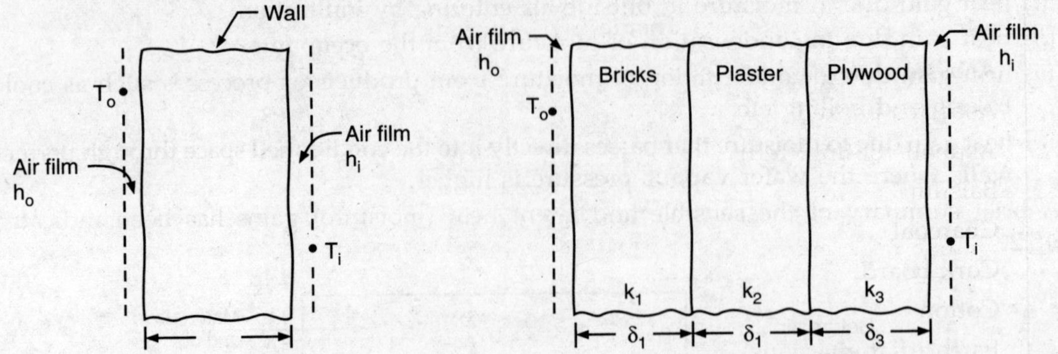

(a) Simple wall (b) Composite wall.

Fig. 15.3.

(*a*) The overall heat transfer coefficient U for a simple wall (Fig. 15.3*a*) is

$$\frac{1}{U} = \frac{1}{h_0} + \frac{x}{k} + \frac{1}{h_i} \quad \text{... (15.2)}$$

where h_0 and h_i are the film coefficient due to outside and inside air film respectively.

(*b*) For a composite wall (Fig. 15.3*b*) of material having conductivities k_1, k_2, k_3, .. and thicknesses $\delta_1, \delta_2, \delta_3, ...,$ the value of U is given by

$$\frac{1}{U} = \frac{1}{h_0} + \frac{\delta_1}{k_1} + \frac{\delta_2}{k_2} + \frac{\delta_3}{k_3} + ... + \frac{1}{h_i}$$

(*c*) For a composite wall with air spaces

$$\frac{1}{U} = \frac{1}{h_0} + \frac{\delta_1}{k_1} + \frac{1}{k_a} + \frac{\delta_2}{k_2} + \frac{\delta_3}{k_3} + ... + \frac{1}{h_i}$$

where k_a is thermal conductance of air space.

The values of thermal conductivity, film or surface coefficients and overall heat transfer coefficients for the commonly used building material under different operating conditions are given in Tables 15.1. to 15.4.

Table 15.1 Thermal conductivity of building and insulating materials

S.No.	Material	Thermal conductivity W/mK
		Building Materials
1.	Asbestos building board	4.70
2.	Brick-low density	8.65
3.	Brick-high density	15.90
4.	Cement mortar	20.82
5.	Cement plaster	20.82
6.	Cinder blocks	8.50
7.	Concrete	20.80
8.	Gypsum plaster	5.72
9.	Lime stone	18.73
10.	Sand and gravel	21.85
11.	Wood	1.72

	Insulating Materials	
1.	Asbestos loose	1.85
2.	Asbestos packed	2.81
3.	Asphalt roofing (felt)	1.21
4.	Balsum	0.60
5.	Charcoal	0.62
6.	Cork board	0.48
7.	Cotton	0.67
8.	Earth (diatomaceous)	0.46
9.	Glass wool (high grade)	0.38
10.	Glass wool (commercial)	0.38 to 0.48
11.	Rock wool	0.44
12.	Rubber	0.31
13.	Saw dust	0.71

Table 15.2. Film or surface coefficient for air film and air spaces

S.No.	Direction of heat flow and air/wind velocity	Surface position	Thermal conductance W/m² K
	Air film (surface)		
1.	Still air – heat flowing up	horizontal	f_i = 9.25
		sloping 45°	9.10
2.	Still air – heat flowing down	horizontal	6.13
		sloping 45°	7.5
3.	Still air – heat flow horizontal	vertical	8.30
4.	Wind at 24 km/hr	Any position	f_0 = 34.0
	Wind at 12 km/hr	Any position	= 22.7
	Air spaces bounded by structural material		
5.	Heat flowing up	19 to 100 mm horizontal	6.7
		19 to 100 mm with 45° slope	6.3
6.	Heat flowing down	19 mm horizontal	5.56
		38 mm horizontal	4.95
		100 mm horizontal	4.60
		19 to 100 mm with 45° slope	5.50
7.	Heat flowing horizontal	19 to 100 mm vertical	5.85

Note : These values of thermal conductance are for a mean temperature of 10° to 32°C.

Table 15.3 Conductance of building materials

Material	Conductance W/K
Plaster board 9 mm thick	21.16
Plaster board 12.5 mm thick	16.15
Roofs	
Asphalt, composition or prepared roofing	36.84
Shingles, asbestos	34.10
Shingles, asphalt	36.64
Shingles, slate	59.00
Metal lath plaster, total thickness 12.5 mm	24.95
Wood lath plaster, total thickness 18 mm	14.19
For sheathing 25 mm and building paper	4.87
For sheathing 25 mm building paper and yellow pine lap siding	2.84
For sheathing 25 mm building paper and stucco	4.64
Hollow clay tile : 10 cm thick	5.66
Hollow clay tile : 15 cm thick	3.65
Hollow clay tile : 20 cm thick	3.39
Hollow clay tile : 25 cm thick	3.30
Hollow clay tile : 30 cm thick	2.27
Hollow clay tile : 40 cm thick	1.76
Concrete block (sand gravel) : 20 cm thick	5.67
Concrete block (sand gravel) : 30 cm thick	4.55
Concrete block (sand gravel) : 20 cm thick	3.41
Concrete block (sand gravel) : 30 cm thick	3.01

Table 15.4 Overall coefficient of heat transmission (U) for structure with a wind velocity 24 km/for outside

Structure	Overall coefficient of heat transfer (U)
Brick Wall	
: 20 cm bare	2.83 W/m^2 K
: 20 cm with plaster one side on brick	2.60
: 20 cm with plaster one side on metal lath-furred	1.80
: 40 cm bare	1.60
: 40 cm with plaster one side on metal lath-furred	1.19
Hollow Tile Stucco Exterior	
: 20 cm bare	2.26 W/m^2 K
: 20 cm with plaster on metal lath-furred	1.60
: 30 cm bare	1.69
: 30 cm with plaster on metal lath-furred	1.26
Cinder Blocks	
: 20 cm bare	2.37 W/m^2 K
: 20 cm with plaster one side on metal lath-furred	1.60
: 20 cm bare (concrete block)	3.19
Wood shingle frame construction with plaster on wood lath	1.41 W/m^2 K
Stucco frame construction with plaster or wood lath	1.69
Brick veneer frame construction with plaster on wood lath	1.54

EXAMPLE 15.1

A double glazed window is made of 2 glass panes of 6 mm thick with an air gap of 6 mm between them. Assuming that the air layer is stagnant and only conduction is involved, determine the overall heat transfer coefficient and heat flow from outside ambient to indoor. Take the following data :

Ambient and inside temperatures = 40°C and 22°C

Outside and inside heat transfer coefficients = 9 W/m² K and 15 W/m² K

Thermal conductivity of glass and air = 1.4 W/mK and 0.025 W/mK.

(b) How the calculated parameters would be affected if the window is made of single glass 12 mm thick.

Solution : Refer Fig. 15.4 and consider unit area.

Total thermal resistance,

$$R_t = \frac{1}{h_0 A} + \frac{\delta_1}{k_g A} + \frac{\delta_2}{k_a A} + \frac{\delta_3}{k_g A} + \frac{1}{h_i A}$$

$$= \frac{1}{15 \times 1} + \frac{0.006}{1.4 \times 1} + \frac{0.006}{0.025 \times 1} + \frac{0.006}{1.4 \times 1} + \frac{1}{9 \times 1}$$

$$= 0.067 + 0.00428 + 0.24 + 0.00428 + 0.111$$

$$= 0.426 \text{ m}^2 \text{ K/W}$$

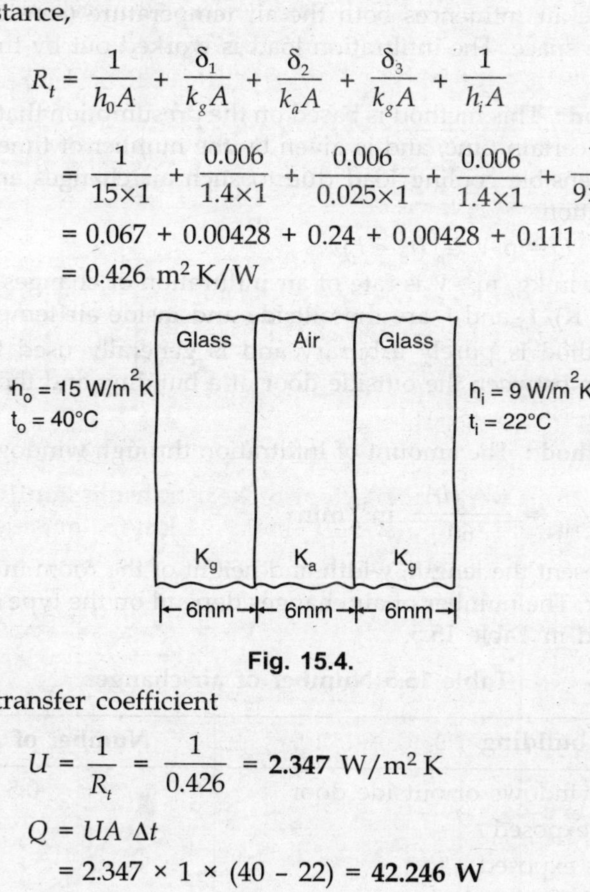

Fig. 15.4.

Then overall heat transfer coefficient

$$U = \frac{1}{R_t} = \frac{1}{0.426} = \mathbf{2.347} \text{ W/m}^2 \text{ K}$$

Heat transfer $Q = UA\,\Delta t$

$$= 2.347 \times 1 \times (40 - 22) = \mathbf{42.246 \text{ W}}$$

$$(b)\ R_t = \frac{1}{15} + \frac{0.012}{1.4} + \frac{1}{9}$$

$$= 0.067 + 0.00857 + 0.111 = 0.1866 \text{ m}^2 \text{ K/W}$$

$$U = \frac{1}{R_t} = \frac{1}{0.1866} = 5.36 \text{ W/m}^2 \text{ K}$$

$$Q = UA\,\Delta T = 5.36 \times 1 \times (40 - 22) = \mathbf{96.48 \text{ W}}$$

15.3. INFILTRATION LOAD

Uncontrolled entry of unconditioned outside air directly into the building

The infiltration load refers to the heat carried by the outside air that enters the air-conditioned space through window cracks and through doors when opened. The infiltration is caused by

(*i*) pressure difference on the two sides of the widows and doors,

(*ii*) wind velocity and its direction,

(*iii*) buoyancy, *i.e.*, difference in densities due to temperature difference between the inside and outside air.

Further, the volumetric flow rate of infiltration depends on the length and width of cracks, quality of construction, age of building and its deterioration.

The entry of outside air influences both the air temperature (sensible load) and humidity level (latent load) in the space. The infiltration load is worked out by the air change and crack length method.

- **Air change method :** This method is based on the presumption that infiltration air replaces the air in the room in a certain time, and is given by the number of times air changes to fill the volume of room. The sensible cooling load due to such air changes and other infiltrations is computed from the relation.

$$Q = \rho\, V\, C_p\, (t_0 - t_i) \qquad \text{... (15.3)}$$

where ρ is density of air in kg/m^3, V is rate of air infiltration or changes in m^3/hr, C_p is specific heat of air (1.025 kJ/kg K), t_0 and t_i are the outside and inside air temperatures.

The air change method is purely arbitrary and is generally used for reception halls and vestibules (enclosed area between the outside door of a building and the inside door) where air changes are frequent.

- **Crack length method :** The amount of infiltration through windows and walls is taken as

$$V_{inf} = \frac{LWHA_c}{60} \text{ m}^3/\text{min} \qquad \text{... (15.4)}$$

where L, W and H represent the length, width and height of the room in metres, and A_c denotes the air changes per hour. The number of air changes depend on the type of room and exposures, and have been indicated in Table 15.5.

Table 15.5 Number of air changes

Type of room or building	Number of air changes per hour
• Room : with no windows or outside door	0.5 to 0.75
: one wall exposed	1.0
: two walls exposed	1.5
: three walls exposed	2.0
: four walls exposed	2.0
• Entrance halls	2.0 to 3.0
• Reception halls	2.0
• Bath rooms	2.0

Further, there also occurs infiltration when a person passes through a door that leads to outside or to an unconditioned space. The amount of this door infiltration is given in Table 15.6 and this has to be added to the infiltration air through windows and walls for obtaining the total building infiltration. The values indicated in Table 15.6 pertain to infiltrated air for 1.8 metres revolving door in m^3 per person per passage. Further, these values are based on the assumption that there is no wind pressure and the swinging doors are in use in one-way only. For a 0.9 m swinging door, the air infiltration is generally taken as 3 m^3/min.

Table 15.6. Air infiltration through a door

Usage of door	Air infiltration for 1.8 m revolving door per person per passage	
	Freely revolving door	Brake equipped door
• Infrequency	2.5 m^3	2.0 m^3
• Average	2.0	1.75
• Heavy	1.5	1.25

EXAMPLE 15.2

A room measuring 20 m × 5 m × 4 m high is to be maintained at dry bulb temperature of 25°C and relative humidity 75%. The ambient air is at dbt = 32°C and φ = 60%. Determine the infiltration load if the infiltration amounts to two air changes per hour.

Solution : Volume of room = 20 × 5 × 4 = 400 m^3

$$\text{Volume of infiltration air } V_{inf} = 2 \times 400 \text{ m}^3/\text{hr}$$

$$= \frac{2 \times 400}{3600} = 0.222 \text{ m}^3/\text{s}$$

Locate on psychrometric chart the state point 1 corresponding to ambient conditions (*dbt* = 32°C and φ = 60%) and state point 2 corresponding to inside conditions (*dbt* = 25°C and φ = 75%). Read therefrom

$$h_1 = 77.5 \text{ kJ/kg of dry air} ; v_1 = 0.885 \text{ m}^3/\text{kg}$$

$$h_2 = 63 \text{ kJ/kg of dry air}$$

Then $$Q_{inf} = \frac{V_{inf}}{v_1}(h_1 - h_2) = \frac{0.222}{0.885}(77.5 - 63) = 3.637 \text{ kJ/s}$$

$$= \mathbf{3.637 \text{ kW}}$$

EXAMPLE 15.3

Calculate the infiltration load for a room of dimensions 15 m × 8 m × 4 m high. The ambient and inside temperatures are 40°C and 22°C respectively. It may be presumed that for the given volume of the room and storage period, the system has 3.1 air changes per 24 hours long. Proceed to calculate the cooling load when the ambient and inside wet bulb temperatures are 17°C and 10°C respectively. Comment on the results.

Assume C_p = 1.025 kJ/kg K and atmospheric pressure = 1.013 bar

Solution : Volume of room = 15 × 8 × 4 = 480 m^3

$$\text{Volume of infliltrated air} = \frac{3.1 \times 480}{24} = 62 \text{ m}^3/\text{hr}$$

$$\text{Density of air } \rho = \frac{p}{RT} = \frac{1.013 \times 10^5}{287 \times (273 + 40)} = 1.128 \text{ kg/m}^3$$

$$\text{Infiltration load } Q_{inf} = \rho V C_p (T_a - T_i)$$

$$= 1.128 \times 62 \times 1.025\ (40 - 22) = 1290 \text{ kJ/hr}$$

(*b*) From psychrometric chart,

At $dbt = 40°C$ and $wbt = 17°C$ (ambient conditions)

$h_a = 48$ kJ/kg and $v_a = 0.885$ m^3/kg

At $dbt = 22°C$ and $wbt = 10°C$ (indoor conditions)

$h_i = 29.5$ kJ/kg

$$\text{Then : } \quad Q_{inf} = \frac{V}{v_a}(h_a - h_i) = \frac{62}{0.885}(48 - 29.5)$$

$$= 12.96 \text{ kJ/hr}$$

Obviously for moist air, the cooling load depends on the moisture content in the air.

15.4. HEAT GAIN DUE TO VENTILATION

Ventilation refers to the supply of fresh outside air to the conditioned space by natural or by mechanical means. Such a provision is necessary for maintaining the indoor environment free from health threatening contaminants such as tobacco smoke, body odour, moisture, carbon dioxide and other undesirable gases.

Natural ventilation is unpowered air flow through open windows, doors and other intentional openings in the building envelope. Forced ventilation is intentional and powered air exchange by a fan or blower and vents that are specially designed and installed for ventilation.

The recommended values for the fresh air required per person per minute for types of applications (buildings and premises) have been indicated in Table 15.7.

Table 15.7. Ventilation loads (outside air)

S.No.	Application (Buildings and Premises)	Smoking	Recommended outside air in m³/min/person
1.	Apartment	Some	0.6
2.	Bank	Occasional	0.3
3.	Bars	Heavy	0.9
4.	Cafeterias	Considerable	0.36
5.	Departmental stores	None	0.3
6.	Drug stores	Some	0.3
7.	Factories	None	0.3
8.	Hospitals	None	0.75
9.	Hotels	Heavy	0.9
10.	Offices	Some or none	0.45 – 0.75
11.	Restaurants	Considerable	0.45
12.	Shops	Occasional	0.3
13.	Theatres	None	0.25

The salient aspects of ventilation loads are :

(*i*) Ventilation air quantity depends on occupancy of smokes and duration of their stay in the building.

(*ii*) Ventilation air should be more than the infiltration air.

(*iii*) The quantity of outside fresh air used for ventilation should provide atleast half of the air changes per hour in the building.

(*iv*) For general applications such as offices, a minimum ventilation rate of 10 litres per person is recommended.

(*v*) Ventilation air is generally introduced at the air handling unit (AHU) rather than directly into the confined space. It than becomes a cooling coil load component instead of a space load component.

(*vi*) The ventilation air adds to sensible as well as latent heat.

The sensible heat gain corresponding to the change in dry bulb temperature Δt for a given air flow is

$$Q_s = 1.20\ (1.005 + 1.84\ \omega)\ V\ \Delta T\ \approx\ 1.23\ V\ \Delta T \qquad \text{... (15.5)}$$

where Q_s is the sensible heat change (*W*), ω is the specific humidity (kg water/kg dry air) and *V* is the air flow rate (m^3/s).

The latent heat gain corresponding to change in specific humidity (Δω) for a given air flow *V* is

$$Q_e = 1.20 \times 2500\ \ Q_s\ \Delta\omega = 3010\ Q_s\ \Delta\omega \qquad \text{... (15.6)}$$

15.5. HEAT LOAD FROM OCCUPANTS

The occupants in a conditioned space give off thermal energy continuously because of certain metabolic reactions that depend on their breathing rate, sex, age and the level and type of activity in which they are engaged. A person seated at rest releases lesser heat than a working person. The occupants contribute towards sensible and latent loads, and the relative proportion of these loads depend on the ambient dry bulb temperature. Lower the dbt, the greater is the heat given out as sensible heat. Table 15.8 shows the heat released (in watts) from occupants doing different activities in specific applications. The listed values are based on 27°C dry bulb temperature.

Table 15.8 Heat gain from occupants (in watts)

S.No.	Degree of activity	Application area	Sensible load	Latent load
1.	Seated-at rest	Theatre	53	44
2.	Seated-very light work	Offices, hotels, apartments	57	67
3.	Standing or walking (light work)	Department store	59	73
4.	Light bench work	Factory	64	155
5.	Heavy work	Factory	136	288
6.	Moderate work	Hotel, restaurant, retail store	70	85
7.	Dancing	Dance hall	72	177
8.	Exercising	Gymnasium	120	248
9.	Bowling	Bowling alley (narrow passage)	136	288

EXAMPLE 15.4

The data given below refers to the different categories of persons working in an office and the nature of work being performed by them. The thermal energy released in kJ/hr per person has also been indicated.

8 typists (normal duty : 500); 2 supervisors (light work : 475)

1 senior supervisor (moderately heavy work : 1050)

2 attendants (standing and light work : 580)

12 persons (moderately active office work : 500)

Calculate the total energy released by all the persons.

Solution : The energy released by the various categories of persons is worked out as

(*i*) Energy from 8 typists each releasing 500 kJ/hr

$$Q_1 = 8 \times 500 = 4000 \text{ kJ/hr}$$

(*ii*) Energy from 2 supervisors each releasing 475 kJ/hr

$$Q_2 = 2 \times 475 = 950 \text{ kJ/hr}$$

(*iii*) Energy from 1 senior supervisor releasing 1050 kJ/hr

$$Q_3 = 1 \times 1050 = 1050 \text{ kJ/hr}$$

(*iv*) Energy from 2 attendants each releasing 580 kJ/hr

$$Q_4 = 2 \times 580 = 1160 \text{ kJ/hr}$$

(*v*) Energy release from other 12 persons each releasing 500 kJ/hr

$$Q_5 = 12 \times 500 = 6000 \text{ kJ/hr}$$

Total energy released by all the persons working in the office,

$$Q = Q_1 + Q_2 + Q_3 + Q_4 + Q_5$$
$$= 4000 + 950 + 1050 + 1160 + 6000$$
$$= \mathbf{13160} \text{ kJ/hr}$$

15.6. PRODUCT HEAT GAIN

The product (fruits and vegetables) are brought into the refrigerated space at a temperature which is higher than the temperature of refrigerated space. Obviously it gives off heat (sensible as well as latent) until it reaches the desired storage temperature. The quantity of heat released by the product depends upon

- state of the product when it enters the refrigerated space
- state attained by the product after storage
- nature and quantity of the product
- specific heat of the product above and below its freezing point
- freezing temperature, and latent heat of fusion.

The computations for product loads at different stages of cooing are made as follows :

(*i*) Cooling load above freezing (Q_1) :

$$Q_1 = \frac{m c_p (t_1 - t_2)}{t_{ch}} \quad \text{... (15.7)}$$

where m is the mass of product, c_p is the mean specific heat of product, t_{ch} is the time of chilling, t_1 and t_2 represent respectively the product temperature and the desired storage temperature.

(*ii*) Cooling load below freezing (Q_2) :

$$Q_2 = \frac{m c_p (t_1' - t_2')}{t_{ch}} \quad \text{... (15.8)}$$

where t_1' is the actual storage temperature of product and t_2' and is the desired freezing temperature.

(*iii*) Freezing load (Q_3) :

$$Q_3 = \frac{mL}{t_f} \quad \text{... (15.9)}$$

where L is the latent heat of freezing of product and t_f is the time of freezing.

(iv) **Respiration load (Q_4) :**

During storage, the product releases heat called the heat of respiration

$$Q_4 = m \times \text{heat evolution rate} \quad \text{... (15.10)}$$

For apple, the heat of respiration is about 1500 W at a storage temperature of 0°C, 2800 W at 4°C and 800 W at 15°C.

Total heat from products $Q = Q_1 + Q_2 + Q_3 + Q_4$

The rate of heat production for different types of fruit and vegetable has been given in Table 15.9.

Table 15.9 Heat production from different foods at different storage temperatures

S.No.	Food Products	Heat Evolved from food products in J/kg per 24 hours		
		Storage temperature 0°C	4.5°C	15.5°C
1.	Apples	600 – 1050	1200 – 1875	4650 – 7000
2.	Cabbage	1250	1750	4250
3.	Carrots	2250	3650	8500
4.	Cauliflowers	—	4650	10500
5.	Cherries	700 – 1850	—	11500 – 14000
6.	Cucumbers	—	—	2250 – 6850
7.	Grape fruits	480	1140	2950
8.	Lemons	600	850	2200
9.	Melons	1350	2050	8850
10.	Onions	700 – 1200	1850 – 2080	at 10°C
11.	Oranges	750 – 975	1160	5275
12.	Peaches	900 – 1450	1450 – 2050	7550 – 9825
13.	Pears	700 – 925	—	—
14.	Peas	8500 – 8700	13500 – 16500	41000 – 4600
15.	Potatoes	450 – 925	1175 – 1850	2325 – 3725
16.	Rasp berries	4650 – 7000 at 2°C	–	16250 – 18600
17.	Straw berries	2900 – 4000	5400 – 7000	16500 – 20000
18.	Tomatoes	600	1150	6575

Note :

(i) The bananas (green and ripe) are generally stored at 20°C and the heat evolution per kg per 24 hours is about 8850 kJ

(ii) The storage temperature for mushrooms is 10°C and at this temperature, they release about 23000 J/kg per 24 hours.

EXAMPLE 15.5

The following data refers to a meat product that enters a chiller at 5°C and is frozen and chilled to a final temperature of – 10°C for storage in 12 hours.

Specific heat above freezing = 3.25 kJ/kg-deg

Specific heat below freezing = 1.65 kJ/kg-deg

Freezing temperature = – 2.5°C

Patent heat of fusion = 167.5 kJ

If the mass of meat products is 2500 kg, calculate the product load in kJ/day of 24 hour.

Solution : The cooling takes place in three stages namely

(*i*) Heat removed to cool from 5°C to freezing temperature of – 2.5°C.

$$Q_1 = m\,c_p\,(t_1 - t_f) = 2500 \times 3.25\,[5 - (-2.5)] = 60937 \text{ kJ}$$

(*ii*) Heat removed to freeze

$$Q_2 = m \times h_{fg} = 2500 \times 167.5 = 418750 \text{ kJ}$$

(*iii*) Heat removed from freezing temperature to storage temperature

$$Q_3 = m \times c_{p_i} \times (t_1 - t_f) = 2500 \times 1.65 \times [-2.5 - (-10)] = 30937 \text{ kJ}$$

Total heat removed from the meat product

$$Q = Q_1 + Q_2 + Q_3 = 60937 + 418750 + 30937 = 510624 \text{ kJ}$$

∴ Product cooling load in kJ/day of 24 hours

$$= \frac{510624}{12} \times 24 = 1021248 \text{ kJ} = \mathbf{1021.24 \text{ MJ}}$$

15.7. HEAT GAIN FROM APPLIANCES

The appliances are daily use equipment used in conditioned spaces. The heat gain from such appliances depends on their size, capacity, power consumption and operation whether electrical, gas fired, oil fired or steam heated. Both sensible and latent heat loads are contributed by most appliances, and the quantum of these loads for some standard appliances has been indicated in Table 15.10.

Table 15.10 : Heat gain from appliances (in watts)

S.No.	Food Products	Electrical		Gas	
		Sensible	Latent	Sensible	Latent
1.	Coffee brewer, 2¼ litres	265	65	395	105
2.	Coffee brewer with tank, 20 litres	1405	350	2110	530
3.	Egg boiler, 2 cups	350	235	–	–
4.	Food warmer with plat warmer, per sq m of top surface	1140	1140	–	–
5.	Food warmer only, per sq m of top surface	650	1140	2750	1400
6.	Fry kettle, per sq m of fry area	1025	1465	1760	1175
7.	Grill, meat, per sq m of fry area	1380	735	2930	735
8.	Grill, sand witch, per sq m of fry area	790	205	–	–
9.	Hair dryer, blower type	675	120	–	–
10.	Hot plate	650	1140	–	–
11.	Stoves	–	–	1230	965
12.	Sterilizer instrument for physicians	190	350	–	–
13.	Toaster, 360 slices per hour	1495	380	2260	970

15.8. HEAT GAIN FROM LIGHTING EQUIPMENT

Various types of lighting equipments are in an air-conditioned space. These contribute heat equivalent to energy consumed. Part of the energy emitted from lights is in the form of convective heat that becomes instantaneous cooing load. The remaining part is radiant heat which is initially absorbed by the walls, floor, furniture and upholstery and then released to the room air after a time lag.

The heat gain from light emitting elements is computed from the relation

$$Q = \text{wattage} \times \text{use factor} \times \text{allowance factor}$$

- Wattage is the power of the light tube or bulb.
- Use factor is the ratio of actual wattage in use to the installed wattage. The use factor is usually unity for residences, commercial stores and shops. For industrial workshops, use factor is 0.5.
- Allowance factor pertains to fluorescent fixture and/or fixtures that are either ventilated or installed so that only a part of the heat goes to the conditioned space.

15.9. HEAT GAIN FROM POWER EQUIPMENT

The instantaneous heat gain from the equipment operated by electrical motors within the conditioned space is computed as

$$Q = \frac{\text{power rating of motor in kW}}{\text{motors efficiency}} \times \text{load factor}$$

The load factor is the fraction of total load at which the motor works. The load factor is applied when the motor operation is intermittent. For conventional applications, it is taken as 1.0.

Fans that circulate air through the air-conditioning systems add energy to the system by the following processes :

- temperature rise in the air stream by fan efficiency
- temperature rise in the air stream as a consequence of static and velocity pressure
- temperature rise from heat generated by motor and drive efficiencies.

When the fan is located before the air-conditioner, its heat is added to the total load. However, if the fan is installed after the air-conditioner, its heat is added to the room sensible heat.

EXAMPLE 15.6

The following data pertains to the electrical appliances fitted inside an air conditioned compartment :

8 tube lights of 40 W each ; 4 fans of 100 W each ; 2 motors of 2 kW each ; one refrigerator of 200 W and 2 blowers each of 500 W.

What will be the total load due to these electrical appliances ?

Solution : The load due to various electrical appliances is worked out as :

(*i*) Load due to 8 tube lights each of 40 W rating

$$Q_1 = (8 \times 40) \times 1.2 = 384 \text{ W}$$

As per the normal practice, a blast factor 1.2 has been assumed while estimating the actual load of tube lights.

(*ii*) Load due to 4 fans each of 100 W rating

$$Q_2 = 4 \times 100 = 400 \text{ W}$$

(*iii*) Load due to 2 motors each of 2 kW power

$$Q_3 = 2 \times 2 = 4 \text{ kW} = 4000 \text{ W}$$

(*iv*) Load due to 1 refrigerator of 2000 W capacity

$$Q_4 = 1 \times 200 = 200 \text{ W}$$

(*v*) Load due to 2 blowers each of 500 W rating

$$Q_5 = 2 \times 500 = 1000 \text{ W}$$

Total load due to electrical appliances,

$$Q = Q_1 + Q_2 + Q_3 + Q_4 + Q_5$$

$$= 384 + 400 + 4000 + 200 + 1000 = \mathbf{5984\ W}$$

15.10. DUCT HEAT GAIN AND LEAKAGE

The heat gain due to supply duct is computed by using the relation

$$Q_d = U\,A_d\left(t_a - \frac{t_i + t_0}{2}\right) \qquad \text{... (15.11)}$$

where U is the overall heat transfer coefficient, A_d is the area of duct, t_a is the temperature of ambient air, t_i and t_0 respectively denote the temperature of air entering and leaving the duct.

In general :

(*i*) The heat gain through a duct is generally taken to be 5 percent of the room sensible heat

(*ii*) The duct heat gain is zero for a duct located in the conditioned space.

(*iii*) There is heat gain as well as condensation for the duct located in unconditioned space. This is precisely the reason for providing insulation to ducts.

Air leakage from duct joints is of the order of 5 – 15%. This outward leakage from supply duct is a direct loss of cooling and dehumidifying capacity and needs to be compensated by increased flow. While designing a duct, the duct air leakage is generally assumed to be 10 percent for long runs and 5 percent for medium rains.

15.11. SOLAR HEAT GAIN (RADIATION FROM SUN)

Solar radiation is the principle variable that has maximum influence on the cooling load of building. Below is presented a brief account of the salient aspects related to solar heat gain :

1. The solar radiations strike the external face of the building and contribute towards sensible heat gain :
2. The solar heat gain depends on :
   (*i*) time of the day and day of the year
   (*ii*) position of building on earth's surface
   (*iii*) clearness of sky
   (*iv*) absorptivity of the exposed surface of the building
   (*v*) temperature of ground and surroundings
   (*vi*) overall heat transfer coefficient of the wall
3. The solar radiations from the outer spaces reach the earth's surface by two ways : direct radiation and diffuse radiation.

Direct radiation refers to the part of solar radiation that travels through the atmosphere and reaches the earth's surface directly. The intensity of such radiations depends on the orientation of the surface. It is maximum for a surface normal to the sun's rays and is zero when the surface is oriented parallel to the sun's rays.

A substantial part of the solar radiation is scattered, reflected back into space and absorbed by moisture, dirt and dust particles present in the atmosphere. **Diffuse radiation** refers to that part of this radiation which is re-radiated and reaches the earth's surface uniformly from all directions. Because of its diffuse nature, it does not change much with the orientation of the surface.

4. The total radiation that reaches a surface equal the sum of direct and diffuse radiation. Mathematically.

$$I_{i\theta} = I_{DN} \quad \cos\theta + I_{d\theta} + I_s \qquad \text{... (15.12)}$$

where $I_{i\theta}$ = total solar irradiation of the surface

I_{DN} = direct radiation from sun

θ = incident angle as shown in Fig. 15.5

$I_{d\theta}$ = diffuse radiation component from sky

I_r = shortwave radiation reflected from other surfaces.

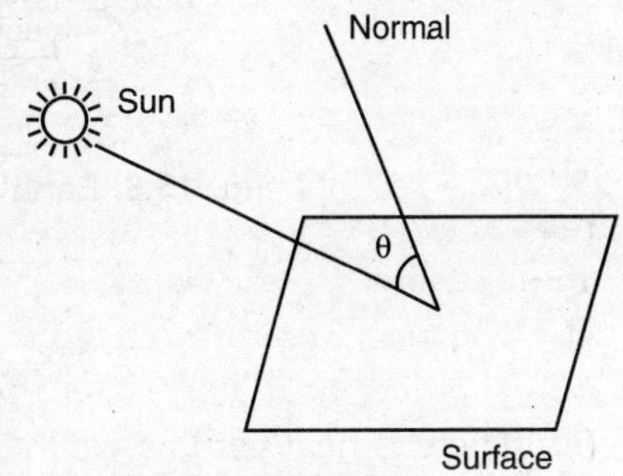

Fig. 15.5

The I_{DN} cos θ term constitutes about 85 percent of the total on the day of clear sky. However, when the sky is hazy or cloudy, there is increase in diffuse radiation and a considerable decrease in direct radiation.

4. The incident angle θ is a direct function of solar geometry and is dependent on the following variables:

- the latitude of location r : the angular distance of the point on the earth measured north or south of the equator.
- the hour angle H : the angle through which the earth must turn to bring the meridian of the point directly in line with the sun's rays. The hour angle expresses the time of the day with respect to the solar noon.

$$\text{One hour of time} = \frac{360}{24} = 15 \text{ degree of hour angle}$$

- the sun's declination δ : the angular distance of sun's rays north or south of the equator.

δ = 0 at the autumnal and vernal equinoxes

= + 23.5° at the summer solastic (21st June)

= – 23.5 at the winter solastic (21st December)

At intervening periods of the year, δ is worked out from the following sinusoidal relation.

$$\delta = 23.47 \sin \frac{360(284 + N)}{365}$$

where N is the day of the year numbered from January 1.

- the solar altitude β : the angle measured from a horizontal plane on the earth upto the sun.
- the solar azimuth ϕ : the angle between the sun's rays and the south.

These angles have been illustrated in Fig. 15.6 and their computations are made from the relations given below

$$\sin\beta = \cos L \cos H \cos\delta + \sin L \sin\delta$$

$$\sin\phi = \frac{\cos\delta \sin H}{\cos\beta} \quad \text{for } \phi \le 90° \qquad \text{...(15.14)}$$

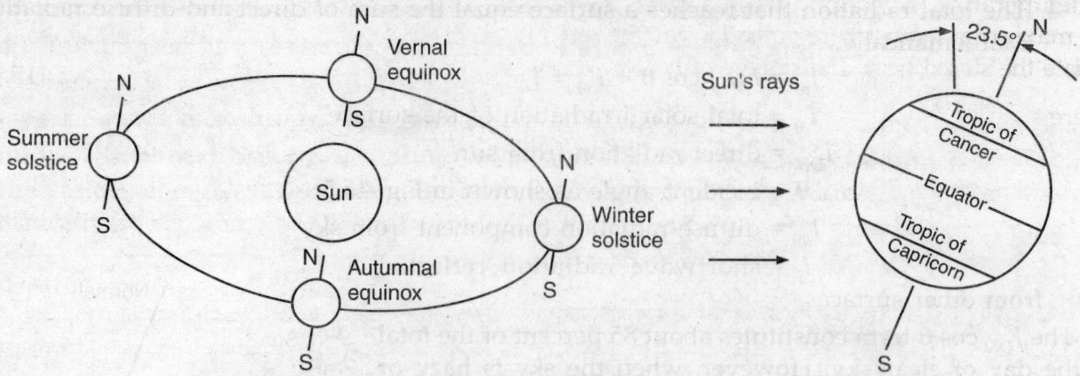

Fig. 15.6. Earth's revolution and relative position of sun's rays

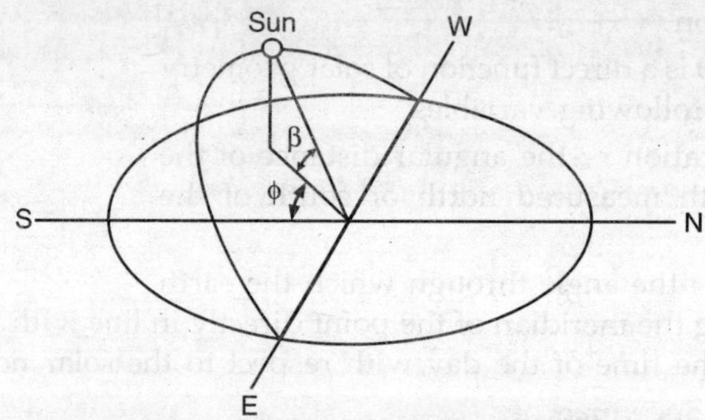

Fig. 15.7. Solar altitude β and solar azimuth ϕ

- the surface azimuth ψ : the angle a vertical plane wall makes with the south.
- the solar-surface azimuth γ: the angle between the solar azimuth and the surface azimuth.

$$\gamma = \phi + \psi \quad \text{when } \phi \text{ and } \psi \text{ are on opposite sides of south}$$
$$= \phi - \psi \quad \text{when } \phi \text{ and } \psi \text{ are an same side of south} \quad ...(15.15)$$

The angle of incidence (the angle between the sun's rays and normal to the surface is then taken as

$$\theta = 90° - \beta \quad \text{for a horizontal surface}$$
$$= \cos^{-1} (\cos \beta \cos \gamma) \text{ for a vertical surface.}$$

5. The direct radiation from the sun (I_{DN}) is worked out from the relation

$$I_{DN} = \frac{A}{\exp(B/\sin\beta)} \quad ...(15.16)$$

The values of parameters A and B depend on the month of year.

$$A = 1230 \text{ W/m}^2 \text{ in December and January}$$
$$= 1080 \text{ W/m}^2 \text{ in mid summer}$$
$$B = \text{ranges from 0.14 in winter to 0.21 in summer}$$

15.11.1. Solar Heat Gain Through Walls

The intensity of solar radiation and the ambient conditions (outside air temperature) change periodically over a cycle of 24 hours. An increases in the outside air temperature during a particular period increases the wall temperature and obviously that causes storage of heat in the wall. The stored heat is subsequently released to the room when there occurs a fall in the outside air temperature. Maximum temperature of day is normally reached between 2 PM to 4 PM, and the maximum temperature or load inside the room is felt in the evening (7 PM to 8 PM) when the walls radiate the stored heat. The storage of heat by the wall thus introduces a lag in the solar load.

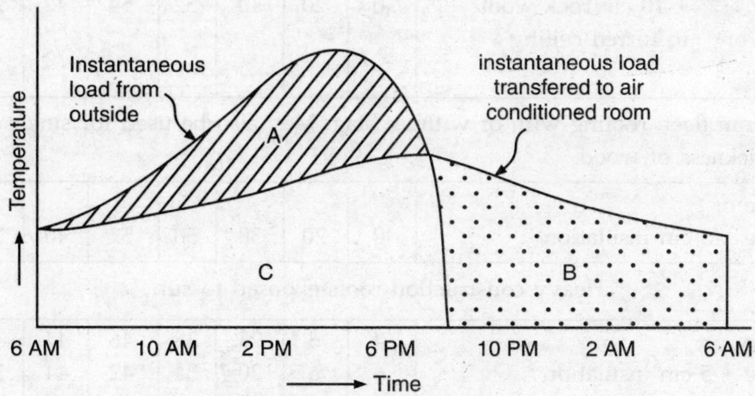

Fig. 15.8. Solar load curves

The curves of intstantaneous load coming from outside and the actual load felt inside are shown in Fig. 15.8. The areas under the two curves are equal and since area *C* is common to both,

heat stored by the wall structure (shaded area *A*)

= heat released by the wall structure (dotted area *B*)

Apparently the transmission of solar radiation through the walls is a case of unsteady heat transfer, and the peak of incoming heat to air-conditioned space is delayed (and also reduced) by the storage nature and capacity of the wall structure.

The lag of solar radiation due to wall storage is taken into account by introducing a term called this ***equivalent temperature differential*** T_e. Then

$$Q = UA\,T_e \qquad \text{... (15.17)}$$

The equivalent temperature differentials for sunlit shaded roofs and walls have been given in Tables 15.11 and 15.12.

Table 15.11 : Total equivalent temperature differentials for calculating heat gain through sunlit and shaded roof

Description of roof construction	Sun-time								
	AM			PM					
	8	10	12	2	24	65	8	10	12
Light construction roofs-exposed to sun									
2.5 cm wood or 2.5 cm wood + 2.5 cm or 5 cm insulation or 5 cm wood	12	38	54	62	50	26	10	4	0
Medium construction roofs-exposed to sun									
5 cm concrete or 5 cm concrete + 2.5 cm or 5 cm insulation 5 cm wood	6	30	48	58	50	32	14	6	2

5 cm gypsum or 5 cm gypsum + 2.5 cm insulation 2.5 cm wood 5 cm wood or + 10 cm rock wool 5 cm concrete or in furred ceiling 5 cm gypsum	0	20	40	52	54	42	20	10	6sss

- Includes 9 mm fleet, roofing with or without stag. May also be used for single roof.
- Nominal thickness of wood.

10 cm concrete or 10 cm concrete + 5 cm insulation	0	20	38	50	52	40	22	12	6
Heavy construction roofs-exposed to sun									
15 cm concrete	4	6	24	38	46	44	32	18	12
15 cm concrete + 5 cm insulation	6	6	20	34	42	44	34	20	14
Roofs covered with water-exposed to sun									
Light construction roof with 2.5 cm water0	4	16	22	18	14	10	2	0	
Heavy construction roof with 2.5 cm water	– 2	– 2	– 4	10	14	16	14	10	6
Any roof with 15 cm water	– 2	0	0	6	10	10	8	4	0
Roof in shade									
Light construction	– 4	0	5	12	14	12	8	2	0
Medium construction	– 4	– 2	2	8	12	12	10	6	2
Heavy construction	– 2	– 2	7	4	8	10	10	8	4

Fig. 15.12. Total equivalent temperature differentials for calculating heat gain through units and shaded walls

North Latitude Wall Facing	Sun-Time																		South Latitude Wall Facing
	AM						PM												
	8		10		12		2		4		6		8		10		12		
	Exterior colour of wall : D-dark, L-light																		
		D	L	D	L	D	L	D	L	D	L	D	L	D	L	D	L	D	
	Frame																		
NE	22	13	24	12	14	10	12	10	14	14	14	14	10	10	6	4	2	2	SE
E	30	14	36	18	32	16	12	12	14	14	14	14	10	10	6	6	2	2	E
SE	13	6	26	16	28	18	24	16	16	14	14	14	10	10	6	4	2	2	NE
S	–4	–4	4	6	22	12	30	20	26	20	16	14	10	10	6	6	2	2	N
SW	–4	–4	0	–2	6	4	26	22	40	28	42	28	24	20	6	4	2	2	NW
W	–4	–4	0	6	6	6	20	12	40	28	48	34	22	22	8	8	2	2	W
NW	–4	–4	0	–2	6	4	12	10	24	20	40	26	34	24	6	4	2	2	SW
N (shades)	–4	–4	–2	–2	4	4	10	10	14	14	2	12	8	8	4	4	0	2	S (shades)

10 cm brick or stone veneer frame

NE	–2	–4	24	12	20	10	10	6	12	10	14	14	12	12	10	10	5	4	SE
E	2	0	30	14	31	17	14	14	12	12	14	14	12	12	10	8	6	6	E
SE	2	–2	20	10	28	16	26	16	18	14	14	14	12	12	10	8	6	6	NE
S	–4	–4	–2	–2	12	6	24	16	26	18	20	16	12	12	8	8	4	4	N
SW	0	–2	0	–2	2	2	12	8	32	22	36	26	34	24	10	8	6	6	NW
W	0	–2	0	0	4	2	10	8	26	18	40	28	42	28	16	14	6	6	W
NW	–4	–4	–2	–2	2	2	8	6	12	12	30	22	34	24	12	10	6	6	SW
N (shade)	–4	–4	–2	–2	0	0	6	6	10	10	12	12	12	12	8	8	4	4	S (shade)

20 cm hollow tile or 20 cm cinder block

NE	0	0	0	0	20	10	16	10	10	6	12	10	14	12	12	10	8	8	SE
E	4	2	12	4	24	12	26	14	20	12	12	10	14	12	14	10	10	8	E
SE	2	0	2	0	16	8	20	12	20	14	14	12	14	12	12	10	8	6	E
S	0	0	0	0	2	0	12	6	24	14	26	16	20	14	12	10	8	9	N
SW	2	0	2	0	2	0	6	4	12	10	26	18	30	20	26	18	8	6	NW
W	4	2	4	2	4	2	6	4	10	8	18	14	30	22	32	22	18	11	W
NW	0	0	0	0	2	0	4	2	8	6	12	10	22	18	30	22	10	14	SW
N (shade)	–2	–2	–2	–2	–2	–2	0	0	6	6	10	11	10	10	10	10	6	6	S (shade)

20 cm brick or 30 cm hollow tile or 30 cm cinder block

NE	2	2	2	2	10	2	16	8	14	8	10	6	10	8	10	10	10	8	SE
E	8	6	8	6	14	8	18	10	18	10	14	8	14	10	14	10	12	10	E
SE	8	4	6	4	6	4	14	1	18	12	16	12	12	10	12	10	12	10	NE
S	4	2	4	2	4	2	4	2	10	6	16	10	16	12	12	10	12	8	N
SW	8	4	6	4	6	14	8	4	10	6	12	8	20	12	24	16	20	14	NW
W	8	4	6	4	6	6	8	6	10	6	14	8	20	16	24	16	20	14	W
NW	2	2	2	2	2	2	4	2	6	4	8	6	10	8	16	14	18	14	SW
N (shade)	0	0	0	0	0	0	0	0	2	2	6	6	8	8	8	8	6	6	S (shade)

30 cm brick

NE	8	6	8	6	8	4	8	4	10	4	12	6	12	6	10	6	10	6	SE
E	12	8	12	8	12	8	10	6	12	8	14	10	14	10	14	8	14	8	E
SE	10	6	10	6	10	6	10	6	10	6	12	8	14	10	14	10	12	8	NE
S	8	6	8	6	6	4	6	4	6	4	8	4	10	6	12	8	12	8	N
SW	10	6	10	6	10	6	10	10	10	6	10	6	10	8	12	8	14	10	NW
W	12	8	12	8	12	8	10	10	10	6	10	6	10	6	12	8	14	10	W
NW	8	6	8	6	8	4	8	4	8	4	8	4	8	6	10	6	10	6	SW
N (shade)	4	4	2	2	2	2	2	2	2	2	2	2	2	2	4	4	4	6	S (shade)

20 cm concrete or Stone or 15 cm or 20 concrete block

NE	4	2	4	0	16	8	14	8	10	6	12	8	12	10	10	8	8	6	SE
E	6	4	14	8	24	12	24	12	18	10	14	10	14	10	12	10	10	8	E
SE	6	2	6	4	16	10	18	12	18	12	14	12	12	10	12	10	10	8	NE
S	2	1	2	1	4	1	12	6	16	12	18	12	14	12	10	8	1	6	E
SW	6	2	4	2	6	2	8	4	14	10	22	16	24	16	22	16	10	3	NW
W	6	4	6	4	6	4	8	6	12	8	20	14	28	18	26	13	14	10	W
NW	4	2	4	0	4	2	4	4	6	6	12	10	20	14	22	16	8	6	SW
N (shade)	0	0	0	0	0	0	2	2	4	4	6	6	8	8	6	6	4	4	S (shade)

100 cm concrete or stone

NE	6	4	6	2	6	14	8	14	8	10	8	10	8	12	12	10	10	8	SE
E	10	6	8	6	10	6	18	10	18	12	16	10	12	10	14	10	14	10	E
SE	8	4	8	4	6	4	14	8	16	10	16	10	14	10	12	10	12	10	NE
S	6	4	4	2	4	2	4	2	10	6	14	10	16	12	14	10	10	8	N
SW	8	4	8	4	6	4	6	4	8	6	10	8	8	14	20	14	18	16	NW
W	10	6	8	6	8	6	10	6	10	6	12	8	16	10	24	14	22	16	W
NW	6	4	6	2	6	2	6	4	6	4	8	6	16	8	18	12	20	14	SW
N (shade)	0	0	0	0	0	0	0	0	2	2	4	4	0	6	8	8	6	8	S (shade)

Sol Air Temperature

There is a need to evolve a temperature that helps to calculate the heat received by the outside surface of a building by the combined effect of convection and radiation.

With reference to Fig. 15.9

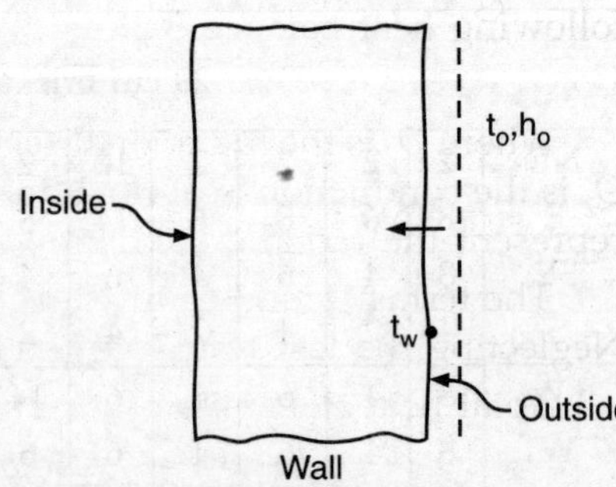

Fig. 15.9

Heat received by outside surface of wall by convection

$$q_c = h_0\ (t_0 - t_w)$$

where h_0 is outside film coefficient, t_0 is temperature of outside ambient air and t_w denotes temperature of outside surface of the wall.

Heat received by outside surface of wall by radiation,

$$q_r = I\alpha$$

where I is the total radiation intensity and α is the absorptivity of the wall surface.

The total heat received by the outside surface of the wall is then given by the sum of heats received by convection and radiation. That is

$$\begin{aligned} q_{\text{total}} &= q_c + q_r \\ &= h_0\ (t_0 - t_w) + I\alpha \\ &= h_0\ t_0 - h_0\ t_w + I\alpha \\ &= h_0 \left[t_0 + \frac{I\alpha}{h_0} \right] - h_0\ t_w \\ &= h_0\ t_e - h_0\ t_w \\ &= h_0\ (t_e - t_w) \end{aligned} \qquad \text{... (15.18)}$$

where $t_e = t_0 + \frac{I\alpha}{h_0}$ is referred to as the ***sol air temperature***.

Apparently the sol-air temperature may be defined as *the temperature of outside air that, in the absence of all radiation changes, gives the same rate of heat entry as would the combination of incident solar radiation and convective heat exchange with outside air*.

15.11.2. Solar Heat Gain Through Transparent Surfaces

The amount of solar heat delivered to an air-conditioned space through glass varies from hour to hour, from day to day and from latitude to latitude. When solar radiation impinge on an unshaded glass windows, they are partly transmitted, partly reflected and partly absorbed.

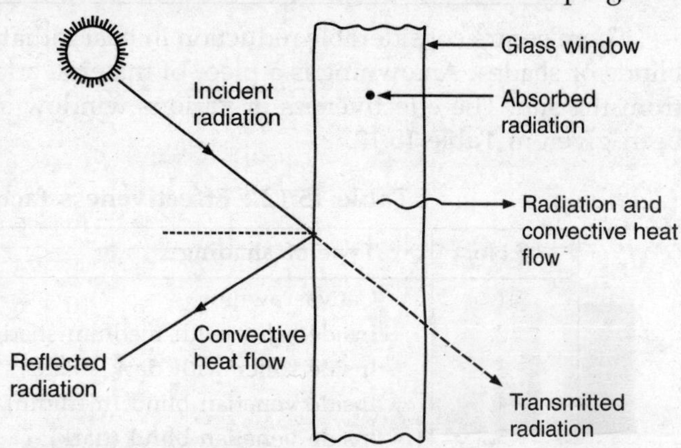

Fig. 15.10. Absorption, reflection and transmission of solar radiation in glass

Due to absorption, there occurs an increase in the temperature of glass. Subsequently the heated glass radiates energy to the inside of room. The heat is also driven off by convection currents from inside the building. There is also energy loss due to interchange of radiant energy between the glass and all the surfaces it can see. Accordingly, the total instantaneous heat gain through the glass window can be computed from the following heat balance.

$$Q_{total} = Q_c + Q_a + Q_c \pm Q_{rc} \quad \text{... (15.19)}$$

where Q_c is the radiation directly transmitted through glass, Q_a is the absorbed solar radiation, Q_c is the conduction heat flow due temperature difference between outdoors and indoor, and Q_{rc} represent the radiation and convective heat exchange between glass and outside surfaces.

The terms Q_c and Q_a are generally grouped together and designated as Q_{shg} (solar heat gain). Neglecting the last term, we can write

Total heat transmission through glass

= solar heat gain + conduction heat gain

Generally, the solar radiation being transmitted through transparent glass is expressed as

$$Q_{sr} = A\,(\tau\, I_t + F\,\alpha\, I_t)$$
$$= A\, I_t\,(\tau + F\alpha) \quad \text{... (15.20)}$$

where α and τ denote total the absorptivity and transmissivity of the glass, I_t is the solar radiation incident on the surface and F is the fraction of absorbed radiation transferred by conduction and convection to inside environment. Under steady conditions, $F = U/h_0$ where U is the overall heat transfer coefficient for the glass and h_0 is the outside heat transfer coefficient : Then

$$Q_{sr} = A\, I_t \left(\tau + \frac{U\alpha}{h_0}\right) = A \times \text{SHGF} \quad \text{... (15.21)}$$

The term $I_t\left(\tau + \frac{U\alpha}{h_0}\right)$ is referred to as solar heat gain factor (SHGF).

The values of solar heat gain for the unprotected and covered glasses has been indicated in Table 15.11.

Table 15.11 : Solar heat gain

Type of glass	Solar heat gain (East exposure)	Solar heat gain (South and West exposure)
Unprotected	1150 kJ/m^2	1610 kJ/m^2
White shades	460	520
Venetian blendes	690	815

There occurs considerable reduction in heat radiations through glass with the use of awnings, blinds or shades. An owning is a piece of material attached to the surface which provides shelter from the sun. The effectiveness of various window shadings in reducing the sun radiation has been given in Table 15.12.

Table 15.12 : Effectiveness factor window shadings

S.No.	Type of shading	Effectiveness factor
1.	Canvas awning	0.35
2.	Inside roller with medium shade	0.60
3.	Inside roller with dark shade	0.80
4.	Inside venetian blind (medium)	0.65
5.	Inside venetian blind (dark)	0.75
6.	Internal curtains	0.85

Further, it is to be noted that direct solar heat gain can be reduced by using different type of glass, glass construction and shades. For example :

(*i*) Double plane glass reduces the solar radiation by 10 to 20%.
(*ii*) Special heat absorbing glass reduces the solar heat by 25%.
(*iii*) Stained glass, depending on its colour, can reduce the solar heat gain upto 65%.
(*iv*) Shading devices installed on the inside of windows reduce sun load upto 15%.
(*v*) Ventilation blinds and curtain shades reduce the solar heat gain by 30 to 30%.

EXAMPLE 15.7

On a particular sunny day, the temperature on the outside and inside of the wall of a room has been noted to be 35°C and 20°C respectively. The solar intensity for the location is 350 kJ/hr-m^2 and the overall heat transfer coefficient for the wall surfaces is 3.8 kJ/m^2 hr K. If the fraction of the solar radiation absorbed by the medium coloured surface of the wall is 0.7, calculate the heat transfer from the surroundings to the room.

Solution : The heat transfer from the surroundings to room is obtained from the relation.

$$Q = FA\, e\, I = 0.23\, U \times A\, e\, I$$

where $F = 0.23\, U$ is an empirical correlation that represents fraction of absorbed solar radiation that is transmitted to the inside wall.

$$\therefore \quad \frac{Q}{A} = (0.23 \times 4) \times 0.7 \times 350 = \mathbf{225.4\ kJ/m^2\ hr}$$

It is to be noted that data about ambient and inner temperature is not needed for making the said calculations.

EXAMPLE 15.8

Set up an expression for the sol-air-temperature and calculated its value for a wall with the following conditions :

Total of direct and diffuse solar radiation = 275 W/m²; absorptivity of surface = 0.85
Outside surface heat transfer coefficient = 25 W/m² K
Outside air temperature = 35°C.

Solution : Sol-air-temperature

$$t_e = t_0 + \frac{I\alpha}{h_0} = 35 + \frac{275 \times 0.85}{25} = \mathbf{44.35°C}$$

EXAMPLE 15.9

(a) State the function of making an estimate of load in the context of summer air-conditioning.
(b) List the variables considered in estimating the cooling load for a building.
(c) Mention the categories in which the loads are generally divided.

Solution : The primary function of heat gain calculations is to estimate the capacity that will be required for the various air-conditioning components necessary to maintain comfort within a space.

(*b*) The following variables need consideration in the estimation of load :

(*i*) nature of construction, material used and age of building
(*ii*) orientation of openings, windows and doors
(*iii*) number of persons, their period of occupancy and the activities in which they are engaged
(*iv*) outside humidity and temperature
(*v*) magnitude and direction of wind velocity

(*c*) The loads are generally categorized into the following groups :

- **Transmission :** Heat gain due to a temperature difference across a building element.
- **Solar :** Heat gain due to transmission of solar energy through a transparent building component or absorption by an opaque building element.
- **Infiltration :** Heat loss or heat gain due to infiltration of outside air into a conditioned space.
- **Internal :** Heat gain due to release of energy within a space (people, light, equipment and appliances etc).

EXAMPLE 15.10

For an office of 6 m × 3 m × 4 m high, the ambient and indoor conditions are stated as :

$$dbt = 40°,\ \ wbt = 26°C \text{ and } dbt = 22°C \text{ and } \phi = 55\%$$

Further, it may be presumed that the office has structural load of 6000 kJ/hr, 5 tube lights each of 40 W rating, 13.5 air changes per 24 hour for the infiltration load, 7.1 × 10⁻³ m³/s per person of ventilation, 10 person occupancy and each releases 500 kJ/hr.

Workout the capacity of a window air-conditioner that will suffice to achieve the desired objective.

Solution : Locate state point 1 (ambient conditions dbt = 40°C and wbt = 26°C) and state point 2 (dbt = 22°C and ϕ = 55%) on psychrometric chart and read therefore :

$$h_1 = 80.5 \text{ kJ/kg},\ \ v_1 = 0.94 \text{ m}^3\text{/kg of dry air and } h_2 = 46 \text{ kJ/kg}$$

The estimate of loads from different sources is :

(*i*) Structure load = 5000 kJ/hr = $\frac{6000}{3600}$ kJ/s = 1.667 kW

(*ii*) Electrical load = (5 × 40) × 1.2 = 240 W = 0.24 kW

As per usual practice, a blast factor of 1.2 has been assumed for estimating the actual load from tube light.

(*iii*) Volume of air infiltration $V_i = \frac{13.5}{24} \times (6 \times 3 \times 4) = 40.5 \text{ m}^3/\text{hr}$

$$\text{Mass of air infiltration} = \frac{V_i}{v_1} = \frac{40.5}{3600 \times 0.94} = 0.012 \text{ kg/s}$$

$$\text{Infiltration load} = m_i (h_1 - h_2) = 0.012 \, (80.5 - 45)$$

$$= 0.426 \text{ kW}$$

(*iv*) Volume of ventilated air $V_v = (7.1 \times 10^{-3}) \times 10 = 0.071 \text{ m}^3/\text{s}$

$$\text{Ventilation load} = m_v (h_1 - h_2) = \frac{V_v}{v_1} (h_1 - h_2)$$

$$= \frac{0.071}{0.94} (80.5 - 45) = 2.64 \text{ kW}$$

(*v*) Occupancy load = heat released from the occupants

= 10 × 500 = 5000 kJ/hr = 1.389 kW

Total load the air conditioner has to meet

= 1.667 + 0.24 + 0.426 + 2.64 + 1.389 = 6.362 kW

Since 1 ton of refrigeration is equivalent to 3.5 kW, we have

Refrigeration capacity of the air conditioner

$$= \frac{6.326}{3.5} = \mathbf{1.807 \text{ tons}}$$

EXAMPLE 15.11

Determine the sensible heat factor and capacity of the refrigeration system to be installed for a bank building to be designed for 100 persons. The pertinent data is stated as :

Outside ambient conditions = 40°C dbt and 26°C wbt

Inside conditions = 22°C dbt and 55% RH

Building size = 20 m × 15 m × 5 m high

Number of changes for infiltration load = 2.8 per 24 hour

Ventilation air = 4.7 × 10⁻³ m³/s per person

Electrical load = 11500 kJ/hr

Latent and sensible heat release per person = 625 kJ/hr and 420 kJ/hr

Overall heat transfer coefficients for wall and ceiling

= 18 kJ/m² hr and 10.5 kJ/m² hr K

Solution : Locate state point 1 (outside conditions *dbt* = 40°C and *wbt* = 26°C) and state point 2 (inside conditions *dbt* = 20°C and ϕ = 55%) on psychrometric chart. Join points 1 and 2 and mark point A where a vertical line through point 1 meets the horizontal line drawn through point 2. Then from psychrometric chart

$$v_1 = 0.94 \text{ m}^3/\text{kg of dry air}$$

$$h_1 = 80.5 \text{ kJ/kg} ; \quad h_2 = 45 \text{ kJ/kg} \quad \text{and} \quad h_A = 64 \text{ kJ/kg}$$

The estimate of load from different sources is :

(*i*) Structure load Q_1

$$\text{Area of walls} = (20 \times 5 + 15 \times 5) \times 2 = 350 \text{ m}^2$$

$$\text{Area of ceiling} = 20 \times 15 = 300 \text{ m}^2$$

Heat transfer through walls and ceilings ($U\,A\,\Delta T$)

$$Q_1 = 18 \times 350 \times (40 - 22) + 10.5 \times 300\ (40 - 22)$$

$$= 113400 + 56700 = 170100 \text{ kJ/hr}$$

(*ii*) Electrical load $Q_2 = 11500$ kJ/hr (Given)

(*iii*) Volume of air infiltration $V_i = \dfrac{2.8}{24}(20 \times 15 \times 5) = 175 \text{ m}^3/\text{hr}$

Volume of air ventilation $V_v = (4.7 \times 10^{-3} \times 100) \times 3600 = 1692 \text{ m}^2/\text{hr}$

Mass of air due to infiltration and ventilation,

$$m_a = \frac{V_i + V_v}{v_1} = \frac{175 + 1692}{0.94} = 1986 \text{ kg/hr}$$

Then sensible heat gain due to infiltration and ventilation

$$= m_a\,(h_a - h_2) = 1986\ (64 - 45) = 37734 \text{ kJ/hr}$$

Latent heat gain due to infiltration and ventilation

$$= m_a(h_1 - h_A) = 1986\ (80.5 - 64) = 32769 \text{ kJ/hr}$$

(*iv*) Sensible heat gain from the persons

$$= \text{number of persons} \times Q_s \text{ per person}$$

$$= 100 \times 420 = 42000 \text{ kJ/hr}$$

Latent heat gain from persons

$$= \text{number of persons} \times Q_l \text{ per person}$$

$$= 100 \times 625 = 62500 \text{ kJ/hr}$$

(*v*) Room sensible heat (*RSH*), *i.e.*, total sensible load

= sensible heat gain through structure (walls and ceiling)
+ electrical load
+ sensible heat gain due to infiltration and ventilation
+ sensible heat gain from persons
= 170100 + 11500 + 37734 + 42000 = 2612334 kJ/hr

Room latent heat (*RLH*), *i.e.*, total latent load

= latent heat gain due to infiltration and ventilation
+ latent heat gain from persons
= 32769 + 62500 = 95269 kJ/hr

$$\therefore \quad Q_{total} = RSH + RLH = 261334 + 95269$$

$$= 356603 \text{ kJ/hr}$$

$$\text{Sensible heat factor } (SHF) = \frac{RSH}{RSH + RLH} = \frac{261334}{356603} = \mathbf{0.733}$$

Since 1 ton of refrigeration is equivalent to 12600 kJ/hr, we get

Required capacity of refrigeration system

$$= \frac{356603}{12600} = \mathbf{28.3 \text{ ton}}$$

EXAMPLE 15.12

For the air-conditioning of a 30 m × 20 m × 4 m high single storey office building with 30 m wall facing north, the following data is available :

Three glass windows, in each of the east, west and north walls ... 2 m × 1.5 m each

Two doors in the south wall ... 2.5 m × 3 m each

Lighting load ... 15 W fluorescent per m^2 of floor area

Infiltration ... one air change

Solar heat gain factors (SHGF) for east, west and north walls ... 50, 350 and 50 W/m^2 respectively

Overall heat coefficients for walls, roof, floor, door and windows ... 2.5, 2, 3, 1.5 and 6 W/m^2 K

Corrected equivalent temperature differences for east, west, north, south walls, roof and floor. ... 12, 17, 12, 15, 20 and 2.5°C

Number of occupants ... 150

Sensible and latent heat loads per person ... 75 and 55 W

Ventilation requirement per person ... 0.3 m^3/min

Outdoor conditions ... dbt 43°C and humidity ratio 0.0277 kg/kg of dry air

Indoor conditions ... dbt 25°C and humidity ratio 0.01 kg/kg of dry air.

Use factor for fluorescent ... 1.25

Make calculations for the room sensible heat load and room latent heat load.

Assumptions may be made as 5% safety factor, 5% for fan power, 2% leakage of supply air and 0.5% heat leakage to supply air duct.

Solution : Refer Fig. 15.11 for the plan of the office building

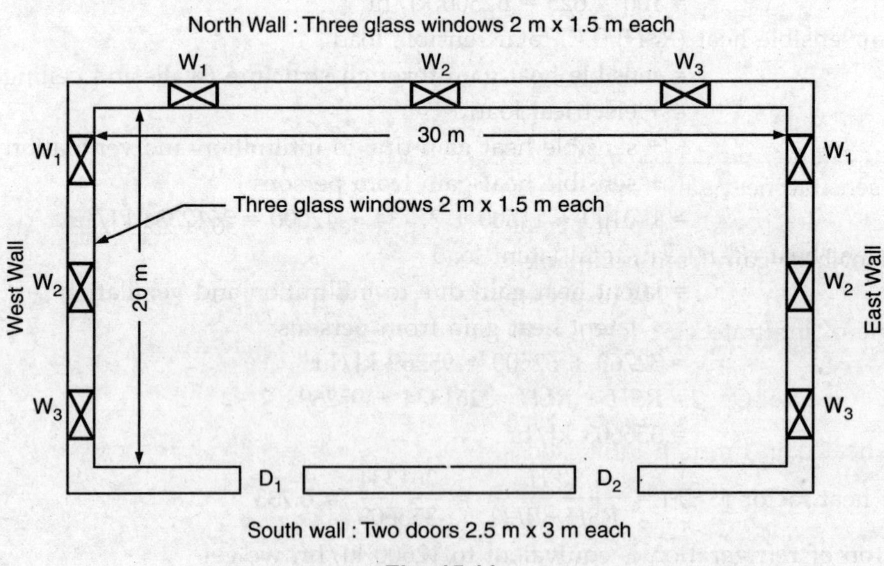

Fig. 15.11

- Sensible heat gain through walls, roof, floor and windows

Area of one door = 2.5 × 3 = 7.5 m^2

Area of one window = 2 × 1.5 = 3.0 m^2

Equivalent temperature difference for glass windows

$$t_e = dbt_1 - dbt_2 = 43 - 25 = 18°C$$

S.No.	Source	Overall heat transfer coefficient U in W/m²K	Area (A) m²	Equivalent temperature difference t_e in °C	Sensible heat gain = $U\,A\,t_e$ in watt
1.	North wall	2.5	30 × 4–3 × 3 = 111	12	2.5 × 111 × 12 = 3330
2.	South wall	2.5	30 × 4–2 × 7.5 = 105	15	2.5 × 105 × 15 = 3937.5
3.	East wall	2.5	20 × 4 – 3 × 3 = 71	12	2.5 × 71 × 12 = 2130
4.	West wall	2.5	20 × 4 – 3 × 3 = 71	17	2.5 × 71 × 17 = 3017.5
5.	Roof	2.0	30 × 20 = 600	20	2.0 × 600 × 20 = 24000
6.	Floor	3.0	30 × 20 = 600	2.5	3.0 × 600 × 2.5 = 4500
7.	Door	1.5	2 × 7.5 = 15	12	1.5 × 15 × 120 = 270
8.	Windows				
	(a) East wall 3 Nos	6	3 × 3 = 9	18	6 × 9 × 18 = 972
	(b) West wall 3 Nos	6	3 × 3 = 9	18	6 × 9 × 18 = 972
	(c) South wall 3 Nos	6	3 × 3 = 9	18	6 × 9 × 18 = 972
					Total = 44101 W = 44.1 kW

* Solar heat gain through glass

Glass	Area m²	SHGF W/m²	Q = area × SHGF W
East	3 × 3 = 9	50	9 × 50 = 450
West	3 × 3 = 9	50	9 × 50 = 450
North	3 × 3 = 9	350	9 × 350 = 3150
			Total = 4050 W = 40.5 kW

- Total sensible heat gain from occupants

$$= \text{No of persons} \times Q_s \text{ per person} = 150 \times 75 = 11250 \text{ W}$$

Total latent heat gain from occupants

$$= \text{No of persons} \times Q_l \text{ per person} = 150 \times 55 = 8250 \text{ W}$$

- Volume of infiltrated air $V = L \times W \times H \times$ number of air changes

$$= \frac{30 \times 20 \times 4 \times 1}{60} = 40 \text{ m}^3/\text{min}$$

Sensible heat gain through infiltrated air

$$= m \times c_p \times \Delta t$$

$$= \rho V \times c_p \times \Delta t = 1.2 \times \frac{40}{60} \times 1.02 \times (43 - 25)$$

$$= 14.7 \text{ kJ/s} = 14.7 \text{ kW}$$

where $\rho = 1.2$ kg/m³ is the density of moist air and $c_p = 1.02$ kJ/kg K is the specific heat of humid air.

Latent heat gain through infiltrated air

$$= m\, h_{fg}\, \Delta\omega$$

$$= \rho V\, h_{fg}\, \Delta\omega = 1.2 \times \frac{40}{60} \times 2500 \times (0.0277 - 0.01)$$

$$= 34.4 \text{ kJ/s} = 34.4 \text{ kW}$$

- Volume of ventilation or outside air

$$V = \text{No of persons} \times \text{ventilation required per person}$$

$$= 150 \times 0.3 = 45 \text{ m}^3/\text{min}$$

Sensible heat gain through ventilation air

$$= \rho\, V\, c_p\, \Delta t = 1.2 \times \frac{45}{60} \times 1.02 \times (43 - 25)$$

$$= 16.55 \text{ kJ/s} = 16.55 \text{ kW}$$

Laten heat gain through ventilation air

$$= \rho V\, h_{fg}\, \Delta\omega = 1.2 \times \frac{45}{60} \times 2500 \times (0.0277 - 0.01)$$

$$= 39.82 \text{ kJ/s} = 39.82 \text{ kW}$$

Sensible heat gain due to lighting

$$= \text{total wattage of lights} \times \text{use factor} \times \text{allowance factor}$$

$$= 15\,(300 \times 20) \times 1 \times 1.25 = 11250 \text{ W} = 11.25 \text{ kW}$$

(*a*) If fan is assumed to the placed before the air conditioner, the factor 5% for fan power can be neglected. Then the total room sensible heat needs to be increased by :

5% safety factor + 2% leakage of supply air + 0.5% of heat leakage to supply air duct.

∴ Total room sensible heat

$$= (1 + 0.05 + 0.02 + 0.005)\,[44.8 + 40.5 + 11.25 + 14.7 + 16.55 + 11.25]$$

$$= 1.075 \times 139.05 = 149.48 \text{ kW}$$

The total room latent heat is to be increased by :

5% safety factor + 2% leakage of supply air

∴ Total room latent heat

$$= (1 + 0.05 + 0.02)\,(8.25 + 35.4 + 39.82)$$

$$= 1.07 \times 83.49 = 89.31 \text{ kW}$$

(*b*) In case, the fan is located after the conditioner, the fan power also adds to room sensible heat. That is total room sensible heat

$$= (1 + 0.05 + 0.02 + 0.005 + 0.05)\,(44.8 + 40.5 + 11.25 + 14.7 + 16.55 + 11.25)$$

$$= 1.125 \times 139.05 = 156.43 \text{ kW}$$

The total room latent heat will, however, remain unaffected.

EXAMPLE 15.13

The following is the design data of the air-conditioning system for a restaurant :
Outside conditions : 35°C dbt and 26°C wbt
Inside design condition : 27°C dbt and 55% RH
Minimum temperature of air supplied to the room : 17°C dbt
Total heat flow through walls, roof and floor : 6.25 kW
Solar heat gain through glass = 2 kW
Sensible and latent heat gain from equipment : 3 kW and 0.7 kW

Seating chairs for dining : 50
Number of employers serving meals : 5
Sensible heat gain per person : 60 W
latent heat gain per sitting person and per employer : 45 W and 75 W
Sensible and latent heat added from meals : 0.16 kW and 0.25 kW
Total infiltration air : 6.65 kg/min
Total amount of fresh air supplied : 26.65 m³/min
Power of motor running the fan : 7.5 kW

Determine the following parameters presuming that the fan is located before the air-conditioner :

(a) amount of air delivered to the room in m³/hr,
(b) percentage of recirculated air,
(c) refrigeration load on the coil, and
(d) by-pass factor and dew point temperature of the coil.

Solution : Refer Fig. 15.12 for the schematic flow diagram of the air-conditioning system and the various processes laid on psychrometric chart.

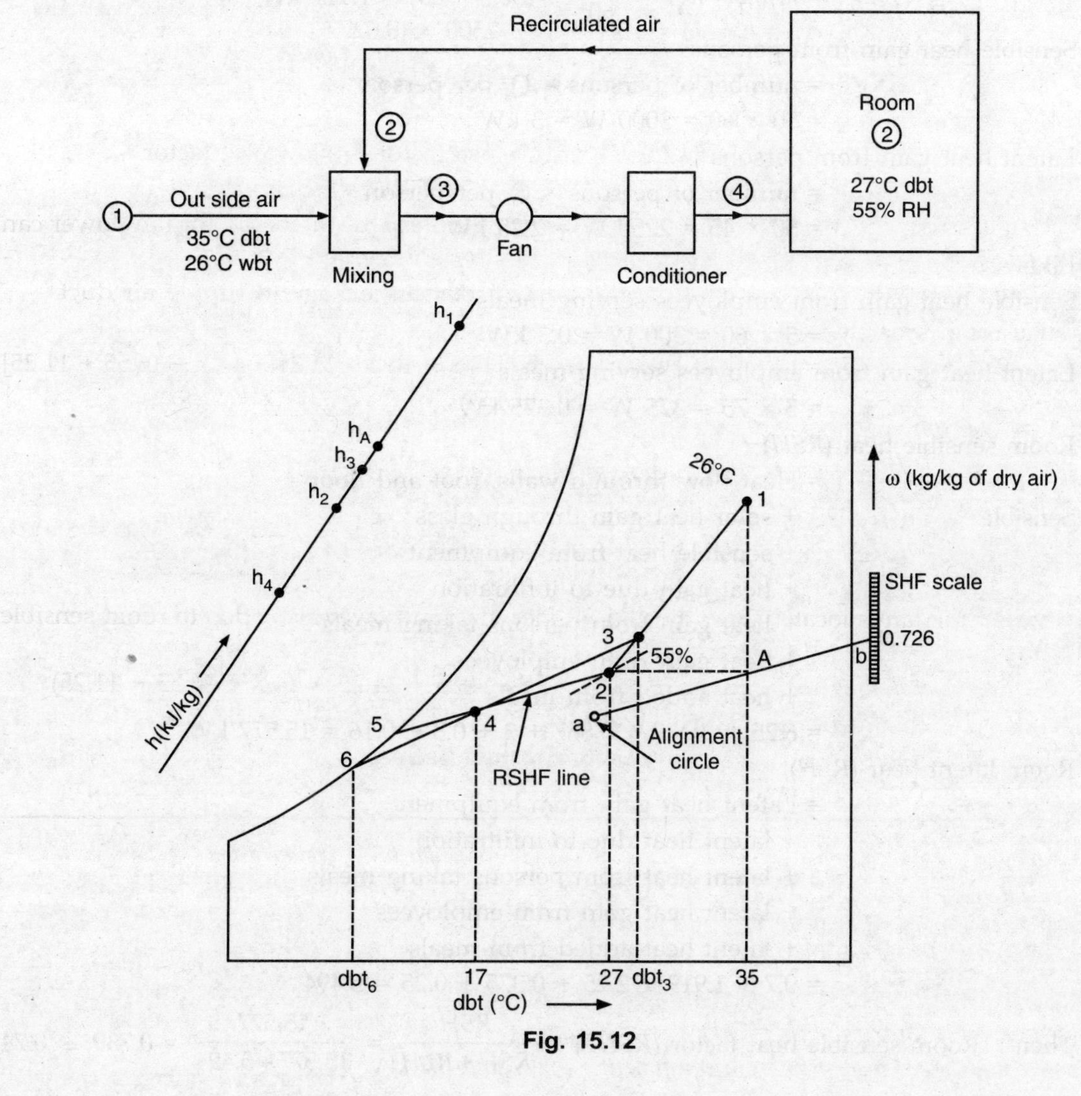

Fig. 15.12

(*i*) Locate state point 1 (outdoor conditions = 35°C *dbt* and 26°C *wbt*) and state point 2 (inside design conditions = 27°C *dbt* and 55% *RH*) on psychrometric chart. Joint points 1 and 2, and mark point A where a vertical line drown through point 1 meets the horizontal line drawn through point 2.

Then from psychrometric chart :

$$v_1 = 0.895 \text{ m}^3/\text{kg of dry air}$$

$$h_1 = 80.5 \text{ kJ/kg} ; \; h_2 = 58.0 \text{ kJ/g and } h_A = 65 \text{ kJ/kg}$$

(*ii*) Mass of infiltration air $m_1 = \dfrac{6.65}{0.895} = 7.43$ kg/min

Sensible heat gain due to infiltration air,

$$= m_1(h_A - h_2) = \frac{7.43}{60} \times (65 - 58) = 0.867 \text{ kW}$$

Latent heat gain due to infiltration air,

$$= m_1(h_1 - h_A) = \frac{7.43}{60} \times (80.5 - 65) = 1.919 \text{ kW}$$

(*iii*) Sensible heat gain from persons

= number of persons × Q_s per person

= 50 × 60 = 3000 W = 3 kW

Latent heat gain from persons

= number of persons × Q_l per person

= 50 × 45 = 2250 W = 2.25 kW

Likewise :

Sensible heat gain from employees serving meals

= 5 × 60 = 300 W = 0.3 kW

Latent heat gain from employees serving meals

= 5 × 75 = 375 W = 0.375 kW

(*iv*) Room sensible heat (*RSH*)

= Heat flow through walls, roof and floor

Sensible + solar heat gain through glass

+ sensible heat from equipment

+ heat gain due to infiltration

+ heat gain from persons taking meals

+ heat gain from employees

+ heat added from meals

= 6.25 + 2 + 3 + 0.867 + 3 + 0.3 + 0.16 = 15.577 kW

Room latent heat (*RLH*)

= Latent heat gain from equipment

+ latent heat due to infiltration

+ latent heat from persons taking meals

+ latent heat gain from employees

+ latent heat added from meals

= 0.7 + 1.919 + 2.25 + 0.375 + 0.25 = 5.494

Then : Room sensible heat factor (*RSHF*) $= \dfrac{RSH}{RSH + RLH} = \dfrac{15.577}{15.577 + 5.494} = 0.739 \simeq 0.74$

(*v*) On psychrometric chart, draw the alignment line. It is the line which joins the 26°C *dbt* and 50% relative humidity point *a* to 0.741 RSHF point *b* on SHF scale. .

From point 2, draw a line 2-5 parallel to line *a-b*. The line 2-5 is referred to as *RSHF* line. Draw a vertical line through 17°C *dbt* such that it intersects *RSHF* line at point 4. Then from the chart

$$v_4 = 0.825 \text{ m}^3/\text{kg} \quad \text{and} \quad h_4 = 45 \text{ kJ/kg}$$

If m_a is the mass of air delivered to room, then

$$m_a \times \text{heat removed} = \text{total room heat}$$

$$m_a \times (h_2 - h_4) = RSH + RLH$$

That gives : $$m_a = \frac{15.577 + 5.494}{58 - 45} = 1.621 \text{ kg/s}$$

Volume of air delivered to room,

$$V_a = m_a \times v_4 = 1.621 \times 0.825 = 1.337 \text{ m}^3/\text{s} = 4813 \text{ m}^3/\text{hr}$$

Mass of fresh air supplied, $$m_f = \frac{V_f}{v_1} = \frac{26.65}{0.895} = 29.78 \text{ kg/min} = 0.496 \text{ kg/s}$$

∴ Mass of recirculated air = 1.621 – 0.496 = 1.125 kg/s

Percentage of recirculated air = $\frac{1.125}{1.621} \times 100 = \mathbf{69.4\%}$

The 69.4% air at condition 2 is mixed with 30.6% air at condition 1. The mixture condition is represented by the point 3 (before entering the cooling coil). The mixing point 3 is so located that length 2-3 = 0.4 × length 1-2.

From the psychrometric chart we note that h_3 = 64.5 kJ/kg

(*vi*) Refrigeration load $= m_a (h_3 - h_4) + \text{motor power}$

$$= 1.621 (64.5 - 45) + 7.5 = 39.11 \text{ kJ/s}$$

Since 1 ton of refrigeration is equivalent to 3.5 kJ/s, we get

$$\text{Refrigeration load} = \frac{39.11}{3.5} = \mathbf{11.17} \text{ TR}$$

(*vii*) Join point 3 with point 4, and extend this line to intersect the saturation curve at point 6. The dew point temperature of the coil is given by the dry bulb temperature corresponding to state point 6. That is

$$dpt = dbt_6 = \mathbf{14.5°C}$$

$$\text{By-pass factor} \quad BPF = \frac{dbt_4 - dbt_6}{dbt_3 - dbt_6} = \frac{17 - 14.5}{29.5 - 14.5} = \mathbf{0.167}$$

EXAMPLE 15.14

A cold storage room is maintained at 20°C dry bulb temperature and 60% relative humidity when outside conditions are 35°C dry bulb temperature and 28°C wet bulb temperature. The other data pertinent to cold storage is :

Quantity of vegetables : 500 tonnes
Water content in vegetables : 76%
Loss of water content from vegetables : 0.01% per hour
People working in cold storage : 25
Sensible heat gain through glass : 11.5 kW

Sensible heat gain through walls and ceilings : 6 kW
Heat from appliances and reaction heat of vegetables : 3.5 kW
Infiltration air and fresh air supplied : 3.35 m³/min and 1.22 m³/min

If the air-conditioning is achieved first by cooling and dehumidifying and then by heating, making calculations for

(a) *amount of recirculated air, if the recirculated air is mixed with fresh air before entering the cooling coil,*
(b) *capacity of the cooling coil and its by-pass factor if the dew point of the coil is 5°C,*
(c) *capacity of the heating coil.*

It may be pressured that :
Sensible and latent heat gain per person working in cold storage = 250 and 210 kJ/hr
Latent heat of evaporation of water = 2500 kJ/kg
Upper limit to temperature of air entering the storage room = 15°C

Solution : Refer Fig. 15.13 for the schematic flow diagram of the air-conditioning system, and the various processes laid on psychrometric chart.

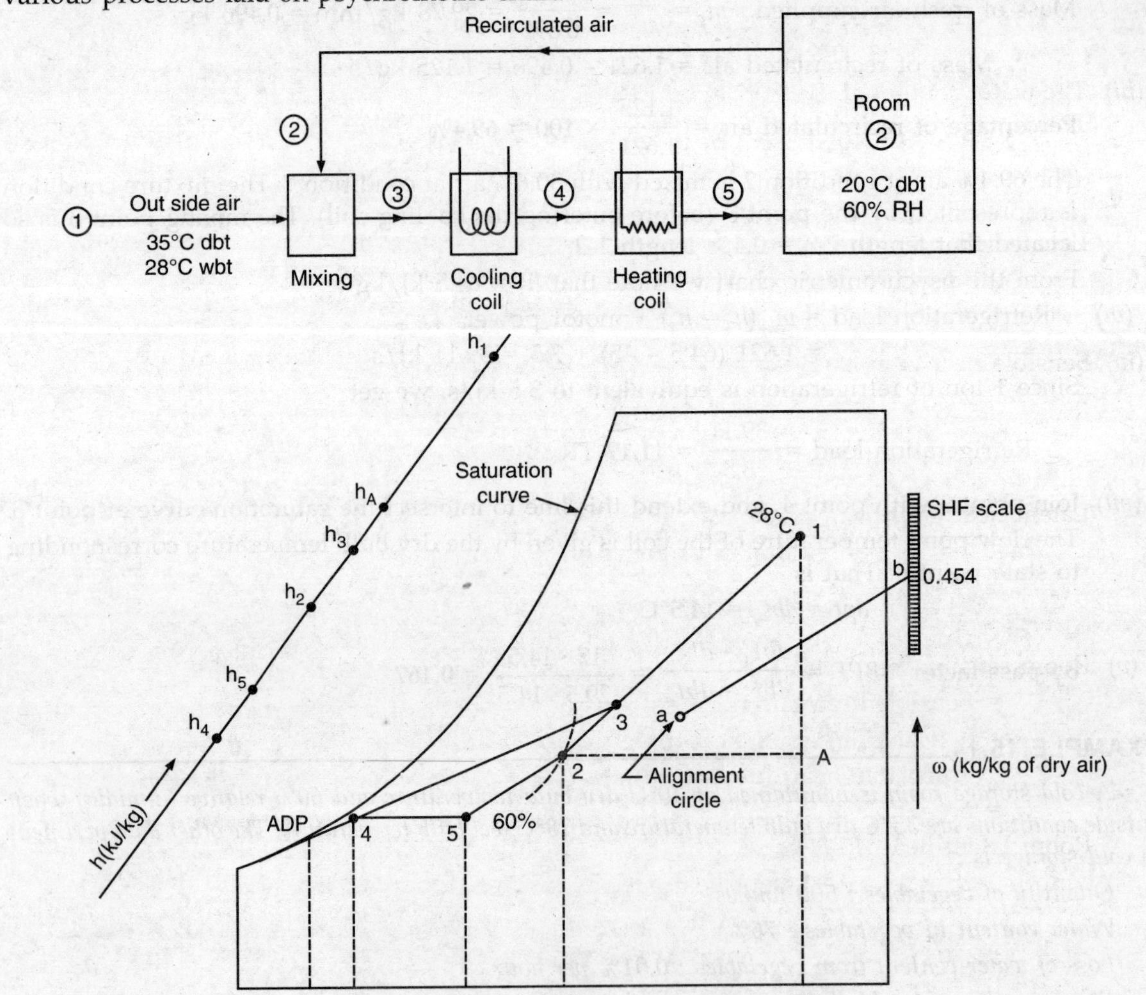

Fig. 15.13

(*i*) Locate state point 1 (outside conditions = 35°C *dbt* and 28°C *wbt*) and state point 2 (inside design conditions = 20°C *dbt* and 60% RH) on psychrometric chart. Join points 1 and 2, and mark point A where a vertical line drawn through point 1 meets the horizontal line drawn through point 2.

Then from psychrometric chart :

$$v_1 = 0.91 \text{ m}^3/\text{kg of dry air}$$

$$h_1 = 90 \text{ kJ/kg} \; ; \; h_2 = 42.5 \text{ kJ/g and } h_A = 58 \text{ kJ/kg}$$

(*ii*) Mass of infiltration air $m_1 = \dfrac{V_1}{v_1} = \dfrac{3.35}{0.91} = 3.68$ kg/ min

Sensible heat gain due to infiltraton air,

$$= m_1(h_A - h_2) = \frac{3.68}{60} \times (58 - 42.5) = 0.95 \text{ kW}$$

Latent heat gain due to infiltration air,

$$= m_1(h_1 - h_A) = \frac{3.68}{60} \times (90 - 58) = 1.96 \text{ kW}$$

(*iii*) The water content of the vegetables is 76% and the loss of water content is 0.01% per hour. Therefore, water loss from vegetables which is carried by air

$$= (500 \times 10^3) \times 0.76 \times \frac{0.01}{100} = 38 \text{ kg/hr}$$

Latent heat gain due to evaporation of water from vegetables

$$= \frac{38}{3600} \times 2500 = 26.39 \text{ kW}$$

(*iv*) Sensible heat gain from persons working in the cold storage

$$= \text{number of persons} \times Q_s \text{ per person}$$

$$= 25 \times \frac{250}{3600} = 1.736 \text{ kW}$$

Latent heat gain from persons

$$= \text{number of persons} \times Q_l \text{ per person}$$

$$= 25 \times \frac{210}{3600} = 1.458 \text{ kW}$$

(*v*) Room sensible heat (*RSH*)

= Sensible heat gain through walls and ceilings + sensible heat gain through glass
+ heat from appliances and reaction heat of vegetables
+ sensible heat due to infiltration air + sensible heat from persons

$$= 6 + 11.5 + 3.5 + 0.95 + 1.736 \simeq 23.69 \text{ kW}$$

Room latent heat (*RLH*)

= Latent heat gain due to infiltration air
+ latent heat due to evaporation of water from vegetables
+ latent heat gain from persons

$$= 1.96 + 26.39 + 1.458 \simeq 29.81 \text{ kW}$$

Then : Room sensible heat factor (*RSHF*) $= \dfrac{RSH}{RSH + RLH}$

$$= \frac{23.69}{23.69 + 29.81} = 0.443$$

(*vi*) On psychrometric chart, draw the alignment line. It is the line which joins the 26°C *dbt* and 50% relative humidity point *a* to 0.741 *RSHF* point *b* on *SHF* scale.

From point 2, draw a line 2-5 parallel to line *a-b*. The line 2-5 is referred to as *RSHF* line. Draw a vertical line through 15°C *dbt* such that it intersects *RSHF* line at point 5. The state point 5 represents the condition of air entering the room. Then from the chart

$$h_5 = 31.6 \text{ kJ/kg}$$

If m_a is the mass of air delivered to room, then

$$m_a \times \text{heat removed} = RSH + RLH$$

$$m_a \times (h_2 - h_5) = RSH + RLH$$

That gives : $m_a = \dfrac{23.69 + 29.81}{42.5 - 31.6} = 4.908 \text{ kg/s}$

Mass of fresh air supplied from outside, $m_f = \dfrac{V_f}{v_1}$

$$= \frac{1.22}{0.91} = 1.341 \text{ kg/s}$$

∴ Mass of recirculated air $m_r = m_a - m_f$

$$= 4.908 - 1.341 = 3.567 \text{ kg/s}$$

Percentage of recirculated air $= \dfrac{3.567}{4.908} \times 100 =$ **72.7%**

Then 72.7% air at condition 2 is mixed with 27.3% air at condition 1. The mixture condition is represented by the point 3 (before entering the cooling coil). The mixing point 3 is so located that

$$\text{length 2-3} = 0.273 \times \text{length 1-2}$$

From the psychrometric chart, we note that:

$$h_3 = 56 \text{ kJ/kg and} \quad dbt_3 = 24.5°\text{C}$$

(*vii*) Draw a line joining point 3 and dew point temperature 5°C of the coil. This line intersects the specific humidity line drawn through point 5 at point 4. The state point 4 represents the conditions of air leaving the cooling coil. Line 4-5 represents the heating process of air. From the chart

$$h_4 = 24 \text{ kJ/kg}, \quad h_5 = 31.5 \text{ kJ/kg} \quad \text{and} \quad dbt_4 = 8°\text{C}$$

(*viii*) Capacity of the cooling coil $= m_a (h_3 - h_4)$

$$= 4.908 (56 - 24) = 157 \text{ kJ/s}$$

Since 1 ton of refrigeration is equivalent to 3.5 kJ/s, we get

Refrigeration load of the cooling coil $= \dfrac{157}{3.5} =$ **44.86 TR**

By-pass factor BPF $= \dfrac{dpt_4 - dpt}{dbt_3 - dpt} = \dfrac{8-5}{24.5-5} =$ **0.1538**

Capacity of heat coil,

$$= m_a (h_5 - h_4)$$

$= 4.908 (31.5 - 24) =$ **36.81 kW**

EXAMPLE 15.15

The following data relates to the administrative office which has a maximum seating capacity for 25 employees :

Outside design conditions ... 35°C dbt and 28°C wbt

Inside design conditions... 24°C dbt and 56% RH

Latent and sensible heat gain per person .... 100 W and 90 W respectively.

Solar and lighting load ... 10 and 2.5 kW respectively

Sensible heat load from other sources .... 12.5 kW

Infiltration rate ...15 m³ per minute

Make calculations for the dew point temperature of the coil and the plant capacity.

You may assume 40% fresh air, 60% of recirculated air passing through the evaporator coil, and a by-pass factor of 0.15. ***(MRIU)***

Solution : Refer Fig. 15.14 for the flow diagram of the air-conditioning system, and the psychrometric chart corresponding to the given data :

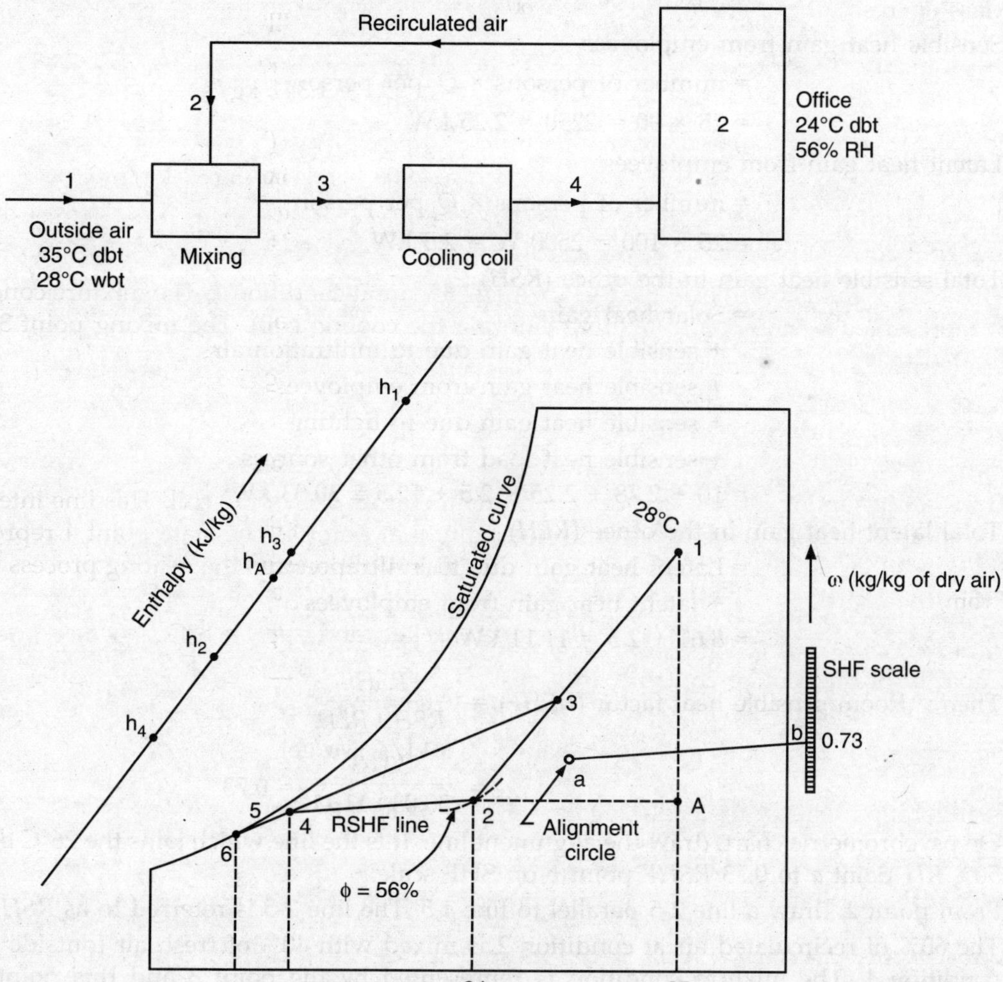

Fig. 15.14

(*i*) Locate state point 1 (outdoor conditions = 35°C *dbt* and 28°C *wbt*) and state point 2 (inside design conditions = 24°C *dbt* and 56% RH). Joint points 1 and 2, and mark point A where a vertical line drawn through point 1 meets the horizontal line drawn through point 2.

Then from psychrometric chart :

$$v_1 = 0.9 \text{ m}^3/\text{kg of dry air}$$

$$h_1 = 89 \text{ kJ/kg} ; \; h_2 = 48 \text{ kJ/g and } h_A = 58 \text{ kJ/kg}$$

(*ii*) Mass of infiltration air $m_1 = \frac{V_1}{v_1} = \frac{15}{0.9} = 16.67$ kg/ min.

Sensible heat gain due to infiltraton air,

$$= m_1(h_A - h_2) = \frac{16.67}{60} (58 - 48) = 2.78 \text{ kW}$$

Latent heat gain due to infiltration air,

$$= m_1(h_1 - h_A) = \frac{16.67}{60} \times (89 - 58) = 8.61 \text{ kW}$$

(*iii*) Sensible heat gain from employees

= number of persons × Q_s per person

= 25 × 90 = 2250 = 2.25 kW

Latent heat gain from employees

= number of persons × Q_l per person

= 25 × 100 = 2500 W = 2.5 kW

(*iv*) Total sensible heat gain in the office (*RSH*)

= Solar heat gain

+ sensible heat gain due to infiltration air

+ sensible heat gain from employees

+ sensible heat gain due to lighting

+ sensible heat load from other sources

= 10 + 2.78 + 2.25 + 2.5 + 12.5 ≡ 30.03 kW

Total latent heat gain in the office (*RLH*)

= Latent heat gain due to infiltration air

+ latent heat gain from employees

= 8.61 + 2.5 = 11.11 kW

Then : Room sensible heat factor (*RSHF*) $= \frac{RSH}{RSH + RLH}$

$$= \frac{30.03}{30.03 + 11.11} = 0.73$$

(*v*) On psychrometric chart, draw the alignment line. It is the line which joins the 26°C *dbt* and 50% *RH* point *a* to 0.73 *RSHF* point *b* on SHF scale.

From point 2, draw a line 2-5 parallel to line *a-b*. The line 2-5 is referred to as *RSHF* line. The 60% of recirculated air at condition 2 is mixed with 40% of fresh air (outside air) at condition 1. The mixture condition is represented by the point 3 and this point is so located that length 2-3 = 0.4 × length 1-2

From the chart : dbt_3 = 28°C and h_3 = 64.5 kJ/kg

(vi) Let dbt_6 be the dew point temperature of the coil. Join points 3 and 6, and let the line 3-6 meet the 2-5 line at point 4.

$$\text{By-pass factor of the coil} = \frac{dbt_4 - dbt_6}{dbt_3 - dbt_6}$$

$$0.15 = \frac{dbt_4 - dbt_6}{28 - dbt_6}$$

We adopt trial and error method and compute dbt_4 and dbt_6 : That gives :

$$dbt_4 = 10.5°C \text{ and } dbt_6 = 7.4°C$$

From the chart : h_4 = 28.5 kJ/kg

(vii) If m_a is the mass of air delivered to room, then

$$m_a \times \text{heat removed} = RSH + RLH$$

$$m_a \times (h_2 - h_4) = RSH + RLH$$

That gives : $$m_a = \frac{28.78 + 11.11}{48 - 28.5} = 2.04 \text{ kg/s}$$

Capacity of the cooling coil = m_a (h_3 - h_4)

= 2.04(64.5 - 28.5) = 73.44 kJ/s

Since 1 ton of refrigeration is equivalent to 3.5 kJ/s, we get

$$\text{Refrigeration load of the cooling coil} = \frac{73.44}{3.5} = \mathbf{20.98\ TR}$$

REVIEW QUESTIONS

A. Conceptual and conventional questions.

1. State the importance of cooling load in case of an air-conditioning system.
2. List the different sources of heat gains in an enclosed space.
3. What is sensible heat load and latent heat load ?
4. Enumerate and describe briefly the components of internal heat gains.
5. Give a brief account of the following loads :
   (a) heat transmission load through building structure
   (b) product load and
   (c) occupancy load.
6. What is meant by the term infiltration ? How an estimate can be made of air leakage through window and door cracks ?
7. (a) How does infiltration load differ from ventilation load ?
   (b) How an estimate is made of the heat gain through ducts ?
8. Present a brief account of the salient aspects related to solar heat gain.
9. Explain the concept of equivalent temperature differential.
10. A classroom with a seating capacity of 25 students is to be air-conditioned for the following summer conditions:

Outdoor conditions	: 34°C *dbt* and 28°C *wbt*
Required comfort conditons	: 24°C *dbt* and 50% RH

Total heat gain through walls, floor and roof	: 407 kW
Solar heat gain through window glass	: 4.4 kW
Sensible and latent heat gain per student	: 85 W and 105 W
Sensible heat gain from other sources	: 11.6 kW
Total infiltrated air	: 14 m^3/min
Internal lighting load	: 15 lamps of 100 W each and 10 fluorescent fixtures of 80 W each
By pass factor of the oil	: 0.35

If the return and outdoor air are adiabatically mixed in the ratio 3 : 2 by mass and then passed through the conditioner coil, determine

(*a*) *dbt* and RH of supply air,

(*b*) dew point temperature of the coil, and

(*c*) capacity of the conditioner in kW. (**Ans.** 13°C and 92.5%; 7°C; 88 kW)

11. The design data for air conditioning of a restaurant hall with seating capacity for 50 persons is as detailed below:

Outdoor conditions	: 35°C *dbt* and 26°C *wbt*
Inside comfort conditions	: 27°C *dbt* and 55% RH
Total heat gain through walls, floor and foof	: 6.15 kW
Solar heat gain through glass	: 2 kW
Equipment sensible and latent heat gain	: 2.9 kW, 0.7 kW
Total amount of infiltrated air	: 1400 m^3/hr
Total amount of fresh air supplied	: 1600 m^3/hr
Sensible and latent heat gain for each person taking the meals	: 58 W and 44 W
Employees serving the meals	: 5
Sensible and latent heat gain for each employee serving the meals	: 58 W and 76 W
Sensible and latent heat added from the meals	: 3.5 W and 5.8 W

If the fan is installed before the conditioner and the air supplied to the room is at a minimum dry bulb temperature of 17°C, make calculations for the following parameters:

(*a*) Quantity of air delivered to the room in m^3/hr,

(*b*) Percentage of recirculated air,

(*c*) Refrigeration load on the coil in tons of refrigeration,

(*d*) Dew point temperature of the cooling coil and its by pass factor.

The motor power connected to the fan equals 8.8 kW

(**Ans.** 5520 kg/hr; 67.8%; 11.5 tons; 14°C and 0.19)

12. The following data pertains to the air-conditioning system of a big shop:

Outside conditions	: 40°C *dbt* and 27 *wbt*
Inside design conditions	: 25°C *dbt* and 50% RH
Sensible and latent heat load	: 58 kW and 14.5 kW

If 70 m^3/min of fresh ventilation air is supplied to the shop and the cooling coil has a by-pass factor of 0.15, determine:

(*a*) Effective sensible heat factor and apparatus dew point,

(*b*) Dehumidified air quantity, and

(*c*) Conditions of air entering and leaving the system.

(**Ans.** 0.78 and 11°C; 215 m^3/min; 30°C *dbt* and 22°C *wbt*; 14°C *dbt* and 13°C *wbt*)

B. Fill in the blanks with appropriate word/words

1. .... is the total heat required to be removed from the space for keeping it at the desired temperature by the air conditioning and refrigeration equipment.
2. .... and .... gains are the two main components of a cooling load imposed on an air conditioning plant operating during summer.
3. The latent heat gain when the outside air is introduced for ventilation purposes is due to difference of .... between the fresh air and air in space.
4. The solar heat reaches any part of the earth's surface in the form of .... and .... radiations.
5. .... is the hypothetical temperature used to calculate the heat received by the outside surface of a building wall by the combined effect of convection and radiation.
6. The sol-air-temperature t_e is prescribed by the relation

$$t_e = t_0 + \frac{I\alpha}{h_0}$$

where I is the .... and α is the .... of the wall surface.

7. .... and .... are the two methods used for estimation of air leakage through window and door cracks.
8. The heat loss or heat gain due to absorptivity infiltration of outside air into the conditioned space is called ....
9. .... is obtained by adding by-passed air load to the room sensible heat.
10. The sensible and latent heat gains due to occupants, lights, appliances, machines and piping etc. within the conditioned space form the components of ....

Answers

1. cooling load 2. sensible heat, latent heat 3. humidity 4. beam (direct), diffused (sky) 5. sol air temperature 6. total radiation intensity 7. air change, crack length 8. infiltration load 9. effective room sensible heat 10. internal heat gain

❑❑❑

Chapter – 16

Ventilation, Filtration and Air Distribution

Learning objectives : A study of the subject matter presented in this chapter will enable the students/readers to have an overview of

- *standard and state of excellence of air in an enclosed space*
- *ventilation, filtration and the duct system needed for air movement and proper distribution in the air conditioning systems*
- *pressure losses in ducts and duct design methods*
- *salient features of centrifugal and axial flow fans used for transmission of air in air conditioning systems*

Mere conditioning of air with regard to the maintenance of acceptable thermal and humidify conditions is not sufficient to provide an environment suitable for the health and comfort of human beings. In addition, the air needs to be purified, filtered and made free from airborne pollutants such as dust, smoke, pollens, odour, allergens, bacteria, metabolic products etc. This implies that a state of excellence of air has to be ensured in the enclosed space. Further, a required volume of treated and conditions air has to be supplied and uniformly distributed in the different enclosures that are to be maintained at specific design conditions.

16.1. INDOOR AIR QUALITY (IAQ)

Indoor air quality refers to the standard and state of excellence of the air in an enclosed space. Broadly it defines the nature of air that affects the health and well being of occupants. It is essentially related to how well indoor air satisfies the three basic requirements for human occupancy :

(*i*) thermal acceptability,
(*ii*) maintenance of normal concentrations of respiratory gases; and
(*iii*) dilution and removal of contaminants to levels health or odour discomfort thresholds.

The concept of IAQ is constantly changing and is governed by many factors that include :

(*i*) sources of pollutants and odours
(*ii*) moisture and humidity
(*iii*) maintenance and operation of building ventilation systems
(*iv*) occupant perceptions and susceptibilities, physical, emotional and mental reactions of the occupants.

Recent findings have demonstrated that indoor air is often more polluted and a greater health hazard than the corresponding outdoor setting. Further, people spend most of their time indoors and the concentration of contaminants may build up in an enclosed space. The poor indoor air quality can have a significant impact on people's lives, especially those who are most vulnerable such as infants and children, pregnant woman, the elderly and those who have chronic illnesses.

Pollutants and their sources : Indoor impurities can originate in the outdoor air, be carried by the outdoor or indoor air or be generated by building materials, building occupants and their activities. Some of the common pollutants are :

(*i*) **Radon :** an invisible radioactive atomic gas that results from radioactive decay of some forms of uranium that may be found in rock formations beneath buildings or in certain building materials.

(*ii*) **Molds and allergens :** a soft, grey, green or blue substance that sometimes forms in spots on damp walls or clothes. Moisture build up inside buildings may arise from water penetration from plumbing leaks or from ground moisture. The molds, in general, contain some toxic compounds having allergenic properties that can trigger a serious respiratory problem in persons that already have asthma.

Indoor mold growth can be inhibited by keeping humidity levels low and by eliminating leaks or moisture condensation and accumulation.

(*iii*) **Carbon monoxide :** an acutely toxic indoor contaminant resulting from tobacco smoke. Each cigarette has the potential to release billions of particles containing smoke, dust, aerosol and vapours. By depriving the brain of oxygen, high levels of carbon monoxide can lead to nausea, unconsciousness and even death.

(*iv*) **Legionellosis :** a disease caused by water borne bacterium that grows best in slow moving or still warm water. A common source of this in commercial buildings is from poorly placed or maintained evaporative cooling towers, which often release aerosolized water that may enter nearby ventilation intakes.

(*v*) **Asbestos fibres :** many common building materials such as floor and ceiling titles, pipe wrap release asbestos when disturbed by cutting, sand and drilling. Inhalation of asbestos fibres over long exposure times is associated with increased incidence of lung cancer.

(*vi*) **Carbon dioxide :** a surrogate for indoor pollutants emitted by humans and correlates with human metabolic activity. Carbon dioxide at levels that are unusually high may cause occupants to grow drowsy, get headaches, or function at low activity levels.

(*vii*) **Dust :** a household containment that contains a variety of biological, organic and inorganic agents including mites, viruses, bacteria, and asbestos fibres. The dust particles act as carriers for microbes and outdoor pollutants.

Effects of indoor air pollution : Successive studies have shown that poor air quality does affect the comfort, health, productivity and morale of the occupants.

Basically the effects of indoor air pollution are categorized as :

(*a*) respiratory, heat and cancer disease
(*b*) chronic obstructive pulmonary disease
(*c*) infectious disease and microbial toxins
(*d*) immunologic disorders
(*e*) irritation phenomenon and odours

Surprisingly, irrtants or odour annoyance and distraction phenomenon are most frequently discussed and documented, even though least risky.

These health effects may be experienced soon after exposure or may show up later after exposure has occurred or only after long or repeated periods of exposure.

Sick building syndrome (SBS) and Building related illness (BRI) : The common symptoms associated with discomfort due to poor indoor air quality are :

- Eyes, nose and throat irritation
- Headache, lethargy, fatigue and dizziness
- Nasal congestion and difficulty in breathing
- Dry skin and rashes

This group of loosely associated symptoms is known as ***sick building syndrome.*** These symptoms are known to occur in more than 20% those exposed and are found to disappear when the person exits the building. SBS are not fundamental health hazards; rather they correlate more with reasonable expectations of comfort and wellness.

In contrast, clinical contributions to indoor air quality have demonstrated a set of so called ***building related illness*** which are the diseases resulting from exposure to indoor air and pollutants. These diseases have defined diagnostic criterion, recognizable causes and defined treatments. Examples of BRI include respiratory problems, humidifier fever, infections (legionnaire's disease) and skin diseases.

How to improve indoor air quality : Using ventilation to dilute contaminants, filtration and source control are the primary methods for improving indoor air quality in most buildings. The samples on air and building surfaces are analyzed for mold, bacteria and chemicals etc. The findings can lead to an understanding of the sources of contaminants and ultimately to stretegies for removing the unwanted elements from air.

The following steps have been suggested to help improve indoor air quality at home, the work place and in other indoor environments :

- ensure good ventilation
- conduct regular inspection of air-conditioning system to verify that there is no internal contamination build up contributing to poor indoor air quality
- change filters and air cleaners according to the manufacturer's directions
- use natural household cleaning products and reduce exposure to potentially toxic airborne substances
- do not smoke or allow smoking in the home, and avoid indoor areas where smoking takes place
- avoid or reduce biological contaminants by proper maintenance of humidifiers, dehumidifiers and air-conditioners

16.2. VENTILATION

Ventilation is the continuous injection of certain amount of outside fresh air into air-conditioned space to offset the contaminants produced by occupants. Ventilation serves to meet the following conditions for the comfort and wellness of the occupants of the enclosed space :

(*i*) freedom form dust, bacteria and noxious fumes

(*ii*) provide oxygen, remove respired carbon dioxide and dilute the body odours

(*iii*) maintain the standard and state of excellence of the air, *i.e.*, provide acceptable indoor air quality

Obviously, ventilation air is necessary to dilute odours and limit the concentration of carbon dioxide and airborne pollutants such as respirable suspended particles and volatile organic compounds. Inadequate ventilation in a densely occupied space can cause the level of CO_2 to increase leading to sleepiness and reduced efficiency at work. It is worthwhile to state that injection of fresh air does not eliminate the contaminants but it dilutes their concentration and this helps to improve indoor air quality.

Ventilation rate : In commercial, industrial and institutional buildings, air is often recirculated. However a fraction of the return air is normally exhausted and replaced by outside fresh air. ASHRAE has provided certain standards which prescribe the rate of ventilation required. The ventilation rate is normally expressed by the flow rate of outside air being intentionally induced into the building. The typical units are m^3/min or litres/sec. Often the ventilation rate is expressed on per person or per unit floor area basis such as m^3/min/person or $m^3/min/m^2$. For residential buildings, the common measure of ventilation rate is the number of times the whole interior volume of air is replaced per hour and is called air changes per hour.

Ventilation methods : The process of the exchange of outdoor air with the air already in the building can be broadly classified into ventilation and infiltration.

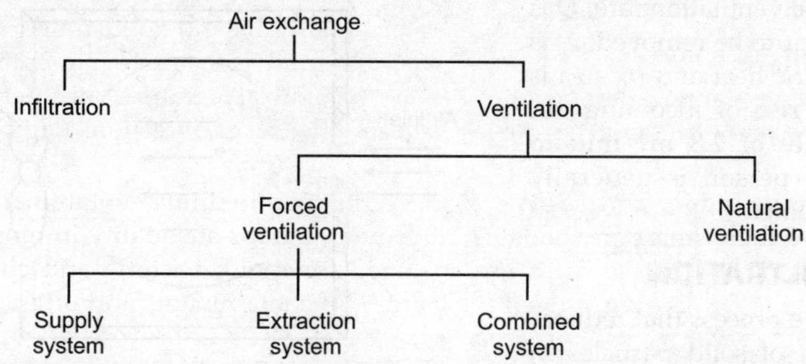

Infiltration is the uncontrolled flow of outdoor air into the building through the building fabric having cracks and other unintentional openings. However, this too serves to provide ventilation air that offsets the contaminants produced by occupants.

Natural ventilation occurs when the air in the enclosed space is changed with outdoor air without the use of any mechanical system such as a fan. The natural ventilation occurs through

- windows, doors, grills, louvers
- roof ventilators, and
- inlet and outlet openings that are specially designed.

The natural ventilation occurring in buildings may be wind driven ventilation or stack ventilation. The pressures generated by buoyancy (also known as stack effect) are quite low while wind pressures are usually far greater. There is no control over the flow and the dust particles from the surroundings may enter the ventilated space.

Mechanical or ***forced ventilation*** is through an air handling unit or direct injection into a building using fans. The fan would also enhance infiltration and natural ventilation, thus increasing the ventilation flow rate. The quantity and direction of flow can be easily controlled by regulation of fan speed.

Mechanical ventilation systems are available in three different arrangements and their schematics are as shown in Fig. 16.1.

- **Supply system :** The fan forces air to the ventilated area. The action is similar to that of a forced draught and the system maintains positive pressure gradient in the conditioned space.
- **Extraction system :** The fan sucks air from the application area. The action is similar to that of induced draught and the effectiveness of the system depends on location of air inlets in the ventilated space.
- **Combined system :** The fans supply air to the conditioned space and also extract air from it. The action is similar to that of balanced drought.

The supply system arrangement is generally favoured in the air-conditioning systems.

The ventilation flow needed to remove a given amount of heat from a building is generally worked out from the relation

$$V = \frac{Q}{c_p(t_0 - t_i)} \quad ...(16.1)$$

where V is the ventilation rate, Q is the amount of heat to be removed, c_p is the average specific heat and $(t_0 - t_i)$ is the temperature rise of incoming air. A ventilation rate of 2.8 m^3/min to 5.7 m^3/min per person is generally considered adequate.

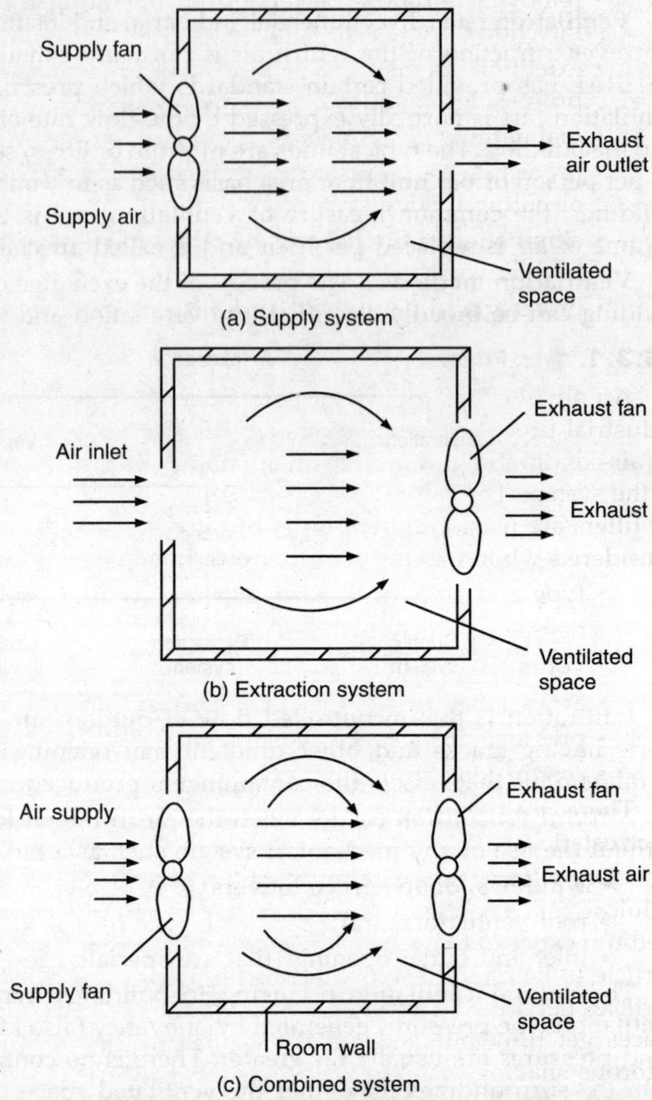

Fig. 16.1. Mechanical ventilation systems

16.3. AIR FILTRATION

Air filtration is the process that reduces the concentration of solid particles in an airstream to a level that can be tolerated in a process or space occupancy. Solid particles in the airstream range in size from 0.01 micrometer to objects that can be caught by ordinary fly screens such as lint, feathers and insects. The particles generally include :

- solid impurities such as soot, ash, smoke, fumes and soil lint etc.
- living organisms such as virus, bacteria and mold spores

The indoor air too gets polluted due to :

- inadequate ventilation
- use of variable air volume (VAV) system of air-conditioning
- improper filteration
- reduced infiltration due to air tight structures, and
- carpets and furnishings etc. made from synthetic materials

Thus both the outside (fresh) air and recirculated air of an air-conditioning system would consist of dust, suspended impurities and vapour particles (tobacco smoke, perspiration) etc. The importance of air filtration extends from the simple task of preventing lint and other debris from plugging heating/cooling coils to remove particles as small as 0.1 micron which could cause a short circuit on a microchip. Air cleaning becomes a necessity for a wide variety of purposes, some of which include

- protecting the general well being, health and comfort of the occupants of the space

- removing airborne bacteria from operating room air to help prevent post operative infection
- extention of shelf life of perishale dairy products by removing airborne mold during processing operations
- maintaining cleanliness of room surface and furnishing
- protecting the contents of occupied spaces including paintings and other items of historic or cultural value
- elimination of fire hazards by removing lint and other materials which might accumulate in duct work
- protection of equipment and air-conditioning machinery

16.3.1. Air Filters

An air filter is a device or component of most systems where good quality air is used for industrial processes, ventilation or comfort air-conditioning. Filters improve the performance of an air-conditioning system by maintaining air purity, improving life and reducing maintenance of the system. Due to wide variety of airborne contaminants and the diversity of systems where air filters are used, different types of filter are commercially available. The following factors are considered while making selection of a filter needed for a particular application :

- type and amount of particular matter in air to be removed
- degree of purity required
- volume of air handled
- arrangement made for the disposal of dust etc.
- pressure drop, space available and initial cost
- cost of maintaining or replacing filters

There are three basic types of air filters in common use : dry, viscous impingement and electrostatic.

- **Dry filters :** The filtering medium consists of fabrics made of cloth, coarse paper, wood or cellulose felt. The medium is supported in pockets of wire frame and that increases the area of medium exposed to air. The intevening spaces between the medium are smaller than the smallest particle to be filtered. When the air flows over the filter in a steady stream, the associated dust particles get arrested and not allowed to move further. The air flow ceases when the intervening spaces get fully plugged up. After collection of full dust load, the filter is either cleaned by vigorous shaking or is replaced by a new one.

- **Wet filters :** The filter incorporates an arrangement for wetting the dust particles by water spray and that makes the heavier particles settle down. The filter is, however, ineffective in removing the greasy particles like pollens present in air. These filters are normally employed in industrial air-conditioning systems.

- **Viscous filters :** The filter is made up of a relatively loosely arranged medium consisting of spun glass fibres, metal screens or layers of crimped expanded metal whose surfaces are coated with oils and greases. In the impingement type viscous filter, the air stream is forced to change direction frequently as it passes through the filter medium, and viscous layer draps dust and other particles from air. In the automatic viscous filter, there is a continuous roll of material coated with oil and is motor driven across the stream. The oil coated roller passes over rollers and is made to move alternately through an oil trough and the air steam. The trough serves to wash off the dirt and recoat the roll fabric with relatively clean oil.

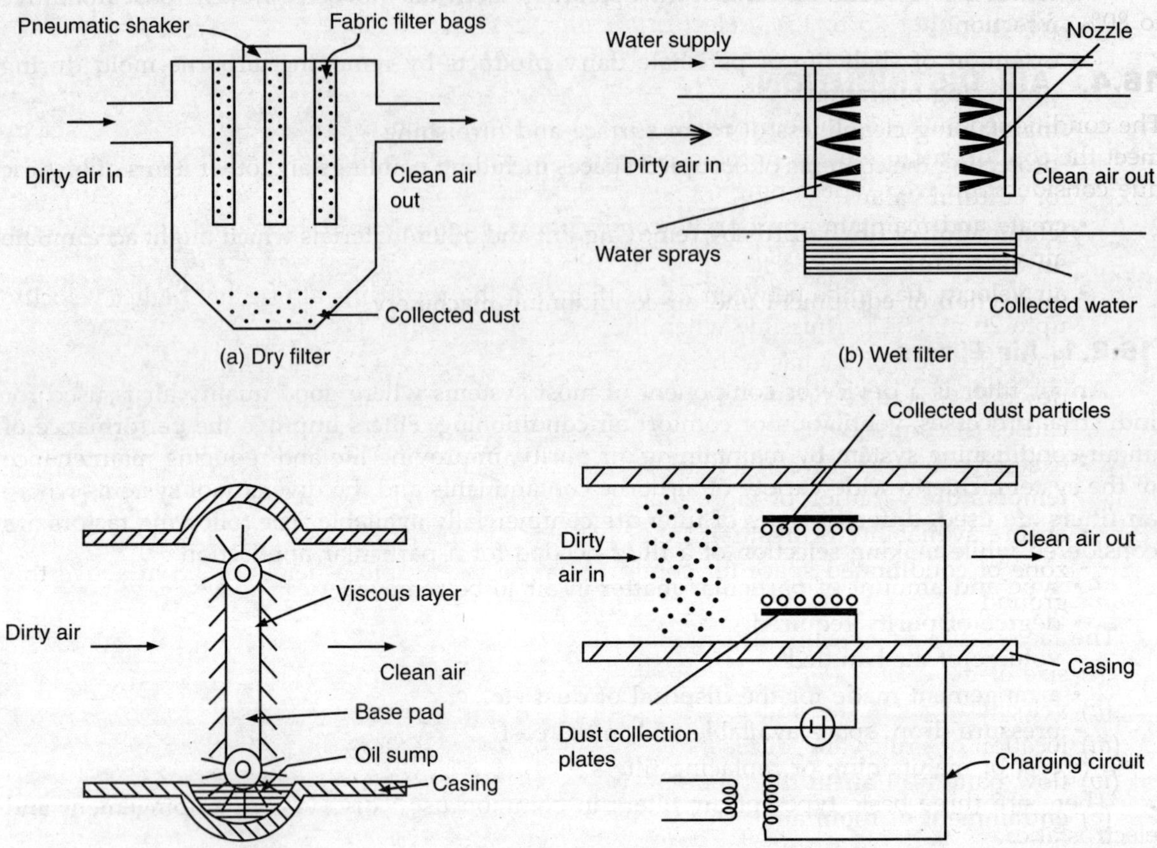

Fig. 16.2. Air-conditioning filters

• **Electrostatic filters :** The electrostatic air cleaner is a particulate collection device that removes particles from flowing air using the force of an induced electrostatic charge. The system operates by charging the particles inductivity with an electric field, then attracting them to collector plates which have been given an opposite charge. From time to time, the collector plates are removed and cleaned from the particles which got attached to it.

Electrostatic filters are highly efficient filtration devices that minimally impede the flow of gases, can easily remove fine particulate matter with a reasonably pressure drop, have the ability to handle large volume of gases and as such have become common in numerous industrial applications.

The overall effectiveness of an air cleaning device depends upon the efficiency of the unit, static pressure drop across it and the amount of air drawn across it.

The efficiency of an air filter is measure of its ability to remove contaminants, and is defined as

$$\eta_f = \frac{W_1 - W_2}{W_1} \times 100 \qquad \text{... (16.2)}$$

where W_1 is weight of dust particles in unit volume of air entering the filter, and W_2 is weight of dust particles in unit volume of air leaving the filter.

The efficiencies of different air filters lie in the ranges : dry filter : 60 to 95% ; wet filter : 75 to 80% ; viscous filter : 5 to 15% ; electrostatic filter : 70 to 90%.

16.4. AIR DISTRIBUTION

The conditioned air needs to be properly and uniformly distributed within the enclosed space to meet the comfort requirements of the occupants. For that, the following points have to be given due consideration while designing the air distribution systems :

- create and maintain appropriate combination of temperature, humidity, air purity and air velocity.
- air velocity has to be between 7.5 to 15 m/s if the occupants are sitting; higher velocity upto 20 m/s is permissible when the occupants are moving.
- flow direction of air has to be towards the faces (not back side) of occupants.
- down flow is to be preferred over upward flow; the upward flow produces draft which causes discomfort.
- excessive temperature fluctuations and excessive air motion has to be avoided; a temperature variation of more than 2°C is not to be allowed.
- ensure availability of fresh air in all parts of the room.
- zone of conditioned space for comfort is to be taken upto a height of 1.8 m above the ground.

The factors affecting the air distribution are :

(*i*) size of the duct
(*ii*) velocity of air in the duct
(*iii*) location of grilles and diffusers
(*iv*) flow pattern of air in the room
(*v*) entrainment of room air

16.4.1. Terms Related to Air Distribution System

Some of the terms related to air-distribution system are :

- ***Intake and outlet*** : Intake is the opening through which air is returned or exhausted from the space. The opening through which air is supplied to the spaced being conditioned is called outlet.
- ***Grille and diffuser :*** A grille is a perforated covering that controls the direction of flow of air and provides a decorative appearance to the end of duct. Grilles consist of a frame and fixed or adjustable parallel bars. The fixed grilles direct the air in one direction only while adjustable grilles can be arranged to direct the air in different directions. Registers are grilles with volume control dampers mounted behind the grille. A diffuser is the outlet grille designed to guide the direction of air. The diffusers ensure proper supply of conditioned air to the room.

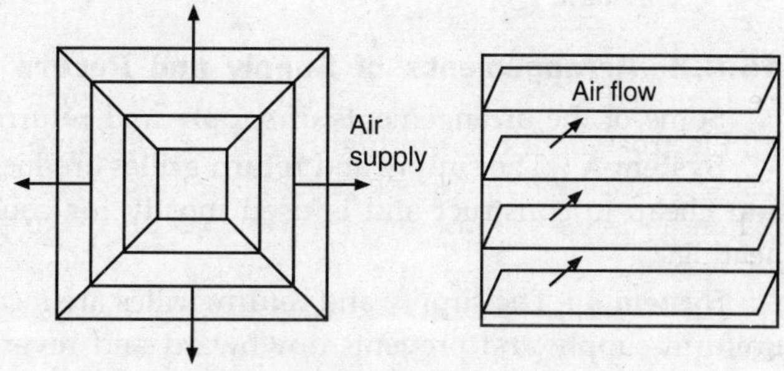

Fig. 16.3. Diffuser and grille

Grilles and diffusers improve the aesthetics of the air-conditioning system. They are usually located at the level of false ceiling which is provided to conceal the duct.

- *Throw, drop, spread :* Blow or throw is distance travelled by the supply air stream in the horizontal direction on leaving the air outlet and reaching a specified low velocity known as terminal velocity. The velocity is measured at 1.8 m above the floor level and is usually taken as 0.25 m/s, 0.5 m/s or 0.75 m/s depending on the application. Drop is the vertical distance the air moves after it leaves the outlet and reaches the end of throw.

  Spread is the angle of divergence of air stream after it leaves the outlet. Straight vanes produce a spread of 14° to 24°. Converging vanes produce the same spread as the straight vanes but are 15 percent longer. Diverging 45° vanes produce a spread of 60°.

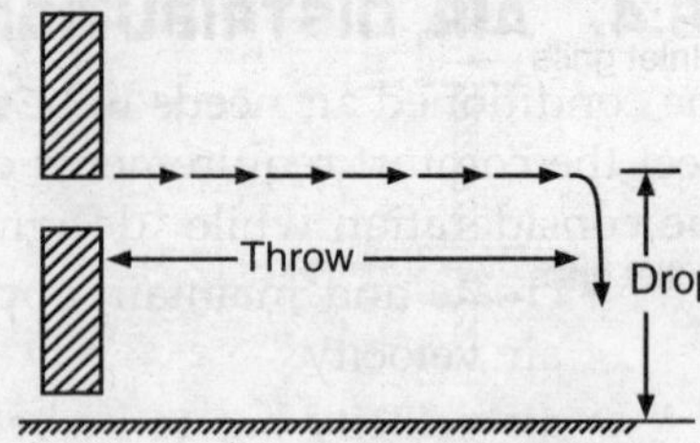

Fig. 16.4. Throw and drap

- *Slot and ceiling diffusers :* A slot diffuser is a long strip shaped outlet with one or more narrow openings. This elongated outlet has an aspect ratio of 25 : 1 and its maximum height is limited to 7.5 cm. The slot diffusers are fitted in high side wall locations or in perimeter installations in floors.

  The ceiling diffusers consists of separated concentric rings or louvres with a collar or neck to connect to duct. they may be round, square, rectangular or in the form of perforated panels. The louvres may be of adjustable pattern or variable area type. These diffusers are mounted in the ceilings.

- *Primary, secondary and total air :* Primary air is the air that comes out of the outlet. The room air picked up by the primary air by entrainment is called secondary air. The sum of primary air and secondary air is called the total air. The induction ratio is defined as the ratio of total air to primary air.

  Blow depends on the initial supply air velocity, temperature difference between the supply and room air and induction ratio.

- *Draft* : Draft is the localised feeling of coolness or warmth of any portion of the body due to both air movement and air temperature, with humidity and radiation considered constant.

16.4.2. Arrangements of Supply and Return Air

Some of the arrangements of supply and return air have been depicted in Fig. 16.5.

System A : The supply and return grilles are located on the same wall. The system is simple and cheap to construct and is used mostly for cooling (summer air-conditioning) rather than heating.

System B : The supply and return grilles are located on different walls. The system provides uniform supply and prevents downward and reverse current. However, there is possibility of creation of a stagnant space below the supply duct.

Another arrangement for imparting still better air distribution is obtained when the inlet and outlet ducts are installed as shown in system C.

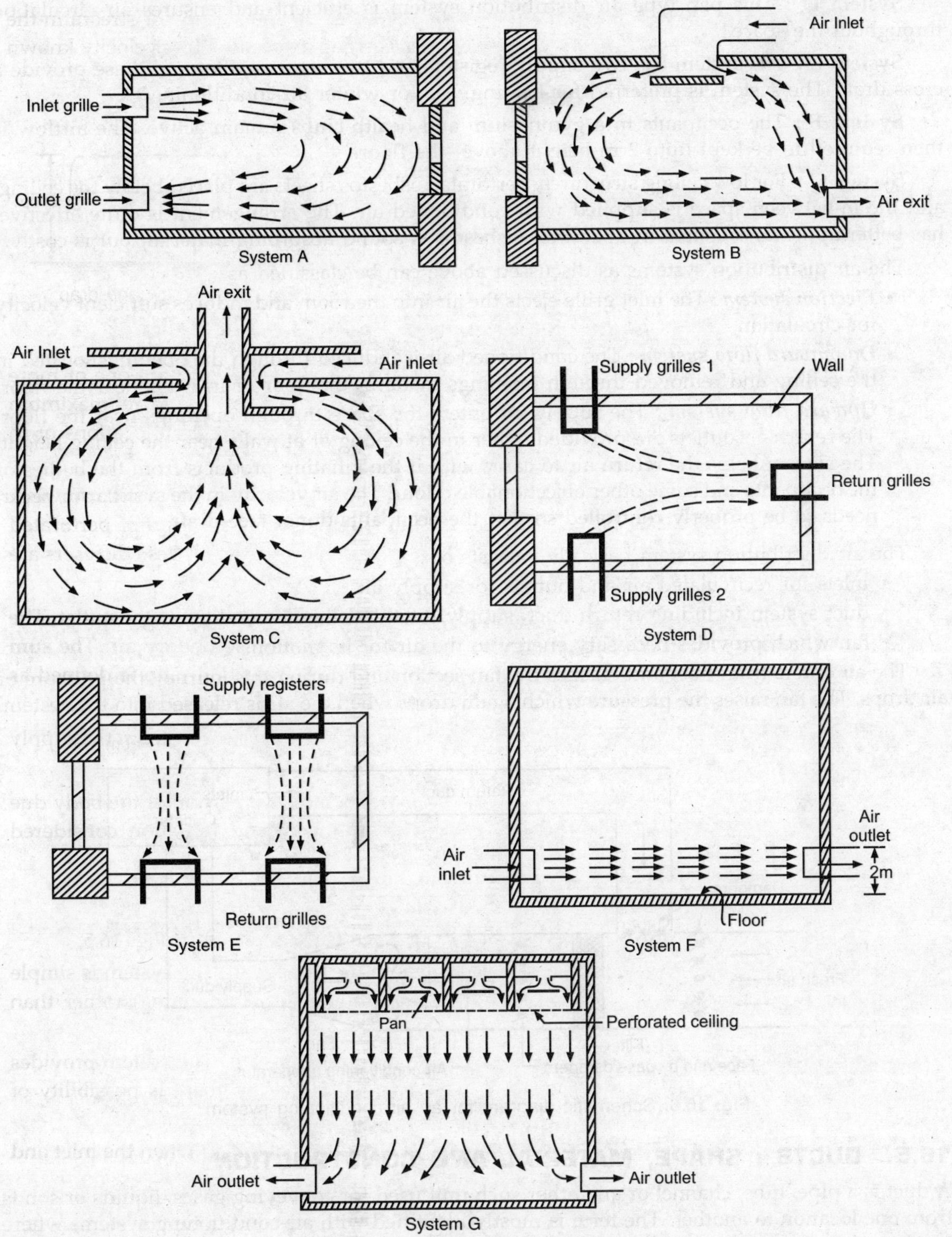

Fig. 16.5. Air distribution combinations

System D : This pan type air distribution system is efficient and ensures air circulation throughout the space.

System E : The system has two supply registers and two return grilles and these provide a cross draft. The system is preferred for heating, *i.e.*, for winter air-conditioning.

System F : The occupants in a gymnasium and health clubs remain active. The airflow is then required to be kept upto 2 m height above the floor.

System G : For low ceiling structures, perforated asbestos sheets are placed below the ceiling, and the in-between space is supplied with conditioned air. The arrangement is quite effective, has better appearance, silent in operation (asbestos is sound absorbing material) but is costly.

The air distribution systems as discussed above can be classified as

- ***Ejection system :*** The inlet grille ejects the air into the room and induces sufficient velocity for circulation.
- ***Downward flow system :*** The conditioned air is induced through the openings located in the ceiling and removed through openings made in the floor or in walls near the floor.
- ***Upward flow system :*** The supply air enters the space through openings near the floor. The return air outlets are provided either in the ceiling or at walls near the ceiling height. The system helps the return air to carry with it the vitiating products from the bodies of the occupants and some other objectionable odour. The air velocity in the system, however, needs to be properly controlled so that the occupants donot face draft.

The air distribution system basically consists of :

- inlets for recirculated air and outlets for supply air
- duct system including return duct, supply duct and the air-conditioning apparatus.
- fan which provides necessary energy to the air for its motion.

The air enters from inlet and reaches the fan section and during this journey the pressure of air drops. The fan raises the pressure which again drops when the air is released into the system.

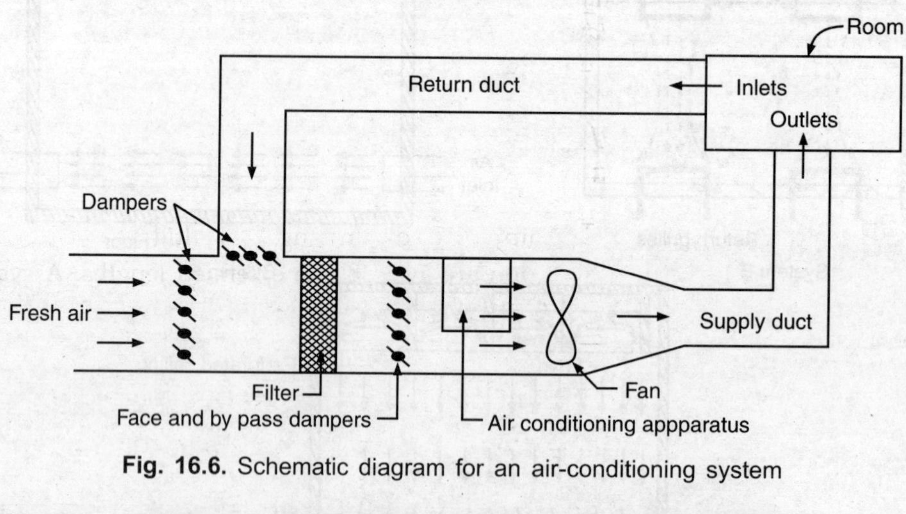

Fig. 16.6. Schematic diagram for an air-conditioning system

16.5. DUCTS : SHAPE, MATERIAL AND CONSTRUCTION

A duct is a pipe, tube, channel or any other such unit used for conveying gases, liquids or solids from one location to another. The term is mostly identified with air-conditioning systems where transfer of air is necessary. In that context, ducts are the air passages that convey the conditioned

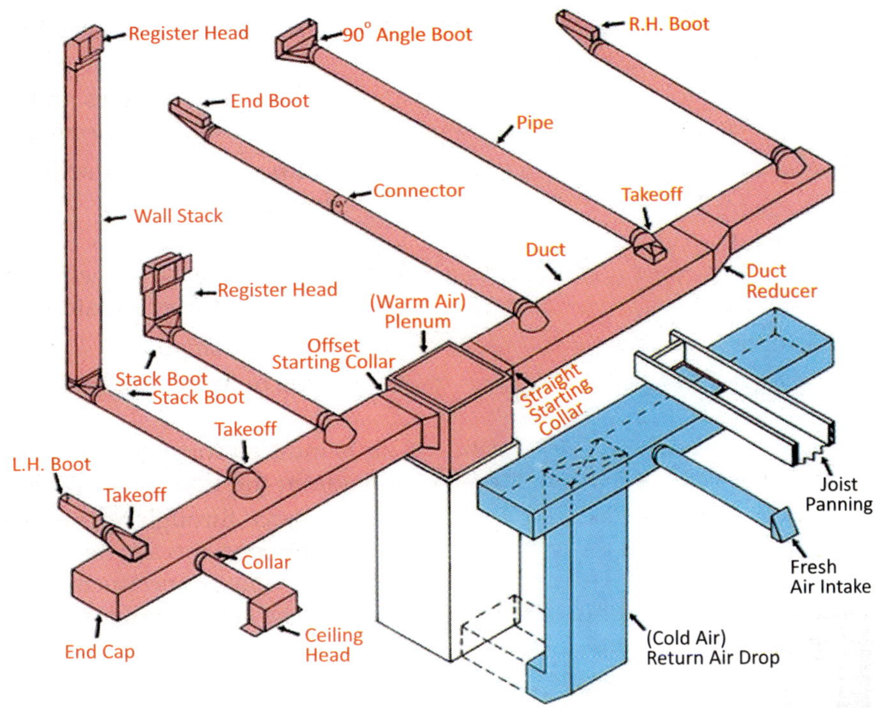

(a)

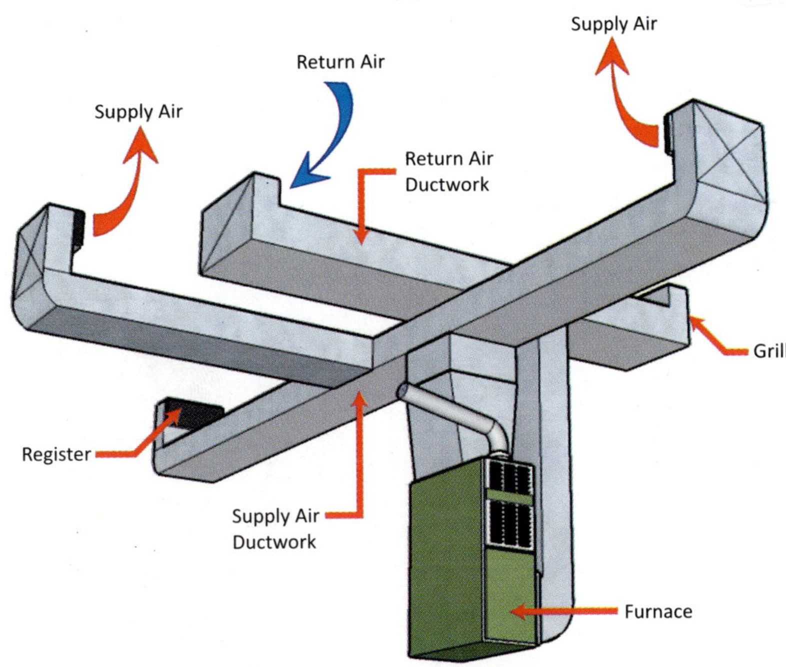

(b)

Duct Systems

air from air handling unit to the air distribution points (air supply outlets in the room) and carry the return air from the room back to air handling unit for reconditioning and recirculation. The installation is referred to as duct work.

16.5.1. Duct Classification

Ducts are generally categorized on the basis of :

(*i*) Air supplied :

- Supply air duct : a duct which supplies the conditioned air from the air-conditioning equipment to the space to be conditioned.
- Return air duct : a duct which carries the recirculation air from the conditioned space back to the air-conditioning equipment.
- Fresh air duct : a duct which supplies outside fresh air to the system.
- Exhaust duct : a duct which exhausts a part of air leaving the conditioned space to the atmosphere.

(*ii*) Pressure in the duct :

- Low pressure duct : the static pressure in the duct is 50 mm of water column or less.
- Medium pressure duct : the static pressure in the duct is from 50 mm to 150 mm of water column.
- High pressure duct : the static pressure in the duct is higher than 150 mm of water column.

(*iii*) Velocity of air in the duct

- Low velocity duct : the velocity of air in the duct is upto 10 m/s
- High velocity duct : the velocity of air in the duct is more than 10 m/s

16.5.2. Duct Shape

Ducts are usually fabricated in round or rectangular shapes.

(*i*) Round ducts have a smaller pressure drop per unit area and are generally the most cost effective; they can carry higher amount of air in less space. The use of round ducts instead of rectangular ducts saves about 20% metal weight. Further, round fittings are commercially available and provide convenient connection to flexible duct; they are easier to insulate and seal. However, the round ducts require more clear height for installation and the larger sizes may cause shipping and handling difficulties.

(*ii*) Round ducts with spiral construction are available in extended lengths. A system with long sections has few joints with low pressure losses and less leakage than a system constructed of many shorter sections.

(*iii*) Recently, there has been development of flat oval duct which is essentially a round spiral duct that has been stretched to an oval shape. These ducts have smaller height requirements and also retain most of the advantages of round ducts. However, their fittings are difficult to fabricate or modify at the site. They also have a tendency to become more round under pressure.

(*iv*) Rectangular ducts provide flat surfaces and are conveniently fabricated. However, they create high pressure drop, use more material for the same air-flow rate as round duct. Further, the joint length is limited to the sheet width, joints are more difficult to seal, and the rectangular transverse joints are more costly to install.

Aspects ratio, the ratio of long side to short side, is an important factor to be considered in the design of rectangular ducts. At higher aspect ratio of the duct, more metal surface is required for the same cross-sectional area. That increases not only the initial cost but also the operating cost by way of increased heat gain or loss. A rectangular duct should be designed for an aspect ratio as low as possible; the duct should be as nearly square as possible.

16.5.3. Duct Material

The selection of duct material depends largely on the nature of environment, required life of the duct and initial investment etc.

(*i*) Galvanized iron sheet is the most commonly used material for duct construction. The sheet thickness varies from 26 gauge (0.55 mm) to 16 gauge (1.6 mm). These sheets are given a coating of zinc to prevent rusting.

(*ii*) Aluminium sheet metal has the benefit of light weight and resistance to moisture.

(*iii*) Fibre glass ducts are used in low velocity applications where thermal insulation or acoustic treatment is required.

(*iv*) Cement asbestos ducts are employed for underground air distribution and for exhausting corrosive material.

(*v*) The wooden ducts are employed in places where moisture content of air is not very large.

16.5.4. Duct Construction

Some of the general considerations in the fabrication/construction of a duct system are :

(*i*) Various types of sheet metal joints have been suggested and are being used in the construction of ducts. Duct hangers are used to support long runs of ducts.

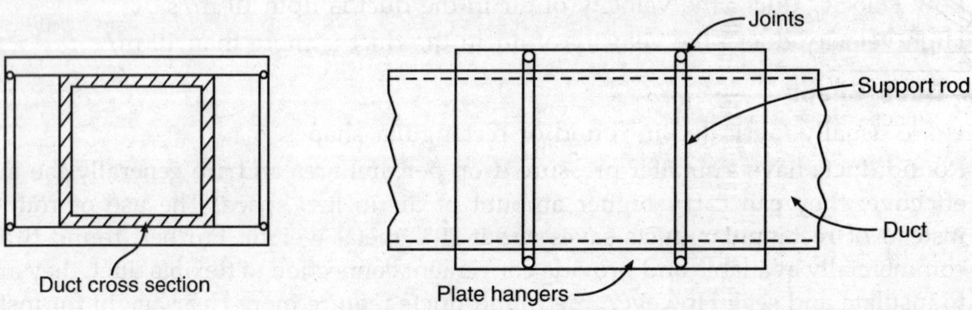

Fig. 16.7. Duct supports

(*ii*) The joints are to be air tight and strong. To obtain additional strength, the joints are often riveted. Further, the joints are made leak proof by sealing them with special tape.

(*iii*) When the ducts are larger and are to operate under high pressure, duct reinforcement is recommended. With reinforcement, the chance of bulging or collapsing of the duct and separating of the seam gets reduced. Duct reinforcing is done by sheet metal itself or by steel or extruded aluminium angles.

(*iv*) The fabric joints are employed when ducts fasten to air-conditioner and also in situations where there is possibility of expansion/contraction of sheet metal due to hearing/cooling. The fabric joints have the advantage of absorbing such movements.

(*v*) Quite often, the duct has to travel through an unconditioned space. In that situation, the outside surface of the duct is provided insulation to reduce the heat gain or heat loss. The insulation material is fastened to the duct surface either with an adhesive or with metal clips.

16.5.5. Duct System

The ducts employed in the air handling system of an air-conditioning plant are categorised as :

- ***Supply ducts :*** These convey air from the air handling unit to the supply air grilles or diffusers for its further distribution in the room.
- ***Return ducts :*** These extract the exhaust gas from the conditioned space and take it back to the air handling unit for recirculation.

The supply air ducts are normally arranged in the following ways :

1. **Loop perimeter duct system :** The arrangement has several feeder ducts which carry the conditioned air from the plenum to a common duct that runs all along the room space to from a continuous closed loop. The air is then supplied to the room through outlets fitted in the common duct. The system has application in concrete slab construction on ground for offsetting down draft and cold floors.

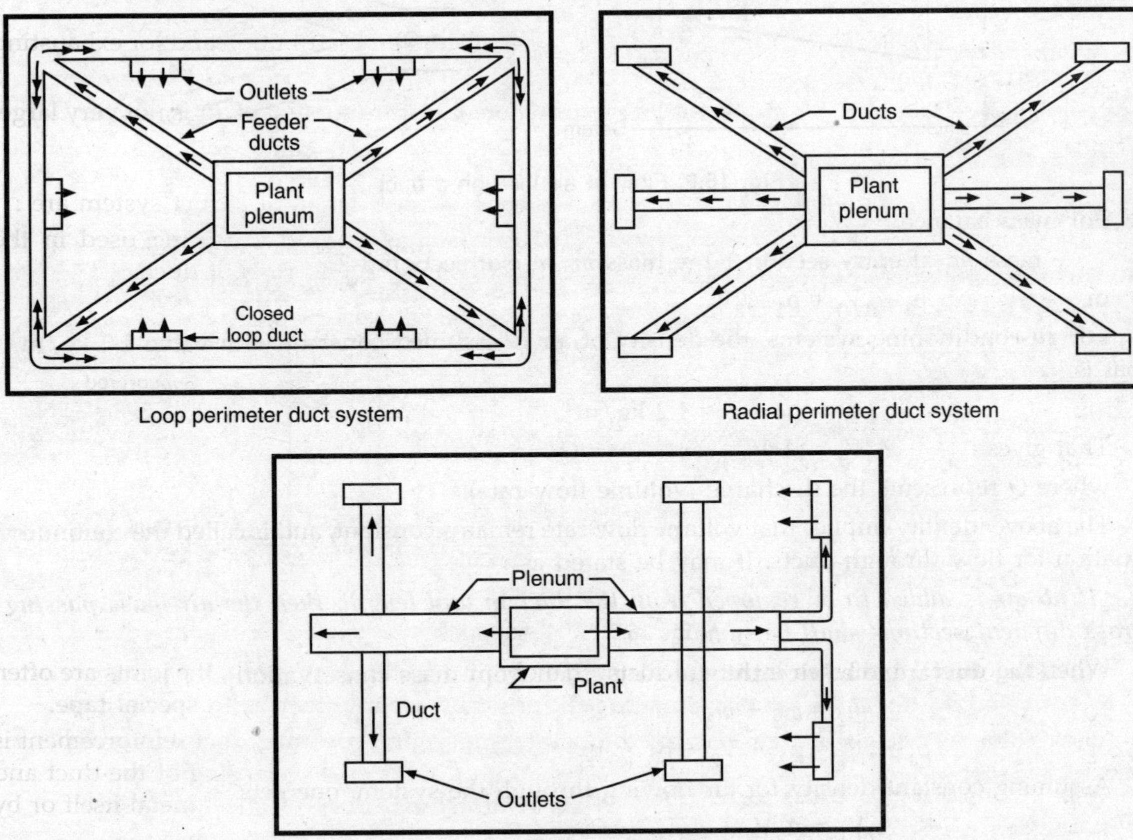

Fig. 16.8. Arrangements for supply ducts

2. **Radial perimeter duct system :** The arrangement has separate/independent ducts which supply conditioned air from the plenum to the respective supply outlets. This arrangement is normally employed for crawl space houses and in slab floors. However, it is not as effective as the loop system in warming the perimeter of floor.

3. **Extended plenum duct system :** The plenum of the unit has extensions provided towards one or both sides of the air-conditioning plant. There is a separate duct for feeding the conditioned air to the supply outlet. The system is adopted for houses with basement or crawl space.

16.6. FLOW THROUGH A DUCT

The salient aspects of flow through a duct system are :

16.6.1. Continuity Equation

Consider the flow of air through a duct between two sections ; inlet section 1-1 and outlet section 2-2.

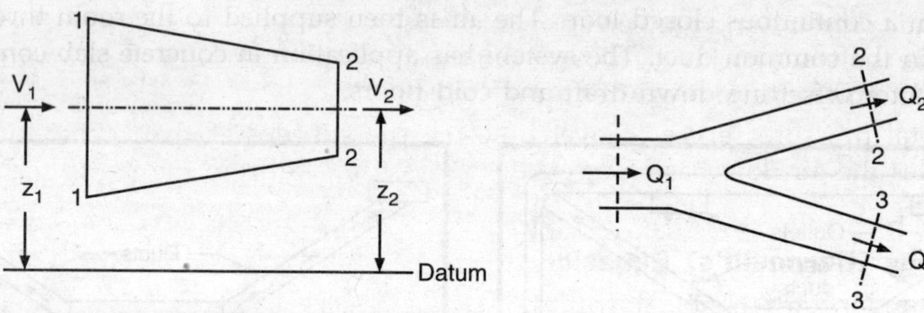

Fig. 16.9. Flow of air through a duct

For mass balance,

mass m_1 at entry section 1-1 = mass m_2 at exit section 2-2.

or $$\rho_1 A_1 V_1 = \rho_1 A_2 V_2$$

For air-conditioning systems, the density of air is assumed constant with value 1.2 kg/m³. That is

$$\rho_1 = \rho_2 = \rho = 1.2 \text{ kg/m}^2$$

That gives : $$A_1 V_1 = A_2 V_2 \quad \text{or} \quad Q_1 = Q_2$$

where Q represents the discharge (volume flow rate).

The above identity implies that volume flow rate remains constant, and is called the continuity equation for flow through ducts. It may be stated as :

"If no air is added to or removed from the duct in any length, then the air mass passing across different sections shall be same."

When the duct 1 branches into two ducts, then from mass conservation

$$m_1 = m_2 + m_3$$

or $$\rho_1 A_1 V_1 = \rho_2 A_2 V_2 + \rho_3 A_3 V_3$$

Assuming constant density for air flowing through the system, one gets

$$A_1 V_1 = A_2 V_2 + A_3 V_3$$

or $$Q_1 = Q_2 + Q_3$$

***(ii)* Pressure in ducts :** The following pressures are involved in a duct system :

The ***static pressure*** (p_s) is the pressure that acts normal to the walls of the duct and always exists; it is independent of air movement. It is essentially a measure of the potential energy at any point in the system. The static pressure has a tendency to burst the duct if it becomes greater than the atmospheric pressure, and tends to collapse the duct when its force is less than that of atmosphere. It provides the force needed to overcome the friction and shock losses that occur when the air flows from the delivery of the fan to the outlet of duct.

The ***velocity pressure*** (p_v), also called the dynamic pressure, manifests when the flow takes place. The dynamic pressure is flow dependent and it equals the drop in static pressure necessary to produce the desired flow velocity. Conversely the dynamic pressure equals the increase in static pressure when the flow velocity becomes zero.

The sum total of static pressure and velocity pressure is called the ***total pressure*** (p_t). Mathematically

$$p_t = p_s + p_v \qquad ...(16.3)$$

It is worthwhile to point out that :

(*i*) The static pressure and total pressure may either be positive or negative, where as velocity pressure is always positive. The term positive/negative means that the pressure is greater/less than atmospheric pressure.

(*ii*) The total pressure serves to overcome the pressure losses that occur due to various obstacles the air-flow has to meet during the passage from the fan to the conditioned space.

16.6.2. Energy (Bernoulli's) Equation

For steady and incompressible flow through a duct, the Bernoulli's equation can be written as

$$\frac{p}{\rho} + \frac{V^2}{2} + gz = \text{constant} \qquad ...(16.4)$$

Each term has the units of

$$\frac{\text{m}^2}{\text{s}^2} = \frac{\text{m (kg m/s}^2)}{\text{kg}} = \frac{\text{Nm}}{\text{kg}} \qquad (\because 1 \text{ N} = 1 \text{ kg m/s}^2)$$

or units of $\frac{\text{Nm}}{\text{kg}}$ and as such each term represents energy per unit kilogram mass.

Dividing equation 16.4 by g and using the relation $w = \rho g$, we get

$$\frac{V^2}{2g} + \frac{p}{\rho g} + z = \text{constant} \qquad ...(16.5)$$

or $$\frac{V^2}{2g} + \frac{p}{w} + z = \text{constant}$$

This is a form of Bernoulli's equation commonly used by hydraulic engineers. Here each term represents energy per weight (Nm/N) and has physical dimensions of height (metre). The quantities are called heads.

$\frac{V^2}{2g}$ = velocity head ; $\frac{p}{w}$ = pressure head or static head and

z = elevation head, position head or geodetic head.

While dealing with air and gases, Bernoulli's equation is recast as

$$\frac{\rho V^2}{2} + p + \rho g y = \text{constant} \qquad ...(16.6)$$

Here each term is in N/m² or Nm/m³, *i.e.*, energy per unit volume. Since elevation changes are small, the term $\rho g y$ can be dropped. Then

$$\frac{\rho V^2}{2} + p = \text{constant} \quad ...(16.7)$$

or $$p_v + p_s = \text{constant}\ ; \quad p_{v1} + p_{s1} = p_{v2} + p_{s2} \quad ...(16.8)$$

Bernaulli's equation for air flow through a duct may then by stated as

In a steady system of frictionless incompressible fluid flow, the sum of velocity pressure and static pressure remains constant at every section.

The above statement is, however, on the assumption that energy is neither added to nor taken away by some external agency. A pressure loss Δp_l takes place along the duct due to friction, and sudden change in area and flow direction. Further, a fan introduced between the two sections does work on air and that has the effect of raising its total pressure. Then

$$p_{v1} + p_{s1} + p_{ft} = p_{v2} + p_{s2} + \Delta p_l$$

where p_{ft} is the pressure rise due to fan work and is called the fan total pressure.

The pressures in the duct are usually expressed in mm of water. If the velocity through the duct is in m/s and ρ = 1.2 kg/m^3 is taken as the standard density of air, then

$$p_v = \frac{\rho V^2}{2} = \frac{1.2 V^2}{2} = 0.6\ V^2\ \text{N/m}^2$$

Since 1 mm of water is equivalent to 9.81 N/m^2,

$$p_v = \frac{0.6 V^2}{9.81} = \left(\frac{V}{4.04}\right)^2 \text{ mm of water} \quad ...(16.9)$$

16.7. PRESSURE LOSSES IN DUCTS

When flow of air continues in a duct, the air has to undergo pressure losses at every component of the duct system. The pressure losses are essentially due to

(*i*) duct friction, and

(*ii*) change of direction and/or velocity

The pressure drop in the duct length due to friction is known as static pressure drop or friction loss. The pressure loss due to velocity and direction changes is called the dynamic loss, and it occurs at

(*i*) Duct fittings such as elbows, bends etc., which cause change in direction

(*ii*) Sudden enlargement/contraction, dampers, suction and discharge openings which cause changes in area or velocity

Computation of various losses is necessary for the proper selection and sizing of fan.

Pressure loss due to duct friction : The pressure loss due to friction in a straight duct is computed by using the D'arcy relation

$$p_f = f\,\frac{\rho l V^2}{2m} \quad ...(16.10)$$

where p_f is pressure loss due to friction in N/m^2; l is the length of duct in metres; V is velocity of air flowing through the duct in m/s; ρ is density of air in kg/m^3; f is a dimensionless friction factor depending upon the surface of duct, and m is hydraulic mean depth in metres.

$$m = \frac{\text{cross-sectional area of the duct } (A)}{\text{wetted perimeter of the duct } (P)}$$

For a circular duct of diameter D,

$$m = \frac{A}{P} = \frac{\frac{\pi}{4}D^2}{\pi D} = \frac{D}{4}$$

For a rectangular duct of sides a and b,

$$m = \frac{A}{P} = \frac{ab}{2(a+b)}$$

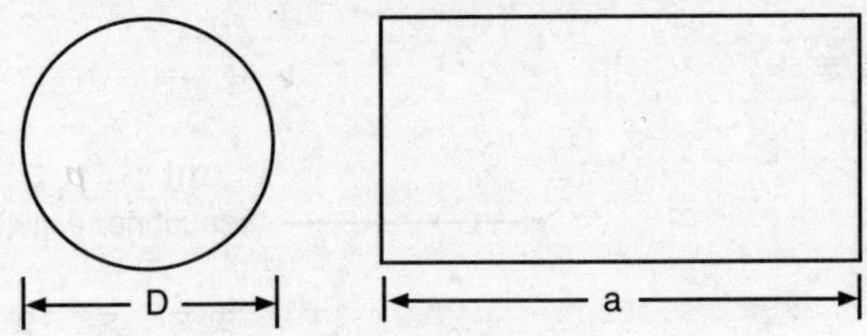

Fig. 16.10. Circular and rectangular duct

The friction factor f is a function of the flow Reynolds number $\left(Re = \frac{\rho V D}{\mu}\right)$ and relative roughness of the duct surface $\left(\frac{\in}{D}\right)$ where $\in$ is the absolute roughness. The friction factor is estimated either through empirical correlations or graphical charts.

The following empirical relations have been suggested for the evaluation of the friction factor:

(*i*) For laminar flow : $f = \frac{64}{Re}$ $\quad$ (Re <2000) ...(16.11)

(*ii*) For turbulent flow : $f = \frac{0.3164}{(Re)^{0.25}}$ ...(16.12)

(*iii*) For rough pipes and ducts : $f = \frac{1}{\left[1.74 - 2\log\left(\frac{2\in}{D}\right)\right]^2}$ ...(16.13)

– The absolute roughness depends upon the quality of duct material.
– The values of friction factor for different Reynolds number and different roughness factors are read directly from the Moody chart shown in Fig. 16.11(*a*).
– The following relation has been suggested by Fritzsche for obtaining the frictional loss in circular ducts.

$$p_f = \frac{0.1422 l V^{1.852}}{D^{1.269}} \text{ N/m}^2$$

where velocity V is in m/s, length l and diameter D are in metres.

$$\text{Mean velocity } V = \frac{\text{volume flow rate } (Q)}{\text{cross-sectional area } (A)} = \frac{4Q}{\pi D^2}$$

Then :

$$p_f = \frac{0.1422 l\,(4Q/\pi D^2)}{D^{1.269}}$$

$$= \frac{0.02243 l Q^{1.852}}{D^{4.973}} \text{ N/m}^2 \qquad ...(16.14)$$

$$= \frac{0.002268\, l Q^{1.852}}{D^{4.973}} \text{ mm of water} \qquad (\because 9.81 \text{ N/m}^2 = 1 \text{ mm of } H_2O)$$

In terms of mean velocity V and discharge Q (by eliminating D),

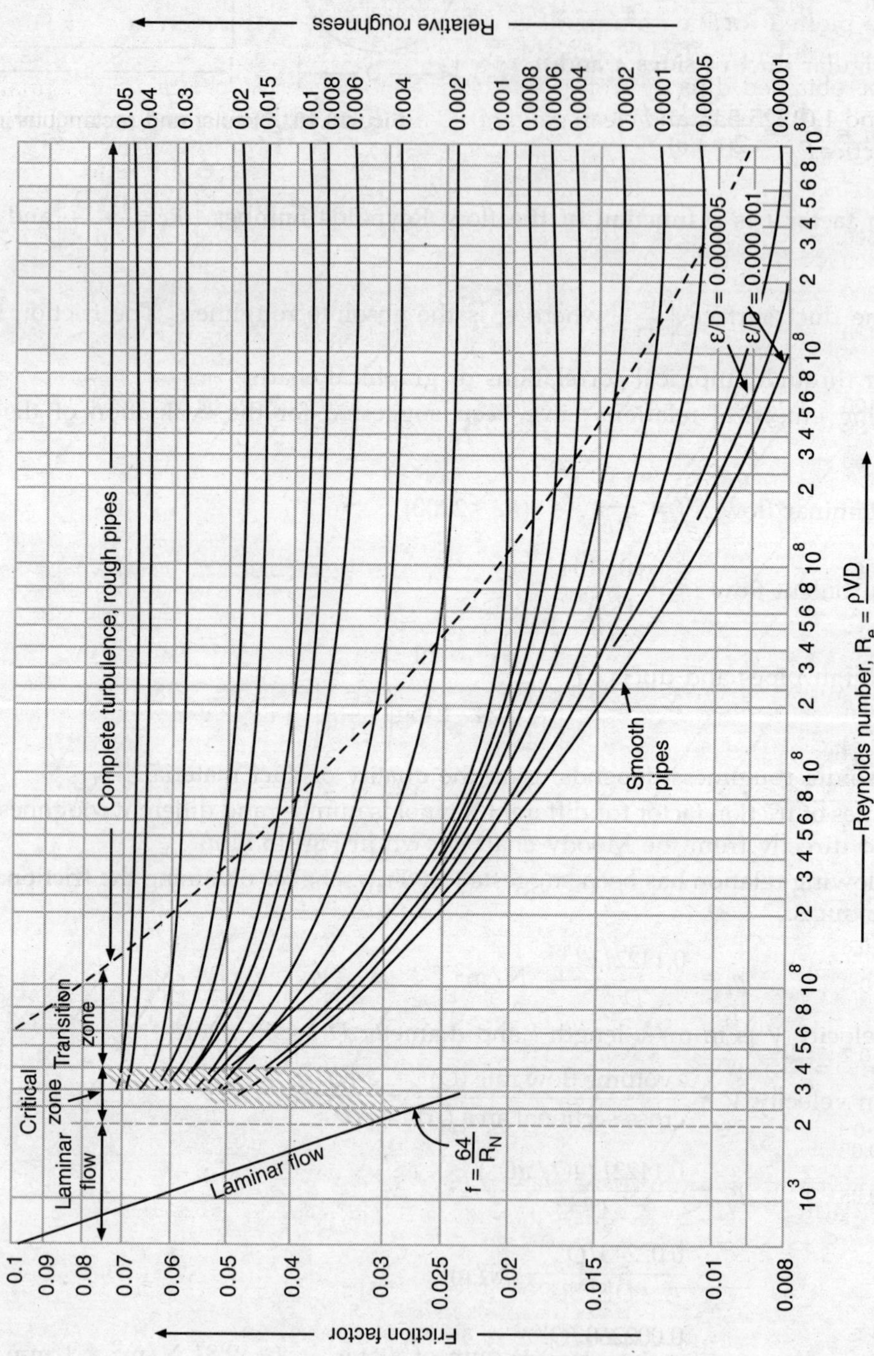

Fig. 16.11(a). Moody chart showing friction factors for fluid flow in circular pipes or ducts

$$p_f = \frac{0.012199\, l\, V^{2.4865}}{Q^{0.6343}} \quad ...(16.15)$$

The above relations are used to develop the duct friction charts as shown in Fig. 16.11(*b*). These charts are plotted for the volume flow rate as a function of friction rate, *i.e.*, the friction pressure drop per unit length. The frictional loss for circular ducts for various velocities and duct diameters can be obtained directly from these charts. However, the charts are essentially valid for air at 20°C and 1.01325 bar, and clean galvanized air ducts with joints and seams having good commercial practice.

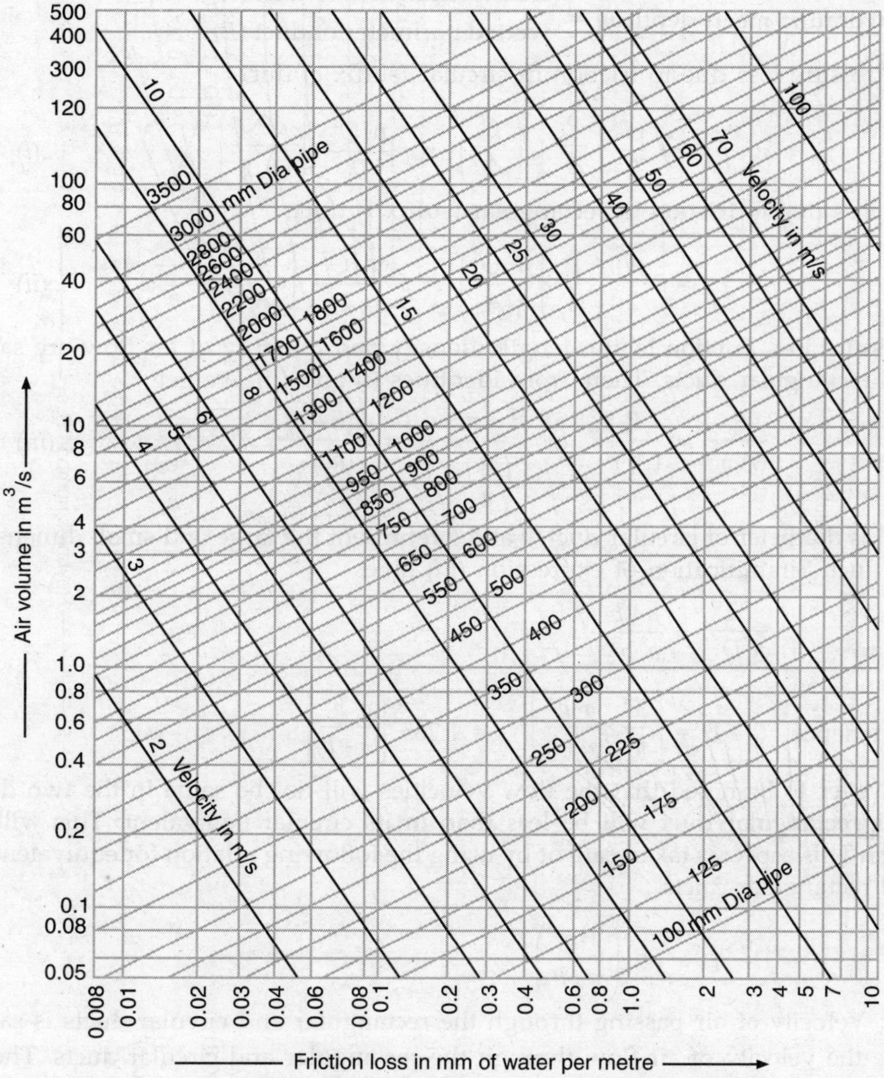

Fig. 16.11(b). Friction chart for circular ducts

16.8. RECTANGULAR EQUIVALENTS OF CIRCULAR DUCTS

The usual practice in the design of air ducts is to first size for round sections. If the requirement is for rectangular ducts, than these are sized to provide the same flow rates and to have the same rate of pressure drop as for round ducts.

Case 1 : Quality of air passing through the rectangular and circular ducts is same.

From Darcy relation, the pressure loss due to friction is given by

$$p_f = f\ \frac{\rho l V^2}{2m} = f\ \frac{\rho l}{2m}\left(\frac{Q}{A}\right)^2 \qquad \left(\because V = \frac{Q}{A}\right)$$

where hydraulic mean depth $m = \dfrac{\text{cross-section area of duct } (A)}{\text{wetted parimeter of duct } (P)}$

Then : Pressure loss due to friction in circular (suffix c) duct,

$$p_{fc} = f\ \frac{\rho l}{2}\left(\frac{P_c}{A_c}\right)\left(\frac{Q}{A_c}\right)^2 = f\ \frac{\rho l Q^2}{2}\left[\frac{P_c}{(A_c)^3}\right] \qquad ...(i)$$

Pressue loss due to friction in rectangular (suffix r) duct,

$$p_{fr} = f\ \frac{\rho l}{2}\left(\frac{P_r}{A_r}\right)\left(\frac{Q}{A_r}\right)^2 = f\ \frac{\rho l Q^2}{2}\left[\frac{P_r}{(A_r)^3}\right] \qquad ...(ii)$$

The pressure loss, friction factor, length, density and quantity of air flow are same for the circular and rectangular ducts. Then, from identities (*i*) and (*ii*), we get

$$\frac{P_c}{(A_c)^3} = \frac{P_r}{(A_r)^3} \quad \text{or} \quad \frac{\pi D}{\left(\frac{\pi}{4}D^2\right)^3} = \frac{2(a+b)}{(ab)^3} \qquad ...(iii)$$

where D is diameter of circular duct, a and b represent the large and small dimensions of the rectangular duct. Simplification of expression (*iii*) gives

$$\frac{32}{\pi^2 D^5} = \frac{a+b}{a^3\, b^3}$$

$$\text{whence} \qquad D = \left[\frac{32}{\pi^2}\times\frac{a^3\, b^3}{a+b}\right]^{1/5} = 1.265\left[\frac{a^3\, b^3}{a+b}\right]^{1/5} \qquad ...(16.16)$$

It is however to be noted that the flow velocities will not be same in the two ducts; mean velocity in a rectangular duct will be less than in its circular equivalent. This will effect the friction factor. This aspect is taken care of by using the following relation for equivalence between circular and rectangular ducts.

$$D = 1.3\left(\frac{a^3\, b^3}{a+b}\right)^{1/5} \qquad ...(16.17)$$

Case 2 : Velocity of air passing through the rectangular and circular ducts is same.

Let V be the velocity of air flow through the rectangular and circular ducts. Then

Pressure loss due to friction for a circular duct,

$$p_{fc} = f\ \frac{\rho l V^2}{2m} = f\ \frac{\rho l V^2}{2}\left(\frac{P_c}{A_c}\right) \qquad ...(a)$$

Pressure loss due to friction for a rectangular duct,

$$p_{fr,} = f\frac{\rho l V^2}{2m} = f\frac{\rho l V^2}{2}\left(\frac{P_r}{A_r}\right) \qquad ...(b)$$

The friction loss, friction factor, length, density and velocity are same for the circular and rectangular ducts. Then from expressions (*a*) and (*b*), we get

$$\frac{P_c}{A_c} = \frac{P_r}{A_r} \quad \text{or} \quad \frac{\pi D}{\frac{\pi}{4}D^2} = \frac{2(a+b)}{ab}$$

where $$D = \frac{2ab}{a+b} = \frac{2a}{\frac{a}{b}+1} = \frac{2a}{AR+1} \qquad ...(16.18)$$

where $AR = \frac{a}{b}$ is known is *aspect ratio.*

16.9. DYNAMIC PRESSURE LOSS

The dynamic pressure loss manifests due to certain local features or disturbances in the flow passage. The disturbances may be caused by changes in the size or shape of the duct; this affects the velocity distribution and results in eddy formation and consequently pressure loss.

- **Sudden expansion :** In case of enlargement of flow passage, the total pressure drops due to friction and turbulence. The pressure loss is

$$\Delta p_l = p_{t1} - p_{t2} = (p_{s1} + p_{v1}) - (p_{s2} + p_{v2})$$
$$= (p_{s1} - p_{s2}) + (p_{v1} - p_{v2})$$

When the area increases, the velocity decreases with rise in pressure. There is conversion of velocity pressure into static pressure and this conversion is called ***static regain.***

$\therefore$ Static regain $\quad (p_{s2} - p_{s1}) = (p_{v1} - p_{v2}) - \Delta p_l \qquad ...(16.19)$

If the enlargement is gradual, there is no loss of pressure and the static regain is 100 percent.

Fig. 16.12 depicts a duct having a sudden enlargement. The air flowing through the duct fails to make an adjustment with the change in direction needed for complete filling of the enlarged section. The flow breaks away from the edges of narrow section; eddies form and the resulting turbulence causes dissipation of energy (pressure loss). The initiation and onset of turbulence is due to air momentum and its inertia.

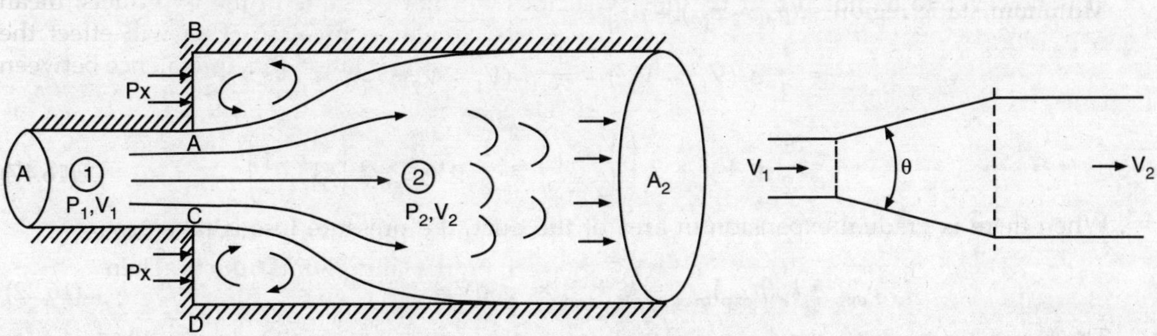

Fig. 16.12. Abrupt expansion in a duct

Consider the cross-sections of the stream : section 1-1 before expansion and section 2-2 which marks the end of the region of turbulence caused by fluid separation. From momentum consideration:

$$p_{s1} A_1 + p_{sx} (A_2 - A_1) - p_{s2} A_2 = \rho Q(V_2 - V_1)$$

The flow velocity at the annular wall AB and CD is very small and so the pressure there can be assumed to be equal to the pressure of incoming fluid, *i.e.*, $p_{sx} = p_{s1}$, and then the momentum equation transforms to

$$(p_{s1} - p_{s2}) A_2 = \rho Q(V_2 - V_1)$$

or $$p_{s1} - p_{s2} = \frac{\rho Q}{A_2}(V_2 - V_1) = \rho V_2(V_2 - V_1) \quad \text{... (i)}$$

Invoking the Bernoulli's equation,

$$p_{s1} + \frac{\rho V_1^2}{2} = p_{s2} + \frac{\rho V_2^2}{2} + p_{exp}$$

where p_{exp} is the pressure loss due to sudden expansion.

$$p_{exp} = (p_{s1} - p_{s2}) + \frac{\rho(V_1^2 - V_2^2)}{2} \quad \text{... (ii)}$$

Combining expressions (*i*) and (*ii*)

$$p_{exp} = \rho V_2(V_2 - V_1) + \frac{\rho(V_1^2 - V_2^2)}{2} = \frac{\rho(V_1 - V_2)^2}{2} \quad \text{...(16.20)}$$

Evidently the pressure loss due to abrupt expansion equals the velocity pressure computed for the velocity difference. Since by continuity $A_1V_1 = A_2V_2$; $V_2 = A_1V_1/A_2$ and therefore equation 16.20 may be alternatively written as

$$p_{exp} = \left[1 - \frac{A_1}{A_2}\right]^2 \frac{\rho V_1^2}{2} = k_{exp} \times \frac{1}{2}\rho V_1^2$$

Thus, the pressure loss is a function of incoming velocity pressure.

The parameter $k_{exp} = \left(1 - \frac{A_1}{A_2}\right)^2$ is known as ***loss coefficient***.

Equation 16.20 represents the maximum possible loss due to an enlargement, and hence corresponds to a minimum value of static regain.

Minimum static region $= (p_{v1} - p_{v2}) - p_{exp}$

$$= \frac{1}{2}\rho(V_1^2 - V_2^2) - \frac{1}{2}\rho(V_1 - V_2)^2$$

$$= \frac{1}{2}\rho V_2^2 \times 2\left(\frac{V_1}{V_2} - 1\right) = \frac{1}{2}\rho V_2^2 \times 2\left(\frac{A_2}{A_1} - 1\right) \quad \text{...(16.21)}$$

When there is gradual expansion in area of the duct, the pressure loss is taken as

$$p_{exp} = k_r (p_{exp})_{max} = k_r k_{exp} \times \frac{1}{2}\rho V_1^2 \quad \text{...(16.22)}$$

The factor k_r represents the ratio of actual loss to the maximum possible loss. Table 16.1 lists the values of k_r as a function of the included angle θ of the sides.

Table 16.1 : Relative loss coefficient for gradual expansion

Included angle θ	5	7	10	20	30	40
Relative loss coefficient k_r	0.17	0.22	0.28	0.45	0.59	0.73

Sudden contraction : Fig. 16.13 depicts a duct in which abrupt contraction occurs.

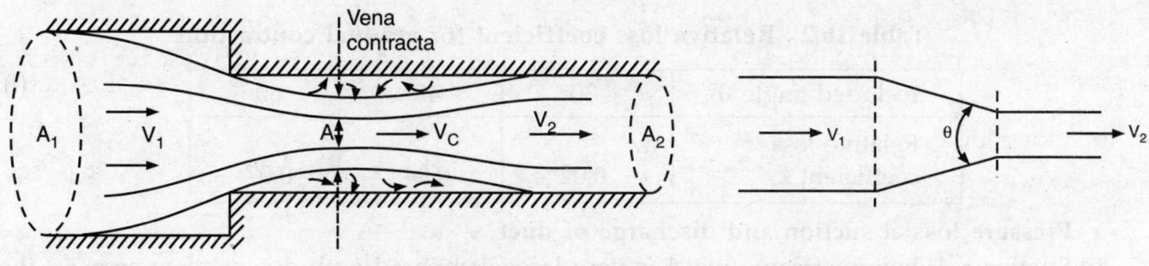

Fig. 16.13. Abrupt contraction in a duct

Inspection of the flow pattern reveals that it exists in two phases :

(*i*) A converging flow wherein the flow stream leaves the surface at the corner of junction and attains a minimum cross-sectional area. The section of least cross-section and hence of maximum contraction is called ***vena-contracta***. At vena-contracta, the effective flow area becomes considerably less than cross-sectional area of the small size duct. The accelerating flow is stable; boundary layer has no chance to separate and consequently there is little loss of pressure between the entrance and the vena-contracta.

(*ii*) A diverging flow downstream from vena-contracta wherein the stream flow expands and ultimately assumes uniform flow over the entire cross-section of the small size duct. During this sudden and unguided expansion, vortices are formed between mainstream and the wall of duct. Eddy formation and the consequent energy dissipation are essentially responsible for most of the pressure loss.

The loss of pressure in duct contraction is thus caused mainly by the turbulence created by abrupt expansion of the flow just after it has passed through vena-contracta; it can be approximated by the abrupt expansion equation. That is

$$p_{con} = \frac{1}{2} \rho (V_c - V_2)^2 \quad ...(i)$$

The cross-sectional area of the jet at vena-contract is defined as :

$A_c = C_c A_2$ where C_c is the coefficient of contraction.

Further, from continuity consideration

$$A_2 V_2 = A_c V_c = C_c A_2 V_c \quad \text{or} \quad V_c = \frac{V_2}{C_c}$$

and hence by substitution in expression (*i*),

$$p_{con} = \frac{1}{2} \rho (V_2/C_c - V_2)^2$$

$$= \frac{1}{2} \rho V_2^2 \left[\frac{1}{C_c} - 1\right]^2 = k_{con} \times \frac{1}{2} \rho V_2^2 \quad ...(16.23)$$

The contraction loss coefficient k_{con} depends essentially on the rate of contraction, *i.e.*, on area ratio A_2/A_1 and to some extent on the flow Reynolds number.

When there is gradual contraction in area of the duct, the pressure loss is taken as

$$p_{con} = k_r \, k_{con} \times \frac{1}{2} \rho \, V_2^2 \quad \text{...(16.24)}$$

where k_r is referred to as relative loss coefficient. Table lists the values of k_r as a function of the included angle θ of the sides.

Table 16.2 : Relative loss coefficient for gradual contraction

Included angle θ°	30	45	60
Relative loss coefficient k_r	0.02	0.04	0.07

- **Pressure loss at suction and discharge of duct**

***(i)* Suction :** When upstream area A is very large compared with downstream area A_2, the situation corresponds entrance to the duct. The pressure loss at suction is taken as

$$p_{suc} = k_{suc} \times \frac{1}{2} \rho \, V_2^2$$

where V is the velocity of air in the duct.

The suction loss coefficient k_{suc} depends on the nature of suction openings. For abrupt suction opening k_{suc} = 0.85. The corresponding values for flanged entrance and formed entrance are 0.34 and 0.03 respectively.

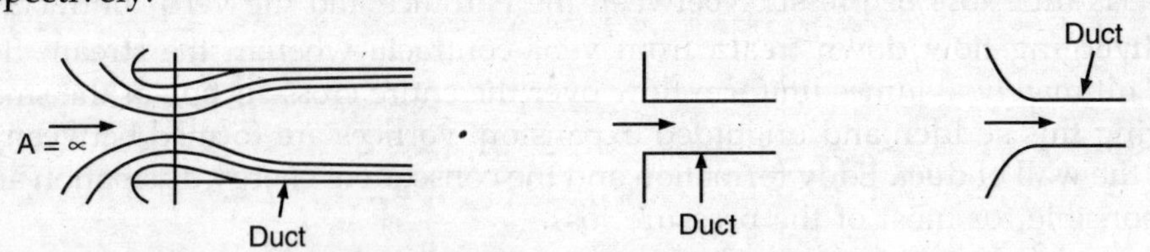

Fig. 16.14. Nature of suction openings

***(ii)* Discharge :** When downstream area A_2 is very large compared to upstream area A_1, the velocity V_2 can be assumed to be zero. Loss of velocity pressure and kinetic energy is then complete. The flow situation corresponds to a duct discharging into atmosphere. The pressure loss is then called the duct exit loss p_{exit} and is given as

$$p_{exit} = \frac{1}{2} \rho \, V^2$$

where V represents the average velocity in the duct.

- **Losses at bends, elbows and tees :** Fig. 16.15 depicts the flow pattern regarding separation and eddying in the region of separation in bends and tees. The resulting pressure loss can be prescribed by the relation $p = k \times \frac{1}{2} \rho \, V^2$ where V is the average flow velocity, and the loss coefficient depends on parameters defining the geometry of the section and the flow. Loss coefficient of a bend is primarily a function of the total angle of bend and the ratio r/d where r is the radius of the bend and d is the duct diameter. Resistance of large sized elbows can be reduced appreciably by splitting the flow into a number of streams by a set of guide vanes.

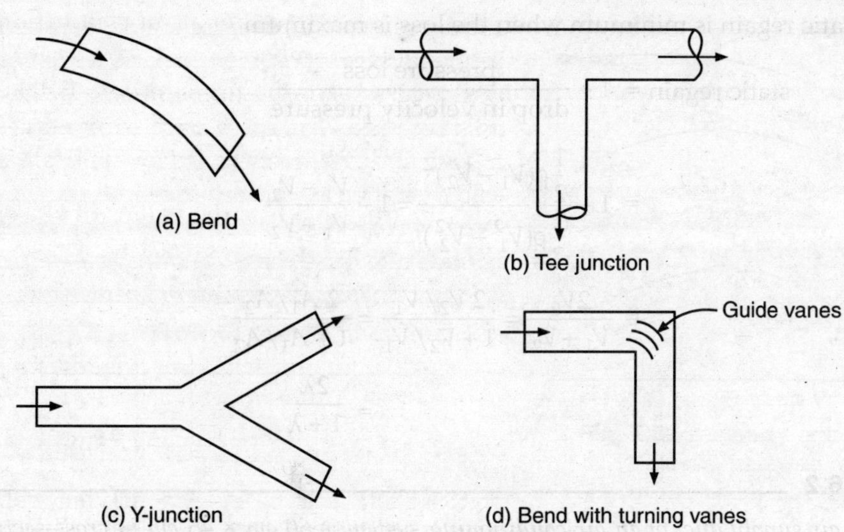

Fig. 16.15. Duct bends, elbows and tees

EXAMPLE 16.1

The loss of pressure in a sudden enlargement of duct is assumed as:

$$P_l = \rho\left(\frac{V_1 - V_2}{2}\right)^2$$

where V_1 is velocity in smaller duct, V_2 is velocity in larger duct and ρ is fluid density.

Show that in a gradual expander, the static pressure down stream is always greater than the static pressure upstrem, and also that the minimum static regain possible is given by,

$$\text{static regain} = \frac{2\lambda}{1+\lambda} \quad \text{where } \lambda \text{ is the area ratio } A_1/A_2.$$

(WBUT)

Solution : For an expander, the loss will be maximum when there is complete detachment of the fluid stream from the boundary wall. The state is referred to as sudden enlargement for which the pressure loss is given by,

$$P_l = \frac{1}{2}\rho(V_1 - V_2)^2$$

Further, velocity pressures are given as $P_{v1} = \frac{1}{2}\rho V_1^2$ and $P_{v2} = \frac{1}{2}\rho V_2^2$

Then from the energy principle,

$$P_{s1} + \frac{1}{2}\rho V_1^2 = P_{s2} + \frac{1}{2}\rho V_2^2 + \frac{1}{2}\rho(V_1 - V_2)^2$$

That gives: $$p_{s2} - p_{s1} = \frac{1}{2}\rho\left[(V_1^2 - V_2^2) - (V_1 - V_2)^2\right] = \rho V_2(V_1 - V_2)$$

Since $V_2 > 0$ and $V_1 > V_2$: $(P_{s2} - P_{s1})$ is +ve and therefore static pressure (P_{s2}) downstream is greater than the static pressure (P_{s1}) upstream

(*b*) The static regain is minimum when the loss is maximum

$$\text{static regain} = 1 - \frac{\text{pressure loss}}{\text{drop in velocity pressure}}$$

$$= 1 - \frac{\frac{1}{2}\rho(V_1 - V_2)^2}{\frac{1}{2}\rho(V_1^2 - V_2^2)} = 1 - \frac{V_1 - V_2}{V_1 + V_2}$$

$$= \frac{2V_2}{V_1 + V_2} = \frac{2\,V_2/V_1}{1 + V_2/V_1} = \frac{2\,A_1/A_2}{1 + A_1/A_2}$$

$$= \frac{2\lambda}{1+\lambda}$$

EXAMPLE 16.2

The main air supply duct of an air-conditioning system is 60 cm × 45 cm in cross-section and carries 4 m³/s of standard air. It branches into two ducts of cross-section 55 cm × 35 cm and 45 cm × 35 cm. If the mean velocity of air in the smaller branch is 12 m/s, make calculations to workout

(a) mean air flow velocity in the main duct and larger branch,

(b) mean velocity pressure in each duct.

Solution : Refer Fig. 16.16 for the duct system. Subscripts 1, 2, 3 refer to main duct, larger branch and smaller branch respectively.

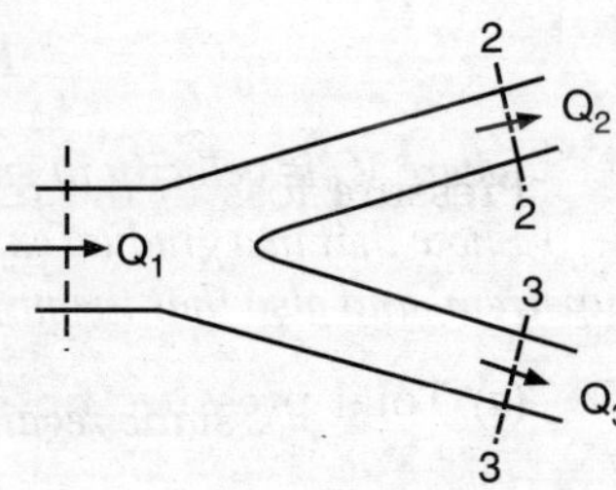

Fig. 16.16

Duct areas (*A*) :

$$A_1 = 0.6 \times 0.45 = 0.27 \text{ m}^2$$

$$A_2 = 0.55 \times 0.35 = 0.1925 \text{ m}^2$$

$$A_3 = 0.45 \times 0.35 = 0.1575 \text{ m}^2$$

Flow rates (*Q*) and air velocities (*V*) :

$$Q_1 = 4 \text{ m}^3/\text{s} \quad \text{and} \quad V_1 = \frac{Q_1}{A_1} = \frac{4}{0.27} = \mathbf{14.81} \text{ m/s}$$

$$V_3 = 12 \text{ m/s} \quad \text{and} \quad Q_3 = A_3 V_3 = 0.1575 \times 12 = 1.89 \text{ m}^3/\text{s}$$

$$Q_2 = Q_1 - Q_3 = 4 - 1.89 = 2.11 \text{ m}^3/\text{s}$$

$$V_2 = \frac{Q_2}{A_2} = \frac{2.11}{0.1925} = \mathbf{10.96} \text{ m/s}$$

Mean velocity pressure (p_v)

$$p_{v1} = \left(\frac{V_1}{4.04}\right)^2 = \left(\frac{14.81}{4.04}\right)^2 = \mathbf{13.44} \text{ mm of water column}$$

$$p_{v2} = \left(\frac{V_2}{4.04}\right)^2 = \left(\frac{10.96}{4.04}\right)^2 = \mathbf{8.82} \text{ mm of water column}$$

$$p_{v3} = \left(\frac{V_2}{4.04}\right)^2 = \left(\frac{12}{4.04}\right)^2 = \mathbf{7.36} \text{ mm of water column}$$

EXAMPLE 16.3

A 100 mm diameter duct converges gradually to 80 mm diameter. The static pressure and the air flow velocity at the inlet are noted to be 25 mm of water and 7.5 cm/s respectively. If the pressure loss in the reducer are estimated to be 12 percent of velocity head in the downstream of the reducer, determine :

(a) total pressure upstream and downstream of the reducer,

(b) pressure indicated by a U-tube water manometer connected differentially to tappings putstream and downstream of the reducer.

Solution : Refer figure 16.17. Subscripts 1 and 2 refer to section just upstream and downstream of the reducer.

From continuity considerations, $A_1V_1 = A_2V_2$

$$V_2 = \frac{A_1}{A_2}V_1 = \left(\frac{D_1}{D_2}\right)^2 V_1 = \left(\frac{100}{80}\right)^2 \times 7.5$$

$$= 11.72 \text{ m/s}$$

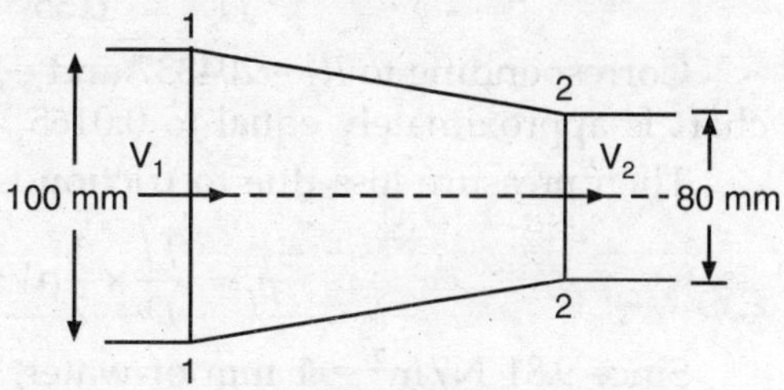

Fig. 16.17

Velocity pressures upstream and downstream of the reducer are :

$$p_{v_1} = \left(\frac{V_1}{4.04}\right)^2 = \left(\frac{7.5}{4.04}\right)^2 = 3.45 \text{ mm of water}$$

$$p_{v_2} = \left(\frac{V_2}{4.04}\right)^2 = \left(\frac{11.72}{4.04}\right)^2 = 8.42 \text{ mm of water}$$

Pressure loss in the reducer

$$= 12\% \text{ of velocity head downstream of reducer}$$

$$p_{\text{loss}} = 0.12 \times 8.42 = 1.01 \text{ mm of water}$$

(*a*) Total pressure upstream of reducer

$$p_{t_1} = p_{s_1} + p_{v_1} = 25 + 3.45 = \mathbf{28.45} \text{ mm of water}$$

From energy considerations, $p_{t_1} = p_{t_2} + p_{loss}$

∴ Total pressure downstream of reducer,

$$p_{t_2} = p_{t_1} - p_{loss} = 28.45 - 1.01 = \mathbf{27.44} \text{ mm of water}$$

(*b*) Static pressure downstream of reducer,

$$p_{s_2} = p_{t_2} - p_{v_2} = 27.44 - 8.42 = 19.02 \text{ mm of water}$$

The pressure indicated by the manometer is the difference between the static pressures at the tapping sections

$$= p_{s_1} - p_{s_2} = 25 - 19.02 = \mathbf{5.98} \text{ mm of water}$$

EXAMPLE 16.4

An air-conditioning duct 35 cm in diameter and 15 m long conveys air at the rate of 50 m³/min. The duct is made of galvanised steel for which the surface roughness is 0.12 mm and for the air being conveyed at specified pressure and temperature, the mass density is 1.2 kg/m³ and the absolute or dynamic viscosity is 17.8 × 10⁻⁶ Ns/m². Determine the pressure loss due to friction in mm of water.

Solution : Velocity of air $V = \frac{Q}{A} = \frac{50/60}{\frac{\pi}{4}(0.35)^2} = 8.66 \text{ m/s}$

Reynolds number $Re = \frac{VD\rho}{\mu} = \frac{8.66 \times 0.35 \times 1.2}{17.8 \times 10^{-6}} = 204337$

Roughness factor $\frac{\in}{D} = \frac{0.00012}{0.35} = 0.000343$

Corresponding to $R_e = 204337$ and $\in/D = 0.000343$, the friction factor as read from the Moody chart is approximately equal to 0.0165

Then pressure loss due to friction is

$$p_f = \frac{fl}{D} \times \frac{1}{2}\rho V^2 = \frac{0.0165 \times 15}{0.35} \times \frac{1}{2} \times 1.2 \times (8.66)^2 = 31.82 \text{ N/m}^2$$

Since 9.81 N/m^2 = 1 mm of water,

$$p_f = \frac{31.82}{9.81} = \mathbf{3.24} \text{ mm of water}$$

Note : If the Moody chart is not available, the friction factor f can be worked out by using the following correlation :

$$f = \frac{1}{\left[1.74 - 2\log\left(\frac{2\in}{D}\right)\right]^2} = \frac{1}{[1.74 - 2\log(2 \times 0.000343)]^2}$$

$$= \frac{1}{[1.74 - (-6.327)]^2} = 0.0154$$

This relation considers the surface roughness and flow to be turbulent. However, the effect of the magnitude of Reynolds number is not taken care of.

EXAMPLE 16.5

A tunnel of length l meter is ventilated by a duct of area A, perimeter P and frictional coefficient f. The fresh air from the duct discharges into tunnel through a continuous flow of air regulators which give uniform distribution q m^3/min per meter run of the tunnel? The total air quantity is Q m^3/min and the frictional pressure drop down ρ_f is in mm of water column. Neglecting resistance of tunnel and taking density of air as ρ kg/m^3, show that

$$\rho_f = \frac{Pf\rho lQ^2}{6A^3}$$ ***(WBUT)***

Solution : Taking Q as total air quantity at duct inlet and q as quantity dischare per metre length, we have

$$q = \frac{Q}{l} \quad \text{or} \quad Q = lq$$

Pressure loss (drop) $\rho_f = f\frac{\rho l V^2}{2m}$ where

$$m = \frac{A}{P} \quad \text{and} \quad V = \frac{Q}{A}$$

Then :

$$P_f = \frac{f}{2} \times \frac{P\rho l}{A}\left(\frac{Q}{A}\right)^2 = \frac{Pf\rho l\, Q^2}{2\,A^3}$$

Consider any section of duct at distance x from the inlet. Then frictional loss over distance dx is,

$$dP_f = \frac{Pf\rho}{2A^3}(Q - xq)^2\, dx$$

and total pressure loss

$$P_f = \frac{Pf\rho}{2A^3}\int_0^l (Q - xq)^2\, dx$$

$$= \frac{Pf\rho}{2A^3}\left[\int_0^l Q^2 dx - 2\int_0^l Qqx\, dx + \int_0^l x^2 q^2 dx\right]$$

$$= \frac{Pf\rho}{2A^3}\left[Q^2 l - \frac{2Qql^2}{2} + \frac{q^2 l^3}{3}\right]$$

$$= \frac{Pf\rho}{2A^3}\left[Q^2 l - \frac{Q^2 l}{3}\right] \qquad (\because \; Q = ql)$$

$$= \frac{Pf\rho l}{6A^3}$$

EXAMPLE 16.6

A rectangular duct of section 500 mm × 300 mm size is used to convey 60 m³/min of air having density 1.12 kg/m³. Determine the equivalent diameter of a circular duct if for both the cases,

(a) the quantity of air carried is same

(b) the velocity of air is same.

Proceed to calculate the pressure loss per 100 mm length of duct if the friction factor f = 0.01 for the sheet metal used in the fabrication of duct.

Solution : (*a*) When the quantity of air carried by rectangular and circular ducts is same, the equivalent diameter of circular duct is given by

$$D = 1.265\left(\frac{a^3 b^3}{a+b}\right)^{1/5} = 1.265 \times \left(\frac{0.5^3 \times 0.3^3}{0.5+0.3}\right)^{1/5}$$

$$= 1.265 \times \left(\frac{0.003375}{0.8}\right)^{1/5} = \mathbf{0.424\ m}$$

(*b*) When the velocity of air passing through the rectangular and circular ducts is same, the equivalent diameter of circular duct is given by

$$D = \frac{2ab}{a+b} = \frac{2 \times 0.5 \times 0.3}{0.5+0.3} = \mathbf{0.375\ m}$$

(c) Velocity of air passing through the rectangular duct

$$V = \frac{Q}{A} = \frac{60}{0.5 \times 0.3} = 400 \text{ m/min} = 6.67 \text{ m/sec}$$

$$\text{Mean hydraulic depth } m = \frac{A}{P} = \frac{ab}{2(a+b)} = \frac{0.5 \times 0.3}{2(0.5+0.3)} = 0.09375 \text{ m}$$

From Darcy relation, the pressure loss due to friction is

$$p_f = f\frac{\rho l V^2}{2m} = 0.01\left[\frac{1.12 \times 100 \times 6.67^2}{2 \times 0.09375}\right] = 265.75 \text{ N/m}^2$$

$$= \frac{265.75}{9.81} = \mathbf{27.09 \text{ mm of water}}$$

EXAMPLE 16.7

A circular duct has been selected to carry air at the rate of 0.95 m³/s to an air-conditioned space. If the velocity of air flow in the duct is limited to 7.25 m/s to keep the noise at desired level, what should be the duct diameter to meet these requirements ?

(b) If this duct is subsequently to be replaced by a rectangular duct of aspect ratio 1.6, determine the dimension of the rectangular duct for equal friction method when :

(i) discharge rate of air in two ducts is same and (ii) velocity of air in two ducts is same.

Solution : Let D be the diameter of the circular duct. Then

$$\frac{\pi}{4} D^2 \times 7.25 = 0.95$$

or $$D = \left(\frac{4 \times 0.95}{7.25\pi}\right)^{1/2} = 0.408 \text{ m} = \mathbf{40.8 \text{ cm}}$$

(b) For the same discharge rate of air,

$$D = 1.265\left(\frac{a^3 b^3}{a+b}\right)^{1/5}$$

where a and b represent the longer and shorter side of the rectangular duct. Also a/b = aspect ratio = 1.6 (given). Then

$$40.8 = 1.265\left[\frac{(1.6b)^3 \times b^3}{1.6b+b}\right]^{1/5} = 1.265 \times 1.095\, b$$

That gives : $$b = \frac{40.8}{1.265 \times 1.095} = 29.45 \text{ cm}$$

and $$a = 1.6 \times 29.45 = 47.13 \text{ cm}$$

∴ Size of duct is 47.13 cm × 29.45 cm

(ii) For the same velocity of air,

$$D = \frac{2ab}{a+b}\ ; \quad 40.8 = \frac{2(1.6b \times b)}{1.6b+b}$$

$$b = \frac{40.8 \times 2.6}{2 \times 1.6} = 33.15 \text{ cm}$$

and $$a = 1.6 \times 33.15 = 53.04 \text{ cm}$$

∴ Size of duct is 53.04 cm × 33.15 cm

EXAMPLE 16.8

A rectangular duct of 12 cm × 10 cm in size and 15 m long carries standard air at the rate of 0.2 m³/s. Determine :

(a) the total pressure required at inlet to the duct in order to maintain this flow, and

(b) the air power required.

Assume the friction factor f = 0.006 for the duct material.

Solution : Velocity of air in the duct $V = \dfrac{Q}{A} = \dfrac{0.2}{0.12 \times 0.1} = 16.67$ m/s

Velocity pressure in the duct $p_v = \dfrac{1}{2}\rho V^2 = \dfrac{1}{2} \times 1.2 \times (16.67)^2 = 166.73$ N/m²

Hydraulic mean depth of duct $m = \dfrac{ab}{2(a+b)} = \dfrac{0.12 \times 0.1}{2(0.12+0.1)} = 0.02727$ m

Using Darcy equation, the pressure loss in the duct due to friction is

$$p_f = f\,\frac{\rho l V^2}{2m} = 0.006 \times \frac{1.2 \times 15 \times (16.67)^2}{2 \times 0.02727} = 550.27 \text{ N/m}^2$$

∴ Total pressure required at inlet to the duct to maintain flow,

$$p_t = p_v + p_f = 550.27 + 166.73 = \mathbf{717\ N/m^2}$$

(*b*) Air power = $Q \times p_t = 0.2 \times 717$

= 143.4 Nm/s = **143.4 W**

EXAMPLE 16.9

A rectangular duct 1.5 m × 1 m in size carrying conditioned air runs 50 m from the fan outlet. Thereafter it branches into two parts each 75 m long and 1.5 m × 1 m size as shown in the figure given below :

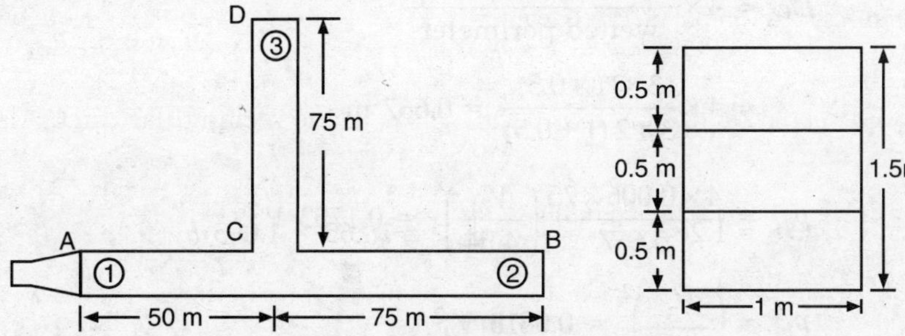

Fig. 16.18

To keep the noise level within prescribed limits, the branch CD has been divided into there segments and lined with an absorbent material. If the air discharge at section B is 16 m³/sec determine the quantity of air discharged at section D and the static pressure of an fan outlet A. Duct friction loss may be calculated from the relation

$$p_f = \frac{4fl}{De}\left(\frac{V}{4.04}\right)^2 \text{ mm of water}$$

where l is length in metres, De is equivalent diameter in metres, V is velocity in m/sec and f is the friction coefficient which is 0.005 for sheet metal and 0.006 for absorbent material.

Solution : The term equivalent diameter is defined as

$$D_e = 4 \times \frac{\text{cross-sectional area}}{\text{wetted perimeter}}$$

Segment CB (subscript 2)

$$V_2 = \frac{Q_2}{A_2} = \frac{16}{1.5 \times 1} = 10.67 \text{ m/s}$$

$$De_2 = 4 \times \frac{1.5 \times 1}{2(1.5+1)} = 1.2 \text{ m}$$

$$p_{f_2} = \frac{4 f_2 l_2}{De_2}\left(\frac{V_2}{4.04}\right)^2$$

$$= \frac{4 \times 0.005 \times 75}{1.2}\left(\frac{10.67}{4.04}\right)^2 = 8.72 \text{ mm of water}$$

$$p_{v_2} = \left(\frac{V_2}{4.04}\right)^2 = \left(\frac{10.67}{4.04}\right)^2 = 6.97 \text{ mm of water}$$

Then, the total pressure at section C is given as

$$p_{t_c} = 8.72 + 6.97 = 15.69 \text{ mm of water}$$

Segment CD (subscript 3)

The pressure p_{t_c} = 15.69 mm of water has to be equal to the sum of velocity pressure and friction loss in segment CD.

Since the duct CD is divided into three parts, each being 1 m × 0.5 m

$$De_3 = 4 \times \frac{\text{cross-sectional area}}{\text{wetted perimeter}}$$

$$= 4 \times \frac{3 \times (1 \times 0.5)}{3 \times 2\,(1 + 0.5)} = 0.667 \text{ m}$$

$$p_{f_3} = \frac{4 \times 0.006 \times 75}{0.667}\left(\frac{V_3}{4.04}\right)^2 = 0.1753\ V_3^2$$

$$p_{v_3} = \left(\frac{V_3}{4.04}\right)^2 = 0.0613\ V_3^2$$

Then : $0.1753\ V_3^2 + 0.0613\ V_3^2 = 15.69$

$$\text{or} \quad V_3 = \left[\frac{15.69}{0.1753 + 0.0613}\right]^{1/2} = 8.14 \text{ m/s}$$

∴ Quantity of air discharged at D

$$Q_3 = (1.5 \times 1) \times 8.14 = 12.21 \text{ m}^3/\text{s}$$

Segment AC (subscript 1)

Quantity of air flowing through segment AC

$$Q_1 = Q_2 + Q_3 = 16 + 12.21 = 28.21 \text{ m}^3/\text{s}$$

$$V_1 = \frac{Q_1}{A_1} = \frac{28.21}{1.5 \times 1} = 18.81 \text{ m/s}$$

$$p_{v_1} = \left(\frac{V_1}{4.04}\right)^2 = \left(\frac{18.81}{4.04}\right)^2 = 21.68 \text{ mm of water}$$

$$De_1 = 4 \times \frac{1.5 \times 1}{2(1.5+1)} = 1.2 \text{ m}$$

$$p_{f_3} = \frac{4 f l_1}{De_1}\left(\frac{V_1}{4.04}\right)^2$$

$$= \frac{4 \times 0.005 \times 50}{1.2}\left(\frac{18.81}{4.04}\right)^2 = 18.07 \text{ mm of water}$$

Total pressure at fan out let *A*,

$$p_{t_A} = p_{t_c} + p_{f_3} = 15.69 + 18.07 = 33.76 \text{ mm of water}$$

Static pressure at fan outlet A

$$p_{s_A} = p_{t_A} - p_{v_1} = 33.76 - 21.68 = \mathbf{12.08} \text{ mm of water}$$

16.10. DUCT DESIGN

The object of duct design is to determine the dimensions of all ducts in the given system. A good duct design optimizes

- duct size (manufacturing costs)
- duct system pressure loss (operation costs)
- duct system acoustics (environmental costs)
- air balance procedure (commissioning costs)

The requirements of a good duct system are :

- capacity to transport a given volume of air with minimum pressure drop, loss noise and vibration
- space occupation to be as less as possible
- easy and inexpensive repair and maintenance
- initial cost to be properly balanced with the operating cost so that total yearly expenditure is minimum

The initial cost is detemined by the cost of duct system which depends on duct sizes. The operating cost is governed by the fan power consumption which depends on the pressure drop in the air-handling unit and the duct system. The pressure drop can be reduced by increasing the duct size but that leads to an increase in the first cost. Obviously the need for a proper balance.

16.10.1. General Aspects

The following are the few general aspects to be considered in the design of ducts :

(*i*) The air supply to the prescribed locations should be as direct as possible. This will economise on power, material and space.

(*ii*) The air velocity in the duct should be within prescribed limits; high enough to reduce the size of ducts but low to reduce the noise and pressure losses to economize power requirements.

(*iii*) Sudden changes in the direction should be avoided. When bends are essential, use splitter vanes to minimize pressure loss.

(*iv*) The area changes must be gradual; for diverging ducts, the angle of divergence should not be more than 20-degree. The converging angle for reducing ducts has upper limit of 60-degree.

(*v*) For smaller pressure drop and cost effectiveness, round ducts are to be preferred. If rectangular ducts are essential, aspects ratio of 4 and less is desirable but it should not be greater than 8 in any case.

(*vi*) Each branch outlet should be provided with a damper for balancing the system.

(*vii*) Ducts should preferably be made of smooth materials such as galvanized iron or aluminium sheet metals.

(*viii*) Avoid obstructions in the path of air flow from the fan to the outlet, *i.e.*, air-conditioned space.

16.10.2. Determination of Duct Size

The design procedure for sizing of duct is based on three methods namely equal friction method, velocity reduction method and static region method. The salient features of these common methods are outlined below :

1. Equal friction method : The ducts are sized to give equal pressure drop or friction loss per unit length in all the ducts. In accordance with permissible noise level for a particular application, selection is made of suitable velocity in the main duct. For example, the maximum duct velocity recommended for residential apartments and theatres is 180 m/min and 240 m/min respectively. Knowing the air flow rate and the selected air velocity, the size and friction loss are determined. The remaining ducts are then sized, maintaining the friction loss per unit length at this value for their respective air-flow rates.

This method finds favour for designing low velocity duct systems and when the duct layout has same length in each run. There would then be equal pressure loss in each branch and no dampers would be required to balance the system. However, when the runs are of different lengths, the shortest run will have minimum loss and consequently high pressure at the outlet. For reducing the high pressure, the shortest duct will need a considerable amount of dampering. Alternatively, the design will be modified to provide higher velocities. That will create objectionable noise and to arrest that, the noise absorbing outlets and fittings will have to be provided.

This method of sizing ducts, automatically, reduces the air velocity in the direction of flow.

2. Velocity reduction method : The duct size is so determined that the velocities in the ducts progressively decrease as the flow proceeds. The starting velocities for the main duct and the branch ducts depend upon the application. For example, the recommended values for the main and branch velocities for residential apartments are 300 and 180 m/min respectively. The corresponding values for theatres are 390 and 300 m/min. Corresponding to the assumed velocities, the pressure drops are calculated for the main duct and respective branches. Then

(*i*) the duct sizes are determined from assumed velocities and the known quantities of air being supplied through ducts

(*ii*) the pressure at the outlet is adjusted by dampers provided in the respective ducts

(*iii*) the fan is designed to overcome the pressure losses along any single run including the losses of main duct, branch duct, elbows, enlargements and contractions

The method is the easiest in sizing ducts and adjustment in velocities can be made to avoid noise. However, the method is not normally adopted unless the designer has considerable practical experience, judgement and knowledge of selecting velocities so as to the make the system optimum in economy and power.

3. Static regain method : The design procedure by static regain method is based on the maxim that drop in pressure during a given length is made up by reducing velocity at the end of the section. This implies that the friction loss in each branch is made equal to the gain in pressure due to reduction in velocity. That is

$$\text{Static region} = R(p_{v1} - p_{v2}) = R \times \frac{1}{2}\,\rho\,(V_1^2 - V_2^2)$$

where R is the static regain factor. The value of R varies from 0.5 to 0.7 as a part of kinetic energy is not converted into pressure energy. When $R = 1$, the region is complete and the static pressure remains about the same throughout the system.

This design method is used for high velocity duct systems which have long runs of ducts with many branches. Advantage of this method lies in :

(*i*) long runs as well as short runs near the fan can be designed for complete regain.

(*ii*) designing the main duct for complete regain and using the same pressure at outlets of the branches.

(*iii*) self-balancing of ducts.

However, the decrease in air velocities increases duct sizes and that affects the economy of the system.

A comparison of these duct design methods is as given below :

S.No.	Equal friction (pressure drop) method	Velocity reduction method	Static regain method
1.	Equal pressure drop per unit length in all ducts.	Velocities in the ducts progressively decrease as the flow proceeds.	Static pressure remains about the same at all outlets.
2.	Friction loss is estimated for the main duct and the branch ducts are then sized maintaining the friction loss per unit length at this value.	Velocities are assumed for the main duct and branches, and the corresponding pressure drops are calculated.	The design is based on velocity pressure which leads to static regain.
3.	Dampers are required to balance pressure drop in various branches of different lengths.	The pressure at the duct outlet can be varied by dampers.	Balancing of ducts is automatic.
4.	Noise is a problem; noise absorbing outlets and fittings are to be provided.	No noise problem	No noise problem
5.	Duct size radius downstream.	Duct size reduces downstream	Duct size increases downstream.
6.	Adopted for large systems and long run of ducts.	Adopted for simple systems and long run ducts.	Adopted for large systems; may not become suitable for long ducts. Increase in duct size due to reduction in velocity may not satisfy the economic criterion.

EXAMPLE 16.10

The main circular duct of an air-conditioning system is 1.2 m in diameter and it conveys air with a velocity of 10 m/s. This duct consists of three branch ducts taking equal air volumes at equal intervals. Each branch duct has a friction loss of 1.25 mm of water and to cope up with this friction loss, a static pressure of 4 mm of water is necessary at each branch. Make calculations for the velocities and diameters of second and third lengths whereby the static pressure regain is sufficient to overcome the friction loss in succeeding length of main duct upto the next branch. A static pressure regain factor of 0.6 may be assumed.

Further proceed to identify the static, velocity and total pressure at the appropriate points of change.

Solution : Refer Fig. 16.19 for the schematic arrangement of the duct system.

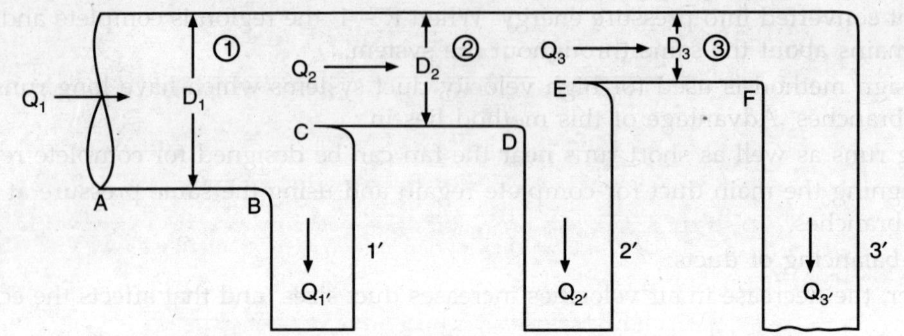

Fig. 16.19

As per the given conditions : $p_{f_1} = p_{f_2} = p_{f_3} = 1.25$ mm of water

and $p_{s_B} = p_{s_D} = p_{s_F} = 4$ mm of water

For duct AB : Discharge $Q_1 = A_1 V_1 = \dfrac{\pi}{4}(1.2)^2 \times 10 = 11.30 \text{ m}^3/\text{s}$

$$\text{Velocity pressure } p_{v_1} = \left(\frac{V_1}{4.04}\right)^2 = \left(\frac{10}{4.04}\right)^2 = 6.13 \text{ mm of water}$$

Quantity of air passing through each branch,

$$Q_{1'} = Q_{2'} = Q_{3'} = \frac{11.30}{3} = 3.77 \text{ m}^3/\text{s}$$

For duct CD : Discharge $Q_2 = Q_1 - Q_{1'} = 11.30 - 3.77 = 7.53 \text{ m}^3/\text{s}$

$$\text{Static pressure regain} = R(p_{v_1} - p_{v_2})$$

Since the static pressure regain equals the friction loss in the duct CD, we have

$$R(p_{v_1} - p_{v_2}) = p_{f_2}$$

or $$0.6\,(6.13 - p_{v_2}) = 1.25$$

That gives : $$p_{v_2} = 6.13 - \frac{1.25}{0.6} = 4.05 \text{ mm of water}$$

Applying the relation $p_{v_2} = \left(\dfrac{V_2}{4.04}\right)^2$, we get

$$V_2 = 4.04 \times \sqrt{p_{v_2}} = 4.04 \times \sqrt{4.05} = 8.13 \text{ m/s}$$

Types of Fans

588 (A)

Then : $\frac{\pi}{4} D_2^2 \times V_2 = Q_2$

$$\therefore \quad D_2 = \left(\frac{4Q_2}{\pi V_2}\right)^{1/2} = \left[\frac{4 \times 7.53}{\pi \times 8.13}\right]^{1/2} = \mathbf{1.086\ m}$$

For duct EF : Discharge $Q_3 = Q_2 - Q_{2'} = 7.53 - 3.77 = 3.76\ m^3/s$

The static pressure regain equals the friction loss, and therefore

$$R(p_{v_2} - p_{v_3}) = p_{f_3}$$

or $\quad 0.6\,(4.05 - p_{v_3}) = 1.25$

That gives : $\quad p_{v_3} = 4.05 - \frac{1.25}{0.6} = 1.97$ mm of water

Applying the relation $p_{v_3} = \left(\frac{V_3}{4.04}\right)^2$, we get

$$V_3 = 4.04 \times \sqrt{p_{v_3}} = 4.04 \times \sqrt{1.97} = 5.67\ m/s$$

Then : $\frac{\pi}{4} D_3^2 \times V_3 = Q_3$

$$\therefore \quad D_3 = \left[\frac{4Q_3}{\pi V_3}\right]^{1/2} = \left[\frac{4 \times 3.76}{\pi \times 5.67}\right]^{1/2} = \mathbf{0.919\ m}$$

The static, velocity and total pressures at the appropriate points of change are as tabulated below :

Pressure mm of water	**Appropriate point of change**				
	B	**C**	**D**	**E**	**F**
• Static pressure (p_s)	4.00	5.25	4.00	5.25	4.00
• Velocity pressure (p_v)	6.13	4.05	4.05	1.97	1.97
• Total pressure (p_t)	10.13	9.30	8.05	7.22	5.97

16.11. FANS

A fan is a mechanical device that induces air flow for cooling and ventilation purposes, and in conjuction with a heat source for heating and drying. The fan can be any revolving vane system that produces air flows with high volume and pressures enough to overcome the resistance offered to air flow by the various components in the entire duct system.

A fan essentially consists of rotating wheel (called impeller) and a stationary surrounding member known as stator or housing. The impeller is driven by a motor and the energy so imparted to air creates the pressure difference which moves the air stream. The air at inlet to the fan is below atmospheric pressure while at outlet, the air is above atmospheric pressure. The air feed into a fan is called induced draft while the air exhaust from the fan is called forced draft.

In air conditioning systems, the fans used for transmission of air are of two types : centrifugal fan and axial flow fan.

16.12. CENTRIFUGAL FANS

The centrifugal fan consists of an impeller or wheel mounted in a scroll type of housing. The impeller is driven by a motor through a belt and pulley mechanism. The air enters the centre of rotating impeller, is accelerated by the blades and is propelled from the periphery of the wheel into the discharge opening of the scroll. The scroll; acts as a diffuser where the velocity attained by the air at the wheel periphery is converted into static pressure by the time the air is discharged from the housing and gets directed towards the outlet.

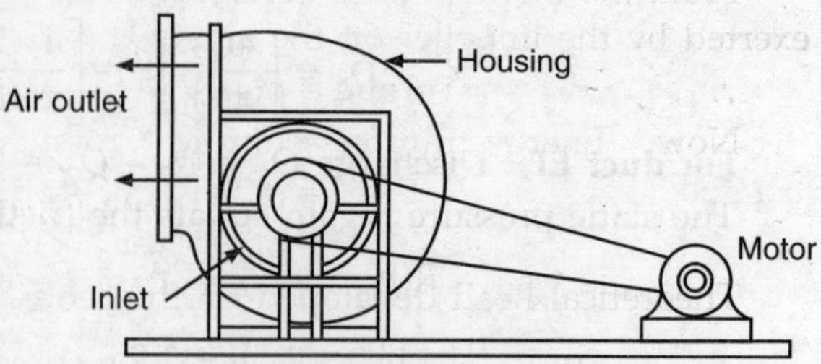

Fig. 16.20. Centrifugal fan

The air enters axially and exits radially at the periphery of impeller blades. Further the fan may have a single or double entry depending upon whether the air enteres from one or both sides.

16.12.1. Velocity Vector Diagram and Work Done

The velocity vector diagram at the inlet and outlet of the impeller of a centrifugal fan are shown in Fig. 16.21.

Following nomenclature has been used while drawing these diagrams :

V = absolute velocity of air; u = blade or peripheral velocity;
V_r = relative air velocity ; V_f = flow component of absolute velocity,
V_u = whirl or tangential component of absolute velocity.

Further, suffix 1 and 2 represent the conditions at inlet and outlet of impeller.

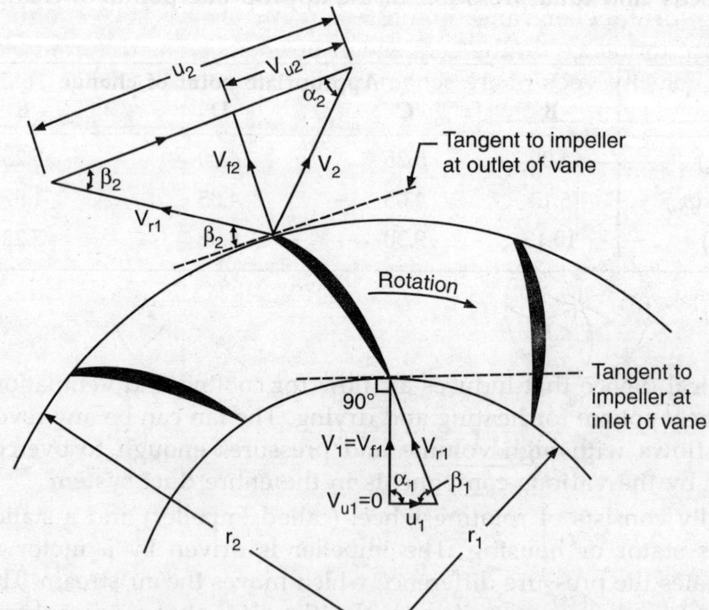

Fig. 16.21. Velocity vector diagram of a centrifugal impeller

Considering the flow of air through a rotor of any shape, the rate of change of angular momentum is given by

$$= m\,(V_{u2}\, r_2 - V_{u1}\, r_1) = \rho Q\,(V_{u2}\, r_2 - V_{u1}\, r_1)$$

where Q is the quantity of air entering and leaving the impeller in a unit time ; r_2 and r_1 are the impeller radii at outlet and inlet respectively.

From the momentum theorem, the rate of change of angular momentum equals the torque exerted by the impeller on the air

$$\therefore \quad \text{Torque} = \rho Q\,(V_{u2}\, r_2 - V_{u1}\, r_1)$$

Now, Energy transfer = torque × rotation speed in radian/sec

$$= \rho Q\,(V_{u2}\, r_2 - V_{u1}\, r_1)\,\omega$$

$$= \rho Q\,(V_{u2}\, u_2 - V_{u1}\, u_1) \qquad (\because u \quad \omega r)$$

Theoretical head developed by the centrifugal fan,

$$H = \frac{\text{work done per second}}{\text{weight of air per second}} = \frac{m(V_{u_2}\, u_2 - V_{u_1}\, u_1)}{mg} \qquad (\because W = mg)$$

$$= \frac{V_{u_2}\, u_2 - V_{u_1}\, u_1}{g} \text{ in m of air}$$

The above expression is referred to as ***Euler's equation***.

When the air enters the blades at right angles, *i.e.*, radially to the direction of motion of blades, the whirl or tangential component of absolute velocity is zero. The term $V_{u1}\, u_1$ then vanishes and the Euler's equation takes the form

$$\text{Work done} = m\, V_{u1}\, u_1 \text{ in Nm or J}$$

and $$\text{Theoretical head developed} = \frac{V_{u_2}\, u_2}{g} \text{ in m of air}$$

16.12.2. Vane Shape

The shape/form of impeller vanes is fundamentally changed depending on blade angle β_2 which has significant influence on the conversion of energy. The different blade configurations and the respective velocity vectors are schematically represented in Fig. 16.22.

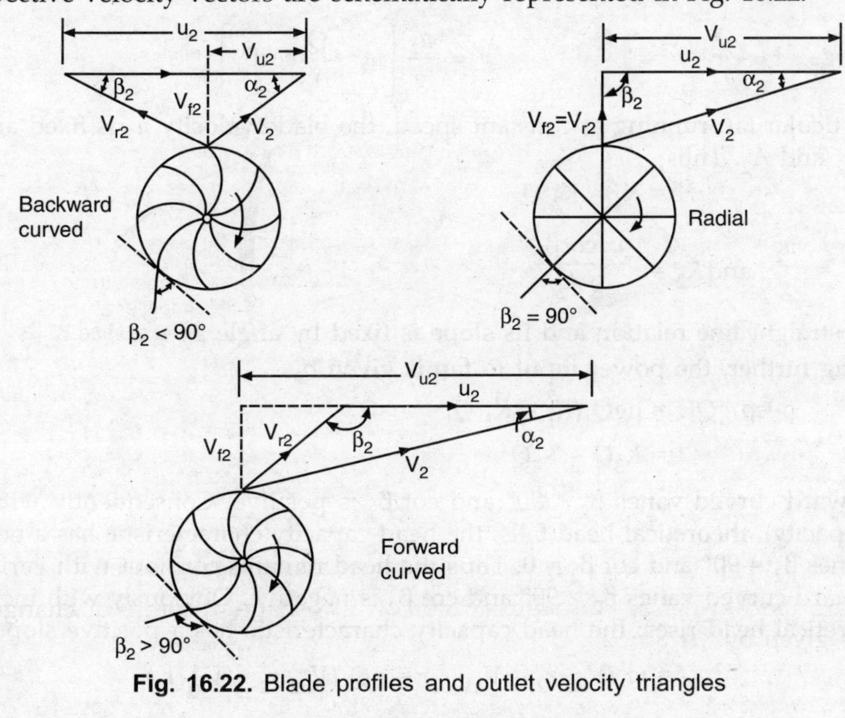

Fig. 16.22. Blade profiles and outlet velocity triangles

• **Backward curved blade :** The outlet tip of the blade is curved backward with respect to the direction of motion of the fan wheel, and the angle between the blade tip and tangent to rotor at exit is acute ($\beta_2 < 90°$)

$$V_{u2} = u_2 - V_{f2} \cot \beta_2 = u_2 - \frac{Q}{A_2} \cot \beta_2$$

• **Radial blade :** The blade has no curvature, is straight and is fixed raidally. The air leaves the blade with relative velocity in radial direction and angle $\beta_2 = 90°$

$$V_{u2} = u_2$$

• **Forward curved blade :** The outlet tip of the blade is inclined forward with respect to the rotation of the fan wheel, and the angle between the blade tip and tangent to rotor at exit is obtuse ($\beta_2 > 90°$)

$$V_{u2} = u_2 + V_{f2} \cot \beta_2 = u_2 + \frac{Q}{A_2} \cot \beta_2$$

The volume flow through the fan is given by

$$Q = \pi\, d_2\, b_2\, V_{f2} = \pi\, d_1\, b_1\, V_{f1}$$

where d is the impeller diameter and b is the width of shrouds. In practice, the flow velocities at the inlet and outlet of the impeller are kept same, and are of the order of

$$V_{f1} = V_{f2} = 0.2\, u_2$$

16.12.3. Head capacity relationships

Recalling the Euler expression for theoretical head produced and using the outlet velocity diagram,

$$\text{Theoretical head developed } H = \frac{V_{u_2}\, u_2}{g} = \frac{u_2}{g}\,[u_2 - V_{f2} \cot \beta_2]$$

$$= \frac{u_2}{g}\left[u_2 - \frac{Q}{A_2}\cot\beta_2\right]$$

For a particular fan running at constant speed, the blade velocity u_2 is fixed and so are the parameters b_2 and A_2. Thus

$$H = K_1 - K_2\, Q$$

where $K_1 = \frac{u_2^2}{g}$ and $K_2 = \frac{u_2 \cot\beta_2}{g A_2}$

This is a straight line relation and its slope is fixed by angle β_2.

Proceeding further, the power input to fan is given by

$$\rho = \rho\, gQH = \rho g Q\,(K_1 - K_2\, Q)$$

$$= K_3 Q - K_4 Q^2$$

For backward curved vanes $\beta_2 < 90°$ and cot β_2 is positive. Consequently with increase in discharge (capacity), theoretical head falls; the head-capacity characteristic has a negative slope. For radial vanes $\beta_2 = 90°$ and cot $\beta_2 = 0$. Thus the head remains constant with variation in flow rate. For forward curved vanes $\beta_2 > 90°$ and cot β_2 is negative. Obviously with increase in flow rate the theoretical head rises; the head capacity characteristic has a positive slope.

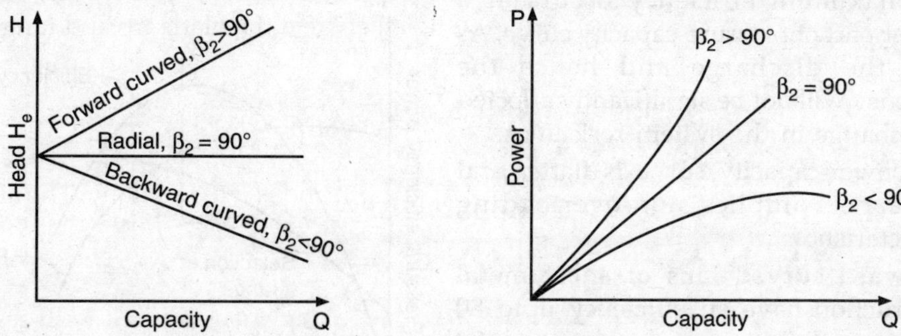

Fig. 16.23. Influence of vane exit angle on head-capacity and power-capacity relationship

Theoretical head-capacity and power-capacity variation for different type of impeller vanes is shown in Fig. 16.23. For forward curved vanes ($\beta_2 > 90°$), the power input is seen to rise steeply with the discharge. A change in the nature of these occurs due to intake losses, friction and eddy losses and the discharge losses in the diffusing section.

Other distinguishing aspects of these category of centrifugal fans are :

(*i*) For backward curved vanes, the tangential; component is much reduced and consequently for a give impeller speed, the fan will have a low energy transfer $\left(H - \dfrac{V_{u2}\,u_2}{g}\right)$

(*ii*) In case of forward curved vanes, V_{u2} is increased and consequently the energy transfer becomes more. However, the absolute velocity V_2 at impeller outlet also increases. A higher value of V_2 is not desirable as its conversion into static pressure cannot be carried out efficiently in the diffuser section.

(*iii*) For the same static pressure to be produced, the backward curved fan has to be run at a speed higher than that for the fan with forward curved vanes.

(*iv*) The fans with forward curved vanes develop the highest pressure for a given diameter and speed. Accordingly for the same fan duty, the forward curved fans are smaller and slow running.

(*v*) The forward blade fans provide maximum air velocity for a given wheel size and speed, and therefore are lighter and smaller for the same volume capacity.

(*vi*) Noise and turbulence level are greater in a forward blade fan. However, they tend to be quieter and cheaper for fan total pressures upto 750 N/m^2 (7.6 mm of water).

(*vii*) Forward curved and radial fans rarely have efficiencies greater than 75 percent. Backward curved fans of sheet metal construction have an efficiency upto 80 percent.

(*viii*) The backward curved fans find favour for use in air-conditioning systems due to better performance characteristics.

A typical set of fan curves for a centrifugal fan with backward curved vanes is shown in Fig. 16.24. These performance characteristics are usually plotted for a particular rotational speed with volume flow rate as the independent variable versus power, static pressure and efficiency as dependent variables. The notable aspects of these curves are :

- Due to steeper pressure-capacity curve, the volume of air delivered remains fairly constant even with change in resistance of the duct system.

- The maximum efficiency occurs at a steeper part of pressure-capacity curve. As such the discharge and hence the efficiency will not be significantly affected by a change in the system resistance.
- The power-capacity curve is flatter and this aspect implies non-overloading characteristic
- Backward curved fans of sheet metal construction have an efficiency upto 80 percent and this can even go upto 90 percent with aerfoil section.

The dark portion of the curves represents the normal selection range. A designer would select the fan for the required operating condition at the highest point fan efficiency, thus minimising fan losses and energy requirements.

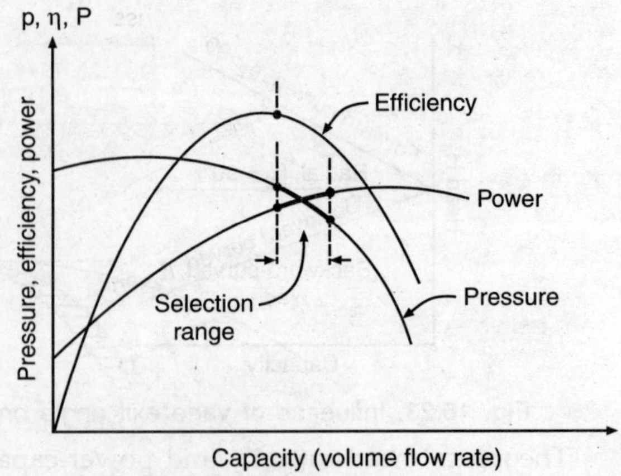

Fig. 16.24. Characteristic curves for centrifugal fan

EXAMPLE 16.11

The following data refers to a centrifugal fan that delivers 120 m³/min. When turning 1000 revolutions per minute. The air enters the impellers with a small whirl component in the direction of impeller rotation but the relative velocity meets the blade tangentially.

Impeller diameter.... 68 cm

Diameter at the blade inlet .... 45 cm

Impeller width at inlet and outlet ... 18 cm and 12 cm

Blade angles for the backward curved blade .... 22.5° and 50° with the tangent at inlet and outlet

Losses at inlet, in the impeller and in the casing ... 70 percent of velocity head at impeller outlet

Velocity head at the fan discharge ... 10 per cent of velocity head at impeller outlet.

Draw the inlet and outlet velocity triangles and determine the theoretical head developed by the impeller. Proceed to calculate the fan static pressure in mm of water.

Assume air density to be 1.2 kg/m³ and neglect the effect of blade thickness and interblade circulation.

Solution : Refer Fig. 12.25 for the velocity vector diagrams and the nomenclature. The subscripts 1 and 2 denote the conditions at inlet and outlet of the impeller.

$$u_1 = \frac{\pi D_1 N}{60} = \frac{\pi \times 0.45 \times 1000}{60} = 23.55 \text{ m/s}$$

$$u_2 = \frac{\pi D_2 N}{60} = \frac{\pi \times 0.68 \times 1000}{60} = 35.59 \text{ m/s}$$

$$V_{f1} = \frac{Q}{\pi D_1 b_1} = \frac{120/60}{\pi \times 0.45 \times 0.18} = 7.86 \text{ m/s}$$

$$V_{f2} = \frac{Q}{\pi D_2 b_2} = \frac{120/60}{\pi \times 0.68 \times 0.12} = 7.80 \text{ m/s}$$

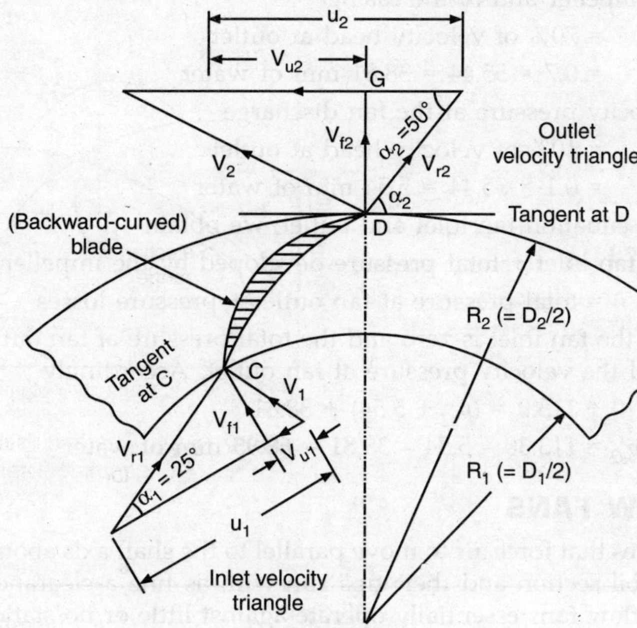

Fig. 16.25

From the initial velocity triangle,

$$u_1 - V_{u1} = V_{f1} \cot 22.5°$$

or $$V_{u1} = u_1 - V_{f1} \cot 22.5° = 23.55 - 7.86 \times 2.414 = 4.57 \text{ m/s}$$

Similarly from the outlet velocity triangle

$$V_{u2} = u_2 - V_{f2} \cot 50° = 35.59 - 7.80 \times 0.8389 = 29.05 \text{ m/s}$$

From Euler relation, the theoretical head developed by the fan impeller is given by

$$H = \frac{V_{u_2} u_2 - V_{u_1} u_1}{g} = \frac{29.05 \times 35.59 - 4.57 \times 23.55}{9.8}$$

$$= \frac{1033.89 - 107.62}{9.81} = \mathbf{94.42 \text{ m}} \text{ of air}$$

(*b*) Total pressure developed by the fan impeller

$$p_{tf} = \rho_a\, g\, H = 1.2 \times 9.81 \times 94.42 = 1111.51 \text{ N/m}^2$$

Since 1 mm of water = 9.81 N/m²,

$$p_{tf} = \frac{1111.51}{9.81} = 113.30 \text{ mm of water}$$

From the outlet velocity triangle, the absolute velocity of air leaving the blade is

$$V_2 = \sqrt{V_{u2}^2 + V_{f2}^2} = \sqrt{29.05^2 + 7.80^2} = \sqrt{843.90 + 60.84} = 30.08 \text{ m/s}$$

Then : Velocity head or velocity pressure at the impeller outlet,

$$p_{v2} = \left(\frac{V_2}{4.04}\right)^2 = \left(\frac{30.08}{4.04}\right)^2 = 55.44 \text{ mm of water}$$

Loss at inlet, in the impeller and in the casing

$$= 70\% \text{ of velocity head at outlet}$$
$$= 0.7 \times 55.44 = 38.81 \text{ mm of water}$$

Velocity head or velocity pressure at the fan discharge

$$= 10\% \text{ of velocity head at outlet}$$
$$= 0.1 \times 55.44 = 5.54 \text{ mm of water}$$

Applying Bernoulli's equation fan inlet and outlet, we obtain

total pressure at fan inlet + total pressure developed by the impeller

= total pressure at fan outlet + pressure losses

The total pressure at the fan inlet is zero and the total pressure at fan outlet equals the sum of fan static pressure and the velocity pressure at fan outlet. Accordingly

$$0 + 113.0 = (p_{s2} + 5.54) + 38.81$$
$$\therefore \quad p_{s2} = 113.30 - 5.54 - 38.81 = \mathbf{68.95} \text{ mm of water}$$

16.13. AXIAL FLOW FANS

Axial flow fans have blades that force air to move parallel to the shaft axis about which the blades rotate. Blades are of airfoil section and their tips run with as fine a clearance as possible in a cylindrical casing. Axial flow fans essentially operate against little or no static pressure and are used chiefly for exhaust and circulation purposes.

The axial flow fans are categorized as propeller fans, tube axial fans and vane axial fans.

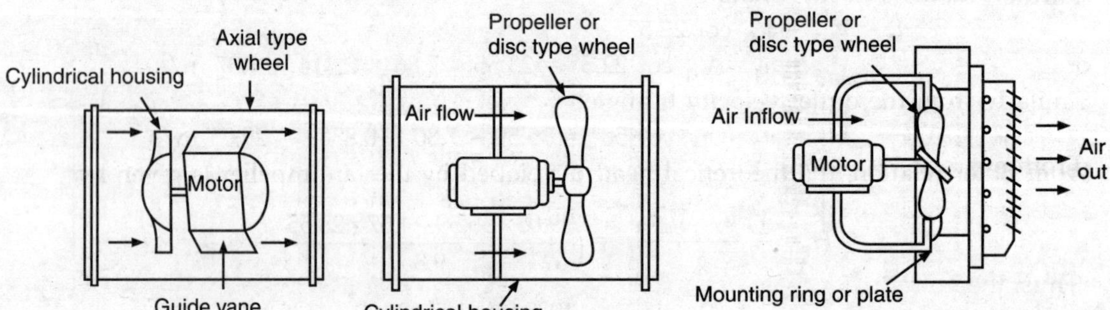

Fig. 16.26. Axial flow fans

• **Propeller fan :** The unit consists of a propeller or disc wheel which works within a mounting ring or plate. The ring surrounding the wheel is so designed that it does not permit the discharged air to come back into the wheel around its periphery.

These are low pressure, high capacity fans and are used where resistance to the movement is small and the situations where little or no duct work is required. The notable applications are :

– ventilation of attic spaces, wash rooms etc.
– removal of cooking odour from kitchen

• **Tube axial fan :** A fan consisting of a propeller or disc type wheel within a cylinder; may be either belt driven or connected directly to a motor. The wheel is formed with a large diameter hub provided with die formed blades. These design aspects minimize re-entry losses and produce a uniform air velocity from hub to tip. The air discharge from the fan follows a spiral path as it leaves the cylindrical casing.

These fans are more efficient than propeller fans, are easy to install, and are able to handle a wide range of air volume against medium pressures.

• **Vane axial fan :** A fan consisting of a disc type wheel mounted in a cylindrical casing with a set of air guide vanes located before or after the wheel. The vanes help to straighten out the spiral flow of air coming out of the fan unit. The fan may be belt-driven or connected directly to a motor.

The vane axial fans have better efficiency and pressure characteristics, and are quieter in operation compared to propeller and tube axial type fans.

16.14. SIMILARITY RELATIONS AND FAN LAWS

While investigating the operating characteristics of homologous (identical) fans, the following assumptions are made :

- the fans are kept geometrically similar and all linear dimensions are proportional to the impeller diameter, *e.g.*, breadth $B \propto$ diameter D.
- there is similarity in the velocity triangles
- all velocities are proportional to square root of pressure head H developed by the fan.
- there is equality of flow Reynolds number through the fans being examined.

(*a*) Discharge $Q \propto$ area of flow × flow velocity

$$\propto \pi D B V_f$$

Since $B \propto D$ and $V_f \propto \sqrt{H}$, we obtain

$$Q \propto D^2 \sqrt{H} \qquad ...(i)$$

Further, the blade velocity or tangential velocity of impeller,

$$u = \frac{\pi DN}{60} \; ; \; D \propto \frac{u}{N} \propto \frac{\sqrt{H}}{N}$$

$$\text{or} \quad \sqrt{H} \propto ND \qquad ...(ii)$$

From expressions (*i*) and (*ii*),

$$Q \propto D^2 \times ND \propto ND^3 \quad \text{or} \quad \frac{Q}{ND^3} = \text{constant} \qquad ...(iii)$$

Thus, the volume flow rate or capacity (Q) of a fan is directly proportional to the fan speed (N) and cube of impeller diameter (D).

The factor $\frac{Q}{ND^3}$ is called the ***flow coefficient***.

(*b*) Relationship $D \propto \frac{u}{N} \propto \frac{\sqrt{H}}{N}$ gives $H \propto N^2 D^2$

The total pressure developed by the fan is given by

$$p_{tf} \propto \rho_a gH \propto \rho H \propto \rho N^2 D^2$$

$$\text{or} \quad \frac{p_{tf}}{\rho_a N^2 D^2} = \text{constant} \qquad ...(iv)$$

That is, the total pressure developed by the fan is directly proportional to the square of fan speed (N^2), square of the impeller diameter (D^2) and density of air (ρ_a).

The parameter $\frac{p_{tf}}{\rho_a N^2 D^2}$ is called the ***pressure*** or ***head coefficient***.

(c) Power $P = Q\, p_{tf} \propto (D^2 \sqrt{H})\ (\rho_a N^2 D^2)$

$$\propto (D^2 \times ND)\ (\rho_a N^2 D^2)$$

$$\propto \rho_a N^3 D^5$$

or $$\frac{P}{\rho_a N^3 D^5} = \text{constant}$$

Thus the power (P) of a fan is directly proportional to cube of fan speed (N^3), fifth power of impeller diameter (D^5) and density of the air (ρ_a).

The parameter $\frac{P}{\rho_a N^3 D^5}$ is called the ***power coefficient***.

If it is merely needed to determine the performance of one particular machine and density of air is assumed to remain constant, then the general non-dimensional parameters get simplified to:

$$\frac{Q}{N} = \text{constant}\ ; \quad \frac{H}{N^2} = \text{constant and } \frac{P}{N^3} = \text{constant}$$

Note : The above laws apply to what is called a constant system; one with no change in duct and fittings.

EXAMPLE 16.12

Test runs at the workshop on a fan indicate that it delivers 5 m³/s of air of density 1.2 kg/m³ against a static pressure of 62.5 mm of water with a static efficiency of 60 percent.

If the fan is driven by a constant speed motor, calculate the static pressure developed and the power consumed at the site of use where the air density is 0.95 kg/m³.

Solution : Invoking the pressure or head coefficient relation,

$$\frac{p_{sf}}{\rho_a N^2 D^2} = \text{constant}$$

For one particular machine running at constant speed (D and N are fixed), the above relation gets modified to

$$\frac{p_{sf_1}}{\rho_{a_1}} = \frac{p_{sf_2}}{\rho_{a_2}}\ ; \qquad p_{sf_2} = \frac{\rho_{a_2}}{\rho_{a_1}} \times p_{sf_1}$$

$$= \frac{0.95}{1.2} \times 62.5 = 49.48 \text{ mm of water}$$

$$= 49.48 \times 9.81 = 485.4 \text{ N/m}^2$$

(b) Invoking the flow coefficient relation, $\frac{Q}{ND^3}$ = constant and modifying it for a particular machine operating at a constant speed,

$$Q_2 = Q_1 = 5 \text{ m}^3/\text{s}$$

$$\text{Static fan power} = Q \times p_{sf} = 5 \times 485.4 = 2427 \text{ Nm/s} = 2427 \text{ W}$$

Since the machine is stated to have static efficiency of 60%,

Shaft power at the site $= \frac{2427}{0.6} = 4045 \text{ W} = \mathbf{4.045\ kW}$

EXAMPLE 16.13

A centrifugal fan of 60 cm impeller diameter discharges 125 m³/min of air at 17°C while developing a total pressure of 725 N/m². A motor drives the fan and the tachometer registers a speed of 1250 rev/min. Taking total efficiency of fan as 85 percent, determine the volume of air delivered, total pressure developed and power consumed under the following set of conditions :

(a) the air temperature is 45°C

(b) the air temperature is 45°C and the fan speed is increased to 1500 rev/min

(c) the air temperature is 45°C, the fan speed is 1500 rev/min and a geometrical similar fan of 50 cm impeller diameter is operated.

Solution : The density ratio at the given temperature is

$$\frac{\rho_{a_2}}{\rho_{a_1}} = \frac{273+t_1}{273+t_2} = \frac{273+17}{273+45} = 0.912$$

The similarity relations for discharge and total pressure developed by the fan are :

$$\frac{Q}{ND^3} = \text{constant} \; ; \quad \frac{p_{tf}}{\rho_a N^2 D^2} = \text{constant} \; ; \quad \frac{P}{\rho_a N^3 D^5} = \text{constant}$$

(*a*) For a particular machine operating at constant speed (no change in diameter and speed), the similarity relations get modified to

$$Q = \text{constant and } \frac{p_{tf}}{\rho_a}$$

$$\therefore \quad Q_2 = Q_1 = 125 \text{ m}^3/\text{min}$$

$$\text{and} \quad p_{tf_2} = \frac{\rho_{a_2}}{\rho_{a_1}} p_{tf_1} = 0.912 \times 725 = \mathbf{661.2 \ N/m^2}$$

$$\text{Power consumed } P_2 = \frac{Q_2 p_{tf_2}}{\eta} = \frac{125 \times 66.12}{0.85}$$

$$= 97235.3 \text{ Nm/min} = 1620.6 \text{ Nm/s} = \mathbf{1620.6 \ W}$$

(*b*) Let subscripts 2 and 3 refer to the running conditions at speeds 1250 rpm and 1500 rpm respectively; both at temperature 45°C (no change in density)

$$Q_3 = Q_2 \times \frac{N_3}{N_2} = 125 \times \frac{1500}{1250} = 150 \text{ m}^3/\text{min}$$

$$p_{tf_3} = p_{tf_2} \left(\frac{N_3}{N_2}\right)^2 = 661.2 \times \left(\frac{1500}{1250}\right)^2 = 952.13 \text{ N/m}^2$$

$$P_3 = P_2 \left(\frac{N_3}{N_2}\right)^3 = 1620.6 \times \left(\frac{1500}{1250}\right)^3 = \mathbf{2800.4 \ W}$$

$$\text{Check :} \quad P_3 = \frac{Q_3 p_{tf_3}}{\eta} = \frac{150 \times 952.13}{0.85} = 168023 \text{ Nm/min} = 2800.4 \text{ W}$$

(*c*) Let subscripts 3 and 4 refer to the running conditions for the given fan (D_3 = 60 cm) and the geometrically similar fan (D_4 = 50 cm). Other parameters namely running speed and temperature (density) of air delivered remain constant. Then

$$Q_4 = Q_3 \times \left(\frac{D_4}{D_3}\right)^3 = 150 \times \left(\frac{0.5}{0.6}\right)^3 = 86.80 \text{ m}^3/\text{min}$$

$$p_{tf_4} = p_{tf_3}\left(\frac{D_4}{D_3}\right)^2 = 952.13 \times \left(\frac{0.5}{0.6}\right)^2 = 661.2 \text{ N/m}^2$$

$$P_4 = P_3\left(\frac{D_4}{D_3}\right)^5 = 2800.4 \times \left(\frac{0.5}{0.6}\right)^5 = \mathbf{1125.42 \text{ W}}$$

Check : $$P_4 = \frac{Q_4 p_{tf_4}}{\eta} = \frac{86.80 \times 661.2}{0.85} = 67520 \text{ Nm/min} = 1125.34 \text{ W}$$

EXAMPLE 16.14

The air supply system for an air-conditioning plant consists of straight inlet and outlet ducts. The inlet duct is 15 m long and is of circular section 80 cm in diameter. A centrifugal fan of outlet 80 cm × 60 cm and discharging 560 m³/min of air is located after the inlet duct and there after is installed a diffuser that connects with the outlet duct 80 cm in diameter and 60 m in length. The pressure loss at entry to the inlet duct is 0.5 times of the velocity pressure there, and the loss of presssure in the outlet duct can be taken as 0.3 times the difference between velocity pressures. Assuming friction factor to be 0.0042 for the inlet duct and 0.0034 for the outlet duct, determine :

(a) fan total pressure

(b) static pressure at the fan inlet and outlet.

Proceed to plot the variation of total pressure and static pressure along the system.

Assume that air is sucked in by the inlet duct and is delivered by the outlet duct at atmospheric pressure.

Solution : Refer figure 16.27 for the duct system

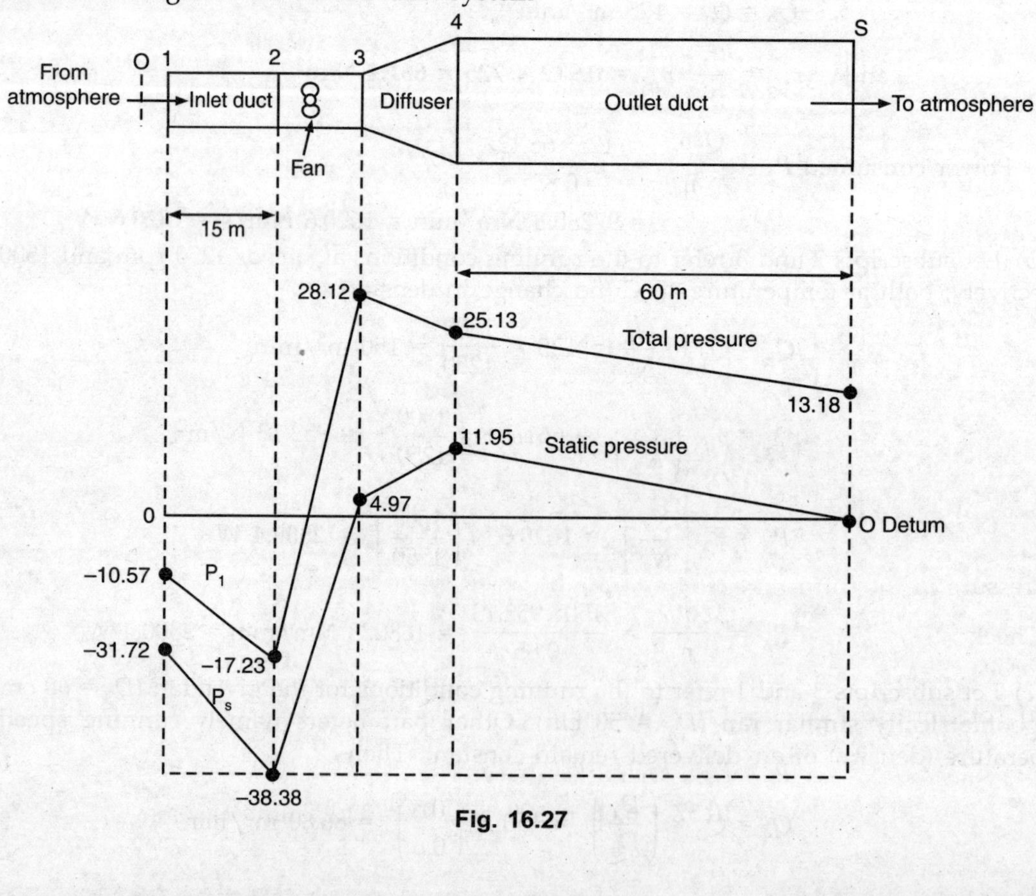

Fig. 16.27

Inlet duct 1-2

$$\text{Velocity } V_1 = V_2 = \frac{Q}{A} = \frac{560/60}{\frac{\pi}{4}(0.8)^2} = 18.58 \text{ m/s}$$

$$\text{Velocity pressure } p_{v_1} = p_{v_2} = \left(\frac{V_1}{4.04}\right)^2 = \left(\frac{18.58}{4.04}\right)^2 = 21.15 \text{ mm of water}$$

Pressure loss due to friction,

$$p_{f_{1-2}} = \frac{f_1 l_1}{m_1} p_{v_1} = \frac{4 f_1 l_1}{D_1} p_{v_1} \qquad \left(\because m_1 = \frac{D_1}{4}\right)$$

$$= \frac{4 \times 0.0042 \times 15}{0.8} \times 21.15 = 6.66 \text{ mm of water}$$

Pressure loss at entry at inlet duct at point 1

$$p_{l_1} = 0.5\, p_{v1} = 0.5 \times 21.15 = 10.57 \text{ mm of water}$$

Fan section 2-3 :

$$A_3 = 0.8 \times 0.6 = 0.48 \text{ m}^2$$

$$V_3 = \frac{Q}{A_3} = \frac{560/60}{0.48} = 19.44 \text{ m/s}$$

$$p_{v3} = \left(\frac{V_3}{4.04}\right)^2 = \left(\frac{19.44}{4.04}\right)^2 = 23.15 \text{ mm of water}$$

Outlet duct 4-5 :

$$A_4 = A_5 = \frac{\pi}{4}(0.9)^2 = 0.686 \text{ m}^2$$

$$V_4 = V_5 = \frac{Q}{A_4} = \frac{560/60}{0.636} = 14.67 \text{ m/s}$$

$$p_{v4} = p_{v5} = \left(\frac{V_4}{4.04}\right)^2 = \left(\frac{14.67}{4.04}\right)^2 = 13.18 \text{ mm of water}$$

$$p_{f4-5} = \frac{f_4 l_4}{m_4} p_{v_4} = \frac{4 f_4 l_4}{D_4} p_{v_4} \qquad \left(\because m_4 - \frac{D_4}{4}\right)$$

$$= \frac{4 \times 0.0034 \times 60}{0.9} \times 13.18 = 11.95 \text{ mm of water}$$

Pressure in the diffuser section 3 – 4

$$p_{l_{3-4}} = 0.3\,(p_{v_3} - p_{v_4}) = 0.3(23.15 - 13.18)$$

$$= 2.99 \text{ mm of water}$$

(*a*) Total pressure loss in the duct system

$$p_l = p_{l_1} + p_{f_{1-2}} + p_{l_{3-4}} + p_{f_{4-5}}$$

$$= 10.57 + 6.66 + 2.99 + 11.95 = 32.17 \text{ mm of water}$$

Fan total pressure $p_{t_f} = p_l + p_{v_4}$ = 32.17 + 13.18

= **45.35** mm of water

(*b*) Let p_{s_2} and p_{s_3} be the static pressures at fan inlet and fan outlet respectively.

Applying Bernoulli's equation to a section at 0 in the atmosphere upstream of the entrance to inlet duct and to section 2 at inlet to fan,

Total pressure at 0 = Total pressure at 2

$$p_{t_0} = p_{t_2} = p_{s_2} + p_{v_2} + p_{l_1} + p_{f_{1-2}}$$

The total pressure in the atmosphere p_{t_0} is zero and accordingly

$$0 = p_{s_2} + 21.15 + 10.57 + 6.66$$

That gives : p_{s_2} = – 38.38 mm of water

Again applying Bernoulli's equation to fan inlet (section 2) and fan outlet (section 3),

$$p_{t_2} = p_{t_3}$$

or $$p_{s_2} + p_{v_2} + p_{t_f} = p_{s_3} + p_{v_3}$$

or $$-38.38 + 21.15 + 45.35 = p_{s_3} + 23.15$$

That gives : p_{s_3} = 4.97 mm of water

(*c*) Variation of total pressure and static pressure along the system

section 1 (entrance to inlet duct) :

$$p_{t_1} = p_{t_0} - p_{l_1} = 0 - 10.57 = -10.57 \text{ mm of water}$$

$$p_{s_1} = p_{t_1} - p_{v_1} = -10.57 - 21.15 = -31.72 \text{ mm of water}$$

section 2 (inlet to fan) :

$$p_{t_2} = p_{t_1} - p_{f_{1-2}} = -10.57 - 6.66 = -17.23 \text{ mm of water}$$

$$p_{s_2} = p_{t_2} - p_{v_2} = -17.23 - 21.15 = -38.38 \text{ mm of water}$$

section 3 (fan outlet or inlet of diffuser)

$$p_{t_3} = p_{t_2} + p_{t_f} = -17.23 + 45.35 = 28.12 \text{ mm of water}$$

$$p_{s_3} = p_{t_3} - p_{v_3} = 28.12 - 23.15 = 4.97 \text{ mm of water}$$

section 4 (outlet of diffuser or entrance to outlet duct)

$$p_{t_4} = p_{t_3} - p_{l_{3-4}} = 28.12 - 2.99 = 25.13 \text{ mm of water}$$

$$p_{s_4} = p_{t_4} - p_{v_4} = 25.13 - 13.18 = 11.95 \text{ mm of water}$$

section 5 (exit of outlet duct)

$$p_{t_5} = p_{t_4} - p_{f_{4-5}} = 25.13 - 11.95 = 13.18 \text{ mm of water}$$

$$p_{s_5} = p_{t_5} - p_{v_5} = 13.18 - 13.18 = 0.0 \text{ mm of water}$$

The variation of total pressure and static pressure along the system has been depicted in Fig. 16.27.

EXAMPLE 16.15

A rectangular fan has a circular inlet duct of 50 cm diameter and a rectangular outlet duct of 50 cm × 40 cm size. Measurement data shows that when the fan delivers 125 m³/min of air, the static pressure at the fan inlet and fan outlet are – 10.5 and 25 mm of water. If input to the fan is 1.1 W, determine :

(a) total pressure at fan inlet and outlet,

(b) fan total pressure and fan static pressure, and

(c) fan total efficiency and fan static efficiency.

Assume standard air density ρ = 1.2 kg/m³ in both ducts and compressibility factor as unity.

Solution : Let suffix 1 and 2 denote the conditions at inlet to and exit from the fan.

At inlet to fan : $A_1 = \frac{\pi}{4} D^2 = \frac{\pi}{4} \times 0.5^2 = 0.19625 \text{ m}^2$

$$V_1 = \frac{125}{0.19625} = 636.94 \text{ m/min} = 10.616 \text{ m/sec}$$

$$p_{v1} = \left(\frac{V_1}{4.04}\right)^2 = \left(\frac{10.616}{4.04}\right)^2 = 6.905 \text{ mm of water}$$

∴ Total pressure at fan inlet,

$$p_{t1} = p_{s1} + p_{v1} = -10.5 + 6.905 = -\mathbf{3.595} \text{ mm of water}$$
$$= -3.595 \times 9.81 = -35.27 \text{ N/m}^2$$

At exit from fan : $A_2 = ab = 0.5 \times 0.4 = 0.2 \text{ m}^2$

$$V_2 = \frac{125}{0.2} = 625 \text{ m/min} = 10.42 \text{ m/sec}$$

$$p_{v2} = \left(\frac{V_2}{4.04}\right)^2 = \left(\frac{10.42}{4.04}\right)^2 = 6.65 \text{ mm of water}$$

∴ Total pressure at fan outlet,

$$p_{t2} = p_{s2} + p_{v2} = 25 + 6.65 = -\mathbf{31.65} \text{ mm of water.}$$
$$= 31.65 \times 9.81 = 310.49 \text{ N/m}^2$$

(*b*) Fan total pressure $p_{tf} = p_{t2} - p_{t1}$

$$= 31.65 - (-3.595) = \mathbf{35.245} \text{ mm of water}$$
$$= 35.245 \times 9.81 = 345.75 \text{ N/m}^2$$

Fan static pressure $p_{sf} = p_{tf} - p_{v2}$

$$= 35.245 - 6.65 = \mathbf{28.595} \text{ mm of water}$$
$$= 28.595 \times 9.81 = 280.52 \text{ N/m}^2$$

(*c*) Total fan air power $= Q\, p_{tf}\, K$

$$P_{tf} = \frac{125}{60} \times 345.75 \times 1 = 720.31 \text{ W} \simeq 0.72 \text{ kW}$$

Static fan air power $= Q\, p_{sf}\, K$

$$P_{sf} = \frac{125}{60} \times 280.52 \times 1 = 584.42 \text{ W} \simeq 0.584 \text{ kW}$$

Fan total efficiency $= \frac{P_{tf}}{\text{power input}} = \frac{0.72}{1.1} = \mathbf{6545 \text{ or } 65.56\%}$

$$\text{Fan static efficiency} = \frac{P_{sf}}{\text{power input}} = \frac{0.584}{1.1} = \mathbf{0.5309 \text{ or } 53.09\%}$$

16.15. FAN PRESSURE, POWERS AND EFFICIENCIES

• The fan total pressure (p_{tf}) is the difference between the total pressure at fan outlet and the total pressure at fan inlet. That is

$$p_{tf} = p_{t2} - p_{t1}$$
$$= (p_{s2} + p_{v2}) - (p_{s1} + p_{v1})$$

The fan total pressure essentially consists of the following :

- the pressure loss of suction opening and friction past the inlet grille,
- the pressure loss of return duct and air-conditioning equipment,
- the pressure loss of supply duct,
- the friction past the outlet grille, and
- the velocity pressure at the outlet.

As such the fan total pressure equals the total pressure drop plus the velocity pressure at the outlet.

The total pressure at a point in the flow stream is the sum of static pressure and velocity pressure at that point. Thus for a fan

fan total pressure = fan static pressure + fan velocity pressure

That is : $$p_{tf} = p_{sf} + p_{vf}$$

The fan velocity pressure is the velocity pressure corresponding to mean velocity of air at the fan outlet based on total area without any deductions for motors and other obstructions ($p_{vf} = p_{v2}$). Then

$$p_{sf} = p_{tf} - p_{v2}$$

The fan static pressure thus represents the difference between the fan total pressure and the velocity pressure at the fan outlet. It essentially represents the pressure increase produced by fan.

• The power output of a fan is expressed in terms of air power and it represents the work done by fan

$$\text{Total fan air power } P_{tf} = Q\, p_{tf}\, K \quad \text{(in watts)}$$

where Q is the total quantity of air flowing at the fan inlet in m^3/s; p_{tf} is the fan total pressure in N/m^2 and K is the compressibility coefficient (normally presumed to be unity)

$$\text{Static fan air power } P_{sf} = Q\, p_{sf}\, K \quad \text{(in watts)}$$

• The ratio of the total fan power and the brake power (driving power required at the fan shaft) is known as total fan efficiency.

$$\eta_{tf} = \frac{\text{total fan air power}}{\text{brake power}} = \frac{P_{tf}}{BP}$$

The fan static efficiency is defined as the ratio of static fan air power to the brake or input power of fan

$$\eta_{tf} = \frac{P_{sf}}{BP}$$

16.16. FANS IN SERIES AND IN PARALLEL

Fans are said to be connected in ***series*** when the outlet of first fan is connected to the inlet of second fan, and outlet of the second fan is connected to the inlet of third fan and so on.

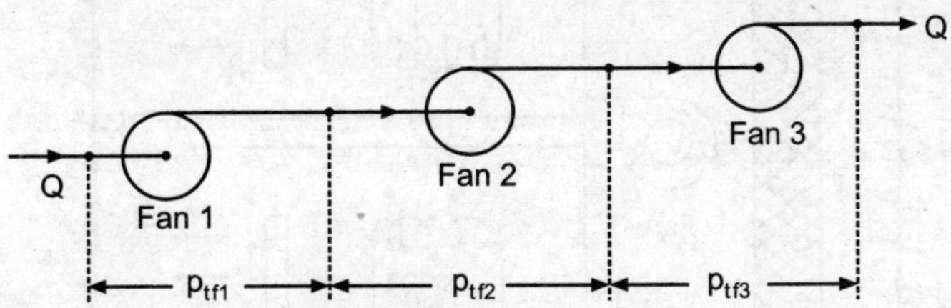

Fig. 16.28. Fans in series

When fans are arranged in series

(*i*) the volume flow rate (Q) through each fan is same. That is

$$Q_1 = Q_2 = Q_3 = Q$$

(*ii*) the overall fan total pressure (p_{TF}) is equal to the sum of fan total pressures developed by individual fans. That is

$$p_{tf} = p_{tf1} + p_{tf2} + p_{tf3}$$

- Fans are said to be connected in ***parallel*** when all the fans have the same inlet and outlet. Each works separately and the discharge from various impellers are connected in a common delivery pipe. The system augments the discharge ; fan total pressure remaining same.

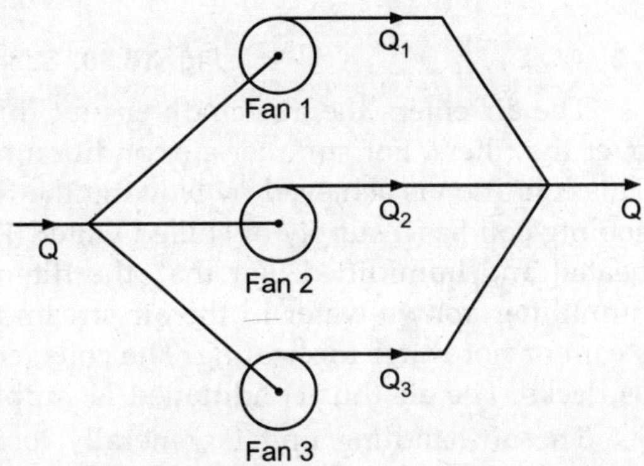

Fig. 16.29. Fans in parallel

When the fans are connected in parallel,

(*i*) the total volume delivered equals the sum of air volumes delivered by individual impellers

$$Q = Q_1 + Q_2 + Q_3$$

(*ii*) the fan total pressure of each fan is same. That is

$$p_{tf1} = p_{tf2} = p_{tf3}$$

The multistage units (series and parallel connections) have a lower efficiency because of the loss associated with the tortuous motion of air between passages.

16.17. AIR HANDLING UNIT (AHU)

The air handling unit is a packaged assembly of air-conditioning equipment that includes :

(*i*) a fan or blower

(*ii*) the heating and/or cooling coils

(*iii*) the humidifying and/or dehumidifying unit

(*iv*) the regulatory controls such as dampers

(*v*) the condensate drain pans,

(*vi*) the air filters and so forth.

These components serve to provide for the treatment of air before it is distributed to the air-conditioned space.

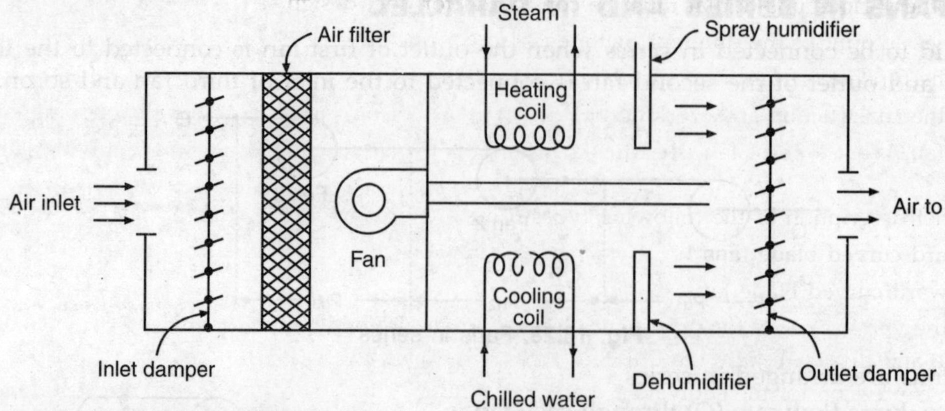

Fig. 16.30. Schematics of air handling unit

The air enters the air handling unit through the inlet damper and gets purified by passing over the filters. For summer air-conditioning, the air is required to be cooled and dehumidified. This condition is achieved by blowing the filtered air over the cooling coil and dehumidifier. The cooling coil has a supply of chilled water. The winter air-conditioning necessitates that the air be heated and humidified. For that, the filtered air is blown over the heating coil and the spray humidifier sprays water in the air stream for humidification. The heating coil has a supply of steam or hot water for heating. The coils (cooling or heating) are usually finned and are referred as decks. The air thus conditioned is supplied to the room through the outlet damper.

The air-handling unit is generally located on vibration proof pads on the roof or in the basement of the building and connected to the duct system. A single air handling unit can supply air to several zones in the building.

REVIEW QUESTIONS

A. Conceptual and conventional questions.

1. State the importance of indoor air quality in a building.
2. Describe briefly the various methods which cause indoor air pollution.
3. What is ventilation ? How ventilation methods are classified ?
4. (*a*) State the function of duct in an air-conditioning system.
   (*b*) Present a brief account of duct shape, duct material and duct construction.
5. (*a*) How ducts are classified ?
   (*b*) Define the duct aspect ratio. Should it have a large or small value ?
6. What is aspect ration and why it is considered an important factor in design of ducts for air distribution in an distribution system?
7. (*a*) Define static pressure, dynamic pressure and total pressure in the context of flow through a duct.
   (*b*) Mention any two factors that cause pressure loss in a duct.
8. Define the following terms as used in air distribution system:
   Grille, register, diffuser, throw and drop
9. List out the characteristics of a good air distribution system.
10. Mention the different methods of air distribution system and point out their salient features.
11. State the principle of air distribution. What is an air handling system ?

12. Mention the various points that must be considered for duct design.
13. Give a brief account of the different methods used for duct design.
14. What is the function of fan in an air-conditioning system ?
15. Describe the construction and working of different types of fans used in an air-conditioning system.
16. Define a fan. What type of fan is particularly suitable for an air conditioning system with elaborate ducting.
17. Give a brief description of the following types of fans and compare thier performance.
    (*a*) forward curved blade fans
    (*b*) backward curved blade fans
18. (*a*) List the various needs for proper air cleaning in an air conditioning system.
    (*b*) Name and discuss briefly the different types of air cleaning devices.
19. Derive an expression for the equivalent diameter of circular duct corresponding to a rectangular duct of sides a and b for the same pressure loss per unit length when the velocity of air flowing through both ducts is same.
    Ans. $\left(\frac{2ab}{a+b}\right)$
20. Prove that the equivalent diameter for a circular duct to replace rectangular duct for the same volume rate of flow of air is : $D_e = .1.265\left(\frac{a^3b^3}{a+b}\right)^{1/5}$
    where, D_e = equivalent diameter for round duct
    a, b = width and breadth respecively of rectangular duct.
21. The main air supply duct of an air-conditioning system is 100 cm × 90 cm in cross-section and conveys 10 m³/s of standard air. It branches off into two ducts of cross-section 80 cm × 80 cm and 80 cm × 60 cm. If the mean velocity of air in the larger branch is 9 m/s, make calculations to work out mean velocity pressure in each duct. (**Ans.** 7.55, 4.96, 4.78 mm of water)
22. An air duct of 60 cm × 40 cm rectangular section supplies 100 m³/min of standard air (density 1.2 kg/m³) to an air-conditioning system. Determine the equivalent diameter of circular duct if (*a*) the velocity of air in two ducts is same, (*b*) the discharge rate of air in two ducts is same.
    Also workout the pressure loss per 100 metre length of duct. The friction factor for the sheet metal used for the fabrication of duct may be taken as 0.015. (**Ans.** 48 cm, 54.4 cm, 6.25 metres of air)
23. A circular duct of 30 cm diameter conveys standard air to an air-conditioned space at a velocity of 6 m/s. This duct is subsequently required to be replaced by a rectangular duct having the same pressure loss per unit length due to friction. Determine the size of rectangular duct if the aspect ratio is 1.5 for (*a*) the same velocity in the two ducts, (*b*) the same discharge rate of air in the two ducts. (**Ans.** 37.5 cm, 25 cm and 33.6 cm × 22.5 cm)
24. A fan delivers 10 m³/s of standard air at a pressure of 700 Pascal at a fan speed of 600 rpm and a power input of 15 kW. What speed, input power and static pressure are required if the flow rate is increased to 15 m³/s by increasing the rotational speed? (**Ans.** 900 rpm, 50.62 kW; 1575 P_a)
25. A 30 m long rectangular duct 20 mm × 160 mm in size carries standard air at the rate of 24 m³/min. Assuming friction factor f = 0.0048, determine,
    (*a*) total pressure required at inlet to the duct to maintain this flow
    (*b*) air power required. (**Ans.** 397.8 N/m², 159.2 W)
26. A rectangualr duct of 500 × 350 mm size carries 75 m³/min of air having density of 1.15 kg/m³. Determine the equivalent diameter of a circular duct if
    (*a*) the quantity of air carried in both cases is same
    (*b*) the velocity of air in both cases is same
    If f = 0.01, find the pressure loss per 100 m length of the duct. (**Ans.** 460 mm, 412 mm, 29.9 mm)

27. A rectangular duct 2 m × 1 m in size carries conditioned air for a straight run of 50 m. Thereafter, as shown in the figure given below, it branches into two segments, Each segment is 2 m × 1 m in cross-section, 80 m long and the discharge at end C is measured to be 1600 m^3/min. Taking a friction factor $f = 0.005$ for the main and branch segments, determine:
   (a) Quantity of air discharged at end point D,
   (b) Static pressure at the fan outlet A, and
   (c) Pressure loss in the duct AB due to friction.
   (Ans. 1600 m^3/min/; 13 mm of water; 32.62 mm of water**)**

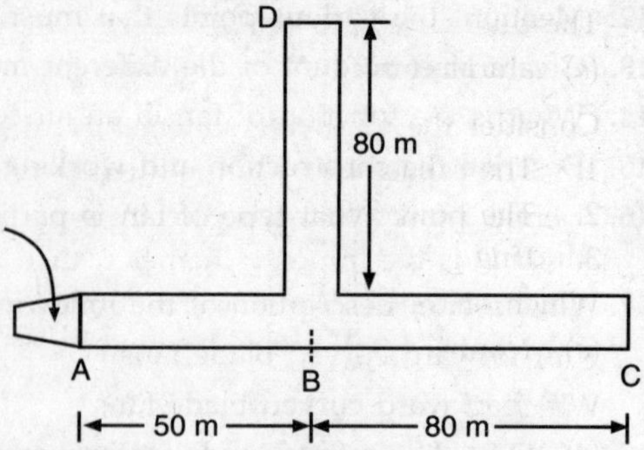

Fig. 16.31

B. Fill in the blanks with appropriate word/words

1. Grills are used for .... in an air conditioning system.
2. .... is the grille provided with a volume control dampler.
3. .... is th horizontal or vertical axial distance that an air stream travels on leaving the outlet.
4. .... is the continuous injection of certain amount of outside fresh air into air conditioned space to offset the contaminants produced by occupants.
5. The localised feeling of coolness or wormth of any portion of the body due to both air movement and air temperature is called ....
6. The three common methods of duct sizing are the velocity reduction method, the equal friction method and the .... method.
7. The static regain method of duct design is based on the principle of maintenance of constant static pressure before each .... and each ....
8. In designing ducts, the .... method is ideal when the system is balanaced.
9. For rectangular ducts, the aspect ratio is equal to the ratio.of ....
10. The primary functions of .... is to produce air movement through heating, ventilating and air conditioning equipment.
11. A .... fan consists of an impeller running in a casing which is normally of volute shape.

Answers

1. air distribution 2. register 3. throw 4. ventilation 5. draft 6. static regain 7. terminal, branch 8. equal friction 9. long side to short size 10. fan 11. centrifugal.

C. Multiple choice questions.

1. The most commonly used method for the design of duct size is the
   (a) velocity reduction method (b) equal friction method
   (c) static regain method (d) dual or double method
2. Equal friction method is the method used for designing
   (a) condenser (b) blower or fan (c) evaporator (d) air distribution duct
3. For rectangular ducts, the aspect ratio is equal to
   (a) ratio of longer to shorter side
   (b) average of the longer and shorter side
   (c) geometric mean of longer and shorter sides
   (d) square root of the ratio of the longer to shorter side
4. If the frictional loss in the air conditioning duct of rectangular section (width a and breadth b) equals the frictional loss in a round duct, then the equivalent diameter d_e of the round duct is
   (a) $d_e = \left(\frac{b}{a+b}\right)^{1/2}$ (b) $d_e = \frac{2ab}{a+b}$ (c) $d_e = \left(\frac{a^2 b^2}{a+b}\right)^{1/2}$ (d) $d_e = \left(\frac{a+b}{ab}\right)^{1/3}$

5. The ducts used for air distribution in an air conditioning system are generally made of
   (*a*) aluminium (*b*) fibre glass (*c*) galvanised iron (*d*) any one of these
6. Consider the following statements :
   1. The capacity of a fan is directly proportional to cube of the impeller diameter
   2. The power of fan is directly proportional to N^3 where N is the fan speed
   3. If D is the impeller diameter, then power of a fan is directly proportional to D^5.

   Which of these is/are correct statement ?
   (*a*) 1 only (*b*) 2 and 3 (*c*) 1 and 2 (*d*) 1, 2 and 3
7. Which of these is a correct statement in the context of axial flow fans ?
   (*a*) the air flows parallel to the axis of the impeller
   (*b*) the air guide vanes may be installed to eliminate spiral flow of discharge air
   (*c*) suitable for handling large volumes of air at relatively low pressure
   (*d*) all of these are correct
8. Identify the term not associated with air distribution in a refrigeration system.
   (*a*) diffuser (*b*) throw (*c*) drop (*d*) sol air

Answers

1. (*c*) 2. (*b*) 3. (*a*) 4. (*b*) 5. (*d*) 6. (*d*) 7. (*d*) 8. (*d*)

Appendix

Thermodynamics Properties of Refrigerants

TABLE 1. SATURATED AMMONIA (NH_3), R-717

Temp °C	Pressure bar	Enthalpy kJ/kg		Entropy kJ/kg K		Specific volume m^3/kg	
		h_f	h_g	s_f	s_g	$v_f \times 10^3$	v_g
–50	0.4103	–25.434	1390.64	0.0926	6.438	1.424	2.616
–45	0.5474	–3.302	1399.07	0.1905	6.337	1.437	1.999
–40	0.7201	18.902	1407.26	0.2865	6.241	1.449	1.547
–35	0.9349	41.188	1414.52	0.3808	6.150	1.462	1.212
–30	1.1990	63.563	1422.86	0.4735	6.064	1.476	0.961
–25	1.5216	86.033	1430.20	0.5646	5.981	1.489	0.770
–20	1.9074	108.60	1437.23	0.6544	5.902	1.504	0.622
–18	2.0826	117.66	1439.94	0.6898	5.872	1.510	0.573
–16	2.2704	126.73	1442.60	0.7251	5.842	1.515	0.528
–15	2.3709	131.27	1443.90	0.7426	5.822	1.518	0.508
–14	2.4714	135.82	1445.20	0.7602	5.812	1.521	0.488
–12	2.6863	144.93	1447.74	0.7950	5.783	1.528	0.451
–10	2.9157	154.06	1450.22	0.8296	5.755	1.534	0.417
–9	3.0360	158.63	1451.44	0.8469	5.741	1.5379	0.402
–8	3.1602	163.20	1452.64	0.8641	5.757	1.540	0.387
–7	3.2884	167.78	1453.83	0.8812	5.713	1.543	0.973
–6	3.4207	172.37	1455.00	0.8983	5.699	1.546	0.359
–5	3.5571	176.96	1456.15	0.9154	5.686	1.550	0.346
–4	3.6977	181.56	1457.29	0.9324	5.672	1.553	0.333
–3	3.8426	186.16	1458.42	0.9494	5.659	1.556	0.322
–2	3.9920	190.77	1459.53	0.9663	5.645	1.559	0.310
–1	4.1458	195.38	1460.62	0.9832	5.632	1.563	0.299
0	4.3043	200.00	1461.70	1.0000	5.619	1.566	0.289
1	4.4674	204.62	1462.76	1.0168	5.606	1.569	0.280
2	4.6353	209.25	1463.80	1.0335	5.593	1.573	0.269
3	4.8081	213.89	1467.83	1.0502	5.580	1.576	0.260
4	4.9859	218.53	1465.84	1.0669	5.567	1.580	0.251
5	5.1687	223.18	1466.84	1.0835	5.554	1.583	0.243
6	5.3567	227.84	1467.82	1.1001	5.542	1.587	0.235

(Contd...)

Table 1 *(Contd...)*

Temp °C	Pressure bar	Enthalpy kJ/kg		Entropy kJ/kg K		Specific volume m³/kg	
		h_f	h_g	s_f	s_g	$v_f \times 10^3$	v_g
7	5.5500	232.50	1468.78	1.1167	5.529	1.590	0.227
8	5.7487	237.17	1469.72	1.1332	5.517	1.594	0.219
9	5.9528	241.85	1470.65	1.1496	5.505	1.597	0.212
10	6.1625	246.53	1471.57	1.1661	5.492	1.601	0.205
11	6.3778	251.22	1472.46	1.1825	5.480	1.604	0.198
12	6.5989	255.92	1473.34	1.1988	5.498	1.608	0.192
13	6.8259	260.62	1474.20	1.2151	5.456	1.612	0.186
14	7.0588	265.34	1475.05	1.2314	5.444	1.616	0.180
15	7.2979	270.05	1475.88	1.2477	5.432	1.619	0.174
16	7.5431	274.78	1476.69	1.2639	5.420	1.623	0.169
17	7.7946	279.51	1477.48	1.2801	5.409	1.627	0.164
18	8.0525	284.25	1478.25	1.2963	5.397	1.631	0.157
19	8.3169	289.00	1479.01	1.3124	5.385	1.634	0.154
20	8.5879	293.78	1479.75	1.3288	5.374	1.639	0.150
21	8.8657	298.53	1480.48	1.3445	5.363	1.642	0.144
22	9.1503	303.30	1481.18	1.3605	5.351	1.647	0.140
23	9.4418	308.08	1481.87	1.3765	5.340	1.651	0.136
24	9.7403	312.87	1482.53	1.3925	5.329	1.655	0.132
25	10.046	317.69	1483.18	1.4084	5.317	1.659	0.128
26	10.359	322.47	1483.81	1.4243	5.306	1.663	0.124
27	10.680	327.28	1484.42	1.4402	5.295	1.667	0.121
28	11.007	332.10	1485.01	1.4560	5.284	1.671	0.118
29	11.343	336.93	1485.59	1.4718	5.273	1.676	0.114
30	11.686	341.77	1486.14	1.4876	5.262	1.680	0.110
31	12.037	343.61	1486.67	1.5034	5.252	1.684	0.107
32	12.396	351.47	1487.18	1.5191	5.2418	1.689	0.107
33	12.763	356.33	1487.66	1.5348	5.230	1.693	0.101
34	13.139	361.19	1488.13	1.5504	5.219	1.698	0.098
35	13.522	366.07	1488.57	1.5660	5.208	1.702	0.096
36	13.915	370.96	1488.99	1.5816	5.198	1.707	0.093
37	14.315	375.85	1489.39	1.5972	5.187	1.711	0.090
38	17.724	380.75	1489.76	1.6127	5.177	1.716	0.088
39	15.143	385.67	1490.10	1.6283	5.166	1.721	0.085
40	15.570	390.59	1490.42	1.6438	5.156	1.726	0.083
42	16.451	400.46	1490.98	1.6747	5.135	1.735	0.079
44	17.370	410.38	1491.41	1.7055	5.114	1.745	0.074
46	18.326	420.36	1491.72	1.7363	5.093	1.755	0.070
48	19.322	430.40	1491.88	1.7671	5.072	1.766	0.065
50	20.359	440.52	1491.89	1.7980	5.051	1.777	0.063

TABLE 2: TRICHLORO MONOFLUORO METHANE (CCl_3F), R-11

Temp °C	Pressure bar	Enthalpy kJ/kg		Entropy kJ/kg K		Specific volume m³/kg	
		h_f	h_g	s_f	s_g	$v_f \times 10^3$	v_g
– 60	0.0129	140.08	350.04	0.7897	1.7747	0.6007	9.9605
– 55	0.0188	153.31	361.45	0.8092	1.7634	0.6045	7.0230
– 50	0.0267	157.54	363.88	0.8284	1.7531	0.6084	5.0461
– 45	0.0373	161.76	366.33	0.8471	1.7138	0.6123	3.6896
– 40	0.0512	165.99	368.79	0.8655	1.7353	0.6164	2.7417
– 35	0.0693	170.22	371.27	0.8834	1.7276	0.6205	2.0681
– 30	0.0924	174.25	373.57	0.9010	1.7207	0.6247	1.5818
– 25	0.1215	178.53	376.11	0.9182	1.7145	0.6289	1.2255
– 20	0.1578	182.81	378.66	0.9352	1.7088	0.6333	0.6909
– 15	0.2025	187.09	381.22	0.9518	1.7038	0.9378	0.7619
– 10	0.2571	191.39	383.77	0.9681	1.6992	0.6423	0.6105
– 8	0.2820	193.11	384.80	0.9746	1.6975	0.6442	0.5602
– 6	0.3088	194.83	385.82	0.9810	1.6959	0.6461	0.5148
– 5	0.3229	195.72	386.36	0.9842	1.6951	0.6470	0.4941
– 4	0.3376	196.55	386.84	0.9874	1.6944	0.6479	0.4739
– 2	0.3686	198.27	387.86	0.9937	1.6929	0.6498	0.4368
0	0.4018	200.00	388.89	1.0000	1.6915	0.6518	0.4031
1	0.4192	200.86	389.40	1.0031	1.6908	0.6527	0.3875
2	0.4373	201.73	389.91	1.0062	1.6902	0.6537	0.3726
3	0.4560	202.59	390.42	1.0094	1.6895	0.6547	0.3584
4	0.4754	203.46	390.93	1.0125	1.6889	0.6556	0.3448
5	0.4953	204.32	391.44	1.0155	1.6883	0.657	0.3319
6	0.5160	205.19	391.95	1.0186	1.6877	0.6577	0.3195
7	0.5373	206.05	392.46	1.0217	1.6871	0.6587	0.3077
8	0.5593	206.92	392.97	1.0248	1.6865	0.6597	0.2964
9	0.5821	207.79	393.47	1.0278	1.6859	0.6607	0.2856
10	0.6055	208.65	393.98	1.0309	1.6854	0.6612	0.2753
11	0.6297	209.52	394.49	1.0339	1.6849	0.6627	0.2655
12	0.6547	210.39	395.00	1.0369	1.6843	0.6633	0.2560
13	0.6804	211.26	395.51	1.0399	1.6838	0.6647	0.2470
14	0.7070	212.13	396.02	1.0429	1.6833	0.6658	0.2384
15	0.7343	213.00	396.52	1.0459	1.6828	0.6668	0.2301
16	0.7625	213.87	397.03	1.0489	1.6823	0.6679	0.2222
17	0.7915	214.74	397.54	1.0519	1.6819	0.6689	0.2146
18	0.8214	215.61	398.04	1.0549	1.6815	0.7000	0.2073
19	0.8521	216.48	398.55	1.0578	1.6810	0.6710	0.2004

(Contd...)

Table 2 *(Contd...)*

Temp °C	Pressure bar	Enthalpy kJ/kg		Entropy kJ/kg K		Specific volume m³/kg	
		h_f	h_g	s_f	s_g	$v_f \times 10^3$	v_g
20	0.8838	217.35	399.05	1.0608	1.6806	0.6721	0.1937
21	0.9164	218.22	399.56	1.0637	1.6802	0.6732	0.1872
22	0.9499	219.10	400.06	4.0667	1.6798	0.6742	0.1811
23	0.9844	219.97	400.57	1.0696	1.6794	0.6753	0.1752
24	1.0198	220.84	401.07	1.0725	1.6790	0.6764	0.1695
25	1.0562	221.72	401.57	1.0754	1.6787	0.6775	0.1640
26	1.0937	222.59	402.07	1.0784	1.6783	0.6786	0.1588
27	1.1321	223.47	402.57	1.0812	1.6780	0.6797	0.1537
28	1.1716	224.34	403.08	1.0841	1.6776	0.6808	0.1489
29	1.2122	225.22	403.58	1.0870	1.6773	0.6819	0.1442
30	1.2538	226.10	404.08	1.0899	1.6470	0.6831	0.1398
32	1.3405	228.85	405.07	1.0956	1.6739	0.6853	0.1313
34	1.4318	229.61	406.07	1.1013	1.6758	0.6876	0.1235
36	1.5278	231.37	407.06	1.1070	1.6753	0.6899	0.1161
38	1.6287	233.13	408.05	1.1126	1.6748	0.6923	0.1094
40	1.7346	234.90	409.04	4.1182	1.67439	0.6948	0.1031
45	2.0228	239.32	411.49	1.1321	1.6732	0.7007	0.0893
50	2.3464	243.75	413.93	1.1458	1.6724	0.7070	0.0776
55	2.7083	248.21	416.34	1.1593	1.6716	0.7135	0.0678
60	3.1110	252.68	418.73	1.1727	1.6711	0.7201	0.0594
70	4.0515	261.68	423.42	1.1990	1.6703	0.7342	0.0462
80	5.1921	270.79	427.98	1.2248	1.6699	0.7494	0.0364
90	6.5584	279.61	432.01	1.2502	1.6698	0.7658	0.0290
100	8.1769	288.96	436.23	1.2752	1.6699	0.7837	0.0233
110	10.0757	298.49	440.25	1.3000	1.6700	0.8035	0.0189
120	12.2844	308.27	444.03	1.3247	1.6701	08.255	0.0154
130	14.8340	318.36	447.52	1.3495	1.6699	0.8504	0.0126
140	17.7576	328.84	450.62	1.3746	1.6694	0.8791	0.0103
150	21.0905	339.84	453.24	1.4002	1.6682	0.8128	0.0085
160	24.8704	351.52	455.19	1.4266	1.6660	0.9540	0.0069
170	29.1384	364.19	456.16	1.4546	1.6621	1.0076	0.0056
180	33.9394	378.38	455.49	1.4851	1.6553	1.0807	0.0044
190	39.3238	395.64	451.27	1.5215	1.6416	1.2075	0.0033

TABLE 3: DICHLORO DIFLUORO METHANE (CCl_2F_2), R-12

Temp °C	Pressure bar	Enthalpy kJ/kg		Entropy kJ/kg K		Specific volume m^3/kg	
		h_f	h_g	s_f	s_g	$v_f \times 10^3$	v_g
– 30	1.0041	172.810	338.143	0.8952	1.57507	0.6720	0.1594
– 28	1.0927	174.593	339.057	0.9024	1.57326	0.6746	0.1473
– 26	1.1872	176.380	339.968	0.9097	1.57152	0.6786	0.1364
– 24	1.2880	178.171	340.876	0.9169	1.56985	0.6800	0.1263
– 22	1.3953	179.965	341.780	0.9240	1.56825	0.6827	0.1172
– 20	15093	181.764	342.682	0.9311	1.56672	0.6855	0.1088
– 18	1.6304	183.567	343.580	0.9382	1.56526	0.6883	0.1012
– 16	1.7589	185.374	344.474	0.9459	1.56385	0.6911	0.0943
– 14	1.8950	187.185	345.365	0.9522	1.56250	0.6941	0.0879
– 12	2.0390	189.001	346.252	0.9591	1.56121	0.6970	0.0820
– 10	2.1912	190.822	347.134	0.9660	1.55997	0.7000	0.0766
– 9	2.2704	191.734	347.574	0.9694	1.55938	0.7016	0.0741
– 8	2.3519	192.647	348.012	0.9729	1.55897	0.7031	0.0717
– 7	2.4355	193.562	348.450	0.9763	1.55822	0.7046	0.0693
– 6	2.5214	194.477	348.886	0.9798	1.55765	0.7062	0.0671
– 5	2.6096	195.395	349.321	09831	1.55710	0.7078	0.0649
– 4	2.7001	196.313	349.755	0.9865	1.55657	0.7039	0.0629
– 3	2.7930	197.233	350.187	0.9899	1.55604	0.7110	0.0609
– 2	2.8882	198.154	350.619	0.9933	1.55552	0.7126	0.0590
– 1	2.9859	199.076	351.049	0.9966	1.55502	0.7142	0.0572
0	3.0861	200.000	351.477	1.0000	1.55452	0.7159	0.0554
1	3.1888	200.925	351.905	1.0033	1.55404	0.7176	0.0537
2	3.2940	201.852	352.331	1.0067	1.55356	0.7192	0.0520
3	3.4019	202.780	352.755	1.0100	1.55310	0.7209	0.0505
4	3.5124	203.710	353.179	1.0134	1.55264	0.7226	0.0489
5	3.6255	204.642	353.600	1.0167	1.55220	0.7244	0.0475
6	3.7414	205.575	354.020	1.0200	1.5518	0.7261	0.0461
7	3.8601	206.509	354.439	1.0233	1.5513	0.7279	0.0447
8	3.9815	207.445	354.856	1.0266	1.5509	0.7297	0.0434
9	4.1058	208.383	355.272	1.0299	1.5505	0.7315	0.0421
10	4.2330	209.323	355.686	1.0332	1.5501	0.7333	0.0409
11	4.3631	210.264	356.098	1.0365	1.5497	0.7351	0.0397
12	4.4962	221.207	356.509	1.0398	1.5493	0.7369	0.0386
13	4.6323	212.152	356.918	1.0430	1.5489	0.7388	0.0375
14	4.7714	213.099	357.325	1.0463	1.5486	0.7407	0.0364
15	4.9137	214.048	357.730	1.0496	1.5482	0.7426	0.0354
16	5.0591	214.998	358.134	1.0528	1.5479	0.7445	0.0344
17	5.2076	215.951	358.535	1.0561	1.5475	0.7465	0.0335

(Contd...)

Table 3 *(Contd...)*

Temp °C	Pressure bar	Enthalpy kJ/kg		Entropy kJ/kg K		Specific volume m^3/kg	
		h_f	h_g	s_f	s_g	$v_f \times 10^3$	v_g
18	5.3594	216.906	358.935	1.0593	1.5471	0.7485	0.0325
19	5.5145	217.863	359.333	1.0626	1.5468	0.7504	0.0316
20	5.6729	218.821	359.729	1.0658	1.5464	0.7525	0.0308
21	5.8347	219.783	360.122	1.0690	1.5461	0.7545	0.0299
22	5.9980	220.746	360.514	1.0723	1.5456	0.7565	0.0291
23	6.1684	221.712	360.904	1.0755	1.5455	0.7586	0.0283
24	6.3405	222.680	361.291	1.0787	1.5451	0.7607	0.0276
25	6.5162	223.650	361.676	1.0819	1.5448	0.7629	0.0268
26	6.6954	224.623	362.059	1.0851	1.5445	0.7650	0.0261
27	6.8782	225.598	362.439	1.0883	1.5442	0.7672	0.0254
28	7.0647	226.576	362.817	1.0915	1.5439	0.7694	0.0248
29	7.2550	227.557	363.193	1.0947	1.5436	0.7716	0.0241
30	7.4490	228.540	363.566	1.0979	1.5433	0.7739	0.0235
31	7.6468	229.526	363.937	1.1011	1.5430	0.7761	0.0229
32	7.8485	230.515	364.305	1.1043	1.5428	0.7784	0.0223
33	8.0541	231.506	364.670	1.1075	1.5425	0.7808	0.0217
34	8.2636	232.501	365.033	1.1107	1.5422	0.7832	0.0212
35	8.4772	233.498	365.392	1.1139	1.5419	0.7856	0.0206
36	8.6948	234.499	365.749	1.1171	1.5416	0.7880	0.0201
37	8.9164	235.503	366.103	1.1203	1.5413	0.7904	0.0196
38	9.1423	236.510	366.454	1.1238	1.5411	0.7929	0.0191
39	9.3723	237.521	366.802	1.1266	1.5408	0.7955	0.0186
40	9.6065	238.535	367.146	1.1298	1.5405	0.7980	0.0182
41	9.8451	239.552	367.487	1.1330	1.5402	0.8006	0.0177
42	10.088	240.574	367.825	1.1362	1.5400	0.8032	0.0173
43	10.335	241.598	368.160	1.1394	1.5397	0.8059	0.0168
44	10.587	242.627	368.491	1.1426	1.5394	0.8086	0.0164
45	10.843	243.659	368.818	1.1457	1.5391	0.8114	0.0160
46	11.369	245.736	369.141	1.1489	1.5388	0.8142	0.0156
47	11.369	245.736	369.461	1.1521	1.5386	0.8170	0.0153
48	11.639	246.781	369.777	1.1553	1.5383	0.8198	0.0149
49	11.914	247.830	370.088	1.1885	1.5380	0.8228	0.0145
50	12.193	248.884	370.396	1.1617	1.5377	0.8257	0.0142
52	12.766	251.004	370.997	1.1681	1.5371	0.8318	0.0135
54	13.359	253.144	371.581	1.1745	1.5365	0.8380	0.0128
56	13.972	255.304	372.145	1.1809	1.5359	0.8445	0.0122
58	14.605	257.486	372.688	1.1874	1.5352	0.8512	0.0117
60	15.259	259.690	373.210	1.1938	1.5346	0.8581	0.0111

TABLE 4: MONOCHLORO DIFLUORO METHANE ($CHClF_2$), R-22

Temp t °C	Pressure p, bar	Enthalpy kJ/kg		Entropy kJ/kg K		Specific volume m³/kg	
		h_f	h_g	s_f	s_g	$v_f \times 10^3$	v_g
– 50	0.6439	144.959	383.921	0.7792	1.8500	0.6953	0.3245
– 45	0.8271	150.153	386.282	0.8022	1.8371	0.7022	0.2570
– 40	1.0495	155.414	388.609	0.8250	1.8250	0.7094	0.2057
– 35	1.3168	160.742	390.896	0.8474	1.8138	0.7168	0.1664
– 30	1.6348	166.140	393.138	0.8698	1.8033	0.7245	0.1358
– 28	1.7776	168.318	394.021	0.8786	1.7993	0.7277	0.1255
– 26	1.9299	170.507	394.896	0.8875	1.7953	0.7309	0.1162
– 24	2.0922	172.708	395.762	0.8964	1.7915	0.7342	0.1077
– 22	2.2648	174.919	396.619	0.9051	1.7878	0.7375	0.0999
– 20	2.4483	177.142	397.467	0.9139	1.7841	0.7409	0.0928
– 18	2.6429	179.376	398.305	0.9226	1.7806	0.7444	0.0863
– 16	2.8493	181.622	399.133	0.9313	1.7771	0.7479	0.0804
– 14	3.0678	183.878	399.951	0.9400	1.7737	0.7514	0.0749
– 12	3.2989	186.147	400.759	0.9486	1.7704	0.7551	0.0699
– 10	3.5430	188.426	401.555	0.9572	1.7671	0.7588	0.0653
– 9	3.6701	189.571	401.949	0.9615	1.7655	0.7606	0.0632
– 8	3.8006	190.718	402.341	0.9658	1.7639	0.7625	0.0611
– 7	3.9347	191.868	402.729	0.9701	1.7624	0.7644	0.0591
– 6	4.0723	193.021	403.114	0.9744	1.7608	0.7664	0.0572
– 5	4.2135	194.335	403.876	0.9830	1.7593	0.7683	0.0537
– 4	4.3584	195.335	403.876	0.9830	1.7577	0.7703	0.0519
– 3	4.5070	196.497	404.252	0.9872	1.7562	0.7723	0.0519
– 2	4.6594	197.662	404.626	0.9915	1.7547	0.7743	0.0487
– 1	4.8157	198.828	404.994	0.9957	1.7533	0.7763	0.0486
0	4.9759	200.000	405.361	1.0000	1.7502	0.7783	0.0471
1	5.1401	201.174	405.724	1.0042	1.7503	0.7840	0.0457
2	5.3083	202.351	406.084	1.0085	1.7488	0.7825	0.0443
3	5.4806	203.530	406.440	1.0127	1.7475	0.7846	0.0430
4	5.6571	204.713	406.793	1.0170	1.7460	0.7867	0.0416
5	5.8379	205.899	407.143	1.0212	1.7446	0.7889	0.0403
6	6.0228	207.089	407.489	1.02547	1.7432	0.7911	0.0391
7	6.2122	208.281	407.831	1.0296	1.7418	0.7933	0.0380
8	6.4059	209.477	408.169	1.0338	1.7405	0.7955	0.0368
9	6.6042	210.675	407.504	1.0380	1.7391	0.7977	0.0358
10	6.8070	211.877	408.835	1.0422	1.7377	0.8000	0.0347
11	7.0144	213.083	409.162	1.0464	1.7364	0.7802	0.0337
12	7.2265	214.291	409.485	1.0506	1.7351	0.8046	0.0327

(Contd...)

Table 4 *(Contd...)*

Temp t	Pressure p, bar	Enthalpy kJ/kg		Entropy kJ/kg K		Specific volume m^3/kg	
°C		h_f	h_g	s_f	s_g	$v_f \times 10^3$	v_g
13	7.4433	215.503	409.804	1.0547	1.7337	0.8070	0.0318
14	7.6650	216.719	410.119	1.0589	1.7324	0.8094	0.0309
15	7.8915	217.937	410.430	1.0631	1.7311	0.8118	0.0300
16	8.1229	219.160	410.736	1.0673	1.7298	0.8142	0.0291
17	8.3593	220.386	411.038	1.0714	1.7285	0.8167	0.0283
18	8.6008	221.615	411.336	1.0756	1.7272	0.8192	0.0275
19	8.8475	222.848	411.629	1.0797	1.7259	0.8217	0.0267
20	9.0993	224.084	411.918	1.0839	1.7246	0.8243	0.0260
21	9.3564	225.324	412.202	1.0880	1.7233	0.8269	0.0253
22	9.6189	226.568	412.481	1.0922	1.7221	0.8294	0.0246
23	9.8867	227.516	412.755	1.0963	1.7208	0.8322	0.0239
24	10.160	229.068	413.028	1.1005	1.7195	0.8349	0.0232
25	10.439	230.324	413.289	1.1046	1.7183	0.8376	0.0226
26	10.723	231.583	413.548	1.1088	1.7170	0.8404	0.0220
27	11.014	232.847	413.802	1.1129	1.7158	0.8432	0.0214
28	11.309	234.115	414.050	1.1170	1.7145	0.8461	0.0208
29	11.611	235.387	414.293	1.1212	1.7132	0.8490	0.0203
30	11.919	236.664	414.530	1.1253	1.7120	0.8519	0.0197
31	12.232	237.944	414.762	1.1294	1.7107	0.8549	0.0192
32	12.552	239.230	414.987	1.1335	1.7095	0.8579	0.0187
33	12.878	240.520	415.207	1.1377	1.7083	0.8610	0.0182
34	13.210	241.814	415.420	1.1418	1.7070	0.8641	0.0177
35	13.548	243.114	415.627	1.1759	1.7058	0.8673	0.0173
36	13.892	244.418	415.828	1.1501	1.7045	0.8738	0.0168
37	14.243	245.727	416.021	1.1542	1.7032	0.8738	0.0164
38	14.601	247.041	416.208	1.1583	1.7020	0.8771	0.0159
39	14.965	248.361	416.388	1.1625	1.7007	0.8805	0.0155
40	15.335	249.686	416.561	1.1666	1.6995	0.8839	0.0151
41	15.712	251.016	416.726	1.1707	1.6982	0.8874	0.0147
42	16.096	252.352	416.883	1.1749	1.6969	0.8910	0.0146
43	16.487	253.694	417.033	1.1790	1.6956	0.8946	0.0140
44	16.885	255.042	417.174	1.1831	1.6943	0.8983	0.0136
45	17.290	256.396	417.308	1.1873	1.6917	0.9059	0.0129
46	17.702	257.756	417.432	1.1914	1.6917	0.9059	0.0129
47	18.121	259.123	417.548	1.1956	1.6904	0.9098	0.0126
48	18.548	260.497	417.655	1.1998	1.6891	0.9137	0.0122
49	18.982	261.877	417.752	1.2039	1.6878	0.9178	0.0120
50	19.423	263.264	417.838	1.2081	1.6864	0.9219	0.1167

TABLE 5 : TETRAFLUOROETHANE (CH_2 FCF_3), R-134(a)

Temp °C	Absolute pressure	Enthalpy kJ/kg		Entropy kJ/(kg K)		Specific volume kJ/(kg K)	
	bar	Liquid h_f	Vapour h_g	Liquid s_f	Vapour s_g	Liquid $v_f \times 10^3$	Vapour v_g
– 40.00	0.5122	148.57	374.16	0.7973	1.7649	0.705	0.3567
– 30.00	0.8436	161.10	380.45	0.8498	1.7519	0.720	0.2240
– 28.00	0.9268	163.62	381.70	0.8601	1 7497	0.723	0.2051
– 26.00	1.0164	166.16	382.94	0.8704	1.7476	0.726	0.1881
– 24.00	1.1127	168.70	384.19	0.8806	1.7455	0.730	0.1727
– 22.00	1.2160	171.26	385.43	0.8908	1.7436	0.733	0.1589
– 20.00	1.3268	173.82	386.66	0.9009	1.7417	0.736	0.1464
–18.00	1.4454	176.39	387.89	0.9110	1.7399	0.739	0.1350
–16.00	1.5721	178.97	389.11	0.9211	1.7383	0.743	0.1247
–14.00	1.7074	181.56	390.33	0.9311	1.7367	0.746	0.1153
–12.00	1.8516	184.16	391.55	0.9410	1.7351	0.750	0.1067
–10.00	2.0052	186.78	392.75	0.9509	1.7337	0.753	0.0989
– 8.00	2.1684	189.40	393.95	0.9608	1.7323	0.757	0.0918
– 6.00	2.3418	192.03	395.15	0.9707	1.7310	0.761	0.0853
– 4.00	2.5257	194.68	396.33	0.9805	1.7297	0.764	0.0794
– 2.00	2.7206	197.33	397.51	0.9903	1.7285	0.768	0.0739
0.00	2.9269	200.00	398.68	1.0000	1.7274	0.772	0.0689
2.00	3.1450	202.68	399.84	1.0097	1.7263	0.776	0.0642
4.00	3.3755	205.37	401.00	1.0194	1.7252	0.780	0.0600
6.00	3.6186	208.08	402.14	1.0291	1.7242	0.784	0.0561
8.00	3.8749	210.80	403.27	1 .0387	1.7233	0.788	0.0525
10.00	4.1449	213.53	404.40	1.0483	1.7224	0.793	0.0491
12.00	4.4289	216.27	405.51	1.0579	1.7215	0.797	0.0460
14.00	4.7276	219.03	406.61	1.0674	1 7207	0.802	0.0432
16.00	5.0143	221.80	407.70	1.0770	1.7199	0.806	0.0405
18.00	5.3706	224.59	408.78	1.0865	1.7191	0.811	0.0381
20.00	5.7159	227.40	409.84	1.0960	1.7183	0.816	0.0358
22.00	6.0777	230.21	410.89	1.1055	1.7176	0.821	0.0336
24.00	6.4566	233.05	411.93	1 1149	1.7169	0.826	0.0317
26.00	6.8531	235.90	412.95	1 1244	1.7162	0.831	0.0298
28.00	7.2678	238.77	413.95	1.1338	1.7155	0.836	0.0281
30.00	7.7008	241.65	414.94	1.1432	1.7149	0.842	0.0265

(Contd...)

Table 5 *(Contd...)*

Temp °C	Absolute pressure bar	Enthalpy kJ/kg		Entropy kJ/(kg K)		Specific volume kJ/(kg K)	
		Liquid h_f	Vapour h_g	Liquid s_f	Vapour s_g	Liquid $v_f \times 10^3$	Vapour v_g
32.00	8.1530	244.55	415.90	1.1527	1.7142	0.847	0.0250
34.00	8.6250	247.47	416.85	1.1621	1.7135	0.853	0.0236
36.00	9.1172	250.41	417.78	1.1715	1.7129	0.859	0.0222
38.00	9.6301	253.37	418.69	1.1809	1.7122	0.865	0.0210
40.00	10.165	256.35	419.58	1.1903	1.7115	0.871	0.0199
42.00	10.721	259.35	420.44	1.1997	1.7108	0.878	0.0188
44.00	11.300	262.38	421.28	1.2091	1.7101	0.885	0.0177
46.00	11.901	265.42	422.09	1.2185	1.7094	0.892	0.0168
48.00	12.527	268.49	422.88	1.2279	1.7086	0.899	0.0159
50.00	13.177	271.59	423.63	1.2373	1.7078	0.906	0.0150
52.00	13.852	274.71	424.35	1.2468	1.7070	0.914	0.0142
54.00	14.553	277.86	425.03	1.2562	1.7061	0.922	0.0134
56.00	15.280	281.04	425.68	1.2657	1.7051	0.931	0.0127
58.00	16.033	284.25	426.29	1.2752	1.7041	0.940	0.0120
60.00	16.815	287.49	426.86	1.2847	1.7031	0.949	0.0114
62.00	17625	290.77	427.37	1.2943	1.7019	0.959	0.0108
64.00	18.464	294.08	427.84	1.3039	1.7007	0.969	0.0102
66.00	19.334	297.44	428.25	1.3136	1.6993	0.979	0.0097
68.00	20.234	300.84	428.61	1.3234	1.6979	0.991	0.0091
70.00	21.165	304.29	428.89	1.3332	1.6963	0.003	0.0086

TABLE 6 : PSYCHROMETRIC PROPERTIES OF WATER & STEAM, AND AIR AT STANDARD BAROMETRIC PRESSURE OF 760 mm Mecury, OR 1.01325 BAR

Temp. t°C	Saturation pressure of water and steam, p_s bar	Enthalpy of saturated water, h_f kJ/kg	Enthalpy of saturated steam, h_g kJ/kg	Specific volume of saturated steam, v_g m³/kg	Specific volume of dry air v_a m³/kg	Enthalpy, of dry air h_a kJ/kg	Humidity ratio kg/kg dry air
0	0.006107	0.00	2501.5	206.31	0.7734	0.000	0.00379
1	0.006566	4.17	2503.4	192.61	0.7762	1.006	0.00407
2	0.007055	8.39	2505.2	179.92	0.7791	2.012	0.00438
3	0.007575	12.0	2507.1	168.17	0.7819	3.018	0.00471
4	0.008130	16.80	2508.9	157.27	0.7848	4.024	0.00505
5	0.008719	21.01	2510.7	147.16	0.7876	5.030	0.00542
6	0.009347	25.21	2512.6	137.79	0.7904	6.036	0.00582
7	0.010013	29.41	2514.4	129.06	0.7933	7.041	0.00624
8	0.010722	33.60	2516.2	120.97	0.7961	8.047	0.00668
9	0.011474	37.80	2518.1	113.44	0.7990	9.053	0.00716
10	0.012272	41.99	2519.9	106.43	0.8018	10.059	0.00766
11	0.013119	46.19	2521.7	99.909	0.8046	11.065	0.00820
12	0.014017	50.38	2523.6	93.835	0.8075	12.071	0.00876
13	0.014969	54.57	2525.4	88.176	0.8103	13.077	0.00937
14	0.015977	58.75	2527.2	82.900	0.8132	14.084	0.01001
15	0.017044	62.94	2529.1	77.978	0.8160	15.090	0.01069
16	0.018173	67.13	2530.9	73.384	0.8188	16.096	0.01141
17	0.019367	71.31	2532.7	69.095	0.8217	17.102	0.01218
18	0.020630	75.50	2534.5	.65.087	0.8245	18.108	0.01299
19	0.021964	79.68	2536.4	61.341	0.8274	19.114	0.01384
20	0.023373	83.86	2538.2	57.838	0.8302	20.121	0.01475
21	0.024861	88.04	2540.0	54.561	0.8330	21.127	0.01572
22	0.026431	92.23	2541.8	51.492	0.8359	22.133	0.01674
23	0.028086	96.41	2543.6	48.619	0.8387	23.140	0.01781
24	0.029832	100.59	2545.5	45.926	0.8416	24.146	0.01896
25	0.031671	104.77	2547.3	43.402	0.8444	25.153	0.02016
26	0.033609	108.95	2549.1	41.034	0.8472	26.159	0.02144
27	0.035649	113.13	2550.9	38.813	0.8501	27.165	0.02279

Table 6 *(Contd...)*

Temp. t°C	Saturation pressure of water and steam, p_s bar	Enthalpy of saturated water, h_f kJ/kg	Enthalpy of saturated steam, h_g kJ/kg	Specific volume of saturated steam, v_g m³/kg	Specific volume of dry air v_a m³/kg	Enthalpy, of dry air h_a kJ/kg	Humidity ratio kg/kg dry air
28	0.037797	117.31	2552.7	36.728	0.8529	28.172	0.02422
29	0.040055	121.48	2554.5	34.769	0.8558	29.179	0.02572
30	0.042431	125.66	2556.4	32.929	0.8586	30.185	0.02732
31	0.044928	129.84	2558.2	31.199	0.8614	31.192	0.02900
32	0.047552	134.02	2560.0	29.572	0.8643	32.198	0.03078
33	0.050308	138.20	2561.8	28.042	0.8671	33.205	0.03266
34	0.053201	142.38	2563.6	26.601	0.8700	34.212	0.03464
35	0.056237	146.56	2565.4	25.245	0.8728	35.219	0.03674
36	0.059423	150.74	2567.2	23.967	0.8756	36.226	0.03895
37	0.062764	154.92	2569.0	22.763	0.8785	37.233	0.04129
38	0.066265	158.09	2570.8	21.627	0.8813	38.239	0.04376
39	0.069935	163.27	2572.6	20.557	0.8842	39.246	0.04636
40	0.073778	167.45	2574.4	19.546	0.8870	40.253	0.04911
41	0.077803	171.63	2576.2	18.592	0.8898	41.261	0.05202
42	0.082016	175.81	2577.9	17.692	0.8927	42.268	0.05509
43	0.086424	179.99	2579.7	16.841	0.8955	43.275	0.05833
44	0.091036	184.17	2581.5	16.036	0.8983	44.282	0.06176

PRESSURE-ENTHALPY DIAGRAMS

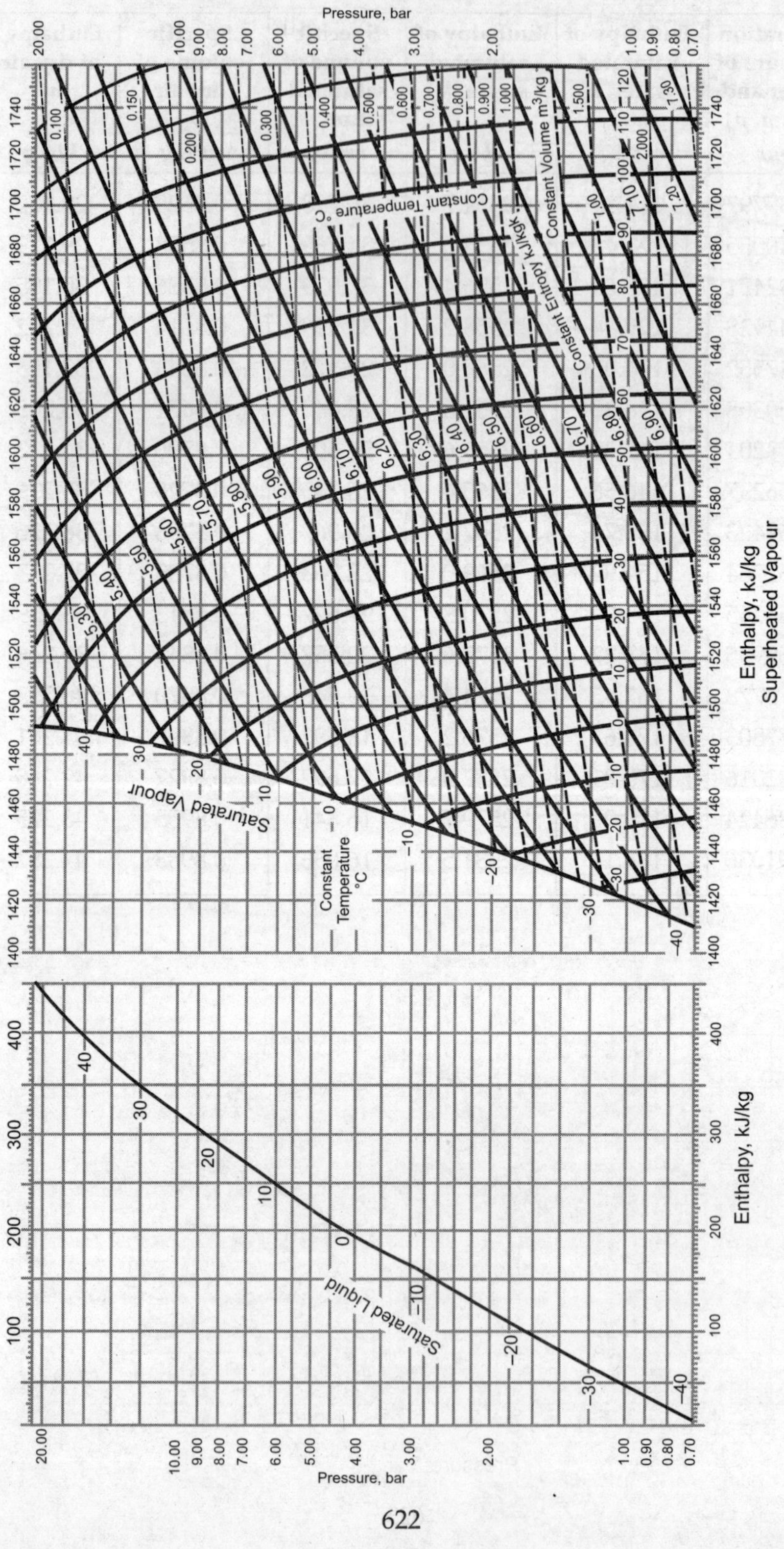

PRESSURE-ENTHALPY DIAGRAM : AMMONIA (R-717)

(Contd...)

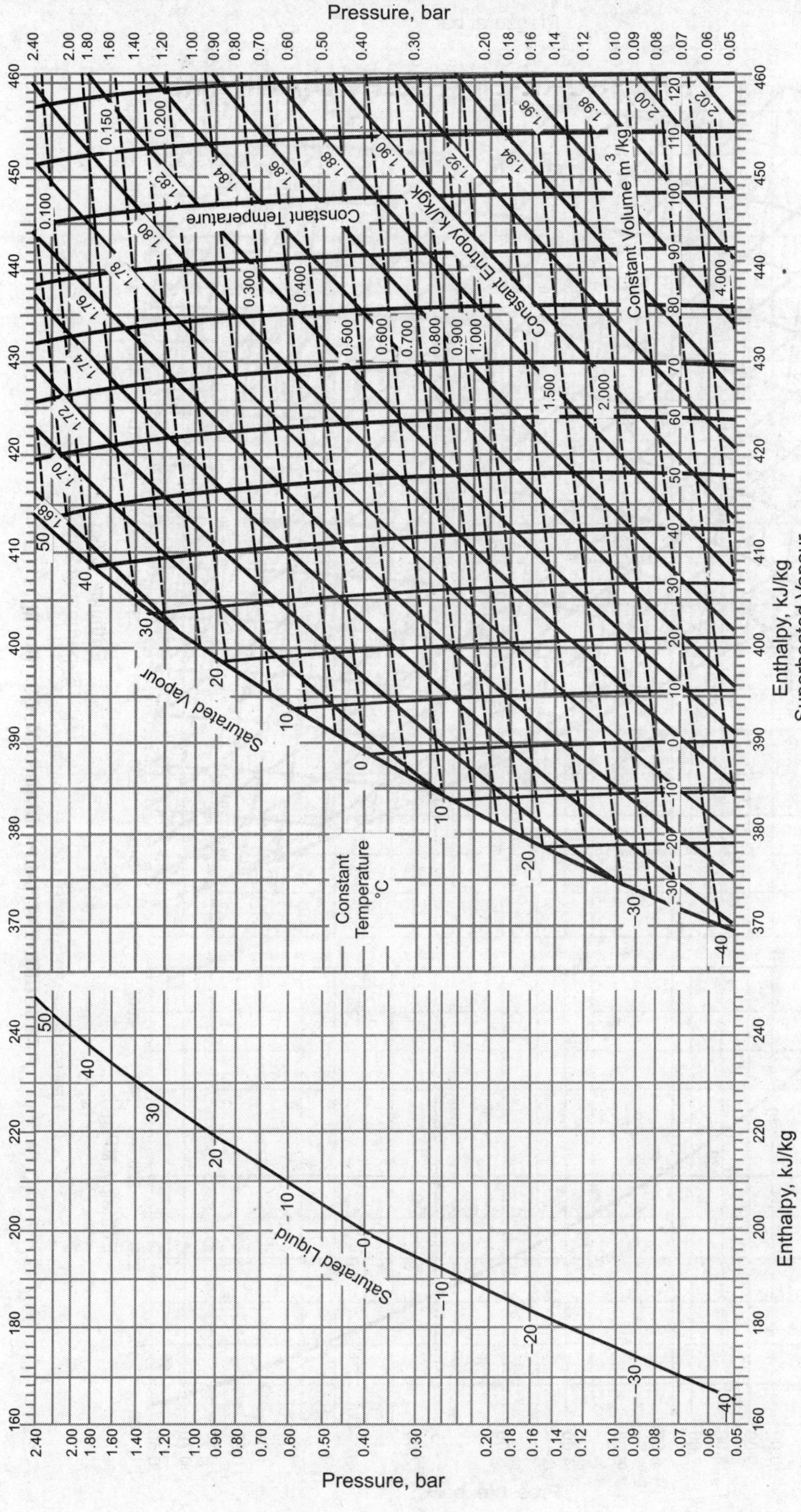

PRESSURE-ENTHALPY DIAGRAM: CCl_2F (R-11)

(Contd...)

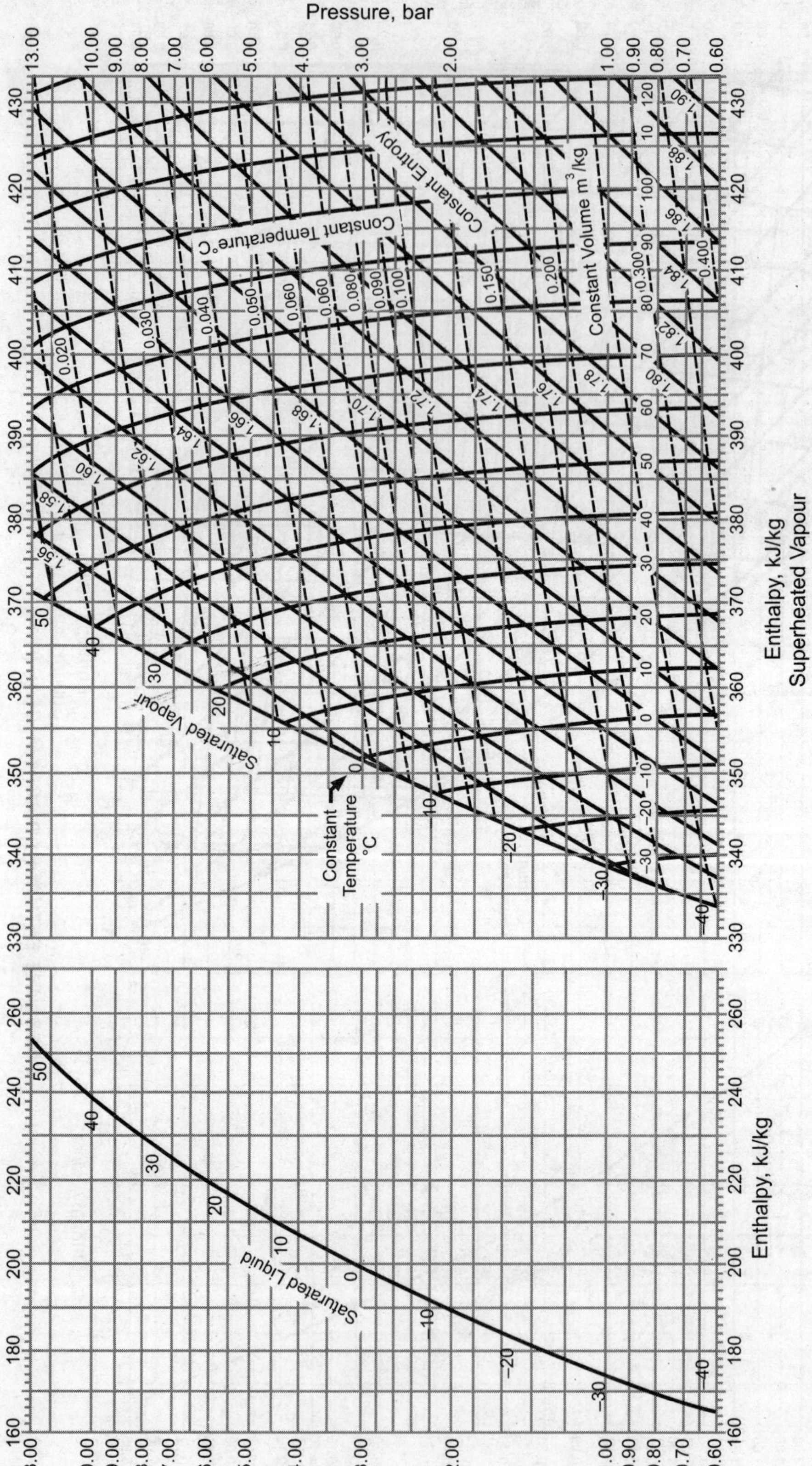

PRESSURE-ENTHALPY DIAGRAM: CCl_2 (R-12)

(Contd...)

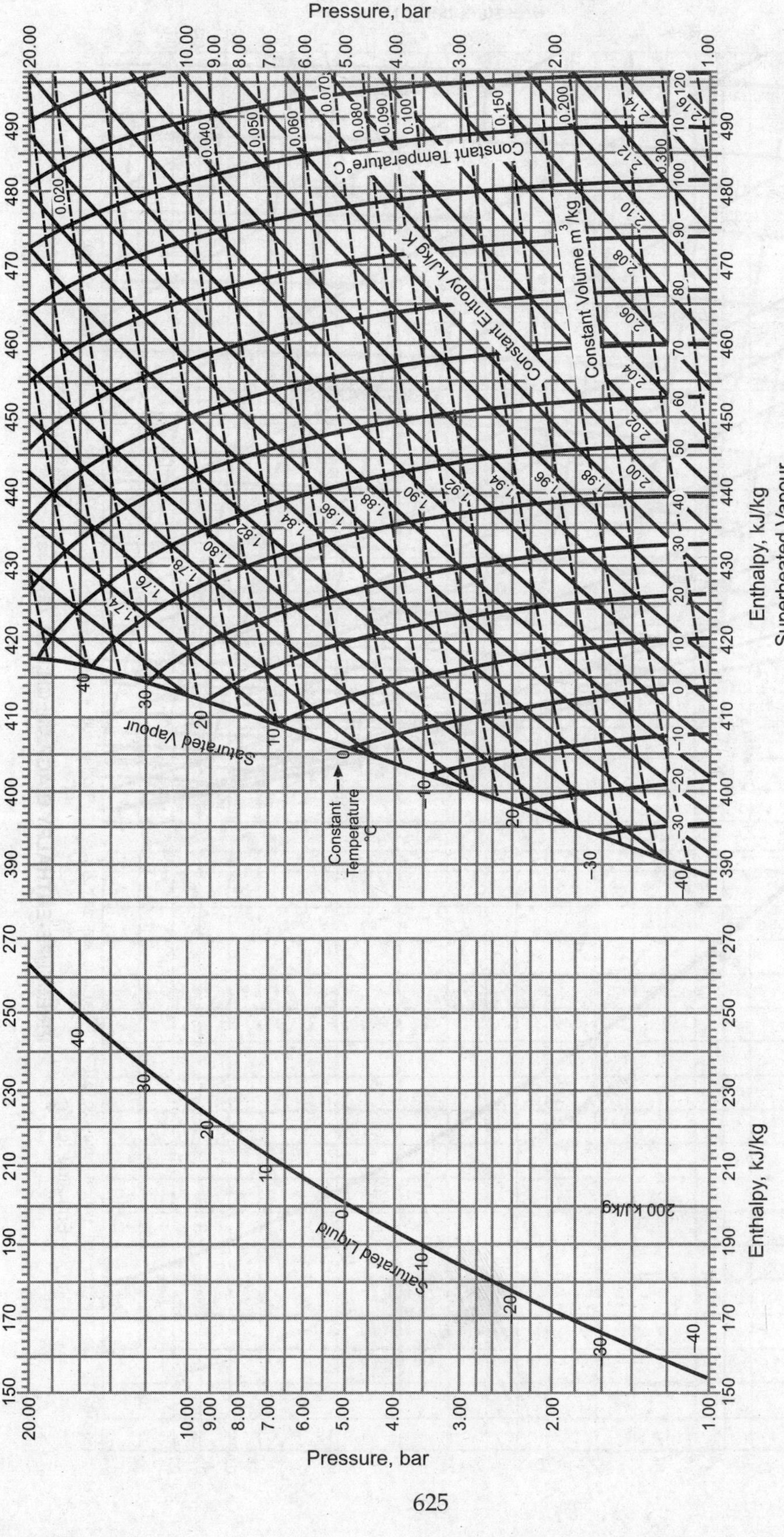

PRESSURE-ENTHALPY DIAGRAM: CHCl F_2 (R-22)

(Contd...)

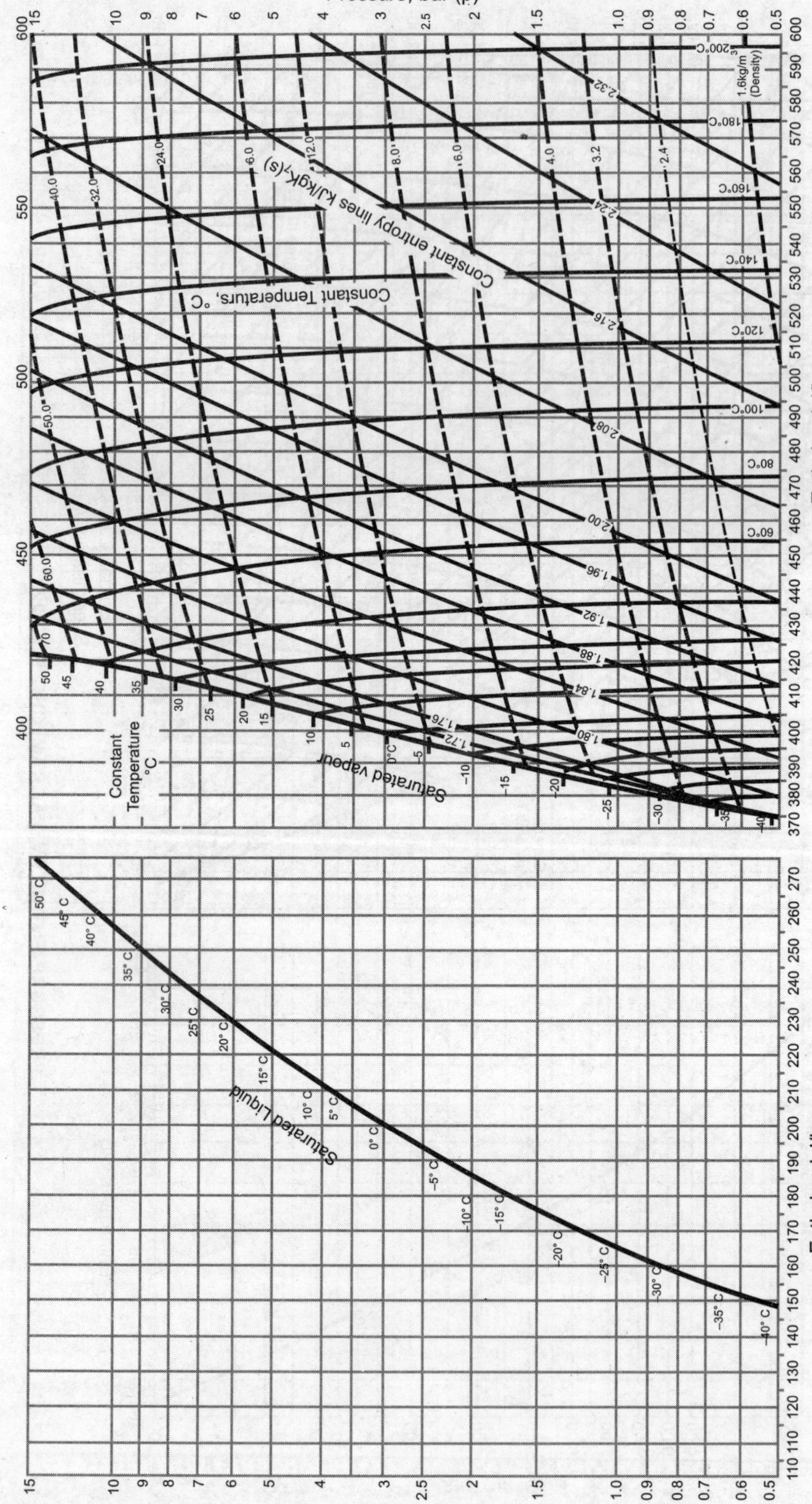

PRESSURE-ENTHALPY DIAGRAM: CH_2FCF_3 (R-134a)

(Contd...)

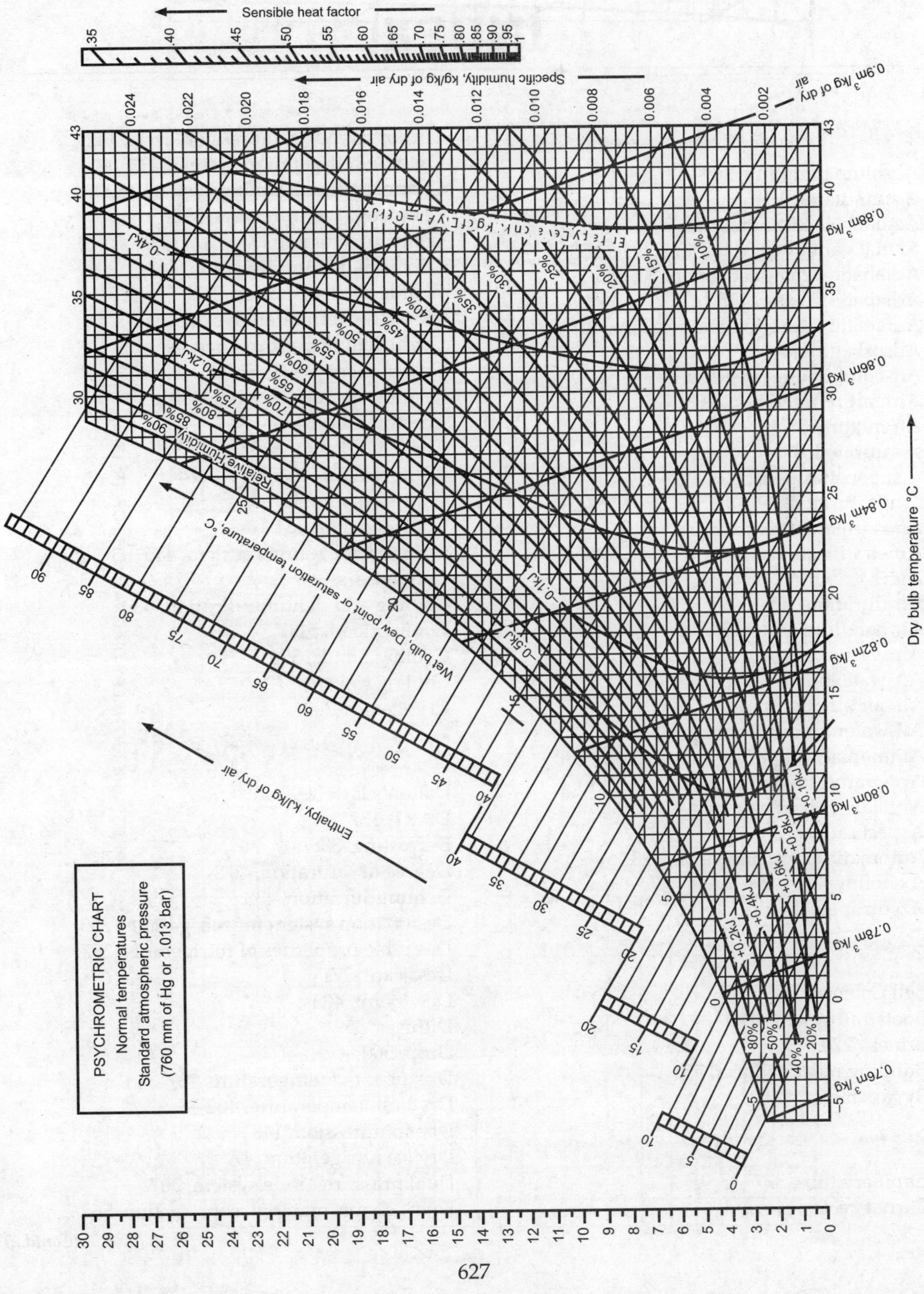
PSYCHROMETRIC CHART
Normal temperatures
Standard atmospheric pressure
(760 mm of Hg or 1.013 bar)
Sensible heat factor
Specific humidity, kg/kg of dry air
Dry bulb temperature °C
Wet bulb, Dew point or saturation temperature, °C
Enthalpy, kJ/kg of dry air
Relative Humidity
0.9m3/kg of dry air
0.88m3/kg
0.86m3/kg
0.84m3/kg
0.82m3/kg
0.80m3/kg
0.78m3/kg
0.76m3/kg

INDEX

□□□

Win Prizes

ATTENTION READERS

Win Prizes

We request for your frank assessment regarding various aspects of the book as given below :

REFRIGERATION AND AIR-CONDITIONING

By: Dr. D.S Kumar

Please Fill up the Sheet in Neat and Clean Writing

(i) What topics of your syllabus, important for your examination point of view, are not covered in this book ?

....

....

....

....

(ii) What are the chapters wherein the treatment of the subject matter is not systematic or organised or updated ?

....

....

....

....

(iii) Have you come across misprints/mistakes/factual inaccuracies in the book? Please specify the chapters and the page numbers.

....

....

....

....

(iv) What other books on the same subject have you found/heard better than the present book? Please specify in terms of price and quality

....

....

....

....

(v) Further suggestions and comments for the improvement of this book.

....

....

....

....

(PS: If need be please attach a separate sheet to write your views)

OTHER DETAILS

(i) Who recommended you this book ? (Please tick (✓) in the box near the option relevant to you.

☐ Teacher ☐ Friends ☐ Bookseller

....

....

....

....

(ii) Name of the recommending teacher, his designation and address :

....

....

....

....

(iii) Name and address of the bookseller you purchased the book from :

....

....

....

....

(iv) Name and address of your institution (Please mention the University or Board as the case may be)

....

....

....

....

(v) Your name and complete postal address :

....

....

....

....

(vi) Course you are enrolled in :

....

....

....

....

The best assessment will be awarded each month: The award will be in the form of our publications as decided by the Editorial Board.

Please mail the filled up coupon at your earliest to :

S.K. KATARIA & SONS®

4885/109, Prakash Mahal, Dr. Subhash Bhargav Lane,

Opposite Delhi Medical Association, Daryaganj, New Delhi–110002 (INDIA)

Phone: +91-11-23243489, 23269324; 43551243 Telefax: +91-11-23243489

e-mail: katariabooks@yahoo.com; katariabook@gmail.com

Website: www.skkatariaandsons.com; *For e-books:* www.katsonbooks.com